Praise for *Wigmore on Alcohol*

"It will certainly be the new Alcohol Bible for decades to come, and be of great value to active practitioners and researchers."

—Dr. Kurt M. Dubowski, George Lynn Cross Distinguished Professor Emeritus of Medicine, University of Oklahoma

"This book is a must for anyone who is working on medicolegal issues involving alcohol. It will be appreciated by lay persons interested in this topic as well as scientific and legal experts involved in alcohol-related medicolegal issues"

—Thomas L. Pazdernik, PhD, University of Kansas Medical Center

"In short, *Wigmore on Alcohol* is a wonderful research tool that will prove a valuable resource in both legal and scientific libraries."

—Justice Libman, Judge of the Ontario Court of Justice

"I regret that *Wigmore on Alcohol* was not available to me during my years of active practice in forensic science"

—Doug Lucas, MSc, PhD (Hons), Retired Director of the Centre of Forensic Sciences, Toronto

"In summary, this work provides scientist and non-scientist, student and expert alike with a source of relevant articles to assist them in learning about, understand or deepening their knowledge of alcohol toxicology."

—Canadian Society of Forensic Science Journal

"I worked in the insurance industry for many, many years and I have attended mediations, discoveries and trials. I have also published articles and presented at seminars. Your book would have been of such a help as a reference had it been available . . . It's certainly one that should be used by the Insurance Industry, the Law Society and Police Departments across the country."

—Carolan Shannon, RN, Claims Consultant

Praise for *Wigmore on Cannabis*

"Medical and legal practitioners will find this an indispensable tome with up-to-date scientific evidence related to cannabis and its use. Academic law libraries will also benefit from having this practical book in their collections, as it emphasizes the depth of analysis that has been undertaken and will continue to be required as new cannabis-related issues arise in our courts."

—Margo Jeske, *Canadian Law Library Review*

"I found *Wigmore on Cannabis* to be an extremely useful, detailed and accurate reference book that assisted me to gain comprehensive knowledge of this complex substance. With over 5,000 peer reviewed scientific studies being assessed meant every possible angle of the drug was covered and especially useful was the information on the effect of cannabis on driving skills. The indexing and short abstract of each article was extremely user-friendly and easy to understand."

—Caro Smit, Founder and Director of South Africans Against Drunk Driving (SADD), Alcohol and Drug Educator and Counselor, Psychiatric Social Worker, Road Safety Expert

"In summary, I believe the reader will find this a very useful book for the breadth and depth of the studies and areas covered, and also for the many tables that elegantly summarize and provide an easier way to understand the take-away messages for a range of studies Jim has categorized in his second *Wigmore On* book.

—Robert M. Langille, PhD, Forensic Toxicologist

"The science of cannabis is a rapidly evolving area of research. This book is a valuable tool for finding published literature and keeping up-to-date on a variety of topics related to cannabis."

—Karen Woodall, PhD, Assistant Professor, Forensic Toxicology, University of Toronto

WIGMORE ON NICOTINE AND ITS DRUG DELIVERY SYSTEMS

Also by James G. Wigmore

Wigmore on Alcohol: Courtroom Alcohol Toxicology for the Medicolegal Professional

The Abridged Wigmore on Alcohol: Courtroom Alcohol Toxicology for the Medicolegal Professional

Wigmore on Cannibis: The Forensic Toxicology of Marijuana for Lawyers and Other Medicolegal Professionals

WIGMORE ON NICOTINE

AND ITS DRUG DELIVERY SYSTEMS

The Medicolegal Aspects of Our Most Addictive and Dangerous Legal Drug

JAMES G. WIGMORE

IRWIN LAW

James G. Wigmore
Forensic Alcohol Toxicologist
455 Danforth Ave
Toronto, ON M4K 1P1
james.wigmore@rogers.com

Published in 2022 by
Irwin Law Inc.
14 Duncan Street, Suite 206
Toronto, ON M5H 3G8
www.irwinlaw.com

ISBN 978-1-55221-647-7 (print) | e-book ISBN 978-1-55221-648-4 (PDF)

Cataloguing in Publication available from Library and Archives Canada

Printed and bound by CPI Group (UK) Ltd, Croydon, CR0 4YY

1 2 3 4 5 26 25 24 23 22

This book is dedicated to my grandchildren Calvin and Alice Spence; Chloe, Simon, and Benji Hearn; may they live in a smoke and nicotine-free world in which no one uses tobacco in any form. And to their grandmother Jacquie Berrow (Baba), who died too young of lung cancer due to scourge of nicotine addiction and cigarettes.

Summary Table of Contents

Detailed Table of Contents

List of Tables

(Reference number in bold parentheses)

- Hazard Ratios (HRs) of Various Adverse Events Comparing Bupropion or Varenicline and Nicotine Replacement Therapy (NRT) **(70225)**
- Stage of the Smoking Epidemic in Various Countries **(70303)**
- Disease/Toxic Constituent Health Warning Labels in Australia, Canada, and Mexico in 2011-2012 **(70307)**
- Changes in Cigarette Packages Design Before and After Increased Regulations in Canada (1999-2017) **(70311)**
- Percent of English Brand Names and Any English on Cigarette Packs in Various LMICs **(70313)**
- Percentage of Smokers in Canada, US, UK and Australia Who Agree with Various Statements About Smoking and the Tobacco Industry **(70322)**
- The Related Federal Court RICO Findings About the Tobacco Companies **(70323)**
- Incidence of Depression and Anhedonia (inability to feel pleasure) by Number of Cigarettes Per Day in College Students **(80103)**
- Hazard Ratio of First Depressive Episode Between 1990-2011 According to Cigarette Smoking History **(80105)**
- Adjusted Odds Ratio of Cigarette, E-Cigarette, Waterpipe, Cigars and Smokeless Tobacco Use and Mental Health Diagnosis (2+ Diagnosis or Depression Scale Score) **(80108)**
- Causes of Housefires in North Carolina, 1985 **(80202)**
- Residential Fire and Smoking Material Related Fire Death Rates Per 100,000 Population **(80204)**
- Odds Ratio (OR) of Death, Severe Casualty and Casualty According to Heat Source **(80205)**
- The Frequency of Body Areas That Were Burned Due to ECs **(80209)**
- Location of E-cigarette Lithium Batteries Explosions **(80210)**
- Smoking Status and Adjusted OR for Invasive Pneumococcal Disease **(80302)**
- Risks of Neurological COVID-19 Infection **(80313)**
- Multivariate Analysis of Risk Factors for COVID-19 Disease Progression **(80314)**
- Comparison of the Percentage of MVCs in Smokers and Non-smokers Before and After the 2009 Law **(80402)**
- Possible Factors for the Increased Risk of Motor Vehicle Collisions in Smokers **(80404)**
- The Odds Ratio (OR) of Unintentional Injury and Smoking Status **(80407)**
- Cigarettes Per Day in the 30 Days Prior to Basic Training and the Hazard Ratio of Injury **(80409)**
- Relative Risk of Mortality from Accidents and Adverse Events According to the Percentage of Adult Smokers in Each US County **(80411)**
- Odds Ratio of Developing Acute Lung Injury in Smoking Blunt Trauma Patients **(80412)**

- The Mean Serum Cotinine Concentration in Non-Hispanic Black and White Smokers and Number of Cigarettes Per Day **(90106)**
- Serum Cotinine Concentrations and Odds Ratio (Risk) of Lung Cancer **(90109)**
- Mean Urine Tobacco Alkaloid Concentrations (ng/mL) in the 24-hour Urine of Cigarette, Cigar, and Pipe Smokers and SLT Users **(90202)**
- Urine Nicotine/Metabolite and Anabasine Concentrations (ng/mL) in 8 Patients Undergoing NRT **(90203)**

Foreword

I have known James Wigmore for over twenty years. We first met when I joined the toxicology section at the Centre of Forensic Sciences. Jim was an excellent mentor who was always happy to pass on his extensive knowledge about the field of alcohol and also willing to share his experience about testifying in court and numerous tips about providing expert testimony. That type of valuable information from a colleague with years of experience is invaluable, something you can't learn from reading a textbook! Following his retirement, Jim used his time to publish a comprehensive review on alcohol based on his over 30 years' experience as a forensic toxicologist. He then applied his skills on reviewing and summarizing scientific literature for his second book, this time about cannabis. Now, Jim turns his attention to the often-overlooked drug nicotine.

I've worked in the field of Forensic Toxicology for more than 20 years but, considering the frequency of nicotine use in society, I (and many other forensic toxicologists) have paid little attention to this drug for many years. The popularity of smoking seemed to me to be diminishing each year and, in my opinion, the general public appeared to be well aware of its negative health consequences.

Even though nicotine has not been a primary focus of forensic toxicologists, considering the prevalence of nicotine use in society and its toxicity, it is surprising that a comprehensive review of the scientific literature about this drug has never been published. Jim fills this gap by continuing his successful series of "Wigmore on" books. He uses the same structure that readers of *Wigmore on Alcohol* and *Wigmore on Cannabis* will be familiar with.

The book is well organized and covers background information in each chapter, containing helpful tables of data and brief summaries of each article. The new quotations about the specific topic at the beginning of each chapter and section are an interesting addition to his "Wigmore on" structure and provide useful insight into people's comprehension and opinions on the various topics.

The book begins with the history of nicotine and also provides basic information about the drug, including its addiction potential and general toxicity. A number of studies have been published regarding the toxic effects observed in farmers working with tobacco plants and there are some interesting citations about Green Tobacco Sickness (GTS), an illness associated with dermal exposure to nicotine from the green, growing tobacco plant. The articles on GTS in this book are a good example of how toxic nicotine can be.

Forensic toxicologists are often questioned about alternative reasons a drug may be present in an individual aside from a typical route of exposure. I was fascinated to learn that although nicotine can be found in naturally occurring plants, including tomatoes and potatoes, it would take the consumption of over 90 kilograms of tomatoes to obtain detectable levels of nicotine.

The range of topics continues with chapters that cover important information about the pharmacology of nicotine and its detection in biological fluids. The author summarizes the multiple ways nicotine is consumed, and in addition to background information on traditional cigarettes, there is up-to-date information on newer delivery systems, including vaping.

In recent years, the introduction of vaping devices has led to an explosion in its popularity, especially amongst youth, so education in this area is extremely important. Information published by the government of Canada states that "Vaping can increase your exposure to chemicals that could harm your health (e.g. cause lung damage). Vaping could also expose you to nicotine, which is addictive. There are also concerns about the appeal of vaping products among youth and their potential to promote tobacco use."[1]

---

1 Government of Canada, "Risks of Vaping" (31 March 2022), online: www.canada.ca/en/health-canada/services/smoking-tobacco/vaping/risks.html.

The articles Jim has summarized about this topic provide interesting and current information on the dangers of vaping, including "popcorn lung," a recently documented risk that arises when the chemical diacetyl is present.

In researching this book, Jim uses his well-established techniques for critically reviewing the scientific literature. He reduces potential bias by focusing his reviews using research studies that have no conflict of interest. The importance of unbiased scientific studies and well-designed research is crucial in any field. He provides a good example of this in one of the chapter quotes: "Studies with tobacco and EC funding are 67X more likely to find no harm with ECs compared to studies without a conflict of interest," he writes, citing a study titled "What are the Respiratory Effects of E-Cigarettes?" (Gotts et al, 2019).

Using his expertise in reviewing and curating articles, Jim has incorporated more than 800 peer-reviewed published scientific studies into this book, which provide useful insight into the science of nicotine. Due to the increase in availability of nicotine products and the rise in popularity of e-cigarettes amongst youth, this is a timely addition to the scientific literature.

In summary, this book is an excellent resource for toxicologists, physicians, lawyers, and anyone with an interest in learning more about one of the most widely consumed drugs in the world.

Karen Woodall, PhD
Assistant Professor, Forensic Toxicology,
University of Toronto, Mississauga (2019–present)

Centre of Forensic Sciences, Toxicology Section (1999–2019)
Canadian Society of Forensic Sciences,
Drugs and Driving Committee (2019–present)

Preface

NICOTINE: THE NEGLECTED MEDICOLEGAL DRUG

> "Due to the widespread prevalence of smoking and therefore the frequency with which nicotine is detected by toxicological screening, there is the potential of its significance to be overlooked when present in overdose."
>
> —Corkery, Button, Vento & Schifano, "Two UK Suicides Using Nicotine Extracted from Tobacco Employing Instructions Available on the Internet" (2010)

After my first two books were published, one on alcohol, the other on tetrahydrocannabinol (THC), a number of people suggested that I write a book on opioids. No one, especially those who were forensic toxicologists or medicolegal professionals, suggested that I write a book on nicotine. This is typical of the neglect of nicotine in forensic science.

When I was a forensic toxicologist in the 1980s and '90s at the Centre of Forensic Sciences in Toronto, routinely during a drug screen involving medicolegal cases (e.g., sudden unexplained death, drug overdose, impaired driving, and other criminal investigations), nicotine would be detected. When I asked senior toxicologists about this nicotine peak on the gas chromatogram, I would get a response that the person was a smoker and nothing more. During that time smoking was much more common, and it seemed to be even more common in medicolegal cases than in the general public. Its presence in these cases was so frequent that nicotine could almost be considered to be an internal standard.

I also was lulled into dismissing nicotine and ignored this alkaloid for most of my career in forensic toxicology.

But nicotine is an extremely poisonous drug, that is involved in homicides, suicides, and accidental deaths. In one case of a suicide due to an overdose of nicotine, the forensic lab initially missed the fatal level of nicotine, perhaps because smoking was so common. But as you can read in Chapter 8, nicotine is not only highly addictive, but it is involved with injuries and motor vehicle collisions, fires and burns, psychiatric disorders, as well as homicides and suicides. Yet the forensic laboratories in general miss these connections.

If the forensic labs became more focused on nicotine and its metabolites as well, it could be determined whether the deceased obtained nicotine as an e-liquid, patch, or from tobacco, how long ago the deceased smoked before death, and perhaps whether the person was undergoing nicotine withdrawal, which would cause impairment of physical and mental abilities.

Hopefully, this book will awaken a renewed interest in the medico-legal aspects of nicotine and not just concentrate on its addictive and health problems, especially now that highly concentrated liquid nicotine is readily available to the general public on the web.

THE TOBACCO PLANT IS A HARSH MISTRESS

> "Warning: Nicotine can kill you, don't drink a tea made from its leaves, always wear gloves, the neurotoxin readily and quickly absorbs through the skin, it kills fish and bees and other animals, and it stinks."
>
> —Harold Roth, *The Witching Herbs: 13 Essential Plants and Herbs for Your Magical Garden* (2017)

A field of tobacco plants (Nicotiana tabacum) from a distance can look quite beautiful, but up close the tobacco plant is not something to admire or to plant in your home garden. Tobacco belongs to the nightshade family of plants, which includes potatoes and tomatoes and uses poisonous nicotine to protect their leaves from insects and animals. For food, we use the nonpoisonous parts of the nightshade family, the tubers and fruits. For the tobacco plant alone, we use the noxious leaves to inhale smoke into our delicate lungs.

Nicotiana tabacum is a thick stem plant that grows about 4 to 6 feet high and has large, rough, sticky leaves. It requires at least 100 frost-free days to grow. It seldom grows in the wild and requires careful cultivation. Unlike the cannabis plant, which grows like a weed, the tobacco plant needs a lot of backbreaking work. This work was initially achieved through slave labor and is now performed in less developed countries, frequently by children and women.

It is a difficult crop to grow. The seeds are extremely small and are generally planted inside before being transplanted to the field in the warmer weather. The tobacco plant requires a lot of fertilizer, especially nitrogen, to enable the plant to produce alkaloids, especially nicotine. The plant accumulates the toxic metals and radioactivity found naturally in fertilizer and soil into the leaves, which results in additional health risks to smokers. Before the invention of fertilizers, the tobacco plant could exhaust the soil in a few years. Thus, the farmers would be required to cultivate newer and newer fields, causing more ecological damage and deforestation.

The tobacco plant also requires a lot of pesticides to control insect infestation. It is ironic that a lot of pesticides are required to grow a pesticide (i.e., nicotine).

The tobacco plant has delicate roots and requires the loosest possible soil, which necessitates a lot of cultivation and weeding. The flowers and suckers of the plant are removed by hand or sprayed with chemicals, in order to divert production of the plant to the leaves and to increase its nicotine content.

A field of tobacco plants can cross-contaminate crops grown near it with nicotine and other alkaloids. This is important because many countries, including the European Union, restrict the nicotine content of food products, as nicotine is a banned pesticide.

It can also poison farm animals such as horses or cows that break into the tobacco field. Tobacco is also known as yellow henbane as it is toxic to fowl as well. Field workers develop Green Tobacco Syndrome by the absorption of nicotine through the skin if they work in the field without protective equipment.

The tobacco farmer receives very little money for all that work. Most of the money is made in the production of tobacco products, especially cigarettes, by advertising, and by the numerous tobacco industry lawyers. At one time lawyers were in every part of the tobacco companies and all communication had to be conducted through them to keep the information from the courts and public, as they claimed lawyer-client privilege.

Fortunately, that ploy did not work, and millions and millions of secret, internal tobacco industry documents have been made available and can be accessed on the internet. But the tobacco industry continues to hide its documents, by moving essential business to countries such as Switzerland, out of the reach of the US courts.

DOWN THE RABBIT HOLE

> "Doubt is our product since it is the best means of competing with the body of fact that exists in the minds of the general public. It is also the means of establishing a controversy."
>
> —Brown and Williamson Tobacco Company memo (1969)

When I decided to complete my trilogy of the forensic and medicolegal aspects of our most common, psychoactive legal drugs, I did not realize when I came to writing about nicotine, that I would be drawn down a rabbit hole. My first book on alcohol was relatively straightforward, as the drug had been studied for many years and there had been numerous publications in both English and German.

The book on cannabis was a bit more difficult, as there were a lot of opinions and fears about the legalization of this drug and the amount of research conducted about the effect of THC on humans was limited due to its widespread prohibition. But I found that by keeping to the peer-reviewed published data I could present a scientifically valid view of cannabis.

This was the most difficult book of the three to write. The science of nicotine and tobacco has been distorted and perverted by moneyed interests, conflict of interests, and the wealthy and deep-pocketed tobacco industry for many years, in order to delay legislative and social action to control and eventually remove this highly addictive drug from the general public and especially youth. Not only did the tobacco industry finance favorable studies, but it also influenced university institutions and whole journals to promote its view that tobacco did not cause cancer.

Due to all the misuse of science to promote the sale of cigarettes and other nicotine products, I have taken the unusual step of removing all studies in which the authors, researchers, or the journal were funded or given research grants by tobacco interests. My screening procedure was relatively simple, as studies that were neutral or positive towards tobacco products almost invariably had links to or were financed by the tobacco companies.

For example, it is reported that the editor-in-chief of the journal *Regulatory Toxicology and Pharmacology* was paid $30,000 to publish an article dismissing the health risks of secondhand smoke.[1]

I initially thought that I would include these scientific papers published by researchers with links to or financed by the tobacco industry and just highlight the article to notify the reader that this paper is suspect. Ultimately, however, I concluded that including those biased and questionable sources would serve only to confuse the information about smoking and nicotine, as they were designed to do.

If I were writing a book about gravity, I would not include papers claiming that gravity does not exist, nor would I include, in a book on vaccines for COVID-19, that vaccines are useless and will change your DNA. Likewise, in my book on nicotine, I have not included studies financed by the tobacco industry, or written by scientists with a conflict of interest, claiming that nicotine is not addictive or that smoking doesn't really cause lung cancer or that vape is just harmless water vapor.

SOCIAL MEDIA AND BOTS

A social bot (robot) is an automated account that uses artificial intelligence to steer discussion and promote products or messages on social media such as Twitter or Facebook. Tobacco interests have peppered social media with bots to minimize the health risks of cigarettes and other tobacco products and especially vaping, and to glamorize their use.

I personally had experience of a bot. I posted an article on Redditt in the Vaping community that warned about the dangers of inhaling propylene glycol and vegetable glycerin, as these chemicals, although generally regarded as safe, have only been tested on ingestion and on vaping. In fact, the manufacturers of propylene glycol have warned that it should not be vaped.

The article, however, did not seem to be getting any traction. Even several weeks later the number of upvotes of my article remained resolutely at zero. I did not realize what was happening until one friendly member messaged me and asked, "Do you have a bot?"

At the time I knew nothing about bots and said that I didn't know. He messaged that he would check for me.

---

1 Jenssen et al, "Heat-not-Burn Tobacco Products: Tobacco Industry Claims No Substitute for Science." *Pediatrics* 141(1): 3pp, 2018.

Several hours later he messaged that yes, I did have a bot. Someone had attached a bot to my slightly critical vaping article that would automatically downvote my article whenever anyone upvoted it. Thus, keeping my article at a zero level of voting and buried in the vaping community.

I have noticed that of the articles about nicotine, alcohol, or THC that I post on social media, those about nicotine always have the lowest interest or views and the lowest number of upvotes or likes, perhaps due in part to manipulation by tobacco interests.

RACKETEER INFLUENCED AND CORRUPT ORGANIZATION (RICO)

US federal Judge Kessler in her 2006 landmark 1,683-page decision ruled, after seven years of trial, that the tobacco companies had violated the RICO Act by engaging in numerous acts of fraud to further a conspiracy to deceive the US public about nicotine addiction and the health effects of cigarettes and secondhand smoke. The judge concluded that the tobacco companies had:

- Violated civil racketeering laws (engaged in an organized conspiracy to commit fraud)
- Committed fraud
- Are likely to continue to commit fraud
- Denied that they control the level of nicotine to create and sustain addiction
- Suppressed and concealed scientific research
- Marketed low tar and light cigarettes as less harmful though they knew they were not
- Marketed cigarettes to young people to replace smokers who die or quit smoking
- Denied that secondhand smoke harms nonsmokers
- Denied that smoking is addictive
- Denied the health consequences of smoking.[2]

The corrective statements ordered by the federal judge are in Appendix 2.

2 *US v Philip Morris USA, Inc*, 449 F.Supp.2d 16.

GOOGLE, GOOGLE SCHOLAR, AND PUBMED

When I searched Google for quotes about tobacco, e-cigarettes, nicotine, and cigarettes, I found numerous critical, realistic quotes about cigarettes, but I could not find any negative comments about ECs. The only quotes I could find in a Google search were ones about how safe ECs are, that vape is mainly water vapor, and that they are less addictive than cigarettes. This shows again how the social media is biased and has been manipulated by tobacco interests.

For that reason, I would recommend that any search about medical issues such as nicotine, should be conducted on Google Scholar which, as the name indicates, basically removes commercial interests and ranking. It lists only published research papers or studies and shows much less bias.

The gold standard still remains the US National Library of Medicine at PubMed. This website has over 32 million citations from the biomedical literature and should not be substantially influenced by the tobacco industry in the listing of the papers, unlike social media. I tried to choose articles that were listed in PubMed and whose pdfs were free, so almost all of the studies I have listed in this book are available at no cost to the reader.

STRUCTURE OF THE "WIGMORE ON ..." BOOKS

As I did for my previously published books on alcohol and cannabis, I initially extensively reviewed and assessed numerous books and more than 2,000 peer-reviewed published studies. From these studies I have selected over 800 for this book.

I organized the studies according to their content into nine main chapters, in a structure similar to that of my other two books. Each study or reference has a unique five-digit number assigned to it. The first digit indicates the chapter, the next two digits indicate the section, and the last two digits indicate the position of the study in the section.

For example, reference number 80406 indicates that it is in Chapter 8 ("Other Effects of Nicotine Products/Nicotine"), section 4 (04) ("Injuries/Motor Vehicle Collisions") and is study number 6. In which you will find:

Reference Number: 80406

HUTCHENS, L., T.M. SENSERRICK, P.E. JAMIESON, D. ROMER, AND F.K. WINSTON. "Teen Driver Crash Risk and Associations with Smoking and Drowsy Driving" *Accident Analysis and Prevention*, 40: 869–876, 2008 (4 tables, 49 references)

Abstract: A telephone survey was conducted of 550 drivers ages 14 to 22 years in the United States in 2005. One hundred and twenty-eight (23.3%) were current smokers and 202 (40.7%) had a crash as a driver. The odds ratio (OR) of crash involvement was 1.8× for driving drowsy alone, 1.3× for the length of licensure, and 2.1× for being a current smoker.

> Possible explanations given for smoking's association with increased odds of injury include distractibility involved in the act of smoking (including one-handed steering), direct toxicity (potentially contributing to reduced night vision, vision performance, and reduced driving performance), and associated risk-taking. Another plausible rationale is that people who risk their health by engaging in smoking may also be more risk-taking drivers; for example, smokers have been found to wear their safety belts less often. Smoking was also significantly associated with sensation seeking and alcohol consumption in our study. However, we found that smoking was associated with higher crash odds despite holding constant general risk-taking tendencies as well as other indicators of risky behavior tendencies, such as drinking alcohol, using marijuana, and driving without a seatbelt. Hence, it appears that the risk conferred by smoking is unique to this form of risky behavior.

The first paragraph of the reference lists the authors, the title of the study in quotation marks, and the journal citation. In addition, I list the number of tables, figures, and references that appear in this study.

The second paragraph is an abstract of the study, which I wrote and which summarizes the particular area of the paper involving the main points I would like to raise.

The third paragraph is in sans-serif font and is a direct excerpt from the study, showing what the researcher(s) actually wrote. I thought it was important to have the researchers' own words to describe their research and give a flavor of the study.

At the beginning of each chapter, I have summarized a broad overview of its content. Each section of the chapter has a more detailed review citing the actual studies. Initially, the reader could just concentrate on the section summaries and not get bogged down in the details of the numerous studies.

In this book I have added a new feature, a quotation about the specific topic at the beginning of each chapter and section, which I thought would add interest and show that the study of nicotine and tobacco is not only of academic interest but affects virtually all aspects of modern life. In fact, one could obtain a lot of information about nicotine by just reading the quotations from each section.

I have also included more tables in this book, which were popular in my previous two books.

This book was designed to offer an extensive scientific overview of the drug nicotine. It is an extensive book, not an intensive book. I do not cite all the papers on a specific topic. If you would like more detail on the topic, I would suggest you read the original study, most of which are available free at PubMed or Google Scholar.

COMPARISON OF ALCOHOL, THC, AND NICOTINE

When I was a junior technologist at the Centre of Forensic Sciences in Toronto, a wise older forensic toxicologist told me, "If it has nitrogen, it is a drug." Perhaps it is an over-generalization but is largely insightful. Most of the drugs that cause addiction and serious health problems such as morphine, heroin, cocaine, and amphetamines contain a nitrogen in their molecular structures.

Of the three drugs that I have written extensively about, only nicotine contains nitrogen and is the most addictive and toxic of the three. Both alcohol and THC do not contain a nitrogen atom in their structure and are much less toxic and addictive than nicotine.

The following table contains a comparison of these three drugs.

	Alcohol	THC	Nicotine
Chemical structure	C_2H_5OH	$C_{21}H_{30}O_2$	$C_{10}H_{14}N_2$
Chemical family	Alcohol	Alcohol	Alkaloid
Drug type	Depressant	Depressant	Stimulant
Boiling point	78°C	157°C	247°C
Toxicity	++	+	+++
Psychosis with use	+	++	+
Impairment of driving ability	+++	++	+

	Alcohol	THC	Nicotine
Addictiveness	++	+	+++
Absorption routes	Oral	Oral Inhalation Sublingual	Inhalation Dermal Sublingual Oral Ocular
Source	Fermentation of sugars by yeast	Buds of unfertilized female cannabis plants (sinsemilla)	Leaves of de-flowered tobacco plant
Metabolites	Acetaldehyde Acetate	11-hydroxy THC 11-carboxy THC	cotinine 3' hydroxycotinine

Acknowledgements

I would like to acknowledge my colleagues Karen Woodall and Rob Langille, who have joined me in retirement to continue the wonderful journey of the discovery of forensic toxicology, especially of alcohol, cannabis, and tobacco. I would also like to acknowledge my dear, wonderful, and amazing spouse Penelope, without whom none of this would be possible.

I would also like to acknowledge the Toronto Public Library system, which is the largest public library system in Canada with 100 local branches throughout the city. It is the largest neighborhood-based library system in the world. To those living or working in Toronto all its services are free. I was also able to give numerous free lectures to the public in its branches about alcohol, cannabis, and nicotine (especially vaping). The lectures were very helpful to me in thinking and questioning about these drugs and their role in the public at large.

Finally, I would also like to acknowledge my publisher, Irwin Law, the only independent legal publisher in Canada, for its many years of support, and its terrific editors and designers, Danna Petersen-Deeprose, Lesley Steeve, and Heather Raven.

"The cigarette industry is peddling a deadly weapon. It is dealing in people's lives for financial gain. The industry we seek to regulate is powerful and resourceful. Each new effort to regulate will bring new ways to evade. Still, we must be equal to the task, for the stakes involved are nothing less than the lives and health of millions of people around the world. But this is a battle that can and will be won."

—Robert Kennedy, address to 1st World Conference on Smoking and Health, 1967

"A cigarette is pinch of tobacco in paper with a fire at one end and a fool at the other."

—George Bernard Shaw

"Tobacco is a legal drug that kills many of its users when used exactly as intended by the manufacturers."

—WHO global report on trends in tobacco smoking, 2015

"We don't smoke that s____. We just sell it. We reserve the right to smoke for the young, the poor, the black and stupid."

—R.J. Reynolds executive's reply when asked why he did not smoke, Times of London, *2 August 1992*

CHAPTER 1

Nicotine: The Drug

> "Smoking is highly addictive. Nicotine is the addictive drug in tobacco. Cigarette companies intentionally designed cigarettes with enough nicotine to create and sustain addiction."
>
> —Court-ordered Corrective Statements, *US v Philip Morris USA Inc*, RICO decision (2006)

> "Indeed, a chemist at Philip Morris would have been among a select few in the country who could appreciate just how nasty nicotine really is. Many at the research center learned the hard way that merely leaving a bottle of the clear liquid open in a warm room brings on a wave of coughing and gagging, followed in short order by dizziness and nausea."
>
> —Dan Zegart, *Civil Warriors the Legal Siege on the Tobacco Industry* (2001)

> "If it's got nitrogen, it is a drug."
>
> —Wan-Ching Sun, Forensic Toxicologist, Centre of Forensic Science (1976)

Nicotine is a much more toxic and addictive drug than either alcohol or tetrahydrocannabinol (THC) and is the only one of the three drugs that originated from the New World. Nicotine was named after an early advocate of the use of this drug and tobacco, Jean Nicot, the French ambassador to Portugal who obtained some of the first tobacco seeds. The name is derived from "Nicot" and "ine," as the drug contained nitrogen (amines) and was an alkaloid. The species of tobacco plant was also named after him ("nicotiana").

The first section on the History and Cultivation of Tobacco does not present a sugar-coated, nostalgic view of tobacco, such as Sir Walter Raleigh's servant rushing into his room with a bucket of water to put out the pipe he was smoking, as the servant thought he was on fire. Instead, it concentrates on the major adverse health effects tobacco has had on humans and on the environment, especially in the cultivation and processing of tobacco.

The other sections in this chapter describe the addictive and poisoning effects of nicotine. Other minor and insignificant sources of nicotine such as tomatoes or potatoes, which are in the same plant family, have been used as a smoke screen by the tobacco industry to cast doubt that the nicotine found in non-smokers was caused by secondhand smoke but by the ingestion of these vegetables.

1.01 HISTORY AND CULTIVATION OF TOBACCO

> "Both tobacco cultivation and curing, then are associated with farming methods that destroy the environment in developing countries, while it can provide profits for farmers and local communities, these societies are impoverished by the loss of resources."
>
> —De Granda-Orive et al, "World Health Organization Positioning. The Impact of Tobacco in the Environment: Cultivation, Curing, Manufacturing, Transport, and Third and Fourth-Hand Smoking" (2018)

> "If tobacco were to be introduced today, it would not be approved for human consumption anywhere in the world."
>
> —Chandra and Govindraju, "Prevalence of Oral Mucosal Lesions Among Tobacco Users" (2012)

Tobacco plants are a member of the nightshade family (Solanaceae), which also includes tomatoes and potatoes. The main tobacco plants (nicotiana tabacum and nicotiana rustica) have been cultivated in the Andes of South American since about 5,000 BCE and spread to North America, where the Indigenous people offered Columbus gifts of tobacco (10101, 10103). At first Columbus threw the tobacco overboard but members of his crew started smoking and its use grew in popularity. At that time smoking was conducted mainly by using various types of pipes. The Maya uniquely did not used pipes but wrapped the dried tobacco in a corn husk or other plant material to form a type of crude cigar.[1]

Nicotiana rustica was the main tobacco grown in North America before colonization. It produced a harsh smoke as it was more basic (i.e., bitter) than Nicotiana tabacum, and so was not inhaled completely into the lungs. Instead, the smoke was kept in the mouth, which allowed the nicotine to be absorbed through the mucous membranes of the oral cavity. This prevented the tobacco from becoming a major cause of lung cancer and addiction among the Indigenous peoples in North America until the production of modern cigarettes.

Many modern Indigenous communities' sacred ceremonies use commercial tobacco, and there is a call to begin using the other traditional types of tobacco plants originally cultivated here instead (10102). Different types of native tobacco plants were also grown by the Indigenous communities in other countries such as Australia (10104).

Tobacco is considered to be a valuable cash crop. But the farmers where this crop is grown are typically poorer and have more health problems than farmers of nontobacco crops (10110, 10111). Any profit seems to benefit the tobacco industry and not the developing countries where it is grown (10114). Globally, approximately 4.2 million hectares of land are under cultivation for tobacco (10108) and approximately 6.25 trillion (6,250,000,000,000!) cigarettes are produced annually (10113).

Nicotiana tabacum concentrates toxic heavy metals, such as cadmium, arsenic, and lead, and radioactive elements from soil, pesticides, and fertilizers into its leaves, which also would pose an additional health risk to tobacco users (10106, 10107, 10109, 10130). The tobacco plant depletes the soil, requiring large amounts of fertilizers and pesticides, which causes substantial environmental degradation (10108, 10113).

About 95% of the alkaloids produced by the tobacco plant is nicotine, which is produced by the roots but accumulates in the leaves. The nicotine content of the tobacco leaves can therefore be reduced by grafting the plant onto eggplant roots, but this would reduce the addictiveness of cigarettes (10115). Nicotine is a powerful stimulant and parasympathetic alkaloid which is readily absorbed through the lungs and the skin (10126–7). This absorption of nicotine causes Green Tobacco Sickness

1 Ian Gately, *La Diva Nicotina—the Story of How Tobacco Seduced the World* (UK: Simon and Schuster, 2001).

(or Syndrome) (GTS) in tobacco workers, especially when harvesting or transporting the green tobacco leaves.

The symptoms of GTS are due to the poisoning effects of nicotine and include dizziness, headache, nausea, vomiting, blurred vision, and abdominal pain (10119, 10120). GTS can be prevented by the use of personal protective equipment (PPE) such as wearing gloves, long sleeve shirts, and masks, and changing wet clothes in the tobacco fields (10122). But the wearing of PPEs can induce heat stroke and dehydration in the tobacco workers in hot countries (10128).

GTS is not the only workplace hazard faced by tobacco workers. The liberal use of pesticides causes additional health problems. Since children less than 15 years of age are more tolerant to the effects of pesticides, they are often employed on tobacco plantations in less developed countries (10123).

Reference Number: 10101

KUNITZ, S.J. "Historical Influences on Contemporary Tobacco Use by Northern Plains and Southwestern American Indians." *American Journal of Public Health*, 106: 246–255, 2016 (3 tables, 97 references)

Abstract: The two types of tobacco that were cultivated in North America before European colonization were Nicotiana rustica and N tabacum. N rustica has a higher nicotine content but is more difficult to inhale. In the early 20th century, roll-your-on cigarettes were widely available, and Bull Durham was the most popular brand. The calumet (pipe) ceremony with tobacco was frequently performed in the Northern and Southern plains but not the Southwest. The rate of smoking is 2.23× greater in Northern Plain Indigenous men than in the Southwest and the lung cancer rate is 5.6× greater.

> Tobacco was, thus, integrated into social relationships with religious sanctions on the Northern Plains, which increased the acceptability of commercial cigarettes in the 20th century. Smoking is, therefore, more deeply embedded in religious practices and social relationships on the Northern Plains than in the Southwest.

Reference Number: 10102

TUSHINGHAM, S., C.M. SNYDER, K.J. BROWNSTEIN, W.J. DAMITO, AND D.R. GANG. "Biomolecular Archeology Reveals Ancient Origins of Indigenous Tobacco

Smoking in North American Plateau." *Proceedings of the National Academy of Sciences of the United States of America*, 115: 11742–11747, 2018 (1 table, 4 figures, 50 references)

Abstract: The tobacco plant (genus Nicotiana) was first domesticated about 6,000 to 8,000 years ago in the Andes of South America. Ultimately the native people developed tobacco species such as Nicotiana rustica and Nicotiana tabacum, which have larger leaves and a higher nicotine concentration than the earlier wild varieties. Domesticated tobacco spread all over North America, except tor the western regions where desert or wild tobacco was used (Nicotiana quadrivalvis, Nicotiana attenuata (coyote tobacco), and Nicotiana obtusifolia). In many parts of the northwest, kinnikinnick or bearberry was the main plant smoked. Tobacco was also used as an offering in a religious context. Twelve archeological pipes from the Columbia River basin dating over a span of 1,300 years were analysed for nicotine by UPLC-MS and it was identified in eight of the pipes.

Tobacco has clear negative health consequences worldwide, yet we still have much to learn about the evolution of the human-tobacco relationship. The attractiveness of psychoactive plants to humans is clearly related to their physiological effects when ingested. Tobacco's powerful pull on humans is due to its intoxicating properties and addictive nicotine content, which is the result of thousands of years of coevolution with people. Historical processes have obscured our understanding of early tobacco use by indigenous people of western North America. One of the greatest puzzles in northwestern North American archeology is the timeline and trajectory of indigenous tobacco consumption by Plateau hunter-gathers before the introduction of domesticated trade tobacco (N. tabacum). This study establishes a pattern of ongoing tobacco smoking in this region from at least 1,200 y age through the historical period. These results conflict with ethnohistoric descriptions of smoking practices on the Columbia Plateau; the sites represented in this study are located securely within a region where only kinnikinnick was smoked and tobacco was a completely introduced trade commodity. Thus, it appears that when the famous leader Chief Joseph stated that French traders introduced tobacco to the Nez Perce in the early 1800s, he was specifically referring to trade tobacco.

Commercially grown species of tobacco have replaced indigenous tobacco and use of other smoke plants among many natives, a process that began several hundred years ago during the earliest days of European colonization of the Americas. The shift from traditional smoking of indigenous tobaccos and other smoke plants to commercial tobacco products has had significant deleterious effects on tribal culture and health. Increased awareness of the history of tobacco and its role in the process of colonialization is critical to future smoking abatement programs, as is distinguishing between commercial tobacco and tobacco used in ceremonial and other traditional manners (often distinguished as sacred tobacco by native peoples). Indeed, few realize the extent to which domesticated strains of commercial tobacco have replaced indigenous species of tobacco and other smoke plants. In many cases, commercial tobacco has become the tobacco used for ceremonial purposes, replacing the many Nicotiana species or other botanicals originally used by Indians who had a religious smoking tradition. This information can be helpful to Tribal efforts toward educating youth about the dangers of commercial tobacco and the sacredness of indigenous tobaccos. Rejuvenation of additive-free indigenous tobacco (e.g., N. quadrivalvus, N. obtusifolia) used in traditional contexts can be part of this process, including sacred use practices in daily or ceremonial contests (e.g., tobacco smudging, tobacco used as gifts and as offerings).

Reference Number: 10103

SIKORSKA-JAROSZYŃSKA, M.H.J., M. MIELNIK-BŁASZCZAK, D. KRAWCZYK, A. NASIŁOWSKA-BARUD, AND J. BLASZCZAK. “Passive Smoking as an Environmental Risk Factor.” *Annals of Agricultural and Environmental Medicine*, 19: 547–550, 2012 (62 references)

Abstract: A general review was conducted on the history of tobacco and secondhand smoke. Tobacco first appeared around 6,000 BCE and was treated in the Americas as a sacred smoke which gave Indigenous people the possibility of reconciliation and eternal peace. Originally tobacco was considered a decorative plant in Europe, and only began to be treated as a medicinal herb by Nicolas Manardes in 1571, who described tobacco as a treatment for 36 diseases including mental illness, fever, headache, and frostbite. Throughout the 18th century, many countries including Russia and the Vatican banned smoking tobacco. Wars and epidemics encouraged the widespread use of tobacco.

In the 1980s and 1990s, due to the results of experimental, clinical, and epidemiological studies revealing the harmfulness of tobacco, there appeared new laws limiting smoking in public places, especially at work. The Framework Convention on Tobacco Control (FCTC) is the first international treaty concerning public health. It was issued after four years of negotiations among the WHO members. It was signed in 2004 by 192 countries, and its provisions were officially introduced in February 2005. The provisions of FCTC mainly concern the tobacco industry, the necessity for the labeling of tobacco products packing, introduction of new taxes on tobacco products, and illegal trade and protection from exposure to tobacco smoke. In 2005, the Oral Health Network of Tobacco Use Prevention and Cessation was established. In 2008, a work group was formed in the European Parliament aimed at the introduction of a complete ban on tobacco trade in the European Union by 2025.

Reference Number: 10104

RATSCH, A.M., A. MASON, L. RIVE, F.E. BOGOSSIAN, AND K.J. STEADMAN. "The Pituri Learning Circle: Central Australian Aboriginal Women's Knowledge and Practices Around the Use of *Nicotiana* spp., as a Chewing Tobacco." *Rural and Remote Health*, 17: 4044 (15 pp), 2017 (1 table, 3 figures, 24 references)

Abstract: Australia has 22 native species and 4 subspecies of Nicotiana, which are primarily an annual plant dependent on the rain. The *Nicotiana gossei* plant is referred to as pituri or mingkulpa by the Aboriginal peoples in Australia and its nicotine levels can vary from 0.6 to 18.46 mg/g dry weight. The pH of pituri is acidic at 5.47 and so nicotine is not readily absorbed through the oral cavity. Aboriginal peoples combine pituri with wood ash, which has a pH of 12.2 to 13.2 and allows the nicotine to become free-base and absorbed more rapidly through the oral cavity.

Table: Signs of Pituri (Mingkulpa) Overdose and Toxicity as Reported by the Aboriginal Population

Mingkulpa Overdose and Toxicity Signs	
• Really weak	• Heart is irregular
• Feel sick	• Confused
• Having chest pain, burning chest	• Vomit
• Headache	• Lie down and feel like you're spinning
• Drowsy	• Dizzy
• Heart is weak	• Really sleepy
• Can't see properly	• Fall on ground and lie there
• My heart got burn	

Source: Adapted from Ratsch et al, 2017

Central Australian Aboriginal women have a firmly established knowledge and understanding of the pharmacological principles related to the content of Nicotiana spp. and the extraction of nicotine from the plant. Widespread use of Nicotiana spp. as a chewing tobacco by Aboriginal population in the southern, central, and western desert regions of Australia is attested to by participants who assert that everyone uses it, with girls in these remote areas commencing use between 5 and 7 years of age. Central Australian Aboriginal people who chew Nicotiana spp do not consider it to be a tobacco plant and will strongly refute that they are tobacco users. Central Australian Aboriginal people do not consider that the Western Health information regarding tobacco (as a smoker product) is applicable or aligned to their use of pituri. Nicotiana spp users will deny tobacco use at health assessment. There is a requirement to develop and provide health information on a broader image of tobacco and nicotine products in ways that are considered credible by the Aboriginal population. Health messages around pituri use need to account for the dominant role that pituri occupies in the contest of central Australian Aboriginal women's lives.

Reference Number: 10105

TASSEW, Z., AND B.S. CHANDRAVANSHI. "Levels of Nicotine in Ethiopian Tobacco Leaves." *Springer Plus*, 4: 649 (6 pp), 2015 (3 tables, 29 references)

Abstract: Tobacco is a valuable cash crop and is the most widely grown non-food crop. Nicotine is a pale yellow to dark brown oily liquid with an unpleasant pungent odor and a sharp, persistent, bitter taste and is soluble in organic liquids such as alcohol and chloroform. In small doses nicotine has a stimulating effect which increases alertness and improves memory, but repeated users only report relief from the symptoms of nicotine withdrawal. Nicotine also increases the heart rate and blood pressure and reduces appetite. It is more toxic than many other alkaloids. The nicotine content of four types of tobacco leaves from four regions of Ethiopia were determined by high pressure liquid chromatography. The average nicotine concentrations ranged from traces to 3.26%.

Table: Mean Nicotine Concentration of Various Tobacco Species Grown in Ethiopia

Tobacco Species	Mean Nicotine Concentration (%)
Virginia (Billate)	3.26
Virginia (Shewa Robit)	2.20
Burley (Awassa)	0.65
Oriental (Awassa)	Traces
Gaya (Wollayita)	1.10

Source: Adapted from Tassew and Chandravanshi, 2015

The level of nicotine in the Ethiopian tobacco leaves were determined by high performance liquid chromatography. The analysis of tobacco leaves was based on their leaf position, places of cultivation and variety of species. Among the variety of species Virginia tobacco leaves were found to have high nicotine level (3.26%) followed by the native tobacco Gaya (1.1%), then Burley tobacco leaves (0.650%). The nicotine level in Oriental tobacco leaves was below the detection limit and its accurate level of nicotine was not determined by this study. Based on their leaf positions, the nicotine level of Ethiopian Virginia tobacco leaves like that of other countries increases from bottom to top (tip) leaves. Furthermore, this study has found that nicotine level of a given tobacco species cultivated in different places is different. The levels of nicotine in Ethiopian tobacco leaves are comparable with that cultivated in the rest of the world.

Reference Number: 10106

O'CONNOR, R.J., L.M. SCHNELLER, R.V. CARUSO, W.E. STEPHENS, Q. LI, J. YUAN, AND G.T. FONG. "Toxic Metal and Nicotine Content of Cigarettes Sold in China, 2009 and 2012." *Tobacco Control*, 24: iv55–iv59, 2015 (1 table, 3 figures, 37 references)

Abstract: Toxic heavy metals can accumulate in the tobacco plant leaves from the soil or fertilizer and contribute to cancers when smokers inhale them. The toxic heavy metal and nicotine concentrations were determined in cigarettes purchased from seven cities in China in 2009 and 2012. Nicotine constitutes up to 95% of the alkaloid content of tobacco. Nitrogen fertilizer can increase the nicotine content of tobacco, but if used improperly can increase the toxic, heavy metal content as well. The mean arsenic, cadmium, and lead levels in Chinese cigarettes ware substantially greater than in US cigarettes.

> These results highlight the sustained exposure among Chinese smokers to high levels of toxic metals. The high levels of Cd and Pb in Chinese cigarettes can be potentially problematic given China's high smoking prevalence and per capita consumption, as a significant fraction of Cd and Pb transfer directly into smoke and are correlated with smoking intensity. Furthermore, Cd has a biological half-life of one to four decades and minimal excretion through the urinary tract indicates the body experiences long-term exposures. Metal contamination of agricultural products due to environmental pollution is a broader public health concern in China, particularly for foodstuffs and recent reports have highlighted high levels of metals in vegetables, rice, and fish. So, Chinese smokers may be adding additional metal burden to existing exposures.
>
> Lastly, nicotine, the addictive component of tobacco is thought to be affected by the plants' supply of nitrogen throughout growth. Given that fertilizers are intended to provide nitrogen as well as being a potential source of heavy metal contamination, the use of fertilizers on tobacco crops may explain the observed correlation between the tobacco rod nicotine and metal content. Therefore, the regulation of fertilizer use may act as a means to decrease the amount of heavy metals found within the Chinese tobacco.

Reference Number: 10107

CHIBA, M., AND R. MASIRONI. "Toxic and Trace Elements in Tobacco and Tobacco Smoke." *Bulletin of the World Health Organization*, 70: 269–275, 1992 (4 tables, 52 references)

Abstract: Tobacco plants have a special affinity to absorb cadmium and nickel from the soil and to accumulate these heavy metals in a high concentration in the leaves. Cadmium has a half-life of 9.4 years once in the body. Cadmium concentrations in various biological tissues are higher in smokers than in non-smokers.

Table: Cadmium Concentrations in Various Postmortem Samples in Non-smokers, Ex-smokers, and Smokers

Body Tissue	Non-smokers	Ex-smokers	Smokers
Lung (μg/g)	0.05	0.22	0.32
Muscle (μg/g)	0.052	0.054	0.061
Prostate (μg/g)	0.06	0.10	0.15
Blood (μg/L)	2.1	2.8	4.2
Urine	2.47	2.78	2.47

Source: Adapted from Chiba and Masironi, 1992

The scientific literature is filled with evidence on the harmful effects of carbon monoxide, nicotine, tar, irritants, and other noxious gases emitted in tobacco smoke. Not enough attention, however, has been paid to the presence of heavy metals and other toxic and trace elements in tobacco smoke and their possible effects on biochemical processes in the human body. Once inhaled through smoking, heavy metals have a long biological half-life. Chronic adverse effects on human health may, therefore, in later years result from prolonged intake of such toxic elements, some of which are powerful carcinogens. Several of them accumulate in bone and may trigger disorders of mineral metabolism, e.g., osteoporosis. The body burden of heavy metals increased as a result of occupational exposure and tobacco smoking enhances the adverse effects of such exposure.

In the tobacco plant, the concentration of elements varies along the stalk, being higher in the older leaves than in the younger, top leaves. In cigarettes, element concentrations vary among brands and even within the same brand. Besides tobacco the main sources of trace and other elements in cigarettes are the paper and filter.

Reference Number: 10108

MASANOTTI, G.M., E. ABBAFATI, E. PETRELLA, S. VINCIGUERRA, AND F. SRACCI. "Intensive Tobacco Cultivations, A Possible Public Health Risk?" *Environmental Science and Pollution Research*, 26: 12616–12621, 2019 (2 tables, 5 figures, 23 references)

Abstract: Tobacco plantations cause environmental degradation due to the large amounts of pesticides, fertilizers, and other chemicals required. The tobacco plant depletes the soil and will cause it to collapse, requiring cultivation to be moved. About 4.2 million hectares of land are used globally for tobacco cultivation. Italy is a major producer of tobacco in Europe and produces 50,000 tons of tobacco, mainly in Campania, Umbria, Veneto, and Tuscany. The nicotine content of 300 lots of medicinal wild plants growing near tobacco fields was tested. Nicotine was detected in 82% of the samples in 2015 and 63% of the samples in 2016.

In this study, the endogenous presence of nicotine as a secondary metabolite was detected in the various cultures examined classified as medicinal plants and wild plants. Considering the bio-accumulation of nicotine in these crops, it can be inferred that nicotine is present in the ecosystem. However, the phenomenon that leads to the contamination of the different environmental matrices remains, still, unidentified.

Reference Number: 10109

PAPASTEFANOU, C. "Radioactivity of Tobacco Leaves and Radiation Dose Induced from Smoking." *International Journal of Environmental Research and Public Health*, 6: 558–567, 2009 (3 tables, 1 figure, 29 references)

Abstract: Tobacco plants incorporate and concentrate natural radioactivity by its roots from the soil and phosphate fertilizers. The trichomes (sticky hairs) on the tobacco leaves may also accumulate radioactivity from small atmospheric particulates. Seventeen different samples of tobacco leaves from different areas of Greece in 1990 were analysed for radioactivity using gamma ray spectrometry. The Chernobyl nuclear reactor accident occurred in May 1986. It is estimated that a daily smoker would have approximately 200× the radioactive dose inhaled into the lungs as a non-smoker.

The radioactivity in tobacco leaves collected from 15 different regions of Greece before cigarette production has been studied in order to find any association between the uptake of naturally occurring primordial radionuclides and radioisotopes of cesium of Chernobyl origin. The activities of the radioisotopes of radium 236Ra and 228Ra in the tobacco leaves would be either through root uptake or through the fertilizers used for cultivation of tobacco plants in the fields, as the air concentration of the radioisotopes of radium are extremely low. Lead-210 originating from air was deposited onto the tobacco leaves and trapped by the trichomes. Potassium-40 in the tobacco leaves was due to root uptake from soils or from fertilizers. The radioisotopes of cesium 137Cs and 134Cs in the tobacco leaves were due to root uptake and not to deposition onto the leaf foliage as they still remined in soil four years after the Chernobyl reactor accident but were absent from the atmosphere because of the rain washout (precipitation) and gravitational settling.

Reference Number: 10110

MINH, H.V., K.B. GIANG, N.M. BICH, AND N.T. HUONG. "Tobacco Farming in Rural Vietnam: Questionable Economic Gain but Evident Health Risks." *BMC Public Health*, 9: 24 (10 pp), 2009 (5 tables, 29 references)

Abstract: Tobacco companies have promoted tobacco growing in many developing countries such as Vietnam, claiming that it will bring unparalleled prosperity to farmers and their communities. A comparison was conducted of two tobacco farming communities and two nontobacco growing communities in Vietnam (a total of 480 households). The mean per capita income in US dollars were $563 and $502 for the tobacco growing communities and $568 and $487 for the nontobacco growing communities, respectively. More tobacco farmers reported health problems such as tiredness/weakness, nausea, pains, and poor appetite than nontobacco farmers.

While the cigarette industry argues that tobacco farming is a major contributor to the country's economy, the seriously damaging health and environmental impacts caused by tobacco farming have been well documented. From the moment the tobacco seed is planted to the time the tobacco plant is harvested and cured, the health of those who cultivate the crop is constantly at risk.

The hazards posed by tobacco cultivation place tobacco workers at increased risk of injury and illness. Children and adults (mainly women) working with tobacco frequently suffer from green tobacco sickness (GTS), which is caused by dermal absorption of nicotine from contact with wet tobacco leaves. GTS is characterized by symptoms that may include nausea, vomiting, weakness, headache, dizziness, abdominal cramps, and difficulty in breathing as well as fluctuations in blood pressure and heart rate. Large and frequent applications of pesticides to protect the plant from insects and diseases can cause poisoning, skin and eye irritations and other disorders of the nervous, respiratory systems as well as kidney damage.

Tobacco growing also causes a lot of damage to the environment. In many developing countries wood is used as fuel to cure tobacco leaves and to construct curing barns. An internationally estimated 200,000 hectares of forests and woodlands are cut down each year because of tobacco farming. Environmental degradation is also caused by the tobacco plant, which leaches nutrients from the soil as well as pollution from pesticides and fertilizers applied to tobacco fields.

Reference Number: 10111

MILLS, S.D., R. KURTZMAN, S.D. GOLDEN, A.Y. KONG, AND K.M. RIBISL. "Cultivating New Directions: The Changing Role of Tobacco in North Carolina's Economy." *North Carolina Medical Journal*, 79: 30–33, 2018 (29 references)

Abstract: In North Carolina, 14,200 people have died from smoking in 2014 and it is estimated that tobacco causes $3.8 billion in annual health care costs and $4.2 billion in productivity loss. The tobacco industry has prevented tobacco control measures and increases in tobacco taxes (NC is fifth lowest tobacco tax in the US) and foster the belief that tobacco is central to NC's economy. But the number of tobacco farms in North Carolina have dwindled from 150,000 in 1954 to 18,000 in 1992 and 1,682 in 2010. This decline in tobacco farms is due in part to the Tobacco Transition Payment Project or tobacco buyout, which provided payments to tobacco farmers from 2005–14 to help them transition from the federally regulated tobacco quota program. The majority of NC total farm income in 2015 comes from dairy, livestock, and poultry. Although tobacco was the leading cash crop ($647 million), it was followed closely by soybeans ($572 million), corn ($491 million), and sweet potatoes ($342 million).

North Carolina is disproportionately burdened by tobacco use and the state lag behind in key areas of tobacco control. Tobacco's role in North Carolina's agriculture and manufacturing sectors have been declining due to changing state and federal regulations as well as automation and outsourcing of tobacco farming. As a result, small tobacco farms are less common in the state and other commodities have become more dominant. Despite tobacco industry claims otherwise, enhancing state tobacco control is unlikely to significantly harm the economy and provides health and economic benefits. Raising the tobacco excise tax and increasing funding for tobacco control programs could raise government revenue and reduce tobacco-related health costs. As industries change in the state, North Carolina should consider the education and training needs of new industry professions and their impact on all North Carolinians.

Reference Number: 10112

HENDLIN, Y.H., AND S.A. BIALOUS. "The Environmental Externalities of Tobacco Manufacturing: A Review of Tobacco Industry Reporting." *Ambio*, 49: 17–34, 2020 (5 tables, 4 figures, 125 references)

Abstract: Tobacco smoke emission alone from the burning of 6.25 trillion cigarettes annually results in an estimated 3,000–6,000 tons of formaldehyde, 12,000–47,000 tons of nicotine, and a large amount of greenhouse gas (CO_2, methane, and nitrous oxides) as well. Manufacturing is the most environmental destroying step of tobacco production and constitutes 43 cents out of every dollar of tobacco cost, whereas only 4 cents goes towards purchasing of tobacco leaf. It is estimated that the CO_2e emission is 5.72 grams per cigarette, resulting in 39.4 million tons of CO_2e, which is similar to the CO_2e of major oil companies (e.g., Chevron emitted 66 million tons of CO_2e in 2016).

Table: Reported CO_2e Emissions from Tobacco Manufacturing

Company	Thousand of tons CO_2e
Altria (2016)	392
British American Tobacco (2016)	862
Imperial (2016)	329
Japan Tobacco Incorporated (2016)	6,513
PMI (2015/2016)	5,690
RAI (2015)	244
Global total	30,958,904

Source: Adapted from Hendlin and Bialous, 2020

As the 2017 WHO report on the environmental impact of tobacco concludes, "the adage 'there is no such thing as a safe cigarette' could be extended to assert that there is no such thing as an environmentally neutral tobacco industry." Especially, if these companies adhered to Trucost accounting which incorporates environmental externalities (water use, air pollution, land degradation, etc.), tobacco would not be a profitable industry. Yet, until the tobacco industry is required to internalize its social and environmental harms, citizens, governments, future generations, and the earth are subsidizing the profits these companies reap. While for some products this trade-off may be judged acceptable in exchange for the goods an industry provides to society, tobacco provides no such social good, and deserves a utility calculus accounting for all of its ranging harms, including environmental ones.

Reference Number: 10113

KULI, M.C., S.A. BIALOUS, S. MUNTHALI, AND W. MAX. "Tobacco Growing and the Sustainable Development Goals, Malawi." *Bulletin of the World Health Organisation*, 95: 362–367, 2017 (50 references)

Abstract: The seven largest tobacco producers in the world are China, Brazil, India, the United States, Indonesia, Zimbabwe, and Malawi. Malawi is the world's largest producer of burley leaf tobacco and is one of the poorest countries in the world.

The tobacco production process has negative health impacts on tobacco workers. Each day, a tobacco worker who plants, cultivates, and harvests tobacco may absorb as much nicotine as found in 50 cigarettes. In consequence, tobacco workers may suffer from the nicotine poisoning known as green tobacco sickness. Given the common physical requirements of tobacco production, it also seems likely that tobacco workers in Malawi suffer similar musculoskeletal injuries to those recorded among tobacco workers in the United States.

Tobacco farming is particularly devastating for the farmers' children. They work in the fields to help their families increase production and are at risk from green tobacco sickness, being injured, and/or missing days of school because of illness or farming needs. Children are particularly vulnerable to green tobacco sickness because of their size relative to the amount of nicotine absorbed. Their poor schooling makes it impossible for them to break out of the cycle of poverty. Although Malawi's Employment Act of 2000 prohibits children younger than 14 years from working, it is not enforced. It has been estimated that 80,000 children work in tobacco farming in Malawi. Poor education is common among children in rural Malawi: although 90% of rural children from all wealth quintiles attend primary school, only 8.8% attend secondary school and the corresponding rates for the lowest wealth quintile are even lower, at 83.2% and 3.2% respectively. Child labour itself is destructive for those concerned, and it has been estimated that the tobacco industry in Malawi annually saved the equivalent of $10.7 million United States dollars (US$) through unpaid child labour from 2000–2010.

Environmental effects

Tobacco cultivation also leads to large-scale environmental degradation. Even though most of Malawi's tobacco is air-cured, the annual cost of tobacco-related deforestation in Malawi has been estimated at US$ 6.4 million—based on the hectares of land cleared for tobacco production. Such costs are incurred despite the tobacco industry's reforestation programmes—which often involve the planting of non-native species to be used in the flue-curing of tobacco. The pesticide use that accompanies tobacco production caused environmental pollution that reduces animal and plant biodiversity and damages farmers' health. The tobacco industry has recommended that Kenyan tobacco farmers make as many as 16 applications of pesticide during the first few months of seedling production.

Reference Number: 10114

LI, V.C., AND S. TANG. "China's New Road for Tobacco Control: Tobacco Crop Substitution [Editorial]." *American Journal of Public Health*, 108: 1316–1317, 2018 (7 references)

Abstract: In 2005, China ratified the Framework Convention of Tobacco Control (FCTC). To increase food security, China has replaced and encouraged the conversion of tobacco acreage to food crops.

China is the first country in the world to have a national policy on tobacco crop substitution. This policy has resulted in the reduction of tobacco acreage nationwide, thereby reducing the supply of tobacco. The alliance between public health and agriculture has ushered in a new era of tobacco control for China. The critical lesson from the Yuxi pilot project is that farmers, many of whom are semiliterate, could learn the knowledge and skills necessary to operate a commercial-industrial entity that gives them a higher income that tobacco farming allows. Training farmers in skills for crop substitution and ensuring them higher earnings than are attainable through growing tobacco, coupled with political will, are essential ingredients for broad-based crop substitution.

When the FCTC was ratified, crop substitution pertained mostly to economic analysis and was yet to be proven an effective strategy. Tobacco crop substitution's time has come. We urge the FCTC to add it to its set of effective strategies for tobacco control. In light of the need for global food security to feed a growing population, phasing out tobacco crops to expand food crops is a win-win.

Reference Number: 10115

REN, M., M. ZHANG, H. YANG, AND H. SHI. "Reducing the Nicotine Content of Tobacco by Grafting with Eggplant." *BMC Plant Biology*, 20: 285 (10 pp), 2020 (4 tables, 2 figures, 42 references)

Abstract: Nicotine is a stimulant and a potent parasympathomimetic alkaloid found in the nightshade family of plants (Solanaceae) and is the main active alkaloid in tobacco leaves. Reducing nicotine concentrations in the tobacco leaves may reduce the addiction of cigarettes. Nicotine is produced in the roots of the tobacco plant, so when the tobacco plant was grafted to eggplant roots, the nicotine concentration decreased by 95% after 65–80 days of growth. There was, however, no decrease in tar or carbon monoxide (CO) in the smoke of the tobacco/eggplant leaves.

> Our present grafting results show that the nicotine levels of grafted tobacco leaves decreased dramatically, while the botanical and agronomic characteristics of the tobacco leaves did not change significantly in the tobacco and eggplant grafting group. The contents of nornicotine and anabasine decreased and the contents of amino acids and precursors of alkaloids increased significantly in grafter tobacco. Higher levels of protein, starch, and total nitrogen and lower levels of sugar contents were also observed. The contents of NNN, NNK, NAT, total TSNAs and the nicotine of mainstream cigarette smoke decreased significantly.

Reference Number: 10116

PARK, S-J., H-S. LIM, K. LEE, AND S-J. YOO. "Green Tobacco Sickness Among Tobacco Harvesters in a Korean Village." *Safety and Health at Work*, 9: 71–74, 2018 (4 tables, 39 references)

Abstract: Green Tobacco Sickness (GTS) is an occupational disease in tobacco harvesters who develop nicotine poisoning from the absorption of nicotine through the skin. In Korea, there are an estimated 11,000 tobacco harvesters. Surveys were conducted on 20 male and 20 female tobacco harvesters and urine cotinine concentrations were measured. GTS occurred in 55% of the women and 20% of the men. The mean urinary cotinine concentration decreased from 500 ng/mL during working to 135 ng/mL after working ceased. GTS occurred mainly in July and August during harvesting. The mean half-life of nicotine is 2 to 2.5 hours after smoking but

4 to 5 hours when absorbed through the skin. Cotinine has a half-life of 18 to 24 hours.

> At present, cotinine has been shown to be the best available biomarker of nicotine exposure. Cotinine is the major nicotine metabolite, and an average of 72% of nicotine was converted to cotinine. The use of urine cotinine is illustrated in several circumstances where smoking status assessment is of interest. Such situations include evaluation of the impact of smoking cessation programs, monitoring of pregnancy and other groups at risk, assessment of occupational exposure to industrial pollutants, validation of phase I clinical trials, and the assessment of life insurance candidates.

Reference Number: 10117

SALEEON, T., W. SIRIWONG, H.L. MALDONADO-PÉREZ, AND M.G. ROBSON. "Green Tobacco Sickness Among Thai Traditional Tobacco Farmers, Thailand." *International Journal of Occupational and Environmental Medicine*, 6: 169–176, 2015 (3 tables, 21 references)

Abstract: The mature leaf of the traditional Thai tobacco (Nicotiana abacus) contains 3–4× the amount of nicotine found in Virginia type tobacco (Nicotiana tabacum) and thus the hazards of Green Tobacco Syndrome among workers who cultivate and harvest this type of tobacco may be increased. GTS occurs because the nicotine in the tobacco leaves readily passes through the skin, resulting in acute nicotine poisoning that can cause nausea, vomiting, headache, abdominal cramps, breathing difficulties, pallor, chills, and fluctuations in blood pressure and heart rate. A survey was conducted of 473 Thai traditional tobacco farmers from rural areas in Nan province. The presence of GTS was 17.9% in men and 27.5% in women. Smoking was protective against GTS, but exposure to environmental tobacco smoke (ETS) or living with a smoker were not.

> In conclusion, dermal exposure is a major risk for GTS. The results of the present study represent the first investigation studying the prevalence of GTS among Thai traditional tobacco farmers in Nan province, northern Thailand. The findings have indicated various contributing factors for GTS including working in wet conditions, doing activities such as watering and working with tobacco plants. Moreover, insufficient use of protective equipment may increase health symptoms related to GTS. For all these reasons, health education programs that discuss health risks exposure reduction are recommended.

Reference Number: 10118

SATORA, L., H. GOSZCZ, E. GOMÓŁKA, AND W. BIEDROŃ. "Green Tobacco Sickness in Poland." *Polskie Archiwum Medycyny Wewnetrznej*, 119: 184–185, 2009 (5 references)

Abstract: Green Tobacco Sickness (GTS) is an illness associated with exposure to nicotine from the green, growing tobacco plant and with the farming of it. Symptoms include vomiting, headache, and dizziness. A 27-year-old man who worked on a tobacco plantation for 14 hours breaking the tops of Virginia bright tobacco with his bare hands developed GTS. He did not use any protective equipment or decontamination. His symptoms appeared within three hours of work completion. At the hospital, his urine cotinine concentration was 869 ng/mL, which declined to 61 ng/mL on the third day, although he was a non-smoker. Rainy and humid days increase the risk of nicotine poisoning. About 96% of nicotine can be removed from the hands by washing.

> Tobacco workers perform numerous activities manually with no use of protective clothes. They do not use skin decontamination methods either. Moreover, the use of herbicide may confuse GTS with herbicide poisoning. Furthermore, if symptoms resolve spontaneously, subjects do not report to the doctor. It seems vital to launch a campaign on the use of protective measures and decontamination rules during work in tobacco farming.

Reference Number: 10119

FASSA, A.G., R.D. MEUCCI, N.S. FIORI, M.L.V. CARRETT, AND N.M.X. FARIA. "Urinary Cotinine in Tobacco Farmers in Southern Brazil." *Revista de Saúde Pública*, 52: 70 (8 pp), 2018 (4 tables, 11 references)

Abstract: Urinary cotinine concentrations were measured by gas chromatography (GC) in 582 tobacco workers (137 reported Green Tobacco Sickness (GTS) and 445 did not). GTS was defined as dizziness or headache together with nausea or vomiting after picking tobacco. Smokers had a higher urine cotinine concentration than non-smokers. The saturation point of urine cotinine concentration is about 15–20 cigarettes per day. The mean urine cotinine concentration was 158.7 ng/mL and 181.5 ng/mL for female and male workers with GTS, respectively, and 134.1 ng/mL and 154.8 ng/mL for female and male workers without GTS.

Table: Percentage of Men and Women Tobacco Workers who Suffered Symptoms after Harvesting Tobacco

Symptom	Male (%)	Female (%)
Weakness	33.1	57.8
Headache	19.4	42.2
Dizziness	14.7	38.9
Nausea	12.2	32.2
Palpitation	9.7	20.0
Vomiting	9.4	16.7
Abdominal pain	9.4	16.7

Source: Adapted from Fassa et al, 2018

The urinary cotinine measures exposure to nicotine up to its saturation point; while green tobacco sickness, affected by tolerance, indicates nicotine poisoning. Strategies to reduce nicotine exposure in tobacco production are needed. Mechanization could be an alternative, as long as it overcame the challenge of irregular terrain and did not affect the quality of the leaf. More studies are needed to evaluate the chronic effects of nicotine exposure.

Reference Number: 10120

DE OLIVEIRA, P.P.V., C.B. SIHLER, L. DE MOURA, D.C. MALTA, M.C. DE ALBUQUERQUE TORRES, S.M.C.P. LIMA, A.L.A. DE LIMA, C.E. LEITE, V.L. DA COSTA-E-SILVA, J. SOBEL, AND T.M. LANZIERI. "First Reported Outbreak of Green Tobacco Sickness in Brazil." *Cadernos Saúde Pública*, 26: 2263–2269, 2010 (3 tables, 1 figure, 21 references)

Abstract: Dermal absorption of nicotine from the tobacco plant (Nicotiana tabacum) can cause Green Tobacco Sickness, which results in nausea, vomiting, dizziness, and headache and afflicts mainly tobacco workers during harvest. Pesticide use is minimal during the harvest season. Brazil is the second largest tobacco producer in the world, with more than 190,000 family tobacco farmers. A comparison was conducted of 107 people with GTS and matched to controls without GTS. Urine samples were collected and urine cotinine concentrations were determined by high pressure liquid chromatography (HPLC). Among smokers there was no significant difference between persons with GTS and controls (811 ng/mL vs 1,293 ng/mL). Among non-smokers, persons with GTS had a urine cotinine concentration of 288 ng/mL compared to 156 ng/mL for non-smoking controls.

Table: Signs and Symptoms of Green Tobacco Sickness in Tobacco Workers in Brazil

Signs and Symptoms	Percent Affected
Dizziness	90%
Weakness	88%
Vomiting	88%
Nausea	82%
Headache	59%
Blurred vision	52%
Dyspnea	31%
Colic	26%
Tremors	20%
Diarrhea	18%
Skin irritation	14%
Fainting	12%

Source: Adapted from de Oliveira et al, 2010

In our study, we have observed that the case-patients were most commonly non-smokers rather than smokers. A study performed by Arcury et al. found that smoking may reduce dermic absorption of nicotine, which could be caused either by the vessel constriction due to tobacco consumption or the metabolic adaptation or acquired nicotine tolerance by tobacco consumers. This tolerance may be explained by the increase in nicotine receptors found in smokers which cause a non-responsive phase after binding of nicotine to receptors. This phenomena is called "taquiphilaxis" which is the introduction of low doses of toxic products protecting the organism from sudden administration of new doses in a short interval. However, if the exposure to nicotine is seriously exceeded as observed in occupational exposure, green tobacco sickness may be observed in smokers as well, in addition to the several effects of smoking on health, such as lung, trachea and bronchial cancer and heart disease.

Reference Number: 10121

ONUKI, M., K. YOKOYAMA, K. KIMURA, H. SATO, R.B. NORDIN, L. NAING, Y. MORITA, T. SAKAI, Y. KOBAYASHI, AND S. ARAKI. "Assessment of Urinary Cotinine as a Marker of Nicotine Absorption from Tobacco Leaves: A Study on Tobacco Farmers in Malaysia." *Journal of Occupational Health*, 45: 140–145, 2003 (5 tables, 1 figure, 39 references)

Abstract: Urinary cotinine concentrations were determined by HPLC in 77 male tobacco growing farmers and 38 healthy male controls in Kelantan, Malaysia. Among non-smokers, the median urinary cotinine concentration was 17.7 ng/mL in tobacco farmers and 0.0 in non-smoking controls. Among smokers who smoked >10 cigarettes a day, the median urinary cotinine concentrations were 1119.2 in tobacco farmers and 1194.5 ng/mL for smoking controls.

It was observed that, among non-smokers, urinary cotinine levels of tobacco farmers were significantly increased, suggesting the absorption of nicotine from tobacco leaves. Furthermore, those subjects with higher levels of urinary cotinine had runny eyes and blurred vision. These symptoms have not been observed in other studies; but the fact that they are reportedly caused by nicotine suggests they are related to GTS. As the urinary cotinine levels of non-smoking farmers were less than those of smokers consuming 1–10 cigarettes per day, these symptoms seem to be caused by very low levels of exposure, considering the short half-lives of nicotine and cotinine: 2 to 3 and 19.7 (range 11 to 37) hours, respectively. In the present study, no significant associations were observed between non-adjusted urinary cotinine levels and symptoms in farmers.

Reference Number: 10122

SALEEON, T., W. SIRIWONG, H.L. MALDONADO-PÉREZ, AND M.G. ROBSON. "Salivary Cotinine Levels as a Biomarker for Green Tobacco Sickness in Dry Tobacco Production Among Thai Traditional Tobacco Farmers." *Roczniki Państwowego Zakładu Higieny*, 67: 121–130, 2016 (6 tables, 28 references)

Abstract: Dry Thai traditional tobacco production involves non-Virginia type tobacco plants, which have 3–4× the nicotine level. The Thai tobacco processing process involves numerous exposures to nicotine for farmers such as picking tobacco leaves, curing the tobacco leaves, removing the stems of the leaves, rolling a bundle of leaves, cutting the leaves with a machine, placing the tobacco leaves on a bamboo rack, and spraying the leaves with a tobacco solution to adjust its color. Farmers are exposed to nicotine through the skin or inhalation and can develop Green Tobacco Syndrome. The salivary cotinine concentration of 20 male and 20 female tobacco workers (mean age 50.2 years), along with 40 control subjects, was determined by Nic-Alert saliva strips biweekly for 14 weeks. There was a strong correlation between headache, nausea, vomiting, and dizziness (common GTS symptoms) and salivary cotinine levels.

Table: Personal Protective Equipment Which Reduced Salivary Cotinine Concentrations in Tobacco Farmers

Personal Protective Equipment (PPE) Use that Lowers Salivary Cotinine Levels
• Long sleeve shirt • Long legged pants • Gloves • Boots • Mask • Dry suit (changing out of wet suit during work)

Source: Adapted from Saleeon et al, 2016

The Thai traditional tobacco process produces a final product of dried tobacco. Almost all of the farmers are located in the northern part of Thailand. The unique method and process are carried out by the Thai traditional tobacco farmers with intensive hard-work and family labor. The demographic characteristics of the study populations are representative of rural areas in the northern part of Thailand. Almost all of the farmers graduated from primary school and followed farming as the traditional occupation. The average age of farmers was quite high which is a big proportion of agriculturalists at the present time in the rural areas.

Reference Number: 10123

RIQUINHO, D. L., AND E.A. HENNINGTON. "Health, Environment and Working Conditions in Tobacco Cultivation: A Review of the Literature." *Revista Ciência & Saúde Coletiva*, 17: 1587–1600, 2012 (3 tables, 71 references)

Abstract: A review was conducted of health, environment, and working conditions in the cultivation of tobacco throughout the world. Between 1970 and 2000, the US output of tobacco declined from 20% to 8% of the global production. Tobacco production in the European Union declined from 14% to 7%, with similar declines in Japan. The estimated number of smokers is 1.3 billion, of which 84% live in developing countries. GTS of tobacco workers involving headache and abdominal cramps was first recorded in 1713 by Bernardino Ramozzini in Italy. Pesticides, which are used extensively in tobacco cultivation, can be absorbed through the mouth and skin or inhaled; because of this, children less than 15 years of age, who are more tolerant to the effects of pesticides, are often employed. In addition, pesticides that are banned in the United States and European Union are used. The tobacco production chain is a complex oligopoly in which a few companies control the farmers, resulting in working conditions that are often inadequate and unhealthy.

Table: Main Health Issues Associated with Tobacco Production

Health Issue	Signs and Symptoms
Green Tobacco Sickness (GTS)	Vomiting, nausea, dizziness, headache, abdominal pain, diarrhea, abnormal blood pressure, and heart rate. Visual complaints
Respiratory problems	Worsening pulmonary function, increased risk of emphysema
Musculoskeletal lesions	Trauma, various injuries
Pesticide poisoning	Altered nerve conduction velocity, postural sway, parkinsonism, and depression. Acute and chronic toxicity due to various products during tobacco growing. Organophosphate poisoning is predominant.

Source: Adapted from Riquinho and Hennington, 2012

This study sought to present and discuss the working conditions and health impact of tobacco cultivation through a review of the literature. A worldwide trend toward standardization of the production process of tobacco was observed, especially concerning the establishment of a direct agreement between the industry and the farmer. Whereas in the U.S. the production of tobacco has been carried out by "others"—European women, African Americans, and more recently male Latino migrants—, in developing countries the workforce is constituted by families. However, with the changes in family structure, especially resulting from the decline in fertility rates, including in rural Brazil, the question remains of how tobacco production will be organized in the future—whether there will be a reduction in this crop or a shift towards the hiring of daily workers with worse working conditions, considering that mechanization is limited and a significant number of workers is required.

Reference Number: 10124

YOSHIOKA, D., H. ISHII, Y. HATANO, K. UMEKI, N. SAKAMOTO, R. SHIRAI, S. FUJIWARA, AND J. KADOTA. "A Case of Eosinophilic Pneumonia in a Tobacco Harvester." *Allergology International*, 60: 551–554, 2011 (3 figures, 12 references)

Abstract: A 60-year-old non-smoking woman presented at hospital after developing a persistent cough, breathlessness, and general malaise after working 2 months sorting tobacco leaves. She always wore waterproof gloves, but no mask. She was diagnosed with eosinophilic pneumonia and treated successfully with corticosteroids.

In conclusion, we herein describe a rare case of eosinophilic pneumonia in a tobacco harvester, which was likely caused by tobacco leaf exposure through the airway while the patient was harvesting and sorting tobacco leaves. Clinicians should be aware of the possibility of eosinophilic pneumonia as an occupational illness of tobacco leaf harvesters.

Reference Number: 10125

FASSA, A.G., N.M.X. FARIA, R.D. MEUCCI, N.S. FIORI, V.I. MIRANDA, AND L.A. FACCHINI. "Green Tobacco Sickness Among Tobacco Farmers in Southern Brazil." *American Journal of Industrial Medicine*, 57: 726–735, 2014 (4 tables, 18 references)

Abstract: Tobacco is cultivated in more than 100 countries, but China, Brazil, India, the United States, Malawi, and Indonesia account for two-thirds of the production. Brazil is the second leading tobacco producing country in the world. A survey was conducted of 1,464 male and 1,005 female tobacco workers who worked at least 15 hours a week in Brazil during 2011. Green Tobacco Sickness was characterized as the occurrence of dizziness, headaches, nausea, or vomiting within 2 days after tobacco harvesting. The GTS prevalence was 6.6% among men and 11.9% among women.

Among men, age, being a non-smoker, hanging tobacco sticks in the barn, harvesting wet leaves, and exposure to physical exertion were risk factors for GTS. Among women, tying bales of tobacco, transporting bales, harvesting wet leaves, having had contact with pesticides, and exposure to physical exertion were positively associated with GTS.

Reference Number: 10126

FIORI, N.S., A.G. FASSA, N.M.X. FARIA, R.D. MEUCCI, V.I. MIRANDA, AND D.C. CHRISTIANI. "Wheezing in Tobacco Farm Workers in Southern Brazil." *American Journal of Industrial Medicine*, 58: 1217–1228, 2015 (4 tables, 42 references)

Abstract: Nicotine is absorbed by the skin and reaches the lungs without the first pass effect that occurs with ingestion, reducing its bioavailability. Physical exercise in hot weather increased nicotine's dermal absorption via vasodilation and increased blood flow to the skin, increasing heart rate and sweating. A cross-sectional study was conducted of wheezing prevalence in 2,469 tobacco family farming growers in southern Brazil. Tobacco leaf harvesting was undertaken by more than 90% of the workers of both genders. Wheezing prevalence was 11% with no difference between the genders. Wheezing was associated with age, smoking, strenuous work, pesticide use, contact with vegetable dust and dried tobacco dust, lifting sticks with tobacco leaves to curing barns, harvesting the lower leaves (which contain the highest nicotine concentration), and GTS.

The findings highlight the importance of broadening perspectives about nicotine exposure considering its multiple source of absorption and variable intensity of exposure. GTS is an indicator of heavy exposure to nicotine, however intense exposure might also course asymptomatic due to tolerance mechanisms.

These results provide an important support to draw actions to accomplish Article 18 of the Framework Convention on Tobacco Control regarding protection of the people and the environment. Mechanization of the harvesting and baling process could reduce tobacco farmers' exposure to nicotine. Often, farmers do not use personal protective equipment (PPE) or use PPE that has lost its effectiveness or is not certified to reduce nicotine exposure. Both developing efficient PPE and also ensuring its use in hot weather conditions in the harvesting season continue to be challenges. Article 17 of the same Convention addresses the need of governments to promote economically viable alternatives to tobacco plantation. However, promoting crop diversification is not enough to protect farmers' health, a sustainable agricultural production model with decreased pesticide use is also necessary.

Reference Number: 10127

YOO, S-J., S-J. PARK, B-S. KIM, K. LEE, H-S. LIM, J-S. KIM, AND I-S. KIM. "Airborne Nicotine Concentrations in the Workplaces of Tobacco Farmers." *Journal of Preventive Medicine and Public Health*, 47: 144–149, 2014 (3 tables, 3 figures, 27 references)

Abstract: Airborne nicotine concentrations were determined at tobacco farms in rural South Korea using XAD-4 sorbent tubes and analysis by GC at the laboratory. The workplace threshold for nicotine is 0.5 mg/m^3. The mean ambient nicotine concentration was 93.3 mg/m^3 in the lower region of the tobacco field and 83.4 mg/m^3 in the upper field. The personal sample of one farmer was 150 mg/m^3.

Table: Ambient Nicotine Concentrations at Various Locations on a Tobacco Farm

Location	Mean Nicotine Concentration (mg/m^3)
Private curing barns	323.7
Workers	121.0
Front yard of curing barns	73.7
Joint curing barn	610.3
Schoolyard (control)	Not detected

Source: Adapted from Yoo et al, 2014

Until recently, GTS has been known globally as a disease that occurs due to the absorption of nicotine through the skin. However, recent studies conducted by us and others have suggested that absorption may also occur through the respiratory system. Airborne nicotine concentrations in a private curing barn as well as in the tobacco field were very high; therefore, these tobacco farmers may have been exposed to nicotine during harvesting and while curing the tobacco leaves.

Reference Number: 10128

MCKNIGHT, R.H., AND H.A. SPILLER. "Green Tobacco Sickness in Children and Adolescents." *Public Health Reports*, 120: 602–605, 2005 (1 table, 27 references)

Abstract: In 2004 some 5.73 million metric tons of tobacco were harvested in 100 countries worldwide. The top five producers were China, Brazil, India, the United States, and Malawi. Green Tobacco Sickness is an occupational poisoning associated with tobacco farming due to nicotine exposure through the skin, lungs, nose, and eyes. The two main farming tasks that lead to GTS are harvesting and topping (removing the buds or flowers of the tobacco plant). Tobacco harvesting in a wet field should be equated to working in a field recently sprayed with pesticides—both should be avoided.

GTS prevention strategies recommended for adults could be adapted for children. Protective clothing such as rain gear and water-tight gloves, measures that have been cited in other studies as precautions adult tobacco workers should adopt, could also be adopted by child workers. An obvious requirement of wearing such impermeable gear, however, would be evaluating the dangers workers face from heat exhaustion and dehydration in regions with high heat and humidity. Additionally, the U.S. Environmental Protection Agency's Worker Protection Standard could be revised to address nicotine toxicity in wet tobacco fields.

Clinicians and public health officials and agencies need to become more knowledgeable about the causes of GTS and its signs and symptoms, as the effects of acute nicotine poisoning might be mistaken for those of pesticide poisoning or heat exhaustion.

Reference Number: 10129

GANGULY, K., B. LEVÄNEN, L. PALMBERG, A. ÅKESSON, AND A. LINDÉN. "Cadmium in Tobacco Smokers: A Neglected Link to Lung Disease?" *European Respiratory Review*, 27: (8 pp), 2018 (1 figure, 60 references)

Abstract: Cadmium is a major concern for public health and this toxic metal is hyper-accumulated by the tobacco plant into the leaves. The cadmium concentration in the tobacco plant has been found to range between 1 and 2 μg/g dry weight, independent of its soil content. This would result in approximately 0.5-1 μg of cadmium per cigarette. Cadmium oxide generated during tobacco smoking is deposited in the lungs and absorbed into the systemic blood circulation. Among active smokers, the mean blood cadmium concentration was 1.04 μg/L in those with obstructive lung disease and 0.70 μg/L in those without obstructive lung disease.

> Local accumulation of cadmium in the lungs appears to be a critical component of predisposition to lung diseases among long-term smokers. This may be particularly important considering that the biological half-life of cadmium in the human body is >25 years, a substantial period of time, suggesting the possibility of significant retention of cadmium in the lungs of long-term smokers. Moreover, it is plausible that retention of cadmium in the lungs and other organs will inevitably influence intracellular signalling. This may in turn result in impaired function of host defence including innate immunity, possibly contributing to the enhanced susceptibility to bacterial colonisation and infections leading to chronic inflammation, fibrosis and emphysema. These events may in turn lead to impaired ventilatory lung function as well as reduced gas exchange, key clinical characteristics in severe COPD among long-term smokers.

1.02 ADDICTION

> "When you smoke, the nicotine changes the brain—that's why quitting is so hard."
>
> —*US v Philip Morris USA Inc*, RICO decision (2006)

Addiction to cigarettes and other tobacco products such as electronic cigarettes or cigars is mainly due to nicotine. If it were not for the rapid delivery of nicotine to the brain by inhalation, no one would be smoking or vaping. Nicotine stimulates the release of dopamine in the mesolimbic system, which is pleasurable, and it changes the neuron structure of the brain after long-term use (10202, 10214). One of the major differences between nicotine and alcohol or THC is the lack of a prolonged overt high or intoxication (10212). This euphoric effect of nicotine is more pronounced after an overnight sleep, when the nicotine levels in the blood are low (10212).

Nicotine causes physical tolerance, withdrawal, loss of control, and craving, as do other addictive drugs such as heroin (10201, 10205, 10206).

The tobacco industry has always sought to minimize the addictiveness of nicotine using various means, including the testimony of tobacco company CEOs under oath that nicotine is not addictive and attempts to influence the American Psychiatric Association (10212). The faster nicotine is absorbed into the brain the more addictive it is. Cigarette smoking allows nicotine to be absorbed faster than even IV injection (10213).

A good illustration about the addictiveness of nicotine is seen in one experiment where mice exposed to smoke in a chamber on the first day refused to enter the exposure chamber on the second day, whereas mice exposed to the same smoke with nicotine added on the first day entered the exposure chamber on the second day without hesitation (10204).

Reference Number: 10201

ROH, S. "Scientific Evidence of the Addictiveness of Tobacco and Smoking Cessation in Tobacco Litigation." *Journal of Preventative Medicine and Public Health*, 51: 1–5, 2018 (23 references)

Abstract: A general review was presented of the addictiveness of nicotine. The WHO reports that tobacco causes 6 million deaths a year globally, including 600,000 non-smokers due to secondhand smoke. The tobacco industry insists that smokers choose to smoke, according to their right to self-determination, and that, with motivation and willpower, they can easily stop smoking. The change in nicotine receptors in the brain due to continuous smoking leads to addiction.

Tobacco causes physical tolerance, withdrawal, loss of control, craving, and compulsive smoking behaviors. Such changes occur proportionally to the amount and duration of smoking. Therefore, the argument made by the tobacco companies that anyone can succeed in smoking with a will to quit and self-confidence is flawed and is not consistent with the scientific evidence on nicotine's influence on the brain. In conclusion, tobacco alters the human brain and causes an addictive disorder called tobacco use disorder. Smokers become dependent on nicotine regardless of their will.

Reference Number: 10202

PICCIOTTO, M.R., AND Y.S. MINEUR. "Molecules and Circuits Involved in Nicotine Addiction: The Many Faces of Smoking." *Neuropharmacology*, 76: (22 pp), 2014 (1 table, 156 references)

Abstract: This article reviews the neural basis of nicotine addiction. Tobacco smoking is one of the most persistent and widespread addictions and is driven by the nicotine in tobacco smoke. Nicotine stimulates dopamine (DA) release from neurons in the mesolimbic system. Smokers use tobacco because it is pleasurable, controls appetite, and helps with mood. Weight control is the primary reason for tobacco use in teenage girls. The rate of smoking in depressed persons is 2× that of the average population. Maternal smoking is associated with ADHD in children.

There is an essential paradox at the heart of smoking behavior. Among humans, tobacco is a highly addictive drug, however, in animal models, the primary addictive component of tobacco, nicotine, is less reinforcing than some other drugs of abuse, such as cocaine. There are likely to be many reasons for this discrepancy. First, tobacco is legal and access is almost universal, so there is more chance of exposure to tobacco than to illicit drugs, and the stigma of use may be lower. Second, there are more than 4,000 constituents of tobacco smoke, and other constituents in addition to nicotine may contribute to tobacco use. Third, and perhaps most important, nicotine has many effects on brain circuits and behavior in addition to its ability to stimulate neuronal systems involved in primary reinforcement, and the complex actions on these systems may contribute to smoking behavior and relapse in human tobacco users.

Reference Number: 10203

CHEN, A., M. MACHIORLATTI, N.M. KREBS, AND J.E. MUSCAT. "Socioeconomic Differences in Nicotine Exposure and Dependence in Adult Daily Smokers." *BMC Public Health*, 19: 375 (10 pp), 2019 (4 tables, 51 references)

Abstract: Socioeconomic status (SES), cigarette use, and nicotine dependence was determined in 352 adult smokers (mean age 37.6 years). Oral fluid concentrations of nicotine metabolites cotinine, 3-hydroxycotinine (3-HC), and total metabolites were measured by GC/MS. The average income level of lighter smokers (<20 cigarettes per day (CPD)) was $10,000 more than that heavier smokers. Higher Fagerstrom Test for Nicotine Dependence scores were associated with lower income and job status, ranging from 5.4 in the unemployed, 4.4 in blue collar, and 3.8 in white collar smokers.

Table: Smoking Variables and Percentage Among the Unemployed, Blue Collar, and White Collar Smokers

Smoking Variable	Unemployed	Blue Collar	White Collar
<20 CPD	69.4%	79.0%	90.6%
>20 CPD	30.6%	21.0%	9.4%
Menthol cigarette smoker	44.4%	43.0%	51.2%
Non-menthol cigarette smoker	55.6%	57.0%	48.8%

Source: Adapted from Chen et al, 2019

In the current study, lower SES was associated with a higher number of cigarettes smoked per day, which is consistent with previous studies. We also found that cigarettes smoked per day was highest among the unemployed. Little is known about the smoking habits of the unemployed. In the Framingham Heart Study Offspring Cohort, men but not women who smoked and became unemployed increased their daily cigarette consumption. These data indicate that unemployment or the stressors and lack of support structures associated with unemployment has an effect on cigarette consumption independent of low income.

Reference Number: 10204

DRAZEN, J.M., S. MORRISSEY, AND E.W. CAMPION. "The Dangerous Flavors of E-Cigarettes." *The New England Journal of Medicine*, 380: 679–680, 2019 (11 references)

Abstract: Nicotine is very addictive; in one experiment, mice exposed to smoke refused to enter the exposure chamber the following day, whereas mice exposed to the same smoke with nicotine entered the exposure chamber without hesitation, almost eagerly. E-cigarettes (EC) are electronic delivery devices for nicotine, and the addition of flavors makes the initial exposure to smoking more pleasurable, especially to teenagers. Altria, a large producer of cigarettes that has nearly 250,000 retail outlets in the United States, has acquired one-third of JUUL, a company that produces a brand of e-cigarettes which is especially popular among adolescents.

We think the FDA should simply ban the sale of flavored nicotine products for use in e-cigarettes. The public health problem that e-cigarettes can help solve—by helping people who are users of combustible tobacco products stop smoking by switching to vaping—is adequately addressed by liquids that are not flavored to appeal to adolescents. We urge the FDA to use its statutory powers in regulating nicotine delivery devices to take the bold step of removing these flavored products from the market.

Reference Number: 10205

HATSUKAMI, D.K., L.F. STEAD, AND P.C. GUPTA. "Tobacco Addiction." *Lancet*, 371: 2027–2038, 2008 (10 tables, 154 references)

Abstract: An overview of tobacco addiction and its treatment is presented. The primary difference between nicotine addiction and addiction to other drugs such as alcohol or cannabis is the lack of an overt intoxication or high. Nicotine addiction should be treated as a chronic disorder for which relapse is a likely occurrence. Cigarette reduction of 50% or greater results in only a 30% reduction in biomarkers for toxicant exposure, as smokers compensate for the reduced number of cigarettes. In addition, a 50% reduction in number of cigarettes does not reduce the risk of fatal and nonfatal myocardial infarction and does not reduce hospitalizations for Chronic Obstructive Pulmonary Disease (COPD). The risk of cancer, however, is reduced by 25%.

Table: Criteria for Nicotine Dependence based on the Diagnostic Statistical Manual Version IV

DSM-IV
A maladaptive pattern of substance use, leading to clinically significant impairment or distress, as manifest by three (or more) of the following, occurring at any time in the same 12-month period. • Tolerance (diminished effect with continued use of same amount) • Withdrawal • Substance often taken in larger amounts over a longer period than intended • Persistent or unsuccessful efforts to cut down or control substance use • Great deal of time spent in activities necessary to obtain the substance or recover from its effects. • Important social, occupational, or recreational activities given up or reduced because of substance use • Substance use continued despite knowledge of having a persistent or recurrent physical or psychological problem likely to have been caused or exacerbated by the substance

Source: Adapted from Hatsukami et al, 2008

The 1988 Surgeon General's report, *The Health Consequences of Smoking: Nicotine Addiction*, described three major conclusions: 1) cigarettes and other forms of tobacco are addictive, 2) nicotine is the drug that causes this addiction, and 3) the pharmacological and behavioral processes that determine tobacco addiction are similar to those that determine addiction to drugs such as heroin and cocaine.

Reference Number: 10206

RUSSELL, M.A., M.J. JARVIS, G. JONES, AND C. FEYERABEND. "Non-Smokers Show Acute Tolerance to Subcutaneous Nicotine." *Psychopharmacology*, 102: 56–58, 1990 (2 figures, 12 references)

Abstract: Four male and two female subjects (ages 29–54 years), including one occasional smoker and two former smokers, were injected with nicotine subcutaneously as a bitartrate salt at a mean dose of 0.75–1.0 mg nicotine. The mean peak plasma nicotine concentration was 8.5 ng/mL at 15 minutes after the injection. The heart rate increased with plasma nicotine concentration by an average of 11 beats per minute at 10 minutes after injection. Five subjects felt light-headed and slightly dizzy for about 10–15 minutes. One subject felt a slight coldness in the extremities and one subject who had never smoked felt nausea.

> Our demonstration of acute tolerance in the three subjects who had never smoked serves to emphasise that this form of tolerance is part of the primary action of nicotine at certain sites. Unlike chronic tolerance, it is not an acquired phenomenon dependent on prior exposures and the development of adaptive changes. Acute tolerance was first demonstrated a century ago in the superior cervical ganglion of the cat, where the initial agonist effect of nicotine at post-synaptic cholinergic receptors was followed by synaptic blockade due to a form of prolonged depolarisation, termed tachyphylaxis. This form of acute tolerance is relatively short-lived and is no longer evident after nicotine has been cleared from the body and hence from nicotinic receptors.

Reference Number: 10207

PROCHASKA, J.J., AND N.L. BENOWITZ. "Current Advances in Research in Treatment and Recovery: Nicotine Addiction." *Science Advances*, 5: (23 pp), 2019 (2 tables, 235 references)

Abstract: Nicotine addiction in the form of cigarette smoking causes more harm to the public than any other drug addiction. More than 90% of smokers try to quit and most make multiple attempts. About one-half of smokers quit smoking long term, although most do not obtain total abstinence until after the age of 30 years. Nearly 90% of cigarette smokers start before the age of 18 years. Persons who begin smoking in their teens are more likely to become long-term smokers. Early onset smoking may predispose the person to depression, anxiety disorders, and schizophrenia. But the tobacco industry funds research into the idea that people with psychiatric disorders need to self-medicate with tobacco in order to lessen their symptoms. All forms of nicotine replacement therapy (NRT) increase quit rates by 50 to 100% compared to behavioral treatment alone.

Compliance with NRT is the greatest with patches, lower with gum and lozenges, and lower still with sprays and inhalers. All NRT products have an alkaline pH to lower faster oral absorption. Coffee, citrus juices, sodas, and alcoholic beverages, which are all acidic, should be avoided for 10 minutes or more before use. Varenicline is the most effective smoking cessation therapy and is a partial agonist of nicotine. It both activates the receptors (50% of the maximal effect of nicotine) and blocks nicotine from the receptors, which reduces the withdrawal symptoms and lessens the satisfaction of smoking.

> Positive psychoactive effects of nicotine include pleasure, stimulation, and mood modulation, with reduced anxiety and stress. A smoker often reports pleasure and stimulation with the first cigarette of the day, stimulation and increased concentration from smoking during repetitive tasks during the day, and relaxation at times of stress and at bedtime. However, tolerance develops to many of nicotine's effects such that even within the day, the pleasure experienced from each cigarette diminishes. As nicotine levels decline, withdrawal symptoms develop, reversing nicotine's positive effects. Thus, an abstinent smoker may feel anxious, irritable, and depressed and have problems concentrating. Hedonic dysregulation (a reduced ability to experience pleasure) may be experienced, presumably related to a relative deficiency of dopaminergic activity. Nicotine increases metabolic rate and suppresses appetite, resulting in smokers, on average, weighing less than nonsmokers. During nicotine withdrawal, smokers typically experience hunger and gain weight.
>
> Some of the perceived benefits of nicotine are mediated by the reduction of adverse effects of nicotine withdrawal (termed negative reinforcement). Thus, the pharmacologic role of nicotine in addiction is a combination of providing positive and negative reinforcement.

Reference Number: 10208

JOHN, U., J. RIEDEL, H-J. RUMPF, U. HAPKE, AND C. MEYER. "Associations of Perceived Work Strain with Nicotine Dependence in a Community Sample." *Occupational and Environmental Medicine*, 63: 207–211, 2006 (3 tables, 38 references)

Abstract: A study was conducted of 2,128 adults (aged 18–64 years) living in Lubeck, Germany, who worked 35+ hours per week and of 421 adults who worked 15–34 hours per week. Self-reported smoking status was never (36.6%), former (23.5%), and current daily smokers (39.2%). Among the current smokers, 26.0% were nicotine dependent. Work strain as defined as a high work demand and low work control was assessed with a questionnaire. Subjects with work strain were 1.6× as likely to be nicotine dependent, although current daily smoking was unrelated to work strain.

> We conclude that nicotine dependence is related to work strain, although no causal relations could be determined because of the limitations of the study design. One of the most plausible reasons for the relation may be that both work strain and nicotine dependence provoke physical arousal which the individual seeks to sedate by smoking. Nicotine dependence might function as an effect modifier in the relation between stress and somatic disease—for example, in cardiovascular disease. In this process, nicotine dependence and work strain may function as mutually reinforcing factors. Nicotine dependence plus work strain might be a risk factor particularly predictive of cardiovascular and other diseases. Consequently, workplace smoking bans should help to avoid such diseases. This confirms the finding that the enactment of a law against smoking in workplaces coincided with a significant reduction of hospital admissions of acute myocardial infarction. Public health interventions at the workplace should consider the work strain–nicotine dependence relation—for example, by including alternative ways to cope with arousal or by adequate smoking cessation aids in addition to more general approaches, particularly workplace smoking bans. Workplace smoking bans may add to giving nicotine dependent smokers the opportunity to detect nicotine withdrawal. Our results show that this subpopulation of smokers may constitute a substantial part of the workforce.

Reference Number: 10209

DIFRANZA, J.R. "Can Tobacco Dependence Provide Insights into Other Drug Addictions?" *BioMed Central Psychiatry*, 16: 365 (11 pp), 2016 (3 tables, 78 references)

Abstract: Nicotine is among the least rewarding of all addictive drugs, as only 20% of first-time users find smoking relaxing. But smoking provides an image and a shared social activity to youth, which keep smokers motivated until nicotine dependence takes over. Smoking two cigarettes per week in adolescence increases the risk of heavy adult smoking by 174×. The three phases of physical dependence of nicotine are wanting, craving, and needing. Unlike alcohol, there is no binge smoking. The number of cigarettes smoked per day is typically remarkably uniform for a regular smoker. Of all other drugs, opiate addiction most closely resembles that of nicotine.

> The natural history of tobacco dependence is now fairly well established, at least in regard to its onset. As tobacco does not produce intoxication, it is clear that intoxication is not an essential quality for an addictive drug. Even without the allure of intoxication, the continuation rates for tobacco use are very high in comparison to other addictive substances: an adolescent would need to smoke only four cigarettes to have a 90% chance of smoking for decades. Whether it is measured using medical or psychiatric criteria, the prevalence of dependence among tobacco users is very high in comparison to that for drugs such as marijuana and alcohol. The high continuation rates for tobacco use likely reflect the rapidity with which symptoms of dependence emerge. Until the earliest symptoms of dependence on other drugs are identified it will not be possible to determine how quickly dependence on other drugs can develop.
>
> No minimum in terms of amount or frequency of tobacco use has been identified that does not carry a risk of dependence for the most vulnerable individuals. Research on other substances might benefit from case studies focused on identifying the amount and frequency of use when users of those substances first felt "hooked." The only way to discover the natural history of the development of dependence is through the experiences of those who have lived through it. In the author's opinion, excessive skepticism regarding subjective symptoms has impeded progress. It was only through listening to the personal experiences of adolescents that the natural history of tobacco dependence was revealed.

Reference Number: 10210

BENOWITZ, N.L. "Nicotine Addiction." *New England Journal of Medicine*, 362: 2295–2303, 2010 (4 figures, 80 references)

Abstract: The chance that a life-long smoker will die prematurely from a tobacco-related illness is approximately 50%. Smoking is a highly efficient form of drug administration. Inhalation of smoke from a cigarette carries nicotine into the lungs which rapidly passes into the bloodstream and then quickly into the brain, where it binds to nicotinic cholinergic receptors which then release various neurotransmitters such as dopamine (which causes pleasure). With repeated nicotine exposure, the number of binding sites on the nicotinic cholinergic receptors increases dramatically and tolerance to some of the effects of nicotine develop. Nicotine withdrawal causes anxiety and stress, both of which are powerful incentives to continue smoking. Nicotine withdrawal can also cause irritability, depressed moods, restlessness, and anhedonia (the feeling that there is little pleasure in life, which also occurs in the withdrawal from other drugs of addiction).

> Nicotine sustains tobacco addiction, a major cause of disability and premature death, by acting on nicotinic cholinergic receptors in the brain to trigger the release of dopamine and other neurotransmitters. Release of dopamine, glutamate, and GABA is particularly important in the development of nicotine dependence, and CRF may play a key role in withdrawal. Neuroadaptation and tolerance involve changes in nicotinic receptors and neural plasticity. Nicotine addiction occurs when smokers come to rely on smoking to modulate mood and arousal, relieve withdrawal symptoms, or both. Light or occasional smokers smoke mainly for positive reinforcement in specific situations. Genetic studies indicate that nicotinic receptor subtypes and the genes involved in neuroplasticity and learning play a part in the development of dependence. People with psychiatric or substance-abuse disorders, who account for a large proportion of current smokers, have an increased susceptibility to tobacco addiction. Nicotine is metabolized primarily by the enzyme CYP2A6, and variation in the rate of nicotine metabolism contributes to differences in vulnerability to tobacco dependence and the response to smoking-cessation treatment.

Reference Number: 10211

HIRSHBEIN, L.D. "Politics, Profit, and Psychiatric Diagnosis. A Case Study of Tobacco Use Disorder." *American Journal of Public Health*, 104: 2076–2084, 2014 (72 references)

Abstract: In May 1976, cigarette giant Philip Morris was alerted of worrying developments in the psychiatric field regarding a newly considered diagnosis, originally called compulsive tobacco use. Such a new psychiatric diagnosis could dramatically affect cigarette sales and perhaps enable smokers to eventually quit more easily, removing more lifelong customers. The American Psychiatric Association (APA) published the Diagnostic and Statistical Manual (DSM)-III in 1980, which listed tobacco dependence as a disorder, despite another tobacco giant RJ Reynolds attempting to prevent tobacco's psychiatric diagnosis by seeking contact with both psychiatrists in tobacco-friendly US states and the president-elect of the APA. The Tobacco Institute and the tobacco industry law firms encouraged physicians and scientists to criticize the DSM-III and then cited them as a group of concerned scientists in their public announcements, creating doubt and controversy.

> As historian Allan Brandt has pointed out, the tobacco industry was a pioneer in what we now recognize as conflict of interest—industry leaders (and their attorneys) used company resources to organize allies and push their own agenda to limit regulation about smoking. But the example of tobacco use disorder is more than just another illustration of the devious actions of the tobacco industry. Instead, it highlights the political and commercial alliances that can promote—or undermine—laudable public health goals. In the case of tobacco dependence, those opposed to the diagnosis were both connected to the tobacco companies and in favor of a modest role for psychiatry in society. Those promoting the diagnosis included psychiatrists who in addition to wanting to help people were also interested in professional status and reimbursement. And the pharmaceutical industry wanted the medical backing to promote an increasingly large (and profitable) range of products.
>
> In the last several decades, the tobacco industry has been much less successful in influencing psychiatrist opinion about smoking. It is probably not an accident, then, that the industry has been more critical of psychiatry. In 1988, the same year that the surgeon general released a massive volume devoted to a review of the research on nicotine addiction, Philip Morris placed an advertisement entitled "Getting the Buttons in the Wrong Button Holes," which cited psychiatry critic Thomas Szasz. Philip Morris criticized Surgeon General C. Everett Koop for saying that smoking was the same as heroin or cocaine, and used an analogy attributed to Szasz in his critique of the psychiatric establishment—that the buttons (the evidence) were in the wrong holes (the conclusions) from the top down. Philip Morris attempted to echo Szasz's well-known criticism of the field for engaging in social control rather than medical treatment. And in the 1990s, when the FDA was contemplating regulation of tobacco products, tobacco industry attorneys convinced University of California, Los Angeles social welfare professor Stuart Kirk to testify on the problems of labeling dependence on nicotine or smoking as a disease. Kirk had written critically about the history of the

construction of the DSM-III, and argued that the contested and unscientific basis of DSM diagnoses meant that the FDA could not use them as objective facts on which to base its actions. Tobacco attorneys exploited psychiatry's critics to support their own arguments against the profession.

Reference Number: 10212

POMERLEAU, C.S., AND O.F. POMERLEAU. "Euphoriant Effects of Nicotine in Smokers." *Psychopharmacology*, 108: 460–465, 1992 (3 figures, 18 references)

Abstract: In a study of around 1,000 drug dependent persons, cigarettes were rated as less pleasurable than alcohol or other drugs. But nicotine has repeatedly been shown to produce electroencephalogram changes of cortical activation, whether delivered through IV or smoking. In one study, a total of 10 male smokers (aged 20–40 years) arrived at the lab at 8:00 am, after not having cigarettes or caffeine since midnight. In the first session, they were given an ultralow nicotine cigarette; in the second session, a high nicotine cigarette; and in the third session, one of their own usual brand cigarettes. In a follow-up study, 12 male smokers were given an ultralow, medium, and high nicotine cigarette over the course of three sessions. The subjects in both studies reported any euphoric sensations they experienced. For about 10% of the time the cigarette was smoked, starting at around 2.5 minutes after lighting up, 19 of the 22 total smokers reported at least one euphoric symptom.

Together, these studies support the existence of a dose-response relationship for nicotine-induced euphoric sensations; suggest that they are more pronounced following overnight abstinence than following minimal deprivation, and in more dependent smokers; and characterize in detail the temporal features of these sensations.

Reference Number: 10213

BENOWITZ, N.L. "Emerging Nicotine Delivery Products. Implications for Public Health." *Annals of the American Thoracic Society*, 11: 231–235, 2014 (34 references)

Abstract: A new oral tobacco product that dissolves in the mouth and requires no spitting has been marketed as Camel, Marlboro, and Skoal sticks, strips, and orbs. Dissolvable tobacco products are finely ground tobacco compressed into sticks, strips, and tablets (orbs). The product contains lower levels of tobacco-specific nitrosamines (TSNAs) than snus contain. The nicotine delivery of US snus and dissolvables is substantially lower than in traditional smokeless tobacco products, result in lower plasma nicotine concentration, and have small effects on the withdrawal symptoms in abstinent smokers. The dissolvable products have not been popular due to poor nicotine delivery, as they do not satisfy the cravings of addicted smokers.

> Nicotine establishes and maintains tobacco addiction by complex actions that affect the neurochemistry of the brain. Nicotine from cigarette smoke is rapidly absorbed in the lungs, from which it rapidly passes into the brain. The rapidity of absorption is an important determinant of the addictiveness of a drug, and cigarette smoking is the most rapid method of nicotine delivery. Nicotine exerts its effects by binding to nicotinic cholinergic receptors in the brain. Nicotine affects a number of neurotransmitter systems, including the dopamine system, which is critical in experiencing pleasure. Nicotine results in positive psychological effects including pleasure, arousal, and mood modulation.
>
> With chronic nicotine exposure, as is the case with addicted smokers, neuroadaptation occurs, such that more nicotine is required to deliver the same neurochemical effect. As the brain becomes tolerant, nicotine is needed to maintain normal brain functioning. In this context, stopping smoking is associated with deficient neurotransmitter release and associated withdrawal symptoms, including irritability, anxiety, problems getting along with family and friends, difficulty concentrating, increased hunger and eating, and weight gain. Nicotine addiction is thus sustained by a combination of positive effects related to pleasure and arousal and the need to continue to take nicotine to avoid the unpleasant effects of nicotine withdrawal.
>
> In addition to the pharmacological aspects of nicotine addiction, conditioning also plays an important role in tobacco addiction. Smoking often becomes associated with specific behaviors, such as drinking a cup of coffee, alcohol consumption, talking on the phone, driving a car, and/or after meals. Through conditioning these behaviors become cues for smoking and contribute to maintained smoking. Smoking also facilitates nicotine dependence through sensorimotor factors associated with the act of smoking. The factors include the smoking process and the smell, taste, and feel of the cigarette smoke. Denicotinized cigarettes can produce some smoking satisfaction and can reduce cigarette craving.

Reference Number: 10214

TIWARI, R.K., V. SHARMA, R.K. PANDEY, AND S.S. SHUKLA. "Nicotine Addiction: Neurobiology and Mechanism." *Journal of Pharmacopuncture*, 23: 1–7, 2020 (2 figures, 52 references)

Abstract: Nicotine is the major phytoconstituent of tobacco responsible for addiction. It was isolated from Nicotiana tabacum by Posselt and Reimann in 1828 and was named after Jean Nicot, who introduced and popularized the tobacco plant to the French court in 1560. The majority of nicotine (70–80%) is metabolized to cotinine. Cotinine is further metabolized to trans 3' hydroxycotinine (3-HC). Nicotine stimulates specific receptors (nAChR) which release acetylcholine and also enhances the release of dopamine, which results in addiction.

Nicotine is a psychoactive substance easily available in all corners of the world. Nicotine as whole serves as a major enforcer in the form of tobacco, cigarettes etc. in the human beings. It shows a complex mechanism involving neuronal system in the brain. Various studies have shown that nicotine increases the activity of dopaminergic system, its release leading to the craving and withdrawal effect. Further the study of nicotine on different brain system such as cholinergic, serotonergic, GABAergic and noradrenergic may also be involved in evoking pleasurable effect of nicotine. Molecular mechanism of nicotine addiction is still a challenging task for the researchers as so many brain systems are involved in it. The combination of molecular modern techniques with behavioral analysis may provide an exact mechanism and knowledge about nicotine addiction.

1.03 OTHER ADVERSE EFFECTS

"The role of nicotine as the major addictive component of tobacco products may have distracted our attention from toxicological effects on cell growth, angiogenesis, and tumor malignancy."

—Sanner and Grimsrud, "Nicotine: Carcinogenicity and Effects on Response to Cancer Treatment—A Review" (2015)

There is a push by the tobacco and vaping industries to downplay the health effects of nicotine because e-liquids do not contain many of the tobacco-related carcinogens found in cigarettes (10307, 10311, 10316). But nicotine itself is a neurotoxin and has been used as an insecticide since the 17th century. Nicotine has now been banned as a pesticide in many countries as it is too toxic to the environment, but has not been banned from cigarettes and other tobacco products (10301, 10302). It is ironic that nicotine is banned from agricultural use but not from inhalation into the lungs (10305).

Although most of the health harm due to tobacco use is due mainly to the thousands of carcinogenic chemicals that are inhaled with smoking (10303), nicotine itself also causes serious adverse health effects including high blood pressure, rapid heart rate, diabetes, COPD, and myocardial infarction, as well as suppressing the immune system and increasing the risk of community acquired pneumonia (10304, 10306, 10309, 10311, 10313, 10317). It also is associated with circadian rhythm changes and sudden infant death syndrome (10305). Nicotine is rated as having similar health risks to acrolein or formaldehyde (10303).

Although nicotine does not currently appear to cause cancer directly, it does contribute to cancer and promote cancer and its spread, such that in smokers, cancer spreads rapidly from the lungs into the brain (10307, 10317). Nicotine also promotes fibrosis in the heart, liver, and lung (10318).

Only 70.9% of current vapers and by 86.1% of adults who have not used any tobacco products view nicotine as definitely harmful to children (10308), even though nicotine is a known promoter of oxidation in neural cells and is especially toxic to the neurodevelopment of fetuses and young children (10320).

Nicotine simulates the Central Nervous System (CNS) and increases the risk of autoimmune diseases such as multiple sclerosis, rheumatoid arthritis, and Crohn's disease (10321). Nicotine also impairs healing after surgery (10322).

Nicotine did not improve the mood of abstinent smokers, contrary to the claim that cigarettes reduce stress and anxiety, and worsened the mood of never-smokers (10323).

Reference Number: 10301

MISHRA, A., P. CHATURVEDI, S. DATTA, S. SINUKUMAR, P. JOSHI, AND A. GARG. "Harmful Effects of Nicotine." *Indian Society of Medical & Paediatric Oncology*, 36: 24–31, 2015 (2 tables, 89 references)

Abstract: Nicotine has been used as an insecticide since the 17th century, but the US Environmental Protection Agency (EPA) has banned its use since January 2014 due to its toxicity to mammals. Nicotine, a strong alkaloid, is a clear liquid that turns brown upon exposure to air. On direct application, nicotine causes irritation and a burning sensation in the mouth and throat, increased salivation, nausea, abdominal pain, vomiting, and diarrhea. Nicotine causes an increase in HR, BP, free fatty acids (FFAs), hyperglycemia, and catecholamines. It is one of the most toxic of all poisons. Brain imaging shows nicotine increases activity in the prefrontal cortex and visual system and causes increased oxidative stress, neuronal apoptosis, DNA damage, reactive oxygen species, and lipid peroxide increase.

> Nicotine is the fundamental cause of addiction among tobacco users. Nicotine adversely affects many organs as shown in human and animal studies. Its biological effects are widespread and extend to all systems of the body including cardiovascular, respiratory, renal and reproductive systems. Nicotine has also been found to be carcinogenic in several studies. It promotes tumorigenesis by affecting cell proliferation, angiogenesis and apoptotic pathways. It causes resistance to the chemotherapeutic agents. Nicotine replacement therapy (NRT) is an effective adjunct in management of withdrawal symptoms and improves the success of cessation programs. Any substantive beneficial effect of nicotine on human body is yet to be proven. Nicotine should be used only under supervision of trained cessation personnel therefore its sale needs to be strictly regulated.

Reference Number: 10302

SANNER, T., AND T.K. GRIMSRUD. "Nicotine: Carcinogenicity and Effects on Response to Cancer Treatment—A Review." *Frontiers in Oncology*, 5: 196 (10 pp), 2015 (1 table, 1 figure, 98 references)

Abstract: A person who smokes 25 cigarettes per day will obtain a blood nicotine concentration ranging from 4 to 72 ng/mL and absorb about 0.43 mg/kg body weight. Nicotine binds to the nicotinic acetylcholine receptors (nAChRs) located in the CNS and parasympathetic autonomic nervous system. It binds to the nAChRs to a greater extent than acetylcholine. Tobacco-specific nitrosamines are formed from alkaloids in tobacco, especially nicotine. TSNAs may be formed in the body from nicotine. Nicotine and its metabolites, cotinine, have been found to promote cancer growth but is it not known whether nicotine is a complete carcinogen.

Table: Urine TSNAs (NNAL) in Tobacco Product Users, Nicotine Replacement Therapy Users, and Non-Smokers Exposed to SHS

Group	Total NNAL (pmol/mg creatinine)
Smokers	2.6
Snuff (old type)	3.3
Swedish snus	1.4
Nicotine Replacement Therapy	0.3
Non smokers exposed to SHS	0.042

Source: Adapted from Sanner and Grimsrud, 2015

Recent studies have shown that nicotine can affect several important steps in the development of cancer, and suggest that it may cause aggravation and recurrence of the disease. TSNA may be formed from nicotine in the body. The role of nicotine as the major addictive component of tobacco products may have distracted our attention from toxicological effects on cell growth, angiogenesis, and tumor malignancy. Effects on cancer disease are important aspects in the evaluation of possible long-term effects from sources of nicotine, such as e-cigarettes and products for nicotine replacement therapy, which both have a potential for life-long use.

Reference Number: 10303

BAUMUNG, C., J. REHM, H. FRANKE, AND D.W. LACHENMEIER. "Comparative Risk Assessment of Tobacco Smoke Constituents Using the Margin of Exposure Approach: The Neglected Contribution of Nicotine." *Scientific Reports*, 6: (9 pp), 2016 (3 tables, 1 figure, 65 references)

Abstract: Tobacco smoke is a complex mix of more than 5,000 compounds and at least 16 are carcinogenic to humans. Nicotine was not among the compounds in tobacco smoke that was evaluated for risk and so a health risk assessment of nicotine was conducted using a margin of error (MOE) approach. Nicotine was found to have a similar health risk as acrolein or formaldehyde. The calculated MOEs ranged from below 1 to up to 7.6, which indicates a considerable consumer risk.

There is immediate need of action concerning the risk of nicotine also with regard to electronic cigarettes and smokeless tobacco.

Reference Number: 10304

TWEED, J.O., S.H. HSIA, K. LUTFY, AND T.C. FRIEDMAN. "The Endocrine Effects of Nicotine and Cigarette Smoke." *Trends in Endocrinology and Metabolism*, 23: 334–342, 2012 (1 table, 1 figure, 117 references)

Abstract: A review was conducted to separate or tease the effects of nicotine from those of cigarette smoke in the endocrine system in humans.

The predominant effects of nicotine in humans include increased release of catecholamines into the bloodstream that increase pulse rate and blood pressure, the release of plasma free fatty acids, and the mobilization of blood glucose. Decreases in skin temperatures, arousal and relaxation are also noted following nicotine administration. At the cellular level, the effects of nicotine include increased synthesis and release of neurotransmitters and hormones, induction of oxidative stress, activation of transcription factors and the catecholamine-synthesizing enzyme tyrosine hydroxylase, as well as prevention of apoptosis. Serving as the fundamental mediator of neurotransmission in the CNS and PNS, the activation of nAChRs has important physiological consequences for multiple organs, including the endocrine system.

Reference Number: 10305

ALKAM, T., AND T. NABESHIMA. "Molecular Mechanisms for Nicotine Intoxication." *Neurochemistry International,* 125: 117–126, 2019 (2 tables, 1 figure, 203 references)

Abstract: Nicotine is a neurotoxin that was once used as a pesticide, although it is now banned in many countries. Nicotine is a pale yellow, oily hygroscopic liquid. Each cigarette contains approximately 8 to 20 milligrams of nicotine. Nicotine was never smoked or used alone until the use of nicotine gum (NRT) and e-cigarettes. Nicotine readily crosses the blood/brain barrier and is distributed throughout the brain, where it binds to the nicotinic cholinergic receptors (nAChRs). This binding or stimulation of nAChRs results in a release of acetylcholine, norepinephrine, dopamine, serotonin, vasopressin, growth hormone, and adrenocorticotropic hormone.

Nicotine, one of the more than 4700 ingredients in tobacco smoke, is a neurotoxin and once used as pesticides in agriculture. Although its use in agriculture is prohibited in many countries, nicotine intoxication is still a problem among the workers in tobacco farms, and young children as well as adults due to the accidental or suicidal ingestions of nicotine products. Understanding the mechanism of nicotine intoxication is important not only for the prevention and treatment but also for the appropriate regulatory approaches. Here, we review pharmacokinetics of nicotine and the molecular mechanisms for acute and chronic intoxication from nicotine that might be relevant to the central and the peripheral nervous system. We include green tobacco sickness, acute intoxication from popular nicotine products, circadian rhythm changes, chronic intoxication from nicotine through prenatal nicotine exposure, newborn behaviors, and sudden infant death syndrome.

Reference Number: 10306

STEWART, P.M., AND J.R. CATTERALL. "Chronic Nicotine Ingestion and Atrial Fibrillation." *British Heart Journal*, 54: 222–223, 1985 (5 references)

Abstract: A 35-year-old man attended hospital as a result of prolonged periods of irregular palpitations. For 18 months he had been trying to stop smoking (40 cigarettes/day addiction) by chewing initially 20 pieces of 2 mg nicotine gum daily and then 30 pieces per day. He was diagnosed as having atrial fibrillation with 150 bpm heart rate. His blood nicotine concentration was 4.8 ng/mL five hours after experiencing the palpitations. Over the next 12 months, he abstained from cigarette smoking and the use of Nicorette gum and had no reoccurrence of the symptoms. His blood nicotine concentration was zero at the time, which confirmed his abstinence.

Nicotine chewing gum was developed as an aid to smoking withdrawal and is usually effective in a dose of eight 2 mg pieces daily. Reports of cardiovascular effects of nicotine gum have been limited to an increase in heart rate, although nicotine has been shown to cause arrhythmias in animals. We report a patient who developed atrial fibrillation while consuming three to four times the usual amount of nicotine chewing gum.

Reference Number: 10307

SAJI, S., S.S. PATIL, M. ALLEYN, R. LOCKEY, AND N. KOLLIPUTI. "Nicotine in E-Cigarette Smoke: Cancer Culprit?" *Journal of Cell Communication and Signaling*, 14: 127–128, 2020 (10 references)

Abstract: Nicotine in tobacco smoke contributes to diseases such as lung cancer, coronary artery disease, and asthma. The derivatives of nicotine in electronic cigarette smoke (ECS), such as tobacco-specific nitrosamines, are carcinogenic. The major DNA repair pathways are nucleotide excision repair (NER) and base excision repair (BER).

However, studies have found that E-cigarette smoke (ECS) not only induces DNA damage but also reduces DNA repair activity via BER and NER pathways. Further research conducted with cells damaged by Ultra-Violet (UV) light or hydrogen peroxide (H_2O_2) indicates that ECS can function as a comutagen; nicotine can amplify mutagenic activity by merging with other mutagens. The downstream metabolites derived from nicotine found in ECS put E-cigarette smokers at a higher risk for developing lung or bladder cancers or heart disease than their non-smoking counterparts. Overall, these findings are instrumental in our understanding of the harmful effects of ECS.

Reference Number: 10308

KEMP, C.B., C.A. SPEARS, T.F. PECHACEK, AND M.P. ERIKSEN. "Adults' Perceptions of Nicotine Harm to Children." *Pediatrics,* 142: (8 pp), 2018 (2 tables, 38 references)

Abstract: A US national survey was conducted between 2015 and 2016 as to whether nicotine is harmful to children. Overall, the majority of adults (83.2%) characterized nicotine use as definitely harmful to children and 2.3% felt nicotine was not harmful.

Table: Percentage of Adults who Thought Nicotine was Definitely, Maybe, Unlikely, or Not Harmful to Children

Characteristic	Definitely Harmful (%)	Maybe (%)	Unlikely (%)	Not (%)
Men	81.5	7.6	2.1	0.6
Women	84.8	5.0	1.3	0.6
Less than high school diploma	75.3	9.1	1.9	1.4
Current cigarette use	73.8	11.9	3.4	1.2
Current vaping	70.9	15.5	3.8	2.4
No tobacco products used	86.1	4.4	1.1	0.4

Source: Adapted from Kemp et al, 2018

We analyzed US adults' perceptions of nicotine harm to children in a nationally representative sample and found that although the majority of the population perceived nicotine to be harmful when used by children, important subgroup variations exist. For example, the finding that adults who use tobacco are significantly more likely to underestimate nicotine's harmfulness to children regardless of the number of different products used is particularly concerning. These individuals have the greatest likelihood of possessing nicotine-containing products, providing potential sources of exposure to children. Results also suggest that some men and individuals from racial and/or ethnic minority backgrounds might benefit from interventions to correct inaccurate risk perceptions.

Authors of previously published research have demonstrated that after cigarettes, ENDS products and chewing tobacco have been the principal sources of unintended nicotine exposure in children since 2012. The increasing use of ENDS and other noncombusted tobacco products in the US adult population presents new sources of nicotine consumption and exposure not only to the users but also to the children who live with them. The food-grade flavorings used in these products include many fruit and candy flavors that are appealing to children. The number of cases of nicotine poisonings resulting from children ingesting e-liquids has risen sharply in recent years.

Reference Number: 10309

BENOWITZ, N.L., AND A.D. BURBANK. "Cardiovascular Toxicity of Nicotine: Implications for Electronic Cigarette Use." Trends in Cardiovascular Medicine, 26: 515–523, 2016 (1 table, 1 figure, 39 references)

Abstract: Cigarette smoke contains more than 9,000 chemicals and more than 69 known carcinogens. Nicotine in smoke is inhaled primarily in the particulate phase followed by rapid diffusion of nicotine into the vapor phase within the lungs. Nicotine reaches the brain within 15 to 20 seconds after a puff. Each puff delivers 100 to 150 μg of nicotine and a cigarette normally delivers 1 to 2 milligrams of nicotine to the systemic circulation. Nicotine releases epinephrine. Nicotine level generally plateaus after 6 to 8 hours of smoking and has a gradual decline during the night. Typical blood concentration is 10 to 40 ng/mL and is 2 to 3× greater in the heart and brain blood. Heart rate and blood pressure increase after nicotine use. Nicotine constricts the blood flow to the skin and coronary vessels and increases blood flow to the skeletal muscle. Nicotine may contribute to impaired wound healing, macular degeneration, and progressive renal disease. Cigarette smoking and smokeless tobacco increase risk of heart failure. Cigarette smoking is also an important risk factor for Type 2 diabetes.

Table: Some Cardiovascular Disorders Caused by Cigarettes and Nicotine

Some Cardiovascular Disorders Caused by Cigarette Smoking and Nicotine
• Accelerated atherosclerosis
• Acute myocardial infarction
• Shorter exercise time to angina
• Coronary spasm
• Stroke
• Aortic Aneurysm
• Peripheral obstructive arterial disease
• Stent thrombosis after PCI
• Graft occlusion after coronary bypass surgery
• Sudden cardiac death
• Atrial fibrillation

Source: Adapted from Benowitz and Burbank, 2016

The cardiovascular safety of nicotine is an important element in the debate on the benefits vs risk of e-cigarette use and related public policy. Based on current knowledge, we believe that the cardiovascular risks of nicotine from e-cigarette use in people without CVD are quite low. We have concerns that nicotine from e-cigarettes could pose some risk for users with CVD. Other constituents of e-cigarette aerosol could theoretically pose some cardiovascular risk, but experimental evidence of such risk is lacking.

Reference Number: 10310

CHEN, E.Y., A. SUN, C-S. CHEN, A.J. MINTZ, AND W-C. CHIN. "Nicotine Alters Mucin Rheological Properties." *Lung Cellular and Molecular Physiology*, 307: L149–L157, 2014 (1 table, 5 figures, 58 references)

Abstract: Cigarette smoking and SHS are major risk factors of developing COPD. It is thought that cigarette smoking induces a chronic airway inflammation, hypersecretion of mucin, and underhydration of the airway surfaces, and impairs mucociliary clearance. The more viscous, poorly hydrated mucin is characteristic of the thick, viscous occlusion commonly found in COPD. An in-vitro human airway A547 was constructed and exposed to nicotine alone and no other constituents of tobacco smoke. Both nicotine and mucin are amphiphilic molecules and perhaps mucin binds to monoprotonated nicotine at neutral pH and becomes viscous and underhydrated.

> Furthermore, our results indicate that mucin gels serve not only as a reservoir but also as a control-release modulator for nicotine. Consequently, the body may experience prolonged exposure due to a sustained nicotine release from airway mucus beyond the period of active smoking and ETS exposures. Our data are supported by the lack of COPD decline among formerly heavy smokers. That chronic inflammation persisted after smoking cessation can be elucidated by the persistent nicotine released from mucus that attracts and activates neutrophils. In addition, gradual release and retention of tobacco carcinogens and nicotine from mucus may be an additional mechanism linking lung carcinogenesis and COPD susceptibility.
>
> This study has provided evidence supporting that nicotine directly retards mucin hydration through interactions with mucins possibly via cross-linking. The fact that low levels of tobacco smoke exposure may lead to nicotine accumulation in secretory granules highlights the hazardous potential of ETS for nonsmokers. Furthermore, airway mucus potentially sustains harmful blood nicotine concentration for extended time after tobacco inhalation. In addition, nicotine is likely to cause significant systemic effects by inducing widespread changes in the release kinetics of other secretory matrices that may contribute to the complex pathology found in tobacco smokers or in individuals chronically exposed to ETS.

Reference Number: 10311

KAPAYA, M., D.V. D'ANGELO, V.T. TONG, L. ENGLAND, N. RUFFO, S. COX, L. WARNER, J. BOMBARD, T. GUTHRIE, A. LAMPKINS, AND B.A. KING. "Use of Electronic Vapor Products Before, During and After Pregnancy Among Women with a Recent Live Birth—Oklahoma and Texas, 2015." *Morbidity and Mortality Weekly Report*, 68: 189–194, 2019 (1 table, 2 figures, 9 references)

Abstract: A study of 1,955 women from Oklahoma and 1,322 women from Texas with a recent live birth were surveyed as to cigarette and electronic vaping products (EVP) use. Approximately 83% had used EVP at some point, although at the time of pregnancy, the use was 7% overall.

> Although aerosol from EVPs contains lower levels of toxicants than does cigarette smoke, EVPs are not safe to use during pregnancy because most contain nicotine. Nicotine, a developmental toxicant, adversely affects

pregnancy and infant outcomes. Although the U.S. Preventive Services Task Force has determined that, currently, insufficient evidence exists to recommend EVPs for tobacco cessation among adults (including pregnant women), many women report using EVPs in an attempt to quit smoking cigarettes around the time of pregnancy.

Reference Number: 10312

HERGENS, M-P., L. ALFREDSSON, G. BOLINDER, M. LAMBE, G. PERSHAGEN, AND W. YE. "Long-Term Use of Swedish Moist Snuff and the Risk of Myocardial Infarction Amongst Men." *Journal of Internal Medicine*, 262: 351–359, 2007 (3 tables, 1 figure, 33 references)

Abstract: The risk of myocardial infarction (MI) was studied in 118,395 non-smoking men between 1978–1993 and 2004 in Sweden. Nearly 30% of the men had used snuff. During follow-ups, 3,651 cases of acute MI were identified, of which 841 were fatal. Swedish moist snuff (snus) does not appear to increase the overall risk of MI but does increase the risk of fatal MI, especially among heavy users.

However, it is well known that nicotine puts additional demands on the heart by increasing the heart rate, cardiac stroke volume and coronary blood flow as well as by vasoconstriction. Animal studies show that nicotine exposure can induce cardiac arrhythmias and increase the vulnerability for ventricular fibrillation following myocardial infarction. Therefore, it is conceivable that high nicotine exposure can increase the severity of myocardial infarction amongst snuff users leading to increased risks for a fatal outcome.

Reference Number: 10313

TAS, D., H. SEVKETBEYOGLU, A.F. AYDIN, K. CELIK, AND M.A. KARACA. "The Relationship Between Nicotine Dependence Level and Community-Acquired Pneumonia in Young Soldiers: A Case Control Study." *Internal Medicine* 47: 2117–2120, 2008 (2 tables, 17 references)

Abstract: Smoking and nicotine dependence (Fagerstrom Test for Nicotine Dependence) were determined in 58 male patients hospitalized at the Izmir Military Hospital with community acquired pneumonia (CAP) and 580 matched controls without CAP. The mean age of the soldiers in both groups was 22.2 years. Pneumonia development risk was 2.19× higher in smokers than non-smokers. In the patients with the highest nicotine dependence (Fagerstrom Test for Nicotine Dependence score of >5), the risk increased to 2.93×.

> Nicotine affects both cellular and humoral immunity, so it is an important immune suppressive agent of smoke as well as other agents. When nicotine dependence increases, it is expected that there should be an increase in infection risk.
>
> Pneumonia is still a major cause of mortality and morbidity all over the world. In respect to community health, it is important to decrease the incidence of CAP together with smoking-related diseases in order to decrease health costs. In this context, the results of this study highlight the importance of campaigns against smoking. In addition, physicians should recommend pneumococcal vaccines to subjects with high NDL although they do not have additional immune suppression.

Reference Number: 10314

XIE, X., Q. LIU, J. WU, AND M. WAKUI. "Impact of Cigarette Smoking in Type 2 Diabetes Development." *Acta Pharmacologica Sinica*, 6: 784–787, 2009 (33 references)

Abstract: Smoking is a well-known risk factor for coronary heart disease, due to pathological changes such as arteriosclerosis and inflammatory responses. Nicotine also increases the blood sugar levels.

> Numerous clinical and experimental reports have described the association between the active cigarette smoking, the development of diabetes, glycemic control, and diabetic complications. Although confounding social factors are sometimes argued rather than a direct effect of nicotine or other cigarette components, it is probable that most of these effects are exerted by nicotine since both prenatal and postnatal exposure to nicotine can directly induce imbalance of metabolic control. Effects of nicotine on insulin action, secretion, and development of diabetes have been widely studied both in animal models and human subjects. Results have indicated nicotine exposure could induce a reduction of insulin release, and negatively affect insulin action, suggesting nicotine could be a cause for development of insulin resistance.

Reference Number: 10315

COOKE, J.P. "New Insights into Tobacco-Induced Vascular Disease: Clinical Ramifications." *Methodist DeBakey Cardiovascular Journal*, 11: 156–159, 2015 (1 figure, 38 references)

Abstract: Nicotine is the highly addictive chemical in tobacco smoke which also causes adverse cardiovascular effects. Nicotine activates the sympathetic nervous system by releasing catecholamines, which elevate blood pressure and heart rate. Long-term effects of elevated HR and BP can be expected to accelerate the progression of atherosclerosis and increase the risk of myocardial infarction.

> Our observations regarding the vascular effects of nicotine are relevant to the long-term safety of nicotine replacement therapies (NRTs) for smokers. Medicinal nicotine is a mainstay in tobacco cessation programs. Whereas short-term use of NRTs has been shown to be safe and effective in tobacco cessation, there are little clinical data on the safety and efficacy of long-term use of NRTs. The preclinical data discussed above reveals that nicotine is a potent angiogenic agent and raises the concern that the angiogenic component of tobacco-related diseases (cancer, age-related macular degeneration, and atherosclerosis) may be promoted by medicinal nicotine. Clearly tobacco smoke, with its potpourri of carcinogens, toxins, and particulate matter, is more hazardous than medicinal nicotine. Furthermore, there is no evidence from observational studies that individuals taking NRTs for greater than the recommended periods are at greater risk for adverse events related to pathological angiogenesis. For this reason, and because of the importance of tobacco cessation, the U.S. Food and Drug Administration has softened its stance on the long-term use of medicinal nicotine. However, in the absence of randomized clinical trials on the long-term use of nicotine, the astute clinician might consider additional approaches toward tobacco cessation so as to reduce the long-term use of medicinal nicotine. Although the effect of secondhand smoke is dependent upon many factors, including the intensity of cigarette use and the air exchange in a room, a recent meta-analysis of secondhand tobacco smoke studies indicated that plasma nicotine levels are often high enough to have biological effects.

Reference Number: 10316

ANTONIEWICZ, L., A. BRYNEDAL, L. HEDMAN, M. LUNDBÄCK, AND J.A. BOSSON. "Acute Effects of Electronic Cigarette Inhalation on the Vasculature and the Conducting Airways." *Cardiovascular Toxicology*, 19: 441–450, 2019 (2 tables, 4 figures, 37 references)

Abstract: Nine female and 6 male occasional cigarette smokers (mean age 26 years) inhaled 30 puffs of a variable mod third generation EC (cVic-VT) with a dual coil nickel atomizer at 230°C with or without nicotine (19 mg/mL). The e-liquid contained 49.4% propylene glycol, 44.4% vegetable nicotine, and 5% ethanol, but no flavorings. Serum cotinine concentrations were measured by enzyme-linked immunosorbent assay (ELISA) to ensure compliance. Blood pressure, hematocrit, arterial stiffness, and respiratory function were determined before the exposure and at both 2 and 4 hours following the 30 minutes of vaping. Respiratory measurements were also determined at 6 hours following exposure. Electronic cigarette aerosol (ECA) with nicotine causes a sudden increase in heart rate and arterial stiffness and a sudden increase in flow resistance. Both ECAs increased the BP.

> There is an ongoing debate on which role nicotine plays in the pathophysiology of atherosclerosis and whether if, and how it may accelerate vascular disease. There are many pathways through which nicotine may cause negative effects on the cardiovascular system. Nicotine has been demonstrated to elicit strong sympathomimetic effects, diminish coronary blood flow, impair endothelial function, enhance inflammation and arteriogenesis, as well as cause insulin resistance. Epidemiological data that correspond with these findings are sparse, as most people tend to use tobacco products as opposed to products containing solely nicotine. Investigating oral snus usage may to some extent reflect long-term nicotine effects, as this product contains high amounts of nicotine and smaller amounts of known health hazardous compounds associated with combustible tobacco. Snus use has been demonstrated to be associated with increased mortality following myocardial infarctions and stroke as well as increased risk of type II diabetes. However, transdermal nicotine replacement therapy seems to be safe in patients with cardiovascular disease. It is not unlikely that the pathophysiological effect of nicotine is determined by pharmacokinetics; nicotine administered in tablet form does not give the very quick and high rise in blood nicotine levels as seen following cigarette smoking, inhalation of ECA, or oral snus use.

Reference Number: 10317

WU, S-Y., F. XING, S. SHARMA, K. WU, A. TYAGI, Y. LIU, D. ZHAO, P.R. DESHPANDE, Y. SHIOZAWA, T. AHMED, W. ZHANG, M. CHAN, J. RUIZ, T.W. LYCAN, A. DOTHARD, AND K. WATABE. "Nicotine Promotes Brain Metastasis by Polarizing Microglia and Suppressing Innate Immune Function." *Journal of Experimental Medicine*, 217: (17 pp), e20191131, 2020 (6 figures, 82 references)

Abstract: Cigarette smoking is one of the major risk factors for lung cancer, as tobacco smoke contains a toxic mix of more than 7,000 chemicals. Nicotine is not considered to be a carcinogen but does promote cancerous tumors by activating the nicotinic acetylcholine receptors (nAChR) in the cells. The nAChR also present in a variety of brain cells including the neurons, astrocytes, and the microglia. Up to 40% of lung cancer patients develop brain metastasis. A study was conducted of 281 patients with lung cancer that metastasized to the brain, with or without a history of smoking. The rate of brain metastasis was 22% for never smokers, 25% for former smokers, and 42% for current smokers. Cell cultures and animal testing were also conducted. Nicotine was found to suppress the innate immune function of the microglia. This can result in a nicotine replacement therapy for tobacco cessation entailing some potential risk, especially in smoking lung cancer victims.

The incidence of brain metastasis among lung cancer patients is 10–36%, although these data are considered to be underestimated. The majority of the lung cancer patients are diagnosed with advanced stage disease, most with stage IV cancer. More than 90% of lung cancer deaths are attributed directly to the metastatic disease and include those brain metastases. Cigarette smoking is a major risk factor for lung cancer, and our previous study showed a negative impact of smoking on lung cancer brain metastasis. In the present study, we have found a significantly higher incidence of brain metastasis among current smokers with stage IV lung cancer.

Nicotine is not a carcinogen, but it causes addiction by activating nAch receptor, which is abundantly expressed by various brain cells including microglia. Direct tumor promoting effects of nicotine have been previously reported.

Reference Number: 10318

JENSEN, K., D. NIZAMUTDINOV, M. GUERRIER, S. AFROZE, D. DOSTAL, AND S. GLASER. "General Mechanisms of Nicotine-Induced Fibrogenesis." *Federation of American Societies for Experimental Biology Journal*, 26: 4778–4787, 2012 (5 figures, 85 references)

Abstract: Nicotine is a powerful biological agent and exerts its action mainly by binding to the nicotinic acetylcholine receptors (nAChRs) in cell membranes. In addition to promoting cancer, nicotine promotes fibrosis in cardiac, liver, lung, and other organs, in part due to causing a dramatic increase in circulating catecholamines and other hormones.

> Nicotine is especially known for its role in addiction and plays a role in other physiological effects of smoking and tobacco use. Recent studies have provided compelling evidence that, in addition to promoting cancer, nicotine also plays a pathogenic role in systems, such as the lung, kidney, heart, and liver. In many organ systems, nicotine modulates fibrosis by altering the functions of fibroblasts. Understanding the processes modulated by nicotine holds therapeutic potential and may guide future clinical and research decisions. This review discusses the role of nicotine in the general fibrogenic process that governs fibrosis and fibrosis-related diseases, focusing on the cellular mechanisms that have implications in multiple organ systems. Potential research directions for the management of nicotine-induced fibrosis, and potential clinical considerations with regard to nicotine-replacement therapy (NRT) are presented.

Reference Number: 10319

MILLQVIST, E., AND M. BENDE. "Capsaicin Cough Sensitivity is Decreased in Smokers." *Respiratory Medicine*, 95: 19–21, 2001 (2 figures, 11 references)

Abstract: Capsaicin is an ingredient found in red peppers and has been used to induce coughing to test the effectiveness of cough medicine. In this study, 10 female and 5 male smokers (mean age 43 years) who smoked on average 12.5 CPD for at least 1 year and 32 female and 7 male non-smokers were tested as to the number of coughs after the inhalation of 1 mL capsaicin spray. As the capsaicin concentration increased, the number of coughs increased. At the highest concentration, the mean number of coughs was 26.0 for non-smokers and 14.3 for smokers.

Table: Number of Coughs after Inhalation of Capsaicin Spray (for 10 minutes) in Smokers and Non-smokers

Capsaicin Concentration	# of coughs in smokers	# of coughs in non-smokers
0.4 μM	0.2	2.0
2 μM	2.1	9.5
10 μM	14.3	26.0

Source: Adapted from Milliqvist and Bende, 2001

The authors conclude that smokers are less sensitive to capsaicin than non-smokers, which supports the hypothesis that nicotine either inhibits the receptors of the C-fibres of the sensory nerves in the respiratory epithelium, or induces a depletion of neuropeptides. Clinically, this may serve to explain increasing airway symptoms that are often seen after cessation of smoking.

Reference Number: 10320

SWAN, G.E., AND C.N. LESSOV-SCHLAGGAR. "The Effects of Tobacco Smoke and Nicotine on Cognition and the Brain." *Neuropsychology Review*, 17: 259–273, 2007 (1 figure, 161 references)

Abstract: Tobacco smoke is probably the single most significant cause of toxic chemical exposure in humans. Cigarette smoke consists of 2 phases—a particulate (tar) phase and a gas phase. The particulate phase contains more than 10^{17} free radicals per gram and the gas phase contains more than 10^{15} free radicals per puff. The radicals associated with the gas phase have a shorter duration (seconds) compared to the particulate phase, which has a duration of hours to months. Ammonium compounds are added to cigarettes to allow for a more complete and faster absorption of nicotine and enhance the impact of nicotine on the nicotinic receptors. Acute administration of nicotine (but not tobacco smoke) increased vigilance, selective attention, and working and verbal memory on adult non-smokers and non-deprived smokers. Nicotine is a promotor of oxidation in neural cells and is especially toxic to the neurodevelopment of fetuses and young children.

Tobacco smoke is a highly complex mixture of compounds, many of which have known toxic effects on the cardiovascular, cerebrovascular, and pulmonary systems. Because of the number of compounds present in tobacco smoke, a complete picture of organ-specific toxicity is not yet available.

However, there is sufficient evidence at the epidemiologic level to support the conclusion that a history of smoking is clearly associated with preclinical changes in the brain (atrophy, silent infarcts, and WMHIs), accelerated cognitive decline (executive function, verbal memory, speed of processing), and increased risk for dementia (AD and VaD). Exposure to sidestream smoke, especially in fetuses and children, is associated with poorer neurocognitive performance. While the precise mechanisms underlying these associations is not known definitively, in part because of the complexity of tobacco smoke, work in animal models and at the cellular level suggest that tobacco smoke increases oxidative stress in the brain and other organs and induces an inflammatory response that may directly or indirectly promote atherosclerosis and neuropathology associated with AD.

Reference Number: 10321

PEŘINOVÁ, P., E. FEKETEOVÁ, D. KEMLINK, P. KOVALSKÁ, K. CHLEBUŠOVÁ, J. NEPOŽITEK, V. IBARBURU, E. KRÁLÍKOVÁ, S. NEVŠÍMALOVÁ, AND K. ŠONKA. "Smoking Prevalence and Its Clinical Correlations in Patients with Narcolepsy-Cataplexy." *Prague Medical Report*, 117: 81–89, 2016 (1 table, 39 references)

Abstract: Narcolepsy-cataplexy (NC) is a chronic neurological disease with a prevalence of approximately 0.045% in North America and Europe. The basis of the disease is a deficiency of neurons in the lateral hypothalamus that produces hypocretin and may be due to an autoimmune attack. The characteristics of NC are excessive daytime sleepiness (EDS), cataplexy, and fragmented sleep. Nicotine stimulates the CNS and increases the risk of autoimmune diseases such as multiple sclerosis, rheumatoid arthritis, and Crohn's disease. An analysis was conducted of 49 female and 38 male adult patients (mean age 46.4 years) with NC. Forty-four percent of the patients were regular smokers, compared with 23.5% of the population. Seventy-two percent of the smokers with NC started smoking before the onset of the disease.

Although narcolepsy symptom relief may be viewed as a benefit of nicotine and may thus be a barrier to smoking cessation in narcoleptics, the act of smoking is itself of concern because of many reasons. First is the risk of falling asleep while smoking resulting in injury and damage and second is the already mentioned risk factor of many other diseases. Third reason is based on the fact that nicotine is a highly addictive substance, and this obviously applies to the narcolepsy population as well.

Reference Number: 10322

FRACOL, M., R. DORFMAN, L. JANES, S. KULKARNI, K. BETHKE, N. HANSEN, AND J. KIM. "The Surgical Impact of E-Cigarettes: A Case Report and Review of the Current Literature." *Archives of Plastic Surgery*, 44: 477–481, 2017 (1 figure, 2 tables, 19 references)

Abstract: A 51-year-old woman with a 25 pack year smoking history and a diagnosis of right breast cancer underwent a right skin sparring mastectomy and immediate tissue expander reconstruction. She had stopped smoking for several months prior to the surgery and reported herself as a non-smoker despite substantial e-cigarette use instead. She developed skin flap necrosis and breast reconstructive failure. Nicotine increases the risk of skin flap necrosis and surgical site infection. Smokers undergoing facelifts had a 13× risk of skin flap necrosis compared to non-smokers.

In conclusion, our case report depicts the potential dangers associated with e-cigarette use in the surgical patient. There is a public misconception that e-cigarettes are healthier than traditional cigarettes and as such their use may go unreported by patients. Early evidence suggests e-cigarettes may induce some of the same physiologic changes as traditional cigarettes and may have a significant deleterious effect on wound healing.

Reference Number: 10323

FOULDS, J., J.A. STAPLETON, N. BELL, J. SWETTENHAM, M.J. JARVIS, AND M.A. RUSSELL. "Mood and Physiological Effects of Subcutaneous Nicotine in Smokers and Never-Smokers." *Drug and Alcohol Dependence*, 44: 105–115, 1997 (2 tables, 4 figures, 39 references)

Abstract: Smokers frequently claim that cigarettes reduce stress and anxiety and calm them down. But nicotine actually increases the heart rate and the plasma adrenaline and noradrenaline concentrations. Eighteen smokers (mean use 21 CPD) and 18 never-smokers were nicotine deprived for 24 hours and then administered 2 doses of placebo (saline), 2 doses of 0.3 mg nicotine, or 2 doses of 0.6 mg nicotine subcutaneously. The 0.6 mg dose would produce approximately one-half the blood nicotine concentration of smoking a cigarette. Various subjective measures of mood and physiological measures were determined. Two never-smokers had to pause testing after receiving the second 0.6 mg nicotine dose due to excessive dizziness and nausea (one vomited).

> The results of this study confirm that abstinence from tobacco worsens mood in smokers, and indicates that when a moderate acute dose of nicotine is administered subcutaneously under double-blind conditions it produces no improvements in mood in abstinent smokers and worsens mood in never-smokers. Acute mood enhancements from nicotine have been hard to detect in laboratory studies, particularly in controlled studies using tobacco-free delivery systems. They might occur only through self-administration by smoking in response to situation-specific mood fluctuations.

1.04 POISONINGS/FATALITIES

> "Nicotine, which is found in tobacco is one of the most toxic of all known poisons."
>
> —Miyauchi et al, "Gastric Lavage Guided by Ultrathin Transnasal Esophagogastroduodenoscopy in a Life-Threatening Case of Tobacco Extract Poisoning: A Case Report" (2013)

Nicotine is one of the most poisonous drugs. There is some controversy as to the fatal dose of nicotine, with recent studies suggesting that it is not 60 mg, the commonly accepted number, but in fact between 500 and 1000 mg (10401). But this is still much lower than 212,000 mg of alcohol typically required for fatal poisonings. Nicotine was added to ground beef by a disgruntled supermarket employee which poisoned 100 persons who ate the meat. The most common symptoms were the burning of the mouth, nausea, vomiting, and dizziness (10402).

Children have been poisoned by nicotine for many years by ingesting cigarettes or cigarette butts (10403–5). New sources of poisoning in children include nicotine patches (10406), chewing tobacco, and snuff (10407). An increasing cause of nicotine poisoning in children is accidental ingestion of e-liquids containing nicotine (10408–13). Children exposed to e-liquids had 5.2× the risk of health care admission and 2.6× the risk of a more severe outcome than those exposed to cigarettes (10410). Standardized packaging, warnings, and child-resistant storage of e-liquids are required to prevent this source of nicotine poisoning (10409).

Fortunately, only about 4% of the children poisoned by nicotine required hospitalization and most were treated in ER. The most common symptoms were nausea, crying, red eye, cough, sleepiness, and unresponsiveness (10411). But at higher doses, nicotine can cause death due to cardiac and respiratory failure (10419), and e-liquids have been used in suicides (10415–18).

In a global study of 26 cases of nicotine e-liquid intoxications, 11 died. The mean plasma nicotine concentration of the survivors was 307 ng/mL and it was 3,360 ng/mL for the deceased (10420). The nicotine in ECs have caused poisoning in children not only by ingestion of the e-liquids but by inhaling the high nicotine concentration in vape (10421). Infants are more susceptible to the poisoning effects of nicotine. A 15-month-old baby girl died after ingesting 5 mL of a nicotine containing e-liquid (10422).

A smoker died of nicotine poisoning when he used nicotine patches while continuing to smoke cigarettes (10423).

It is estimated that a fatal nicotine dose is 500–1,000 mg for adults but only 19 mg for infants (10424). All e-liquids containing nicotine should be marked with warning labels and in child resistant containers (10424). An EC modified with a stronger coil that resulted in a doubling of the coil temperature and thus a much higher nicotine concentration in the vape caused nicotine intoxication in two enlisted military men (10425).

Nicotine can also poison cattle via dermal absorption (10426).

Reference Number: 10401

MAYER, B. "How Much Nicotine Kills a Human? Tracing Back the Generally Accepted Lethal Dose to Dubious Self-Experiments in the Nineteenth Century." *Archives of Toxicology* 88: 5–7, 2014 (21 references)

Abstract: Smoking a cigarette results in an uptake of about 2 mg of nicotine and in an arterial plasma concentration of 30 ng/mL. An oral dose of 60 mg of nicotine (often cited as the lethal dose) would result in a plasma nicotine concentration of 180 ng/mL. Recent studies have indicated that the lower limit of lethal blood nicotine concentration is 200 ng/mL, which suggests the lethal dose of nicotine to be 500 to 1,000 mg.

The discrepancy between the 60-mg dose and published cases of nicotine intoxication has been noted previously, but nonetheless, this value is still accepted without scrutiny and taken as the basis for worldwide safety regulations of tobacco and other nicotine-containing products. Nicotine is a toxic compound that should be handled with care, but the frequent warnings of potential fatalities caused by ingestion of small amounts of tobacco products or diluted nicotine-containing solutions are unjustified and need to be revised in light of overwhelming data indicating that more than 0.5 g of oral nicotine is required to kill an adult.

Reference Number: 10402

BOULTON, M., M. STANBURY, D. WADE, J. TILDEN, D. BRYAN, J. PAYNE, AND B. EISENGA. "Nicotine Poisoning After Ingestion of Contaminated Ground Beef—Michigan, 2003." *Morbidity and Mortality Weekly Report*, 52: 413–416, 2003 (9 references)

Abstract: A recall was conducted on approximately 1,700 pounds of ground beef as a result of about 100 persons becoming ill after eating it. The symptoms were burning of the mouth, nausea, vomiting, and dizziness. Two persons were treated at hospital, one of whom had atrial fibrillation. An employee of the supermarket was charged with poisoning 200 pounds of ground beef using an insecticide (Black Leaf), which consists mainly of nicotine.

Acute nicotine toxicity is associated with overstimulation of nicotinic receptors. Burning in the throat with nausea and vomiting occurs quickly after ingestion. More toxic manifestations include cardiac tachyarrhythmias, seizures, and hypertensive crisis.... Nicotine is used in a limited number of pesticides because of its toxic properties; nine nicotine containing pesticides are registered for use in Michigan and note of the product labels list nicotine at more than 14%. Black Leaf 40 contained 40% nicotine and the EPA cancelled its product registration because of its toxicity.

Reference Number: 10403

PETRIDOU, E., N. KOURI, A. POLYCHRONOPOULOU, K. SIAFAS, M. STOIKIDOU, AND D. TRICHOPOULOS. "Risk Factors for Childhood Poisoning: A Case-Control Study in Greece." *Injury Prevention*, 2: 208–211, 1996 (4 tables, 29 references)

Abstract: One hundred consecutive childhood accidental poisoning cases brought to the ERs of two Athenian hospitals in 1995 were studied. For each poisoning, two children brought to the same hospital for other reasons were used as a control. Chewing and swallowing cigarettes was the second most common type of accidental poisoning (15%). The odds ratio (OR) for unintentional tobacco poisoning was 3.2x with one smoking parent and 7.9x with two smoking parents compared with non-smoking parents.

Table: Percent of Accidental Childhood Poisonings According to the Poison Involved

Poisoning Agent	Percent of Poisonings
Medication	58.0%
Cigarettes	15.0%
Pesticides, Herbicides	12.0%
Petroleum Products	5.0%
House Cleaning Products	3.0%
Other	7.0%

Source: Adapted from Petridou et al, 1996

Chewing or swallowing cigarettes or cigarette butts was the second most common type of poisoning in this study, responsible for 15% of all cases. Because most types of childhood poisoning depend on the accessibility of the respective products, this type of poisoning is likely to be common in any population with a high prevalence of smoking. Pointing out to care givers the potential of tobacco products to cause poisoning may enhance their awareness. This is a necessary, although hardly sufficient, condition for reducing poisoning by tobacco products.

Reference Number: 10404

RIORDAN, M., G. RYLANCE, AND K. BERRY. "Poisoning in Children 5: Rare and Dangerous Poisons." *Archives of Disease in Childhood*, 87: 407–410, 2002 (1 table, 21 references)

Abstract: A short review was conducted on accidental poisoning in children. Among the poisons discussed were carbon monoxide, ethylene glycol, and nicotine.

Nicotine is a toxic alkaloid found in a range of plants, most notably tobacco. Nicotine is also used as a pesticide. Nicotine stimulates nicotinic acetylcholine receptors throughout the central and autonomic nervous systems. In severe poisoning confusion and convulsions are accompanied by arrhythmias and parasympathetic over stimulation (see organophosphate poisoning).

Children most frequently encounter nicotine in the form of cigarette or cigarette butt ingestion. Vomiting is the commonest symptom following cigarette ingestion and tends to limit toxicity. Asymptomatic children who have ingested less than one cigarette or two cigarette butts should be observed for two hours. No specific treatment in required. Children who have consumed more than this amount, should receive activated charcoal.

Reference Number: 10405

LEWANDER, W., H. WINE, R. CARNEVALE, J. LINDENMAYER, E. HARVEY, C. HALL-WALKER, L. LAMBRIGHT, AND E. MANZO, "Ingestion of Cigarettes and Cigarette Butts by Children—Rhode Island, January 1994–July 1996." *Morbidity and Mortality Weekly Report*, 46: 125–128, 1997 (1 table, 9 references)

Abstract: Ingestion of products containing nicotine was reported in 146 children (<6 years of age) in Rhode Island between 1994 and 1996. Follow-up information was available for 90 children. Fifty (56%) had ingested cigarettes and 40 (44%) had ingested cigarette butts. About 72% of the children ingested less than a whole cigarette and 55% had ingested less than a whole cigarette butt. Children in homes where smoking occurred in front of them were 7.8× more likely to consume cigarettes or butts. Children in homes with ashtrays were 5.9× as likely to ingest tobacco products.

Symptoms were reported in 30 (33.3%) of the children and included spontaneous vomiting (up to four episodes) (26 {87%}), nausea (two {7%}), pale or flushed appearance (two {7%}), lethargy (one {3%}), and gagging (one {3%}). Thirteen (14%) of the children had been taken to a health-care facility. All 30 children recovered fully within 12 hours.

Reference Number: 10406

WAIN, A.A., AND J. MARTIN. "Can Transdermal Nicotine Patch Cause Acute Intoxication in a Child? A Case Report and Review of Literature." *The Ulster Medical Journal*, 73: 65–66, 2004 (7 references)

Abstract: Transdermal nicotine patches (TNPs) are an efficient method of drug delivery to help adults avoid tobacco smoking. A 11-year-old boy who wore his mother's nicotine patch attended hospital as a result of vomiting, being unable to stand, head and stomach ache, and cold and clammy skin. Atropine sulphate (0.02 mg/kg) was administered IM and he was asymptomatic after 4 hours. The transdermal route may cause serious toxicity due to the lack of the first pass effect. TNPs are available without prescription in Ireland.

> Therapeutic use has been associated with a variety of adverse effects including skin rashes, allergic skin reactions, nausea and vomiting, sleep disturbances, headache, chest pain. Symptoms following oral ingestion in children include gastrointestinal symptoms, increased salivation, pallor, weakness, and dizziness.

Reference Number: 10407

WANG, B., AND B. ROSTRON. "Tobacco-Related Poison Events Involving Young Children in the US, 2001–2016." *Tobacco Regulatory Science*, 3: 525–535, 2017 (3 tables, 2 figures, 47 references)

Abstract: Between 2001 and 2016, Poison Control Centers across the United States received 123,876 tobacco-related exposure calls involving children less than 5 years of age. Most calls involved cigarettes, which remained constant at about 5,000 calls/year, followed by chewing tobacco (568 calls in 2001, which increased to 1,302 calls in 2015). EC related calls increased from 7 in 2010 to 84 in 2011 to 2,558 in 2015. Two hundred and seventy-eight children were admitted to intensive care in hospital and 497 children were admitted to non-intensive care in hospital.

Table: Route of Exposure of Tobacco-Related Poison Events Among Children (<5 years of age) by Product Type

Type of Tobacco Product	Dermal (%)	Ingestion (%)	Inhalation (%)	Ocular (%)
Cigarettes	1.0	98.9	0.05	0.1
Cigarette butts	2.2	97.7	0.10	0.0
Chewing tobacco	2.3	96.7	0.06	0.9
Snuff	2.8	95.6	0.20	1.4
Dissolvable tobacco	0.0	100.0	0.00	0.0
Cigars	0.8	98.9	0.2	0.08
E-cigarette device	10.9	83.0	3.7	2.40
E-cigarette liquid	12.1	85.9	0.2	0.08
Hookah	4.3	73.9	17.4	4.30

Source: Adapted from Wang and Rostron, 2017

Our study results indicate that these poison events continue to affect thousands of young Americans each year and remain a serious public health issue. Moreover, the youngest children are particularly at risk, with more than 80% of tobacco-related poison exposure calls involving children aged 5 years or younger. Nicotine has been identified as the toxic substance most commonly reported as causing symptoms such as nausea and vomiting in low doses, and more extensive neurological symptoms with higher doses among children who have consumed cigarettes or butts. The incidence of such harmful events could be reduced by eliminating young children's access to tobacco products, given that previous research has shown that cigarettes and butts were most commonly ingested by young children in homes where smoking occurred in the presence of children and cigarettes and cigarette butts were accessible. The rapid increase in e-cigarette use and associated poison events in recent years raise additional public health concerns. National data indicate that current e-cigarette use has increased dramatically among US youth and adults. Currently, e-cigarette use in the US is still much lower than cigarette smoking, both in terms of prevalence in adults (3.7% vs 16.8% based on 2014 National Health Interview Survey data) and sales (an estimated $2 billion vs $80 billion in 2013), but e-cigarettes now account for a disproportionate share of tobacco-related poison exposure calls. Further monitoring of this situation is warranted given the increasing prevalence of e-cigarette use and the growing number of associated poisoning events. On August 8, 2016, the FDA began the implementation of the Deeming Final Rule, which extends the Agency's authority to all tobacco products, including e-cigarettes.

Reference Number: 10408

ORDONEZ, J.E., K.C. KLEINSCHMIDT, AND M.B. FORRESTER. "Electronic Cigarette Exposures Reported to Texas Poison Centers." *Nicotine and Tobacco Research*, 25: 209–211, 2014 (22 references)

Abstract: A total of 225 exposures to EC liquids were reported to Texas poison centers between 2009 and 2014. Fifty-three percent of those exposed were children less than 15 years of age. The genders were evenly divided. The routes of exposure were ingestion (78%), inhalation (9%), dermal (8%), ocular (4%), and mixed (10%). The clinical symptoms most reported were vomiting (20%), nausea (10%), dizziness (5%), and headache (4%). During the same time period, there were 2,888 exposures to other nicotine/tobacco products including cigarettes (1,893), snuff (272), chewing tobacco (267), cigars (56), and filter tips/butts (49). If ingested, propyl glycol is metabolized to lactate and propionic acid.

The number of e-cig exposures reported to TPCN, albeit currently small, is increasing. Fifty three percent were less than 5-years of age; 41% were ≥20-years of age. The exposures tended to be unintentional, occurred through ingestion, and occurred at the patients' homes. While serious exposures were uncommon, because of the high concentrations of the nicotine solutions and the increasing use of these products, there is still the potential for dangerous toxicity in the future and close surveillance is warranted.

Reference Number: 10409

CHOI, A., M. LE, T. RAHIM, C. ROSE, AND T. KOSATSKY. "Electronic Cigarette Exposures Reported to the British Columbia Drug and Poison Information Centre: An Observational Case Series." *Canadian Medical Association Journal Open*, 7: E462–E471, 2019 (6 tables, 2 figures, 41 references)

Abstract: Between 2012 and 2017, there were 243 calls recorded at the BC Drug and Poisoning Information Centre, of which 186 were unique exposure to ECs and e-juice. Of the 186 unique exposures, there were 108 males (58.1%) with a median age of 3 years. Most nicotine exposures occurred in the place of residence. Almost all exposures in children less than 5 years of age were accidental (95%) and involved the ingestion of bottled e-juice (53%). In adults aged 25 years or over, the exposure to nicotine was due to spills (13%), EC malfunction, or mistaking e-liquids for other substances such as eye drops (13%).

Table: Symptoms Present in Cases with Dermal, Ingestion, and Inhalation Exposures in Children

Symptoms Present	Dermal	Ingestion	Inhalation
Yes	54.5%	31.8%	89.3%
No	31.8%	49.2%	7.1%
Not Recorded	13.6%	19.7%	3.6%

Source: Adapted from Choi et al, 2019

The availability of poison centre data presented a timely opportunity to study a rapidly changing, relatively novel exposure through detailed exposure records obtained within minutes to hours of the exposure. British Columbia experienced a sixfold increase in ENDS-related calls to the provincial poison centre between 2012 and 2017, driven by ingestions in young children. Electronic cigarettes are relatively novel, and it is important that both the public and health care providers be aware of potential health effects. Nicotine is the primary known hazard, and the amount of nicotine present in e-juices can exceed the labelled concentrations. High-concentration products may present a unique hazard to young children given their low body weight. Moreover, e-juice bottles are sometimes mistaken for other substances. Clear, standardized labelling and child-resistant packaging should be required. As rates of exposure are not declining, increased surveillance (including surveillance across multiple poison centres) may elucidate the ongoing epidemiologic features of exposure.

Reference Number: 10410

KAMBOJ, A., H.A. SPILLER, M.J. CASAVANT, T. CHOUNTHIRATH, AND G.A. SMITH. "Pediatric Exposure to E-Cigarettes, Nicotine, and Tobacco Products in the United States." *Pediatrics*, 137: (9 pp), 2016 (3 tables, 2 figures, 30 references)

Abstract: Between 2012 and 2014 there were 29,141 calls to the US National Poison Data Systems for exposure to nicotine and tobacco products among children less than 6 years of age. Cigarettes accounted for 60.1% of the exposures, ECs for 14.2% ,and other tobacco products for 16.4%. The monthly number of exposures due to ECs increased by 1,492.9% over that period of time. Children exposed to ECs had 5.2× OR of health care admission and 2.6× OR of a more severe outcome than with cigarettes. There was one pediatric death due to EC nicotine exposure.

Table: How Children Obtained the Tobacco Product in Poisoning Cases

Access to Product	E-cigarettes	Cigarettes	Other
Stored within sight of the child	41.2%	44.5%	48.2%
Product always left out	6.2%	10.3%	12.7%
Product temporarily left out and in use	12.3%	6.6%	11.4%
Child access product from purse	3.3%	11.5%	–
Product stored inappropriately	7.8%	5.0%	8.2%
Other	29.2%	22.1%	19.6%

Source: Adapted from Kamboj et al, 2016

Young children are at risk for ingestion because of their curiosity and tendency for oral exploration. As they get older and their taste discrimination matures, they may be more attracted to flavored tobacco products. Under the Family Smoking Prevention and Tobacco Control Act, a ban was implemented in 2009 on cigarettes with characterizing flavors other than menthol and tobacco, including fruit- and candy-flavored cigarettes. Currently, no such ban exists on e-cigarettes and other tobacco products. The plethora of added flavors, colorful packaging, and candylike appearance associated with e-cigarette devices and their liquid refill products, as well as other tobacco products, may make them more attractive to young children. With increasing mobility and dexterity, toddlers are also able to access and open liquid nicotine refill containers, which currently generally lack child-resistant closures and often contain potentially lethal amounts of nicotine for a toddler. These developmental changes in taste discrimination, mobility, and dexterity may help explain the slightly greater median age of exposure for e-cigarettes (2.0 years) and other tobacco products (1.3 years) compared with cigarettes (1.1 years).

Reference Number: 10411

CHANG, J.T., B. WANG, C.M. CHANG, AND B.K. AMBROSE. "National Estimates of Poisoning Events Related to Liquid Nicotine in Young Children Treated in US Hospital Emergency Departments, 2013–2017." *Injury Epidemiology*, 6: 10 (6 pp), 2019 (3 tables, 20 references)

Abstract: An estimated 2 million youths and 7.9 million adults in the US vape nicotine. Between 2013 and 2017, it is estimated that there were 4,745 nicotine poisoning cases in children under the age of 5 years treated at hospital. The most common route of exposure was by ingestion (96.9%) followed by dermal exposure (2.6%). Many US states do not require e-liquids to have child-resistant packaging. Most of the poisonings were treated at ER and 4% were admitted to hospital. The most common symptoms were vomiting and nausea, crying, red eyes, cough, sleepiness, and unresponsiveness.

Table: E-Liquid Poisonings in Children <5 Years of Age

ENDS Liquid Unit of Measurement	Minimum	Maximum	Mean
mg	1.8	100	3
mL	0.2	118.3	16.8
Bottles	0.5	1	0.875

Source: Adapted from Chang et al, 2019

In conclusion, our study demonstrated that thousands of young children experienced ENDS-related nicotine poisoning and were treated in hospital EDs. Some of these children developed serious clinical symptoms that could be life-threatening. While federal legislation now requires special packaging for liquid nicotine containers such as child-resistant packaging, to prevent child exposure to liquid nicotine, additional efforts, such as public education, could complement the existing child-resistant packaging law to further lower the burden of unintended nicotine exposure to ENDS e-liquids in children.

Reference Number: 10412

GOVINDARAJAN, P., H.A. SPILLER, M.J. CASAVANT, T. CHOUNTHIRATH, AND G.A. SMITH. "E-Cigarette and Liquid Nicotine Exposures Among Young Children." *Pediatrics*, 141: (8 pp), 2018 (1 table, 3 figures, 38 references)

Abstract: There were 8,269 liquid nicotine exposures to children less than 6 years of age reported to US poison control centers between 2012 and 2017. At least 2 young children have already died of liquid nicotine exposure. On 26 July 2016, federal law required child-resistant liquid nicotine packaging. In the US states without a childproof nicotine law, the number of exposures decreased 4.4% after the new law went into effect. Fatalities could be prevented by limiting the volume and concentration of nicotine to a sublethal dose for an average child as was successful with aspirin poisoning.

Liquid nicotine can pose a serious poisoning risk for young children. The frequency of liquid nicotine exposures among children <6 years old has decreased since January 2015, which may, in part, be attributable to adoption of child-resistant packaging laws for liquid nicotine containers and greater public awareness of the risks associated with these products. Additional measures, including child-resistant e-cigarette devices, use of flow restrictors on liquid nicotine containers, and regulations on e-cigarette liquid flavoring, labeling, and concentrations, could further reduce the incidence of these exposures and the likelihood of serious medical outcomes when exposures do occur.

Reference Number: 10413

NOBLE, M.J., B. LONGSTREET, R.G. HENDRICKSON, AND R. GERONA. "Unintentional Pediatric Ingestion of Electronic Cigarette Nicotine Refill Liquid Necessitating Intubation." *Annals of Emergency Medicine*, 69: 94–97, 2017 (2 figures, 13 references)

Abstract: Nicotine e-liquids used in ECs are highly concentrated, unreliably packaged, and poorly regulated. A 6-year-old girl was administered 10 mL of a nicotine e-liquid from a bottle that looked similar to children's Motrin. It caused an immediate burning sensation in her mouth and throat and she rapidly lost consciousness. The patient arrived at the hospital 25 minutes after exposure. She fluctuated between agitation and unresponsiveness. A serum sample taken 60 minutes after exposure showed 348 ng/mL nicotine, 742 ng/mL cotinine, and 32 ng/mL 3-HC. The e-liquid was analyzed by LC/MS and was found to contain 70.3 mg/mL nicotine rather than the labelled 60 mg/mL. The patient was intubated and given supportive therapy. She recovered and was released the next morning.

With supportive care, this patient survived an acute ingestion of 703 mg, or 35 mg/kg, of liquid nicotine. The patient had virtually immediate onset of symptoms consistent with nicotine toxicity, including initial sympathetic (agitation) and neuromuscular effects (increased neuromuscular activity, myoclonic movements, and fasciculations), followed shortly by parasympathetic and cholinergic effects (bradycardia, obtundation, vomiting, diaphoresis, and copious secretions). Within 30 minutes of the exposure, the patient required intubation in the ED for airway protection. She also received activated charcoal, which is a reasonable approach for a patient whose airway is protected, given data to support adsorption of nicotine. The patient's clinical course was thereafter uneventful, and she was extubated within approximately 12 hours of the ingestion.

Reference Number: 10414

GILL, N., G. SANGHA, N. POONAI, AND R. LIM. "E-Cigarette Liquid Nicotine Ingestion in a Child: Case Report and Discussion." *Canadian Journal of Emergency Medicine*, 17: 699–703, 2015 (2 tables, 2 figures, 12 references)

Abstract: A 2-year-old girl was crying and vomiting with purple discoloration of her lips and with an open 60 mL bottle of e-liquid (24 ng/mL: nicotine) with 15 mL missing. Liquid had spilled on her shirt. She recovered uneventfully.

Table: The Nicotine Content of Various Tobacco Products

Tobacco Product	Nicotine Content
Cigarette butt	4–5 mg/butt
Cigarette	13–30 mg/cigarette
Used nicotine patch	83 mg/patch
EC liquid	6–72 mg/mL

Source: Adapted from Gill et al, 2015

The commonly seen initial vomiting can help to reduce absorption and is usually self-limited; treatment with anti-emetics is not recommended, but maintaining adequate hydration in the form of intravenous fluids may be necessary in severe cases. The onset of symptoms is usually within two hours, with the exception of delayed-release products such as patches.

Overstimulation of the nicotinic receptors can cause receptor inhibition after the initial stimulatory effects. More severe toxicity can result in seizures, confusion, weakness, bradycardia, hypotension, and respiratory muscle paralysis. Initial hypertension, tachycardia, and tachypnea can be followed by hypotension, bradycardia, and bradypnea.

The treatment of nicotine toxicity is primarily symptomatic and supportive. In cases of dermal exposure to a nicotine solution such as the one in this case report, contaminated clothing should be removed and affected areas should be washed thoroughly with mild soap and copious amounts of cold water to reduce absorption.

Reference Number: 10415

DAVIES, P., S. LEVY, A. PAHARI, AND D. MARTINEZ. "Acute Nicotine Poisoning Associated with a Traditional Remedy for Eczema." *Archives of Disease in Childhood*, 85: 500–502, 2001 (2 figures, 6 references)

Abstract: An 8-year-old boy with moderate eczema had a traditional herbal paste spread on his affected parts of his body. The paste was made from a ground mixture of tobacco leaves, lime, and freeze-dried coffee. After 30 minutes he felt dizzy, vomited, and became agitated and then unconscious. His serum nicotine and cotinine concentrations 12 hours after the onset of symptoms were 89 ng/mL and 1,430 ng/mL, respectively. The boy improved with supportive care and was discharged 5 days later.

This is the first report of nicotine poisoning secondary to dermal absorption from a traditional remedy. Nicotine is a highly toxic substance in overdose and is rapidly absorbed from the skin. We recommend that parents are educated about the potential toxicity of seemingly innocuous substances used in traditional remedies, and that extra care needs to be taken when taking a history through an interpreter. Health care professionals should be aware of the symptoms and signs of nicotine poisoning and that rapid absorption of the drug is possible through the skin. Treatment of nicotine poisoning is essentially supportive but atropine can specifically treat muscarinic symptoms such as bradycardia, salivation, and wheezing.

Reference Number: 10416

THORNTON, S.L., L. OLLER, AND T. SAWYER. "Fatal Intravenous Injection of Electronic Nicotine Delivery System Refilling Solution." *Journal of Medical Toxicology*, 10: 202–204, 2014 (24 references)

Abstract: A 29-year-old man with a history of depression was found unresponsive beside a suicide note indicating that he had injected himself with an e-liquid refill. He was a smoker and had been attempting to quit using EC. He developed multiple seizures and despite treatment was declared brain dead after 5 days. His admission serum nicotine concentration was 2,000 ng/mL (normal range 5 to 90 ng/mL) and cotinine was 2,100 ng/mL (normal range 100 to 1,200 ng/mL). No injection site was found upon examination and it is possible the victim consumed the e-liquid orally.

Finally, there are multiple other chemicals in these eLiquid solutions, including propylene glycol. It is possible that this might have caused or contributed to his cardiovascular collapse.

As the use of ENDS continues to grow, exposures to eLiquids and other associated nicotine-containing solutions may increase. This case highlights the potentially fatal results.

Reference Number: 10417

SCHIPPER, E.M., L.C.G. DE GRAAFF, B.C.P. KOCH, Z. BRKIC, E.B. WILMS, J. ALSMA, AND S.C.E. SCHUIT. "A New Challenge: Suicide Attempt Using Nicotine Fillings for Electronic Cigarettes." *British Journal of Clinical Pharmacology*, 78: 1469–1471, 2014 (1 figure, 11 references)

Abstract: A 27-year-old man with a borderline personality disorder presented at a hospital one hour after ingesting 5 e-liquid refills (a total of 420 mg nicotine) and alcohol. He vomited 3 times before arrival. His only symptoms were tachycardia and excessive salivation. He was administered activated charcoal twice. His serum nicotine concentration as measured by LC/MS/MS was 50 ng/mL at 2 hours and not detectable at 4 hours. The corresponding cotinine concentrations were 250 ng/mL and 180 ng/mL, respectively. His BAC was 390 mg/dL.

> In conclusion, ingestion of e-liquid can lead to potentially lethal nicotine poisoning. Given the increasing popularity of electronic cigarettes, the frequency of nicotine intoxications is likely to rise in the future. Physicians working in the emergency department and general practice should become familiar with this intoxication.

Reference Number: 10418

HOLOYDA, B., K. FRITZ, AND G.L. XIONG. "A Case of Nicotine Overdose-Induced Seizure on an Inpatient Psychiatric Unit." *Annals of Clinical Psychiatry*, 28: 66–67, 2016

Abstract: A 39-year-old non-smoking psychiatric patient developed a seizure and dazed and confused behavior after hording other patients' nicotine patches and chewing them.

> The behavior had occurred over a period of days, making it unclear how much nicotine she absorbed in the 24 hours preceding her seizure event, the ED physician did not draw nicotine levels.

Reference Number: 10419

PAIK, J.H., S. KANG, A. DUREY, J.H. KIM, AND A.J. KIM. "Symptomatic Bradycardia due to Nicotine Intoxication." *Revista Brasileira de terapia intensiva*, 30: 121–126, 2018 (1 table, 1 figure, 17 references)

Abstract: Nicotine is a dangerous drug extracted from tobacco leaves and when absorbed in excessive amounts can lead to respiratory and cardiac failure. E-cig liquids are a new method of nicotine poisoning. Severe nicotine poisoning is biphasic, initially excitatory (salivation, nausea, tachycardia, hypertension, anxiety, muscle spasms, and seizures) followed by paradoxical inhibitory such as bradycardia, lethargy, hypotension, and paralysis. The nicotine dose that could result in death in an adult was estimated to be 40–60 milligrams. A 53-year-old man was found with impaired consciousness, diarrhea, and vomiting after consuming 3 mL of Pure Nicotine (150 mg/mL) in a suicide attempt. His blood cotinine concentration was 1,296 ng/mL. He was administered atropine, dopamine and saline IV and recovered. He was discharged after 3 days.

Each country has different laws and regulations regarding e-cigarettes; in the United States, the Food and Drug Administration manages cigarette products, including e-cigarettes. In Korea, cigarette-related legislation is currently under the jurisdiction of the Ministry of Strategy and Finance. We believe that this limits profound consideration of the public health risks of cigarettes. Allowing the general public to purchase high-concentrations of liquid nicotine for use in e-cigarettes should be recognized as increasing the risk of acute, life-threatening toxicity. There is a need for additional, multi-faceted regulations that account for the unique risks of e-cigarettes, as they differ from commercial cigarettes.

Reference Number: 10420

MAESSEN, G.C., A.M. WIJNHOVEN, R.L. NEIJZEN, M.C. PAULUS, D.A.M. VAN HEEL, B.H.A. BOMERS, L.E. BOERSMA, B. KONYA, AND M.A.G. VAN DER HEYDEN. "Nicotine Intoxication by E-Cigarette Liquids: A Study of Case Reports and Pathophysiology." *Clinical Toxicology*, 58: (8 pp), 2019 (3 tables, 3 figures, 71 references)

Abstract: The symptoms most frequently seen in nicotine intoxication are vomiting, agitation, pallor, hypertension, tachycardia, and headache. PG intoxication symptoms are lactic acidosis, renal failure, and CNS depression. Vegetable glycerin (VG) related symptoms of intoxication are headache, nausea, diuresis, and hyperglycemia. A review was conducted of 26 cases of e-liquid nicotine intoxication involving 31 patients. Nicotine binds to a variety of nicotinic cholinergic receptors throughout the body and acute intoxication can induce both stimulation and depressant effects and can result in biphasic effects. For the patients up to 10 years of age, all e-liquid ingestions were accidental. But between 13 to 53 years of age, nearly all cases were intentional—probably suicide attempts. Eleven of 31 cases resulted in death. The mean plasma nicotine concentration among survivors was 307 ng/mL and 3,360 ng/mL in patients who died. The main symptoms of nicotine intoxication for those who died included tachycardia, cardiac arrest, seizure, and hypoxic brain injury.

Table: Major Symptoms Observed for Nicotine, Propylene Glycol and Vegetable Glycerin Overdoses

Drug	Symptoms of Overdose
Nicotine	Vomiting, agitation, pallor, hypertension, tachycardia, headache
Propylene Glycol	Hyperosmolality, lactic acidosis, hemolysis, renal failure, CNS depression
Vegetable Glycerin	Headache, nausea, diuresis, hyperglycemia, dehydration

Source: Adapted from Maessen, Wijnhoven, Neijzen, et al, 2019

> As e-liquids are very easily accessible and purchased by many people, we urge to raise more awareness about the dangers of these substances when ingested or injected. We expect this will especially prevent more cases of accidental ingestion in young children. On a global scale, stricter regulation is necessary to ensure accurate descriptions of contents on e-liquid cartridges. This is also important for healthcare professionals, who need to know the amount of substances ingested by patients. As for further research, we suggest examining possible PG/VG-nicotine interactions.

Reference Number: 10421

RICHMOND, S.A., I. PIKE, J.L. MAGUIRE, AND A. MACPHERSON. "E-Cigarettes: A New Hazard for Children and Adolescents." *Paediatrics and Child Health*, 23: 255–259, 2018 (1 table, 36 references)

Abstract: A survey was conducted of practicing pediatricians and showed 220 cases of children and adolescents with adverse effects due to inhalation or ingestion of ECs and e-liquids. There were 135 reported cases (ages 15 to 19 years) due to inhalation, with mainly cough/nausea treated at an outpatient clinic or office. The ages of the ingestion cases were between 1 and 4 years and involved mainly nausea, vomiting, and coughing treated at the ER.

Table: Comparison of Inhalation and Ingestion Cases Involving E-cigarettes in Canada

	Ingestion Cases	Inhalation Cases
Age	1–4 years	15–19 years
Most Reported Symptom	Nausea, vomiting, cough	Cough, nausea, vomiting
Setting Patient Treated	ED	Outpatient clinic, office
E-Cigarette Access	Home	Mall kiosk, store

Source: Adapted from Richmond et al, 2018

E-cigarettes, recently introduced into the Canadian market are causing harm to children and adolescents. Results of this study highlight the spectrum of injuries resulting from exposure to e-cigarettes and e-fluid. Children and youth reported in this survey presented to paediatricians with injuries including cough, nausea/vomiting, throat and respiratory irritation or acute nicotine toxicity. Improving education for parents and caregivers about the risk of injury that can result from exposure to e-cigarettes and e-fluid may reduce the risk. However, more stringent legislation around the use, access, and control of sale and purchase of e-cigarettes in Canada and the USA is needed. Policy recommendations from the Canadian Paediatric Society related to the federal Tobacco Control Act include: regulation of the marketing, packaging (including child resistant packaging on all e-fluid containers) and labelling of devices (i.e., enforcement of a maximum dose of nicotine in e-fluids together with accurate labelling; all packaging to include information of the potential and known harmful effects of use; banning advertising, sponsorship and marketing at events and activities where children and youth are involved); restriction of e-cigarette manufacturers or sellers from making positive health claims related to the use of the products; restrictions on the sale of e-cigarettes and products over the internet to identified adults with direct mailing options; and enactment of legislation, across all provinces and territories, making it illegal for children and youth (by provincial legal age determinants) to use, purchase or have any form of e-cigarette device.

Reference Number: 10422

SEO, A.D., D.C. KIM, H.J. YU, AND M.J. KANG. "Accidental Ingestion of E-Cigarette Liquid Nicotine in a 15-Month-Old Child: An Infant Mortality Case of Nicotine Intoxication." *Korean Journal of Pediatrics*, 59: 490–493, 2016 (2 tables, 4 figures, 13 references)

Abstract: A 15-month-old baby girl consumed 5 mL of a liquid nicotine e-liquid containing 10 mg/mL of nicotine. The liquid was mistaken for cough medicine. She vomited immediately and soon lost consciousness and was taken to hospital by ambulance. At hospital she was unresponsive and both pupils were fixed and dilated. She weighed 12 kg. Despite treatment she died 44 hours later. It is estimated that she consumed 50 mg of liquid nicotine, which corresponds to a dose of 3 mg/kg- higher than the estimated fatal nicotine dose in children of 1 mg/kg. Treatment of nicotine poisoning is usually supportive. Ingestion of a high dose of nicotine leads to arrhythmia, bradycardia, convulsions, coma, respiratory failure, and rapid progression to hypotension and death.

> In conclusion, the use of e-cigarettes is increasing with little awareness of liquid nicotine's toxicity. Even though we attempted all means to maintain the child's vital signs, hypoxic brain damage was fatal and irreversible. Although serious events or fatal overdoses of this agent have rarely been reported, even a small dose which is fatal to children can lead to death, as in our case. Therefore, efforts should be made to increase public awareness nicotine toxicity to prevent such accidents.

Reference Number: 10423

SANCHEZ, P., J-L DUCASSÉ, M. LAPEYRE-MESTRE, O. MARTINET, P. ROUGÉ, M.F. JORDA, AND B. CATHALA. "Nicotine Poisoning as a Cause of Cardiac Arrest?" *Clinical Toxicology,* 34: 475–476, 1996 (9 references)

Abstract: A 33-year-old male smoker who used multiple nicotine patches, but did not decrease his smoking, was found unconscious and covered in vomit. He suffered a cardiac arrest and was treated in hospital and in ICU but died 4 days later. His serum nicotine concentration 11 hours after his cardiac arrest was 3,700 ng/mL, as determined by HPLC and MS. A second plasma nicotine concentration was 1,200 ng/mL 57 hours after his cardiac arrest.

> The clinical manifestations of nicotine poisoning are not specific: nausea, vomiting, diarrhea, increased salivation, pale skin, dizziness, confusion, headache, vision or hearing changes. There can be an initial elevated blood pressure and pulse, followed by hypotension and bradycardia, which can lead to shock. Cardiac arrhythmia can appear. The neurological effects are a general cortical stimulation with seizures followed by general inhibition.

Reference Number: 10424

DAVIS, B., A. RAZO, E. NOTHNAGEL, M. CHEN, AND P. TALBOT. "Unexpected Nicotine in Do-It-Yourself Electronic Cigarette Flavourings." *Tobacco Control*, 25 (E1): e67–e68, 2016 (1 figure, 6 references)

Abstract: EC users can create DIY refill liquids by blending bottled nicotine/propylene glycol/glycerol with a flavoring solution. The flavorings are assumed to be nicotine-free, but of 30 different flavoring solutions tested for nicotine by HPLC and GC/MS, 4 were positive. Two had quantifiable nicotine concentrations of 14.2 and 95.4 mg/mL. The total nicotine content in these 5 mL bottles were 71 mg and 477 mg, respectively. Nicotine can be fatal at a dose of 500–1000 mg in adults and 19 mg in children.

> These products, which are presented to the consumer as "nicotine free", could lead to unwanted addiction, poisoning, or even death. The simplest solution to this problem would be, at minimum, to require that any products containing nicotine be clearly labelled with the term "nicotine" and an accurate concentration. Consumers who wish to use 0% nicotine products could then avoid unwanted exposure and EC users could protect their children from accidental ingestion of nicotine. The demonstration of nicotine in presumably nicotine-free DIY flavouring solutions should be valuable information for regulatory agencies, physicians, EC users and poison control centres.

Reference Number: 10425

BENDEL, G.S., H.M. HILLER, AND A. RALSTON. "Nicotine Toxicity Secondary to Aftermarket Modifications to a Vaping Device." *Military Medicine*, usab223: (4 pp), 2021 (1 figure, 25 references)

Abstract: In 2019, vaporizer use among enlisted soldiers approached 20%, but in October 2019, EC sales were suspended at all army, navy, and air force exchanges in the United States due to EVALI (e-cigarette or vaping product associated lung injury). Two enlisted men (ages 19 and 20 years) presented to EMS due to altered mental state after taking a hit from a friend's vaping device. A urine drug screen only detected cotinine (a metabolite of nicotine) at concentrations of 3,252 and 2,337 ng/mL respectively. The EC had been modified by replacing the 100 W coil with a 150–200 W coil, nearly doubling the amount of heat and nicotine produced.

Treatment modalities in nicotine toxicity, regardless of the route of ingestion, is largely supportive. If worrisome inhibitory effects occur, atropine for bradycardia and supplemental oxygen for hypoxia have shown success in limited cases. Activated charcoal can be administered in both ingestion and inhalation exposures as both result and sustain themselves by enterohepatic circulation.

It is likely that nicotine toxicity from modified vape devices become a worsening issue as these devices become more refined and available on the Internet. Coupled with the high rates of vaping and e-cigarette use in the younger armed forces population, awareness of the potential dangers of this supposedly safe cigarette alternative is paramount not only to soldiers themselves and their leadership, but to Army physicians. Recently, the DOD has taken steps to control the situation by discontinuing all sales of e-cigarette and vaporizer products on military bases, but there is still more to be done. Policy statements and education all the way down to the unit level are required to demonstrate that these devices are not safer than traditional cigarettes.

Reference Number: 10426

ANONYMOUS. "Nicotine Poisoning in Calf." *Canadian Journal of Comparative Medicine and Veterinary Science*, 7: 250, 1943

Abstract: During World War II there was a shortage of rotenone used for the treatment of warbles in cattle. A nicotine sulphate dressing was applied to the affected cattle instead. Unfortunately, nicotine is well absorbed through the skin and caused poisoning of the cattle.

The second case concerned a similar outbreak, after the use of the Ministry of Agriculture nicotine sulphate formula for warbles. Jones and John were convinced in this case that there was no question of the dressing having been licked off the back and that the symptoms were caused by absorption of the nicotine through the warble holes. Four of the cattle were slaughtered because of the effects of the poison and these were the beasts with the greatest numbers of warbles and the most perforations. They are quite definitely of the opinion that the dressing is highly dangerous.

1.05 OTHER SOURCES OF NICOTINE

"... Unless subjects eat 90 kg tomatoes a day."

—Jarvis, "Dietary nicotine ....
Unless Subjects Eat 90 kg Tomatoes a Day" (1994)

... or unless you eat a lot of potato peels (10501–2). Nicotine will be present in a number of plants related to the nicotine (nightshade) family, such as tomatoes and potatoes, although in much lower concentrations than in tobacco. Nicotine is concentrated in the leaves of these plants in order to repel insects, not in the tuber or fruit. Thus, potatoes and tomatoes contain far less nicotine than the tobacco leaves that are smoked. The tobacco industry, as is its wont, attempted to confuse the issue of obtaining nicotine via second and thirdhand smoke by stating that the nicotine detected in non-smokers was due to dietary sources and not tobacco smoke. However, significant urine nicotine concentrations would only occur, however, if the person ate at least 90 kilograms of tomatoes a day (10501–2).

Reference Number: 10501

JARVIS, M.J. "Dietary nicotine.... Unless Subjects Eat 90 kg Tomatoes a Day." *British Medical Journal*, 308: 62, 1994 (5 references)

Abstract: Low nicotine concentrations have been detected in some vegetables such as tomatoes and potatoes, which are in the same plant family as tobacco (Solanaceae). These potential sources of nicotine have been used to discredit research into secondhand smoke (SHS), as the cotinine levels detected in non-smokers may be due not to passive inhalation of cigarette smoke but to food. It is estimated that one would have to consume approximately 90 kg of tomatoes a day to obtain detectable nicotine/cotinine concentrations. Dietary nicotine has curiosity value but is irrelevant for the study of SHS.

> The notion that nicotine from dietary sources might contribute sufficiently to measured cotinine concentrations in non-smokers as to make objective assessment of exposure to environmental tobacco smoke unreliable or impossible has been suggested more than once, most recently by Domino et al. This report received widespread publicity in the media, including both the BMJ and the Lancet. If the claim could be substantiated there would be important implications for investigations of the health effects of

passive smoking, which have increasingly used cotinine as a quantitative marker of exposure.

Reference Number: 10502

SHEEN, S.J. "Detection of Nicotine in Foods and Plant Materials." *Journal of Food Science*, 53: 1572–1573, 1988 (1 table, 9 references)

Abstract: Nicotine is present naturally in plants from the Solanaceae family with the highest concentrations occurring in the Nicotiana genus (N tabacum (2-4%), N rustica (up to 9% nicotine)). The detection of nicotine in plants other than tobacco was first recorded in decayed potatoes in 1852. In this study, the nicotine content of various fruits, vegetables, and processed foods was determined by GC. Nicotine was not detected in non-Solanaceae fruits and vegetables and not detected in processed foods such as ketchup.

Table: Nicotine Content in Various Food and Plant Material

Food/Plant Material	Nicotine Detected (ppm)
Tomato fruit	2.31
Tomato root	3.45
Potato peel	14.80
Peeled potato	Not detected
Eggplant	2.65
Green pepper	3.15

Source: Adapted from Sheen, 1988

Potato peel had a high nicotine concentration, while nicotine was not detected in potato storage tissue. One may speculate that nicotine in potato skin could relate to defense mechanisms against pathogenic fungi and bacteria. There was no detectable nicotine in the non-Solunaceae fruit and vegetables analyzed.

Processed tomato products had no detectable level of nicotine. This may be due to enzymatic oxidation and other chemical reactions of nicotine during processing.

Reference Number: 10503

EL-RAMADY, H., N. ABDALLA, T. ALSHAAL, A. ELHENAWY, M. AMER, A. SHEHATA, J. PAULSEN, M. KLEINWÄCHTER, D. SLEMAR, AND H. TAHA. "Contamination of Plant Foods with Nicotine: An Overview." *Report of the DBG, Kommission VIII*, Munich: (4 pp), 2015 (1 figure, 20 references)

Abstract: Nicotine is the major alkaloid of tobacco and other tobacco species and is highly toxic. It is a stimulant like cocaine or amphetamine and speeds up the rate of the central nervous system. Between 1997 and 2005, there has been a statistically significant increase in the nicotine concentration of US cigarettes by 1.1% (0.019 mg/cigarette) per year. The roots of the tobacco plant can synthesize a large amount of nicotine, which is then transported to the shoot and leaves under adverse conditions such as wounding or insect attack.

> Thus, it could be concluded that, there are generally three nicotine sources for contamination of plant: the first source belongs the plants, which synthesize the endogenous nicotine under certain conditions (e.g. stress), the second source represents nicotine results form contaminations by smokers (smoke of cigarettes / nicotine residues at the fingers of harvesters) and finally nicotine, which is taken up from the soil (e.g. resulting from butts of cigarettes thrown away). These previous sources for contamination of plants with nicotine did not include the illegal usage of nicotine in insecticides.

Reference Number: 10504

SELMAR, D., U.H. ENGELHARDT, S. HÄNSEL, C. THRÄNE, M. NOWAK, AND M KLEINWÄCHTER. "Nicotine Uptake by Peppermint Plants as a Possible Source of Nicotine in Plant-Derived Products." *Agronomy for Sustainable Development*, 35: 1185–1190, 2015 (1 table, 4 figures, 10 references)

Abstract: Nicotine was frequently used as a pesticide in Europe but was banned in the European Union in 2009 due to toxicological concerns. However, traces (contamination) of nicotine have still been found in numerous plant-derived commodities. This study determined the nicotine uptake of the peppermint plants (*Mentha x piperita*) from the soil and the air. In one experiment, cigarette tobacco was mulched around the plant and in the other, cigarettes were smoked in a greenhouse where peppermint plants were growing. The nicotine concentrations were determined by GC/MS.

Table: Occurrence and Frequency of Nicotine Contaminations in Some Plant-Derived Commodities

Commodity	Maximum Nicotine Contamination Detected (µg/g)	Sample Containing Nicotine (%)
Chamomile flowers	4.08	64
Linden flowers	1.08	98
Black tea	0.87	82
Sage leaves	0.35	100
Peppermint leaves	0.11	58

Source: Adapted from Selmar et al, 2015

For the first time, we have shown that peppermint plants take up nicotine from contaminated soils, and the resulting nicotine concentrations are several times higher than the maximum residue level set by the European authorities. Moreover, tremendously elevated nicotine levels were also detected after fumigation with cigarette smoke.

These results suggest that the widespread occurrence of nicotine in medicinal, spice and food plants may, at least in part, be due to exogenous nicotine sources. However, all the peppermint plants analysed in this study, including the untreated controls, contained minor amounts of the alkaloid, which suggested that peppermint plants may be able to synthesise nicotine endogenously. This assumption is supported by our finding that the Mentha plants are also able to metabolise the exogenous applied alkaloid, a process that is most likely based on a putative turnover of endogenously synthesised compounds.

In addition to the nutritional aspects, these results also demonstrate that a certain alkaloid could be translocated from one plant after its death to another plant species, even though it has no direct functional relevance.

Reference Number: 10505

IKKA, T., H. YAMASHITA, I. KURITA, Y. TANAKA, F. TANIGUCHI, A. OGINO, K. TAKEDA, N. HORIE, H. HOJO, F. NANJO, AND A. MORITA. "Quantitative Validation of Nicotine Production in Tea (Camellia sinensis L.)." *Public Library of Science One*, 13: (13 pp), 2018 (5 tables, 38 references)

Abstract: Tea is one of the world's most popular beverages. Green tea is produced mainly in China, Vietnam, and Japan. Since nicotine is a banned pesticide, the Maximum Residue Level (MRL) of nicotine in tea has been set to 0.01 μg/g dry weight in Japan. The nicotine content of 32 green tea, 79 black tea, and one oolong tea samples were determined by LC/MS.

> In this study, all purchased tea samples and samples of roots, stems, flower buds and seeds contained nicotine ranging from 0.003 to 0.694 μg/g DW. In addition, tea cells cultured in B5 medium without nicotine contained nicotine at a content of 0.033 μg/g DW. These results suggest that tea plants contain nicotine that is not derived from exogenous sources. In addition, nicotine contents in tea leaves changed seasonally and differed among cultivars. Furthermore, a difference in nicotine contents was observed among tea plant-growing regions. All these results strongly suggest that tea plants, similar to members of the Solanaceae, contain endogenous nicotine. Although low, the levels are higher than allowed MRLs. Despite these findings, the possibility of contamination of commercial tea by exogenous nicotine cannot completely be excluded.
>
> In this study, all purchased tea samples and samples of roots, stems, flower buds and seeds contained nicotine ranging from 0.003 to 0.694 μg g–1 DW. In addition, tea cells cultured in B5 medium without nicotine contained nicotine at a content of 0.033 μg g–1 DW. These results suggest that tea plants contain nicotine that is not derived from exogenous sources. In addition, nicotine contents in tea leaves changed seasonally and differed among cultivars. Furthermore, a difference in nicotine contents was observed among tea plant-growing regions. All these results strongly suggest that tea plants, similar to members of the Solanaceae, contain endogenous nicotine. Although low, the levels are higher than allowed MRLs. Despite these findings, the possibility of contamination of commercial tea by exogenous nicotine cannot be completely excluded.

Reference Number: 10506

CAVALIERI, C., L. BOLZONI, AND M. BANDINI. "Nicotine Determination in Mushrooms by LC-MS/MS with Preliminary Studies on the Impact of Drying on Nicotine Formation." *Food, Additives and Contaminants*, 27: 473–477, 2010 (2 tables, 1 figure, 7 references)

Abstract: Significant amounts of nicotine were found in dried mushrooms imported from China to the European Union, where nicotine is considered a pesticide residue. A method was developed to determine the nicotine content of fresh, frozen, and dried mushrooms by LC/MS/MS. The nicotine concentration of fresh and frozen mushrooms ranged from 0.01 to 0.03 mg/kg, whereas in dried mushrooms a nicotine concentration of up to 4.5 mg/kg was detected. Preliminary studies suggest that drying does not increase its nicotine concentration.

> No clear reasons has thus far been established for this unexpected presence of nicotine that could be of various origin, but also results from a combination of different factors: 1) pesticide use; in organic farming nicotine is not allowed to be used as pesticide and in general in European countries the use of plant protection products containing nicotine will phase out by the latest in June 2010, but its use in Third Countries may continue and may lead to residues of nicotine in food; 2) inherent presence of nicotine in wild mushrooms and its eventually arise due to an increase of its biosynthesis caused by some not properly controlled conditions of storage during the drying phase; 3) cross contamination due to bad practices in the storage/drying and packaging process (smoking people handling the mushrooms, storage in rooms that have been disinfected with nicotine, simultaneous drying of tobacco and mushrooms in the same room).

Reference Number: 10507

FURER, V., M. HERSCH, N. SILVETZKI, G.S. BREUER, AND S. ZEVIN. "Nicotiana Glauca (Tree Tobacco) Intoxication—Two Cases in One Family." *Journal of Medical Toxicology*, 7: 47–51, 2011 (1 table, 2 figures, 24 references)

Abstract: Two cases of poisoning that occurred in one family as the result of the ingestion of the cooked leaves of the tree tobacco plant are described. The main toxic component of Tree Tobacco is anatabine ($C_{10}H_{14}N_2$), which is similar in structure and effects to nicotine but is less potent. The Tree Tobacco was mistaken for spinach. A 73-year-old woman collapsed and was found unconscious with extreme bradycardia. Despite medical treatment, she died 20 days later. Her 18-year-old grandson, who had a smaller amount of Tree Tobacco, survived. A table is presented of N gluca poisonings.

> Ingestion of N. glauca leaves or their extract has a toxic potential. The toxicity is mediated by nicotine-like alkaloid, anabasine. This poisoning should be included in the differential diagnosis of sudden onset of muscle paralysis and respiratory arrest.

CHAPTER 2

Absorption, Distribution, and Elimination of Nicotine

> "There is good evidence that nicotine plays a role in tobacco dependence. Consequently, it is important to understand the quantity and time course of nicotine in the body. Following smoking, or an intravenous infusion, nicotine distributes rapidly into tissues, after which it is eliminated with a terminal half-life of about 2 hours."
>
> —Jacob III et al, "Recent Studies of Nicotine Metabolism in Humans" (1988)

The absorption, distribution, and elimination of a drug by the body is called pharmacokinetics. Nicotine is a highly water-soluble drug similar to alcohol but is also more fat (lipid) soluble than alcohol. The absorption, distribution, and elimination patterns are a mixture of alcohol and tetrahydrocannabinol (THC) pharmacokinetics. As discussed in the first section, nicotine is absorbed readily through the skin, unlike alcohol or THC, which can cause poisoning when nicotine e-liquids are spilled on the skin.

When a person smokes, nicotine is distributed to the brain rapidly, within a minute, causing a rapid rise in nicotine concentration. This leads to a greater addiction in contrast to alcohol, which is more slowly distributed to the brain after ingestion due to its oral ingestion.

Nicotine concentration, like that of THC, decreases rapidly after inhalation as the drug is distributed throughout the body by the blood. After the initial rapid drop in blood nicotine concentration caused by redistribution, nicotine is eliminated mainly by the liver at a half-life of about 2 hours. This requires smokers to smoke numerous cigarettes to maintain a blood nicotine level and prevent withdrawal symptoms.

2.01 ABSORPTION (INHALATION, IV, DERMAL, ORAL, GASTRIC)

> "Cigarette companies control the impact and delivery of nicotine in many ways, including designing filters and selecting cigarette paper to maximize the ingestion of nicotine, adding ammonia to make the cigarette taste less harsh, and controlling the physical and chemical make-up of the tobacco blend."
>
> —Court-ordered Corrective Statements,
> *US v Philip Morris USA Inc*, RICO decision (2006)

Nicotine absorption can occur through the oral cavity, lung, skin, eyes, urinary bladder, and stomach. However, nicotine is poorly absorbed through the stomach due to the acidity of the gastric fluids (20101). The rate of nicotine absorption is based on pH. Cigarette smoke is mainly acidic (pH <7), and is mainly in the ionized form, but the pH of nicotine rapidly approaches physiological pH of >7 in the lungs and is rapidly absorbed (20102–3). The absorption of nicotine from cigarette smoke is similar to IV injection (20103).

The pH of pipe and cigar tobacco is typically >7 and so nicotine can readily be absorbed through the oral cavity and does not need to be inhaled into the lungs (20102).

One basic principle is that the faster a psychoactive drug is absorbed the more addictive it is. For example, the absorption of nicotine is very rapid when inhaled in cigarettes and vaping devices, but when nicotine is absorbed slowly, such as dermally (nicotine patch), it becomes much less addictive (20110).

Smokers tend to regulate their puffing pattern to maintain a similar blood nicotine concentration no matter what the nicotine concentration of the tobacco they smoke (20104). One indicator of recent smoking is the blood carbon monoxide concentration (COHb), as burning tobacco produces this toxic gas (20105–6). This indicator is of limited value in e-cigarettes, which produce virtually no carbon monoxide (CO) gas in their vapor.

Before laws banning smoking indoors and in numerous other public spaces, it was virtually impossible for a non-smoking person to avoid absorbing nicotine (20106).

Electronic cigarettes have been developed since the early cig-a-likes, which were similar in size to cigarettes, to increase the absorption of nicotine to the speed of regular cigarettes and thus increase its addictiveness (20107–8). Tobacco and vaping companies adjusted the pH of their products, in part by adding flavors, in order to increase nicotine absorption rate (20109–11). Nicotine can also be absorbed if the e-liquid is spilled on the skin (20112).

The dermal absorption of nicotine is employed in patches or transdermal nicotine systems, to allow smokers to avoid the worse symptoms of nicotine withdrawal when quitting cigarettes or other tobacco products (20113–4). A slight increase in blood nicotine concentrations can occur due to increased dermal blood flow during exercise (20115).

Reference Number: 20101

KARACONJI, I.B. "Facts About Nicotine Toxicity." *Arhiv za higijenu rada i toksikologiju*, 56: 363–371, 2005 (1 table, 57 references)

Abstract: Nicotine is an alkaloid obtained from the leaves of the tobacco plant. It was first isolated in 1828. Nicotine absorption can occur through the oral cavity, lungs, skin, urinary bladder, and gastrointestinal tract. Nicotine is absorbed poorly through the stomach, due in part to the acidity of the gastric fluids. The respiratory absorption of nicotine is 60–80% and is the most rapid route of absorption. Nicotine is excreted through the urine, feces, bile, saliva, gastric fluids, sweat, and breast milk.

> Cigarette smoking is the major cause of chronic obstructive lung disease. Nicotine, which is readily absorbed from the lung and distributed to tissue, including bone marrow, increases the expression of the elastase gene, leading to increased elastase protein concentration per cell, suggesting a pathophysiologic mechanism for emphysema. Inhaled nicotine produces a concentration-dependent cough and airway obstruction in healthy subjects, probably because of stimulation of afferent nerve endings in the bronchial mucosa and mediated through parasympathetic cholinergic pathways.

Reference Number: 20102

TUTKA, P., J. MOSIEWICZ, AND M. WIELOSZ. "Pharmacokinetics and Metabolism of Nicotine." *Pharmacological Reports*: 57: 143–153, 2005 (1 table, 1 figure, 95 references)

Abstract: The widespread use of cigarettes and other tobacco products is mainly caused by addiction to nicotine. Pulmonary absorption of nicotine (i.e., smoking or vaping) is extremely rapid and occurs at a rate similar to IV administration. At the acidic pH of a cigarette, nicotine is mostly ionized and does not rapidly cross cell membranes until it enters the lungs and is brought to a physiological pH >7. The elimination half-life of nicotine is 2 hours (range 1–4 hours), with a long terminal half-life due to the slow release of nicotine from the tissues. Nicotine protein binding is about 5%. Renal excretion of nicotine is pH dependent and increases at pH <5, since at higher pH the nicotine is unionized and crosses back into the bladder. Saliva is another important route for excretion. The main organ metabolizing nicotine is the liver (about 80%). In addition to cotinine, another metabolite of nicotine is nornicotine, and up to 40% of any nornicotine detected in the blood may be due to smoking tobacco. Newborns have a prolonged elimination half-life of 3–4× that of adults. Nicotine is less rapidly metabolized in pregnant women.

Table: Nicotine and Cotinine Half-Life and Volume of Distribution in Smokers and Non-smokers

	Nicotine Half-Life	Cotinine Half-Life	Nicotine Volume of Distribution	Cotinine Volume of Distribution
Smokers	157 minutes	1047 minutes	196 L	54 L
Non-smokers	122 minutes	1012 minutes	185 L	58 L

Source: Adapted from Tutka et al, 2005

> There are no detailed data concerning the pharmacokinetics and metabolism of NIC in older patients. A high degree of individual variability in plasma NIC concentrations during smoking or NIC replacement therapy does not significantly change with advancing age. On the other hand, it is possible that aging could decrease the volume of distribution of NIC because it is primarily distributed in lean body mass, which decreases with age. It has been suggested that in older patients, the changes in volume of distribution could influence the plasma NIC concentrations after absorption from rapid NIC delivery systems, such as NIC gum or nasal spray.

Reference Number: 20103

ARMITAGE, A.K., C.T. DOLLERY, C.F. GEORGE, T.H. HOUSEMAN, P.J. LEWIS, AND D.M. TURNER. "Absorption and Metabolism of Nicotine from Cigarettes." *British Medical Journal*, 4: 313–316, 1975 (3 tables, 2 figures, 18 references)

Abstract: Eight male subjects (5 smokers, 3 non-smokers, ages 29 to 51 years) smoked a cigarette containing radioactively labelled nicotine (9 to 11 puffs) and had frequent arterial blood samples collected. Radioactivity in all biological samples collected was determined, as were nicotine and cotinine concentrations. In addition, two subjects were administered radioactively labelled nicotine by IV. Four subjects who retained mainstream smoke the most had a mean plasma nicotine concentration of 34.5 ng/mL, whereas the 3 other subjects had a nicotine concentration of 2.5 to 8.0 ng/mL. The IV and smoking plasma nicotine concentrations were similar. All smokers who inhaled had a higher heart rate, as occurred after IV administration of nicotine, and the heart rate decreased with decreasing nicotine concentration. By 60 minutes after smoking, the plasma cotinine concentrations generally exceeded the nicotine concentrations.

These results confirm that no more than 25% of the total nicotine content of a cigarette is likely to appear in the mainstream smoke. Most of the nicotine is lost into the surrounding air and sidestream smoke or is retained in the butt. Nevertheless, smokers who inhale may absorb up to 90% of the nicotine in the mainstream smoke drawn into their mouths. Our data suggest that this efficient extraction is due to inhalation of the smoke into the lungs. Non-smokers who do not inhale absorb much less nicotine.

Reference Number: 20104

FEYERABEND, C., R.M. INGS, AND M.A. RUSSELL. "Nicotine Pharmacokinetics and Its Application to Intake from Smoking." *British Journal of Clinical Pharmacokinetics*, 19: 239–247, 1985 (1 table, 4 figures, 13 references)

Abstract: Five male smokers (ages 24–51 years) were administered 0.25 μg/kg nicotine IV over 1 minute before and after smoking 6 cigarettes. Plasma nicotine concentrations were measured by gas chromatography (GC). Salivary concentrations were on average 5.8× higher than plasma nicotine concentrations and parallel plasma concentrations for the first 30 minutes. Plasma nicotine concentration decreased in a biphasic manner with an initial half-life of 9 minutes and a terminal half-life of 133 minutes. Both the plasma clearance and volume of distribution of nicotine were high at 915 mL/min and 173 L respectively. The average dose of nicotine obtained from a cigarette was 1.06 mg. Nicotine intake varied widely between subjects and had little relation to the standard machine-smoked yields of their cigarettes.

These findings confirm previous work showing that smokers tend to regulate their nicotine intake by modifying their pattern of puffing and inhalation. In this way they tend to maintain a fairly constant intake whether the cigarettes they smoke have high or low nicotine yields.

Reference Number: 20105

RUSSELL, M.A., M. JARVIS, R. IYER, AND C. FEYERABEND. "Relation of Nicotine Yield of Cigarettes to Blood Nicotine Concentrations in Smokers." *British Medical Journal*, 280: 972–976, 1980 (5 tables, 2 figures, 28 references)

Abstract: Blood nicotine and carboxyhemoglobin concentrations were measured in 206 female and 124 male smokers by GC and IL CO-oximeter, respectively. The blood sample was collected within 2 minutes after the subject finished smoking a cigarette. The blood nicotine concentration varied from 4–72 ng/mL (mean 33 ng/mL) and was the same in men and women smokers, although men generally smoke more cigarettes. Despite the large difference in the nicotine yield of cigarettes (0.8–1.9 mg), the blood nicotine concentration varied only slightly.

Table: Mean Blood Nicotine and COHb Concentrations in Male Smokers of Plain, Medium Nicotine, and Low Nicotine Cigarettes

	Plain	Filter (unventilated)	Filter (ventilated)
Mean blood nicotine concentration (ng/mL)	35.7	33.3	29.1
Mean COHb (%)	6.3	8.0	7.8

Source: Adapted from Russell et al, 1980

This study showed that cigarette consumption, tar and nicotine yield of cigarettes, and the type of cigarette made little difference and accounted for little of the variance in blood nicotine and COHb concentrations. This suggests that the smoke intake of smokers is largely determined by their individual pattern of puffing and inhaling and that the assumed health advantage of smoking cigarettes with lower machine-smoked tar and nicotine yields is mostly offset by the tendency of smokers to modify their smoking pattern to regulate their intake to a fairly constant level. Moreover, the small tendency for smokers of lower-nicotine cigarettes to have lower blood nicotine concentrations might partly have occurred because the subjects selected their own brand of cigarettes. For instance, the smokers who had chosen lower-nicotine brands might possibly have tended to be those with slightly lower blood nicotine concentrations before they changed brands; and this factor, rather than any reduction in intake after changing, might have contributed to the low correlations observed between blood nicotine concentration and nicotine yield of cigarette. Owing to the high correlation of tar and nicotine yields we cannot say from this study whether the self-regulation of smoke intake is mediated by a desire for a constant intake of nicotine or tar or by some other factor. Whatever its cause, our results suggest that switching to cigarettes with lower tar and nicotine yields may not reduce tar and nicotine intake at all, or at most by a disappointingly small degree.

Reference Number: 20106

JARVIS, M.J., M.A. RUSSELL, AND C. FEYERABEND. "Absorption of Nicotine and Carbon Monoxide from Passive Smoking Under Natural Conditions of Exposure." *Thorax*, 38: 829–833, 1983 (3 tables, 4 figures, 11 references)

Abstract: Blood, breath, urine, and saliva samples were collected from three male and four female non-smokers at 11:30 a.m. (baseline) and at 7:45 p.m., after 2 hours exposure to secondhand smoke (SHS) in a pub. Nicotine and cotinine concentrations were determined in blood, urine, and saliva by GC, and CO in breath was determined by IR. The mean serum nicotine and cotinine concentrations increased from 0.76 to 2.49 ng/mL and 1.07 to 7.53 ng/mL, respectively. The exhaled breath CO concentration increased from 4.71 ppm to 10.57 ppm. The concentration of breath CO in non-smokers exposed to SHS was 15.7% of that found in heavy smokers and the blood nicotine concentration was 7.5% of that found in heavy smokers.

The results show that when non-smokers keep away from smokers and try to avoid exposure to tobacco smoke they still have small amounts of nicotine and its metabolite cotinine in their body fluids. This indicates that for a city dweller it is virtually impossible to avoid exposure completely. After a normal day at work in an office, followed by two hours socialising with smokers in a pub, all of the seven non-smokers showed increases in the concentrations of nicotine and cotinine in their body fluids and higher concentrations of carbon monoxide in their expired air. The findings are clear, unequivocal, and statistically significant.

Reference Number: 20107

YINGST, J.M., J. FOULDS, S. VELDHEER, S. HRABOVSKY, N. TRUSHIN, T.T. EISSENBERG, J. WILLIAMS, J.P. RICHIE, T.T. NICHOLS, S.J. WILSON, AND A.L. HOBKIRK. "Nicotine Absorption During Electronic Cigarette Use Among Regular Users." *Public Library of Science One*, 14: e0220300 (15 pp), 2019 (3 tables, 4 figures, 41 references)

Abstract: Early e-cigarettes (ECs) were classified into two types. Cig-a-likes were the first-generation ECs, which were similar in size and shape to cigarettes. Second generation devices were larger, with a greater volume of e-liquids, more powerful batteries, and a manual button to operate the heating of the coil before inhalation. Eight male and six female EC users vaped their own ECs every 20 seconds for 10 minutes. Frequent blood samples were collected via an indwelling catheter and serum nicotine concentrations were determined by LC/MS. Seven male and three female smokers also smoked one cigarette according to the same procedure. The mean maximum serum nicotine concentration was 25.9 ng/mL for cigarettes, 11.5 ng/mL for advanced ECs, and 2.8 ng/mL for cig-a-like devices. However, one subject who used a second-generation EC, an Aspire CF W+, obtained a peak serum nicotine concentration of 35.5 ng/mL, even higher than the cigarette smokers.

Table: Serum Nicotine Concentrations (ng/mL) after 10 minutes Puffing a Cigarette, and Second- and First-Generation E-cigarettes

Time from first puff (minutes)	Cigarette (increase in serum nicotine concentration)	Second Generation E-cigarettes (serum nicotine concentration)	First Generation E-cigarettes (serum nicotine concentration)
4	9 ng/mL	5 ng/mL	1 ng/mL
6	15 ng/mL	7 ng/mL	2 ng/mL
10	10 ng/mL	10 ng/mL	2 ng/mL
12	8 ng/mL	10 ng/mL	2 ng/mL
15	7 ng/mL	9 ng/mL	2 ng/mL

Source: Adapted from Yingst et al, 2019

Finally, this study included participants who used a nicotine concentration of at least 12mg/ml and asked participants to vape on an intensive puffing schedule. While it is now known that e-cigarette users can absorb a sizeable amount of nicotine using lower nicotine concentrations and a less intensive puffing schedule, when this study was designed, it was hypothesized that e-cigarettes did not deliver very high doses of nicotine. We chose to include only those with a higher nicotine concentration and to use an intensive puffing schedule to ensure that participants absorbed some nicotine during use. Despite this intensive puffing schedule, first-generation users absorbed only a very small amount of nicotine (boost of 2.8 ng/ml).

An important implication of our study results is that the first-generation e-cigarettes many smokers first try only deliver small amounts of nicotine, even after very intensive puffing (30 puffs over 10 minutes). It is therefore not surprising that these types of devices do not result in impressive rates of smoking cessation in randomized trials or in cohort studies. The finding that most e-cigarettes deliver significantly lower blood nicotine concentrations is consistent with the evidence that e-cigarette users rate themselves as less addicted to their product than cigarette smokers.

Reference Number: 20108

HAJEK, P., M.L. GONIEWICZ, A. PHILLIPS, K.M. SMITH, O. WEST, AND H. MCROBBIE. "Nicotine Intake from Electronic Cigarettes on Initial Use and After 4 Weeks of Regular Use." Nicotine and Tobacco Research, 17: 175–179, 2015 (1 table, 1 figure, 25 references)

Abstract: Frequent blood samples were collected from one male and five female smokers (mean age 52 years) using a first-generation cig-a-like (Green Smoke EC) with a 2.4% nicotine cartridge, first after initial use and then after 4 weeks of use. The cig-a-like provided 9 mg of nicotine after 300 puffs. Frequent blood samples were collected and plasma nicotine concentrations were determined by LC/MS/MS. The mean peak plasma nicotine concentration increased after 4 weeks of use from 4.6 ng/mL to 5.7 ng/mL. The peak plasma nicotine concentration occurred within 5 minutes, which indicates pulmonary absorption.

In summary, in some EC users, nicotine intake from EC increased with practice. Smokers trying EC for the first time should be informed that nicotine delivery from the device may increase as they learn how to use it. The first-generation EC model used in this study delivered only low levels of nicotine, but did so very quickly compared with the speed of nicotine delivery from nicotine replacement therapy. Fast nicotine absorption is considered essential for smokers' satisfaction. Our finding may explain some of the EC appeal to smokers, and it suggests that if EC manage to provide higher doses of nicotine as well, they may have the potential to replace conventional cigarettes.

Reference Number: 20109

ST. HELEN, G., D.A. DEMPSEY, C.M. HAVEL, P. JACOB III, AND N.L. BENOWITZ. "Impact of E-Liquid Flavors on Nicotine Intake and Pharmacology of E-Cigarettes." *Drug and Alcohol Dependence*, 178: 391–398, 2017 (5 tables, 2 figures, 35 references)

Abstract: Fourteen EC users (age 19–59 years) smoked e-liquids containing strawberry flavor (pH 8.29), tobacco flavor (pH 9.10), or their usual brand (mean pH 6.80). The nicotine concentration was 18 mg/mL for the strawberry and tobacco flavored liquids and 3 to 18 mg/mL for their usual flavor. The subjects used a Kanger Tech mini Pro Tank 3 at 9.1 watts. Each session consisted of 15 puffs, followed by 4 hours of abstinence and 90 minutes of ad libitum use. The actual measured nicotine concentrations were 19.9 mg/mL for the flavored e-liquids and 7.4 mg/mL for the subjects' usual brands. The subjects consumed 43% more with the strawberry flavor e-liquid for the ad libitum session than they did of the tobacco flavor. Blood samples were collected and plasma nicotine concentrations were determined by GC/MS. The peak plasma concentrations after 15 puffs were 6.2 ng/mL (usual flavor), 12.1 ng/mL for strawberry, and 9.5 ng/mL for tobacco flavor.

The tobacco e-liquid (pH, 9.10) was more basic than the strawberry (pH, 8.29) e-liquid, which could have contributed to slower rate of nicotine absorption compared to the strawberry e-liquid. In general, high nicotine-containing e-liquids have higher pH due to the alkalinity of nicotine but flavor additives, which may be slightly acidic, can influence the resulting e-liquid pH. At higher pH, more of the nicotine in the tobacco e-liquid is expected to be in the free-base form (unionized) compared to the strawberry e-liquid. Due to higher volatility of free nicotine, more free nicotine from the tobacco e-liquid would leave the e-liquid solution in the vapor phase, possibly resulting in greater deposition of nicotine in the upper airways. This would lead to slower absorption and lower C_{max} compared to nicotine absorbed through the lungs. The greater deposition of vapor phase nicotine in the upper airways would however, lead to increased sensory effects in the upper airways. Indeed, average rating of sensation in the throat was higher for the tobacco e-liquid than the strawberry e-liquid.

This is consistent with reports of greater upper airway sensory effects from smoking cigarettes with basic additives such as ammonia added. We also found slower rates of absorption (AUCs normalized for retained nicotine dose) and C_{max} normalized for retained nicotine dose when we compared the basic usual e-liquids to the acidic usual e-liquids. Surprisingly, even though nicotine intake from the basic usual e-liquids was double that of the acidic usual e-liquids, max increase in heart rate was, on average, 3 bpm lower with the basic usual e-liquids. This raises questions about the role e-liquid pH plays on e-cigarette physiologic effects.

Reference Number: 20110

TOMAR, S.L., AND J.E. HENNINGFIELD. "Review of the Evidence that pH is a Determinant of Nicotine Dosage from Oral Use of Smokeless Tobacco." *Tobacco Control*, 6: 219–225, 1997 (56 references)

Abstract: Nicotine is an addictive drug that causes an increase or growth of nicotinic receptors in the nervous system, which is associated with the development of a tolerance and physical dependence. The speed of absorption is a major determinant of addiction. A few milligrams of nicotine delivered rapidly via smoking, IV injection, or smokeless tobacco with a high pH produce greater addiction than 20 mg of nicotine absorbed slowly via nicotine patch. The rate of absorption is dependent on the pH. At a high alkaline pH (>7), the nicotine is more rapidly absorbed than at a more acidic or lower pH (<7), when nicotine is in the ionized form. Nicotine occurs naturally in tobacco in the acidic form and is thus slow to be absorbed unless buffered. Saliva has a pH buffering agent which promotes absorption. Epidemiologically, persons who used Kodak or Copenhagen moist snuff brands are 2× as likely to report more symptoms of nicotine withdrawal than users of Skoal Bandits at a lower pH.

Internal documents from British-American Tobacco (BAT) that recently became publicly available provide evidence that the tobacco industry conducted research on pH buffering and nicotine absorption as early as 1968. In a series of in-vitro experiments, BAT researchers found that the pH of the solution controlled the rate at which nicotine diffused through selective synthetic and natural membranes. The author concluded: "The rate at which nicotine is absorbed into the blood stream from the mouth appears to be largely controlled by the pH of the saliva in the mouth." In reviewing several published studies, another BAT internal report noted that nicotine in cigar smoke (pH 8.5) was absorbed more readily through the oral mucous membranes than was nicotine in cigarette smoke (pH 5.3). They attributed this finding to the higher concentration of un-ionised nicotine in the relatively alkaline cigar smoke. They also noted that dogs absorbed much more nicotine administered orally when the nicotine solution had a pH of 10.2 than when its pH was 7.4.

In a document describing the additives used in snuff manufacturing, the Swedish Tobacco Company has indicated that it controls nicotine dosage by controlling the pH level. They describe one product additive as "sodium carbonate (Na_2CO_3) which increases the pH-level, which makes nicotine more easily released from the tobacco and subsequently facilitates the uptake of nicotine through the mucous membranes of the mouth. The sodium carbonate is altered in the snuff into bicarbonate".

Reference Number: 20111

DUELL, A.K., J.F. PANKOW, AND D.H. PEYTON. "Nicotine in Tobacco Product Aerosols: 'It's Déjà Vu All Over Again.'" *Tobacco Control*, 29: 656–662, 2019 (2 tables, 3 figures, 35 references)

Abstract: In 1619, the tobacco that was produced in the Jamestown colony was air-cured tobacco (also known as dark, brown, or black tobacco). It was more basic (i.e., protonated) and harsh, but since the nicotine was freebased, it could be absorbed more readily through the oral cavity. Air drying took 6–8 weeks, which destroyed most of the natural sugars in the tobacco leaf. Flue-cured (or bright) tobacco was developed in the 1850s, in which rapid drying with heat produced a bright yellow leaf with a milder smoke that makes it easier to inhale. Flue-cured tobacco is high in leaf sugars or organic acids, which lower the pH and therefore lower the harshness. Initially, e-liquids were similar to air-cured tobacco, with a high amount of free-base nicotine. JUUL and other ECs developed the addition of benzoic acid or other organic acids to lower the pH and de-free base, making it easier to inhale and to deliver a high dose of nicotine, similar to flue-cured tobacco.

The trajectory in figure 3 for smoked tobacco allowed cigarettes to become much more addictive, abused, and deadly than would have been the case if smoked tobacco remained of an air-cured type. The evolution of e-cigarettes has followed a similar overall trajectory. It is undoubtedly true that this evolution has made e-cigarettes more effective as substitutes to get smokers off combustibles. However, exactly as occurred with smoked tobacco, this evolution has made e-cigarette products vastly more addictive for never-smokers. The full public health implications of widely prevalent e-cigarette use will only become fully apparent perhaps a decade hence.

Reference Number: 20112

FRASCH, H.F., AND A.M. BARBERO. "In Vitro Human Epidermal Permeation of Nicotine from Electronic Cigarette Refill Liquids and Implications for Dermal Exposure Assessment." *Journal of Exposure Science & Environmental Epidemiology*, 27: 618–624, 2017 (2 tables, 5 figures, 34 references)

Abstract: Nicotine can be readily absorbed through the human skin, as shown by nicotine patches used in smoking cessation. E-liquids contain nicotine dissolved in propylene glycol and/or glycerol. Some contain flavoring chemicals such as the terpenes limonene and menthol, which can increase dermal absorption. Data was collected regarding the nicotine permeation on human skin in vitro of a nicotine solution and various e-liquids (24 mg/mL nicotine). The data were then applied to three potential scenarios to predict systemic uptake and plasma concentrations from dermal exposure. In scenario 1, a worker mixing e-liquids has all fingertips or else thumb and index fingers exposed to neat nicotine over a 4-hour shift. Scenario 2 was based on a reported case in which a person developed cardiac problems after an e-cigarette (i.e., pod) broke in her bed and led to 1 mL of e-liquid having contact with the skin. In scenario 3, either 1 teaspoon (5 mL) or 1 tablespoon (15 mL) of e-liquid is spread over the skin.

Table: Calculated Maximum Plasma Nicotine Concentrations According to Dermal Exposure

Scenario	Nicotine Formulation	Exposure	Area (cm^2)	Calculated Maximum Plasma Nicotine Concentration (ng/mL)
1	Neat nicotine	All fingers, immersion	73	184
1	Lemon-Lime eJuice	All fingers, immersion	73	24.9
2	Lemon-Lime eJuice	1 mL spread over skin	154	52.6
3	Lemon-Lime eJuice	5 mL spread over skin	769	263
3	Lemon-Lime eJuice	15 mL spread over skin	2307	789

Source: Adapted from Frasch and Barbero, 2017

The potential for substantial nicotine uptake from dermal contact with e-cigarette refill liquids and neat nicotine has been demonstrated. This hazard is particularly acute in occupational settings in which workers mix e-liquids from neat nicotine and other constituents, and fill cartridges from tanks of e-liquid. The use of personal protective equipment, including adequate gloves, is essential. The same applies to individuals who mix their own e-liquids for personal consumption. Neat nicotine is particularly hazardous, with a steady-state flux over seven times greater than that of the higher permeating e-liquid, Lemon-Lime eJuice.

Casual contact by end users, on the other hand, appears to be less hazardous. A worst-case scenario, in which the entire contents of one e-cigarette cartridge are spread over the body in a thin film, suggests less than 8 mg of nicotine absorption, over a time course exceeding 24 h. However, intentional misuse of e-liquids could lead to serious toxic effects.

Reference Number: 20113

GUPTA, S.K., N.L. BENOWITZ, P. JACOB III, C.N. ROLF, AND J. GORSLINE. "Bioavailability and Absorption Kinetics of Nicotine Following Application of a Transdermal System." *British Journal of Clinical Pharmacology*, 36: 221–227, 1993 (3 tables, 3 figures, 18 references)

Abstract: Thirteen smokers (mean age 32.7 years) with a Fagerstrom nicotine dependency score of 8.7 who smoked at least 1 pack of cigarettes a day were administered a transdermal nicotine system (NTS) once a day for 7 days to different parts of the upper arm, back, and chest. The NTS employed was NicoDerm, which has a surface area of 22 cm^2 and releases 36 mg of nicotine over a 24-hour period (about 1.5 mg nicotine/hour). The subjects were also administered an IV solution of 20 mg of radioactively labelled nicotine bitartrate dihydrate over 24 hours. Frequent blood samples were collected, and plasma nicotine and cotinine concentrations were measured by GC/MS. The subjects were abstinent for 48 hours prior to the testing and a positive plasma nicotine concentration was detected in only 1 subject. The mean plasma concentration after use of the patch was about 10.2 to 12.7 ng/mL through the 24 hours. After removal of the nicotine patch, the mean elimination half-life was 2.8 hours. The labelled plasma nicotine concentration for the 24-hour infusion was 15 ng/mL and the mean half-life was 2 hours. The longer half-life after the use of the dermal patch was due to the residual nicotine remaining on the skin (about 1.9 mg nicotine) after the patch was removed. About 9.2 mg of nicotine was lost from the patch due to evaporation of highly volatile nicotine from the exposed edges.

No significant changes were seen in the pharmacokinetics of nicotine between single and multiple applications of NTS. As expected from the higher total plasma nicotine concentrations, the incidence of adverse effects was higher following simultaneous intravenous and transdermal administration of nicotine. The most frequently reported systemic side effects were nervousness and headache: mild itching was the most frequent topical effect.

Reference Number: 20114

HURT, R.D., L.C. DALE, K.P. OFFORD, G.G. LAUGER, L.B. BASKIN, G.M. LAWSON, N-S. JIANG, AND P.J. HAURI. "Serum Nicotine and Cotinine Levels During Nicotine-Patch Therapy." *Clinical Pharmacology and Therapeutics*, 54: 98–106, 1993 (2 tables, 6 figures, 26 references)

Abstract: Eighteen female and six male smokers with severe nicotine dependence were admitted to a clinical research center and treated with 22 mg nicotine transdermal nicotine patches daily in an intensive smoking cessation program. Daily blood samples were collected over 14 days and plasma nicotine and cotinine concentrations were determined by GC/MS and high pressure liquid chromatography (HPLC), respectively. Withdrawal symptoms and quality of sleep were assessed. All the smokers abstained except for one who smoked half a cigarette in a washroom and one who smoked attending an outside function. The mean plasma nicotine and cotinine concentrations were inversely related to the severity of withdrawal symptoms. A steady state level of nicotine and cotinine concentrations was obtained after 3 days. The plasma nicotine concentrations after patch use were about 40–50% of that upon admission. After 14 days, the mean weight increased by 0.5 kg, the mean blood pressure (BP) decreased by 10%, and the mean heart rate (HR) declined by 15%.

A fixed dose of transdermal nicotine will not be effective for all smokers. Individualization of therapy should be based on objective biologic measures such as serum cotinine and subjective assessment of withdrawal relief.

Reference Number: 20115

KLEMSDAL, T.O., K. GJESDAL, AND K. ZAHLSEN. "Physical Exercise Increases Plasma Concentration of Nicotine During Treatment with a Nicotine Patch." *British Journal of Pharmacology*, 39: 677–679, 1995 (1 table, 1 figure, 11 references)

Abstract: Plasma nicotine concentrations were determined by GC/MS in five female and three male smokers (mean age 38 years) after applying a nicotine patch (14 mg/day) to the hairless chest or shoulder area before and after 20 minutes of either bicycle exercise or rest. The mean plasma nicotine concentration increased from 9.8 to 11.0 ng/mL after exercise and decreased from 10.5 ng/mL to 10.2 ng/mL after rest. The increase in drug concentration is due to an increase in the cutaneous blood flow during exercise and has also been observed with local heating of the transdermal patch and after a sauna.

> The clinical importance of a moderate increase in nicotine concentrations is probably minor. However, a longer period of exercise or heat excess might cause a cumulative increase, and possibly dose-related side effects.

2.02 DISTRIBUTION

> "The rapid rise of nicotine in human brain that we observed after a single smoked puff was comparable to the rise time reported for a single puff of cocaine in the brains of cocaine addicts . . ."
>
> —Berodge et al, "Smoking Produces Rapid Rise of [11C] Nicotine in Human Brain" (2010)

Since nicotine is 95% water soluble, it would distribute like alcohol to the various parts of the body, according to water content. However, when smoked or vaped, nicotine's rapid absorption causes the arterial blood nicotine concentration to be higher than the venous blood nicotine concentration, resulting in a rapid distribution to the brain (20201–3). The rapid distribution of nicotine to the brain is comparable to that of cocaine (20204).

After distribution of nicotine reaches equilibrium, the arterial and venous blood nicotine concentrations are the same (20204). But the initial arterial/venous differences may explain in part why nicotine is so addictive.

Reference Number: 20201

HENNINGFIELD, J.E., J.M. STAPLETON, N.L. BENOWITZ, R.F. GRAYSON, AND E.D. LONDON. "Higher Levels of Nicotine in Arterial Than in Venous Blood After Cigarette Smoking." *Drug and Alcohol Dependence*, 33: 23–29, 1993 (1 table, 2 figures, 21 references)

Abstract: Arterial and venous plasma nicotine concentrations were determined by GC in eight male subjects (mean age 33.6 years) at baseline and at 1, 3, 5, and 10 minutes after smoking.

Table: Arterial and Venous Plasma Nicotine Concentration and Time Since Smoking in Two Subjects

Time since smoking	Arterial (venous) plasma nicotine concentrations (ng/mL)	
(minutes)	Subject 1	Subject 2
1	42.9 (9.3)	64.4 (6.4)
3	78.0 (12.2)	54.0 (14.1)
5	92.5 (12.7)	49.8 (16.0)
10	50.9 (16.6)	24.5 (13.3)

Source: Adapted from Henningfield et al 1993

Because of the rapid equilibrium of drug concentrations between arterial blood and the brain, post-inhalation arterial drug levels should better reflect brain concentrations than peripheral venous levels, which reflect effluent concentrations primarily from skeletal muscle. The present findings suggest that the inhalation route results in exposure of the brain and heart to much greater drug concentrations than would be predicted based on venous plasma measurements. Such information should be useful by providing researchers using in vitro assays with more accurate information on the concentrations of drug to which biologic tissues are exposed. Conversely, venous blood levels of smoked drugs may greatly underestimate the concentrations to which tissue is exposed.

Reference Number: 20202

JACOB III, P., N.L. BENOWITZ, AND A.T. SHULGIN. "Recent Studies of Nicotine Metabolism in Humans." *Pharmacology, Biochemistry, and Behavior*, 30: 249–253, 1988 (1 table, 6 figures, 20 references)

Abstract: Various clinical studies of nicotine and its metabolites were conducted, and the concentrations thereof were determined in the blood and urine by GC and labelled nicotine by GC/MS. The relatively long terminal half-life of nicotine, in spite of a high rate of clearance, is due to its large volume of distribution (183L). Trans 3-hydroxycotinine is a nicotine metabolite that is present in the urine. The best measures of nicotine intake by smoking are the blood nicotine concentration, eCO (or COHB), and urinary cotinine excretion.

> There is good evidence that nicotine plays a role in tobacco dependence. Consequently, it is important to understand the quantity and time course of nicotine in the body. Following smoking or an intravenous infusion, nicotine distributes rapidly into tissues, after which it is eliminated, with a terminal half-life of about 2 hr. Total clearance averages about 1300 ml/min and is highly variable between individuals. The relatively long terminal half-life in the presence of rapid clearance can be explained by the large volume of distribution, which averaged 183 liters.
>
> The amount of a drug in the body is a function of both intake (dose) and rate of elimination. Nicotine differs from therapeutic drugs in that the dose is not known, and varies widely depending upon the habits of the individual tobacco user. In addition, the rate of nicotine elimination is highly variable, with clearance rates varying as much as 4-fold among cigarette smokers. Consequently, a fast metabolizer may consume up to four times the dose of nicotine as a slow metabolizer to produce equivalent blood concentrations.

Reference Number: 20203

GUTHRIE, S.K., J.K. ZUBIETA, L. OHL, L. NI, R.A. KOEPPE, S. MINOSHIMA, AND E.F. DOMINO. "Arterial/Venous Plasma Nicotine Concentrations Following Nicotine Nasal Spray." *European Journal of Clinical Pharmacology*, 55: 639–643, 1999 (2 tables, 3 figures, 16 references)

Abstract: Venous blood drug concentrations are a valid index of arterial blood concentration when distribution equilibrium has occurred. Nine female and nine male smokers (ages 18–52 years) with an eCO of <10 ppm used a nasal nicotine spray (1–2.5 mg nicotine) after an overnight abstinence from smoking and had arterial and venous blood samples collected. Plasma nicotine and cotinine concentrations were determined by HPLC. A positron emission tomography (PET) scan was also conducted.

Nasal nicotine administration produced greater A than V plasma levels. There were no gender differences in A/V nicotine concentrations, disproving the above hypothesis, suggesting that other physiochemical factors besides skeletal muscle mass must be involved. Heart rate increases correlated well with arterial plasma nicotine levels (r = 0.77). Males had less variance than females in the expected increase in arterial plasma nicotine concentrations with increased number of nasal sprays. Although there was considerable overlap, mean A cotinine concentrations were consistently slightly larger than V concentrations.

Reference Number: 20204

BERRIDGE, M.S., S.M. APANA, K.K. NAGANO, C.E. BERRIDGE, G.P. LEISURE, AND M.V. BOSWELL. "Smoking Produces Rapid Rise of [11C] Nicotine in Human Brain." *Psychopharmacology*, 209: 383–394, 2010 (1 table, 8 figures, 59 references)

Abstract: Five female and seven male smokers (mean age 26 years) smoked a 10 mm length of a cigarette that was labelled with carbon 11 nicotine bolus in one inhalation. A PET scan was conducted in the brain and lung to determine the speed of distribution of nicotine. In addition, a previous MRI was conducted. The PET brain images were then aligned with the MRI brain scans. The nicotine distributed throughout the lung similar to how a neutral gas would distribute, followed by a rapid washout into the blood. Arterial blood levels of nicotine rose rapidly within 4 seconds of the single puff, whereas venous blood nicotine concentrations rose only slightly, if at all. The rise time of nicotine in the brain was defined as the time required to reach 80% of the peak from 20% of the peak. Nicotine begins to be detected approximately 5 seconds after the start of inhalation and the rise time is usually less than 15 seconds. Smokers report an immediate hedonic effect or rush from their first puff on a cigarette.

The rapid rise time of nicotine in human brain that we have observed after a single smoked puff was comparable to the rise time reported for a single puff of cocaine in the brains of cocaine addicts and comparable also to rise times reported to lead to maximal impacts of nicotine and other drugs. Therefore, we expect that a rapid rate of rise of nicotine in the brain can be considered as a potential contributing factor to addiction of some smokers to nicotine.

2.03 ELIMINATION

> "Our findings suggest that the rate of nicotine metabolism may play a role in the development of nicotine addiction among adolescent smokers. Specifically, faster metabolizers of nicotine experienced greater symptoms of withdrawal following 24 hours of abstinence compared with slower metabolizers, despite smoking similar numbers of cigarettes."
>
> —Rubenstein et al, "Rates of Nicotine Metabolism and Withdrawal Symptoms in Adolescent Light Smokers" (2008)

Nicotine is eliminated in the liver and converted into cotinine. The cotinine is then metabolized into 3"-hydroxycotinine (3-HC) as follows:

Nicotine → Cotinine → 3-HC

The speed of nicotine elimination can be determined by comparing the serum or saliva cotinine and 3-HC concentrations. It is known as the nicotine metabolite ratio (NMR) and is the 3-HC concentration divided by the cotinine concentration (20301). The higher the ratio, the faster the nicotine elimination rate. For example, NMR of greater than 0.42 can be considered as fast nicotine eliminators (20303).

Faster eliminators remove nicotine from the blood more quickly. Smokers compensate for the more rapid elimination of nicotine in the blood by smoking more cigarettes per day than slower eliminators who do not need to smoke as much to maintain their blood nicotine concentrations (20304). Fast eliminators of nicotine tend to have more difficulty in successfully quitting smoking and suffer more withdrawal symptoms (20302).

The NMR depends on a number of factors, but in general men tend to be faster eliminators than women, and White/European smokers tend to be faster eliminators than Hawaiian or Japanese smokers (20305).

The consumption of alcohol tends to increase the MNR, and chronic alcohol users tend to have a higher nicotine elimination rate, which could explain in part why alcoholics have a poor smoking cessation rate (20306–7).

Reference Number: 20301

SCHNOLL, R.A., F. PATTERSON, E.P. WILEYTO, R.F. TYNDALE, N. BENOWITZ, AND C. LERMAN. "Nicotine Metabolite Rate Predicts Successful Cessation with Transdermal Nicotine: A Validation Study." *Pharmacology, Biochemistry, and Behavior*, 92: 6–11, 2009 (1 table, 4 figures, 28 references)

Abstract: Nicotine is metabolized mainly to cotinine, which is further metabolized to 3-HC, by the cytochrome P450 2A6 (CYP2A6) in the liver. The ratio of plasma 3-HC to cotinine is called the nicotine metabolite ratio (NMR) and provides a marker of the rate of nicotine metabolism. A faster nicotine metabolism (i.e., the greater the NMR) predicts lower plasma nicotine concentration during quitting, a greater amount of craving, and a lower successful quitting rate. In this study, the initial NMR was determined in 568 smokers (mean age 45 years) enrolled in a smoking cessation program that provided counseling and 8 weeks of 21 mg/day nicotine patch. Smokers who were in the upper three quartiles of the NMR (fast metabolizers) were about 50% less likely to be abstinent after 8 weeks than smokers in the first quartile (slow metabolizers).

Nevertheless, this study underscores the potential value of assessing pretreatment nicotine metabolism rate from cigarette smoking when considering use of a standard dose of transdermal nicotine. Perhaps only slow metabolizers (i.e., the first quartile for 3-HC/cotinine ratio) should be considered good candidates for this dose of transdermal nicotine. Recent data suggest that faster metabolizers (those in the 4th quartile) achieve significant benefit from bupropion for smoking cessation, while slow metabolizers quit at similar rates on bupropion and placebo (Patterson et al., 2008). Since bupropion is not metabolized by CYP2A6, this finding suggests that the rate of nicotine metabolism influences the level of nicotine dependence and the need for additional pharmacotherapy. Additional randomized clinical trials are needed to determine whether other non-nicotine medications, such as varenicline, are also more efficacious for faster metabolizers, and to identify the optimal treatments for smokers with nicotine metabolism rates in the intermediate range.

Reference Number: 20302

RUBENSTEIN, M.L., N.L. BENOWITZ, G.M. AUERBACK, AND A-B. MOSCIKI. "Rates of Nicotine Metabolism and Withdrawal Symptoms in Adolescent Light Smokers." *Pediatrics*, 122: e643–647, 2008 (2 tables, 1 figure)

Abstract: Nicotine is largely converted into its primary metabolite (cotinine) by the liver enzyme (CYP2A6) and then the cotinine is converted by the same enzyme into 3-HC. The Nicotine Metabolic Ratio (NMR), the ratio of 3-HC to cotinine in the blood, can measure the speed of nicotine elimination from the blood. The serum concentration of 3-HC and cotinine were measured by LC/MS/MS in 20 adolescent smokers (ages 13–17 years) who were then abstinent for 24 hours. Exhaled CO concentration and withdrawal symptoms were measured as well. Five smokers were classified as faster metabolizers (NMR > 0.5) and 15 were slower metabolizers (NMR < 0.5). Eighty percent of the faster metabolizers were female adolescent smokers compared to 47% of the slower metabolisers.

Table: Withdrawal Symptoms Reported at 24 Hours Abstinent in Faster Versus Slower Nicotine Metabolizers (the higher the number the more severe the symptom)

Withdrawal Symptom	Faster Metabolizer Score	Slower Metabolizer Score
Sad	1.60	0.73
Insomnia	3.00	1.27
Angry	2.80	1.13
Impatient	3.80	2.13
Hungry	2.40	1.73
Craving	3.75	2.31

Source: Adapted from Rubenstein et al, 2008

> Our findings suggest that the rate of nicotine metabolism may play a role in the development of nicotine addiction among adolescent smokers. Specifically, faster metabolizers of nicotine experienced significantly greater symptoms of withdrawal following 24 hours of abstinence compared with slower metabolizers, despite smoking similar numbers of cigarettes. Nicotine withdrawal symptoms are an important feature of nicotine addiction and may represent a major obstacle to smoking cessation. Faster metabolism of nicotine would be associated with a more rapid decline in blood and brain nicotine concentrations after smoking a cigarette, which would be expected to produce more intense withdrawal symptoms, compared to a smoker with slower metabolism. More severe withdrawal symptoms in faster metabolizers could result in more persistent smoking behavior to prevent such symptoms from occurring.
>
> In addition to the total withdrawal score, certain symptoms of withdrawal appeared to be more pronounced in the faster metabolizers; namely insomnia, anger, difficulty concentrating and impatience. These factors may be more important in light smokers and future studies may want to focus on a more in depth assessment of these individual items from the withdrawal scale.

Reference Number: 20302

In addition to increased withdrawal symptoms, faster metabolizers described themselves as more highly addicted compared with the slower metabolizers.

Reference Number: 20303

MERCINCAVAGE, M., K. LOCHBUEHLER, E.P. WILEYTO, N.L. BENOWITZ, R.F. TYNDALE, C. LERMAN, AND A.A. STRASSER. "Association of Reduced Nicotine Content Cigarettes with Smoking Behaviors and Biomarkers of Exposure Among Slow and Fast Nicotine Metabolizers. A Nonrandomized Clinical Trial." *JAMA Network Open*, 1: 1346 (17 pp), 2018 (2 tables, 3 figures, 54 references)

Abstract: The association of reduced nicotine cigarettes content (RNC) cigarettes with smoking behavior and biomarkers of exposure was determined in 59 fast and 41 slow metabolizers of nicotine between 2013 and 2015. The NMR was determined in blood and an NMR of 0.42 or greater was classified as a fast metabolizer and an NMR of 0.26 or less was classified as a slow metabolizer. Black Americans on average have a slower nicotine metabolism than White Americans. Fast metabolizers puff their cigarettes more intensely, smoke more cigarettes, extract more nicotine per day, have higher dependencies scores and lung cancer rates, and have lower cessation rates with transdermal patches than varenicline use. The smoking behavior of subjects smoking their own cigarettes (41% used Marlboros, 13% Camels, and 11% Mavericks) and nicotine reduced cigarettes (RNC 4000 5.2 mg/g and RNC 200, 1.3 mg/g nicotine). RNC were associated with decreased puffing behavior and urinary tobacco biomarkers related to the smoker's preferred tobacco brand, but these associations did not differ significantly between fast and slow nicotine metabolizers.

Table: Characteristic of Slow and Fast Metabolizers of Nicotine

Characteristic	Slow Metabolizers	Fast Metabolizers
Male	82.9%	66.1%
Female	17.1%	33.9%
White	51.2%	89.8%
Black	31.7%	8.5%

Characteristic	Slow Metabolizers	Fast Metabolizers
Cigarettes per day (baseline)	14.11	15.69
Total puff volume (mL)	761.03	767.73
Onset CO level (ppm)	19.16	20.65

Source: Adapted from Mercincavage et al, 2018

This study suggests that fast nicotine metabolizers may not be at greater risk of being negatively affected should the US Food and Drug Administration mandate a reduced nicotine product standard similar to the nicotine levels tested.

Reference Number: 20304

ROSS, K.C., N.R. GUBNER, R.F. TYNDALE, L.W. HAWK JR., C. LERMAN, T.P. GEORGE, P. CINCIRIPINI, R.A. SCHNOLL, AND N. BENOWITZ. "Racial Differences in the Relationship Between the Rate of Nicotine Metabolism and Nicotine Intake from Cigarette Smoking." *Pharmacology, Biochemistry, and Behavior*, 148: 1–7, 2016 (1 table, 3 figures, 32 references)

Abstract: Smokers manipulate the number of cigarettes they consume per day (CPD) to titrate the desired rewards and to prevent nicotine withdrawal symptoms. The differences in the rates of nicotine elimination also affects the CPD. Approximately 80% of the nicotine is metabolized to cotinine by the liver enzyme CYP2A6, which metabolizes cotinine into 3-HC. The ratio of 3-HC/Cotinine is termed the NMR and the higher the ratio, the faster the elimination of nicotine. In general, Black American smokers have a slower rate of nicotine elimination and higher cotinine levels than White American smokers. Menthol has been found to have anesthetizing effects and enable the smoker to inhale deeper and absorb more nicotine per cigarette. Menthol also inhibits the CYP2A6 enzyme and decreases the rate of nicotine metabolism. In addition, there are two general types of smokers; peak-seekers who are motivated by the positive effects of nicotine and the trough-maintainers who are motivated to avoid the nicotine withdrawal symptoms. Black American smokers are more likely to be peak-seekers than White American smokers, who smoke more frequently and are more likely to be trough-maintainers. The NMR and other parameters such as CPD were determined in 1,102 white and 598 Black smokers. The average CPD was 16 for Black smokers and 20 for White smokers. Overall, a higher NMR was associated with a higher CPD.

- Rate of nicotine metabolism (NMR) is an important factor influencing nicotine intake
- We found that NMR influences nicotine intake more in Whites vs. African American smokers
- Whites appear to titrate their nicotine intake based on their NMR
- NMR was unrelated to nicotine intake in African Americans, suggesting no titration.

Reference Number: 20305

DERBY, K.S., K. CUTHRELL, C. CABERTO, S.G. CARMELLA, A.A. FRANKE, S.S. HECHT, S.E. MURPHY, AND L. LE MARCHAND. "Nicotine Metabolism in Three Ethnic/Racial Groups with Different Risks of Lung Cancer." *Cancer Epidemiology, Biomarkers and Prevention*, 17: 3526–3535, 2008 (5 tables, 48 references)

Abstract: Approximately 80% of nicotine in the body is metabolized by the liver enzyme CYP2A6 into cotinine, which is further metabolized by the same enzyme into 3-HC. The nicotine metabolite ratio (3-HC/cotinine) can indicate how rapidly nicotine is metabolized (the higher the ratio, the faster the metabolism). A faster nicotine metabolism rate would be expected to result in more cigarettes being smoked. Between 1993 and 1996, the urinary concentration of nicotine, 3-HC, cotinine, caffeine, metabolite (another indicator of CYP2A6 activity), and numerous tobacco smoke carcinogens were measured by GC/MS in 192 native Hawaiian, 198 White American, and 195 Japanese American smokers.

> We measured urinary nicotine and its metabolites and the extent of nicotine metabolism in a large sample of Native Hawaiian, Japanese and white smokers in Hawaii. Our hypothesis was that ethnic/racial differences in nicotine metabolism may affect smoking behavior in a way that could result in greater carcinogen exposure, adjusted for average daily cigarette consumption, and, thereby, may help explain the ethnic/racial differences in lung cancer risk that we have documented in this population. Indeed, case-control and prospective data in Hawaii have demonstrated that Native Hawaiian smokers have a greater risk, and Japanese American smokers a lower risk, compared to white smokers, even after taking into account smoking dose (specifically, CPD) and duration, and other potential confounders.

Reference Number: 20306

GUBNER, N.R., A. KOZAR-KONIECZNA, I. SZOLTYSEK-BOLDYS, E. SLODCZYK-MANKOWSKI, J. GONIEWICZ, A. SOBCZAK, P. JACOB III, N.L. BENOWITZ, AND M.L. GONIEWICZ. "Cessation of Alcohol Consumption Decreases Rate of Nicotine Metabolism in Male Alcohol-Dependent Smokers." *Drug and Alcohol Dependence*, 163: 157–164, 2016 (2 tables, 3 figures, 50 references)

Abstract: The NMR and total nicotine equivalents (TNE) were determined in urine samples collected from 22 male smokers undergoing alcohol detoxification at weeks 1, 4, and 7. About 75% of the nicotine is metabolized to cotinine, which is further metabolized to 3-HC. The higher the ratio of 3-HC/cotinine in the urine, the faster the rate of nicotine metabolism. Other metabolites of 3-HC indicate TNE. Urine samples were analyzed for nicotine and other metabolites by LC-MS/MS. Urine creatinine was also measured. There was a 50% decrease in NMR over 7 weeks of alcohol abstinence, but no change in TNE, indicating no change in smoking behavior. Low smoking cessation rates are found in alcohol-dependent persons but not in former alcoholics. An increase is the NMR could result in the poorer cessation rates in alcoholics.

> In conclusion, our results indicate that chronic alcohol abuse may increase the rate of nicotine metabolism, which then decreases after alcohol cessation. This research may help to inform future smoking cessation interventions in this population.

Reference Number: 20307

CHENOWETH, M.J., M. NOVALEN, L.W. HAWK JR., R.A. SCHNOLL, T.P. GEORGE, P.M. CINCIRIPINI, C. LERMAN, AND R.F. TYNDALE. "Known and Novel Sources of Variability in the Nicotine Metabolite Ratio in a Large Sample of Treatment-Seeking Smokers." *Cancer Epidemiology, Biomarkers and Prevention*, 23: 1773–1782, 2014 (3 tables, 3 figures, 56 references)

Abstract: A higher NMR indicates a faster elimination of nicotine and is associated with heavier smoking (to maintain plasma nicotine levels), lower cessation rates, and difficulties in stopping smoking. A study was conducted of the NMR (as determined by LC/MS/MS in whole blood) on 1,672 smokers. Alcohol use and smoking were associated with a higher NMR. The NMR was 16% lower in mentholated cigarette users and decreased with increasing BMI. The NMR was higher in White than Black Americans and higher in women than men. Women taking oral contraceptives had an increase in NMR by 19.5%.

> While these factors significantly affected NMR, they contributed little (together <8%; each ≤2%) to total NMR variation.
>
> *Impact*
> Thus when using NMR, for example to prospectively guide smoking cessation therapy, these sources of variation are unlikely to cause NMR misclassification.

2.04 GENERAL PHARMACOKINETICS

> "However, consistent with the half-life of two hours, nicotine accumulates in the body over six to nine hours of regular smoking. Thus, smoking results not in intermittent and transient exposure to nicotine, but in an exposure that lasts 24 hours per day. Arterio-venous differences in nicotine concentration during cigarette smoking are substantial, with arterial levels exceeding venous levels up to tenfold. The persistence of nicotine in the brain throughout the day and night results in changes in the structure and function of nicotinic receptors and in intracellular processes of neuroadaptation as mentioned previously."
>
> —Benowitz, "Pharmacology of Nicotine: Addiction, Smoking-Induced Disease and Therapeutics" (2009)

The absorption, distribution, and elimination of nicotine combine to determine the blood nicotine concentration or its general pharmacokinetics.

Since nicotine is a weak base, its absorption is faster at pH >7. Smoking or vaping allows for the most rapid absorption of nicotine and allows for the highest blood nicotine concentration, because it means that first-pass metabolism or initial elimination by the liver is avoided. It only requires about 10 to 20 seconds after a puff on a cigarette for nicotine in enter the brain, which can be faster than intravenous injection (20401).

There are 6 major factors that affect the elimination of nicotine (20401):

- Diet and meals
- Age greater than 65 years
- Sleep

- Gender
- Smoking
- Race

As with many addictive drugs, tolerance to a lot of the effects of nicotine develops, in this case due to an increase in the number of nicotinic receptor sites in the brain or upregulation (20302–4).

Reference Number: 20401

BENOWITZ, N.L., J. HUKKANEN, AND P. JACOB III. "Nicotine Chemistry, Metabolism, Kinetics and Biomarkers." *Handbook Experimental Pharmacology*, 192: 29–60, 2009 (2 tables, 4 figures, 144 references)

Abstract: A major review of the pharmacokinetics of nicotine was conducted. Nicotine is a natural ingredient of the tobacco leaf and acts as a botanical insecticide. Nicotine is the major alkaloid of the tobacco leaf (95%) and makes up about 1.5% by weight of the usual commercial cigarette. An average cigarette contains 10–14 mg of nicotine and about 1–1.5 mg is absorbed by the body when smoking. Oral snuff and pipe tobacco contain about the same amount of nicotine as cigarettes, whereas cigars and chewing tobacco contains half the amount. The most common alkaloids after nicotine are nornicotine, anatabine, and anabasine. Nicotine is a weak base with a pKa of 8.0. When nicotine is in an acidic environment, it is mainly ionized and does not rapidly cross the cell membrane. The pH of flue-cured tobacco (cigarettes) smoke is acidic (pH 5.5–6.0). The smoker can manipulate nicotine dose puff by puff and intake depends on puff volume, depth of inhalation, extent of dilution with room air, rate, and intensity of smoking. After a puff, high levels of nicotine reach the brain in 10 to 20 seconds, faster than IV administration. Nicotine replacement therapy includes gum, patch, nasal spray, and tablets, which are buffered to an alkaline pH to facilitate nicotine absorption. After absorption, nicotine enters the bloodstream where at a pH of 7.4 about 69% is ionized and 31% unionized. Plasma binding to protein is less than 5%. The steady state volume of distribution is 2.6L/kg. The highest affinity for nicotine is the liver, spleen, and lung, and the lowest is adipose tissue. The breast milk/plasma ratio for nicotine is 2.9×. About 70–80% of the nicotine is converted to cotinine. About 3–5% of the nicotine is converted to nicotine glucuronide and excreted in the urine. There are six major factors affecting nicotine metabolism:

1) Diet and meals. Menthol inhibits nicotine elimination and meals increase liver blood flow by 30%. Nicotine clearance increases by 40% after a meal.
2) Age. Elimination is slower in smokers over 65 years of age.
3) Sleep. During sleep, hepatic blood flow decreases and nicotine elimination decreases by 17%.
4) Gender. Women eliminate nicotine faster than men by 13% and women on oral contraceptive steroids are 28% faster than men. Pregnancy increases nicotine clearance by 60%.
5) Abstinence. Smoking inhibits nicotine clearance, and after 4 days of smoking abstinence, elimination increased by 14%. After 7 days, it is increased by 36%. There are substances in cigarettes that inhibit nicotine metabolism,
6) Race. Black smokers eliminate nicotine more slowly than White smokers and nicotine elimination by Chinese smokers is lowest. Renal clearance by kidney is about 5% of nicotine dose. Nicotine and cotinine bind to melanin, and therefore dark hair binds much more nicotine than blonde or white hair.

Nicotine and cotinine can be detected in the toenails as well. The optimal plasma or serum cotinine concentration of 15 ng/mL, indicating cigarette use since the 1980s has been reduced to 3 ng/mL due to less exposure to secondhand smoke.

> The current optimal plasma cotinine cut-point to distinguish smokers from non-smokers in the general US population is 3 ng ml–1. This cut-point is much lower than that established 20 years ago, reflecting less secondhand smoke exposure due to clear air policies and more light or occasional smoking.

Reference Number: 20402

BENOWITZ, N.L. "Pharmacology of Nicotine: Addiction, Smoking-Induced Disease, and Therapeutics." *Annual Review of Pharmacology and Toxicology*, 49: 57–71, 2009 (88 references)

Abstract: Nicotine is a weak base and absorption through the mucous membranes is the fastest at alkaline pH (i.e., greater than 7.0). Smoking is the most efficient form of drug administration, as the nicotine enters the circulation rapidly and avoids the liver and first-pass metabolism. When a person inhales smoke from a cigarette, the nicotine is distilled from the tobacco and is carried on the smoke particles into the lungs, where it rapidly passes into the brain. Nicotine binds to the nicotinic cholinergic receptors (nAChR) in the brain, which eventually releases the neurotransmitter, dopamine, common to all addictive drugs. Repeated exposure of nicotine causes tolerance to some, but not all, of the effects of nicotine and causes an increase in the number of nAChR binding sites in the brain. Nicotine is metabolized by the CYP2A6 enzyme in the liver and converted into cotinine and eventually to 3-HC. Slow metabolizers of nicotine appear to be less dependent and more likely to quit smoking. Asian smokers generally have a slower rate of elimination than White smokers and have a lower risk of lung cancer. Nicotine is not a direct carcinogen, but it does promote tumors by impairing the killing of cancer cells and promoting angiogenesis (the formation of new blood vessels in the tumor).

> There is considerable peak to trough oscillation in blood levels from cigarette to cigarette. However, consistent with the half-life of two hours, nicotine accumulates in the body over six to nine hours of regular smoking. Thus, smoking results not in intermittent and transient exposure to nicotine, but in an exposure that lasts 24 hours per day. Arterio-venous differences in nicotine concentration during cigarette smoking are substantial, with arterial levels exceeding venous levels up to tenfold. The persistence of nicotine in the brain throughout the day and night results in changes in the structure and function of nicotinic receptors and in intracellular processes of neuroadaptation, as mentioned previously.

Reference Number: 20403

BENOWITZ, N.L., P. JACOB III, R.T. JONES, AND J. ROSENBERG. "Interindividual Variability in the Metabolism and Cardiovascular Effects of Nicotine in Man." *Journal of Pharmacology and Experimental Therapeutics*, 221: 368–372, 1982 (2 tables, 4 figures, 23 references)

Abstract: Fourteen healthy male smokers (aged 21–34 years) were administered 1–2 μg/kg of nicotine bitartrate in saline IV for 30 minutes. Frequent blood samples were collected, and plasma cotinine and nicotine concentrations were measured by GC. HR, BP, skin temperature, and other parameters were measured. During the infusion of nicotine, the mean HR increased 16.4 beats per minute (bpm), the systolic BP increased 8.8 mm Hg, and the skin temperature decreased 5.8°C. The steady state volume of distribution was large (183 L), which, despite a high metabolic clearance of 1089 mL/min, accounted for a long terminal half-life of 119 minutes. Renal clearance accounted for 15.7% of the total clearance. The plasma binding of nicotine was on average 4.9%. Cigarette smoking accelerates the metabolism of many drugs, including nicotine.

> Our observations have potential clinical importance. That low concentrations of nicotine increase heart rate to a maximum suggests the heart rate response to cigarette smoking, after the day's first few cigarettes, will not vary with the amount of nicotine delivered, a finding, in fact, observed. The possibility that vasoconstriction is more related to blood concentrations of nicotine, at least within a particular individual, suggests that the extent and duration of vascular responses might be sensitive to the amount of nicotine consumed and persist as long as nicotine circulates. Indeed, cutaneous vasoconstriction, as assessed by thermography of the hands and feet, lasts as long as 90 min after a single cigarette. We would predict that vasoconstrictor effects would persist in cigarette smokers overnight because nicotine remains in the circulation over that period of time.

Reference Number: 20404

POMPERLEAU, O.F., M. HARIHARAN, C.S. POMERLEAU, O.G. CAMERON, AND S.K. GUTHRIE. "Differences Between Smokers and Never-Smokers in Sensitivity to Nicotine: A Preliminary Report." *Addiction*, 88: 113–118, 1993 (1 table, 2 figures, 17 references)

Abstract: Five male and five female never-smokers and five male and five female smokers (mean age 28 years) were administered nicotine via a nasal spray, with 1.50 mg nicotine for smokers and 0.5 or 0.25 mg for never-smokers. HR, BP, and plasma nicotine and cotinine concentrations were measured by HPLC.

Table: Mean Plasma Nicotine Concentration, HR, and Systolic BP Increases in Smokers and Never-Smokers

Smoking History and Nicotine Dose	Mean Maximum Plasma Nicotine Increase (ng/mL)	Mean Maximum HR Increase (bpm)	Mean Maximum Systolic Blood Pressure Increase (mm Hg)
Smokers (1.5 mg)	11.5	+16.4	+17.4
Never-smokers (0.5 mg)	12.0	+37.3	+24.3
Never-smokers (0.25 mg)	7.5	+17.1	+21.6

Source: Adapted from Pomerleau et al, 1993

Never-smokers exhibited a plasma nicotine pattern of rise and decay similar to that of smokers, but the magnitude of the peak (expressed as plasma level per unit dose administered) was nearly four times as large. This disparity in plasma nicotine accumulation between never-smokers and smokers is consistent with a report by Srivastava et al. (1991), who observed plasma nicotine concentrations in never-smokers that were twice those in smokers following application of 30 mg transdermal nicotine patches (presumably having identical bioavailability).

CHAPTER 3

Nicotine in Blood, Serum/Plasma, Urine, Sweat, Saliva, Breast Milk, Hair, and Postmortem Tissues

"Determination of nicotine and metabolite concentrations in biological fluids at low LOQs is of increasing interest in nicotine cessation and clinical investigations. Highly sensitive assays are required to detect environmental tobacco exposure and to extend the window of drug detection in nicotine cessation programs. Full analytical method validation is required for reliable and accurate drug quantification."

—Skhakleyz and Huestis, "Optimization and Validation of a Liquid Chromatography-Tandem Mass Spectrometry Method for the Simultaneous Quantification of Nicotine, Cotinine, and Trans-3'-Hydroxycotinine and Norcotinine in Human Oral Fluid" (2009)

Nicotine and its main metabolite, cotinine, distribute throughout the entire body and so may be detected in blood, serum, urine, oral fluids, and hair. The detection times are of longer duration in hair >urine >oral fluids >blood (serum).

3.01 METHODS OF ANALYSIS

"Of the different body fluids, saliva is the matrix of choice for assessing the presence of nicotine and its metabolites in humans exposed to ETS."

—Dhar, "Measuring Tobacco Smoke Exposure: Quantifying Nicotine/Cotinine Concentration in Biological Samples by Colorimetry, Chromatography and Immunoassay Methods" (2004)

There are numerous methods to determine the concentration of nicotine and its major metabolite, cotinine, in various biological fluids. The methods range from simple enzymatic tests in the urine or saliva to sophisticated LC/MS/MS methods in blood or plasma (30101–4). A sensitive method was developed to measure nicotine in bronchoalveolar fluid from EVALI (e-cigarette or vaping product associated lung injury) patients to determine recent use of an inhalational nicotine product (30105).

Reference Number: 30101

TANNER, J-A., M. NOVALEN, P. JATLOW, M.A. HUESTIS, S.E. MURPHY, J. KAPRIO, A. KANKAANPÄÄ, L. GALANTI, C. STEFAN, T.P. GEORGE, L. BENOWITZ, C. LERMAN, AND R.F. TYNDALE. "Nicotine Metabolite Ratio (3-hydroxycontinine/cotinine) in Plasma and Urine by Different Analytical Methods and Laboratories: Implications for Clinical Implementation." *Cancer Epidemiology, Biomarkers & Prevention*, 24: 1239–1246, 2015 (2 tables, 1 figure, 23 references)

Abstract: Thirty-five plasma and 25 urine samples were spiked with nicotine and trans 3'-hydroxycotinine (3-HC) and were sent to eight different laboratories for analysis. The labs all employed LC/MS/MS, GC/MS, or high-performance liquid chromatography with ultraviolet detection (urine only) methods to determine the 3-HC and cotinine concentrations. The results showed better agreement in plasma than urine.

> We found that plasma NMR measurements are in good agreement and strongly correlated between methods and sites. Our results provide greater confidence that plasma NMR measures, regardless of method used here, are consistent. These findings reduce the need for differential interpretation of plasma NMR results based on approach used to conduct the analysis.

Reference Number: 30102

HAMILTON, D.A., M.C. MAHONEY, M. NOVALEN, M.J. CHENOWETH, D.F. HEITJAN, C. LERMAN, R.F. TYNDALE, AND L.W. HAWK JR. "Test-Retest Reliability and Stability of the Nicotine Metabolite Ratio Among Treatment-Seeking Smokers." *Nicotine and Tobacco Research*, 17: 1505–1509, 2015 (1 figure, 1 table)

Abstract: The nicotine metabolite ratio (NMR) is the ratio of 3-HC to cotinine and is an indicator of slow versus fast metabolizers of nicotine.

Nicotine is metabolized to cotinine and then to 3-HC by the cytochrome P450 2A6 (CYP2A6) enzyme system. The NMR is not affected by the time of the day sampling or storage at room temperature. The NMR was determined in 72 adult regular smokers (10+CPD) in blood samples analysed by GC/MS/MS, collected two to three weeks apart.

> The results of the present study indicate that the NMR is stable and exhibits excellent short-term test-retest reliability. These findings are consistent with several recent studies of ad libitum smokers; however, the present study makes an important extension of earlier work by providing the first NMR reliability data from treatment-seeking smokers. The time frame in which test-retest analyses were conducted parallels the time frame of a planned quit attempt in the real world (ie, 2–4 weeks).

Reference Number: 30103

SHAIK, F.B., G. NAGAJOTHI, K. SWARNALATHA, C.S. KUMAR, AND N. MADDU. "Quantification of Nicotine and Cotinine in Plasma, Saliva, and Urine by HPLC Method in Chewing Tobacco Users." *Asian Pacific Journal of Cancer Prevention*, 20: 3617–3623, 2019 (5 tables, 39 references)

Abstract: Blood (plasma), urine, and saliva (by expectorating) were collected and analyzed for nicotine and cotinine by high pressure liquid chromatography (HPLC) in 30 controls and 30 smokeless tobacco (SLT) users (aged 18–30 years) in India. Various other plasma variables were analyzed (such as cholesterol and triglycerides) by an auto-analyzer. Minimal cotinine and nicotine concentrations were found in the controls due to environmental and food exposure.

> In conclusion, chewing tobacco users revealed a significant increase in levels of nicotine and cotinine in plasma, saliva, and urine. The significantly higher levels of nitric oxide, peroxynitrites, protein carbonyls, and malondialdehyde act as indicators of development of oxidative stress in response to nicotine and tobacco-specific N-nitrosamines. The chewing tobacco users communicated the normal range of plasma biochemical profiles compared to other smokeless tobacco products (Panmasala with tobacco, gutkha, and khaini).

Reference Number: 30104

DHAR, P. "Measuring Tobacco Smoke Exposure: Quantifying Nicotine/Cotinine Concentration in Biological Samples by Colorimetry,

Chromatography and Immunoassay Methods." *Journal of Pharmaceutical and Biomedical Analysis*, 35: 155–168, 2004 (5 tables, 1 figure, 189 references)

Abstract: The study discusses various analytical methods to measure nicotine and cotinine concentrations in saliva, blood and urine samples, including gas chromatography (GC), HPLC, GC/MS, radioimmunoassay (RIA), enzyme-linked immunosorbent assay (ELISA), and colorimetry, though the author acknowledges that the latter lacks specificity.

Of the different body fluids, saliva is the matrix of choice for assessing the presence of nicotine and its metabolites in humans exposed to ETS. Of the different biomarkers, cotinine appears to be the analyte of choice, as it fulfills the prerequisites of specificity and retention time and is found in detectable concentrations in all three matrices. GC–MS is preferred for analyzing nicotine and its metabolites in smokers whereas GLC is preferred for studies related to passive smokers.

Reference Number: 30105

XIA B., B.C. BLOUNT, AND L. WANG. "Sensitive Quantification of Nicotine in Bronchoalveolar Lavage Fluid by Acetone Precipitation Combined with Isotope-Dilution Chromatography-Tandem Mass Spectrometry." *ACS Omega*, 6: 13962–13969, 2021 (5 tables, 4 figures, 35 references)

Abstract: A simple and sensitive method was developed to measure nicotine in the bronchoalveolar lavage (BAL) fluid to aid in the investigation of EVALI patients. The nicotine was extracted from the BAL fluid by acetone precipitation and was analyzed by LC/MS/MS. The limit of detection was 0.050 ng/mL for a sample volume of 40 uL. Nicotine was detected in 35% of the 43 BAL fluid samples from EVALI patients, indicating recent use of an inhalational nicotine product.

We developed a sensitive method for quantifying nicotine in BAL fluids using acetone precipitation combined with isotope-dilution LC-MS/MS. Because of its high sensitivity, precision, accuracy, and throughput, this assay required only 40 μL of samples for detecting nicotine exposure in the BAL fluid matrix. The method has an LOD of 0.050 ng/mL, which is approximately 10 times more sensitive than the LODs of previously reported methods for urine, serum, or plasma in the literature. This method is useful for understanding the recent inhaled-nicotine exposure as it relates to nicotine uptake as well as forensic investigations to link inhaled product use with acute lung injuries such as those observed in EVALI case patients.

3.02 BLOOD/PLASMA/SERUM/BREAST MILK

"The rapid transfer of nicotine from serum into milk and the short half-lives of nicotine in milk indicate, that mothers who cannot refrain from smoking during the nursing period, should attempt to prolong the time between the last cigarette smoked and breast feeding to minimize exposure of the nursing infant by this substance."

—Luck and Nau, "Nicotine and Cotinine Concentrations in Serum and Milk of Nursing Mothers" (1984)

Whole blood contains red blood cells and plasma. Plasma is the clear liquid that remains in the blood after it is separated from the red blood cells, usually by centrifuging. Serum is the same as plasma except that it contains no clotting factors. The plasma also contains drug protein binding factors which can affect the blood/plasma drug concentrations. In drugs that are highly lipophilic (fat-loving), such as tetrahydrocannabinol (THC), the plasma or serum concentration is about 200% higher than the whole blood concentration. In drugs that are water soluble, such as water and nicotine, the plasma concentration is only approximately 20% higher.

There is a rapid transfer of nicotine from the serum into the breast milk of nursing smokers. The mean breast milk:serum ratios are 2.92× for nicotine and 0.78× for cotinine. The lower the pH of the beast milk, the greater the nicotine concentration (30203).

Reference Number: 30201

BENOWITZ, N.L., AND P. JACOB III. "Trans-3-Hydroxycotinine: Disposition Kinetics, Effects and Plasma Levels During Cigarette Smoking." *British Journal of Clinical Pharmacology*, 51: 53–59, 2001 (3 tables, 4 figures, 26 references)

Abstract: Trans 3'-hydroxycotinine (3-HC) is a major metabolite of nicotine formed from cotinine. It is the most abundant metabolite of nicotine found in urine, accounting for approximately 40% of the nicotine dose. Five male and three female cigarette smokers (mean age 41 years) on a clinical research ward were studied. After 5 days of supervised non-smoking, the subjects were administered 4 μg/kg of 3-HC IV. Blood and urine samples were collected and 3-HC concentrations were measured by GC/MS. Glucuronide conjugated 3-HC (3-HC-Glu) was measured in urine after the addition of beta-glucuronidase. On average, 87% of the dose of 3-HC was recovered in the urine and 29% was recovered as 3-HC-Glu. There was no significant change in blood pressure or heart rate after the infusion of 3-HC. The plasma half-life of 3-HC was 6.6 hours. Plasma levels of 3-HC, nicotine, and cotinine were measured in 62 smokers on up to three occasions.

Table: Mean CPDs and Plasma Concentrations of Nicotine, Cotinine, and 3-HC in 62 Smoking Subjects

Parameter	Mean
Number of cigarettes per day	13.1
Plasma nicotine (ng/mL)	7.2
Plasma cotinine (ng/mL)	146
Plasma 3-HC (ng/mL)	44
Ratio 3-HC/cotinine	0.32
Plasma cotinine/CPD	13.6

Source: Adapted from Benowitz and Jacob III, 2001

These findings extend our understanding of the quantitative nature of nicotine metabolism. Such data may be of use in quantitating human exposure to nicotine from tobacco and in studying individual variability in nicotine metabolism.

Reference Number: 30202

WIENCEK, J.R., E.A. GEHRIE, A.M. KEISER, P.C. SZKLARSKI, K.L. JOHNSON-DAVIS, AND G.S. BOOTH. "Detection of Nicotine and Nicotine Metabolites in Units of Banked Blood." *American Journal of Clinical Pathology*, 151: 516–521, 2019 (3 figures, 28 references)

Abstract: Nicotine, cotinine, and 3-HC concentrations were measured by GC/MS/MS in 105 blood units from volunteers in Tennessee. Twenty blood samples (19%) contained detectable concentrations of nicotine, cotinine, or 3-HC and 19 (18.1%) contained concentrations consistent with the use of a nicotine-containing product within 48 hours of blood donation.

Table: Mean and 95% Confidence Range of Nicotine, Cotinine, and 3-HC in 105 Blood Samples from Blood Donors in Tennessee

Nicotine/Metabolite	Mean (ng/mL)	95% Range (ng/mL)
Nicotine	5.8	3.8–7.8
Cotinine	114.1	83.4–145.0
3-HC	34.0	24.8–43.3

Source: Adapted from Wiencek et al, 2019

It is important to remember that the measurements presented in this study indicate exposure to nicotine, which should not be strictly considered a surrogate marker for smoking. It is possible that the prevalence of cigarette smoking is, in fact, lower among donors than the national average and that a significant source of nicotine was from smoking cessation aids such as nicotine gum, nicotine patches, or even e-cigarettes (which many misconstrue as a "healthier alternative" to cigarettes). However, even though other smoking-related chemicals may be absent in these products, nicotine is believed to impair critical aspects of fetal development, including retinoic acid signaling pathways.

Reference Number: 30203

LUCK, W., AND H. NAU. "Nicotine and Cotinine Concentrations in Serum and Milk of Nursing Smokers." *British Journal of Clinical Pharmacology*, 18: 9–15, 1984 (3 tables, 3 figures, 31 references)

Abstract: A total of 44 milk and 44 serum samples were collected from 23 smokers who were currently nursing and analyzed for nicotine and cotinine by GC. The milk nicotine and cotinine concentrations ranged from 1–62 ng/mL and 12–222 ng/mL, respectively. The serum nicotine and cotinine ranged from 1–28 ng/mL and 16–330 ng/mL, respectively. The mean milk:serum ratios were 2.92× for nicotine and 0.78× for cotinine. The half-life of nicotine in milk was 97 minutes. The milk pH ranged from 6.4 to 7.4, compared to 7.4 ±0.03 in serum. The lower the milk pH, the greater the concentration of nicotine in milk and the higher the ratio.

The rapid transfer of nicotine from serum into milk and the short half-lives of nicotine in milk indicate, that mothers who cannot refrain from smoking during the nursing period, should attempt to prolong the time between the last cigarette smoked and breast feeding to minimize exposure of the nursing infant by this substance.

3.03 URINE

"Dietary intake of nicotine containing food was not identified as a source of nicotine and cotinine excretion in urine in the German adult population. Tobacco smoke was the only source for these biomarkers."

—Heinrich, Holscher, Steiwert, et al, "Nicotine and Cotinine in Adults' Urine: The German Environmental Survey 1998" (2005)

Urine nicotine and cotinine concentrations are detected in smokers longer than the blood concentrations of these compounds. The mean urine nicotine concentration was 703.5 ng/mL for cigarette smokers and 55.0 ng/mL for non-smokers in India (30301). In Germany, the mean urinary cotinine concentration was 856 ng/mL for smokers, 7.1 ng/mL for non-smokers exposed to secondhand smoke (SHS) at home, and 2.8 ng/mL for non-smokers not exposed to SHS at home indicating that SHS exposure of nicotine is greater in India (30302).

In a case of tobacco poisoning, the urine nicotine concentration was over 60× the cotinine concentration, indicating recent massive absorption of nicotine (30303).

Reference Number: 30301

BEHERA, D., R. UPPAL, AND S. MAJUMDAR. "Urinary Levels of Nicotine and Cotinine in Tobacco Users." *Indian Journal of Medical Research*, 118: 129–113, 2003 (1 table, 29 references)

Abstract: A 24-hour urine sample was collected from 30 cigarette smokers, 30 bidi smokers, 10 hookah smokers, 20 tobacco chewers, 20 passive smokers, and 20 non-smokers and analyzed for nicotine and cotinine by HPLC. There was no effect of age or gender on nicotine or cotinine excretion. Bidis contain a larger amount of nicotine compared to cigarettes when matched gram for gram.

Table: Mean Urinary Nicotine and Cotinine Concentrations According to Type of Smoker

Type of Tobacco User	Mean Urinary Nicotine Concentration (ng/mL)	Mean Urinary Cotinine Concentration (ng/mL)
Cigarette smokers	703.5	2736.2
Bidi smokers	268.5	562.6
Hookah smokers	548.0	2379.0

Type of Tobacco User	Mean Urinary Nicotine Concentration (ng/mL)	Mean Urinary Cotinine Concentration (ng/mL)
Tobacco chewers	445.8	2178.3
Passive smokers	109.8	280.8
Non-smokers	55.0	7.3

Source: Adapted from Behera et al, 2003

All forms of tobacco users had significantly higher values compared to passive smokers and nonusers. Thus, cotinine and nicotine levels in urine may be considered as good indicators to assess the exposure to tobacco in our population.

Reference Number: 30302

HEINRICH, J., B. HÖLSCHER, M. SEIWERT, C.L. CARTY, G. MERKEL, AND C. SCHULZ. "Nicotine and Cotinine in Adults' Urine: The German Environmental Survey 1998." *Journal of Exposure Analysis and Environmental Epidemiology*, 15: 74–80, 2005 (4 tables, 24 references)

Abstract: The urine nicotine and cotinine concentrations were determined by HPLC in 1580 smokers and 3125 non-smokers (aged 18–69 years) in Germany from 1997 to 1999. More than 20% of the non-smokers were exposed to SHS at home or at work. The mean urinary nicotine concentration was 367 ng/mL for smokers, 4.7 ng/mL for non-smokers exposed to SHS at home, and 1.4 ng/mL for non-smokers with no smokers at home. The mean urinary cotinine concentration was 856 ng/mL for smokers, 7.1 ng/mL for smokers exposed to SHS at home and 2.8 ng/mL for those with no SHS exposure at home.

Table: Odds Ratio of Detecting Nicotine or Cotinine in the Urine of Non-smokers Exposed to SHS at Home, Workplace, or Other

SHS Exposure	OR of detecting nicotine in urine	OR of detecting cotinine in urine
1 smoker at home	4.06×	3.98×
>1 smoker at home	6.17×	5.25×
Workplace exposure	1.89×	2.06×
Exposure elsewhere	1.32×	1.32×

Source: Adapted from Heinrich et al, 2005

Dietary intake of nicotine containing food was not identified as a source of nicotine and cotinine excretion in urine in the German adult population. Tobacco smoke was the only source for these biomarkers.

Exposure to ETS in the workplace and in public places is still an important public health issue in Germany. While a new federal bill to diminish ETS exposure in the workplace was passed in 2002, regulation in other public places could further decrease ETS exposure. However, protection of nonsmokers from ETS at home also needs more attention.

Reference Number: 30303

MIYAUCHI, M., M. HAYASHIDA, K. HIRATA, K. HIRATA, AND H. YOKOTA. "Gastric Lavage Guided by Ultrathin Transnasal Esophagogastroduodenoscopy in a Life-Threatening Case of Tobacco Extract Poisoning: A Case Report." *Journal of Nippon Medical School*, 80: 307–311, 2013 (1 table, 2 figures, 17 references)

Abstract: Nicotine is one of the most lethal of all known poisons. A 31-year-old female non-smoker attempted suicide by swallowing an extract derived from 20 cigarettes mixed with alcohol. The cigarettes employed were Kent and Lucky Strike, which contain 11.2 mg of nicotine in each cigarette. The maximum nicotine dose would therefore be 224 mg. She attended the emergency department 90 minutes after ingestion of the tobacco extract. The patient had not vomited. Gastric lavage was conducted, and a yellow brownish liquid was obtained which appeared to contain tobacco. She was also administered a single dose of activated charcoal. She was discharged the following day. Serum, urine, and gastric content nicotine concentrations were determined by GC/MS from samples collected approximately 2 hours after ingestion. Nicotine is a weak base and is poorly absorbed through the stomach.

Table: Serum, Urine, and Gastric Content Nicotine and Cotinine in a Case of Ingestion of a Tobacco Extract

	Serum Concentration	Urine Concentration	Gastric Content Concentration
Nicotine (ng/mL)	8	532	14,050
Cotinine (ng/mL)	902	316	1,200

Source: Adapted from Miyauchi et al 2013

The absorption of nicotine can easily be affected by several factors, such as gastric pH and variations in gastric juice. In the present case, we thought that antipsychotic medication, food, and alcohol ingested by the patient in addition to the tobacco extract had likely delayed absorption. However, because the nicotine concentration in the stomach contents remained high, we were concerned that the serum nicotine concentration might continue to increase.

3.04 SALIVA (ORAL FLUID)/SWEAT

> "Oral fluid is an important alternative matrix to blood and urine for monitoring drug and tobacco exposure; collection is simple and noninvasive and can be performed by nonmedical personnel. The matrix is relatively clean and offers measurement of unbound, pharmacologically active drug concentrations."
>
> —Skhakleyz and Huestis, "Optimization and Validation of a Liquid Chromatography-Tandem Mass Spectrometry Method for the Simultaneous Quantification of Nicotine, Cotinine, and Trans-3'-Hydroxycotinine and Norcotinine in Human Oral Fluid" (2009)

Oral fluid drug testing is becoming more prevalent than urine drug testing, as it is easier to collect and has a higher nicotine and cotinine concentration than plasma (30402–4). Nicotine can also pass through the skin and appear in the sweat, which can be analyzed to determine heavy or light cigarette and other tobacco products use (30405).

Reference Number: 30401

SHAKLEYA, D.M., AND M.A. HUESTIS. "Optimization and Validation of a Liquid Chromatography-Tandem Mass Spectrometry Method for the Simultaneous Quantification of Nicotine, Cotinine, Trans-3'-Hydroxycotinine and Norcotinine in Human Oral Fluid." *Analytical and Bioanalytical Chemistry*, 395: 2349–2357, 2009 (5 tables, 1 figure, 26 references)

Abstract: A LC/MS/MS method was developed to identify and quantitate nicotine, cotinine, 3-HC, and norcotinine in 0.5 mL of oral fluid. The OF samples were collected using the Quantisal oral fluid (OF) collection device. The Quantisal OF device collected 1 mL of oral fluid and also contained 3 mL of buffer. Two milliters of the sample (equivalent to 0.5 mL OF) was vortexed with 50 μL of internal standard and 2 mL of sodium phosphate. The samples were placed through a solid phase extraction using a CleanScreen DUA SPE column, and 20 μL was injected into the LC/MS/MS. The linearity of the method ranged from 1 to 2,000 ng/mL for nicotine and norcotinine, from 0.5 to 2,000 ng/mL for 3-HC, and from 0.2 to 2,000 ng/mL for cotinine. Analytical recovery was greater than 91%.

> Determination of nicotine and metabolite concentrations in biological fluids at low LOQs is of increasing interest in nicotine cessation and clinical investigations. Highly sensitive assays are required to detect environmental tobacco exposure and to extend the window of drug detection in nicotine cessation programs. Full analytical method validation is required for reliable and accurate drug quantification. This LCMSMS procedure is simple and easy to perform and should be useful for routine monitoring of nicotine and metabolites in oral fluid.

Reference Number: 30402

FENG, S., O. CUMMINGS, AND G. MCINTIRE. "Nicotine and Cotinine in Oral Fluid: Passive Exposure vs Active Smoking." *Practical Laboratory Medicine*, 12: (8 pp), 2018 (4 tables, 4 figures, 19 references)

Abstract: Oral fluid testing is becoming popular, as the sample can be collected easily and without the need for specialized professionals or dedicated facilities. Oral fluid nicotine and cotinine concentrations were determined by LC/MS/MS in 46 non-smokers, 36 non-smokers exposed to secondhand smoke, 44 light smokers (< 10 CPD), and 46 heavy smokers (>10 CPD). A reporting cut-off of 10 ng/mL for nicotine would detect 88% of active smokers and the same OF cotinine cut-off would detect 98% of active smokers.

Table: Median (Range) Oral Fluid Concentrations After Application of the 7, 14, and 21 mg Transdermal Nicotine Patch

Dose (mg nicotine)	Nicotine, Median (Range) ng/mL		Cotinine, Median (Range) ng/mL	
7	57	(<1–472)	140	(10–150)
14	129	(24–2278)	202	(5–823)
21	151	(6–1016)	199	(3–823)

Source: Adapted from Feng et al, 2018

In summary, nicotine and cotinine levels from transdermal application do not differ significantly from those observed from smoking. Thus, oral fluid nicotine testing cannot differentiate between patch application and active smoking indicating that oral fluid testing for nicotine is inadequate for determining patient adherence (to their prescribed transdermal patch). The normalization and transformation of nicotine data from oral fluid samples is consistent with earlier clinical results which indicated that a reporting cutoff of 10ng/mL is appropriate to differentiate active smoking from passive exposure.

Reference Number: 30403

MILLER, E.I., H-R.K. NORRIS, D.E. ROLLINS, S.T. TIFFANY, C.M. MOORE, M.J. VINCENT, A. AGRAWAL, AND D.G. WILKINS. "Identification and Quantification of Nicotine Biomarkers in Human Oral Fluid from Individuals Receiving Low-Dose Transdermal Nicotine: A Preliminary Study." *Journal of Analytical Toxicology*, 34: 357–366, 2010 (4 tables, 1 figure, 39 references)

Abstract: Oral fluid samples were collected using the Quantisal OF collection device, before, 0.5 hours after, and 0.75 hours after the removal of a 7 mg transdermal nicotine patch in 10 nicotine-free subjects. The patch was applied for 4 hours. Plasma samples were also collected simultaneously. The samples were analyzed for nicotine, cotinine, and other metabolites, such as nornicotine, anatabine, and anabasine, by solid phase extraction (SPE) LC/MS/MS. An enzymatic method (ELISA) was used as a screen for cotinine as well. The average OF nicotine and cotinine concentrations at both times were 25 ng/mL and 21 ng/mL, respectively. The mean oral fluid:plasma ratios were 6.4 for nicotine and 3.3 for cotinine.

An ELISA method was also successfully applied as a screening tool for these samples in conjunction with LC–MS–MS confirmation. An ELISA cut-off concentration of 5.0 ng/mL provided excellent sensitivity for discrimination of COT "positive" post-exposure oral fluid samples collected after low-level transdermal NIC exposure. The findings of this study have shown that both LC–MS–MS and ELISA methods can be complementary, demonstrating excellent sensitivity and specificity for oral fluid samples collected from this particular population as part of a clinical study investigating NIC pharmacokinetics and potential biomarkers of NIC exposure in plasma and oral fluid of low-level smokers.

Reference Number: 30404

TENEGGI, V., L. SQUASSANTE, L. IAVARONE, S. MILLERI, A. BYE, AND R. GOMENI. "Correlation and Predictive Performances of Saliva and Plasma Nicotine Concentration on Tobacco-Withdrawal Induced Craving." *British Journal of Clinical Pharmacology*, 54: 407–414, 2002 (3 tables, 6 figures, 29 references)

Abstract: Plasma and saliva nicotine concentrations were determined in seven female and six male smokers (mean age 32.5 years) with and without abstinence from smoking by LC/MS. A good correlation was found between plasma and saliva nicotine concentrations.

> Analysis of saliva definitely offers a potentially more attractive way to assess nicotine concentration values, as samples can be collected easily and noninvasively. In addition, saliva sampling avoids the pain and discomfort involved in venepuncture. In studies that assess psychological measures, such as subjective mood, blood collection could present a possible confounding factor because of the anxiety and pain that accompanies it. For these reasons saliva can reasonably be considered as the ideal sampling site for all clinical studies conducted for the evaluation of the potential activity of drugs on nicotine deprivation symptoms.

Reference Number: 30405

MEHMETI, E., T. KILIC, C. LAUR, AND S. CARRARA. "Electrochemical Determination of Nicotine in Smokers' Sweat." *Microchemical Journal*, 158: (7 pp), 2020 (3 tables, 6 figures, 41 references)

Abstract: As nicotine is mainly water soluble, it can pass though the skin when applied as a nicotine patch. Likewise, nicotine can pass from the smoker's body into the sweat and be detected by various methods. In this study, a simple, inexpensive, and very sensitive electrochemical sensor for the quantification of nicotine in smokers' sweat is described. Both artificial sweat and sweat collected from sweat patches from smoking and non-smoking subjects who performed strong physical exercise for 1 hour were analyzed.

> In this work, we report the demonstration of the capability of electrochemical sensors to provide measurements directly on the sweat of volunteers to successfully discriminate heavy smokers by light ones. Since its importance as pharmaceutical treatment-agent for nicotine replacement

therapy for purposes of tobacco cessation, and since that is the key compound for the development of reduced-risk cigarettes by tobacco industry, nicotine is one of those markers that are of interest for monitoring in human sweat. Especially in therapies for purposes of smoking cessation, easy methods for an easy measure of nicotine on the patients is highly required for a fine adjustment of the nicotine dosage as release, e.g., through therapeutic patches.

3.05 HAIR

"In this study, nicotine levels in children's hair were related to the number of smokers in the household of the child and the total number of cigarettes smoked in the home."

—Al-Delaimy and Woodward, "Questionnaire and Hair Measurement of Exposure to Tobacco Smoke" (2000)

Hair grows at a rate of approximately 1 cm per month and about 90% of the total hair on the scalp is in the growth phase (30501). Hair is the sample which will detect nicotine exposure for the longest period of time. Hair nicotine has been found in mummified human remains from Northern Chile who died over 2,000 years ago (30509). It is also a sample that is easily collected especially from children (30502–5). Hair nicotine analysis has been conducted to determine exposure to secondhand smoke, exposure to nicotine during pregnancy, the effectiveness of smoke-free laws, and abstinence from smoking (30503–7).

The concentration of nicotine in hair is 10× the cotinine concentration, which is different from urine, where the cotinine concentration is typically much greater than nicotine (30501).

Hair nicotine concentration detected in the dogs of smoking owners is similar to that detected in children (30508).

Reference Number: 30501

SAVANT, S.C., H.G. SHAH, V. SODANI, AND D. BAJAJ. "Nicotine Hair Test—A Potential Biomarker for Smoke Analysis." *Bhavnagar University Journal of Dentistry*, 3: 56-60, 2013 (17 references)

Abstract: Hair grows at approximately 1 cm per month and this growth continues for 2-6 years. When the hair attains full growth, it stops growing for 2-3 months and then is shed and a new hair grows in its place.

Approximately 10% of the total hair on the scalp is in the resting phase and 90% is in the growth phase. Even though the serum concentration is higher than nicotine in tobacco users, the nicotine concentration in hair is typically 10× the hair cotinine concentration.

Hair nicotine provides reliable information on long term smoke exposure than biomarker measures in urine, saliva, or serum because of the shorter half life of the latter biomarkers. The long term (up to several months) exposure is the usual exposure of interest in health-related studies. The absence of drug metabolism in hair and its fairly uniform growth rate for a given location in the body may provide an historical account of smoke exposure, provided that the concentrations measured in a given hair segment are related to their distance from the scalp (usually the proximal 1–2 cm).

Reference Number: 30502

KIM, W.H., A. NAVAS-ACIEN, F. DOMINICI, E. AVILA-TNAG, G. ONICESCU, P. BREYSSE, AND J.M. SAMET. "Determinants of Hair Nicotine Concentration in Nonsmoking Women and Children: A Multicounty Study of Secondhand Smoke Exposure in Homes." *Cancer Epidemiology, Biomarkers & Prevention*, 18: 3407–3414, 2009 (3 tables, 1 figure, 45 references)

Abstract: Hair samples were collected from 852 non-smoking women and 1,017 children (<11 years of age) living in households with smokers from 31 countries between 2005 and 2006. Hair nicotine concentrations were determined by GC/MS. Nicotine air concentrations were determined using a passive air sample, by GC. The median number of cigarettes smoked daily was 12. The rate was highest in Europe and the Middle East (20 CPD) and lowest in South America (8 CPD). The median hair nicotine concentration for women and children in smoking households was 0.44 ng/mg and 0.80 ng/mg, respectively.

This study showed a positive association between SHS exposure and hair nicotine concentrations in a setting outside of the laboratory and showed that home exposure to SHS was an important determinant of hair nicotine among nonsmoking women and children from 31 countries around the world. Under global efforts to reduce the burden of tobacco use led by the Framework Convention on Tobacco Control, a large number of countries have implemented smoking bans in public places including workplaces, restaurants, schools, and hospitals. Private microenvironments, such as homes, cars, and child care in private residences, are outside the jurisdiction of imposed smoking bans, although banning smoking on private property may be preferred by the owners of homes or cars. As SHS in the home can spread quickly throughout the closed microenvironment, parents should implement complete home-smoking bans to make the environment around children smoke free. This analysis, along with previous evidence, provides adequate evidence to support educational and policy initiatives aimed at encouraging home smoking bans and reducing exposure to SHS, especially among young children.

Reference Number: 30503

AL-DELAIMY, W.K., AND A. WOODWARD. "Questionnaire and Hair Measurement of Exposure to Tobacco Smoke." *Journal of Exposure Analysis and Environmental Epidemiology*, 10: 378–384, 2000 (2 tables, 2 figures, 43 references)

Abstract: The most proximal 1–2 cm of hair samples from 112 children and 76 mothers in Wellington, New Zealand were analyzed for nicotine and cotinine through an enzymatic method. The children's hair nicotine concentration ranged from undetectable to 24.7 ng/mg hair and cotinine concentration ranged from undetectable to 1.5 ng/mg hair. The mother's hair nicotine and cotinine concentrations were higher than the children's (up to 70 ng/mg and 5.7 ng/mg hair, respectively) as none of the children were active smokers and 50% of the mothers were. The strong association between hair nicotine and its metabolite cotinine indicate that the hair nicotine is mainly derived from the bloodstream rather than from SHS, as cotinine is only formed inside the body after the metabolism of nicotine.

Table: The Number of Smokers in the House and the Median Hair Nicotine Concentration of Children in the House

Number of smokers in the house	Median hair nicotine concentration (ng/mL)
0	Undetectable (<0.1)
1	0.275
2	1.46
>2	2.02

Source: Adapted from Al-Delaimy and Woodward, 2000

In this study, nicotine levels in children's hair were related to the number of smokers in the household of the child and the total number of cigarettes smoked in the home. However, among active smokers, we did not observe a relationship between number of cigarettes smoked and hair nicotine levels. Our findings indicate that nicotine in hair is a more sensitive measure of ETS exposure than cotinine in hair. This method of measuring ETS exposure appears to offer some advantages; because of the slow hair growth rate, hair nicotine levels provide an integrated measure of exposure over a period of weeks to months rather than a measure of a point in time, and is sensitive and specific to tobacco smoke.

Reference Number: 30504

PICHINI, S., O. GARCIA-ALGAR, L. MUÑOZ, O. VALL, R. PACIFICI, C. FIGUEROA, J.A. PASCUAL, D. DIAZ, AND J. SUNYER. "Assessment of Chronic Exposure to Cigarette Smoke and Its Change During Pregnancy by Segmental Analysis of Maternal Hair Nicotine." *Journal of Exposure Analysis and Environmental Epidemiology*, 13: 144–151, 2003 (5 tables, 1 figure, 29 references)

Abstract: Hair nicotine concentrations were measured in 150 mothers (in two subsequent samples accounting for the first and last months of pregnancy) and in the newborn in Barcelona by GC. Before the analysis, the hair samples were extensively washed. Smoking habit questionnaires were also completed by the mothers in their third trimester and day of delivery. Hair nicotine could be detected in almost all of the mother's samples but only 31% of the newborn's hair. The mean median hair nicotine concentration was 3.75 ng/mg for non-exposed non-smoking mothers, 5.44 ng/mg for mothers exposed to SHS, and 14.40 ng/mg for smoking mothers. Newborn hair nicotine concentrations were only detected in active smoking mothers and not those with solely SHS exposure.

Table: Maternal Hair Nicotine Concentrations and Birth Weight and Head Circumference of Newborns

Hair Nicotine (ng/mg hair)	Percent of Mothers	Birth Weight (g)	Head circumference (cm)
<3	23.6	3292.3	34.34
3 to 18	60.8	3218.6	34.14
19+	15.5	3045.2	33.60

Source: Adapted from Pichini et al, 2003

In conclusion, maternal hair nicotine appeared to be an adequate biomarker of chronic exposure to cigarette smoking during pregnancy because of its ability to be associated with levels of active smoking, level of passive maternal exposure to ETS, and to neonatal effects such as birth weight and head circumference. In agreement with other authors, these somatometric effects, together with the risk of small-for-gestational-age and preterm delivery, were associated with the exposure in the last trimester of pregnancy.

Reference Number: 30505

UEMATSU, T., A. MIZUNO, S. NAGASHIMA, A. OSHIMA, AND M. NAKAMURA. "The Axial Distribution of Nicotine Content Along Hair Shaft as an Indicator of Changes in Smoking Behaviour: Evaluation in a Smoking-Cessation Programme with or Without the Aid of Nicotine Chewing Gum." *British Journal of Clinical Pharmacology*, 39: 665–669, 1995 (1 table, 1 figure, 18 references)

Abstract: Twenty-four male and two female smokers (aged 34–64 years) participated in a 6-month smoking cessation program with (n=16) or without (n=10) the aid of nicotine gum. The axial distribution of nicotine along the hair shaft as measured by GC was compared to the self-reported smoking behavior and breath CO measured at 1-month intervals.

In general, the axial distribution of nicotine concentration along the hair shaft mirrored the monthly profile of self-reported number of daily cigarettes in both the nicotine chewing-gum treated and untreated groups. On the other hand, the monthly profile of CO expiration was less consistent. Since hair needs only to be sampled once at the end of an abstinence programme, its analysis may have some advantage over repeated measurements of CO expiration, as an index of smoking continuity, although there is considerable merit in the repeated measurements of CO content in expired air because the result can be immediately used as a basis for counselling the patient by nursing staff or physician.

As indicated by the difference in nicotine content in black and white hairs in the same subjects, nicotine appears to have a high affinity for melanin. Thus, a slow dissociation of nicotine from melanin, resulting in a prolonged transfer of nicotine from hair bulb cells to the hair matrix, may account for the distribution of nicotine along the hair shafts despite an immediate cessation of smoking.

Reference Number: 30506

HAHN, E.J., M.K. RAYENS, N. YORK, C.T.C. OKOLI, M. ZHANG, M. DIGNAN, AND W.K. AL-DELAIMY. "Effects of a Smoke-Free Law on Hair Nicotine and Respiratory Symptoms of Restaurant and Bar Workers." *Journal of Occupational and Environmental Medicine*, 48: 906–913, 2006 (2 tables, 2 figures, 57 references)

Abstract: The SHS exposure of 105 bar and restaurant workers was determined before, 3 months after, and 6 months after the implementation of a smoke-free ordnance in Lexington, Kentucky, using hair samples. Thirty-eight percent of the workers were current smokers. Restaurant and bar workers experienced a 56% decline in hair nicotine concentration 3 months after implementation. Bar workers had a significantly greater decline in HNC than restaurant workers. Regardless of smoking status, respiratory symptoms also improved following implementation of the law.

Stricter workplace smoke-free laws could protect patrons' and employees' health while also being a cost-effective measure for employers. Having smoke-free establishments may improve employee productivity while decreasing absenteeism and turnover. These same restrictions have been shown to lead to a significant reduction in smokers' daily cigarette use and to promote smokers' cessation efforts. Studies have also documented that smoke-free environments for adolescent workers deter smoking, because teenagers in such environments are 32% less likely to smoke than those who are employed in a workplace without smoking restrictions.

Reference Number: 30507

KIM, S., B.J. APELBERG, E. AVILA-TANG, L. HEPP, D. YUN, J.M. SAMET, AND P.N. BREYSSE. "Utility and Cutoff Value of Hair Nicotine as a Biomarker of Long-Term Tobacco Smoke Exposure, Compared to Salivary Cotinine." *International Journal of Environmental Research and Public Health*, 11: 8368–8382, 2014 (2 tables, 3 figures, 32 references)

Abstract: Cotinine has a much longer half-life than nicotine and is the most widely used biomarker of nicotine or tobacco smoke exposure. The saliva to serum cotinine ratio is reported to be from 1.1 to 1.4 and the urine to serum cotinine ratio is reported as 5. Hair growth is reported to be about 1 cm per month. Approximately 30 to 50 strands of hair, 3 cm from the root end (representing 3 months of growth), were collected from the back of the scalp (where the hair growth is most uniform) from 77 non-smokers, 105 passive smokers (exposed to SHS), and 107 active smokers. Saliva samples were collected by having the subjects spit into a test tube. Hair nicotine and saliva cotinine concentrations were measured by GC/MS, and GC/MS/MS. Both hair nicotine and salivary cotinine concentrations correlated with the number of cigarettes smoked per day. A cutoff concentration of 2.77 ng/mg for hair nicotine and 2.89 ng/mL for saliva cotinine was determined to distinguish a smoker from a non-smoker.

Table: Median Hair Nicotine and Salivary Cotinine Concentrations in Non-Smokers, Passive Smokers, and Active Smokers

	Non-Smokers	Passive Smokers (SHS)	Active Smokers
Median hair nicotine concentration (ng/mg)	0.23	0.36	16.2
Median salivary cotinine concentration (ng/mL)	0.27	0.27	181.0

Source: Adapted from Kim et al, 2014

Salivary cotinine is a well-established and excellent biomarker of recent tobacco exposure. Our results showed that the hair nicotine values could be used to distinguish active or heavy passive smokers from non-SHS exposed nonsmokers. Hair nicotine is a useful biomarker for the assessment of long-term exposure to tobacco smoke.

Reference Number: 30508

KNOTTENBELT, C.M., S. BAWAZEER, J. HAMMOND, D. MELLOR, AND D.G. WATSON. "Nicotine Hair Concentration in Dogs Exposed to Environmental Tobacco Smoke: A Pilot Study." *Journal of Small Animal Practice*, 53: 623–626, 2012 (1 table, 1 figure, 20 references)

Abstract: In humans, hair nicotine concentration (HNC) is a useful biomarker of SHS exposure. Dogs grow hair at approximately the same rate as humans (1 cm/month) except for poodles and other breeds with a non-shedding hair coat. Hair was collected from 23 dogs exposed to SHS and 15 dogs not exposed to SHS. The hair (30 mg) was washed with 2 mL of methanol to determine the surface nicotine concentration. The hair was then treated with sodium hydroxide and 1 μg of 2H4-nicotine and analyzed by a GC/MS. The mean HNC was 0.57 ng/mg for dogs regularly exposed to SHS, 0.09 ng/mg to dogs occasionally exposed to SHS, and 0.01 ng/mg to dogs not exposed to SHS. Low HNCs were found in 11 of the 15 dogs reportedly not exposed to SHS.

Table: Location of Dogs Exposed to SHS

Reported Location of SHS Exposure	Percent of the Dogs Exposed to SHS
Close contact with owner	74
In pet's home	87
During car travel	35
In home of family or friends	39
Outdoors	74

Source: Adapted from Knottenbelt et al, 2012

> Dog hair nicotine concentration appears to be strongly associated with reported exposure to environmental tobacco smoke. The range and median of hair nicotine concentration in dogs exposed to environmental tobacco smoke was similar to those reported in children. This suggests that dog hair could provide a useful method of determining the amount of environmental tobacco smoke exposure in all environments common to pets and children.

Reference Number: 30509

ECHEVERRÍA, J., AND H.M. NIEMEYER. "Nicotine in the Hair of Mummies from San Pedro de Atacama (Northern Chile)." *Journal of Archeological Science*, 40: 3561–3568, 2013 (2 tables, 2 figures, 62 references)

Abstract: Hair strands were collected from 56 mummies in Northern Chile who died between 100 BCE and 1450 CE and were analyzed for nicotine by GC/MS. A positive hair nicotine concentration was found in 35 (65%) of the mummies.

> In summary, the present results show the continuous consumption of nicotine in SPA [San Pedro de Atacama], from the Late Formative period to the Late Intermediate period, in spite of profound changes in the administration technology, which changed from smoking pipes to snuffing trays. The results also support the idea that consumption of psychoactive alkaloids in SPA was performed by members of the society at large, irrespective of their social and wealth status.

3.06 POSTMORTEM

> "High doses of nicotine inhibit the central nervous system, inducing hypotension and bradycardia followed by coma and respiratory paralysis, leading to death by asphyxia."
>
> —Tjoncke et al, "Death Related to Nicotine Replacement Therapy: A Case Report" (2020)

Nicotine causes death mainly by respiratory failure. Nicotine has caused death from using NicoDerm patches, intravenous injection, and oral ingestion, or a combination of these methods (30601–4). The distribution of cotinine and nicotine through the various postmortem tissues depends on how the nicotine is absorbed and the speed of death (30601, 30602, 30604).

Higher postmortem nicotine concentrations than cotinine concentrations can indicate death occurred due to acute poisoning (30602).

Reference Number: 30601

KEMP, P.M., G.S. SNEED, C.E. GEORGE, AND R.E. DISTEFANO. "Postmortem Distribution of Nicotine and Cotinine from a Case Involving the Simultaneous Administration of Multiple Nicotine Transdermal Systems." *Journal of Analytical Toxicology*, 21: 310–313, 1997 (2 tables, 11 references)

Abstract: A 31-year-old woman was found dead by her husband with a plastic bag taped over her nose and mouth and 18 NicoDerm patches (7 mg each) duct taped to her chest. A suicide note was found near the body. The postmortem heart blood and femoral blood nicotine concentrations were determined by GC and were 1,400 ng/mL and 460 ng/mL, respectively. The cause of death was ruled as asphyxia.

Table: Postmortem Nicotine and Cotinine Concentrations (ng/mL or µg/g)

Sample	Nicotine Concentration	Cotinine Concentration
Heart blood	1,400	1,300
Femoral blood	460	–
Urine	2,900	7,600
Vitreous humor	270	–
Liver	2,000	520
Gastric content	80	4,000

Source: Adapted from Kemp et al, 1997

To our knowledge, this report is the first presentation of data from a case of attempted suicide with topical administration of multiple nicotine transdermal systems. The advent of this type of drug-delivery system has provided patients with a convenient method of drug administration and has been used for a number of therapeutic purposes, including the relief of pain, the prevention of motion sickness, and, as with this case, an aid for the cessation of smoking. This case has indicated the need for further study into the pharmacokinetics and potential toxicity from the administration of multiple transdermal systems.

Reference Number: 30602

AOKI, Y., T. IKEDA, N. TANI, A. SHIDA, S. ORITANI, AND T. ISHIKAWA. "Evaluation of the Distribution of Nicotine Intravenous Injection: An Adult Autopsy Case Report with a Review of Literature." *International Journal of Legal Medicine*, 134: 243–249 2019 (3 tables, 2 figures, 46 references)

Abstract: A 19-year-old woman was brought to hospital in cardiorespiratory arrest after being found unconscious in a hotel bathroom by her husband and was subsequently declared dead. A syringe and a bottle of e-liquid containing nicotine were found in the bathroom. Injection marks were found on her left and right arms. Cotinine and nicotine concentrations were determined in numerous blood and tissue samples by GC/MS and the nicotine ranged from 15,023 ng/mL in the upper right arm near the injection site to 3,109 ng/mL in the peripheral blood. The cause of death was determined to be intravenous nicotine poisoning.

Table: Postmortem Nicotine and Cotinine Concentrations (ng/mL or µg/mg) in a Fatal Nicotine Injection Case

Sample	Nicotine Concentration	Cotinine Concentration
Heart blood (right)	3,157	2
Heart blood (left)	1,529	19
Peripheral blood	3,109	15
Vitreous humor	6,226	48
Brain (temporal lobe)	11,637	125
Liver	3,704	280
Injection site (right arm)	15,023	5,495

Source: Adapted from Aoki et al, 2019

In general, the blood concentration of nicotine reaches approximately 10 ng/mL within a few minutes of smoking one cigarette; in the present case, the nicotine concentration in the blood samples was above that level. Intravenous injection of nicotine likely caused the death in the present case. The moment nicotine enters the body, it is metabolized into cotinine, which reaches a concentration that is 2–4 times higher than that of nicotine within an hour. In the present case, the higher levels of nicotine than cotinine in all samples suggested an acute death due to the high concentration injected nicotine.

Reference Number: 30603

SOLARINO, B., F. ROSENBAUM, B. RIESSELMANN, C.T. BUSCHMANN., AND M. TSOKOS. "Death Due to Ingestion of Nicotine-Containing Solution: Case Report and Review of the Literature." *Forensic Science International*, 195: e19–e22, 2010 (2 tables, 1 figure, 40 references)

Abstract: A 42-year-old man was found dead on the floor of his garage surrounded by beer and vodka bottles and a half-empty glass bottle labelled as Pure Nicotine. The femoral venous blood nicotine and cotinine concentrations were measured by GC/MS and both were 2,200 ng/mL, respectively. The femoral blood ethanol concentration as determined by headspace GC was 0.220 g/100mL. The quantity of nicotine detected in the stomach was 1,000 mg. A typical cigarette contains between 8 and 20 mg, but only about 12 mg of nicotine is absorbed into the blood. The most common mechanisms of death due to nicotine intoxication are respiratory failure and to peripheral neuromuscular blockade and cardiovascular arrest. Although nicotine is one of the most toxic commonly used drugs, it rarely causes fatalities.

Thus, the femoral venous blood concentration measured in the present case is much higher than the mean value reported in the literature and can be regarded as an indicator of fatal nicotine poisoning. In addition, comparison with previously reported non-fatal cases of nicotine poisoning shows that the blood nicotine concentration of 2.2 µg/mL detected here was about 30 times higher than the usual level at which people start showing neurological symptoms of nicotine poisoning.

Reference Number: 30604

TJONCKE, J-A., R. GONCALVES, N. CASTAING, M. MOLIMARD, F. TOVAGLIARO, AND K. TITIER. "Death Related to Nicotine Replacement Therapy: A Case Report." *Forensic Science International*, 309: (9 pp), 2020 (2 tables, 22 references)

Abstract: A 41-year-old man was discovered dead by his cellmate, laying on his right side in bed. He had seven nicotine patches on his body. His garbage bin contained 14 nicotine patches (21 mg nicotine) and 5 empty blisters of nicotine tablets (Nicopass, 1.5 mg). Various postmortem samples were collected at autopsy and analyzed for nicotine and cotinine using GC/MS. A drug screen was also conducted. The femoral blood nicotine and cotinine concentrations were 2,239 ng/mL and 1,230 ng/mL and the cardiac blood concentrations were 1,344 and 1,090 ng/mL, respectively. Therapeutic levels of prescription drugs and traces of THC were also detected.

Table: Postmortem Nicotine and Cotinine Concentrations Detected in Various Samples

Postmortem Sample	Nicotine (ng/mL)	Cotinine (ng/mL)
Cardiac blood	1344	1090
Femoral blood	2239	1230
Urine	–	2508
Gastric content	936	–
Vitreous humor	188	578

Source: Adapted from Tjoncke et al, 2020

High doses of nicotine inhibit the central nervous system, inducing hypotension and bradycardia followed by coma and respiratory muscle paralysis, leading to death by asphyxia. The first evidences of police investigation and autopsy findings suggested nicotine poisoning. Indeed, the autopsy revealed signs of asphyxia syndrome (congestion of the face, conjunctival injection, reddish mucus in the nose and mouth, cyanosis of the lips, multi-organ congestion and considerable pulmonary oedema).

The above hypothesis was confirmed by the results of the toxicological analyses.

CHAPTER 4

Cigarettes

"The cigarette is the deadliest artifact in the history of human civilization. It is also a defective product. The cigarette is the seatbeltless car, the lead-painted cribs, the car with faulty brakes, the open unguarded manhole, the rickety ladder, the fouled maggoty meat, the Daikon shield, the smokestack spewing fumes . . ."

—Procter, R.N. "Golden Holocaust, Origins of the Cigarette Catastrophe and the Case for Abolition" (2011)

"In our opinion, it is not sufficient for the tobacco industry to merely concede the obvious point that smoking is a cause of disease when it is evident that decades of misinformation has resulted in a public that is massively ignorant about the risks of smoking low-tar cigarettes, nicotine addiction, and second-hand smoke exposure."

—Cummings et al, "The Cigarette Controversy" 2007

Cigarettes did not become the most addictive and destructive human-made artifacts by accident, they were created that way in order to make obscene amounts of money.

4.01 HISTORY

"[Smoking is] hatefull to the Nose, harmefull to the braine, dangerous to the Lungs."

—King James I of England, "A Counterblaste to Tobacco" (1604)

> "I'll tell you why I like the cigarette business. It costs a penny to make. Sell it for a dollar. It's addictive. And there's fantastic brand loyalty"
>
> —Warren Buffett, quoted in Burrough, Bryan, and John Helyar, *Barbarians at the Gate: The Fall of RJR Nabisco* (1989)

On 5 November 2005, Philip Morris celebrated the 50th anniversary of the Marlboro cigarette by having invitation-only parties in 49 US cities. It is estimated that Marlboro cigarettes alone have caused the deaths of 2.3 million Americans since 1955 and are expected to kill another 1.6 million over the next 10 years (40101–2).

This section on the history of cigarettes does not focus on the quaint stories of the servants who rushed in with a jug of water because they thought that Walter Raleigh was on fire when he was smoking, or the invention of the Bonsack machine, which could make the same number of cigarettes as 30–40 workers, or the marketing genius of advertising agencies. The history of cigarettes especially in the 19th, 20th, and 21st centuries, is one of lies, deceit, coverups, greed, misinformation, perversion of science, racism, and fake news by the tobacco industry to maintain its huge profits, by encouraging the use of one of the most dangerous products ever invented by humans. This strategy is still being used by politicians and petroleum companies who deny the science of climate change. In fact, a recent documentary has shown that some of the "experts" who testified for the tobacco industry in court have reappeared to testify for the petrochemical industry.[1]

Documents have shown that the tobacco industry considers the World Health Organization (WHO) to be one of its leading enemies (40115), so it is not surprising that the tobacco industry showed its political muscle by getting the US President to withdraw funding from the WHO in May 2020, which fortunately has been reversed by a newly elected President.

The numerous court cases against the tobacco industry have resulted in the release of over 9 million internal documents which show *inter alia* that tobacco companies:

- Knew in the 1950s that cigarette smoking caused cancer but vigorously denied it (40104).

---

1 Oreskes, Naomi, & Erik M. Conway Oreskes, *Merchants of Doubt: How a Handful of Scientists Obscured the Truth on Issues from Tobacco Smoke to Global Warming* (London: Bloomsbury Press, 2010).

- Altered the pH of the cigarette to allow faster absorption of (and hence addiction to) nicotine (40101, 40103, 40121).
- Funded research institutions, scientists, and scientific journals to create doubt and controversy about health harms of cigarettes (40104, 40109, 40115, 40122, 40124).
- Funded research to show that type A behavior, and not cigarette smoking, causes cancer and lung disease (40111) and that smoking prevents Alzheimer's disease (40112).
- Encouraged the view among smokers that low tar cigarettes were safer (40113).
- Used movies, and now social media, to glamorize nicotine addiction and vaping/smoking (40107, 40108, 40114, 40118).
- Distorted the history of Nazi Germany and tobacco control (40110) and promoted the view that smoking is free will (40119, 40123).
- Encouraged cigarette smoking in the military by giving out free cigarettes and other promotions (40120).
- Promoted the excuse that air pollution kills more people than cigarettes and that air pollution is more toxic than cigarettes (40126).
- Manipulated research to promote its interests and suppressed research showing the harms of cigarette smoking (40127).

The tobacco industry has also continued the creation of doubt and spread misinformation even to 2021, during the worst pandemic in 100 years:

- Promoted the studies (later withdrawn) that nicotine and smoking helped prevent COVID-19, even though a study of 17.3 million adults in the UK found that smoking increased the risk of COVID-19 related death by 14% (40128–29).

This again shows why my decision of not to include any published studies with tobacco/nicotine/vaping authors or journals, or those with a conflict of interest, was appropriate to this industry and the drug they provide.

Although the Master Settlement Agreement (MSA) between the United States and tobacco companies on 1 February 1999 cost the tobacco industry $206 billion over 25 years, Big Tobacco advertised more after the MSA and more than made up for loss of money by encouraging smoking in China and less developed countries (40117).

Reference Number: 40101

BLUM, A. "The Tobacco Reliquary." *Canadian Medical Association Journal*, 187: 133–134, 2015 (6 figures, 4 references)

Abstract: On 11 January 1964, Dr. Luther L. Terry, the Surgeon General of the US Public Health Service, released a report showing that cigarette smoking is casually related to smoking in men.

> Surgeon General Terry's indictment of cigarettes in 1964 should have marked the beginning of the end of the Marlboro Man. Yet, far from riding off into the sunset, the tobacco industry is still riding high in the saddle.
>
> Right up until the 2000s, cigarette manufacturers publicly disputed the evidence about the harmfulness of smoking and sought to allay consumer anxiety by implying that new filtered, low-tar and "light" brands were not harmful. This corporate-level denial is ongoing in the United States and Canada, as evidenced by testimony in individual and class-action litigation. The result: 50 years after the surgeon general's landmark report, the health and economic toll of smoking remains devastating. Thirty percent of deaths from cancer in the US are attributable to cigarette smoking. Although the percentage of American and Canadian adults who smoke cigarettes has declined from about 40% in 1964 to 18% and 16% respectively, the number of smokers is not much less than in 1964—and the cohort may well be younger than ever.

Reference Number: 40102

HYLAND, A., R. GOLDSTEIN, A. BROWN, R. O'CONNOR, AND K.M. CUMMINGS. "Happy Birthday Marlboro: The Cigarette Whose Taste Outlasts Its Customers." *Tobacco Control*, 15: 75–77, 2006 (1 table, 8 references)

Abstract: On 5 November 2005, Philip Morris celebrated the 50th anniversary of the Marlboro cigarette by having invitation-only parties at 49 US cities. The market share for Marlboro had increased from 1.7% in 1955 to over 10% by 1971, over 20% in 1983, and over 40% in 2005. It is estimated that at the time of writing, Marlboro cigarettes alone had killed more than 2.3 million Americans since 1955 and were expected to kill another 1.6 million in the next 10 years.

Table: Marlboro Market Share and Estimated Number of Marlboro Smokers Dying in the US Assuming a 10 Year Lag Period

Year	Marlboro Market Share (%)	Annual Deaths Attributable to Marlboro
1965	5.3	2,856
1975	15.4	13,455
1985	22.4	60,242
1995	30.1	99,089
2005	40.5	133,213
2015 (est)	40.5	179,171

Source: Adapted from Hyland et al, 2006

More Americans have died or will die because of Marlboro than US soldiers who have died in all foreign wars combined. However, merely considering the number of deaths attributable to Marlboro in the United States over the past half century is surely a huge underestimate of the total death toll from Marlboro, given the brands position as the world's best selling cigarette. According to a recent Philip Morris International report, Marlboro is considered the greatest asset of the company with projected strong growth predicted for the future worldwide. With the death toll from tobacco estimated to reach 10 million annually by 2020, it is likely that Marlboro's next 50 years could prove to be its most deadly.

Reference Number: 40103

WAYNE, G.F., G.N. CONNOLLY, AND J.E. HENNINGFIELD. "Brand Differences of Free-Base Nicotine Delivery in Cigarette Smoke: The View of the Tobacco Industry Documents." *Tobacco Control*, 15: 189–198, 2006 (5 tables, 1 figure, 110 references)

Abstract: More than 7 million tobacco industry documents have been disclosed by the tobacco industry and were made available as a result of the 1998 Master Settlement. Nicotine can be found in both the gas and the particulate matter (PM) of cigarette smoke. It thought that most of the nicotine is in the PM and only 1% is in the gas phase of the smoke. In PM, nicotine can exist as diprotonated, monoprotonated, and unprotonated (free base) forms and occur with increasing pH. The fraction of free base nicotine increases with increasing pH. Smoke pH was used by the tobacco industry as a relative measure of free base nicotine. According to the tobacco industry documents, the more free base nicotine delivered to the lungs, the greater the initial dose strength and the greater its addiction potential. Several methods were developed by the tobacco industry to measure the percentage of free base nicotine including the extractable nicotine, volatile nicotine, and headspace nicotine methods.

> The public release of internal tobacco industry documents has led to increasing sophistication by independent scientists in the characterisation of the design and function of tobacco products. One critical area of research is the role of smoke chemistry in determining the delivery, absorption, and effects of smoke constituents, especially harm producing or pharmacologically active compounds. The adverse health effects of smoking are a function of the toxicity of smoke constituents, as well as the amount and duration of exposure to those toxins, in combination with individual differences in metabolism of toxic compounds and susceptibility. Conventional measurements of smoke delivery, such as ISO measures, have generally focused on "tar" or nicotine in smoke PM as collected from the Cambridge filter using standard machine smoking protocols. Recent findings suggest that these measures fail to account for or describe important design-based differences in smoke chemistry, which may alter exposure or toxicological impact.

Reference Number: 40104

CUMMINGS, K.M., A. BROWN, AND R. O'CONNOR. "The Cigarette Controversy." *Cancer Epidemiology, Biomarkers & Prevention*, 16: 1070–1076, 2007 (2 tables, 76 references)

Abstract: A search was conducted of over 41 million pages of internal documents of the tobacco industry in the US at the Legacy Library and over 9 million documents from the Tobacco Documents online from the 1920s to the present day. The internal industry documents show that the tobacco companies knew cigarette smoking was a cause of cancer by the late 1950s. The tobacco documents show how the tobacco industry worked together to create a pro-cigarette public relation campaign to mislead the public and to continue to market cigarettes for as long as possible. Several research institutes and scientists were funded by the tobacco companies to create doubt and controversary about the harm of smoking. In the 2006 election, the tobacco industry spent $100 million opposing state-initiated proposals to limit smoking in public places and raise cigarette taxes.

Table: Selected Comments from the Tobacco Industry About Smoking and Health

Date	Source	Statement
4 June 1954	"A Frank Statement to Cigarette Smokers" published in major US newspapers	"Distinguished authorities point out: 1. That medical research of recent years indicates many possible causes of lung cancer. 2. That there is no agreement among the authorities regarding what the cause is. 3. That there is no proof that cigarette smoking is one of the causes. 4. That statistics purporting to link cigarette smoking with the disease could apply with equal force to any one of many other aspects of modern life. Indeed, the validity of the statistics themselves is questioned by numerous scientists."
30 March 1954	"Facts Versus Fancy," speech to the National Association of Tobacco Distributors by George Weissman, VP Philip Morris	"[I]f we had any thought or knowledge that in any way we were selling a product that was harmful to consumers, we would stop business tomorrow."
23 April 1968	Letter to an elementary school teacher from R.J. Reynolds Tobacco Company	"[M]edical science has been unable to establish that smoking has a direct causal link with any human disease. . . ."
11 January 1990	Letter to elementary school teacher from R.J. Reynolds Tobacco Company	"Despite all the research going on, the simple and unfortunate fact is that scientists do not know the cause or causes of chronic diseases reported to be associated with smoking."

Source: Adapted from Cummings et al, 2007, and the primary documents

In our opinion, it is not sufficient for the tobacco industry to merely concede the obvious point that smoking is a cause of disease when it is evident that decades of misinformation has resulted in a public that is massively ignorant about the risks of smoking low-tar cigarettes, nicotine addiction, and secondhand smoke exposure. Moreover, claims by tobacco companies that they are involved in sponsoring programs to help smokers to quit and discourage youth from taking up smoking must be seriously questioned in light of recent findings that show that these programs have no beneficial effect and may potentially be iatrogenic. There remains a need for public education efforts to correct consumer misperceptions about the risks of smoking along with government oversight to ensure that industry is not permitted to use its vast marketing resources to continue to mislead the public. Universities should also consider adopting policies that prohibit their faculty from accepting funding from tobacco companies.

Reference Number: 40105

GIVELBER, D., AND L. STRICKLER. "Junking Good Science: Undoing Daubert v Merrill Dow Through Cross-Examination and Argument." *American Journal of Public Health*, 96: 33–37, 2007 (11 references)

Abstract: For more than 40 years the tobacco industry, which was a defendant in many legal actions, prevailed by using the very successful defence that "everybody knew, nobody knows." Everybody knew smoking was harmful, but nobody knows what causes cancer. The tobacco companies' expert witnesses over the years of litigation tended to be aging and depleted, while the injured smoker's medical experts became younger and stronger every year. In the scientific and medical community, there are probably 1,000 who believe cigarette smoking may lead to cancer for every 1 who has doubts. Yet the jury will not be exposed to 1,000 witnesses for the plaintiff to 1 witness for the tobacco companies, but will most likely hear 2 expert witnesses who say smoking causes cancer and 2 who say we do not know.

Although from the outset of tobacco litigation in the 1950s, some courts agreed that the plaintiff's evidence was sufficient to show that the defendant tobacco company behaved in a manner that could render it liable, no individual plaintiff in a tobacco case collected a judgment against a tobacco company until the late 1990s. The explanations include the following: (a) most plaintiffs abandoned their suits as the tobacco companies refused to settle and expended vast amounts of time and money, overwhelming the resources of individual plaintiffs and their lawyers; (b) when a case did go to trial, the tobacco companies successfully transformed the dispute into a morality play about the plaintiff and his or her failure to quit; (c) the tobacco companies persuaded jurors that the plaintiff knew as much about the evils of smoking as the tobacco companies, hence it was unfair to hold the company responsible for the plaintiff's choice; and (d) the tobacco companies persuaded some jurors that whether cigarettes caused disease was debatable and whether smoking caused the particular plaintiff's disease was doubtful. The tobacco lawyers were able to keep alive a debate that had raged and been resolved 40 years ago.

Reference Number: 40106

CARTER, S.M., AND S. CHAPMAN. "Smoking, Disease, and Obdurate Denial: The Australian Tobacco Industry in the 1980s." *Tobacco Control*, 12: iii23–iii30, 2003 (113 references)

Abstract: This study presents an examination of the Australian tobacco industry internal documents. In March 1978, the global British American Tobacco company had a research and development conference in Sydney, Australia which concluded that smoking did cause some diseases. In the 1980s, the employees and affiliates of the Australian tobacco industry at all levels were indoctrinated that nobody knows if smoking causes disease. In 1983, the chief of Philip Morris of Australia gave a speech to journalists where he encouraged smokers in the room to take heart as their lungs were as pink and pretty as the non-smoker next to you. The industry's aggressive public communication campaign, however, did not change public opinion that smoking was bad for you (agreement with that statement increased from 75% to 83% of Australian adults).

The industry's cynicism towards smokers became apparent in the 1990s, as product liability litigation in Australia became a serious cause for concern. Overnight, the industry stopped arguing that there was no proof that smokers were at risk, and instead advised smokers that they should have listened to the public health authorities, and the responsibility for their illness was theirs. The industry's primary motivation in Australia in the 1980s, as ever, was not the pursuit of truth or science, or the wellbeing of its customers, but its own financial survival. Despite privately accepting that the case for smoking causing disease had long been established, the industry vigorously denied this via its public affairs activities. The revelations we have summarised provide important evidence for litigation by smokers exposed to such industry campaigns, and for advocacy against industry mendacity.

Reference Number: 40107

CASALDELLI-MAIA, J.M., A. VENTRIGLIO, AND D. BHUGRA. "Tobacco Smoking: From Glamour to Stigma. A Comprehensive Review." *Psychiatry and Clinical Neurosciences*, 70: 24–33, 2015 (95 references)

Abstract: In the 20th century, movies and the cinemas played an important role in encouraging cigarette smoking by subtle advertising and product placement. Cinema helped create a culture of cigarette smoking and glamor. In November 1998, the MSA banned payments for the placing of tobacco brands in film. There was a sharp decrease in the number of adult films showing cigarettes after the MSA, but no significant decline for movies directed at teenagers. In India and Turkey, movie scenes in which an actor is smoking, a note appears on the screen that states that smoking can cause cancer. In Thailand, the smoking characters are pixilated. Anti-smoking campaigns and public smoking bans have changed the perception of smoking as positive and glamorous to negative and having public stigma.

> The history of tobacco is intertwined with the history of humanity itself, having been one of the many products that were incorporated from the indigenous by the Europeans. It came to represent the luxury of the aristocracy for a while. After that, the dispute for tobacco seeds was one of the most striking facts of the 17th century, and many people became rich due to the mercantile nature of the product. Discovering the fast drying process, mass advertising and increasing mass consumption was simply a matter of time. Its evolution as an extremely successful product of marketing that involved film (the most consumed form of art in the world during the 20th century) meant that cigarette smoking penetrated all environments, presenting its misdeeds in smokers and non-smokers. In the current century, we have inaugurated a new phase, with the ban of smoking in public and even in private settings and with the creation of anti-smoking social pressure. Nowadays, we note the stigma associated with smoking in those who continue to smoke and who are criticized over their inability to give up smoking. Health professionals must be aware of this, as current smokers may well hide their habit for fear of prejudice. The new forms of smoking are nothing more than alternatives (unfortunately mostly ineffective) for humans to continue smoking without the pain caused by the extensive anti-smoking campaign.

Reference Number: 40108

LIANG, Y., X. ZHENG, D.D. ZENG, X. ZHOU, S.J. LEISCHOW, AND W. CHUNG. "Exploring How the Tobacco Industry Presents and Promotes Itself in Social Media." *Journal of Medical Internet Research*, 17: e24 (18 pp), 2015 (6 tables, 3 figures, 30 references)

Abstract: This study examined the presence of the top 70 cigarette brands on Facebook, Wikipedia, and YouTube between 3 and 10 March 2014.

Forty-three cigarette brands had created 238 Facebook fan pages (1,189,976 page likes and 19,022 posts), 44 brands were identified on Wikipedia, and there were over 120,000 pro-tobacco videos on YouTube created by 61 brands.

Table: Promotional Strategies used by Tobacco Companies in Social Media

Promotional Strategies	Description
Brand promotion	Company website links embedded in the textual data or video clip; content about the history and culture of cigarette brands.
Sales promotion	Promotion of tobacco sales through price discounts, tobacco coupons, free shipping, no tax, and embedded URLs for tobacco shops.
Fresh imagery	Images of young men and women smoking; smoking sexual fetish scenarios; smoking animals or cartoon characters, etc.
Sponsorship	Provision of funds for sports matches, festivals, racing, etc. For example, Formula One, tennis matches, music bands, and festivals are common places to see logos or brands names from the tobacco industry. In social media, lots of pictures and videos related to social events are funded by cigarette brands.
Misleading information	Slogans to smooth or blur the side effects of smoking. For example, low tar content is emphasized to demonstrate the products are healthier.

Source: Adapted from Liang et al, 2015

With the popularity of social media such as Facebook, YouTube, and Wikipedia, a huge number of users are exposed to pro-tobacco content and even interact with such information. Obviously, the tobacco industry utilizes social media to raise the visibility of tobacco products and promote tobacco use. As exposure to tobacco promotion is positively associated with a higher smoking prevalence and may encourage smoking initiation, regulations about tobacco control in social media should be enforced as soon as possible by decision makers.

Potential Methods

Through the analysis of Facebook fan pages, we found 61% of brands (43/70) are dedicated to tobacco promotion, especially for international cigarette brands such as Pall Mall and Mild Seven. Furthermore, a surging number of Facebook users have accessed and interacted with these fan pages.

Reference Number: 40109

VELICER, C., G. ST. HELEN, AND S.A. GLANTZ. "Tobacco Papers and Tobacco Industry Ties in Regulator Toxicology and Pharmacology." *Journal of Public Health Policy*, 39: 34–48, 2018 (4 tables, 81 references)

Abstract: Tobacco companies have a long history of funding and conducting research that supports their business and cigarette use. In 1989, the tobacco industry tried to discredit the risk of secondhand smoke (SHS) by publishing the journal "Indoor and Built Environment." The relationship between the tobacco industry and another journal, "Regulatory Toxicology and Pharmacology," was assessed using the Truth Tobacco Industry Documents Library, as well as certain internet sources. The editor, 57% of the associated editors, and 37% of the editorial board members had worked or consulted for the tobacco industry. Of the 50 papers published between 2013 and 2015, 76% drew conclusions positive for the industry and none drew negative conclusions.

Table: Author Affiliation with the Tobacco Industry and the Number of Papers with a Positive, Neutral, or Negative Conclusion

Author Affiliation	Positive Conclusion	Neutral Conclusion	Negative Conclusions	Total number of papers
Affiliated with tobacco industry	38	12	0	50
Unaffiliated/unclear	0	0	2	2
Total	38	12	2	52

Source: Adapted from Velicer et al, 2018

The fact that the tobacco industry is heavily represented in the editorial leadership of *Regulatory Toxicology and Pharmacology* and that almost all the peer-reviewed papers on tobacco topics were written by individuals connected with tobacco companies and found positive results calls into question the confidence that members of the scientific community and tobacco product regulators can have in the conclusions of papers published there.

Reference Number: 40110

BACHINGER, E., M. MCKEE, AND A. GILMORE. "Tobacco Policies in Nazi Germany: Not as Simple as it Seems." *Public Health*, 122: 497–505, 2008 (2 tables, 32 references)

Abstract: The tobacco industry and its supporters have compared current tobacco measures to the Nazi campaign against tobacco in Germany. Tobacco control measures such as not smoking in bars and restaurants and advertising restrictions have been vilified as an unimaginable totalitarian attack on individual liberty comparable to the 1930s and 1940s fascist regime in Germany. A more detailed analysis of the Nazi anti-smoking campaigns has shown that these policies were poorly enforced and ambivalent at the highest levels of the Nazi party. Hermann Goering and other high-ranking Nazi officials were smokers and often smoked in public. Taxes on tobacco were increased, but soldiers at the front and others engaged in military service were given tobacco rations. German cigarette consumption did not to begin to decline until 1942, at which point the consumption had peaked at 80 billion cigarettes. There were no Reich-wide anti-smoking campaigns or persecution of smokers.

> In conclusion, the widespread use of Nazi imagery by pro-smoking groups to attack those seeking to limit the harm caused by tobacco is a distortion of history that cannot be justified.

Reference Number: 40111

PETTICREW, M.P., K. LEE, AND M. MCKEE. "Type A Behavior Pattern and Coronary Heart Disease: Philip Morris's 'Crown Jewel.'" *American Journal of Public Health*, 102: 2018–2025, 2012 (112 references)

Abstract: Type A behavior pattern (TABP), which characterizes individuals who are highly competitive and aggressive, and its link to coronary heart disease (CHD) and cancer has been the subject of research for over 50 years. The authors searched the Legacy Tobacco Document Library for type A behavior pattern and found 66 internal tobacco documents about TABP. Philip Morris and RJ Reynolds were major funders of research that it was TABP that caused coronary heart disease, rather than cigarettes. A 1968 film by the tobacco institute showed research on TABP and concluded that smoking was a symptom, not a cause, of cancer. It was argued that psychological stress, not smoking, was a key contributing cause of cancer. Currently, TABP research has been characterized as "zombie science," i.e., research that continues to be published in spite of continuing negative findings.

> TABP has been the subject of hundreds of studies and reports in the popular media since the 1950s, but the close relationship between TABP and the tobacco industry has not been previously described. The characteristics of industry funding of TABP research were similar to those of the industry's influence on other areas of tobacco and health research, such as review by industry lawyers and support of epidemiological and social science research explicitly to generate scientific controversy. The tobacco industry's support of TABP research allowed the industry to extend its strategy to undermine public health policy by arguing that the causes of cancer and heart disease were multifactorial and that any observed relationships with smoking could be a result of confounding with TABP.
>
> Although it proved nonproductive, Philip Morris's substantial investment in TABP research is notable. By the early 1980s, the few early positive findings were being outweighed by numerous negative studies, and TABP and hostility have rarely been shown to be implicated in either the etiology or the prognosis of CHD.

Reference Number: 40112

CATALDO, J.K., J.J. PROCHASKA, AND S.A. GLANTZ. "Cigarette Smoking is a Risk Factor for Alzheimer's Disease: An Analysis Controlling for Tobacco Industry Affiliation." *Journal of Alzheimer's Disease*, 19: 465–480, 2010 (4 tables, 2 figures, 120 references)

Abstract: In 2006, there were 26.6 million Americans with Alzheimer's disease (AD). Some articles in the media have stated that smoking may protect people from developing AD and as of October 2008, over 100 websites promoted smoking as protective against it. Such confusion may discourage cessation attempts of older smokers and allow for continued cigarette sales. In 1976, the tobacco industry started investing in AD research. A total of 43 original research studies on AD and smoking were analyzed in this review. Eleven of the 43 studies (26%) were conducted by tobacco affiliated investigators, but only 3 of the 11 studies disclosed tobacco affiliation. Tobacco affiliation was determined by searching the Legacy Tobacco Documents Library for the names of the authors. In 14 independent cohort studies, the relative risk (RR) of AD from smoking was 1.45×, whereas of the 3 cohort studies with tobacco affiliation, the RR was 0.60×.

Table: Some Conclusions of Smoking and Alzheimer's Disease Reviews and Tobacco Affiliation

Conclusion	Protective Effect of Smoking?	Tobacco Industry Affiliation?
"A statistically significant inverse association between smoking and Alzheimer's disease was observed at all levels of analysis, with a trend towards decreasing risk with increasing consumption."	YES	YES
"[T]he odds ratio associated with family history of dementia tended to be lower for those with a positive smoking history . . ."	YES	YES
"[T]he negative association is consistent with other data suggesting nicotine protects against AD."	YES	YES
"[A] negative association for AD is controversial . . . not a clear effect."	Ambiguous	NO
"[E]lderly smokers have increased risks of dementia and cognitive decline."	NO	NO

Source: Adapted from Cataldo et al, 2010

For the last two decades, the tobacco industry has been actively funding research that supports the position that cigarette smoking protects against AD, and for the past two decades, the scientific literature has reported conflicting results as to the direction of the association between smoking and AD. Consequently, older smokers and their health care providers have been unaware that smoking is a modifiable risk factor for AD. There is an association between tobacco industry affiliation and the conclusions of individual studies and, probably, review papers of AD. Controlling for industry affiliation, study design and other factors, smoking is not protective against AD; it is a significant and substantial risk factor for Alzheimer's disease.

Reference Number: 40113

HAMMOND, D., M. CHAITON, A. LEE, AND N. COLLISHAW. "Destroyed Documents: Uncovering the Science That Imperial Tobacco Sought to Conceal." *Canadian Medical Association Journal*, 181: 691–698, 2009 (1 table, 89 references)

Abstract: In 1954, the Canadian Medical Association issued its first public warnings about the risks of smoking and in 1963, the Minister of Health announced in the House of Commons that smoking was harmful to the health. In 1992, British American Tobacco had its affiliate Imperial Tobacco Canada destroy 60 internal documents including 47 original research studies about the harms of SHS, low tar cigarettes, addictiveness of nicotine, and carcinogens found in cigarette smoke.

> Efforts to conceal research findings on "low-tar" cigarettes have been particularly damaging to health policy and government regulation. Canadian manufacturers began producing and marketing "lower-tar" cigarettes during the 1960s and 70s as a direct response to growing health concerns among consumers. Without the benefit of independent evidence on the potential benefits of "lower-tar" cigarettes, many leading health agencies supported this industry strategy. For example, in the 1970s, Health and Welfare officials urged the Canadian manufacturers to reduce tar levels further and to print tar numbers on packages to help guide consumers. The destroyed documents indicate that British American Tobacco had collected evidence that cast doubt on the potential health benefits of "low-delivery" products. In vivo studies indicated that human smokers were compensating for "low-tar" cigarettes by smoking them more intensely, while biological testing from Projects Janus and Rio revealed negligible differences in the carcinogenity and mutagenicity of cigarette brands. Without access to this evidence, several decades elapsed before regulators and health agencies reached consensus that "low-tar" products did not reduce risk. In the meantime, millions of smokers switched to these brands, many under the assumption of reduced harm as an alternative to quitting. The legacy of low-tar cigarettes continues today: most Canadian smokers continue to believe that lower-tar cigarettes are less harmful, and misleading tar numbers continue to be printed on Canadian packages.

Reference Number: 40114

FROST, K., E. FRANK, AND E. MAIBACH. "Relative Risk in the News Media: A Quantification of Misrepresentation." *American Journal of Public Health*, 87: 842–845, 1997 (2 figures, 19 references)

Abstract: The amount of print that the US national news media devoted to the 11 major causes and risk factors of death was compared to the proportion of actual death causes and risks in 1990.

Table: Relative Risk of New Media Overrepresenting Actual Risk Factors of Mortality

Risk Factors for Mortality	RR of Overreporting Risk
Tobacco	0.23×
Alcohol	1.10×
Firearms	1.30×
Motor vehicle collisions	12.80×
Illicit use of drugs	17.44×

Source: Adapted from Frost and Maibach, 1997

News media over- and underemphasize certain causes of death for a variety of reasons, including competition for viewers and commercial interests. This pattern has been particularly well documented in the case of the leading risk factor for death, tobacco use, where the relative lack of news coverage has been attributed to the influence that tobacco companies, with their enormous advertising budgets, have over media organizations. News reporting is also driven by rarity, novelty, commercial viability, and drama more than by concerns about relative risk. Yet the unusual, novel, lucrative, or dramatic report must be placed in a broader context to help the reader better interpret the story's implications.

Studies of cognition indicate that people judge the frequency or probability of an occurrence by the ease with which they can retrieve relevant instances from memory or imagination. This bias of cognition is referred to as the availability heuristic. The media's overemphasis on certain causes of death helps distort public risk perception by making these causes more available and thus more easily retrieved. The print news media's distorted representations of mortality likely skew the public's perceptions of risk through a second cognitive mechanism as well. Studies of risk perception indicate that threats that are perceived to be externally imposed loom larger than self-imposed threats. Thus, not only do the news media emphasize relatively rarer causes of and risk factors for death, but those causes emphasized are those that are instinctively overestimated.

Reference Number: 40115

ONG, E.K., AND S.A. GLANTZ. "Constructing 'Sound Science' and 'Good Epidemiology': Tobacco, Lawyers, and Public Relations Firms." *American Journal of Public Health*, 91: 1749–1757, 2001 (1 table, 151 references)

Abstract: The tobacco industry (mainly Philip Morris (PM)) began its sound science and good epidemiology practice (GEP) program in 1993 to minimize the harms of secondhand smoke in both the United States and Europe. From 1994 to 2000, seemingly independent seminars on GEP were conducted by several scientific organizations throughout the world, when in fact all were financed by PM. PM wanted to discredit epidemiologic results with relative risks of less than 2, but no epidemiology body would agree to this standard.

The industry's strategy for dealing with secondhand smoke may be shifting again. In 2000, PM released a Web site that acknowledges that many scientific studies report a small increased risk in disease with secondhand smoke exposure, but also claims that "exposure levels outside the home are lower than regulators have generally assumed."

These claims of inconsequential exposures are based on tobacco industry–funded studies of dubious accuracy that were carried out within the same construct as the GEP efforts and other efforts to subvert the scientific process. Indeed, the pilot studies for these exposure studies have been demonstrated in government proceedings to be unreliable.

As greater understanding of the tobacco industry's subversive operations accumulates, scientists and policymakers face the question of whether or not to include the tobacco industry in their discourses. Pros and cons of publishing any material funded by the tobacco industry have been debated, and some journals and organizations have rules to exclude those who are funded by the tobacco industry.

The World Health Organization's tobacco control effort is the most prominent global case to observe in the near future, as the WHO deals with both the scientific and policy-making worlds and has been targeted by the tobacco industry. The WHO recently published a special inquiry on the tobacco documents, which concluded that the tobacco industry views the WHO as one of its leading enemies and strategized to "contain, neutralise, reorient" WHO's tobacco control initiatives.

Reference Number: 40116

MAX, W., AND T. TSOUKALAS. "Economics on Trial: The Use and Abuse of Economic Methods in Third Party Tobacco Litigation." *Tobacco Control*, 15: iv77–iv83, 2006 (2 tables, 44 references)

Abstract: The study analyzed transcripts from third party litigation since 1994 in the United States, with particular emphasis on the tobacco industry's response.

When the tobacco industry found itself in 1994 facing a new and unprecedented wave of anti-tobacco litigation from third party payers, they tried to adapt strategies that had been used successfully in the past—creation of unfounded doubt and confusion, and manipulation of the discovery process to force plaintiffs to withdraw or concede defeat. However, these strategies proved to be of limited use at best, as indicated by the $206 billion Master Settlement Agreement reached by industry representatives and the attorneys general in 1998.

The industry's previously successful strategies failed for several reasons. First, the plaintiff's attorneys were able to commit significant resources and willing to undertake substantial financial risk to defend their new legal approaches. Second, previous arguments related to individual responsibility were deemed irrelevant in the new wave of litigation. Instead of individual plaintiffs seeking to recover individual losses, third party payers were seeking to recover costs they had incurred for the tens of thousands of individual health care recipients they served. Third, economic models of the health care costs of smoking had been developed that were able to quantify damages to a large group of health care recipients.

Reference Number: 40117

JONES, W.J., AND G.A. SILVESTRI. "The Master Settlement Agreement and Its Impact on Tobacco Use 10 Years Later: Lessons for Physicians about Health Policy Making." *Chest*, 137: 692–700, 2010 (1 table, 62 references)

Abstract: The MSA between United States and the tobacco industry was finalized on 1 February 1999. It involved the payment of $206 billion to the United States over 25 years. After a decade of declining cigarettes rates in the United States in the 1990s, the smoking rates leveled off between 2004 and 2007. Tobacco companies advertised more after the MSA than before and increases in cigarette sales in underdeveloped countries and China more than made up for lost sales in the United States.

Table: Highlights of the Master Settlement Agreement

MSA Settlement Highlights
• A payment of $206 billion to the United States, spread out over a 25-year period. • A $1.5 billion payment over 10 years to support state antismoking measures. • A $250 million payment to fund research into reducing youth smoking. • Permanent limitations on cigarette advertising. • A ban on the use of cartoon characters (such as Joe Camel) in advertising. • A ban on cigarette-branded merchandise. • Limits on tobacco industry sponsorship of sporting events (such as Virginia Slims tennis tournament). • The dissolution of tobacco trade organizations.

Source: Adapted from Jones and Silvestri, 2010

In addition, many state antitobacco initiatives were countered by continued (and effective) efforts by the tobacco industry to market and advertise their products to minors. Cigarette marketing to teens through magazine advertising in Massachusetts actually increased after the MSA took effect, although subsequent public and political pressure forced tobacco companies to reverse this. There was also a post-MSA increase in externally visible cigarette advertising on retail storefronts in Massachusetts, an increase that has not been reversed, except in areas near schools. Nationally, the tobacco industry reacted to the MSA with more aggressive marketing, including more expenditures on advertising, especially point-of-purchase cigarette advertising and promotions. Direct marketing was accelerated via direct mail, coupons, sweepstakes, brand loyalty programs, event sponsorship, and tobacco industry magazines, avenues that were generally not bound by MSA guidelines.

One particularly controversial practice was product placement of cigarettes and their use in movies, which often have high proportions of young viewers. A recent study suggested that approximately 390,000 teens start smoking each year because of smoking behavior they have seen in movies.

Reference Number: 40118

KANAMORI, Y., AND R.E. MALONE. "Conveying Misinformation: Top-Ranked Japanese Books on Tobacco." *Tobacco Induced Diseases*, 9: 3 (6 pp), 2011 (1 table, 30 references)

Abstract: Tobacco control efforts have lagged in Japan due in part to the fact that the Japanese government is a majority owner of Japan Tobacco

Incorporated (JT) and that many Japanese citizens have regarded smoking as an issue of good manners and not as a health risk. This study analyzed 17 books available from Amazon Japan and determined that 9 were pro-smoking, 7 were anti-smoking and 1 was neutral. The main misinformation put forth by the top five pro-smoking Japanese books were (1) creating doubt about the science and causes of lung cancer; (2) suggesting that smoking increases health, virility, and happiness; (3) trivializing the effects of tobacco use; (4) attacking public health advocates as liars and fascists (Hitler-like); and (5) linking tobacco with civil rights or Indigenous American origin.

Table: Some Quotes from Top-Rated Japanese Books on Tobacco

Misinformation from Top-Rated Japanese Books on Tobacco
• "Most importantly, tobacco is a tasty, precious gift from God for us (the smokers) to a better life." • "As the rate of smoking among Japanese men decreases, so has the birth rate." • "Men who quit smoking are not only losing its medicinal value, but are also viewed as men of low ambition who have let go of their masculinity" • "Anti-smokers are the greatest evils. Adolf Hitler is known for disliking tobacco." • "As Japan Tobacco Inc. (JT) states, the correlation between tobacco and cancer is not yet fully clear. Therefore, as journalists, one should report/write from a more objective standpoint." • "It is clear that elderly people who love to smoke in nursing homes live much longer." • "As stated by BAT and JT, it is difficult to scientifically assess the effect of secondhand smoke in chronic illnesses. The current assessment of the effect of secondhand smoke is overrated."

Source: Adapted from Kanamori and Malone, 2011

Japanese readers who use the Amazon.co.jp website to search for books about tobacco may be misled by texts such as these. While our search also retrieved anti-smoking books, the positioning of these pro-smoking authors as "scientific" or "medical" authorities claiming that smoking was more beneficial than harmful could influence beliefs and decisions about tobacco use and the desirability of smoking cessation. Public health advocates in Japan (and globally) should expose and counter such misinformation.

More than 40 years ago, a U.S. tobacco company internal document asserted: "Doubt is our product since it is the best means of competing with the "body of fact" that exists in the mind of the general public. It is also the means of establishing a controversy." While we lack direct evidence that JT influenced these books, at least one was authored by a former JT employee, another uses a well-known JT advertising slogan, and several echo JT claims, suggesting continuation of a long-discredited strategy first perfected by U.S. tobacco companies.

There are, however, hopeful signs. Kanagawa prefecture recently implemented the nation's first smoke-free restaurant ordinance. In February 2010, the Ministry of Health, Labor and Welfare issued "guidelines" on smoke-free restaurants and workplaces and made clear that the disease effects of SHS have been scientifically established. In addition, in October 2010, the tobacco tax was increased, raising the price of a pack of cigarettes by approximately $1. While cigarette sales in September rose 88% over a year ago due to last-minute buying to stock up, it appears that many smokers decided to quit smoking due to the tax increase; tobacco sales declined in October after the increase, and demand for antismoking medications is reportedly higher than expected. Perhaps these developments, and continued global efforts to denormalize smoking, will help counter misleading claims made in books such as those we reviewed. Meanwhile, tobacco control advocates in Japan may wish to "name and shame" those authors who promulgate such misinformation.

Reference Number: 40119

MILBERGER, S., R.M. DAVIS, C.E. DOUGLAS, J.K. BEASLEY, D. BURNS, T. HOUSTON, AND D. SHOPLAND. "Tobacco Manufacturers' Defence Against Plaintiffs' Claim of Cancer Causation: Throwing Mud at the Wall and Hoping Some of it Will Stick." *Tobacco Control*, 15: iv17–iv26, 2006 (7 tables, 58 references)

Abstract: An examination was conducted about how tobacco manufacturers have defended themselves against charges that their tobacco product caused cancer in the plaintiff in 34 personal injury lawsuits between 1986-2003. There were 7 main themes that the tobacco companies used ranging from that there is no proof that cigarette smoking causes lung cancer, to our brand did not cause the cancer, to the plaintiff knew of the health risks but exercised free will and decided to continue smoking anyway.

Table: Trends in the Tobacco Companies' Defence Arguments Over Time

Defence Argument	1996 and earlier	1997–2002	2003
Smoking does not cause cancer	89%	65%	25%
Cancer was caused by other factors	11%	33%	50%
Smoking is free will	56%	71%	100%

Source: Adapted from Milberger et al, 2006

Despite the vast body of literature showing that cigarette smoking causes cancer, and despite tobacco companies' recent admissions that smoking causes cancer, defendants used numerous arguments in these cases to deny that their products had caused cancer in plaintiffs. The cigarette companies, through their public admissions and courtroom arguments, seem to be saying, "Yes, smoking causes lung cancer, but not in people who sue us".

Reference Number: 40120

SMITH, E.A., AND R.E. MALONE. "'Everywhere the Soldier Will Be': Wartime Tobacco Promotion in the US Military." *American Journal of Public Health*, 99: 1595–1602, 2009 (120 references)

Abstract: During the first Gulf War (1990-1991) smoking prevalence among deployed US Naval and Military personnel increased and consumption by smokers also rose. Internal tobacco industry documents between 1986 and 2001 were analyzed as to their marketing of cigarettes to the military during the first Gulf War. During the conflict, the tobacco companies targeted the US troops with free cigarettes, direct advertising, branded items, free communications with family members (via the Marlboro Voice Card) and welcome home events.

Smoking rates in the military are not declining. In part, this may be because military personnel smoke as a response to boredom, anxiety, and stress, which are likely exacerbated for deployed troops. Deployment is associated with smoking uptake and recidivism; one recent survey of a battalion deployed to combat operations in Iraq found that 52% smoked. Thus, the current deployment of thousands of young Americans to Iraq and Afghanistan in military engagements places them at increased risk from the hazards of war, and also increased risk of addiction and disease from tobacco.

As a hierarchical institution, the military communicates important messages through orders. The areas in which the military has met health objectives are those where regulations enforce healthy behaviors, such as exercise and seat belt use. Tobacco remains culturally accepted, and a mixed message prevails: tobacco use is discouraged (e.g., donations are forbidden), but accommodated (e.g., purchase is facilitated). DOD regulations are not adequately supporting tobacco use prevention and cessation.

Reference Number: 40121

STEVENSON, T., AND R.N. PROCTOR. "The Secret and Soul of Marlboro. Philip Morris and the Origins, Spread, and Denial of Nicotine Freebasing." *American Journal of Public Health*, 98: 1184–1194, 2008 (2 figures, 97 references)

Abstract: Ammonia occurs naturally in the cured tobacco leaf and ranges in concentration from near 0% to 1% (in high quality cigar tobacco). In the early 1960s, however, Philip Morris's scientists discovered that ammonia could also be used to increase the free nicotine (freebase) content and provide a more rapid delivery of nicotine than the typical milder, lower pH cigarette tobacco. It was determined that the best concentration of ammonia was 0.25–0.50%. In the 1970s, puffed tobacco was manufactured as a cost-saving measure, as it reduced the amount of tobacco required per cigarette and encouraged smokers to smoke more cigarettes to obtain the same amount of nicotine. The most common method to expand or puff tobacco is the dry ice expanded tobacco (DIET) process which uses dry ice (solid carbon dioxide) to freeze dry and puff-up tobacco. Reconstituted tobacco is another cost-cutting measure, which allows the use of the stems and mid-ribs, which were normally discarded, by using diammonium phosphate (DAP) to mask the unpleasant tastes of these parts of the tobacco leaf. PM patented the diammonium phosphate-blended leaf (DAP-BL) process in 1967. Marlboro cigarettes began in the 1920s as a woman's cigarette that was advertised as "mild as May." It was taken off the market during World War II due to poor sales. PM reintroduced Marlboros in the 1950s with a new and improved filter, a flip-top box, and a new masculine image, and it sold surprisingly well, which was due to DAP-BL used in this brand. Diammonium phosphate breaks down to ammonia when smoked, giving it a smooth chocolate-like taste. Diammonium phosphate tobacco products delivered 0.57 mg of nicotine per cigarette, compared to 0.44 mg of nicotine in untreated tobacco.

In the mid-1990s, FDA Commissioner Kessler was using the tobacco companies' recently discovered manipulation of nicotine to argue for the regulation of tobacco. The industry knew that any admission of using ammonia to increase the availability of nicotine would be an admission that nicotine is the addictive sine qua non of smoking, and tobacco manufacturers did not want to have cigarettes compared to crack cocaine. (According to Farone, the industry made a conscious effort in the 1980s and 1990s to avoid the term freebasing in any documents or statements because it didn't want to be associated with cocaine.) With so much at stake, it is not surprising that the industry attempted to cover up its use of ammonia as a freebasing reagent.

It is important to keep in mind that the ammonia cover-up is only one of dozens of similar cover-ups, part of the industry's calculated disregard for human health. The industry has tried to efface the truth in depositions and even in internal documents, but the facts are clear: Philip Morris by the early 1960s had realized ammonia's ability to augment the potency of nicotine and then used this knowledge to develop cigarettes that were nominally low in tar and nicotine while still having a powerful nicotine kick. Sales of Philip Morris products skyrocketed as a result, spiking the company to the top of the market charts. And the other manufacturers took notice: RJ Reynolds, American Tobacco, Brown and Williamson, Liggett, and Lorillard all worked hard to figure out Marlboro's secret and by the 1970s and 1980s had developed their own versions of ammonia technology.

Reference Number: 40122

BRANDT, A.M. "Inventing Conflicts of Interest: A History of Tobacco Industry Tactics." *American Journal of Public Health*, 102: 63–71, 2012 (43 references)

Abstract: This study evaluates the tactics of the tobacco industry in response to the increasing scientific evidence of the harms of cigarette smoking in the 1950s and 1960s. The industry used the leading public relations firm, Hill & Knowlton, which advised the industry to create an industry-sponsored research entity. It also advised the industry to fund university-based scientists to enlist their support and dependence and make academic institutions partners with the tobacco industry.

By the early 1960s—despite categorical research findings indicating the harms of smoking—a significant "controversy" had arisen (at the behest of the tobacco industry) over the validity and meaning of these findings. Indeed, given the widespread acceptance of the conclusion, especially among those who had analyzed and evaluated the research most closely, the persistence of debate about the harms of smoking is a striking demonstration of the powerful impact of the tobacco industry's public relations campaign. The industry insistence, at the direction of Hill & Knowlton, on the notion of no proof and the need for more research was an inspired if cynical manipulation of the natural tendencies within science to encourage skepticism and seek more complete answers to important questions.

Hill & Knowlton had served its tobacco clients with commitment and fidelity and with great success. But the firm had also taken its clients across a critical moral barrier that would have 2 important effects on American science and society. Trust in science, confidence in the media, and the social responsibility of the corporate enterprise were all substantially harmed by Hill & Knowlton's efforts on behalf of the tobacco industry. By making science fair game in the battle of public relations, the tobacco industry set a destructive precedent that would affect future debates on subjects ranging from global warming to food and pharmaceuticals. In addition, by insinuating itself so significantly into the practice of journalism, Hill & Knowlton compromised the legitimacy and authority of the very instruments upon which public relations depended.

The tobacco industry's public relations campaign permanently changed industry–science relationships and public culture. Their disinformation campaign, built on a foundation of conflicts of interest, demonstrates a series of problems that continue to evolve regarding the relationship between medical science and industrial influence.

Reference Number: 40123

MEIJA, P., L. DOOFMAN, A. CHEYNE, L. NIXON, L. FRIEDMAN, M. GOTTLIEB, AND R. DAYNARD. "The Origins of Personal Responsibility Rhetoric in News Coverage of the Tobacco Industry." *American Journal of Public Health*, 104: 1048–1051, 2014 (1 figure, 35 references)

Abstract: The study used news coverage of the tobacco industry between 1966 and 1991 to assess the use of personal responsibility rhetoric by the tobacco industry representatives. It analyzed a total of 793 published news articles that included arguments from the Tobacco Institute about responsibility, of which 210 (26%) referred to personal responsibility. The first personal responsibility news article relating to tobacco-related diseases occurred in 1977.

The tobacco industry has parlayed the powerful American notion of personal responsibility to frame tobacco problems as a matter for individuals to solve. Not until the late 1980s, after it had fully exercised its courtroom strategy, did the tobacco industry frame smoking primarily as a "choice" in news coverage, implying the consumer's accountability.

Previously, we found that personal responsibility framing was not prominent in the national debate over cigarettes and health from 1952 to 1965. After FCLAA, however, the tobacco industry could claim that consumers were now informed of risks, so anyone who still smoked should be held responsible, a successful courtroom strategy that allowed the tobacco industry to prevail in litigation throughout the end of the *Cipollone* case.

This study identified the genesis of what has become a pillar of the tobacco industry's strategy to deflect regulation and derail litigation. The emergence of the industry's public use of personal responsibility rhetoric in the late 1970s illustrates that its tactics preceded the ascension of personal responsibility rhetoric commonly associated with the "Reagan revolution" in the 1980s. Indeed, the industry used the personal responsibility frame regularly throughout most of Reagan's presidency, most intensely during the Cipollone trial. This chronology also supports research documenting decades-long support from the industry in fomenting the rhetorical origins of the conservative movement that became the Tea Party in the late 2000s.

Reference Number 40124

IIDA, K., AND R.N. PROCTOR. "'The Industry Must be Inconspicuous': Japan Tobacco's Corruption of Science and Health Policy via the Smoking Research Foundation." *Tobacco Control*, 27: e3–e11, 2018 (3 tables, 95 references)

Abstract: The Smoking Research Foundation (SRF) was formed under the protection of the Ministry of Finance in 1986.

In the USA, institutions closest to SRF would be the Council for Tobacco Research (CTR) and the Center for Indoor Air Research, both of which were disbanded with the Master Settlement Agreement of 1998, following recognition of their crucial role in the broader denial campaign. SRF and these American institutions were similar insofar as both supported basic biomedical research, and both were used to question whether cigarettes really cause harm, promulgating a false controversy. Both were used to create "a stable of experts" willing to serve as authors, speakers or witnesses in cigarette-friendly symposia and lawsuits and to influence policy. Both helped create "goodwill" for the industry, conveying an impression of a responsible industry selling a controversial product. CTR and SRF both lent credibility to their funders. CTR funded over a hundred different academic institutions; SRF funded up to 90 in 1 year.

The SRF must also be considered an instrument used by global cigarette makers to transmit the denialist campaign into Asia. In the USA, the formal campaign to deny smoking's harms began in December 1953, with a meeting of American cigarette CEOs at Manhattan's Plaza Hotel. That campaign was first extended into Canada, and then to Europe in 1977 following the establishment of the International Committee on Smoking Issues in England.

Reference Number: 40125

FALLIN, A., AND S.A. GLANTZ. "Tobacco-Control Policies in Tobacco-Growing States: Where Tobacco Was King." *The Milbank Quarterly*, 93: 319–358, 2015 (2 figures, 124 references)

Abstract: A detailed historical review of tobacco control policies was conducted in five major tobacco growing US states (Kentucky, North Carolina, South Carolina, Tennessee, and Virginia). The 1998 MSA, the 2004 tobacco quota buyout, and increasing use of cheaper foreign tobacco led to a rift between the tobacco companies and the tobacco farmers. Since 1960, tobacco production has declined by 50–80% in all five states. The first comprehensive smoke-free local law in a tobacco growing state was passed in 2003.

The tobacco companies were particularly concerned about blocking tobacco-control policies in the tobacco-growing states by promoting a pro-tobacco culture, beginning in the late 1960s. Nevertheless, since 2003, there has been rapid progress in the tobacco-growing states' passage of smoke-free laws. This progress came after the alliance between the tobacco companies and the tobacco farmers fractured and hospitality organizations stopped opposing smoke-free laws. In addition, infrastructure built by National Cancer Institute research projects (COMMIT and ASSIST) led to long-standing tobacco-control coalitions that capitalized on these changes. Although tobacco production has dramatically fallen in these states, pro-tobacco sentiment still hinders tobacco-control policies in the major tobacco-growing states.

Conclusions
The environment has changed in the tobacco-growing states, following a fracture of the alliance between the tobacco companies and their former allies (tobacco growers and hospitality organizations). To continue this progress, health advocates should educate the public and policymakers on the changing reality in the tobacco-growing states, notably the great reduction in the number of tobacco farmers as well as in the volume of tobacco produced.

Reference Number: 40126

GAOP, W., M. SANNA, M. HEFLER, AND C.P. WEN. "Air Pollution is not the 'New Smoking': Comparing the Disease Burden of Air Pollution and Smoking Across the Globe, 1990-2017." *Tobacco Control*, 29: (4 pp), 2019 (1 figure, 28 references)

Abstract: Smoking remains one of the top preventable causes of death and disability despite decades of tobacco control measures. Between 2007 and 2017, the estimated number of deaths due to outdoor particulate matter (PM2.5) increased from 2.42 million to 2.94 million, compared to 8.3 million deaths in 2017 due to tobacco use and SHS. Overall, the life expectancy of smokers is 10 years less than non-smokers. By contrast, air pollution in mild to moderately polluted cities reduced life expectancies by an estimated 0.8–1.6 years. Even in China, which has the greatest air pollution disease burden in the world, 2.5× as many deaths are caused by smoking than air pollution.

If air pollution really is the new smoking, some smokers may conversely think that breathing air is equally as bad as smoking. Among the more than one billion current smokers worldwide, this equivalence may lead some to wonder why they should even bother to quit. Paradoxically, some nonsmokers may consider smoking since they believe that breathing urban air is the same.

The tobacco industry would likely welcome the comparison of air pollution and smoking, as this will create a perception that using their products is "only as bad as breathing polluted air". The ubiquitous presence of outdoor air pollution is a common justification used by adolescents who have begun smoking and by smokers who are uncertain about quitting. However, breathing polluted air is not highly addictive nor is it heavily marketed by a notorious industry. No industry deliberately produces toxic and addictive ambient air as a commodity, in stark contrast to the tobacco industry, which produces and aggressively markets a toxic and addictive commodity—cigarettes. The success of the tobacco industry depends on generating as many nicotine-dependent customers as possible. In addition, unlike smoking, measures to improve air quality have tremendous public support, in part because people are not addicted to toxic air in the same way that smokers are dependent on nicotine.

Reference Number: 40127

BERO, L.A. "Tobacco Industry Manipulation of Research." *Public Health Reports*, 120: 200–208, 2005 (1 table, 86 references)

Abstract: An interest group is an organized group such as the tobacco industry with a narrowly defined viewpoint organized to protect its position or profits no matter what the cost. As a result of the 1998 Master Settlement agreement, millions of internal tobacco industry documents showed that for decades the industry has tried to generate controversy and doubt over the health risks of tobacco. The tobacco industry funded law firms to fund research into other causes of cancer such as a low-protein diet, and minimize the harms of secondhand smoke, to deflect attention from tobacco and win lawsuits.

Table: Tobacco Industry Strategies to Manipulate Scientific Findings and Research

Tobacco Industry Strategies to Manipulate Research
• Fund research that supports the interest group position.
• Publish research that supports the interest group position.
• Suppress research that does not support the interest group position.
• Criticize research that does not support the interest group position.
• Disseminate interest group data or interpretation of risk in the lay press.
• Disseminate group data or interpretation of risk directly to policy makers.

Source: adapted from Bero, 2005

The tobacco industry has had a longstanding strategy of funding research and disseminating it through sponsored, non-peer-reviewed publications. These strategies have remained relatively constant as the industry has progressed from refuting research on active smoking to refuting research on secondhand smoke. Tobacco industry lawyers and executives, rather than scientists, have been in control of the design, conduct, and dissemination of this research. Despite the questionable conduct of much of this research, the tobacco industry has widely disseminated it to lay journalists and policy makers. In addition, the tobacco industry has a record of suppressing and criticizing research that is unfavorable to its position.

Reference Number: 40128

EUROPEAN RESPIRATORY JOURNAL EDITORS. "Retraction Notice for: 'Characteristics and Risk Factors for COVID-19 Diagnosis and Adverse Outcomes in Mexico: An Analysis of 89,756 Laboratory-Confirmed COVID-19 Cases.'": Theodoros V. Giannouchos, Roberto A. Sussman, José M. Mier, Konstantinos Poulas and Konstantinos Farsalinos. *Eur Respir J* 2020; In Press." *European Respiratory Journal*, 57: (1 p), 2021

Abstract: The *European Respiratory Journal* (publishers), the editors, and society agreed that the above-noted article should be withdrawn due to undisclosed conflict of interest with tobacco/vaping interests. The paper claimed that SARS-CoV-2 infection is less prevalent in smokers or tobacco users.

The European Respiratory Society, as a leading medical organisation in the respiratory field whose mission is to promote lung health and alleviate suffering from respiratory disease, has bylaws in place that do not permit individuals with ongoing relationships with the tobacco industry to participate in its activities: as well as society membership, and participation in congresses and meetings, publication in the society's journals is also considered an activity covered by these rules.

Reference Number: 40129

HOREL, S., AND T. KEYZER. "Covid 19: How Harm Reduction Advocates and the Tobacco Industry Capitalised on the Pandemic to Promote Nicotine." *British Medical Journal*, 373: n1303, 2021 (33 references)

Abstract: Several studies that showed nicotine/smoking protects against COVID-19 have been discredited and withdrawn.

It has since been roundly disproved that smoking protects against covid-19. Among other studies, the OpenSafely dataset, based on the primary care records of 17.3 million adults in the UK, found that smoking, when adjusted for age and sex, was associated with a 14% increased chance of covid-19 related death.

The BMJ can today also report on undisclosed financial links between certain scientific authors and the tobacco and e-cigarette industry in a number of covid research papers. This follows the high profile retraction of one such paper in the European Respiratory Journal last month, after two authors failed to disclose conflicts of interest.

4.02 FILTERS/LIGHT CIGARETTES

"Smokers today have a greater risk of developing lung cancer than did smokers in 1964. . . . Evidence suggests that ventilated filters may have contributed to higher risks of lung cancer by enabling smokers to inhale more vigorously, thereby drawing carcinogens contained in cigarette smoke more deeply into lung tissue."

—Office of the Surgeon General, "Health Consequences of Smoking, Surgeon General Fact Sheet" (2014)

The white fibers found in the filter tips of over 90% of commercial cigarettes are made by treating wood pulp with acetic anhydride and then dissolving it into acetone to create solid strands of cellulose acetate fiber (40201). A typical cigarette filter contains more than 12,000 of these plastic-like fibers and includes a delustrant (titanium oxide) and a plasticizer (glycerate triacetate) to bind them (40203). As expected with so many fibers, some become loose in the manufacture, packaging, and transport of cigarettes and can be inhaled into the lung (40203–4). Tobacco tar binds to almost all surfaces, including glass, plastic, cotton, and cellulose acetate, but the smoker is not aware that tar and toxic coated biopersistent filter elements can be inhaled or ingested during smoking (40215).

To counter the science showing the severe health harms of cigarette smoking and prevent people from quitting, the tobacco companies promoted misinformation that filter cigarettes were safer than unfiltered cigarettes and smokers could reduce their risk by switching brands, rather than quitting completely (40214–6). This argument is used over and over again by the tobacco industry, but of course virtually anything is safer than cigarettes, the most destructive artifact ever made by humans.

In fact, the filters were designed to circumvent machine testing for smoking toxins, which do not reflect real-world cigarette smoking and thus highly exaggerated the safety of filter cigarettes (40202, 40205–6, 40210–11), again enticing smokers to switch but not quit. Filter cigarette smokers were able to more deeply inhale than unfiltered cigarette smokers, which resulted in a 23–35% higher carbon monoxide content in the blood (40208).

Filter cigarettes did not reduce the incidence of lung cancer, they merely switched the main type of cancer from squamous cell carcinoma to adenocarcinoma (40201–7).

Reference Number: 40201

HARRIS, B. "The Intractable Cigarette 'Filter Problem.'" *Tobacco Control*, 20 (suppl 1): i10–i16, 2011 (36 references)

Abstract: The study assessed the early history of cigarette filter research in the 1950–1960s using the Legacy Tobacco Documents Library online. Cellulose acetate fibers in cigarette filters are made by treating raw cellulose (from wood pulp) with acetic anhydride and dissolving it into acetone, which results in a viscous liquid that is spun to form the fibres.

> Even though filtered brands currently constitute over 90% of the entire cigarette market, as early as the mid-1960s the Surgeon General judged cigarette filters to be useless in reducing harm to the average smoker. Indeed, epidemiological data gathered between the 1970s and early 2000s continue to support the conclusion that the nearly universal adoption of cigarette filters has done little to protect smokers. In 2010, a team of investigators from the USA and Japan asserted in the International Journal of Cancer that "the shift from nonfilter to filter cigarettes appears to have merely altered the most frequent type of lung cancer, from squamous cell carcinoma to adenocarcinoma" Likewise, there is still no compelling evidence that cigarette filters have done anything to mitigate other health hazards of smoking such as heart disease.

Reference Number: 40202

SONG, M-A., N.L. BENOWITZ, M. BERMAN, T.M. BRASKY, K.M. CUMMINGS, D.K. HATSUKAMI, C. MARIAN, R. O'CONNOR, V.W. REES, C. WOROSZYLO, AND P.G. SHIELDS. "Cigarette Filter Ventilation and its Relationship to Increasing Rates of Lung Adenocarcinoma." *Journal of the National Cancer Institute*, 109: (18 pp), 2017 (1 table, 7 figures, 244 references)

Abstract: Cigarette smoke is a major cause of lung cancer, as it contains numerous carcinogens, mutagens, and other toxicants. The most common subtype of lung cancer was squamous cell during the 1950–1960s, when lung cancer rates increased, and then rates decreased over the next 40 years as cigarette smoking decreased dramatically. The rate of lung adenocarcinomas during the 1990s, however, did not decrease and this subtype of lung cancer now accounts for 60% of non–small cell lung cancers. Lung adenocarcinoma develops in the glandular cells in the lung, which secrete mucous, and in the smaller airways such as the alveoli. The 2016 Surgeon General's report on smoking and health concluded that changing cigarette designs including cigarette filter ventilation caused an increase in adenocarcinomas. The uses of holes in the cigarette filters, the use of less tobacco, more reconstituted tobacco, and increasing paper porosity have all been employed by the tobacco industry since the 1950s to reduce the smoking machine–measured tar yields of cigarettes, so they could be promoted as healthier. Many smokers believed this, as the filter ventilation reduced the harshness of the smoke and reduced the resistance to draw while puffing the cigarette. This review was conducted on 3,284 studies on the effect of filter ventilation on public health. Filter ventilation was mainly adopted in the mid 1960s, but the ventilation alters tobacco combustion, which increases smoke toxicants such as TSNAs, and smokers inhale more smoke to maintain their nicotine levels. Filter ventilation causes less air flow through the burning tip of the cigarette, which results in more incomplete combustion and allows the cigarette to smolder longer, producing more toxic constituents. Filer ventilation also allows smokers to have higher puff volumes and more frequent puffs.

Table: Changes in the Chemical Composition of the Smoke and Peripheral Lung Exposure from Filter Ventilation

Changes in Chemical Composition of the Smoke	Peripheral Lung Exposure
• Increased mutagens	• Bigger puffs
• Changes in proportion of chemicals	• Deeper inhalation
• Increased TSNAs	• Larger particles
	• Deeper delivery
	• More cigarettes per day

Source: Adapted from Song et al, 2017

In conclusion, the use of ventilation in the filters of cigarettes has failed to make cigarettes safer, and more than likely has made them more harmful.

Reference Number: 40202

There is no demonstrated public health benefit, and smokers perceive these as less harmful, which in turn encourages smoking and causes harm. The FDA now has the authority to require the elimination of filter ventilation because ventilation does not serve any public health purpose and instead provides a false promise of reduced risk. This single action for banning filter ventilation by the FDA is scientifically justified and within its mandate to improve the public health. Based on these weight-of-evidence reviews, the FDA should embark on a regulatory process of data evaluation and consider regulation(s) for the use of ventilation in filters, up to and including a ban on their use.

Reference Number: 40203

PAULY, J.L., A.B. MEPANI, J.D. LESSES, K.M. CUMMINGS, AND R.J. STRECK. "Cigarettes with Defective Filters Marketed for 40 Years: What Philip Morris Never Told Smokers." *Tobacco Control*, 11; i51–i61, 2002 (1 table, 122 references)

Abstract: Over 90% of the cigarettes sold globally have a filter, which typically consists of more than 12,000 plastic-like cellulose acetate fibers. Filters were introduced to prevent loose tobacco flakes from entering the mouth. In 1950, only 1% of the cigarettes in the United States had filters, which increased to 95% in the United States in 1986, in part due to the Surgeon General's Report 1964 report on cancer and smoking. The cellulose acetate fibers also contain a delustrant, titanium dioxide, and a plasticizer (glycerate triacetate) is applied to bond the fibers. During the high-speed multi-step manufacturing and boxing procedure, cellulose acetate and charcoal granules (used in smoke filters) break loose and are inhaled into the smoker's lung. The loose fragments from the cigarette filters were referred to by Philips Morris as fall-out and have been known by the company since 1961. In 1998, inhaled cellulose and plastic fibers were found in the human lung.

Almost all cigarettes sold worldwide have a filter; thus, the filter is an integral part of today's cigarette. Furthermore, most filters are made of a cylindrical rod or plug that consists of thousands of plastic-like cellulose acetate fibres. We, however, do not know what studies have been conducted by the tobacco companies to: (a) address filter defects that may arise during cigarette manufacturing whereby loose debris of different forms (for example, fibres, fragments, particles, crystals, and fines) contaminates the cut surface of the filter and is released during normal puffing; and (b) assess the toxicology and human health risks of these diverse filter elements that are discharged into mainstream smoke.

This paper discloses that Philip Morris, Inc has known and concealed for approximately 40 years that fibres and particles "fall-out" of the filter of cigarettes during smoking. Other companies have assessed also the discharge of filter fibres. In addressing this filter defect, the tobacco industry's response has been variable, ranging from denial of the discharge of filter fibres to the development of innovative technologies for correcting and preventing the problem. Consumers have not been informed of the filter defect. Further, there is no indication that existing corrective technologies and invention have been uniformly implemented. Summarily, the tobacco industry has been: (a) derelict in concealing information of filter defects; (b) negligent in implementing technologies available to prevent or reduce the emission of filter elements; and (c) wrongful in not investigating the toxicology and harm associated with defective filters of today's cigarettes that are being marketed worldwide.

Reference Number: 40204

PAULY, J.L., S.J. STEGMEIER, A.G. MAYER, J.D. LESSES, AND R.J. STRECK. "Release of Carbon Granules from Cigarettes with Charcoal Filters." *Tobacco Control*, 6: 33–40, 1997 (2 figures, 32 references)

Abstract: More than 95% of cigarettes sold in the United States have filters, which are generally perceived by smokers as safer than unfiltered cigarettes. A study was conducted on 400 Lark cigarettes in 20 different packs. Lark cigarettes have a triple granular filter composed of an activated charcoal layer surrounded by two layers of cellulose acetate fibers. The amount of charcoal in the middle chamber of the filter varied between 25.5 and 78.8 mg and the chamber was on average 60% empty. Charcoal granules were present in the cellulose acetate filters of most (80%) Lark cigarettes. There were often gaps between the tipping paper that surrounds the filter, which would allow for the migration of charcoal granules. On average, 16.2 charcoal granules were released when the Lark cigarette was smoked through a modified cigarette holder by smoking volunteers. The majority of cigarette toxins are not in the particulate (tar) phase, but in the gas phase, where the smooth nonporous nature of the cellulose acetate fibers have an inadequate surface area to absorb them. Cellulose acetate will also allow hydrogen cyanide and hydrogen sulphide to pass through it.

Charcoal granules are incorporated into cigarette filters to aid in removing toxins in cigarette smoke. In studies of Lark, a popular American cigarette with a charcoal filter, charcoal granules were observed on the filter surface, and were released from the filter when the cigarettes were smoked. During smoking, the toxin-containing charcoal granules are inhaled or ingested. The specific adverse health effects of inhaling or ingesting carbon granules have not been addressed; nevertheless, the smoker, as an educated consumer, should be informed of the possible health risks.

Reference Number: 40205

KARZLOWSKI, L.T., N.A. DRESCHEL, S.D. STELLMAN, J. WILKENFELD, E.B. WEISSS, AND M.E. GOLDBERG. "An Extremely Compensatible Cigarette by Design: Documentary Evidence on Industry Awareness and Reactions to the Barclay Filter Design Cheating the Tar Testing System." *Tobacco Control*, 14: 64–70, 2005 (2 tables, 2 figures, 97 references)

Abstract: In 1980, Brown and Wilkinson Tobacco company (B+W), an affiliate of BAT, launched a $340 million marketing and advertising campaign of an ultra low tar cigarette—the Barclay cigarette, which was touted as having only 1 mg tar in King Size. The Barclay cigarette achieved the low tar rating by having a grooved filter system that would pass the mechanical smoking machine tests but would be easily compromised by the pressure exerted by the human smoker's lips and fingers, causing the air dilution channels to collapse. The FTC/International Organization for Standardization (ISO) testing procedure helped cause the public health tragedy of the popular low tar cigarette.

The Barclay cigarette with a novel grooved filter was introduced in the USA in 1980 and was the subject of the most expensive launch in US history. This cigarette design ultimately became infamous around the world because of findings that it could not be properly analysed by official tar testing systems. The filter design became dysfunctional in smoker's mouth and hands, but not in the smoking machine.

We analysed documentary evidence on the manufacturer's awareness of this problem during development and on responses by competitors when Barclay was first sold and before governmental agencies became involved. We found evidence that: (1) the manufacturer was aware of the design problems and planned to respond to criticism before the product was released; and (2) four other companies readily detected the design problems and most considered developing their own deceptive filter. This episode provides insight into the relation between standard mainstream smoke yield and the design of cigarettes to satisfy consumers. The episode also argues for implementation of revised cigarette testing procedures in which cigarette sales trigger more detailed product testing, including exposure studies in human smokers.

Reference Number: 40206

KOZLOWSKI, L.T., R.C. FRECKER, V. KHOUW, AND M.A. POPE. "The Misuse of 'Less Hazardous' Cigarettes and Its Detection: Hole-Blocking of Ventilated Filters." *American Journal of Public Health*, 70: 1202–1203, 1980 (1 table, 1 figure, 10 references)

Abstract: In 1979, about one-quarter of all cigarettes sold in the United States had ventilated filters (i.e., holes or perforations in the filter), which allow for the dilution of the cigarette smoke with the outside air to lower tar and nicotine concentrations. The ventilation filters worked well in smoking machines, but could easily be blocked by smokers using their fingers, lips, or a piece of tape. The blocking of the holes leaves a specific pattern in the spent filter, in which the heavy smoke deposit is found in the entire diameter of the filter, rather than just the inside diameter. A smoking machine analysis was conducted using a popular low tar cigarette with the ventilated filter not blocked, half blocked, and fully blocked.

Table: The Effects of Blocking a Ventilated Filter on Nicotine, Tar, and CO Yield Per Cigarette and Number of Puffs

	Not blocked	Half-blocked	Fully blocked
Nicotine (mg)	0.45	0.73	0.98
Tar (mg)	4.40	7.03	12.60
CO (mg)	4.50	7.80	17.70
Number of puffs	11.10	10.50	9.20

Source: Adapted from Kozlowski et al, 1980

Health researchers and tobacco scientists have been interested in ventilated filters primarily as a way to make less toxic cigarettes; however, smoking behavior can act to sabotage the benefits of these filters.

Reference Number: 40207

GRAY, N. "The Consequences of the Unregulated Cigarette." *Tobacco Control*, 15: 405–408, 2006 (53 references)

Abstract: Substantial efforts have been made to regulate cigarettes. But cigarette design, the constituents of the tobacco, and additives are still within the control of the cigarette manufacturer. Ventilated filters allow deeper inhalations more easily and expose the periphery of the lung to a greater volume of smoke, which delivers nicotine and carcinogens with maximum efficiency. The modern cigarette is both easier to smoke and easier to learn to smoke, and delivers more TSNA. The tobacco industry is not a normal commercial enterprise but should be treated as a drug manufacturer.

It seems, from the above, that there is not one epidemic of lung cancer but (at least) two. One (squamous) has declined but the other (adenocarcinoma) has increased, very often in the face of declining levels of smoking prevalence, tar content and per capita consumption, for which the evidence is strongest in US males. This strongly suggests a different causal basis, and it is extremely likely that this lies within the qualitative differences between modern cigarettes and their predecessors. The adenocarcinoma epidemic can reasonably be described as unnecessary.

Reference Number: 40208

WALD, N.J., M. IDLE, J. BOREHAM, AND A. BAILEY. "Inhaling Habits Among Smokers of Different Types of Cigarette." *Thorax*, 35: 925–928, 1980 (1 table, 1 figure, 14 references)

Abstract: The blood COHb concentration was determined in 110 male smokers of plain cigarettes, 331 of whom smoked ventilated filter cigarettes and 875 of whom smoked unventilated filters, in London between 1975 and 1977. The number of cigarettes smoked per day and the brand of cigarettes were determined. CPD was similar across the three types of cigarette smokers (mean approximately 24 CPD). The COHb was standardized for the number of cigarettes reported to have been smoked before the collection of the blood sample. The standardized COHb levels were 23% higher for ventilated filter smokers and 35% higher for unventilated filter smokers than for smokers of plain cigarettes.

Men who smoked filter cigarettes inhaled substantially more than smokers of plain cigarettes although their cigarette consumption was similar. As a result any benefit to health associated with smoking filter cigarettes, rather than plain ones, is likely to be less than expected on the basis of their relative tar yields. A number of studies have suggested that smokers of low tar cigarettes have a lower risk of lung cancer than smokers of high tar cigarettes, and in the UK, associated with a change to filter cigarettes, there has been a decrease in lung cancer mortality among men aged less than 60 years. The results of our present study suggest that the reduction in tar yield may only partially explain these epidemiological observations.

Reference Number: 40209

KOZLOWSKI, L.T., N.Y. MEHTA, C.T. SWEENEY, S.S. SCHWARTZ, G.P. VOGLER, M.J. JARVIS, AND R.J. WEST. "Filter Ventilation and Nicotine Content of Tobacco in Cigarettes from Canada, the United Kingdom, and the United States." *Tobacco Control*, 7: 369–375, 1998 (8 tables, 25 references)

Abstract: The filter ventilation and nicotine content of tobacco and smoking machine yields were determined in 32 US, 23 Canadian, and 37 British brands of cigarettes. Ninety-four percent of the US brands, 91% of the Canadian brands, and 79% of the British brands were ventilated. Filter vents started at 11–15 mm from the smoker's end of the cigarette and are only detected when the filter covering is taken off and held up to a bright light and examined with a magnifying glass. None of the cigarette packages were labelled as ventilated or not. Recent internal cigarette company documents have shown that invisible vents were a research and development strategy for some manufacturers.

Table: Smoking Machine Yields of Tar, Nicotine, and CO, the Distance of the Closest Vents, and Percent Ventilation of Selected Canadian Cigarettes

Brand	Tar (mg)	Nicotine (mg)	CO (mg)	Closest Vents (mm)	Percent Ventilation
Medallion Ultra mild	1	0.2	2	11.5	70.1%
Craven Ultra light	6	0.6	6	13.0	50.6%
Du Maurier Extra light	9	0.9	9	12.0	40.7%
Export Light	12	1.1	12	12.0	40.0%
Players L Smooth	13	1.3	14	13.0	14.8%
Players Full Flavor	16	1.4	17	–	–

Source: Adapted from Kozlowski et al, 1998

Machine-smoked yields of tar and nicotine have little relevance for smokers' actual intakes, and through brand identifiers, such as "Light" and "Ultralight" may be actively misleading. Our data show that the combination of ample nicotine content with filter vents that are easily, and often unconsciously defeated by smokers provides a readily available means for smokers to satisfy their addiction to nicotine from even low-yield brands. Numerous studies indicate extensive nicotine compensation in smokers of low nicotine cigarettes.

Filter ventilation appears to be the predominant method for reducing machine-smoked yields of tar, nicotine, and CO in three countries. Some cigarettes contain about twice as much nicotine (total content or percent nicotine) as do other brands, indicating that tobacco types or blends and tobacco casings can substantially manipulate nicotine content of cigarettes. It is lamentable that consumers or scientists routinely know so little about so common a consumer "good" that causes so much death and disability. The more the cigarette is allowed to be a "black box"—even to government chemists—the easier it would be to fool most of the people most of the time, if an unscrupulous manufacturer chose to do so.

Reference Number: 40210

PAUWELS, C.G.G.M., W.N.M. KLERX, J.L.A. PENNINGS, A.W. BOOTS, F.J. VAN SCHOOTEN, A. OPPERHUIZEN, AND T. TALHOUT. "Cigarette Filter Ventilation and Smoking Protocol Influence Aldehyde Smoke Yields." *Chemical Research in Toxicology*, 31: 462–471, 2018 (2 tables, 3 figures, 48 tables)

Abstract: The levels of 12 aldehydes, tar, nicotine, and carbon monoxide (TNCO) in mainstream smoke were determined in 11 Dutch cigarette brands in a smoking machine using the International Standards Organization (ISO) and Health Canada Intensive (HCI) protocols with the filter ventilation taped (blocked) or not taped (unblocked). The HCI and taping is more representative as to how smokers actually smoke cigarettes.

Table: Smoking Machine Parameters Using the ISO and Health Canada Intensive (HCI) Protocols

Smoking Machine Parameter	ISO Protocol	HCI Protocol
Puff volume	35 mL	55 mL
Puff duration	2 seconds	2 seconds
Puff interval	60 seconds	30 seconds

Source: Adapted from Pauwels et al, 2018

In all cigarette brands, the highest yields of aldehydes in the cigarette smoke are measured when the filter ventilation holes of the cigarettes are taped during machine smoking. In addition to preventing dilution by sidestream air via filter vents, closing filter vents causes an increased burn rate of the cigarette, leading to more of the tobacco rod being consumed with every puff, thereby reducing the total burn time of the cigarette, as well as the number of puffs needed to completely smoke the cigarette. Aldehyde yields are increased during harsh burning conditions, which are gained by increasing puff volume or blocking ventilation. This explains why the effect of blocking ventilation holes is more pronounced for the ISO regime than the HCI regime. For the HCI regime, the harsh burning conditions due to intense smoking already led to an increase in aldehyde yields. The addition of blocking ventilation for the HCI regime leads to a less distinct increase in aldehyde yields, compared with the ISO regime. This also explains why the impact of ventilation blocking on aldehyde yields was the largest for those cigarettes with the highest initial ventilation (1R5F and Phillip Morris One).

Reference Number: 40211

GENDREAU, P.L., AND F. VIATRO. "The Unbearable Lightness of 'Light' Cigarettes: a Comparison of Smoke Yields in Six Varieties of Canadian 'Light' Cigarettes." *Canadian Journal of Public Health*, 96: 167–172, 2005 (6 figures, 33 references)

Abstract: The tobacco industry developed and heavily promoted light or mild cigarettes to reassure smokers that these cigarettes were safer. Light or mild cigarettes gained popularity and in 2002, 61% of Canadian smokers consumed them. Large numbers of smokers of light or mild cigarettes or extra/ultra light cigarettes believed they were safer, which reduced the smoker's incentive to quit smoking. The amounts of 41 toxic and/or carcinogenic smoke chemicals were determined in 6 varieties of light cigarettes sold in Canada (Player's Light, Players' Light Smooth, Player's Extra Light, Du Maurier Light, Du Maurier Extra Light, and Du Maurier Ultra Light) and were compared to 3 varieties of regular cigarettes. A modified ISO protocol was employed to sample the cigarettes in a smoking machine that increased the puff volume from 35 to 56 mL and decreased the puff interval from 60 to 26 seconds and also blocked all the ventilation holes in the filter tips. The yields of nicotine of light cigarettes were on average 5% higher and the yields of tar were only reduced by on average 16% for the light cigarettes tested in comparison to the regular cigarettes. There were also increased amounts of nickel, chromium, benzene, mercury, and toluene detected in light cigarettes compared to regular cigarettes.

> The present analysis showed no evidence that the different varieties of "light" cigarettes examined here differed substantially from "regular" cigarettes in terms of smoke constituents. Thus we substantiate previous studies investigating the effects of vent blocking on yields of tar, nicotine and CO. In fact, tar was never reduced by 25% or more, and nicotine yields tended to be slightly higher. The only substantial reduction was for CO (30%), and for only one variety of "light" cigarettes. In addition, only one out of four of all other reported yields was reduced by 25% or more in "light" cigarettes compared to "regular" or "full flavour" cigarettes. The lack of sizeable differences between "light" and "regular" cigarettes in laboratory conditions more similar to human smoking behaviour may explain why there has been no reduction in the overall disease burden caused by tobacco smoking since the early marketing of light cigarettes and the subsequent consumer shift from regular to light cigarettes.
>
> Despite anti-tobacco lobbying efforts to uncover the tobacco industry's deceptive marketing strategies, the belief that light cigarettes are safer remains. Enhanced educational tactics and public health policies are necessary to promote the correct notion that toxic exposure from cigarette smoke intake will only be reduced by reducing cigarette consumption, not by smoking "light" cigarettes.

Reference Number: 40212

KAISERMAN, M.J., AND W.S. RICKERT. "Handmade Cigarettes: It's the Tube that Counts." *American Journal of Public Health*, 82: 107–109, 1992 (2 tables, 10 references)

Abstract: Thirty-one brands of roll your own (RYO) tobacco available in Canada were analyzed for tar, nicotine, and CO. The overall mean results were approximately 15.5, 1.1, and 17.7 mg per cigarette for tar, nicotine, and CO, respectively. There were no substantial differences in RYO tobacco as the brands mainly used flue cured tobacco grown in Ontario. TNCO concentrations were determined using 17 commercially available paper cigarette tubes filled with 900 mg of RYO tobacco. The tar delivery varied by 60% depending on the tube used. It is the tube and filter combination, not the tobacco, which controls the delivery of toxic compounds to RYO smokers.

Cigarettes are composed of three major components: tobacco, filters, and the papers used in wrapping the tobacco and filter. Cigarette manufacturers are able to adjust the tar, nicotine, and CO deliveries by manipulating any or all of these components. Most often this involves using porous papers and ventilated filters, which dilute the mainstream smoke and thereby reduce the yields of toxic constituents.

Reference Number: 40213

TALHOUT, R., P.A. RICHTER, I. STEPANOV, C.V. WATSON, AND C.H. WATSON, "Cigarette Design Features: Effects on Emission Levels, User Perception, and Behavior." *Tobacco Regulatory Science*, 4: 592–604, 2018 (1 table, 102 references)

Abstract: A literature review of studies published up to 2015 was conducted on the effect of cigarette design features, such as filter ventilation and additives, new filter designs, and slim cigarettes, on emission levels and user perception and behavior. The amount of filter ventilation ranges from 10–80% depending mainly on the number of ventilation holes in the filter. Higher ventilated filter cigarettes have less complete combustion due to the reduced oxygen flow to the smoldering coal. Increased paper porosity decreases the coal temperature and causes the cigarette to burn faster. Smoke pH and free nicotine increases with increasing tip ventilation. Crimped cellulose acetate fibers are used in 90% of all cigarette filters. Another recent feature of filters are flavor granules, capsules, and flavor threads, which appeal more to young people.

Table: Some Effects of Physical Design Changes in Cigarettes

Design Characteristic	Smoke Emissions	Smoking Behaviors
Increased filter ventilation and paper porosity	Reduced per cigarette machine-generated emissions but less complete combustion	Compensatory smoking behavior: more intense smoking behavior resulting in similar or higher exposures to toxic and carcinogenic emissions
Filter additives like charcoal	Reduced emissions of some but not all smoke chemicals	Compensatory smoking behavior
Filter flavor capsules and flavor threads	Increased emissions of several gas phase smoke components	Increased initiation of smoking for youth
Slim cigarettes	Reduced emissions of some but not all smoke chemicals	Compensatory smoking behavior

Source: Adapted from Talhout et al, 2018

Whereas filter ventilation results in lower machine-generated emissions, it also leads to perceptions of lighter taste and relative safety in smokers who can unwittingly employ more intense smoking behavior to obtain the desired amount of nicotine and sensory appeal. Filter additives that modify smoke emissions can also modify sensory cues, resulting in changes in smoking behavior. Flavor capsules increase the cigarette's appeal and novelty, and lead to misperceptions of reduced harm. Slim cigarettes have lower yields of some smoke emissions, but smoking behavior can be more intense than with standard cigarettes.

Conclusions
Physical design features significantly impact machine-measured emission yields in cigarette smoke, product appeal, smoking behaviors, and exposures in smokers. The influence of current and emerging design features is important in understanding the effectiveness of regulatory actions to reduce smoking-related harm.

Reference Number: 40214

TINDLE, H.A., S. SHIFFMAN, A.M. HARTMAN, AND J.E. BOST. "Switching to 'Lighter' Cigarettes and Quitting Smoking." *Tobacco Control*, 18: 485–490, 2009 (2 tables, 36 references)

Abstract: A survey was conducted of 30,800 ever-smokers who had smoked in 2002–2003 in the United States. There were 24,918 daily and 4,115 occasional smokers, and 38% were cigarette switchers. The three main reasons for switching were to quit smoking, harm reduction, and flavor. The odds ratio (OR) for quitting cigarettes was 0.54×, 0.99×, and 0.45× for each reason, respectively. The so-called light cigarettes deliver amounts of tar and nicotine comparable to regular cigarettes and do not lead to a reduction in tobacco-related illness.

In conclusion, compared to non-switchers, switchers were more likely to make a quit attempt between 2002 and 2003, but were less likely to have recently quit smoking during this period. The magnitude of this association varied by reason for switching. Our analyses support the position that switching should be discouraged and that switching as a step towards quitting should be particularly discouraged. As neither "light" cigarettes nor any other kind of cigarettes have demonstrated convincing clinical evidence of reduced harm, smokers should be encouraged to stop smoking, rather than just change what cigarettes they smoke.

Reference Number: 40215

PAULY, J.L., H.J. LEE, E.L. HURLEY, K.M. CUMMINGS, J.D. LESSES, AND R.J. STRECK. "Glass Fiber Contamination of Cigarette Filters: An Additional Health Risk to the Smoker?" *Cancer Epidemiology, Biomarkers, and Prevention*, 7: 967–979, 1998 (7 figures, 50 references)

Abstract: In 1987, RJ Reynolds announced a new cigarette prototype that was Heat not Burn, called Premier, which did not sell well as it was difficult to light and had poor taste. In June 1996, a reconfigured cigarette design called Eclipse was released. Eclipse contained a central carbon fuel rod, surrounded by insulated mats of glass fibers, interior laminate aluminum foil, and reconstituted tobacco. The purpose of the glass mats is to secure the position of the central fuel rod, insulate the burning rod, and direct heat from the rod to the adjacent tobacco column. Unlike conventional cigarettes Eclipse burns very little paper. When smoked, the Eclipse does not burn to the filter but retains its shape. The ash appearing at the tip of the Eclipse consists mostly of the insulating mat of glass fiber.

> Today's smoker is unaware that cigarette filter elements of different origin and size are being released into their mouths during smoking. These consumers are uninformed that tar- and toxin-coated biopersistent filter components will be inhaled and/or ingested during normal smoking behavior. Battery by these filter elements is not part of the smoking experience.
>
> Today, both conventional and nonconventional smoking articles are being marketed as having a greater consumer benefit, including reduced health risks and decreased environmental smoke. Paradoxically, the filter was designed to absorb and trap tar in cigarette smoke. Tobacco tar binds rapidly and tenaciously to most surfaces (e.g., glass, plastic, cotton, and paper). The binding of tobacco tar to glass fibers used in prototype cigarette filters has been known for many years. Likewise, the use of glass fiber filters (e.g., Cambridge filters) in smoking machines, for the quantitative assessment of tar, as well as glass flasks, used for the condensation of tobacco smoke for extracting different tar components, is known widely. Furthermore, it is widely accepted that the adsorption of tobacco tar onto glass fibers is nonspecific, efficient, and rapid. In this context, we have observed that tobacco tar binds to the Eclipse glass fibers.
>
> The enhanced health hazard associated with cigarette smoking and fiber/dust inhalation has been discussed by many scientists. The synergistic effects of tobacco smoking and inhaled asbestos have been well

documented. We hold the opinion that an increased health risk may arise from the inhalation of tar-coated glass and/or cellulose acetate fibers and particles.

Reference Number: 40216

POLLAY, R.W., AND T. DEWHIRST. "A Premiere Example of the Illusion of Harm Reduction Cigarettes in the 1990s." *Tobacco Control*, 12: 322–332, 2003 (5 tables, 5 figures, 34 references)

Abstract: The 2002 Quebec Superior Court trial on the constitutionality of restricting cigarette advertising resulted in new internal documents from Imperial Tobacco Limited (ITL) on the marketing of the less irritating (i.e., less harmful) Player's Premiere cigarette. ITL is owned by British American Tobacco (BAT) and has a 70% market share in Canada. The promotional promise of being less irritating preceded the development of Premiere. The cigarette was neither new nor improved, and it was only a marketing ploy, offering no real consumer benefit. Future new tobacco products (e.g., vaping and HTPs) that are touted to reduce risk should be met with much scepticism.

The use of the filter's appearance to gain unwarranted "tangible credibility" was described as a "gimmick" in trial testimony, for which the trial judge volunteered the French translation of "ruse". It is clearly not consistent with ITL's claim to "develop products and communicate them honestly and straightforwardly" and to inform consumers. This gimmick, or ruse, might also be termed a gambit, device, stratagem, artifice, contrivance, trick, scam or fraud. In the trial court's judgment upholding the constitutionality of Canada's Tobacco Act, it was described as "nothing more than a massive marketing ploy". The evidence is clearly consistent with the mushroom model of consumer cultivation mentioned earlier, with its crude but apt analogy of cigarette advertising to bullshit. Whether the language is polite or crude, English or French, as Shakespeare might have said, a ruse by any other name would smell as foul.

Advertising effectiveness

The total communications mix employed by ITL generated "considerable curiosity", relatively high degrees of general awareness of the presence of "a unique filter" (even if little understanding of how, why, and to what extent it worked), and precipitated consumer trial. Significantly, the promotional communication also influenced some users' judgments about product performance, apparently convincing about 30% of users who might otherwise have reported no difference to report less irritation.

4.03 MENTHOL

> "The cooling and local anesthetic effects of menthol make mentholated cigarettes easier to smoke than non-mentholated cigarettes. Menthol, therefore, facilitates smoking and contributes to smoking initiation among inexperienced and uninitiated smokers."
>
> —Yerger, "Menthol's Potential Effects on Nicotine Dependence: A Tobacco Industry Perspective" (2011)

Menthol is a type of alcohol ($C_{10}H_{20}O$) that was originally extracted from the peppermint plant (Mentha arvensis) but is now produced synthetically. Menthol is a constituent of about 90% of the cigarettes sold in the United States, even if the cigarette is not labelled as mentholated (40302–3, 40310).

Menthol has anti-tussive, anti-irritant, and cooling properties which are ideal to assist in initiating and maintaining smoking (40301–3). It is not a simple flavorant but also increases nicotine addiction (40308), which makes it more difficult for menthol cigarette smokers to stop (40301–3, 40305, 40307).

The tobacco industry has long targeted Black communities and other racialized groups as well as women and lower socioeconomic status smokers for menthol cigarettes (40301, 40304, 40311). Menthol cigarettes are smoked by 83% of Black American smokers and 36% of women smokers (40307, 40312).

In countries where menthol cigarettes are banned, tobacco companies have been circumventing this regulation by adding a separate menthol crush ball or capsule to the package (40311), which shows how much the tobacco companies value menthol to sell cigarettes.

A promising development has been in San Francisco, where in spite of a major $12 million campaign by RJR Tobacco, voters decided to ban flavored cigarettes (40312).

Reference Number: 40301

WICKHAM, R.J. "The Biological Impact of Menthol on Tobacco Dependence." *Nicotine and Tobacco Research*, 20: 1–9, 2019 (101 references)

Abstract: Menthol provides the smoker with a pleasant mint flavor and a cooling sensation in the mouth, throat, and lungs, providing apparent relief from the irritation of tobacco smoke. Menthol has anti-tussive, anti-irritant, and cooling properties ideal to assist in smoking. In the 1920s, tobacco companies began marketing mentholated cigarettes as a healthier cigarette and have targeted Black communities, women, and lower socioeconomic status smokers. The US FDA has called for a ban on mentholated cigarettes but not on e-cigarettes. Menthol smokers who are in smoking cessation therapy in general have poorer outcomes, and pharmacotherapies such as bupropion are less effective in mentholated cigarette smokers.

In contrast to the dubious health claims by tobacco companies, mentholated cigarettes are in fact more addictive. Smokers of mentholated cigarettes have lower successful quit rates and in some cases are resistant to both behavioral and pharmacological treatment strategies. There is now considerable evidence, especially in the last 5 years, that suggest menthol might influence the addictive potential of nicotine-containing tobacco products via biological mechanisms. First, menthol alters the expression, stoichiometry, and function of nicotinic receptors. Second, menthol's chemosensory properties operate to mask aversive properties of using tobacco products. Third, menthol's chemosensory properties aid in serving as a conditioned cue that can both enhance nicotine intake and drive relapse. Fourth, menthol alters nicotine metabolism, increasing its bioavailability. This review discusses emerging evidence for these mechanisms, with an emphasis on preclinical findings that may shed light on why menthol smokers exhibit greater dependence.

Reference Number 40302

WICKHAM, R.J. "How Menthol Alters Tobacco-Smoking Behavior: A Biological Perspective." *Yale Journal of Biology and Medicine*, 88: 279–287, 2015 (2 figures, 72 references)

Abstract: Nicotine is a natural insecticide for the tobacco plant and is the primary psychoactive and addictive alkaloid of the plant as well. Nicotine dependence is the most common form of chemical dependence in the United States. About 90% of tobacco cigarettes contain menthol and most are not labelled as such. Menthol cigarettes are smoked at higher rates than non-menthol in adolescent compared to adult smokers. Smokers of menthol cigarettes have lower success rates of quitting. More menthol smokers smoke their first cigarette of the day sooner than non-menthol cigarette smokers. Menthol smokers tend to smoke fewer cigarettes per day, suggesting that menthol cigarettes are more reinforcing.

> The profound effects of mentholation on smoking behavior suggest that menthol may be more than just a flavor added to cigarettes. Indeed, it recently has been shown that menthol has other biological effects that may explain how mentholation of tobacco cigarettes can promote nicotine dependence. Menthol has four biological mechanisms that likely contribute to altering smoking behavior. First, menthol's unique sensory properties mask the aversive properties of smoking tobacco cigarettes, such as unappealing taste and burning sensations in the throat and lungs. Second, these same sensory properties serve as cues that can serve as potent triggers of smoking craving and relapse. Third, menthol serves as a negative allosteric modulator of nicotinic acetylcholine receptors (nAChRs), the principle site of action that mediates nicotine reinforcement. Fourth, evidence suggests that menthol can alter nicotine metabolism, increasing nicotine's bioavailability. This review will outline these four major biological mechanisms in the context of smoking behavior.

Reference Number 40303

YERGER, V.B. "Menthol's Potential Effects on Nicotine Dependence: A Tobacco Industry Perspective." *Tobacco Control*, 20: ii29–ii36, 2011 (1 table, 2 figures, 135 references)

Abstract: Menthol, which is mainly extracted from the peppermint plant (Mentha arvensis), is present in 90% of all cigarettes sold in the United States, whether or not they are labelled as mentholated. Menthol cigarettes are smoked by 83% of Black Americans compared to 24% of White American smokers. Despite smoking fewer cigarettes per day, menthol smokers show greater nicotine dependence, such as a shorter time before the first cigarette in the morning. Menthol's cooling action depresses respiratory activity, resulting in breath holding, which leads to a greater exposure of nicotine in the lungs. Internal tobacco documents show that menthol was not just considered to be a flavor additive but also intended to mask the harshness of tobacco and have some nicotine-like sensory effects.

Table: The Transfer Efficiency of Menthol and Percent Menthol Added to the Cigarette Depends on Filtration and Ventilation Rate

Tar Range (more ventilated filters)	% Transfer Efficiency of Menthol	% Menthol Added to Cigarette
Full Flavor	15–16	0.35–0.45
Milds	12–13	0.45–0.55
Lights	8–10	0.60–0.80
Ultras	1–5	0.90–1.25

Source: Adapted from Yerger, 2011

Tobacco companies did not appear to conduct in-house studies on menthol to assess how menthol may affect dependence or exposure measures. Instead, tobacco companies, concerned about how to increase their share of the cigarette market, focused on testing new products and manipulating the menthol and nicotine levels in cigarettes. Internal industry research indicates that menthol has some nicotine-like effects, interacts directly with nicotine to produce cigarettes that are easier to smoke and makes low nicotine delivery products more acceptable to consumers. The cooling and local anaesthetic effects of menthol make mentholated cigarettes easier to smoke than non-mentholated cigarettes. Menthol, therefore, facilitates smoking and contributes to smoking initiation among inexperienced and uninitiated smokers. Because of its effects on smokers, who thus are exposed to the addictive power of nicotine, menthol contributes to the overall burden of tobacco-related disease. Banning the use of menthol in all cigarettes will lead to fewer people starting to smoke and more people quitting, which will have a positive impact on the public health.

Reference Number: 40304

CARABALLO, R.S., AND K. ASMAN. "Epidemiology of Menthol Cigarette Use in the United States." *Tobacco Induced Diseases*, 9: (14 pp), 2011 (6 tables, 14 figures, 39 references)

Abstract: Between 2004 to 2008, about half of adolescent smokers (aged 12 to 17 years) in the United States reported smoking menthol cigarettes. About 70% of adolescent Black Americans smoked menthol cigarettes. There is no gender difference in adolescent smokers of menthol cigarettes. The three top brands of cigarettes smoked by adolescents are Marlboro, Newport, and Camel. Menthol cigarette brands are important for smoking initiation, the younger the age, the more likely menthol cigarettes will be smoked. About 30% of adult smokers use methanol cigarettes. Older adults are less likely to smoke menthol cigarettes than younger adults.

> There is evidence that menthol cigarettes are disproportionately smoked by U.S. adolescents, blacks/African Americans, adult females, those living in the Northeast of the United States, and those with lower family incomes than their counterparts. It is likely that other disparities in menthol cigarette use exist that we have not covered or have not been studied. Based on self-reports of menthol cigarette use, the use of menthol cigarettes among smokers have increased in recent years among adolescents aged 12-17 years, young adults aged 18-25 years, adults aged 26 years or older, white and Hispanic men.

Reference Number: 40305

VILLANTI, A.C., L.K. COLLINS, R.S. NIAURA, S.Y. GAGOSIAN, AND D.B. ABRAMS. "Menthol Cigarettes and the Public Health Standard: A Systematic Review." *BMC Public Health*, 17: 983 (13 pp), 2017 (1 figure, 111 references)

Abstract: A review was conducted of 82 studies on the effect of menthol cigarettes on public health. The US Tobacco Control Act banned all candy and fruit flavored cigarettes but did not include menthol in the ban. Menthol cigarette use among youth has not decreased in the past decade, despite decreases in non-menthol cigarette use.

> This review of the scientific evidence demonstrates that there is more than sufficient evidence to establish a positive relationship between menthol cigarettes and (1) increased youth smoking initiation, (2) increased nicotine dependence, and (3) decreased adult cessation. The weight of the evidence from studies published through 2017 supports that removal of menthol from cigarettes would, in the words of the Tobacco Control Act, decrease the likelihood that those who do not use tobacco products will start using such products and increase the likelihood that existing users of tobacco products will stop using such products.

Reference Number: 40306

MÍGUEZ-BURBANO, M.J., M. VARGAS, C. QUIROS, J.E. LEWIS, L. ESPINOZA, AND D. ASTHANA. "Menthol Cigarettes and the Cardiovascular Risks of People

Living with HIV." *Journal of the Association of Nurses AIDS Care*, 25: 427–435, 2014 (3 tables, 1 figure, 21 references)

Abstract: Various parameters were determined in 153 non-smokers (65% with HIV), 70 regular cigarette smokers (52% HIV), and 170 menthol cigarette smokers (57% HIV) in southern Florida between 2011 and 2013. More people living with HIV (PLWH) smoked menthol cigarettes (70%) compared to people living without HIV (PLWOH) (60%). Users of menthol cigarettes had a higher body mass index (BMI), had a higher rate of abdominal obesity, and were 2× as likely to have hypertension.

The dangers of tobacco products and the excessive rates of menthol use are disquieting when considering that PLWH age at a rapid pace and experience diseases associated with aging 10 to 20 years earlier than in their non-infected peers. We provide additional evidence regarding the risks of menthol cigarettes. Collectively, these findings further justify the need for more studies to closely monitor this public health threat.

While tobacco companies have portrayed menthol as causing innocuous cold sensations, our findings indicated that use of menthol-flavored cigarettes was associated with overall higher prevalence of hypertension. Increases in blood pressure readings were likely the result, at least in part, of excess nicotine levels found in menthol users. In addition to its direct effects on the sympathetic system, the identification of transient receptor potential cation channel subfamily M member 8 (TRPM8; also known as the cold and menthol receptor 1) as a regulator of Ca2+-permeable channel expressed in the vasculature has opened a new pathway to explain the vasopressor effects of menthols.

Reference Number: 40307

GUBNER, N.R., D.D. WILLIAMS, A. PAGANO, B.K. CAMPBELL, AND J. GUYDISH. "Menthol Cigarette Smoking Among Individuals in Treatment for Substance Use Disorder." *Addictive Behaviors*, 80: 135–141, 2018 (3 tables, 52 references)

Abstract: Approximately 30% of all cigarette smokers in the United States use menthol cigarettes, but 83% of Black Americans and 36% of women smoke menthol cigarettes. A study was conducted of 863 current smokers (mean age 37.7 years) enrolled in substance use disorder (SUD) programs from 24 treatment centres. Overall, 53% of all the smokers with SUD used menthol cigarettes. Menthol smokers were more likely to be female and

Black and less likely to have a college degree. Menthol smokers were 3.3× more likely to use cannabis rather than alcohol as their primary drug, even after adjusting for demographics. Regulation targeting the manufacturing, marketing, and sale of menthol cigarettes may benefit smokers with SUD.

> Menthol may intensify tobacco use and dependence both through its flavorant effect (reducing the harshness of cigarettes, and serving as a sensory cue) and through pharmacological interactions with nicotine. Specifically, menthol reduces some of the aversive taste and irritant effects of tobacco smoke through activation of transient receptor potential Ankyrin 1. In a cross-over study, smoking mentholated cigarettes was found to inhibit nicotine metabolism resulting in slower clearance of nicotine, an effect that was also observed using in vitro cell studies; and mice. In addition, studies found greater upregulation of nicotinic acetylcholine receptors (nAChRs) in the brains of menthol versus non-menthol smokers, which the authors attribute to higher nicotine exposure in the former group. However, in a preclinical mouse model, menthol alone, or in combination with nicotine, resulted in upregulation of nAChRs located on midbrain dopamine neurons, involved in reward-related behavior. These findings suggest that menthol, in addition to being a flavorant, may contribute to tobacco use and dependence.

Reference Number 40308

BRODY, A.L., A.G. MUKHIN, J. LA CHARITE, K. TA, J. FARAHI, C.A. SUGAR, M.S. MAMOUN, E. VELLIOS, M. ARCHIE, M. KOZMAN, J. PHUONG, F. ARLONIO, AND M.A. MANDELKERN. "Up-Regulation of Nicotinic Acetylcholine Receptors in Menthol Cigarette Smokers." *International Journal of Neuropsychopharmacology*, 16: 957–966, 2013 (2 tables, 2 figures, 57 references)

Abstract: Cigarette smoking and nicotine lead to the upregulation or large increase of nicotinic acetylcholine receptors (nAChRs) in the brain. Chronic smokers have more receptors than non-smokers, and former smokers have nAChRs densities similar to non-smokers. Approximately one-third of smokers use menthol cigarettes, which leads to elevated serum nicotine concentrations and greater addiction. A positron emission tomography (PET) brain scan was conducted on 22 menthol smokers, 41 non-menthol cigarette smokers, and 51 non-smokers. An exhaled CO test was conducted to verify smoking status (> 8ppm was an active smoker; < 4 ppm was a non-smoker).

Table: Mean Total Binding Volumes of Distribution (nAChR density) in Brain Regions of Non-Smokers, Non-Menthol Cigarette Smokers, and Menthol Cigarette Smokers

Brain Region	Non-Smokers Receptor Density	Non-Menthol Cigarette Smokers Receptor Density	Menthol Cigarette Smokers Receptor Density
Thalamus	16.0	15.7	17.3
Prefrontal Cortex	7.1	7.9	8.6
Brainstem	9.6	11.1	12.7
Cerebellum	8.2	9.4	10.8
Corpus Callosum	6.0	6.6	7.6

Source: Adapted from Brody et al, 2013

Study results demonstrate that menthol smokers have greater up-regulation of nAChRs than non-menthol smokers. This difference is presumably related to higher nicotine exposure in menthol smokers, although other mechanisms for menthol influencing receptor density are possible. These results provide additional information about the severity of menthol cigarette use and may help explain why these smokers have more trouble quitting in standard treatment programmes.

Reference Number 40309

ANTONIO, A.M.M., P. FAGAN, F.D. HAMAMURA, I.J.N. LAGUA, J. LIU, D.J. PARK, P. POKHREL, T.A. HERZOG, I. PAGANO, K. CASSEL, A. SY, D. JORGENSEN, T. LYNCH, C. KAWAMOTO, C.J. BOUSHEY, A. FRANKE, M.S. CLANTON, E.T. MOOLCHAN, AND L.A. ALEXANDER. "Menthol Cigarette Smoking and Obesity in Young Adult Daily Smokers in Hawaii." *Preventive Medicine Reports*, 2: 946–952, 2015 (3 tables, 1 figure, 54 references)

Abstract: The cooling, soothing, anesthetic, and analgesic effect of menthol are attractive to young smokers (18–34 years), as it makes smoking less harsh and more enjoyable. A study was conducted of 120 menthol and 55 non-menthol daily smokers (aged 18–35 years) in Hawaii. BMI was determined and a saliva sample was collected during a one-hour laboratory visit to the Hawaii Cancer Center. Salivary cotinine and 3-HC were measured by HPLC/MS/MS and the nicotine metabolite ratio (NMR) were determined. Over one-third of all daily smokers (40.8% of menthol and 9.1% of non-menthol smokers) were obese. The NMR was significantly lower for overweight menthol daily smokers compared to non-menthol smokers.

This study suggests that the relationship between menthol smoking and obesity, the nation's two leading causes of chronic disease, be further examined. By 2050, menthol smoking is estimated to result in over 300,000 cumulative excess deaths. Future cross-sectional, longitudinal and mechanistic studies are needed to determine how menthol smoking and obesity are related and whether or not these two leading chronic disease risk factors lead to excess disease risk in menthol compared with non-menthol smokers.

Reference Number 40310

LEE, Y.O., AND S.A. GLANTZ. "Menthol: Putting the Pieces Together." *Tobacco Control*, 20: ii1–ii7, 2011 (67 references)

Abstract: On 22 June 2009, President Barack Obama signed the Family Smoking Prevention and Tobacco Control Act, giving the FDA the authority to regulate cigarettes, including banning candy, fruit, and spice flavors but not menthol. The first menthol cigarette was created by Lloyd Spud Hughes who, while suffering from a cold, stored mentholated crystals with his tobacco and discovered the next day that he had created mentholated cigarettes. Today, menthol is added to 90% of cigarettes sold in the United States, whether or not they are labelled mentholated. Internal tobacco industry documents were examined as to what they knew about menthol. Menthol activates and increases smoking, which causes death and disease.

The results of both internal tobacco company research and studies done in the independent scientific literature are remarkably consistent in showing that menthol influences much more than the taste of cigarette smoke. In addition to its candy-like appeal, menthol's sensory effects and ability to mask irritation make menthol cigarettes appealing as a likely starter product. Youths disproportionately smoke menthols, in part because of these sensory characteristics and targeted marketing. Menthol marketing has also been targeted at groups (youths, women and African Americans) that have disproportionate percentages of menthol smokers. Menthol's sensory qualities may also make low-tar cigarettes more palatable and make it easier for inexperienced smokers to progress into increased exposure and, thus, reinforce dependence on nicotine.

Menthol's use for promoting cigarette smoking goes beyond its use as a flavouring, highlighted by the point that tobacco companies knew as early as 1972 that menthol enhances the perception of smoothness even at unidentifiable levels and in non-menthol cigarettes. The ubiquitous addition of menthol by tobacco companies to over 90% of all tobacco products, whether labelled "menthol" or not, demonstrates that menthol is not simply a flavour or brand. Menthol imparts sensory characteristics to cigarettes and has a complex interaction with nicotine that affects smoking behaviour whether it is perceived or not, or whether cigarettes containing menthol are marketed as "menthol" or not.

Reference Number: 40311

HISCOCK, R., K. SILVER, M. ZATOŃSKI, AND A.B. GILMORE. "Tobacco Industry Tactics to Circumvent and Undermine the Menthol Cigarette Ban in the UK." *Tobacco Control*, 29: e38–e42, 2020 (3 figures, 56 references)

Abstract: Menthol is a type of alcohol ($C_{10}H_{20}O$) originally derived from the mint plant but which now can be manufactured synthetically. It has local anaesthetic properties that will increase overall cigarette intake and mask symptoms of early respiratory diseases, which reduce the chances of quitting. Menthol increases smoking rates among youth, Black Americans, and women, and should be banned as an additive to all tobacco products. The latest innovation for adding menthol or other flavors is a small capsule or crush ball which is inserted into the cigarette filter. The flavor is released by crushing the capsule before smoking. A UK (and EU) ban on menthol includes cigarettes and RYO tobacco, but cigars, cigarillos, and pipe tobacco are exempt. Menthol accessories that are not sold in the same cigarette packaging are allowed. A leaked Philip Morris International (PMI) memo revealed that it opposes flavor bans.

Table: Percent of Market Share of Menthol and Capsule Cigarettes in the UK and EU

Countries	2005	2012	2015	2018
UK	7%	10%	16%	21%
EU	4%	7%	9%	5%

Source: Adapted from Hiscock et al, 2020

The tobacco industry has introduced diverse new products to the UK that can get around the ban: menthol accessories sold separately from cigarette packs, and RYO pouches and cigarillos with a capsule. It has also used the ban to promote more profitable heated tobacco products, which unlike quitting are not risk-free. Despite claims of a commitment to reducing harm, this provides yet further evidence that the tobacco industry will circumvent and undermine measures with genuine potential to reduce harm.

Cigarillos are likely to be particularly attractive to the tobacco industry because they are exempt from EU and UK standardised pack legislation (with the exception of large pictorial warnings), are subject to lower taxes, and their definition is such that they can be sold looking similar to conventional cigarettes except for a brown leaf rather than white stick covering. Thus, the tobacco industry may use cigarillos to reintroduce cigarette branding and cheap products which had been removed by standardised packaging.

The EU/UK menthol ban was associated with preceding falls in menthol sales in the EU, but in the UK sales grew up to implementation. This suggests not only that a long implementation period was not needed but that it was misused: menthol sales, which may be particularly harmful, were driven up in the UK. The yearlong standardised packaging implementation period in the UK was similarly exploited by the tobacco industry.

Reference Number: 40312

YANG, Y.T., AND S. GLANTZ. "San Francisco Voters End the Sale of Flavored Tobacco Products Despite Strong Industry Opposition." *Annals of Internal of Internal Medicine*, 169: 708–709, 2018 (10 references)

Abstract: RJR Tobacco manufactures Newport, the best-selling menthol cigarette brand and the second most popular cigarette brand among smokers. The tobacco industry had targeted Black Americans, resulting in 85% of Black smokers using menthol cigarettes compared to 29% of White smokers. In 2013, 81% of youth (aged 12–17 years) initiated tobacco use with a flavored product. RJR launched a $12 million campaign against a ban on flavored tobacco products in San Francisco, but lost.

Despite challenges from tobacco companies, states and localities have clear authority to restrict the sale of flavored tobacco products to reduce tobacco use and its harms to citizens. At least 2 states and more than 100 localities have passed restrictions on the sale of flavored tobacco products, although laws differ in their applicability to specific products and store types. However, Proposition E represents a major step forward because it is the nation's first comprehensive ban on the sale of flavored tobacco, and it has already inspired others to follow.

San Francisco voters showed that with appropriate support for public health campaigns, it is possible to overcome more heavily financed efforts by Big Tobacco if there is enough funding to spread the message. RJR's attempt to roll back the law illustrates how critical flavors are to the tobacco industry's efforts to hook children and keep adults smoking, a lesson that public health advocates should take to heart. In the absence of FDA action, other cities should prioritize passing similar legislation in order to safeguard against addiction and preventable death, especially among vulnerable and minority populations.

4.04 OTHER ADDITIVES/FLAVORS/TOXINS

"RJ Reynolds' addition of beta-carotene to cigarettes suggests that adverse health effects can occur even when a seemingly benign additive is used and points to the need for regulation by the Food and Drug Administration."

—Rabinoff et al, "Pharmacological and Chemical Effects of Cigarette Additives" (2007)

"All cigarettes cause cancer, lung cancer, heart attacks and premature death—lights, low tar, ultra-lights and naturals. There is no safe cigarette."

—Court-ordered Corrective Statements, *US v Philip Morris USA Inc*, RICO decision (2006)

Tobacco is a complex mixture of 6–15% cellulose, 10–15% pectin, 2% lignin, and variable amounts of sugars, mainly glucose, fructose, and sucrose (40401). Additional sugars, cocoa, and licorice extract are also added to the tobacco, as smokers prefer these types of cigarettes; these compounds when consumed orally are relatively harmless, but when burned and inhaled, they produce harmful chemicals such as formaldehyde, acetone, and acrolein (40402–3, 40406).

Cigarettes have been found to contain bacteria, mold, and fungi, which contribute to lung diseases when smoked (40404).

Tobacco also contains polonium 210 (PO-210), the same radioactive poison that was used by the KGB to kill Alexander V. Litvinenko. The majority of PO-210 comes from the high phosphate fertilizer that is repeatedly applied to the soil and absorbed and concentrated in the leaves by the tobacco plant. Polonium 210 emits alpha radiation, and it is estimated that smokers of 1.5 packs of cigarettes a day are exposed to as much radiation as 300 chest X-rays (40405).

Cigarette companies have also added 599 additives to their products (40407). Some compounds such as magnesium oxide and aluminum sulfate are used to mask the odor and visibility of secondhand smoke (40407–8). In 1989, RJR introduced the first cigarette that smells good, New Chelsea, that was mainly aimed at women smokers (40408). Other additives that allow for a smoother, less irritating smoke may mask symptoms of smoking-related illnesses (40407).

Although tobacco companies have insisted additives were not harmful, cigarettes with additives have been found to increase tar and particulate matter, which are harmful to the lung tissue (40406, 40408). Natural cigarettes such as American Spirit and roll your own are as harmful as commercial cigarettes (40409–11).

Methods of curing tobacco, such as direct heat, can increase toxic and carcinogenic tobacco-specific nitrosamines (TSNAs) (40412).

The majority of smokers were susceptible to brand switching if they learned that their cigarettes had a lot more chemicals such as lead, formaldehyde, arsenic, and ammonia but not nicotine (40413). Smokers tend to have positive images about their brand but negative images about tobacco companies in general, referring to them as "killers, liars, greedy, cancer, addiction and big money" (40413).

Reference Number: 40401

BANOŽIĆ, M., S. JOKIĆ, Đ. AČKAR, M. BLAŽIĆ, AND D. ŠUBARIĆ. "Carbohydrates–Key Players in Tobacco Aroma Formation and Quality Determination." *Molecules*, 25: 1734 (13 pp), 2020 (1 table, 3 figures, 66 references)

Abstract: Tobacco is the most widely cultivated non-food crop in the world. Tobacco is a complex mixture that consists of 6–15% cellulose, 10–15% pectin, 2% lignin, and variable amounts of sugars. The most abundant naturally present sugars in the tobacco leaves are glucose, fructose, and sucrose. The sugar content of tobacco depends on the variety, the harvesting, and especially the curing conditions. Flue and sun-dried tobacco result in higher sugar concentrations than tobacco air dried at lower temperatures. Tobacco producers add sugars to tobacco, as high sugar tobaccos (8–30%) are usually preferred by smokers. High contents of carbohydrates in the tobacco leaves, however, can lead to the formation of acetone, acrolein, 2-funfural, acetaldehyde, and formaldehyde in smoke. The increase of toxic compounds in high sugar tobaccos can be 150% compared to low sugar tobaccos. Besides the simple sugars, other carbohydrates found in tobacco include oligosaccharides, cellulose, starch, and pectin.

Table: Percentage of Various Chemicals in Burley, Virginia, and Oriental Tobaccos

Chemical	Burley Tobacco	Virginia Tobacco	Oriental Tobacco
Nicotine (%)	2.07	2.26	1.18
Sugars	1.2	11.1	10.7
Proteins	10.1	7.8	7.8
Chlorine	0.51	0.53	0.84
Crude Ashes	18.4	12.8	16.4

Source: Adapted from Banožić et al, 2020

Degradation of polysaccharides results in a higher yield of simple sugars, but at the same time reduces sugars oxidization and transfer into carbon dioxide and water. Loss of sugar producers will compensate with added sugars, to cover undesirable aroma properties and achieve a better, pleasant taste during smoking. However, tobacco carbohydrates can be precursors for many harmful compounds, including formaldehyde and 5-hydroxymethylfurfural. Keeping in mind that added sugars in tobacco production are unavoidable, it is important to understand all changes in carbohydrates from harvesting to consuming in order to achieve better product properties and avoid the formation of harmful compounds.

Reference Number: 40402

SEIDENBERG, A.B., C.L. JO, AND K.M. RIBISL. "Knowledge and Awareness of Added Sugar in Cigarettes." *Nicotine and Tobacco Research*, 21: 1689–1694, 2019 (3 tables, 31 references)

Abstract: Sugars occur naturally on the tobacco leaf but are also added to cigarettes by the tobacco industry. Flue-cured cigarettes have a higher amount of endogenous sugars, as the higher temperatures of the process prevent fermentation or degradation of the sugars. Up to 4% of the cigarette's weight can be added sugar. Burning sugar generates organic acids, which lower the pH of smoke and make it more palatable. Sugars can also mask the bitter taste of tobacco. Combusting sugars in cigarettes also produce toxins such as acetaldehyde (which enhances the addictiveness of nicotine), formaldehyde, acrolein, and 2-furfural. A survey was conducted of 2,041 male and 2,343 female smokers as to their knowledge of the addition of sugars to cigarettes. Only 5.5% of the smokers knew that sugar is added to the cigarettes and only 3.8% knew that added sugar increases the toxins in cigarette smoke.

> Given the low knowledge of added sugar in cigarettes by smokers and the expressed desire to learn more about added sugar by several participants, FDA and other tobacco control organizations that educate the public on tobacco product risks should consider including information about added sugar in media campaigns. Because of the public's concern about added sugar in processed foods and beverages and the effects of sugars on cigarette smoke, educating the public about the presence of added sugar in cigarettes could be an effective new angle for campaigns to discourage smoking.

Reference Number: 40403

SOKOL, N.A., R.D. KENNEDY, AND G.N. CONNOLLY. "The Role of Cocoa as a Cigarette Additive: Opportunities for Product Regulation." *Nicotine and Tobacco Research*, 16: 984–991, 2014 (1 table, 73 references)

Abstract: US cigarettes consist of Virginia, Burley, Oriental, and reconstituted tobacco (or sheets). Additives can constitute up to 10% of the cigarette weight. Cocoa is one of the oldest additives to cigarettes dating back to at least 1932. Cocoa contains sugar and humectants and is generally applied to air-cured tobacco, as it contains less sugar and is more absorbent than flue-dried tobacco. Cocoa and/or chocolate is widely used in cigarettes (approximately 80% of the brands) at between 0.5-1.5% of the tobacco weight. Cocoa is often used in light cigarettes, as the lowered Tar/Nicotine ratio makes the smoke harsher and more bitter.

> Cocoa has been widely used in American cigarettes. At least two tobacco companies continue to use cocoa as a cigarette additive at levels similar to those recorded in their internal documents. Documents revealed that cocoa has been used in cigarettes for its influence on sensory qualities. The tobacco industry has used cocoa to impact gustatory and olfactory perceptions by enhancing and sweetening tobacco flavor and aroma, and tactile perceptions by reducing smoke irritation. Documents showed the importance of cocoa in improving light cigarettes' acceptability by enhancing smoke flavor lost through ventilation, and smoothing smoke made harsh through lowered T/N ratios. Results from hedonic research indicated

that cigarettes with higher levels of cocoa in their casing are more appealing and acceptable to smokers; however, studies also showed that consumers and researchers felt that when present at a level that would create a recognizable gustatory chocolate/cocoa flavor, the cigarette was no longer appealing for regular use. Although one document suggested that cocoa be investigated for use for physiological effects in future products, documents showed no indication that this project was carried out or that cocoa has any physiological effects at levels used in cigarettes.

Reference Number: 40404

PAULY, J.L., AND G. PASZKIEWICZ. "Cigarette Smoke, Bacteria, Mold, Microbial Toxins and Chronic Lung Inflammation." *Journal of Oncology*, 2011: Article 819129 (13 pp), 2011 (1 table, 1 figure, 119 references)

Abstract: Cigarette smoke induces chronic inflammation of the lung and this inflammation is associated with malignant transformation and tumor cell proliferation. Smoking also impairs immunity (both innate and adaptive). Tobacco-specific nitrosamines arise primarily from the microbial degradation of nicotine in tobacco. Toxic fungi, bacillus spores, bacteria endotoxins, and other microorganisms had been identified in tobacco products.

The results of this literature review have documented that the tobacco microflora has been the subject of many studies by investigators of tobacco industry and academic communities. During the last 50 years, there has been an imbalance, however, in the attention devoted to addressing the identification and propensity of the harm of tobacco- and tobacco-smoke-associated chemicals and in the attention devoted to characterizing microbes and microbial-derived factors.

Ample information has accumulated to suggest that microbes and microbial-derived factors may contribute to the health risks of smoking and smokeless tobacco products. Moreover, the microbes may facilitate microbial colonization of the mouth and airway, the induction of chronic inflammation through the activation of diverse leukocyte subsets, alteration of the tissue microenvironment, and microbial-toxin-induced pathologies. The current health concerns recently expressed by investigators of various disciplines, and with different research interests, in peer-reviewed published research articles are reasonable and validate that additional investigation of the microbiology of tobacco is warranted.

Reference Number: 40405

MUGGLI, M.E., J.O. EBBERT, C. ROBERTSON, AND R.D. HURT. "Waking a Sleeping Giant: The Tobacco Industry's Response to the Polonium-210 Issue." *American Journal of Public Health*, 98: 1643–1650, 2008 (98 references)

Abstract: It has been known since 1964 that tobacco smoke contains polonium 210 (PO-210). In 2006, former KGB agent Alexander V. Litvinenko was poisoned using PO-210. Polonium 210 emits alpha radiation, and it is estimated that smokers of 1.5 packs of cigarettes a day are exposed to as much radiation as 300 chest X-rays. The majority of PO-201 in tobacco plants comes from the high phosphate fertilizer repeatedly applied to the tobacco crop. PO-210 is absorbed from the soil to the leaves via the roots and airborne PO-210 particles stick to the plant trichomes. The tobacco industry initially tried to remove PO-210 by washing tobacco leaves, genetic engineering, and other approaches, all of which failed. This has been revealed in internal British American Tobacco documents released during litigation.

The tobacco industry was clearly concerned about the effect that a widespread debate about the presence of PO-210 in cigarette smoke would have on public perception. In contrast to the general approach of "creating doubt about the health charges without having to deny it," the industry's strategy for handling the PO-210 controversy was to a take a vow of silence and avoid "waking a sleeping giant."

Failed efforts to remove PO-210 from tobacco smoke convinced industry officials that it would never be technologically feasible on a production level. Although some tobacco company researchers genuinely believed that the amount of PO-210 in cigarette smoke was too small to be a major risk factor for lung cancer induction, this opinion was not shared by company lawyers. In their quest to decrease the companies' legal exposure, the lawyers advocated suppressing data that contested published reports on the amount of PO-210 in cigarette smoke, despite the favorable effect publication of such data might have had on the tobacco industry. The internal debate carried on for the better part of a decade, involved most cigarette manufacturers, and pitted tobacco researchers against tobacco lawyers. The lawyers prevailed.

The tobacco industry remains silent on the PO-210 issue, suggesting that it continues to fear public reaction. We found no mention of radioactive particles in tobacco and tobacco smoke on any of the major transnational tobacco companies' current Web sites that present smoking and health information aimed at consumers. We support a proposed new warning label on cigarettes: "Surgeon General's Warning: Cigarettes are a Major Source of Radiation Exposure." This wording would capitalize on public concern over radiation exposure and increase the impact of cigarette warning labels. Health messages highlighting PO-210 as one of the known carcinogens in cigarettes could aid tobacco control efforts.

Reference Number: 40406

BRAUN, M., E-L. FROMM, A. GERBER, D. KLINGELHÖFER, R. MÜLLER, AND D.A. GRONEBERG. "Particulate Matter Emissions of Four Types of One Cigarette Brand with and Without Additives: A Laser Spectrometric Particulate Matter Analysis of Secondhand Smoke." *BMJ Open*, 9: e024400 (7 pp), 2019 (3 tables, 2 figures, 61 references)

Abstract: Particulate matter (PM) causes major adverse health effects such as airway inflammation and cardiovascular and respiratory diseases. It is estimated that secondhand smoke causes 890,000 deaths globally per year. PM from Lucky Strike Red (original and straight), Lucky Strike Blue (original and straight), and a reference cigarette (3R4F) were measured with a laser aerosol spectrometer in an automatic environmental tobacco smoke emitter. The straight form of Lucky Strike were cigarettes without additives. The lowest mean PM_1 was 1762 $\mu g/m^3$ for 3R4F, increasing to 2145 $\mu g/m^3$ for straight blue, 2402 $\mu g/m^3$ for original red, 2530 $\mu g/m^3$ for straight red, and 2865$\mu g/m^3$ for the original blue cigarette brands.

Table: Chemical Constituents and Additives of Various Lucky Strike Cigarettes

Ingredients	3R4F (standard)	Original Red (additives)	Original Blue (additives)	Straight Red (no additives)	Straight Blue (no additives)
Tar (mg)	9.4	10	7	10	7
Nicotine (mg)	0.73	0.8	0.6	0.9	0.6
Glycerin (humectant)	Yes	Yes	Yes	No	No
Sugar	Yes	Yes	Yes	No	No
Cocoa powder	Nil	Yes	Yes	No	No
Licorice extract	Nil	Yes	Yes	No	No
Flavors (<0.1% w/w)	Nil	Yes	Yes	No	No

Source: Adapted from Braun et al, 2019

The findings of this study show the massive increase of PM amount by smoking cigarettes in enclosed spaces and suggest that additives in tobacco products increase the PM amount in SHS.

Reference Number: 40407

RABINOFF, M., N. CASKEY, A. RISSLING, AND C. PARK. "Pharmacological and Chemical Effects of Cigarette Additives." *American Journal of Public Health*, 97: 1981–1991, 2007 (1 table, 1 figure, 117 references)

Abstract: More than 100 of the 599 documented cigarette additives have potential pharmacological actions and are used to camouflage the odor of SHS, enhance nicotine delivery and addictiveness, and mask symptoms (such as coughing) so the smoker will continue smoking.

The tobacco industry has stated that additives are used primarily for flavoring and "smoothing" the smoker's experience. However, a review of botanical medicine sources indicates that many botanical and phyto-chemical additives have other properties, including anesthetic, antibacterial, anticancer, anti-inflammatory, antifungal, and antiviral properties. Industry documents show awareness of and interest in these additional properties.

Unregulated botanical and chemical additives might have "multiple use" purposes, such as enhancing flavor and providing for a "smoother" smoking experience as well as preventing or masking symptoms associated with illnesses induced by smoking. Because inclusion of botanical and chemical additives could reduce, mask, or prevent smokers' awareness of the adverse symptoms caused by smoking (e.g., cough), smokers might continue to smoke even when they are ill, preventing reductions in cigarette consumption and sales revenues.

RJ Reynolds's addition of beta-carotene to cigarettes suggests that adverse health effects can occur even when a seemingly benign additive is used and points to the need for regulation by the Food and Drug Administration. Although the actions of beta-carotene and other additives may have decreased the carcinogenicity of cigarettes, their use may have unintentionally increased the risk for and rate of lung cancer in smokers.

A 1994 study concluded that there was no reduction in the incidence of lung cancer among male smokers after 5 to 8 years of oral supplementation with alpha-tocopherol or beta-carotene. That study raised the possibility that oral beta-carotene supplements might actually have harmful effects in smokers and might increase lung cancer rates. A newer study also has documented the possible adverse effects of oral beta-carotene on lung cancer. This is an example of the potential occurrence of unwanted and unanticipated dangerous effects if appropriate regulatory agencies do not monitor the use of additives.

Reference Number: 40408

CONNOLLY, G.N., G.D. WAYNE, D. LYMPERIS, AND M.C. DOHERTY. "How Cigarette Additives Are Used to Mask Environmental Tobacco Smoke." *Tobacco Control*, 9: 283–291, 2000 (4 tables, 1 figure, 67 references)

Abstract: In 1984, the US federal government required cigarette manufacturers to disclose annually a list of additives placed in the cigarette tobacco (rods). However, the companies are not required to list the chemicals that are added to the wrapper paper or filter. Internal documents from the major US tobacco companies were searched to determine their research and projects into additives to reduce the visibility, odor, irritation, and emission of secondhand smoke. Magnesium oxide and calcium carbonate fillers and wrappers reduce the visibility of sidestream smoke. Aluminum sulfate was found to reduce SHS irritation. Other chemicals would change the odor of cigarette smoke. In 1989, RJ Reynolds introduced the first cigarette (New Chelsea), that smelled good, which was marketed to women smokers. The ad for this cigarette featured a scratch 'n' sniff strip. With the addition of film forming agents such as ammonium alginate or potassium citrate, the permeability of the paper is reduced, which prevents sidestream smoke from escaping during cigarette smoldering.

This review of the tobacco industry documents shows how manufacturers were concerned about the lack of "social acceptability" of their cigarettes caused by negative perceptions of secondhand smoke among both smokers and non-smokers. Cigarette additives were developed along with other product design changes to reduce or mask the aroma, visibility, and irritation of sidestream smoke, and less frequently, to lower actual smoke emissions. Many of these additives were incorporated into special "brand projects" for testing, some of which later entered the marketplace, such as Chelsea/Horizon and Virginia Super Slims. Very little research was found in this review regarding the effect of these additives on the toxicity of mainstream and sidestream smoke or the levels of individual smoke constituents. In addition, there was little evidence to suggest that tobacco manufacturers sought to fully inform consumers of the product changes they were designing and implementing. These are important considerations, since the use of sidestream altering additives could increase non-smokers' involuntary exposure to ETS by reducing the normal warning signs of exposure to smoke toxins. Another major concern is that women of childbearing age are a prime target group for these brand modifications. Thus, women smokers with young children may increase their child's risk of developing respiratory diseases through increased ETS exposure if they themselves are not bothered by the smoke.

Reference Number: 40409

IFFLAND, Y., R. MÜLLER, D. GRONEBERG, AND A. GERBER. "High Particulate Matter Emission from Additive-Free Natural American Spirit Cigarettes." *Springer Plus*, 5: 1958 (7 pp), 2016 (2 tables, 3 figures, 29 references)

Abstract: Combined Mainstream and Sidestream Smoke (CMSS) or secondhand smoke is one of the major sources of indoor pollution. It consists of particulate matter as solid or liquid droplets floating around in the air that can be inhaled into the lungs. The smaller the diameter of the PM, the deeper it can be inhaled into the lungs. PM_{10} is coarse particles with a diameter of 2.5–10.0 μm. $PM_{2.5}$ are fine particles with a diameter of 2.5 μm or less and PM_1 are ultrafine particles with a diameter of less than 1 μm. An automatic environmental tobacco smoke emitter was used to smoke 20 cigarettes each of Natural American Spirit cigarettes (orange, blue, green, and yellow brands) and a reference cigarette 3R4F. The PM was measured with a laser aerosol spectrometer. CMSS is not predicted by CO, tar, or nicotine content of the cigarette but by the origin of the tobacco and design features of the cigarette.

Table: Mean Particulate Matter Concentration (μg/m3) in the Reference Cigarette (3R4F) and Various Natural American Spirit Cigarette Brands

Cigarette Brand	Ultra fine particulate matter (PM_1)	Fine particulate matter ($PM_{2.5}$)
3R4F	33.47	1.11
Orange	97.66	8.04
Blue	73.55	5.90
Green	51.52	2.79
Yellow	53.18	2.70

Source: Adapted from Iffland et al, 2016

Besides various additives and the well known toxic and carcinogenic compounds in tobacco products, CMSS-associated PM constitutes an independent health hazard. The smaller the particles are, passive smokers are exposed to, the more deeply they can be inhaled and even penetrate into the blood stream. Depending on their surface and solubility, particles absorb gases and vapors (e.g. carcinogenic benzopyrenes), permitting their transport in distal lung areas and even to the gas exchanging regions of the lung. Thus, even if we are not able to conduct a toxicological and chemical analysis of SS with our technical possibilities, it can be assumed that significant differences in PM emissions matter. The release of PM by Natural American Spirit cigarettes was shown to be significantly (2–3 times) higher than that of 3R4F reference cigarettes, when smoked in the same standardized way, using our protocol.

Even if Santa Fe foregoes using additives during the production process of their cigarettes, their tobacco products still evoke high PM emission in SS, which reveals a health hazard to passive smokers. We want to illuminate the impact of these differences on the burden of PM-exposure on passive smokers.

Reference Number: 40410

EPPERSON, A.E., T.O. JOHNSTON, N.C. SCHLEICHER, AND L. HENRIKSEN. "The Price of Natural American Spirit Relative to Other Cigarette Brands." *Nicotine and Tobacco Research*, 21: 1715–1720, 2019 (2 tables, 31 references)

Abstract: More than 2.5 million US smokers believe that their cigarettes are less harmful than other brands. American Spirit (branded as organic, without additives) smokers are more than 20× more likely to believe this falsehood than smokers of other brands. The price of American Spirit and other cigarettes were determined in 1,277 tobacco retailers in California in 2017. American Spirit was sold in 75% of the stores at a price 12–34% higher than other cigarettes, but also cost significantly less in tobacco retailers in neighborhoods with a higher proportion of school-age children.

It is important for future research to consider how the ultra-premium price of American Spirit may contribute to misperceptions that the brand is safer than other cigarettes. American Spirit and Pall Mall are the only two brands that increased market share between 2009 and 2013, even as cigarette smoking was declining in the United States. Examining relative price and variation by neighborhood demography may shed further light on brand popularity among population subgroups. Cheaper prices for American Spirit and other cigarette brands in pharmacies compared to other store types reinforces state and local tobacco control objectives to establish tobacco-free pharmacies, as well as for advocacy efforts to divert prescriptions to tobacco-free pharmacies. Even more concerning are lower prices of American Spirit and other cigarettes in neighborhoods with more youth, as evidence shows that lower prices encourage smoking use among this price-sensitive population and among racial/ethnic minorities. With previous research indicating that younger consumers perceive American Spirit to be less harmful than other cigarettes, the current study suggests a need for stronger regulation with respect to tobacco sales in stores near schools and/or colleges. The ultra-premium prices may exacerbate misperceptions that the brand is healthier and organic. Establishing a minimum price on cigarettes is recommended as a policy intervention to regulate the tobacco industry's use of price strategies as a marketing tool, which is relevant to American Spirit and other cigarette brands.

Reference Number: 40411

KOSZOWSKI, B., Z.R. ROSENBERRY, L.C. VIRAY, J.L. POTTS, AND W.B. PICKWORTH. "Make Your Own Cigarettes: Toxicant Exposure, Smoking Topography, and Subjective Effects." *Cancer Epidemiology, Biomarkers and Prevention*, 23: 1793–1803, 2014 (5 tables, 1 figure, 59 references)

Abstract: Due to the increases in the price and restrictions on the sale of conventional cigarettes, and the perceptions that make your own (MYO) cigarettes are safer, the use of MYO has increased significantly. MYO consists of 2 types; roll your own (RYO) or personal machine made (PMM). A study was conducted on 77 daily smokers who smoked RYO, PMM, or factory made (FM) cigarettes, under three conditions: after confirmed overnight abstinence, in an intense smoking procedure, or ad libitum. Blood samples were collected before and 2 minutes after smoking began and plasma nicotine concentrations were determined by GC. Exhaled CO was determined before and within 10 minutes after smoking ended. Puff topography or smoking behavior was observed. There were no significant differences in puff topography between the three different types of cigarettes, except that the puffs per cigarettes and the time to smoke were significantly lower for RYOs. Regardless of the cigarettes, the smokers inhaled the first 3 puffs more vigorously than the last 3 puffs. MYO cigarettes deliver toxicants similar to FM cigarettes and contradict smoker's perceptions that MYO cigarettes are safer.

Table: Increase in Plasma Nicotine Concentration (ng/mL) in 3 Types of Cigarettes After 3 Different Smoking Conditions

Cigarette Type	Smoking Freely	After Overnight Abstinence	Intensive Smoking
Roll your own	14.3	19.4	32.0
Personal machine made	19.8	17.6	36.0
Factory made	18.8	23.5	33.0

Source: Adapted from Koszowski et al, 2014

The study demonstrated that despite differences in their size, tobacco weight and design, RYO, PMM and FM cigarettes delivered similar amounts of nicotine, CO and presumably other toxicants. Toxicant delivery and smoking patterns were similar across a wide range of experimentally manipulated conditions of tobacco craving. Participants of the present study were exclusive RYO, PMM or FM smokers; puffing profiles, toxicant exposure and subjective responses may differ among people who do not exclusively use these products.

Reference Number: 40412

COLLISHAW, N. "Blowing Smoke: The History of Tobacco-Specific Nitrosamines in Canadian Tobacco." *Tobacco Control*, 26: 365–370, 2017 (1 table, 3 figures, 29 references)

Abstract: Recent documents released during a class-action trial of tobacco companies in Montreal showed a change made in the 1960–1970s from indirect heating to direct heating for flue-cured tobacco. Indirect heating had the flue from a wood or gas furnace snaked around the tobacco barn before being vented outside. Direct heating in the 1960–1970s removed the flue and had the tobacco exposed directly to smoke, which increased tobacco-specific nitrosamines (TSNAs). A downdraft kiln also pulled fumes down through the tobacco leaves. In 2000, indirect heating was reinstated, involving use of a heat exchanger. In 2000 and 2001, Canadian cigarettes (100% flue cured) had lower levels of TSNAs than US cigarettes (flue-cured and sun-dried burley tobacco) but higher levels of carcinogens. TSNAs are known inducers of adenocarcinomas.

> It is important to note that, even though the overall rates of death from lung cancer have decreased, due largely to widespread smoking cessation, the risk of death to continuing smokers actually increased, and this increase has been attributed largely to changes in cigarette design and possibly cigarette curing methods that have occurred in both countries.

Reference Number: 40413

MENDEL, J.R., S.A. BAIG, M.G. HALL, M. JEONG, M.J. BYRON, J.C. MORGAN, S.M. NOAR, K.M. RIBISL, AND N.T. BREWER. "Brand Switching and Toxic Chemicals in Cigarette Smoke: A National Study." *Public Library of Science One*, 13: (10 pp), 2018 (4 tables, 34 references)

Abstract: A survey was conducted of 1,151 US smokers (mean age 43 years) by telephone and 1,561 US smokers (mean age 35 years) online as to cigarette brand smoking. The majority (>60%) smoked every day and about 70% had ever used an EC. In the phone survey, about 43% reported to have switched cigarette brands or styles in the past to reduce health risks. The majority of smokers were susceptible to brand switching if they learned that their cigarettes had a lot more chemicals such as lead, formaldehyde, arsenic, and ammonia but not nicotine. Smokers who were female, over 25 years of age, and current light cigarette users were more likely to have switched brands to reduce health risks.

Table: The Odds Ratio (OR) of Switching to a Brand with Less of the Listed Chemicals

Chemical in Current Brand of Cigarette	OR of Switching to a brand with less chemical
Nicotine	1.00×
Carbon Monoxide	2.96×
Lead	3.77×
Formaldehyde	3.00×
Arsenic	4.61×
Ammonia	2.81×

Source: Adapted from Mendel et al, 2018

Brand switching due to inaccurate perceptions of comparative risk may lead to less quitting and worse public health outcomes. Our study shows that many smokers have previously switched cigarette brands with the goal of reducing health risks. The majority of smokers would switch brands in the future if they learned their cigarettes had more of a harmful chemical than other brands. Communications about the harmful chemicals in cigarettes should work to reduce previous misperceptions and prevent new ones.

Reference Number: 40414

JOHNSON, S.E., B.N. COLEMAN, AND C.L. SCHMIDTT. "It's Complicated: Examining Smokers' Relationships with Their Cigarette Brands." *Psychology of Addictive Behaviors*, 30: 887–894, 2016 (3 tables, 46 references)

Abstract: Focus groups were conducted on 64 female and 79 male smokers regarding cigarette smoking and the brands they smoked. In general, smokers' attitude towards their particular brand was positive, whereas their attitudes towards the tobacco companies were largely negative. The strong brand loyalty may be disrupted by requiring plain packaging for cigarettes as the cigarette pack is the primary venue for brand interaction.

Table: Top Word/Phrases Used by Smokers to Describe Tobacco Companies

Top Word/Phrases Used to Describe Tobacco Companies
• Rich/filthy rich
• Money maker/making money/has money/a lot of income
• Greedy
• Money
• Big business/huge
• Killers/killing people/killing people over time to get rich
• Forever/been around forever/historic/long history in this country
• Addiction/addictive products
• Big money
• Cancer
• Liar/fact bender/lies about the danger of smoking

Source: Adapted from Johnson et al, 2016

In sum, these findings reveal that smokers' attitudes toward their cigarette brands are largely positive. For many smokers, thinking about their cigarette brand automatically conjured thoughts about smoking and their enjoyment of smoking, indicating that the satisfaction and withdrawal relief they have come to depend on from smoking is highly associated with their brand of cigarettes. On the other hand, when prompted to think about the tobacco industry, the tone shifted and participants expressed strongly negative attitudes. These contradictory attitudes suggest a paradox for smokers: simultaneously embracing their cigarettes and rejecting the industry that makes them. Taken together, they suggest smokers maintain largely positive brand relationships—and perhaps they do so by diverting their negative feelings about smoking toward the tobacco industry.

4.05 HEALTH RISKS

"Yes, smoking causes lung cancer, but not in people who sue us."

—Milberger et al, "Tobacco Manufacturers' Defence Against Plaintiffs' Claim of Cancer Causation: Throwing Mud at the Wall and Hoping Some of it Will Stick" (2006)

"The scientific evidence is incontrovertible: inhaling tobacco smoke, particularly from cigarettes, is deadly. Since the first Surgeon General's Report in 1964, evidence has linked smoking to diseases of nearly all organs of the body."

—Office of the Surgeon General, "Health Consequences of Smoking, Surgeon General Fact Sheet" (2014)

In 1910, the surgeon Alton Ochsner was a young medical student at Washington University and was eager to participate in an autopsy of a patient who died of lung cancer, thinking this may be the only lung cancer death he would see for the rest of his career (40501). In the 1920s, lung cancer was still quite rare, only accounting for 1% of all cancer deaths, but as the smoking rate increased, so did lung cancer rates, although with about a 10-year lag.

It is now well known that smoking cigarettes causes lung cancer, but as the Surgeon General of the United States reported in 2014, smoking has been linked to diseases of nearly all organs of the body. This section will show some of the lesser-known health harms and diseases, such as:

- Rheumatoid arthritis (40502–6)
- Diabetes (40506–10)
- Tuberculosis, pneumonia, influenza (40511–14)

- Reduced brain grey matter (40515)
- Pancreatitis (40516)
- Asthma (40517–19)
- Multiple Sclerosis (40520–21)
- Vanishing Lung Syndrome (40522–23)
- Nonalcoholic fatty liver disease (40524–25)
- Kidney Disease (40526)
- Crohn's Disease (40527)
- Glaucoma (40528–30)
- Alzheimer's disease (40530)
- Skin diseases, wrinkles, grey skin (40531)
- Reduced Quality of Life (40532)
- COPD (40533)
- Heart disease/cardiovascular disease (40534–38)

In addition to lung cancer, cigarette smoking is associated with over 30% of all cancers, including bladder and prostate cancers and lymphoma (40539–42). Smoking cessation, but not merely smoking reduction or brand switching, will greatly reduce the risks of these diseases (40543–45).

Reference Number: 40501

GRZYWA-CELIŃSKA, A., I. DROGOŃ, J. EMERYK-MAKSYMIUK, I. CHMIELEWSKA, AND J. MILANOWSKI. "Not Only Cigarettes—Other Culprits of Lung Cancer." *Annals of Agricultural and Environmental Medicine*, 26: 661–664, 2019 (41 references)

Abstract: Lung cancer was an extremely rare disease before mass cigarette consumption was promoted and marketed by large and highly profitable tobacco companies. In the 1920s, lung cancer was diagnosed in only 1% of all cancers. The surgeon Alton Ochsner recalls that in 1910, as a medical student at Washington University, he was eager to participate in an autopsy of a patient who died of lung cancer, thinking that he may not see another case for the rest of his career. The first scientific paper linking smoking with premature death was published in 1938, by researcher Raymond Pear. The consumption of cigarettes in the mid-1960s in the United States had increased from 100 cigarettes to 4,400 cigarettes per person per year. Since the Surgeon General's report in 1964, over 20 million people have died prematurely of tobacco-related disease in the United States.

Lung cancer constitutes a huge epidemiological, diagnostic and socio-economic challenge, both in Poland and worldwide. As outlined above, lung carcinoma is the most common disease caused by cigarette smoking; however, there are many other factors that may affect the development of this medical condition, such as smoking other tobacco products discussed above. The growing popularity of e-cigarettes in Poland, which are often advertised as safe substitutes for traditional tobacco products, generates the need to supply reliable clinical data on the subject. The increasing use of marijuana (especially among young adults) requires epidemiological surveillance, because, as of today, the long-term effects of marijuana smoking are not known, and its connection with lung cancer has not been explicitly confirmed (or excluded). The shifting trends in the use of tobacco products, as well as the smoking of their nicotine substitutes, necessitate additional research into the impact of these factors on the risk of lung cancer occurrence. Awareness of the existence of modifiable threats constitutes the first step in an effective prevention of this deadly disease.

Reference Number: 40502

RUIZ-ESQUIDE, V., AND R. SAMNARTÍ. "Tobacco and Other Environmental Risk Factors in Rheumatoid Arthritis." *Reumatologia Clinica*, 8: 342–350, 2012 (3 tables, 123 references)

Abstract: Rheumatoid arthritis (RA) is a chronic inflammatory joint disease that affects up to 1% of the general population and causes joint destruction, disability, and reduced life expectancy. Genetic susceptibility contributes to 50–60% of the development of the disease. Tobacco smoking is also another major risk factor. Current smokers are 1.3–3.8× and former smokers are 1.5–2.6× as likely of developing RA as non-smokers and have an earlier onset of the disease.

The use of tobacco affects multiple organs including the respiratory and cardiovascular systems, but also affects the immune system by producing an inflammatory response. It has been observed that tobacco affects both the humoral and cellular immune response and could have both pro- and anti-inflammatory effects as immunosuppressants through diverse mechanisms.

Reference Number: 40503

MA, Y., H. TONG, X. ZHANG, M. WANG, J. YANG, M. WU, R. HAN, M. CHEN, X. HU, Y. YUAN, G. PAN, Y. ZOU, S. XU, AND F. PAN. "Chronic Obstructive Pulmonary Disease in Rheumatoid Arthritis: A Systematic Review and Meta-Analysis." *Respiratory Research*, 20: 144 (8 pp), 2019 (4 tables, 1 figure, 48 references)

Abstract: RA is a common chronic auto-immune disease with joints destruction, erosion, pain, and mobility limitations, which can cause early unemployment. About half of RA patients also develop lung diseases and about 20% will develop COPD. An evaluation was conducted of 102,205 RA patients. The presence of COPD was positively associated with smoking in RA patients.

> Third, RA patients usually accompanied with systemic inflammation affecting multiple organs. Long-term chronic inflammation was reported to damage the endothelial cell and cause diverse diseases including atherosclerosis. Likewise, diffuse chronic inflammation in lung of RA may cause lasting destruction of normal construction of pulmonary alveoli and increase the susceptibility to COPD. Therefore, we have reasons to hypothesize that COPD and RA share similar pathogenesis of self reactive immune response, and human comorbidity with RA would increase the risk of concurrent COPD, even smoking may actor as a confounding.

Reference Number: 40504

LIU, Y., A.G. WHEATON, L.B. MURPHY, F. XU, J.B. CROFT, AND K.J. GREENLUND. "Chronic Obstructive Pulmonary Disease and Arthritis Among US Adults, 2016." *Preventing Chronic Disease*, 16: E93 (10 pp), 2019 (2 tables, 3 figures, 26 references)

Abstract: More than 54 million US adults have arthritis and more than 15 million report COPD. A study was conducted of 408,774 adults (aged 18+ years). Overall, the prevalence was 6.4% for COPD and 25.2% for arthritis. Respondents with arthritis were more likely to be women aged 45 years or older, retired, to be a former or current smoker, to be obese, and to have current asthma or COPD. Among the never-smokers, the prevalence of COPD was 7.6% for those with arthritis and 1.7% for those without arthritis. After adjusting for numerous factors, those with arthritis were 1.5× as likely to have COPD.

The reason for a positive association between COPD and arthritis is not clear. First, prolonged systemic inflammation may mediate the relationship between arthritis and COPD. Second, smoking is a common risk factor for both arthritis and COPD, and it may exacerbate the 2 underlying conditions. Therefore, smoking cessation is important to preventing and managing arthritis and COPD. Third, stiff joints and joint pain, weak muscles, and poor balance are common symptoms of arthritis. These symptoms can limit a person's physical activity and further impair respiratory function in people who also have COPD. Conversely, dyspnea or shortness of breath are common symptoms of COPD and can make arthritis worse if patients become more sedentary. Therefore, physical inactivity may contribute to the relationship between COPD and arthritis. However, aerobic activity could reduce symptoms of both COPD and arthritis. Therefore, aerobic physical activity interventions or pulmonary rehabilitation may improve long-term health outcomes for patients with COPD and/or arthritis. Finally, follow-up care and treatment of one condition may lead health care providers to detect and diagnose comorbid conditions.

Reference Number: 40505

INTRIAGO, M., G. MALDONADO, J. CÁRDENAS, AND C. RÍOS. "Clinical Characteristics in Patients with Rheumatoid Arthritis; Difference Between Genders." *The Scientific World Journal*: (6 pp), 2019 (1 table, 3 figures, 33 references)

Abstract: This study determined the clinical characteristics of 50 men and 50 women with RA in Guayaquil, Ecuador. The mean age of the patients was 49 years for men and 47 years for women. Smoking occurred in 34% of the men and 8% of the women. The average age of the onset of the disease was 40 years. The severity of RA is inversely related to androgen levels. Testosterone interacts with the immune system suppressing the humoral and cellular response.

It is evident that women have greater activity of the disease than men, represented in poorer quality of life and associated comorbidities such as depression. Due to this, it is important to implement screening tests and perform multidisciplinary and integral management. When considering gender differences, one has to take into account the fact that disease activity measures themselves may be influenced by gender.

Reference Number: 40506

OLSSON, A.R., T. SKOGH, AND G. WINGREN. "Comorbidity and Lifestyle, Reproductive Factors, and Environmental Exposures Associated with Rheumatoid Arthritis." *Annals of the Rheumatic Diseases*, 60: 934–939, 2001 (4 tables, 36 references)

Abstract: Between 1980 and 1995, the risk factors of RA were determined in 281 cases with RA (mean age 62 years) and 507 referents without RA (mean age 58 years) at the University hospital in Linkoping, Sweden. RA has features of an organ specific autoimmune disease with a T helper cell 1 dominated immune response. Current and previous smoking were associated with an increased risk of RA in both men and women. Men had a higher risk of RA if exposed to mold or farm animals. Women had a higher risk with a prior joint injury or prolonged exposure to hair dye.

Table: Odds Ratio of Rheumatoid Arthritis with Pack Years of Smoking in Men and Women

Pack Years of Smoking	OR of RA for Men	OR of RA for Women
Non-smoker	1.0×	1.0×
<5	0.4×	1.8×
5–9	1.4×	1.5×
10–19	2.4×	1.5×
>20	2.7×	1.8×

Source: Adapted from Olsson et al, 2001

There was an increased risk for RA in both current and previous smokers, and in men there was a dose-response relationship which further strengthens the credibility of a true cause-effect relationship. These results are in agreement with previous findings. The tendency for higher ORs in seropositive cases is also in agreement with previous findings of an association between smoking and RA severity.

Reference Number: 40507

LAURIA, V.T., E.F. SPERANDIO, T.L.W. DE SOUSA, W. DE OLIVEIRA VIEIRA, M. ROMITI, A.R. DE TOLEDO GAGLIARDI, R.L. ARANTES, AND V.Z. DOURADO. "Evaluation of Dose-Response Relationship Between Smoking Load and Cardiopulmonary Fitness in Adult Smokers: A Cross-Sectional Study."

Revista Portuguesa de Pneumologia, 23: 79–84, 2017 (2 tables, 2 figures, 27 references)

Abstract: Cardiopulmonary fitness was determined in 35 male and 60 female smokers (mean age 45 years) by conducting cardiopulmonary exercise testing (CPET) on a treadmill. There was no dose-response relationship between smoking load (pack years) and cardiopulmonary fitness. Even very low tobacco use must be avoided to prevent cardiopulmonary fitness decline in smokers.

Table: Characteristics of Subjects with Smoking Loads of <3 Pack Years and >32 Pack Years

Characteristics	Smoking Load (<3 pack years)	Smoking load (>32 pack years)
Mean age (years)	40	57
Diabetes Mellitus	12.5%	31.8%
Dyslipidemia	8.3%	45.5%
Obesity	54.2%	40.9%
Physical inactivity	70.8%	90.0%

Source: Adapted from Lauria et al, 2017

Based on our finding after adjusting the analysis for age and cardiovascular risk, there appears to be no dose-response relationship between SL and cardiopulmonary fitness in adult smokers with preserved pulmonary function. Therefore, smoking cessation might be more useful as the primary strategy to prevent cardiopulmonary fitness decline in smokers, regardless of smoking load. Thus, even a very low dose of tobacco use must be avoided in preventative strategies.

Reference Number: 40508

MADDATU, J., E. ANDERSON-BAUCUM, AND C. EVANS-MOILINA. "Smoking and Risk of Type 2 Diabetes." *Translational Research: The Journal of Laboratory and Clinical Medicine*, 184: 101–107, 2017 (63 references)

Abstract: Diabetes mellitus affects nearly 420 million persons globally and is responsible for 1.5 million deaths directly and 17.5 million deaths indirectly. Type 2 diabetes (T2D) accounts for 90–95% of diabetes cases and is caused by insulin resistance and progressive loss of pancreatic beta cell function and mass. Smoking also worsens glucose tolerance and insulin sensitivity. Numerous epidemiological studies have shown the increased risk of T2D with smoking. According to the 2014 US Surgeon General's Report, smoking increases the risk of T2D by 30–40% for active smokers. In another study of 40,000 US male health professionals between 1986–1992, men who smoked 25 or more cigarettes a day were 1.94× as likely to develop incident diabetes.

> The clear association between T2D risk and disease severity underscore the importance of promoting smoking cessation in individuals with T2D. However, many persons with T2D continue to smoke, despite these contraindications. Increasing rodent and human evidence indicates that T2D may increase the rewarding effects of dopamine, as disrupted insulin signaling during T2D suppresses dopamine signaling in the mesolimbic reward pathway.

Reference Number: 40509

AUNE, D., L.J. VATTEN, AND P. BOFFETTA. "Tobacco Smoking and the Risk of Gallbladder Disease." *European Journal of Epidemiology*, 31: 643–653, 2016 (2 tables, 4 figures, 38 references)

Abstract: Gallbladder disease is a major cause of morbidity in the United States and Europe, affecting approximately 2% of the population. Gallbladder disease is the most frequent and costly of digestive diseases that require hospitalization. Tobacco smoking affects the hepatobiliary system and has been associated with an increased risk of liver and gallbladder cancer. A systematic review and meta-analysis were conducted of 10 studies involving 59,530 gallbladder disease cases among 4.2 million participants.

> To our knowledge this is the first meta-analysis of observational studies of smoking and risk of gallbladder disease and our results confirm a 19 % increased relative risk among current smokers, with a dose–response relationship of increasing risk with increasing number of cigarettes smoked per day. In addition, a 10 and 15 % increase in the relative risk was observed for former and ever smokers as well.
>
> Little is known about the biological mechanisms that could explain the adverse effect of smoking on risk of gallbladder disease. Smoking has been shown to increase the risk of type 2 diabetes which is a risk factor for gallbladder disease. Cigarette smoking has also been associated with increased risk of gallbladder cancer, although the exact mechanism is not known. In addition, smoking may increase gallbladder disease risk

by reducing plasma high density lipoprotein cholesterol concentrations as higher levels of HDL cholesterol are associated with lower gallbladder disease risk. Tobacco smoke contains several dozens of toxic chemicals that may have detrimental effects on the gallbladder by as yet unidentified mechanisms.

Reference Number: 40510

AKTER, S., A. GOTO, AND T. MIZOUE. "Smoking and the Risk of Type 2 Diabetes in Japan: A Systematic Review and Meta-Analysis." *Journal of Epidemiology*, 27: 553–561, 2017 (2 tables, 5 figures, 61 references)

Abstract: A systematic review and meta-analysis were conducted on cigarette smoking and type 2 diabetes mellitus (T2DM) of 22 studies involving 343,573 subjects and 16,383 patients with T2DM. A dose response relationship was found among smokers with the risk of T2DM increasing 16% for each increase of 10 cigarettes smoked per day. The RR of T2DM was 1.38× for current smokers and 1.19× for former smokers. The risk of T2DM decreased with increasing duration of smoking cessation and was not significantly elevated after 10 years of cessation.

In conclusion, the results of this meta-analysis suggest that current smoking is associated with an increased risk of T2DM in a dose-dependent manner among the Japanese. Although the risk of diabetes remains high after short-term smoking cessation, the risk decreases substantially in the long run. Tobacco smoking accounts for 18.8% of T2DM cases among men and 5.4% of T2DM cases among women in Japan. These findings greatly strengthen the evidence on the association between smoking and T2DM and provide an additional rationale for the intensified implementation of tobacco control programs, especially in countries, like Japan, where tobacco smoking is still prevalent.

Reference Number: 40511

KOLAPPAN, C., AND P.G. GOPI. "Tobacco Smoking and Pulmonary Tuberculosis." *Thorax*, 57: 964–966, 2002 (4 tables, 6 references)

Abstract: The prevalence of tuberculosis in men in India is 2–4× higher than that for women, and many men in India smoke tobacco, while few women do. A survey was conducted in 30 villages in southern India of 85 men with bacteriological tuberculosis and 459 control matched men without tuberculosis. In total, 58% of the men were smokers, and 42% were non-smokers. Bidis were the most common tobacco product smoked. There was an overall 2.48× risk of tuberculosis for tobacco smokers than for non-smokers. This risk increased with the number of cigarettes smoked per day and the years of smoking.

Table: OR of Tuberculosis with Years of Smoking Duration and Number of Cigarettes Smoked Per Day

Smoking Duration (years)	OR of Tuberculosis	Number of Cigarettes per Day	OR of Tuberculosis
<10	1.72×	Mild (1–10 CPD)	1.75×
11–20	2.45×	Moderate (11–20 CPD)	3.17×
>20	3.23×	Heavy (>20 CPD)	3.68×

Source: Adapted from Kolappan and Gopi, 2002

> In conclusion, using a case-control design and appropriate analysis, we have shown that there is an association between tobacco smoking and the development of pulmonary tuberculosis which is dose dependent. It is possible that tobacco smoking is a potential risk factor for developing pulmonary tuberculosis. Further studies using different study designs including a follow up period are needed to measure the incidence rate ratios between smokers and non-smokers which will strengthen the evidence for a causal relationship between tobacco smoking and pulmonary tuberculosis.

Reference Number: 40512

SITAS, F., M. URBAN, D. BRADSHAW, D. KIELOWSKI, S. BAH, AND R. PETO. "Tobacco Attributable Deaths in South Africa." *Tobacco Control*, 13: 396–399, 2004 (2 tables, 19 references)

Abstract: A case control study was conducted in South Africa in 1998 of 4,295 people who died of smoking-related diseases (cases) and compared to 1,952 controls who died of infectious diseases or other non-tobacco related diseases. The smoking status of smokers within the last 5 years was obtained from a family member and recorded in the death certificates.

Table: Adjusted OR of Death Due to Smoking in South Africa

Cause of Death	Adjusted Odds Ratio for Death Due to Smoking
Tuberculosis	1.61×
COPD	2.53×
Lung cancer	4.79×
Oral, pharynx, larynx, esophagus, sinus cancer	3.04×
Ischemic heart disease	1.69×

Source: Adapted from Sitas et al, 2004

If smokers had the same death rate as non-smokers, 58% of lung cancer deaths, 37% of COPD deaths, 20% of tuberculosis deaths, and 23% of vascular deaths would have been avoided. About 8% of all adult deaths in South Africa (more than 20 000 deaths a year) were caused by smoking.

Reference Number: 40513

HUTTUNEN, R., T. HEIKKINEN, AND J. SYRJÄNEN. "Smoking and the Outcome of Infection." *Journal of Internal Medicine*, 269: 258–269, 2011 (2 tables, 136 references)

Abstract: The World Health Organization reported that smoking causes 9% of all deaths worldwide. Tobacco smoke contains more than 4,500 chemical compounds and has a substantial effect on the respiratory tract, the immune system, the skin, and soft tissue, which can increase the risk of infectious disease and its severity. Smokers are at an increased risk of infection such as invasive pneumococcal disease, community acquired pneumonia (CAP), influenza, periodontitis, and meningococcal disease.

Large, nationwide case-control studies and estimates of smoking-attributable deaths indicate adverse effects of smoking on survival in individuals with infectious diseases. However, the impact of smoking on infection outcome has not been established for the vast majority of viral and bacterial infections. Available data indicate that ETS is a major factor affecting the course of infectious diseases in children.

Reference Number: 40514

WARD, K.A., P.J. SPOKES, AND J.M. MCANULTY. "Case-Control Study of Risk Factors for Hospitalization Caused by Pandemic (H1N1) 2009." *Emerging Infectious Diseases*, 17: 1409–1416, 2011 (4 tables, 31 references)

Abstract: The risk factors for hospitalization from the 2009 H1N1 pandemic in Sydney, Australia were determined in 302 case-patients and 603 controls. The OR for hospitalization was 2.0× for current or former smokers.

Table: Risk Factors for Hospitalization from Pandemic (H1N1) 2009 Influenza

Patient Characteristic	Adjusted OR
Current smoker	2.0×
Former smoker	2.0×
Asthma (regular medication)	4.3×
Lung disease	6.6×
Pregnancy	22.4×

Source: Adapted from Ward et al, 2011

> Our study highlights the increased risk of moderate to severe illness from pandemic (H1N1) 2009 for pregnant women and introduces smoking as an independent risk factor for hospitalization from pandemic (H1N1) 2009. In addition, our study provides evidence to support the continuation of influenza prevention efforts (including vaccination) targeted to persons with lung disease, immune suppression, asthma, diabetes, and heart disease. Although Aboriginal status and obesity may not be independent risk factors for severe disease, they indicate the likely presence of other risk factors, and so prevention messages should continue to be directed to these groups.

Reference Number: 40515

VŇUKOVÁ, M., R. PTÁČEK, J. RABOCH, AND G.B. STEFANO. "Decreased Central Nervous System Grey Matter Volume (GMV) in Smokers Affects Cognitive Abilities: A Systematic Review." *Medical Science Monitor*, 23: 1907–1915, 2017 (6 tables, 27 references)

Abstract: Although cigarette smoking is the leading cause of preventable mortality, cigarettes are consumed by only 22% of the adult population worldwide. Smoking is a risk factor of cardiovascular disease (CVD) and affects brain processing. A review of four studies showed that the brain matter volume (GMV) decreases in smokers compared to non-smokers.

Overall, smoking causes differences in GMV in various brain areas, and these differences help explain the cognitive impairment and emotional dysregulation in smokers compared with non-smokers. There are considerable differences, not only between males and females, but also between younger and older smokers, and any therapeutic treatment must take this into account. To summarize, smoking decreases GMV in most brain areas, and this decrease is believed to be responsible for the cognitive impairment and difficulties with emotional regulation in smokers.

Reference Number: 40516

ALEXANDRE, M., S.J. PANDOL, F.S. GORELICK, AND E.C. THROWER. "The Emerging Role of Smoking in the Development of Pancreatitis." *Pancreatology*, 11: 469–474, 2011 (1 table, 1 figure, 36 references)

Abstract: Cigarette smoking increases the risk of developing pancreatic cancer and has recently been found to increase the risk of pancreatitis as well. In a study of 535 patients with chronic pancreatitis, 71% reported smoking, yet their physicians listed smoking as a risk factor in only 45% of the patients.

Smoking may affect the risk of developing chronic pancreatitis or its progression. Smoking may also affect the risk for developing acute pancreatitis. Its effects in pancreatitis appear to be dose dependent and its effects may be alcohol independent but synergize with alcohol.

Conclusion

Specific constituents of cigarette smoke, including nicotine and its metabolites, could mediate effects on the pancreas.

Reference Number: 40517

STAPLETON, M., A. HOWARD-THOMPSON, C. GEORGE, R.M. HOOVER, AND T.H. SELF. "Smoking and Asthma." *Journal of the American Board of Family Medicine*, 24: 313–322, 2011 (4 tables, 111 references)

Abstract: Asthma is an inflammatory airway disease that affects 22 million Americans. Approximately 25–35% of asthmatic Americans are current smokers. Current and former smokers are 1.33–1.49× as likely to be asthmatic and have poorer disease control than non-smokers. Young asthmatics (20–34 years) exposed to SHS were 7.1× more likely to be intubated than those in smoke-free homes.

The evidence reviewed in this article regarding the multiple negative effects of smoking and SHS on patients with asthma should underscore the obligation of primary care physicians and other health care professionals to assist patients and those they live with to quit smoking. Specific instructions for family members, caregivers, and friends of patients with asthma should include never smoking in the home, car, or workplace. Simply smoking in another room of the house is not sufficient, nor are any forms of air cleaners effective in reducing SHS exposure.

Reference Number: 40518

CROGHAN, I.T., J.O. EBBERT, J.T. HAYS, D.R. SCHROEDER, A.M. CHAMBERLAIN, V.L. ROGERS, AND R.D. HURT. "Impact of a Countywide Smoke-Free Workplace Law on Emergency Department Visits for Respiratory Diseases: A Retrospective Cohort Study." *BMC Pulmonary Medicine*, 15: 6 (8 pp), 2015 (2 tables, 4 figures, 43 references)

Abstract: Minnesota enacted a smoke-free law (i.e., Freedom to Breathe Act) in workplaces, restaurants, and bars in 2007. The rates of emergency department (ED) visits in Olmstead County MN for asthma and COPD were determined for 2 ½ years before and 2 ½ years after the law was enacted. There was a decrease in ER visits for asthma after the law (RR = 0.849× for adults and RR = 0.751× for children) but no change in hospital attendances due to COPD.

The findings that asthma-related ED visits were significantly reduced after the smoke-free law implementation, but the COPD-related visits were not significantly reduced, is not surprising. The lack of findings with COPD-related visits confirms findings from a recent meta-analysis. Typically COPD is associated with irreversible or minimally reversible airway obstruction, unlike the bronchospasms in asthma, which can be severe but may be completely reversible. Exposure to triggers such as SHS may precipitate acute bronchospasms in asthmatics and is more likely to increase respiratory symptoms but not acute bronchospasms in people with COPD. Acute exacerbations of asthma often require emergency treatment or hospitalization to reverse. In communities with strong smoke-free laws, the changes in COPD-related hospital and ED admissions may be seen 12 months or more after implementation of the law. This likely reflects the slower improvement seen in COPD symptoms and exacerbations as SHS smoke exposure declines.

Reference Number: 40519

MAKADIA, L.D., P.J. ROPER, J.O. ANDREWS, AND M.S. TINGEN. "Tobacco Use and Smoke Exposure in Children: New Trends, Harm, and Strategies to Improve Health Outcomes." *Current Allergy and Asthma Reports*, 17: 55 (15 pp), 2017 (3 tables, 1 figure, 138 references)

Abstract: More than 50 years have passed since the first US Surgeon General's report on the harms of smoking, yet nearly 4,000 adolescents (12–17 years of age) begin smoking in the United States every day. Cigarette smoking accounts for nearly 480,000 adult deaths annually and an additional 41,000 deaths due to secondhand smoke. SHS leads to a rapid and sharp increase in endothelial dysfunction and inflammation, which are implicated in acute cardiovascular events and thrombosis.

Table: Percent of US High School Students Who Used Tobacco Products in 2015

Tobacco Product	Overall (%)	Females (%)	Males (%)
Any tobacco product	25.3	20.3	30.0
Electronic cigarettes	16.0	12.8	19.0
Cigarettes	9.3	7.7	10.7
Cigars	8.6	5.6	11.5
Waterpipes	7.2	6.9	7.4
Smokeless tobacco	6.0	1.8	10.0
Pipes	1.0	0.7	1.4
Bidis	0.6	0.4	0.9

Source: Adapted from Makadia et al, 2017

Twelve–20% of women continue to smoke during pregnancy. New research reveals cognitive differences and behavior-control disorders are seen in elementary school children from prenatal and postnatal exposures. Traditional cigarette smoking has decreased in adolescents; novel and appealing tobacco products have captured their attention, particularly electronic cigarettes, and rates double and often triple from middle to high school. Children with asthma and those living in multi-housing units have higher rates of secondhand smoke exposure than non-asthmatics and children living in single-home dwellings.

Summary

There is no "safe or risk-free" level of tobacco use or exposure. Tobacco use and exposure in childhood and adolescence must be decreased using evidence-based strategies to improve child health.

Reference Number: 40520

HEDSTRÖM, A.K., J. HILLERT, T. OLSSON, AND L. ALFREDSSON. "Smoking and Multiple Sclerosis Susceptibility." *European Journal of Epidemiology*, 28: 867–874, 2013 (4 tables, 16 references)

Abstract: Smoking is one of the most established risk factors for multiple sclerosis (MS). Cigarette smoking causes oxidative stress and inflammatory response in the cells of the lungs and may cause an autoimmune response with central nervous system (CNS) antigens. A study was conducted of the MS risk of 7,883 cases and 9,264 controls in Sweden with differing smoking habits.

Table: Odds Ratio of Multiple Sclerosis According to Smoking Habit

Smoking Habits	OR of MS
Never smoker	1.0×
Current smoker	1.6×
Quit 5 years prior	1.3×
Quit 5–10 years prior	1.2×
Quit >10 years prior	1.0×

Source: Adapted from Hedström et al, 2013

We observed a clear dose–response relationship between cumulative dose of smoking and MS risk. Both duration and intensity of smoking, but not age at debut of smoking, seemed to affect the association between smoking and increased MS risk. Even smoking less than 5 cigarettes per day for many years implied a two-fold increased risk of developing MS. However, the detrimental effect of smoking abates a decade after smoking cessation regardless of cumulative dose of smoking and age at exposure.

Reference Number: 40521

MANOUCHEHRINIA, A., C.R. TENCH, J. MAXTED, R.H. BIBANI, J. BRITTON, AND C.S. CONSTANTINESCU. "Tobacco Smoking and Disability Progression in

Multiple Sclerosis: United Kingdom Cohort Study." *Brain*, 136: 2298–2304, 2013 (2 tables, 2 figures, 47 references)

Abstract: MS is a chronic inflammatory demyelinating autoimmune disease within the brain and CNS. Tobacco smoking has been long associated with a 50% increased risk of MS. The effects of smoking on the disease progression and its severity were determined on 270 male and 625 female MS patients (mean age 49 years). Forty-nine percent of the patients were regular smokers at the time of onset or diagnosis of MS (ever-smokers). The average disease severity was greater in ever-smokers than non-smokers. Current smokers showed a 1.64× risk of reaching the Expanded Disability Status Scale score of 4. Ex-smokers had a reduced risk compared to current smokers.

> There are some likely biological explanations for a mechanistic pathway between smoking and disability accumulation in multiple sclerosis. Exposure to tobacco smoke has been shown to alter the innate and adaptive immune cells. Increased risk of cancer, cardiovascular and other chronic diseases amongst smokers may possibly be related to smoking-induced changes in the immune system. Further work is needed to elucidate the mechanism by which smoking increases the risk of progression in multiple sclerosis.
>
> *Conclusions and implications*
> In summary, we found that ever-smokers with multiple sclerosis accumulate more disability over a shorter period of time and suffer from more severe disease than never-smokers. Our findings point toward the beneficial effect of smoking cessation even after the disease onset in patients with multiple sclerosis. Measures to prevent and reduce smoking may lead to improved outcomes in multiple sclerosis.

Reference Number: 40522

SOOD, N., AND N. SOOD. "A Rare Case of Vanishing Lung Syndrome." *Case Reports in Pulmonology*, 2011: (2 pp), 2011 (2 figures, 11 references)

Abstract: A 33-year-old man who smoked 1 pack of cigarettes a day for 15 years and had 10 years of marijuana abuse attended hospital as a result of Giant Bullous Emphysema or vanishing lung syndrome (VLS).

Vanishing Lung Syndrome is a rare condition which becomes clinically evident in a much advanced stage. Patients should be strongly counseled against any further tobacco and marijuana abuse.

Reference Number: 40523

CLARK, K.D., N. WARDROBE-WONG, J.J. ELLIOTT, P.T. GILL, N.P. TAIT, AND P.D. SNASHALL. "Cigarette Smoke Inhalation and Lung Damage in Smoking Volunteers." *European Respiratory Journal*, 12: 395–399, 1998 (2 tables, 2 figures, 26 references)

Abstract: A high-resolution tomography scan was taken of 80 current cigarette smokers (mean age 51 years) and 20 non-smoking controls (mean age 50 years). Various other lung parameters were determined as well as breath CO concentration (measure by an IR Smokerlyzer) before and after smoking one cigarette. The mean forced expired volume in 1 second (FEV1) was significantly lower in smokers (3.00L) compared to non-smokers (3.43 L). The serum cotinine concentration was 37% higher in smokers with a productive cough (n=16) than those without. The smoker's exhaled breath CO concentration was 8.4 ppm pre-smoking and 11.7 ppm post-smoking.

Emphysema is associated with high alveolar smoke exposure as measured by CO boost. Productive coughing is associated with high nicotine uptake, probably from airway smoke particle deposition.

Reference Number: 40524

OKAMOTO, M., T. MIYAKE, K. KITAI, S. FURUKAWA, S. YAMAMOTO, H. SENBA, S. KANZAKI, A. DEGUCHI, M. KOIZUMI, T. ISHIHARA, H. MIYAOKA, O. YOSHIDA, M. HIROOKA, T. KUMAGI, M. ABE, B. MATSUURA, AND Y. HIASA. "Cigarette Smoking is a Risk Factor for the Onset of Fatty Liver Disease in Non-drinkers: A Longitudinal Cohort Study." *Public Library of Science One*, 13: e0195147 (13 pp), 2018 (5 tables, 2 figures, 26 references)

Abstract: Non-alcoholic fatty liver disease (NAFLD) is a common life-style-related disease that can lead to cirrhosis and liver failure. A ten-year longitudinal study was conducted on the effect of smoking on NAFLD in 1,512 men and 2,348 women. Approximately 20.6% of the current smokers developed NAFLD, compared to 8.5% of the non-smokers. Smoking did not increase the risk of NAFLD in the subjects who had low alcohol consumption (<210 grams/week for men and <140 grams per week for women). If the number of cigarettes smoked per day was 25 or more, the incidence of NAFLD increased to 35% in non-drinkers.

> The findings from this longitudinal study, which excluded subjects whose smoking habits changed during the observation period, demonstrated that cigarette smoking was a significant risk factor associated with the onset of fatty liver disease in individuals who did not consume alcohol, and that this significant association persisted after adjusting for the potential confounders. Moreover, the incidence of fatty liver disease in the non-drinker group increased significantly as the number of cigarettes smoked increased. However, cigarette smoking was not a risk factor for the onset of fatty liver disease in subjects who consumed low amounts of alcohol.

Reference Number: 40525

LUI, Y., M. DAI, Y. BI, M. XU, Y. XU, M. LI, T. WANG, F. HUANG, B. XU, J. ZHANG, X. LI, W. WANG, AND G. NING. "Active Smoking, Passive Smoking, and Risk of Nonalcoholic Fatty Liver Disease (NAFLD): A Population-Based Study in China." *Journal of Epidemiology*, 32: 115–121, 2013 (3 tables, 1 figure, 28 references)

Abstract: Non-alcoholic fatty liver disease (NAFLD) is the most common chronic liver disease and is associated with T2D and CVD. Tobacco use has been associated with an increase in fatty deposits in the liver. A survey was conducted of 2,691 men and 4,889 women (age 40+ years) in Shanghai, China in 2010 regarding smoking history of NAFLD (determined by ultrasound testing and serum liver enzymes). NAFLD prevalence was 29.4% in never-smokers, 34.2% in former smokers, 27.8% in light smokers (<19 CPD), 30.8% in moderate smokers (20–39 CPD), and 43.5% in heavy smokers (40+ CPD). In non-smoking women exposed to SHS in both childhood and adulthood, the OR of NAFLD was 1.25×, but no increase in OR was observed in men exposed to SHS.

Table: Adjusted Odds Ratio of Non-alcoholic Fatty Liver Disease and Smoking Status in Normal Weight and Overweight/Obese Subjects

Smoking Status	Normal Weight	Overweight/Obese
Never-smokers	1.00×	2.69×
Former smokers	1.36×	3.65×

Smoking Status	Normal Weight	Overweight/Obese
Light smokers	1.04×	3.06×
Moderate smokers	1.16×	3.04×
Heavy smokers	2.40×	8.58×

Source: Adapted from Lui et al, 2013

> We observed a positive association between heavy active smoking and NAFLD in a large, community-based population in China. In addition, tobacco smoking and BMI had a synergistic effect on NAFLD prevalence. Finally, exposure to passive smoke during both childhood and adulthood was associated with NAFLD prevalence in never-smoking women. To our knowledge, this is the first clinical study of the relationship between passive smoking and NAFLD.

Reference Number: 40526

LHOTTA, K., H.J. RUMPELT, P. KÖNIG, G. MAYER, AND F. KRONENBERG. "Cigarette Smoking and Vascular Pathology in Renal Biopsies." *Kidney International*, 61: 648–654, 2002 (7 tables, 32 references)

Abstract: Cigarette smoking is highly associated with the reduction in renal function in elderly people and end-stage renal disease. Kidney biopsies and smoking history were determined in 67 male and 40 female patients (mean age 40 years). Forty-eight percent were non-smokers, 17% were ex-smokers, and 36% were current smokers. Renal biopsies showed glomerular disease in 92 patients (86%).

> In patients with renal, especially glomerular disease, cigarette smoking exhibits its deleterious effect on the kidneys primarily through damage of small interlobular arteries.

Reference Number: 40527

TO, N., D.J. GRACIE, AND A.C. FORD. "Systematic Review with Meta-Analysis: The Adverse Effects of Tobacco Smoking on the Natural History of Crohn's Disease." *Alimentary Pharmacology and Therapeutics*, 43: 549–561, 2016 (2 tables, 7 figures, 81 references)

Abstract: Crohn's disease is a chronic inflammatory disorder of the gastrointestinal tract with a prevalence of 130–200 per 100,000 population. A systematic review and meta-analysis were conducted on the effect of tobacco smoking on Crohn's disease. The deleterious effects of smoking on Crohn's disease may be due to the altering of the intestinal microbiota, impairment of mononuclear cells, altered immune response due to nicotine, and inflammation of intestinal mucosa.

> In summary, this meta-analysis has demonstrated that smokers, compared with nonsmokers, have 55–85% higher rates of flares of disease activity, clinical recurrence rates after surgery that are twofold higher, between 54% and 68% higher rates of need for first surgery, and are twice as likely to need a second operation. In the studies we identified, quitting smoking appeared to have a beneficial effect on disease course, particularly for flare of disease activity or need for a second operation. This study, therefore, provides summary estimates of the adverse effect of smoking on the natural history of Crohn's disease that can be communicated easily to patients with Crohn's disease who smoke and used to encourage smoking cessation as a management strategy. The magnitude of the increase in odds of flare of disease activity or need for surgery seen in this meta-analysis among smokers, compared with nonsmokers and ex-smokers, suggests that there is a need for trials of smoking cessation strategies as a primary medical intervention among patients with Crohn's disease who smoke.

Reference Number: 40528

PÉREZ-DE-ARCELUS, M., E. TOLEDO, M.Á. MARTÍNEZ-GONZÁLEZ, N. MARTÍN-CALVO, A. FERNÁNDEZ-MONTERO, AND J. MORENO-MONTAÑÉS. "Smoking and Incidence of Glaucoma." *Medicine*, 96: (6 pp), 2017 (3 tables, 1 figure, 50 references)

Abstract: Glaucoma is a group of disorders which are characterized by a progressive deterioration of the optic nerve associated with a loss of field of vision and is the major cause of irreversible blindness. A study was conducted of 16,797 Spanish undergraduate students without glaucoma who were followed over a median of 8.5 years. Over that time, 184 of the participants developed glaucoma. Current smokers had an adjusted hazard ratio of 1.88× of developing glaucoma compared to never-smokers. Tobacco smoke may increase glaucoma risk by causing an inflammatory response, apoptosis (i.e., cell death), or from its deleterious effect in decreasing blood flow to the optic nerve head.

> In conclusion, in this prospective study we found that current smoking status is associated with a higher risk of developing glaucoma, even

though the same association for former smokers or for passive smokers is not clear. In addition, we found a dose–response relationship that demonstrates that the risk of glaucoma increases as the number of cigarette pack-years increases.

Reference Number: 40529

ZHANG, X., J. KAHENDE, A.Z. FAN, L. BARKER, T.J. THOMPSON, A.H. MOKDAD, Y. LI, AND J.B. SAADDINE. "Smoking and Visual Impairment Among Older Adults with Age-Related Eye Diseases." *Preventing Chronic Disease*, 8: (8 pp), 2011 (3 tables, 34 references)

Abstract: The association between cigarette smoking and age-related eye diseases such as glaucoma, cataracts, age-related macular degeneration, and diabetic retinopathy was determined in 4,517 current smokers, 14,165 former smokers, and 17,651 never-smokers between 2005 and 2008 in the United States.

Table: Adjusted OR of Age-Related Eye Disease in Current, Former, and Never-Smokers

Type of Smoker	Adjusted OR of ARED
Current	1.16×
Former	1.04×
Never	1.0× (reference)

Source: Adapted from Zhang et al, 2011

In summary, smoking is a major modifiable risk factor for AMD, and smoking is the most important modifiable risk factor for primary and secondary prevention of cataracts. A lack of awareness may exist among health care providers and patients about the risks of developing eye diseases and vision loss from smoking. Of an estimated 61 million adults in the United States who are at high risk for serious vision loss due to aging, diabetes, or vision or eye problems, only half visit an eye care provider annually, making the situation even worse. Given the effect of smoking on a person's overall health, and especially on sight, comprehensive tobacco control interventions, as recommended by the Centers for Disease Control and Prevention (CDC), are needed. These interventions include health care provider counseling, telephone quit lines, insurance coverage for cessation therapies, legislation of clean indoor air, and increased tobacco taxes.

Reference Number: 40530

PETERS, R., R. POULTER, J. WARNER, N. BECKETT, L. LURCH, AND C. BULPITT. "Smoking, Dementia and Cognitive Decline in the Elderly, a Systematic Review." *BMC Geriatrics*, 8: 36 (7 pp), 2008 (4 figures, 50 references)

Abstract: A systematic review and meta-analysis were conducted on 28 studies of the effect of current smoking on Alzheimer's disease and other dementias. Cardiovascular risk factors have been linked to an increased risk of dementia.

Current smoking increases risk of Alzheimer's disease and may increase risk of other dementias. This reinforces need for smoking cessation, particularly aged 65 and over. Nicotine alone needs further investigation.

Reference Number: 40531

THOMSEN, S.F., AND L.T. SØRENSEN, "Smoking and Skin Disease." *Skin Therapy Letter*, 15: 4–7, 2010 (1 table, 28 references)

Abstract: The common features of a smoker's face are lines or wrinkles radiating at right angles from the upper and lower lips or corners of the eyes, deep lines in the cheeks, or numerous shallow lines on the checks and lower jaw, a subtle gauntness of the facial features with prominence of the underlying bone structures, an atrophic slightly grey appearance of the skin, and a plethoric, slightly orange, purple, and red complexion. The nails of smokers can show a yellow discoloration and in heavy smokers who suddenly quit smoking (e.g., due to a sudden illness) can develop a sharp demarcation line, which is referred to as Harlequin nails or quitter's nails.

Table: Skin Diseases Associated with Tobacco Smoking

Skin Diseases Associated with Tobacco Smoking
• Basal cell carcinoma
• Cutaneous squamous cell carcinoma
• Genital warts and HPV infection
• Hidradenitis suppurative
• Discoid lupus erythematosus
• Systemic lupus erythematosus
• Oral cancers
• Palmoplantar pustulosis
• Psoriasis

Source: Adapted from Thomsen and Sørensen, 2010

Tobacco smoking induces oxidative stress, which has immunomodulatory effects by changing inflammatory cell function and releasing proteolytic enzymes; the latter alters connective tissue turnover and degrades skin connective tissue. From the literature, it is reasonable to presume an etiological role for tobacco smoking in certain dermatologic conditions, particularly SLE, psoriasis, PPP, cutaneous SCC, hidradenitis suppurativa, and genital warts.

Reference Number: 40532

DE CASTRO, M.R.P., T. MATSUO, AND S.O.V. NUNES. "Clinical Characteristics and Quality of Life of Smokers at a Referral Center for Smoking Cessation." *Jornal brasileiro de pneumologia*, 36: 67–74, 2010 (5 tables, 30 references)

Abstract: A comparison was made between 167 smokers enrolled in a smoking cessation program and 272 never-smoking blood donors in Brazil between 2006 and 2008. Approximately 50% of smokers had some form of depression, compared to 12% of non-smokers. The quality of life (QoL) was higher for non-smokers (16.6) compared to smokers (13.0). The BMI for smokers (24.7) was lower than for non-smokers (26.5).

Table: Clinical Characteristics of Smokers and Never-Smokers

Clinical Characteristics	Smokers (%)	Never-Smokers (%)
Hospitalization in last month	16.8	4.8
Diabetes	10.2	1.5
Hypertension	22.8	12.9
Respiratory disease	31.1	7.0
Heart disease	17.4	3.3

Source: Adapted from de Castro et al, 2010

Smoking prevention programs should focus on the benefits of not smoking, in order to maintain health, improve the QoL and reduce the morbidity and impaired functionality resulting from tobacco consumption, as well as to reduce public health spending.

Reference Number: 40533

SUFFREDINI, D.A., AND R.M. REED. "At the Twisted Heart of Nicotine Addiction." *BMJ Case Reports* (4 pp), 2012 (1 table, 3 figures, 19 references)

Abstract: A male heavy smoker in his late 60s attended ER with a severe shortness of breath. He recently quit smoking only because he became breathless while walking and could no longer buy cigarettes at the store. His dyspnea was due to decompensated severe COPD complicated by pulmonary hypertension. Severe COPD was associated with lung hyperinflation that depresses the diaphragm into the abdomen and causes a shifting of the heart in a clockwise rotation.

- The ECG manifestations can be characteristic but are not specific for chronic obstructive pulmonary disease (COPD) and may be seen in other disease states such as pulmonary hypertension and congenital heart disease.
- Distinct ECG changes seen with COPD are a result of lung hyperinflation causing rotation and medialisation of the heart.
- Recognising how these changes evolve may provide a better understanding of the presentation of simultaneous disease processes such as myocardial ischaemia in a patient with COPD and dyspnoea.

Reference Number: 40534

STÖRK, T., H. EICHSTÄDT, M. MÖCKEL, R. BORTFELDT, R. MÜLLER, AND H. HOCHREIN. "Changes of Diastolic Function Induced by Cigarette Smoking: An Echocardiographic Study in Patients with Coronary Artery Disease." *Clinical Cardiology*, 15: 80–86, 1992 (1 table, 4 figures, 42 references)

Abstract: Eleven female and 17 male chronic smokers (mean age 53 years) with coronary artery disease had left ventricle (LV) relaxation and filling behavior assessed by echocardiography before and after smoking one cigarette (0.9 mg nicotine). The mean heart rate increased from 67 to 81 beats per minute.

> The strong association between long-term cigarette smoking and the incidence of coronary artery disease and consecutive myocardial infarction has been underlined by several epidemiological studies. In addition, a positive correlation between chronic nicotine uptake and the prevalence of sudden cardiac death has been reported on. Furthermore, chronic smoking is known to cause severe changes of lipid metabolism, serum levels of clotting factors, blood flow, and pliability of red blood cells. The acute effect of smoking on myocardial oxygen demand and supply, coronary artery tone and coronary blood flow has been investigated invasively.

The present study shows that smoking significantly affects LV diastolic function in coronary patients by prolonging relaxation time and by inducing a shift of mitral blood flow from early (E wave) to late (A wave) diastole.

Reference Number: 40535

GULLU, H., M. CALISKAN, O. CIFTCI, D. ERDOGAN, S. TOPCU, E. YILDIRIM, E. YILDIRIR, AND H. MUDERRISOGLU. "Light Cigarette Smoking Impairs Coronary Microvascular Functions as Severely as Smoking Regular Cigarettes." *Heart*, 93: 1274–1277, 2007 (2 tables, 19 references)

Abstract: The coronary flow velocity reserve (CFVR) was measured in 20 regular cigarette smokers, 20 light cigarette smokers, and 22 non-smokers. The regular cigarettes contained 12 mg of tar, 0.9 mg of nicotine, and 12 mg CO. The light cigarettes contained 8 mg tar, 0.6 mg of nicotine, and 9 mg of CO. CFVR measures were conducted after 12 hours of cigarette abstinence and again 2 days later, after each subject smoked their normal cigarette.

Our study suggest that reducing the nicotine and tar yield is not sufficient for a cigarette to be called less hazardous, and other noxious compounds in cigarettes continue to compromise human health. Smoking low-tar, low-nicotine cigarettes seems to have the same unfavourable effect on the coronary microvascular functions as smoking regular cigarettes. Action should be taken to prohibit misleading terminology such as "light".

Reference Number: 40536

AUNE, D., S. SCHLESINGER, T. NORAT, AND E. RIBOLI. "Tobacco Smoking and the Risk of Abdominal Aortic Aneurysm: A Systematic Review and Meta Analysis of Prospective Studies." *Scientific Reports*, 8: 4786 (9 pp), 2018 (1 table, 5 figures, 56 references)

Abstract: Aortic aneurysms are localized dilations or enlargements of the aorta that exceed the normal diameter by 50% or more or by more than 3 cm. Rupture of an aortic aneurysm can cause massive internal bleeding and is usually fatal. Most aortic aneurysms are abdominal and are most common in men. A systematic review and meta-analysis were conducted on 22 studies of smoking and the risk of abdominal aortic aneurysm.

There was a 5-fold and 2-fold increase in the risk of abdominal aortic aneurysms among current and former smokers compared to never smokers, respectively. A positive dose-response relationship was observed between increasing number of cigarettes smoked per day and pack-years smoked and the risk of abdominal aortic aneurysm, while there was a reduced risk with increasing duration of smoking cessation with a risk similar to that of never smokers by 25 years of smoking cessation. This together with supportive experimental data provides strong evidence for a causal relationship between smoking and abdominal aortic aneurysms. The findings provide further support for interventions and policies to curb the global tobacco epidemic.

Reference Number: 40537

BANKS, E., G. JOSHY, R.J. KORDA, B. STRAVRESKI, K. SOGA, S. EGGER, C. DAY, N.E. CLARKE, S. LEWINGTON, AND A.D. LOPEZ. "Tobacco Smoking and Risk of 36 Cardiovascular Disease Subtypes: Fatal and Non-Fatal Outcomes in a Large Prospective Australian Study." *BMC Medicine*, 17: 128 (18 pp), 2019 (2 tables, 6 figures, 40 references)

Abstract: A prospective study was conducted in Australia of 188,167 CVD and cancer-free adults (>44 years of age) during 2006–2009 and of their hospitalization and death data up to the end of 2015. The hazard ratios (HR) for hospitalization and mortality for CVD was determined in current and past smokers compared to non-smokers. At baseline 8% of the participants were smokers and 34% were past smokers. During the follow-up time (mean 7.4 years), there were 27,511 major CVD events. The HR for major CVD events increased with increasing intensity of smoking among current smokers.

Table: Hazard Ratio of Combined Fatal and Non-Fatal CVD Outcomes (Compared to 1.0× for Non-smokers)

Cardiovascular Disease	Past Smoker	Current Smoker
Peripheral arterial disease	1.78×	5.06×
Acute myocardial infarction	1.21×	2.45×
Heart failure	1.46×	2.23×
Ischemic heart disease	1.16×	1.57×
Any major CVD	1.16×	1.63×

Source: Adapted from Banks et al, 2019

In conclusion, smoking increases the risk of virtually all CVD subtypes, including that of the newly identified risk of paroxysmal tachycardia. Quitting reduces the risk. Smoking accounts for a substantial proportion of premature CVD events in the setting of an established smoking epidemic with low current smoking prevalence.

Reference Number: 40538

NEGRI, E., M.G. FRANZOSI, C. LA VECCHIA, L. SANTORO, A. NOBILI, AND G. TOGNONI. "Tar Yield of Cigarettes and Risk of Acute Myocardial Infarction." *British Medical Journal*, 306: 1567–1570, 1993 (4 tables, 19 references)

Abstract: The relationship between the tar and nicotine yield and the risk of MI was determined in 916 patients with acute MI and 1,106 control patients without MI in 80 Italian Critical Care Units between 1988 and 1989. Since smokers tend to be admitted to hospital more often than non-smokers, the control group may have had more smokers and therefore underrepresented the risk of MI. The tar and nicotine yield of the cigarette brands by current smokers were determined from <10 mg tar and <0.6 mg nicotine to >20 mg tar and 1.1–1.6 mg nicotine.

Table: Adjusted Relative Risks of Acute Myocardial Infarction According to the Tar and Nicotine Yields of the Brand of Cigarettes Used by Current Smokers

Tar and Nicotine Yields for Cigarettes Currently Smoked	Adjusted Relative Risk of Acute MI
<10 mg tar, <0.6 mg nicotine	3.8×
10–15 mg tar, 0.5–0.9 mg nicotine	4.3×
15-20 mg tar, 0.8–1.3 mg nicotine	3.2×
>20 mg tar, 1.1–1.6 mg nicotine	3.7×

Source: Adapted from Negri et al, 1993

In conclusion, therefore, shifting to lower yield cigarettes is not an effective means of reducing tobacco related morbidity from myocardial infarction.

Reference Number: 40539

JACOBS, E.J., C.C. NEWTON, B.D. CARTER, D. FESKANICH, N.D. FREEDMAN, R.L. PRENTICE, AND W.D. FLANDERS. "What Proportion of Cancer Deaths in the Contemporary United States is Attributable to Cigarette Smoking?" *Annals of Epidemiology*, 25: 179–182, 2015 (3 tables, 17 references)

Abstract: The age- and sex-adjusted population attributable fraction (PAF) for cigarette smoking and cancer mortality in the United States in 2010 was estimated. In 1981, about 30% of all cancer deaths were due to cigarette smoking. In 2010, it was estimated that 31.7% of all cancer deaths were due to smoking.

Despite important declines in smoking prevalence, the PAF for smoking and cancer mortality estimated for 2010 in this analysis is similar to the 30% estimated by Doll and Peto more than 30 years ago. Declines in smoking prevalence have undoubtedly contributed substantially to decreasing both the rate of cancer mortality attributable to smoking and the PAF. Other factors, however, have contributed to increasing the PAF estimate, including the addition of new cancers to the list of those counted as caused by smoking, increases over time in the RR of lung cancer mortality among smokers, and decreases in cancer mortality for reasons unrelated to smoking. Our estimate of the PAF therefore does not indicate a lack of progress in reducing cancer mortality due to smoking. However, our estimate of the PAF, the proportion of cancer deaths currently caused by cigarette smoking, is relevant for prioritizing future public health action.

Our results indicate that cigarette smoking causes about three in 10 cancer deaths in the contemporary United States. Reducing smoking prevalence as rapidly as possible should be a top priority for US public health efforts to prevent future cancer deaths.

Reference Number: 40540

MITHANI, M.H., S.A. KHAN, S. EL KHALID, R. SIDDIQUE, H. HUMAYUN, AND A.S. AWAN. "Awareness of Risk Factors and Fallacies Associated with Urinary Bladder Cancer in Our Population: A Prospective Survey." *Journal of the Pakistan Medical Association*, 68: 55–58, 2018 (3 tables, 1 figure, 21 references)

Abstract: A survey was conducted of 593 male and 406 female patients (mean age 35 years) at the Kidney Centre in Karachi between January and February 2015 regarding the knowledge of the increased risks of bladder disease with smoking. Bladder cancer is the ninth most common cancer worldwide and tobacco smoking is associated with cancer in 50–60% of the male smokers. The most common fallacies about the cause of bladder cancer were due to the low intake of water or having multiple sexual partners.

Table: Percent of Awareness of Smoking as a Risk Factor and Level of Education

Disease	Uneducated	Secondary	Graduation
Heart diseases	45.5%	56.1%	85.2%
Lung cancer	67.3%	81.1%	87.2%
Bladder cancer	9.1%	9.7%	35.2%

Source: Adapted from Mithani et al, 2018

The overall awareness was low regarding the risk factors of bladder cancer. Most patients were unaware that tobacco smoking was linked to the development of bladder cancers. Several false beliefs were also associated with bladder cancer. It is crucial to have awareness programmes to educate the public regarding smoking leading to cancers which will help in preventing smoking-related cancers.

Reference Number: 40541

GIOVANNUCCI, E., E.B. RIMM, A. ASCHERIO, G.A. COLDITZ, D. SPIEGELMAN, M.J. STAMPFER, AND W.C. WILLETT. "Smoking and Risk of Total and Fatal Prostate Cancer in United States Health Professionals." *Cancer Epidemiology, Biomarkers and Prevention*, 8: 277–282, 1999 (3 tables, 51 references)

Abstract: The relationship between smoking and prostate cancer was determined in 47,781 male health professionals in the United States between 1986 and 1994. One hundred and fifty-two men had distant metastatic prostate cancer and 103 had fatal prostate cancer, out of a total of 1,369 male health professionals with prostate cancer.

The higher rate of fatal prostate cancer among smokers, even accounting for screening differences, indicates that tobacco may enhance the aggressive behavior of prostate cancer or cause highly aggressive tumors. Presently, no tissue marker can reliably predict a distinct set of cancers with this aggressive phenotype, but such a group is suggested clinically by cancers that are rapidly progressive and fatal. The associations between recent past and current smoking were particularly strong for prostate cancers that caused death within several years of a negative digital rectal examination. Although it is likely that clones of premalignant cells, or even a small, undetectable cancer, were present at the time of the digital rectal examination, the observation of widespread metastasis and death within a short time period of apparent health suggests a specific "late" event that markedly enhances tumor aggressivity.

Reference Number: 40542

TOABORELLI, M., M. MONTELLA, M. LIBRA, R. TEDESCHI, A. CRISPO, M. GRIMALDI, L. DAL MASO, D. SERRAONO, AND J. POLESEL. "The Dose-Response Relationship Between Tobacco Smoking and the Risk of Lymphomas: A Case-Control Study." *BMC Cancer*, 17: 421 (9 pp), 2017 (4 tables, 2 figures, 35 references)

Abstract: In Europe, approximately 93,500 cases of non-Hodgkin lymphoma (NHL) and 17,500 cases of Hodgkin lymphoma (HL) were diagnosed in 2012 and these types of lymphoma are the 8th most commonly diagnosed cancer. The association of smoking with these cancers was determined in 571 patients with NHL, 1,888 patients with HL, and 1,004 cancer-free controls in Italy between 1999 and 2004. Compared to a non-smoking person, persons who smoked 15+ CPD were 1.42× as likely to develop NHL and 2.47× as likely to develop HL. Former smokers and smokers who smoked less than 15 CPD did not have a significantly increased risk. Smoking duration was significantly associated with HL (OR 2.15×).

These findings support a role of tobacco smoking in the etiology of both NHL and HL, providing evidence of a direct association of risk with smoking intensity.

Reference Number: 40543

O'MALLEY, S.S., R. WU, S.T. MAYNE, AND P.I. JATLOW. "Smoking Cessation is Followed by Increases in Serum Bilirubin, an Endogenous Antioxidant Associated with Lower Risk of Lung Cancer and Cardiovascular Disease." *Nicotine and Tobacco Research*, 16: 1145–1149, 2014 (1 table, 1 figure, 29 references)

Abstract: Unconjugated bilirubin, the primary form of bilirubin in the serum, is a powerful antioxidant and is associated with a lower risk of lung cancer and cardiovascular disease. The serum bilirubin concentration was determined for 155 smokers who were abstinent for 6 weeks and

193 who were not. The mean serum bilirubin increased to 0.06 mg/100mL after smoking cessation compared, to a mean of 0.02 mg/100mL for those who did not quit smoking.

> In conclusion, our study is the first to document that smoking cessation leads to increases in bilirubin concentrations using a longitudinal design. This finding is consistent with cross-sectional studies suggesting a direct role for smoking cessation in increasing bilirubin concentrations. Moreover, we demonstrate that these changes occur shortly after quitting and posit that these modest increases in bilirubin, an endogenous antioxidant, may contribute to some of the early benefits of quitting smoking.

Reference Number: 40544

CHOSI, S.H., J.E. TERRELL, C.R. BRADFORD, T. GHANEM, M.E. SPECTOR, G.T. WOLF, I.M. LIPKUS, AND S.A. DUFFY. "Does Quitting Smoking Make a Difference Among Newly Diagnosed Head and Neck Cancer Patients?" *Nicotine and Tobacco Research*, 18: 2216–2224, 2016 (3 tables, 1 figure, 70 references)

Abstract: The 8-year overall mortality and cancer-specific mortality was determined in 590 new diagnosed head and neck cancer patients. The patients were divided into never-smokers (23.2%), continuing smokers (24.8%), quitters (16.8%), and former smokers (35.3%). Compared to never smokers, continuing smokers had a HR of 2.71× of dying, quitters had 2.38× the risk, and former smokers had 1.68× the risk of dying.

> Compared to never-smokers, continuing smokers have the highest HR of overall mortality followed by quitters and former smokers, which indicates that smoking cessation, even after a cancer diagnosis, may improve overall mortality among newly diagnosed head and neck cancer patients. Health care providers should consider incorporating smoking cessation interventions into standard cancer treatment to improve survival among this population.

Reference Number: 40545

STEIN, J.H., M. BUSHARA, K. BUSHARA, P.E. MCBRIDE, D.E. JORENBY, AND M.C. FIORE. "Smoking Cessation, but Not Smoking Reduction, Reduces Plasma Homocysteine Levels." *Clinical Cardiology*, 25: 23–26, 2002 (2 tables, 1 figure, 18 references)

Abstract: Elevated total plasma homocysteine (Hcy) levels are a risk factor for atherosclerotic vascular disease. Fifty-one daily smokers (mean 35.9 CPD) were randomized into smoking cessation (counselling and nicotine inhaler), reduced smoking (4-8 CPD), or continue smoking groups. After 12 weeks, the subjects in the quit smoking group had a mean plasma Hcy reduced of 11.6%. No significant plasma Hcy reduction occurred in either the subjects who reduced their cigarette consumption or those who maintained it.

In young to middle-aged smokers with normal to mildly elevated Hcy levels, smoking cessation, but not smoking reduction, reduced homocysteine levels. These findings suggest that a "harm reduction" strategy of merely reducing cigarette use may not be sufficient for reducing the vascular risk associated with smoking cigarettes.

4.06 SECONDHAND/PASSIVE SMOKE

"Scientific evidence indicates there is no risk-free level of exposure to secondhand smoke. Breathing even a little secondhand smoke can be harmful to your health."

—Office of the Surgeon General, "There is no Risk-free Level of Exposure to Secondhand Smoke," *Surgeon General's Report* (2006)

Secondhand smoke is the combination of smoke from the burning end of a tobacco product (sidestream smoke) and the smoke that is inhaled and exhaled by smokers (mainstream smoke). Secondhand smoking was initially called involuntary smoking, which in some ways is a better descriptor (40603). It is also known as environmental tobacco smoking (ETS). About 85% of secondhand smoke is sidestream smoke, which is more toxic than mainstream smoke (40601–2).

It is estimated that secondhand smoke causes 53,000 deaths in the United States every year, or approximately one non-smoker death for every eight smoker deaths (40601). SHS has been associated with numerous health problems, including:

- Breast cancer (40616)
- Lung cancer (40610)
- Lung disease/COPD/pneumonia (40609, 40612, 40614)
- Heart attacks/strokes (40604, 40615, 40619)
- Peripheral artery disease/CVD (40604, 40610–11, 40617)

Reference Number: 40545

- Type 2 diabetes (40618)
- Low birthweight (40608)
- Childhood neurological deficits/ADHD (40606–7)
- Periodontal disease (40613)
- Adolescent smoking initiation (40620)

The tobacco industry of course vehemently denied any link between SHS and health problems and labelled it as junk science. "Experts" were hired to testify to this in court (40604). However, this position was no longer tenable, as the evidence mounted and smoking restrictions were imposed on aircraft, bars, restaurants, and finally casinos. When smoke-free laws were introduced, there was a decrease in hospitalization for acute coronary syndrome (40622), hospital admissions for myocardial infarction (40622), and ambulance calls (40623).

Studies on involuntary smoking (i.e., SHS) at workplaces, restaurants, hotels, prisons, and casinos have increased understanding of the risks of SHS and its mitigation (40624–34). The addition of indoor fans and indoor ventilation smoking rooms do not prevent the risk of SHS to non-smokers (40633, 40636). Even outdoor smoking areas expose non-smokers to carcinogenic chemicals in SHS (40635).

Multiunit housing and home visits by health care workers remain problematic (40636–37). In addition, senior non-smokers who were exposed to SHS were 5.4× as likely to be frail as non-smokers not exposed to SHS (40638). This illustrates that not only children, but older adults are more adversely affected by secondhand smoke.

Some good news is that smoke-free laws have dramatically reduced the exposure of non-smokers to carcinogenic tobacco smoke and as a result there has been a 46% decrease in smoking-related cancer mortality among non-smokers (40639).

Reference Number: 40601

SCHICK, S., AND S. GLANTZ. "Philip Morris Toxicological Experiments with Fresh Sidestream Smoke: More Toxic Than Mainstream Smoke." *Tobacco Control*, 14: 396–404, 2005 (4 tables, 92 references)

Abstract: Sidestream smoke is that which arises from the tip of a burning cigarette between puffs. Mainstream smoke is the smoke that arises when the smoker inhales and exhales. Approximately 85% of SHS is sidestream smoke. It is estimated that SHS causes about 53,000 deaths in the United

States every year, which is approximately one non-smoker death for every eight smoker deaths. The tobacco industry has vigorously challenged the negative health effects of SHS. However, according to disclosed internal tobacco documents, Philip Morris conducted 115 studies into the effect of sidestream smoke on laboratory rats at a research institute (INBIFO) in Germany between 1981 and 1989. It was found that sidestream smoke was 4× more toxic per gram of total particulate matter (TPM) than mainstream tobacco smoke.

> Fresh sidestream smoke can cause damage to the respiratory epithelium at low levels, and damage to the respiratory epithelium increases with longer exposures. The number, variety, and results of the fundamental toxicological experiments done by Philip Morris at INBIFO are without parallel in the open scientific literature. These studies were neither published nor revealed to the government in rule making hearings by the US Occupational Safety and Health Administration or in risk assessments by the US EPA and CalEPA. The unpublished research on sidestream smoke condensates and freshly generated cigarette sidestream smoke that Philip Morris did supports the institution of smoke-free policies in workplaces and public places as the only practical way to protect the public from the known risks of lung cancer and other diseases associated with cigarette sidestream smoke.

Reference Number: 40602

ADAMS, J.D., K.J. O'MARA-ADAMS, AND D. HOFFMANN. "Toxic and Carcinogenic Agents in Undiluted Mainstream Smoke and Sidestream Smoke of Different Types of Cigarettes." *Carcinogenesis*, 8: 729–731, 1987 (2 tables, 26 references)

Abstract: Tar, nicotine CO, ammonia, and other toxic chemicals were determined in the mainstream smoke (MS) and sidestream smoke (SS) of four different types of US cigarettes, including one with no filter. The analyses were conducted mainly by GC with a capillary column.

Table: Concentration of Various Toxic Agents in Mainstream and Sidestream Smoke of Filter Cigarette B

	Mainstream Smoke	Sidestream Smoke
Tar (mg)	15.6	24.4
Nicotine (mg)	1.50	4.14
CO (mg)	13.7	36.6
Ammonia (μg)	19.4	893
NNK (ng)	180	752

Source: Adapted from Adams et al, 1987

In summary, we have presented data on the MS/SS distribution of several constituents in the smoke of four US commercial cigarettes. Sidestream deliveries of tar, nicotine, CO and CO_2 are in all cases greater than mainstream deliveries, with the highest SS/MS ratio observed for the ultra-low-yield cigarette. The sidestream deliveries of the carcinogenic, volatile *N*-nitrosamines can be up to 170-fold above mainstream levels which is in line with earlier published data. In the cigarettes tested, NNN was formed primarily during puff-drawing, while NNK was preferentially formed during smouldering. The formation of catechol, BaP and ammonia was also favored during smouldering thus yielding higher levels in SS. These data indicate that the availability of cigarettes with highly reduced mainstream smoke carcinogens has little bearing on the emission of SS carcinogens.

Reference Number: 40603

FRANCIS, J.A., A.K. SHEA, AND J.M. SAMET. "Challenging the Epidemiologic Evidence on Passive Smoking: Tactics of Tobacco Industry Expert Witnesses." *Tobacco Control*, 15: iv68–iv76, 2006 (5 tables, 103 references)

Abstract: An analysis was conducted of seven expert witnesses hired by the tobacco companies to minimize the effects of SHS on health. The experts provided 38 dispositions and testimonies in the 1990s in various legal cases. The witnesses challenged the causation of adverse health effects by SHS and used similar tactics and strategies used in tobacco industry–funded scientific studies.

Just as the tobacco industry developed a strategy to protect its own interests by labelling the epidemiology as "junk science" or "soft science", with the goal of minimising the impact on public policy and public opinion, these general themes are played out in other venues. Unfortunately, epidemiology has been given this pejorative label and the strategy of using the "junk science" label has been used by other parties with financial interests threatened by epidemiological evidence. Criticisms based on the possibility of uncontrolled confounding and the limited utility of epidemiology for "weak effects" have been made in reference to studies of other environmental agents, including air pollution and radiation. In the example of secondhand smoke, these tactics have not remained tenable over time as the scientific evidence continues to mount. Recent reports from the US Surgeon General and the California Environmental Protection Agency have greatly lengthened the list of diseases caused by involuntary smoking.

Reference Number: 40604

MCNEELY, E., I. MORDUKHOVICH, S. STAFFA, S. TIDEMAN, AND B. COULL. "Legacy Health Effects Among Never Smokers Exposed to Occupational Secondhand Smoke." *Public Library of Science One* 14: (12 pp), 2019 (2 tables, 3 figures, 40 references)

Abstract: Secondhand tobacco smoke (SHTS) is a major public health hazard with adverse effects, including lung cancer, CV, and impaired respiratory health. In the 1960–1990s, flight attendants were exposed to high levels of SHTS due to the then high prevalence of inflight smoking, recirculated air, and poor ventilation. In the United States, smoking bans on flights shorter than 2 hours were implemented in 1988 and in 1990 extended to domestic flights of up to 6 hours. Finally, in 1998 nearly all flights to and from the United States were smoke free. The legacy health effects of SHTS were determined using the 2014–2015 Flight Attendant Health Study participants born before 1971, of whom 1,294 were exposed and 1,721 were not exposed to SHTS. No evidence was found of a legacy effect of SHTS for increased rates of cancer. But the OR was elevated for myocardial infarction (1.40×), peripheral artery disease (1.27×), and repeated pneumonia (1.06×).

We report associations between legacy SHTS exposure going back decades and several severe cardiovascular and respiratory outcomes, including MI, PAD, and repeated pneumonia. SHTS is a preventable and modifiable exposure that continues to be a substantial cause of morbidity and premature death worldwide. Only 28 U.S. states have comprehensive public smoking bans and a large percentage of the world's population is exposed to SHTS in the home, community, and workplace, with racial and ethnic minorities and those of lower socioeconomic status disproportionately affected. This represents a tremendous public health burden in light of severe and commonly occurring health outcomes, no known safe level of exposure, and the fact that we observed associations despite decades elapsing since implementation of in-flight smoking bans. Our findings, if confirmed, have implications for informing smoking cessation efforts, and clarifying health education and clinical guidelines among patients with SHTS exposure histories, even in the distant past.

Reference Number: 40605

HOLM, A.L., AND R.M. DAVIS. "Clearing the Airways: Advocacy and Regulation for Smoke-Free Airlines." *Tobacco Control*, 13: i30–i36, 2004 (3 figures, 71 references)

Abstract: The first US national non-smokers rights campaign to make commercial airlines smoke free began in 1969. Progress was slow and between 1969 and 1984, only the creation of smoking and non-smoking sections in the airplane and a ban on pipes and cigars were implemented. Not until 1986, after the publication of the National Academy of Science Report on airplane cabin environment recommended banning smoking on all commercial flights, was significant progress made. In 1988 smoking was banned on all flights of less than 2 hours, and in 1990 it was extended to 6 hours.

The movement to ban smoking in aircraft represents a case study in effective advocacy for smoke-free workplaces. Health advocates, with crucial assistance from flight attendants, used an incremental advocacy process to push for smoking and non-smoking sections on US commercial flights, then for smoking bans on short domestic flights, and finally for completely smoke-free domestic and international flights. Through the course of the battle, advocates from all quarters of tobacco control presented a unified message, exhibited remarkable focus on an attainable goal, and effectively leveraged their relationships with champions in both government and the private sector.

Reference Number: 40606

YERAMANENI, S., K.N. DIETRICH, K. YOLTON, P.J. PARSONS, K.M. ALDOUS, AND E.N. HAYNES. "Secondhand Tobacco Smoke Exposure and Neuromotor

Function in Rural Children." Journal of Pediatrics, 167: 253–259, 2015 (4 tables, 57 references)

Abstract: SHS has been shown to increase the risk of learning disabilities, ADHD, behavior and conduct disorders, and cognition and academic achievement deficit in children. Neuromotor functions of 404 children aged 7–9 years old in Ohio and West Virginia were determined between 2009 and 2013. The children's exposure to SHS were assessed using the parent's self-report of smoking and the child's serum cotinine concentration. A child was considered exposed to SHS if the parent reported smoking more than 1 cigarette per day or the child had a serum cotinine concentration of 0.05 ng/mL or greater. Thirty-seven percent of the children were exposed to a median of 20 cigarettes per day and 50% had a positive serum cotinine concentration.

> Overall, the findings of this study indicate that SHS exposure may play a role in the development of childhood neuromotor function on an individual and population level. Children exposed to SHS have poorer fine motor, gross motor, and visuo-motor skills. Future research is needed to confirm these findings and if these neuromotor function deficits continue to persist in adolescence. Results from our study reinforce the importance of continued interventions to reduce childhood exposure to SHS.

Reference Number: 40607

JOO, H., M-H. LIM, M. HA, H-J. KWON, S.J. YOO, K-H. CHOI, AND K-C. PAIK. "Secondhand Smoke Exposure and Low Blood Lead Levels with Attention-Deficit Hyperactivity Disorder and Its Symptom Domain in Children: A Community-Based Case-Control Study." *Nicotine and Tobacco Research*, 94–101, 2017 (3 tables, 1 figure, 45 references)

Abstract: Urine and blood levels of cotinine and lead respectively were determined in 214 children (aged 6–10 years) with ADHD and 214 matched children without ADHD in South Korea. Low blood lead levels (1.65 μg/dL) were related to ADHD, particularly inattention (OR 1.67×). SHS exposure was related to hyperactivity/impulsivity. In the group with high blood lead levels and SHS, the ORs were 3.16× and 17.06×, respectively. Urinary cotinine was significantly associated with ADHD.

Table: Association Between SHS Exposure (Pre- and Post-Natal) and ADHD

	Prenatal SHS	Postnatal SHS
All ADHD	1.69×	1.98×
Inattention	1.60×	1.50×
Hyperactivity/impulsivity	1.86×	3.85×

Source: Adapted from Joo et al, 2017

In conclusion, low BLLs were associated with ADHD, with inattention symptoms being dominant, and prenatal maternal and/or postnatal SHS exposure in children was associated with ADHD, particularly hyperactivity/impulsivity symptoms. The combined exposure to lead and SHS increased the risk of ADHD synergistically in both symptom domains. In order to protect children from exposure to environmental risk factors for ADHD, further reductions in exposure to lead, even in children whose BLLs are extremely low, as well as cessation of smoking in the home and in children's play areas, are necessary to decrease SHS exposure in pregnant women and children.

Reference Number: 40608

OWILI, P.O., M.A. MUGA, AND H-W. KUO. "Gender Differences in the Association Between Environmental Tobacco Smoke and Birth Weight in Africa." *International Journal of Environmental Research and Public Health*, 15: 1409 (10 pp), 2018 (3 tables, 36 references)

Abstract: Smoking prevalence is increasing in some African countries due to a shift in the tobacco markets from high- to low-income countries. The effect of Environmental Tobacco Smoke (ETS) on birthweight was determined in 23 African countries on 208,207 newborns between 2010 and 2014. Twenty-two percent of mothers were exposed to ETS and the majority (82%) were exposed daily. The adjusted OR of low birthweight was 1.06×, in total, and 1.08× for male newborns.

Exposure to ETS in the household is significantly associated with low birthweight in Africa, particularly among male newborns. Our results find that gender could possibly be a modifier rather than a confounder. More research to explore the biological plausibility of our findings is appropriate. Public health promotions disseminating general behavioral recommendations is likely to have a positive impact on maternal, newborn, and child health.

Reference Number: 40609

SLOWIK, N., S. MA, J. HE, Y.Y. LIN, O.P. SOLDIN, R.A. ROBBINS, AND G.M. TURINO. "The Effect of Secondhand Smoke Exposure on Markers of Elastin Degradation." *Chest*, 140: 946–953, 2011 (2 tables, 3 figures, 19 references)

Abstract: The plasma desmosine and isodesmosine (D/I) ratios (markers of elastin degradation) and cotinine (marker of nicotine exposure) were determined in two groups of passive, active, and non-smokers (n=120) by LC/MS/MS. Elastin is found in many tissues of the body but is of great importance for the structural and functional elements of the alveoli, bronchi, and pleura of the lungs. In patients with COPD, elastin fibers have been shown to be degraded and disorganized and is of low content in the lungs. A higher D/I ratio indicates that there is greater elastin degradation.

Table: Mean D/I and Plasma Cotinine Concentrations According to Cigarette Exposure

Cigarette Exposure	Mean D/I Ratio	Mean Plasma Cotinine Concentration (ng/mL)
Non-exposed	0.22	0.04
SHS	0.29	37.17
Smokers	0.35	75.32

Source: Adapted from Slowik et al, 2011

This study illustrates that SHS exposure may be as dangerous and harmful as active smoking on tissue matrix injury and possibly lung parenchyma. This study demonstrates that D/I can be used to detect early changes of elastin degradation, before clinically significant symptoms occur, and possibly to indicate progression of the disease. Thus, D/I can serve as biomarkers of exposure. It would be significant to determine if increased levels of D/I can be correlated with CT findings of significant parenchymal destruction, before clinical symptoms occur. Overall, having a tissue matrix component affected by SHS and detectable by chemical analysis is a useful adjunct to evaluating the clinical and physiologic consequences of environmental smoke exposure.

Reference Number: 40610

POPE III, C.A., R.T. BURNETT, M.C. TURNER, A. COHEN, D. KREWSKI, M. JERRETT, S.M. GAPSTUR, AND M.J. THUN. "Lung Cancer and Cardiovascular Disease Mortality Associated with Ambient Air Pollution and Cigarette Smoke: Shape of the Exposure-Response Relationships." *Environmental Health Perspectives*, 119: 1616–1621, 2011 (2 tables, 1 figure, 51 references)

Abstract: SHS and exposure to particulate matter less than 2.5 μm in diameter ($PM_{2.5}$) is a cause of both lung cancer and cardiovascular disease (CVD). Data from 1.2 million adults, collected by the American Cancer Society, were analyzed and the relative risks were calculated according to the estimated daily exposure to $PM_{2.5}$ due to smoking, SHS, and pollution. Lung cancer mortality risk increased linearly with increasing $PM_{2.5}$ exposure to a maximum RR of 40 among long term heavy smokers. The CVD mortality risks, on the other hand, increased steeply at low $PM_{2.5}$ exposure and leveled off at about 2–3 for heavy cigarette smoking.

Table: Exposure to $PM_{2.5}$ and Adjusted Relative Risk of Lung Cancer and Cardiovascular Disease

Exposure	Estimated Daily Dose $PM_{2.5}$ (mg)	Adjusted RR of Lung Cancer	Adjusted RR of CVD
Living with smoking spouse	0.54	1.21×	–
Moderate–high SHS exposure	0.90	–	1.26×
Smoking ≤ 3 cigarettes per day	18	10.44×	1.58×
Smoking 20 cigarettes per day	240	19.88×	2.42×
Smoking 30 cigarettes per day	360	26.82×	2.17×
Smoking 45 cigarettes per day	540	39.16×	2.17×

Source: Adapted from Pope III et al, 2011

The exposure–response relationship associated with $PM_{2.5}$ is qualitatively different for lung cancer versus cardiovascular mortality. At low exposure levels, cardiovascular deaths are projected to account for most of the burden of disease, whereas at high levels of $PM_{2.5}$, lung cancer becomes proportionately more important.

Reference Number: 40611

YANKELEVITZ, D.F., C.I. HENSCHKE, R. YIP, P. BOFFETTA, J. SHEMESH, M.D. CHAM, J. NARULA, AND H.S. HECHT. "Second-Hand Tobacco Smoke in Never

Smokers Is a Significant Risk Factor for Coronary Artery Calcification." *JACC: Cardiovascular Imaging*, 6: 651–657, 2013 (4 tables, 1 figure, 31 references)

Abstract: The relationship between SHTS and coronary artery calcification (CAC) was determined in 3,098 never-smokers (ages 40–80 years) enrolled in a Flight Attendants Study between 2005 and 2012. The overall prevalence of CAC was 24.3%, 26.4% in those exposed to SHTS compared to 18.5% in those with minimal exposure. The adjusted OR for CAC was 1.54× for low, 1.60× for moderate, and 1.93× for high SHTS exposure.

With the increasing body of evidence linking SHTS exposure to CAD, consideration should be given to its official recognition as an important independent risk factor. Standard medical history taking should be broadened to include SHTS exposure. It is critical to recognize that it is an eminently avoidable risk factor, as evidenced by the reduction in acute myocardial infarction after implementation of smoke-free laws. The ability to document increased cardiovascular risk in never smokers exposed to SHTS from a low-dose screening CT scan offers an opportunity for combined early detection and treatment of lung cancer, emphysema, and cardiovascular disease, the 3 major diseases attributable to SHTS exposure. Ongoing progress in the development of gated CT scans at low-dose radiation is the next step in the routine inclusion of CAC assessment while screening for lung cancer.

Reference Number: 40612

WU, C-F., N-H. FENG, I-W. CHONG, K-Y. WU, C-H. LEE, J-J. HWANG, C-T. HUANG, C-Y. LEE, S-T. CHOU, D.C. CHRISTIANI, AND M-T. WU. "Second-Hand Smoke and Chronic Bronchitis in Taiwanese Women: A Health-Care Based Study." *BMC Public Health*, 10: 44 (10 pp), 2010 (4 tables, 1 figure, 42 references)

Abstract: The smoking and SHS exposure of 33 women with chronic bronchitis, 182 with probable bronchitis, and 205 with no chronic bronchitis in Taiwan were determined. The urinary cotinine concentration in the women was calculated by LC/MS/MS to determine SHS exposure. The rate of cigarette smoking is 55–60% among Taiwanese men and only 3–4% among women. Women who smoked were 24.8× more likely to have chronic bronchitis and women exposed to SHS were 3.7× as likely to develop bronchitis.

In conclusion, we found a dose-response trend on the effect of lifetime exposure to SHS and risk for chronic bronchitis, with SHS exposure accounting for 47.3% of the attributable risks of having chronic bronchitis. Our findings add additional evidence to the growing body of knowledge supporting the great need for health policies to ensure that a smoke-free environment is created and maintained for the public to decrease the risk of adverse health consequences by non-smokers exposed to SHS.

Reference Number: 40613

ARBES JR., S.J., H. AGÚSTSDÓTTIR, AND G.D. SLADE. "Environmental Tobacco Smoke and Periodontal Disease in the United States." *American Journal of Public Health*, 91: 253–257, 2001 (2 tables, 49 references)

Abstract: Environmental Tobacco Smoke contains more than 4,000 chemicals including nicotine and at least 40 known carcinogens. Approximately 50% of periodontitis in persons younger than 30 years is due to cigarette smoking. A study was conducted on 6,611 persons aged 18 years or greater from the 1988–1994 National Health and Nutrition Examination Survey (NHANES III) who were assessed for periodontal disease and ETS exposure.

The major finding of this study was that among persons in the United States who had never smoked cigarettes (or used pipes, cigars, or smokeless tobacco), the odds of having periodontal disease were 1.6 times greater for persons exposed to ETS than for persons not exposed to ETS, after control for known risk factors for periodontal disease.

Reference Number: 40614

EISNER, M.D., C. IRIBARREN, E.H. YELIN, S. SIDNEY, P.P. KATZ, G. SANCHEZ, AND P.D. BLANC. "The Impact of SHS Exposure on Health Status and Exacerbations Among Patients with COPD." *International Journal of COPD*, 4: 169–176, 2009 (4 tables, 43 references)

Abstract: SHS is a major source of indoor pollution and its effect on COPD was determined in 809 never-smoking patients with COPD. Nearly half of the current non-smokers reported SHS exposure during the last week. Of those, 55% had a lower level and 45% had a higher level of SHS exposure. Higher levels of SHS exposure was associated with poorer physical health status, disease-specific health related quality of life (HRQL), and less distance walked in the Six Minute Walk Test.

> In sum, SHS exposure appears to increase the risk of adverse health outcomes, including poorer physical health status, HRQL, exercise capacity, and disease exacerbation resulting in emergency health care utilization. These poor health outcomes were observed at low levels of SHS exposure. Patients with COPD may comprise a vulnerable population who are at especially high risk of the respiratory health effects of SHS exposure. Promotion of smoke-free indoor environments, including the home and public places, will likely benefit persons with COPD and other chronic respiratory diseases.

Reference Number: 40615

BONITA, R., J. DUNCAN, T. TRUELSEN, R.T. JACKSON, AND R. BEAGLEHOLE. "Passive Smoking as Well as Active Smoking Increases the Risk of Acute Stroke." *Tobacco Control*, 8: 156–160, 1999 (4 tables, 1 figure, 27 references)

Abstract: A comparison was made between 521 (279 male and 242 female) patients with first ever acute stroke and 1,851 community controls (aged 35–74 years) in Auckland, New Zealand between 1991 and 1992. The adjusted OR of stroke was 1.82× in non-smokers exposed to SHS and 4.27× in current smokers. The OR of stroke in current smokers increased to 6.33× when the community controls exposed to SHS were removed as a comparison group.

> This study is one of the few to investigate the association between passive smoking and the risk of acute stroke. We found a significantly increased risk of stroke in men and in women. This study also confirms the higher risk of stroke in men and women who smoke cigarettes compared with non-smokers. The stroke risk increases further when those who have been exposed to ETS are excluded from the non-smoking reference group. These findings also suggest that studies investigating the adverse effects of smoking will underestimate the risk if exposure to ETS is not taken into account.

Reference Number: 40616

WADA, K., T. KAWACHI, A. HORI, N. TAKEYAMA, S. TANABASHI, S. MATSUSHITA, N. TOKIMIOTSU, AND C. NAGATA. "Husband's Smoking Status and Breast

Cancer Risk in Japan: From the Takayama Study." *Cancer Science*, 106: 455–460, 2015 (3 tables, 33 references)

Abstract: A population based prospective study was conducted in Japan of 15,719 women (aged 35+ years) between 1992 and 2008. Former smokers were generally the oldest and current smokers were generally the youngest of the groups studied. The women who had ever smoked were likely to drink more and have fewer childbirths. The adjusted OR for breast cancer was 1.28× for former smokers, 1.71× for never-smoking women exposed to low SHS from their husbands, and 2.03× for never-smoking women exposed to high SHS from their husbands.

> In conclusion, this prospective study of Japanese women demonstrated the increased risk of breast cancer among women exposed to secondhand cigarette smoke from their husband, with a dose-response relationship. The increased risks for women with a smoking husband were stronger among those who do not habitually consume alcohol. These results suggest that exposure to smoke from husbands is a potential risk factor for breast cancer, and, considering the limited data currently available, the modifying effects of alcohol on increased breast cancer risk from passive smoking need to be confirmed in further studies.

Reference Number: 40617

PEINEMANN, F., S. MOEBUS, N. DRAGANO, S. MÖHLENKAMP, N. LEHMANN, H. ZEEB, R. ERBEL, K-H. JÖCKEL, AND B. HOFFMANN. "Secondhand Smoke Exposure and Coronary Artery Calcification Among Nonsmoking Participants of a Population-Based Cohort." *Environmental Health Perspectives*, 119: 1556–1561, 2011 (3 tables, 2 figures, 31 references)

Abstract: Secondhand smoke is a combination of smoke emitted from the burning end of the cigarette and smoke exhaled by the smoker. It consists of fine particulate matter, carcinogens, and various toxins. The association of exposure to SHS and coronary artery calcification (CAC) was determined in 1,766 never-smokers (aged 45–70 years) without clinically manifested coronary heart disease in Germany between 2000 and 2003. Frequent exposure to SHS was reported by 21.5% of the participants. CAC is a measure of subclinical atherosclerosis and is highly predictive of future cardiovascular events. SHS exposure increased the overall risk of having detectable CAC by 1.38×.

The observed association between SHS and CAC is primarily based on the contrast between no calcification and any detectable calcification. Different biological mechanisms could be responsible for the development of plaque, which encompasses earlier stages of atherosclerotic disease, than those leading to the progression of already existing calcified plaque in advanced atherosclerotic disease. Animal experiments support the role of SHS in early stages of atherosclerosis, showing a causal role of SHS in the development of atherosclerotic lesions. Exposure to SHS is associated with permanent inflammation and lipid accumulation in blood vessels in mice and in utero exposure results in an increased and earlier adult atherosclerotic lesion formation. Human studies have shown that the pathophysiology of SHS-induced cardiovascular toxicity includes arterial endothelial dysfunction, platelet activation, oxidation of low-density lipoprotein cholesterol, and increased insulin resistance, which are risk factors for atherosclerosis at early stages of the disease. The association between exposure to SHS and atherosclerosis might be initiated as early as during gestation. Increased thickening of the coronary walls has been observed in autopsies of fetuses of smoking mothers and of infants of smoking parents.

Reference Number: 40618

LAJOUS, M., L. TONDEUR, G. FAGHERAZZI, B. DE LAUZON-GUILLAIN, M-C. BOUTRON-RUAUALT, AND F. CLAVEL-CHAPELON. "Childhood and Adult Secondhand Smoke and Type 2 Diabetes in Women." *Diabetes Care,* 36: 2720–2725, 2013 (2 tables, 1 figure, 28 references)

Abstract: A prospective study was conducted in France of 37,343 women who never smoked and were free of type 2 diabetes, cancer, or cardiovascular disease at baseline in 1992. Between 1992 and 2007, there were 795 cases of type 2 diabetes identified in this group. Childhood SHS was defined as having at least one parent who smoked. Adult SHS was defined if they were currently living with someone who smoked. The subjects also estimated the number of hours they were exposed to SHS at home or outside the home. Women who were exposed to SHS in childhood were 1.8× as likely to have type 2 diabetes. Adult exposure to SHS resulted in being 1.36× as likely to develop type 2 diabetes.

In conclusion, this prospective analysis suggests that secondhand smoke exposure in childhood and adulthood is associated with a higher rate of type 2 diabetes. Smoke-free environments have proven to be relatively easy to implement and effective in the control of cardiovascular disease. Curbing the worldwide type 2 diabetes epidemic requires extensive and lasting changes in public policy. Limiting secondhand smoke exposure by providing smoke-free environments and improving compliance with smoking bans may be an important strategy.

Reference Number: 40619

WHINCUP, P.H., J.A. GILG, J.R. EMBERSON, M.J. JARVIS, C. FEYERABEND, A. BRYANT, M. WALKER, AND D.G. COOK. "Passive Smoking and Risk of Coronary Heart Disease and Stroke: Prospective Study with Cotinine Measurement." *British Medical Journal*, 329: 200–205, 2004 (4 tables, 2 figures, 26 references)

Abstract: The serum cotinine concentrations were determined in 2,015 men in the UK who stated that they did not smoke between 1978 and 1980. Major coronary heart disease and stroke events (fatal and nonfatal) were determined every 5 years for 20 years.

Table: Initial Serum Cotinine Concentration and Adjusted Odds Ratio of Coronary Heart Disease Over 20 Years

Nicotine Cotinine Concentration (ng/mL)	Adjusted OR of CHD Over 20 Years
≤0.7	1.0×
0.8–1.4	1.54×
1.5–2.7	1.89×
2.8–14	1.57×
Active smokers	2.05×

Source: Adapted from Whincup et al, 2004

High overall exposure to passive smoking seems to be associated with a greater excess risk of CHD than partner smoking and is widespread in non-smokers, suggesting that the effects of passive smoking may have been underestimated in earlier studies. Further prospective studies of the association between cotinine (or similar biomarkers) and risk of CHD will help to assess the effects of passive smoking on cardiovascular disease with greater precision. In the meantime, our results add to the weight of evidence suggesting that exposure to passive smoking is a public health hazard and should be minimised.

Reference Number: 40620

MCGRATH, J.J., S. RACICOT, C.T.C. OKOLI, S.K. HAMMOND, AND J. O'LOUGHLIN. "Airborne Nicotine, Secondhand Smoke, and Precursors to Adolescent Smoking." *Pediatrics*, 141: S63-S73, 2018 (3 tables, 1 figure, 62 references)

Abstract: When exposed to SHS, non-smokers inhale 60–80% of the nicotine and absorb concentrations similar to active smokers. Concentrations of nicotine in hair samples of bar and restaurant workers exposed to SHS were the same for both smokers and non-smokers. The hair nicotine concentration of non-smokers whose spouses smoked outside of the home were elevated compared to those whose spouses were non-smokers. Hair nicotine, salivary cotinine, and passive nicotine markers (after 1 week of wearing) were analyzed from 338 adolescents (mean age 13 years). The adolescents were also surveyed as to their beliefs about smoking.

> Airborne nicotine exposure via SHS is a plausible risk factor for smoking initiation during adolescence. Public health implications include limiting airborne nicotine through smoking bans in homes and cars, in addition to stringent restrictions for e-cigarettes.

Reference Number: 40621

GUPTA, R., J. LUO, R.H. ANDERSON, AND A. RAY. "Clean Indoor Air Regulation and Incidence of Hospital Admissions for Acute Coronary Syndrome in Kanawha County, West Virginia." *Preventing Chronic Disease* 8: A77 (8 pp), 2011 (2 tables, 3 figures, 25 references)

Abstract: An analysis was conducted of 14,245 hospital patients admitted for acute coronary syndrome (ACS) in Kanawha County between 2000 and 2008. A modest Clean Indoor Air Regulation (CIAR) was passed in Kanawha County in 1995 and revised with increasing enforcement and restrictions in 2000 and 2003. The cumulative decrease in the age-adjusted ACS hospitalization rate between 2000 and 2008 was a 36% decline for men and 39% in women, but the decline was significant only among non-smokers. Smokers and diabetics did not have a significant decline in ACS.

> To our knowledge, this is the first study to describe a pattern of consistent decline in the rates of hospital admissions for ACS during an 8-year period (2000-2008) in a community with an existing CIAR. During this

period, a significant drop in ACS hospital admission rates was observed for nonsmokers, people without diabetes, and women. However, this benefit did not reach significance for smokers. Additionally, effective January 1, 2004, when the Kanawha County regulation was further strengthened by making restaurants completely smoke-free (by removal of smoking areas in restaurants), male smokers were able to attain similar gains. Our findings indicate that by eliminating smoking areas at restaurants, bars, airports, and other public places and making them completely smoke-free, we may be able to decrease additionally the number of acute coronary events among male smokers. As expected, our study reveals that people who do not have diabetes benefit more from reduced ACS hospital admission rates regardless of sex and smoking status.

Reference Number: 40622

SARGENT, R.P., R.M. SHEPARD, AND S.A. GLANTZ. "Reduced Incidence of Admissions for Myocardial Infarction Associated with Public Smoking Ban: Before and After Study." *British Medical Journal*, 328: 977–980, 2004 (1 table, 1 figure, 24 references)

Abstract: The number of patients admitted for acute myocardial infarction (MI) was determined in Helena, Montana between 1997 and 2003. Helena is a geographically isolated community of 68,140 with one hospital. Between June and November 2002, Helena (but not the surrounding area) was smoke-free. The number of MI admissions in Helena for 6 months before the ban was 40 and decreased to 24 for the 6 smoke-free months. In the surrounding area, the number of MI admissions was 12 and 18 for the same time periods.

Researchers have predicted that smoke-free laws would be associated with a reduced incidence of acute myocardial infarction through a combination of reduced exposure to secondhand smoke and encouraging smokers to quit (38% of the patients with acute myocardial infarction in the study were current smokers, 29% were former smokers, and 33% had never smoked at the time of admission). While both of these effects are probably occurring, we do not have large enough sample size to estimate their relative contribution to our results. Several mechanisms, including increased platelet activation, inhibition of vascular endothelium, impairment of coronary artery dilatation capacity, decreases in antioxidant substances especially ascorbic acid, aortic stiffening, and impaired heart rate variability, all of which could increase the likelihood of an acute coronary event, have been measured within minutes to hours of exposure to secondhand smoke. Even occasional exposure to secondhand smoke has been associated with an increased risk for acute coronary syndromes.

Reference Number: 40623

GLANTZ, S.A., AND E. GIBBS. "Changes in Ambulance Calls After Implementation of a Smoke-Free Law and Its Extension to Casinos." *Circulation*, 128: 811-813, 2013 (6 tables, 28 references)

Abstract: The number of ambulance calls in Gilpin County, Colorado between 2000 and 2012 was determined both before and after the smoke-free workplace law in Colorado (2006) was extended to casinos in 2008. Gilpin County contains 26 casinos and is a major tourist destination. Initial implementation of the smoke-free laws in 2006 resulted in a 22.8% drop in ambulance calls statewide but no drop in ambulance calls to casinos. When the smoke-free law was extended to casinos in 2008, there was a 19.1% drop in ambulance calls from casinos.

These results suggest that exempting casinos from smoke-free laws means that more people will suffer medical emergencies as a result. As of April 5, 2013, only 19 states plus Puerto Rico had laws making all state-regulated gambling smoke-free. In addition, although some tribal casinos are smoke-free, most are not. Our results suggest that applying smoke-free laws to casinos not only would prevent medical emergencies but also would reduce costs to the state, which may incentivize the passing of smoke-free casino laws.

Reference Number: 40624

SU, C-P., G. SYAMLAL, S. TAMERS, J. LI, AND S.E. LUCKHAUPT. "Workplace Secondhand Tobacco Smoke Exposure Among U.S. Nonsmoking Workers, 2015." *Morbidity and Mortality Weekly Report*, 68: 604–607, 2019 (1 table, 1 figure, 10 references)

Abstract: Between 2000 and 2015, the number of US states with smoke-free laws that prohibited smoking in the indoor areas of worksites, restaurants, and bars increased from none to 27. In 2015, about 20% of non-smoking workers reported any exposure to secondhand smoke (SHS). In US states with smoke-free laws, frequent exposure to SHS by workers was 8.6%, compared to 11.0% for workers in those states without smoke-free laws. Workplace SHS exposure is one of the top occupational hazards that contribute to occupational cancer.

Table: Percentage of Non-smokers Exposed to Secondhand Smoke at Work, United States, 2015

Work Location	Percent Exposed
Commercial and industrial machinery repair and maintenance	65.1%
Forestry and logging	52.9%
Beverage and tobacco manufacturing	48.7%
Taxis/limousines	43.6%
Museums	47.6%
Construction	41.9%
Performing arts/sports events	36.5%

Source: Adapted from Su et al, 2019

Workplace SHS exposure is harmful for workers' health. In this study, nonsmoking workers residing in states without comprehensive smoke-free laws and those employed in certain industries were more likely to be frequently exposed to workplace SHS. NIOSH encourages employers, especially those in industries with high prevalences of SHS exposure, to implement workplace-specific smoke-free policies to complement state and local smoke-free laws to help reduce SHS exposure among workers and protect workers' health

Reference Number: 40625

JAMROZIK, K. "Estimate of Deaths Attributable to Passive Smoking Among UK Adults: Database Analysis." *British Medical Journal*, 330: 812 (5 pp), 2005 (3 tables, 30 references)

Abstract: Deaths from SHS in employees of the hospitality industry, general workforce, and general population were estimated in the United Kingdom in 2003. Approximately 320,262 persons worked in pubs, bars, and nightclubs and 829,401 worked in hotels and restaurants. It was estimated that the prevalence of SHS in homes was 42%. At the time, nearly all bars, pubs, hotels, and restaurants permitted smoking.

Passive smoking is accepted as a cause of fatal diseases in adults. Existing studies have shown these risks for both domestic exposure and exposure at work. . . . Passive exposure to tobacco smoke at work may cause more than 600 deaths each year in the UK, including over 50 in people employed in the hospitality industry. Exposure at home may account for 2700 deaths in those aged 20–64 and 8000 in those aged ≥65.

Reference Number: 40626

JONES, M.R., H. WIPFLI, S. SHAHRIR, E. AVILA-TANG, J.M. SAMET, P.N. BREYSSE, AND A. NAVAS-ACIEN. "Secondhand Tobacco Smoke: An Occupational Hazard for Smoking and Non-smoking Bar and Nightclub Employees." *Tobacco Control*, 22: 308–314, 2013 (3 tables, 1 figure, 45 references)

Abstract: As of 2011, only 11% of the world's population was protected by smoke-free policies that included bars, restaurants, and nightclubs. Air nicotine concentrations were measured in 238 bars and nightclubs in 24 large cities throughout the world. Hair nicotine concentrations were also measure in 625 non-smoking and 311 smoking employees. Each centimeter of hair is equivalent to about one month's exposure to cigarette smoke. Eighteen percent of the venues were smoke-free. The mean air nicotine concentration was 3.5 μg/m^3 for smoking and 0.2 μg/m^3 for smoke-free locations. A 2× increase in air nicotine concentration is associated with a 30% increase in hair nicotine in non-smoking staff and 10% increase in smoking staff.

Table. Median Air Nicotine Concentrations (μg/m^3) in Smoking and Smoke-Free Bars and Nightclubs Globally

Smoking Status in Venue	Americas	Eastern Europe	Asia	Africa
Smoking Allowed	3.0	8.5	3.4	1.60
Smoke-Free	0.2	0.1	0.9	0.03

Source: Adapted from Jones et al, 2013

In conclusion, the high levels of airborne nicotine found in bars and nightclubs and the contribution of this exposure to employee hair nicotine concentrations support the need for legislation that regulates smoking in these environments and provides complete protection from secondhand smoke for all employees. This is an opportune moment for countries to honour their commitments under the FCTC and expand the number of people protected from secondhand smoke worldwide. In countries with comprehensive legislation, efforts are needed to ensure complete enforcement. By eliminating secondhand tobacco smoke in socializing and hospitality venues, smoke-free legislations can reduce the burden of disease related to secondhand smoke exposure.

Reference Number: 40627

NEBOT, M., M.J. LÓPEZ, C. ARIZA, M. PÉREZ-RÍOS, M. FU, A. SCHIAFFINO, G. MUÑOZ, E. SALTÓ, AND E. FERNÁNDEZ. "Impact of the Spanish Smoking Law on Exposure to Secondhand Smoke in Offices and Hospitality Venues: Before-and-After Study." *Environmental Health Perspectives,* 117: 344–347, 2009 (3 tables, 39 references)

Abstract: Vapor phase nicotine concentrations were measured using SHS passive samples in 398 premises, including offices, universities, bars, and restaurants throughout Spain before and after a no smoking law came into force in January 2006. However, smoking was banned only in some hospitality venues (about 10–20%). Airborne nicotine concentrations were reduced markedly in smoke-free offices (by 60–97%) and smoke-free restaurants (97%) but not in premises that still allowed smoking.

Table: Median Airborne Nicotine Concentrations in Various Premises in Spain Before and After the Enforcement of a Smoke-free Law

	Median Airborne Nicotine Concentration		
Premises	Before the Law (μg/m^3)	12 Months After the Law (μg/m^3)	Percent Change
Public administration	0.20	0.08	–60.0
Universities	0.21	0.07	–66.7
Private sector	0.39	0.01	–97.4
Bars/restaurants where smoking is banned	2.71	0.09	–96.7
Designated smoking area	5.58	8.89	+37.20
Pubs/discotheques where smoking is allowed	33.31	15.06	–54.79

Source: Adapted from Nebot et al, 2009

Overall, this study shows the positive impact of the law on reducing SHS in indoor workplaces. However, SHS was substantially reduced only in bars and restaurants that became smoke-free. Most hospitality workers continue to be exposed to very high levels of SHS. Therefore, a 100% smoke-free policy for all hospitality venues is required.

Reference Number: 40628

BINSWANGER, I.A., E.A. CARSON, P.M. KRUEGER, S.R. MUELLER, J.F. STEINER, AND W.J. SABOL. "Prison Tobacco Control Policies and Deaths from Smoking in United States Prisons: Population Based Retrospective Analysis." *British Medical Journal*, 349: g4542 (19 pp), 2014 (5 tables, 1 figure, 42 references)

Abstract: Smoking tobacco accounts for approximately 6 million premature deaths a year due to lung cancer, ischemic heart disease, COPD, and cerebrovascular disease. In 1993, the US Supreme Court ruled that exposure of prisoners to environmental tobacco smoke in prison can be considered a cruel and unusual punishment in violation of the Eighth Amendment. Smoking contributes to substantial excess mortality in prisons. Prisons that implemented smoking bans had a 9% reduction in smoking-related deaths.

> Our results suggest that smoking bans in prisons can reduce the risk of death in prison. Smoking bans do not directly reduce mortality but act through effects on smoking and exposure to secondhand smoke. The institution of prison smoking bans has been associated with a reduction in exposure to secondhand smoke and improvements in air quality. Smoking bans can also reduce the initiation of smoking in prison, improve health among correctional officers and other prison staff, improve fire safety, and reduce costs associated with liability insurance, facility maintenance, and healthcare.
>
> On the other hand, complete (indoor and outdoor) smoking and tobacco bans in prisons strongly limit individual autonomy and might not prepare individuals with tobacco use disorders for life in the community, where they can purchase and consume tobacco. Internationally, various types of incomplete bans are being implemented, such as bans that prohibit smoking in public prison spaces, in cells, or in all spaces except for designated smoking areas. Some prisons have also prohibited correctional facilities from forcing people who do not smoke to share cells with inmates who smoke. In addition, smoking bans do not lead to sustained long term cessation. Many people resume smoking after release from prison, and others might be transferred to or re-incarcerated in jails, halfway houses, and drug treatment facilities without bans. Finally, some inmates smoke despite bans, which can lead to correctional sanctions and negative health outcomes for people with tobacco use disorders. Reduced smoking and

smoke exposure, however, can have positive health benefits even if there is not complete adherence with a ban.

Reference Number: 40629

JAYES, L.R., E. RATSCHEN, R.L. MURRAY, S. DYMOND-WHITE, AND J. BRITTON. "Second-Hand Smoke in Four English Prisons: An Air Quality Monitoring Study." *BMC Public Health*, 16: 119 (8 pp), 2016 (2 tables, 5 figures, 19 references)

Abstract: Prisons in England are allowed to have smoke-free landings and/or wings in prisons, but allow prisoners to smoke in single cells or in cells shared with other smokers. About 80% of the 85,000 prisoners detained in England and Wales smoke. TSI SidePak AM510 Personal Aerosol Monitors were used to measure the particulate matter of less than 2.5 µm in diameter at various smoking and non-smoking areas in four prisons. WHO recommends the $PM_{2.5}$ should not exceed 25 µg/m^3 as a 24-hour mean. High concentrations of $PM_{2.5}$ were found in five smoker's cells and ranged from 62.3 to 434.7 µg/m^3. Concentrations of $PM_{2.5}$ in non-smoker's cells were on average 17.0 µg/m^3.

Prisoners in England who want to avoid SHS exposure are entitled to request a non-smoking cell, but our findings suggest that being in a non-smoking cell does not necessarily offer protection against SHS, especially for those on wings with closed narrow corridors. Staff members are also able to opt to work in smoke-free areas of the prison, but such opportunities are relatively rare, resulting in significant exposure for many staff. SHS exposure of pregnant women is also a significant potential hazard for both prisoners and staff members; at the time that this study was carried out, pregnant prisoners were not usually transferred to a smoke-free environment until they have given birth. During data collection at the female closed prison there were 18 pregnant prisoners living on main prison locations, though their smoking status was not known.

Our findings thus provide strong evidence that smoking in prisons in England is a source of high SHS exposure for both staff and prisoners and therefore the current PSI relating to smoking in English and Welsh prisons requires revision. It is likely that our findings are also representative of exposures in similar prison systems in other countries. It is self-evident that this exposure would be reduced by promoting smoking cessation amongst staff and prisoners, increasing the amount of voluntary smoke-free wings and ultimately prevented by making prisons comprehensively smoke-free.

Reference Number: 40630

SEMPLE, S., H. SWEETING, E. DEMOU, G. LOGAN, R. O'DONNELL, AND K. HUNT. "Characterising the Exposure of Prison Staff to Second-Hand Tobacco Smoke." *Annals of Work Exposures and Health*, 61: 809–821, 2017 (3 tables, 1 figure, 35 references)

Abstract: The WHO states that there is no safe level of exposure to SHS. $PM_{2.5}$ and airborne nicotine concentrations were determined in 15 Scottish prisons over a 6-day period. Salivary cotinine concentrations were also measured in non-smoking staff. The median exposure to SHS of $PM_{2.5}$ was 20–30 μg/m^3. Overall, 138 of 149 non-smoking prison staff had a median salivary cotinine concentration increase of +0.138 ng/mL by the end of the shift.

Table: Mean $PM_{2.5}$ Concentrations (μg/m^3) in 15 Prisons

Location	Mean $PM_{2.5}$ (μg/m^3)
Reception	2.4
Teaching area	4.1
Health care/gym	4.7
Outdoor	5.9
Staff office	18.8
Workshops	19.1
Residential corridor/landing	37.5

Source: Adapted from Semple et al, 2017

SHS is harmful and the vast majority of people in the UK, and many other countries, are now protected from exposure to this hazard while at work. Prison staff are aware of the protection afforded to other worker groups and see governmental campaigns and health advice on the importance of living in a smoke-free environment. There is an urgent need to find suitable and effective policies that can help protect prison workers from exposure to SHS given the high proportion of prisoners who smoke.

Reference Number: 40631

HEFLER, M., R. HOPKINS, AND D.P. THOMAS. "Successes and Unintended Consequences of the Northern Territory's Smoke-Free Prisons Policy: Results from a Process Evaluation." *Public Health Research and Practice*, 26: (8 pp), 2016 (2 tables, 26 references)

Abstract: The Northern Territory was the first Australian jurisdiction to introduce a smoke-free policy for all correctional facilities. Prisoners received free nicotine replacement therapy for 6 to 12 months upon admission and free, unmonitored access to a telephone Quitline. Fifty-nine prisoners, 4 visiting family members, and 24 prison staff were interviewed post-ban. Misuse of nicotine patches was a problem initially.

This study found that a comprehensive communication and awareness campaign with a long lead-in time was a vital component of introducing a smoke-free policy in NT correctional services without riots or other major incidents. It allowed informal problem solving among prisoners and staff, which provided a mechanism for potential tensions to dissipate before implementation and generated widespread acceptance of the policy. This collaborative approach, based on building support among staff and prisoners, aligns with findings from similar evaluations of smoke-free prisons in New Zealand and the US, and the introduction of smoke-free policies on US navy submarines in 2010. It also takes into account lessons learnt from unsuccessful implementation in a mental health inpatient unit.

The evaluation highlights the need to understand prisoners' preferences regarding provision of NRT to assist with abrupt, involuntary cessation. Social cues are a relapse risk factor for smokers who are attempting to, or have recently, quit; residing in a facility where these cues are absent or minimal may reduce the need for NRT. Further research in the Australian context should focus on identifying effective support mechanisms for motivating and supporting prisoners to maintain abstinence following their release from smoke-free facilities that have relatively high compliance.

Reference Number: 40632

ROBINSON, A., H. SWEETING, AND K. HUNT. "UK News Media Representations of Smoking, Smoking Policies and Tobacco Bans in Prisons." *Tobacco Control*, 27: 622–630, 2018 (2 tables, 1 figure, 43 references)

Abstract: Smoking rates among prisoners in jails are 2–4× that of the general population. Smoking is seen as a way of helping prisoners with boredom and isolation and thought to be an integral part of prison culture. Many jurisdictions have total or partial prison smoking bans due to SHS, health, safety, and insurance cost issues. An analysis was conducted of 21 newspapers in the United Kingdom and Scotland regarding coverage of the implementation of smoke-free prisons between 2015–2016.

Reporting largely reinforced a negative view of prisoners in discussions of rights to a smoke-free environment, prison smoking cultures and SHS. The general lack of explicit discussion of health impacts, particularly for prisoners, was notable. Reporting in anticipation of smoke-free prisons emphasised problems, despite lack of evidence of disorder resulting from smoking bans; the only actual incidents covered were an Australian riot (repeatedly covered) and a suicide in Wales, attributed to lack of cessation support in a smoke-free prison. Anticipated negative outcomes, particularly violence and unrest, were consistently presented alongside discussion of future smoke-free policies, and when these were countered with more positive statements, references were vague and much shorter. The only representation of a largely trouble-free implementation of smoke-free prison policy was presented as unusual.

As in previous media analyses, stories were episodic, focused mainly on individuals or events, rather than broader public health or societal issues. Opportunities for health gains, and tackling smoking in one of the most disadvantaged populations in whom smoking remains high, were rarely presented, nor was health protection highlighted as a rationale for smoke-free prisons. Discussion of health inequalities in relation to smoking and the challenges of supporting smoking cessation in this population group were also missing.

Reference Number: 40633

ZHOU, Z., D. BOHAC, AND R.G. BOYLE. "Continuous Weeklong Measurements of Indoor Particle Levels in a Minnesota Tribal Casino Resort." *BMC Public Health*, 16: 870 (6 pp), 2016 (1 table, 3 figures, 17 references)

Abstract: Fine particulate matter ($PM_{2.5}$, i.e., <2.5 μm in diameter) was measured at various locations in a casino (both smoking and non-smoking) and outdoors. On average, active smokers accounted for about 15% of the casino patrons in the smoking area. Indoor $PM_{2.5}$ was nearly 13×

the concurrent $PM_{2.5}$ level. Gaming floor $PM_{2.5}$ was highest on Saturday nights and average 62.9 μg/m^3. The highest $PM_{2.5}$ concentration was in the smoking-permitted employee breakroom (600 μg/m^3). $PM_{2.5}$ results in the non-smoking area were in synch with the $PM_{2.5}$ of the smoking area.

Our collaboration with Shooting Star casino enabled us to explore temporal and spatial variability of SHS exposure inside a casino and dining facilities. The results suggest that indoor concentration of PM2.5 is substantially higher than the outdoor level, posing health risks to casino workers and patrons. The results also demonstrate that SHS can migrate into adjacent non-smoking areas very quickly, especially when there is no physical barrier. The casino's ventilation system did not fully eliminate SHS.

Reference Number: 40634

TIMBERLAKE, D.S., J. WU, AND W.K. AL-DELAIMY. "Tribal Casinos in California: The Last Vestige of Indoor Smoking." *BMC Public Health*, 12: 144 (8 pp), 2012 (2 tables, 3 figures, 28 references)

Abstract: The state of California is a pioneer of the tobacco control movement in the United States with smoke-free workplace law in effect since 1994. The smoke-free ban was opposed by the tobacco industry, which sponsored restaurant and casino opposition. The passage of the 1988 Indian Gaming Regulatory Act allowed smoking in casinos. The characteristics of 3,361 persons who had visited a casino within the past year were determined and compared with 7,036 persons who had not visited a casino. The prevalence of current smoking was 17.6% for casino patrons and 10.4% for non-patrons. A majority of casino patrons (60.8%) attempted to avoid secondhand smoke in the casino by moving around. Only 8.5% of patrons (mainly smokers) stated they would have had a shorter stay if there had been a smoking ban.

The findings of this study have important implications for the support of a smoking ban in tribal casinos in California. The data indicates that such a ban would increase casino visitation, possibly resulting in greater revenue and greater customer satisfaction. At the moment, the public health community should continue to support initiatives such as California's Clean Air Project, whose mission is to provide technical assistance for tribal nations' voluntary adoption of smoke-free policies.

Reference Number: 40635

ST. HELEN, G., J.T. BERNET, D.B. HALL, C.S. SOSNOFF, Y. XIA, J.R. BALMES, J.E. VENA, J-S. WANG, N.T. HOLLAND, AND L.P. NAEHER. "Exposure to Secondhand Smoke Outside of a Bar and a Restaurant and Tobacco Exposure Biomarkers in Nonsmokers." *Environmental Health Perspectives*, 7: 1010–1016, 2012 (4 tables, 2 figures, 35 references)

Abstract: Saliva and urine samples were collected from 18 female and 10 male non-smokers (aged 21–37 years) before and after a 3-hour visit to a bar and a restaurant where smoking is allowed, and also in an open air site without any smokers (control). Saliva and urine samples were also collected the morning after exposure. Salivary cotinine was measured by LC/MS/MS and creatinine and total NNAL were measured in the morning through urine samples. Mean salivary cotinine concentration increased by 0.115 ng/mL at the bar and 0.030 ng/mL at the restaurant, after a 3-hour visit. The mean next morning NNAL concentration increased as well.

> Our results indicate that both salivary cotinine and urinary NNAL increased significantly following exposure to SHS outside of restaurants and bars in Athens, Georgia. Although urinary total NNAL concentrations were relatively low, they indicate that nonsmokers exposed to brief periods of SHS in outdoor locations may be exposed to measurable concentrations of carcinogenic tobacco-specific nitrosamines such as NNK. Because of the relatively long half-life of NNAL and much lower NNK exposure in passive smokers than in active smokers, future studies that investigate outdoor SHS exposure should consider measuring NNAL in urine of nonsmokers at ≥24 hr postexposure.

Reference Number: 40636

KING, B.A., S.D. BABB, M.A. TYNAN, AND R.B. GERZOFF. "National and State Estimates of Secondhand Smoke Infiltration Among US Multiunit Housing Residents." *Nicotine Tobacco Research*, 15: 1316-1321, 2013 (4 tables, 41 references)

Abstract: One-quarter of US residents (79 million) live in multi-unit housing (MUH) and are particularly vulnerable to involuntary SHS exposure. Among the MUH residents who are smoke-free at home, an estimated 29 million have experienced SHS infiltration in their homes.

Separating smokers from nonsmokers, cleaning the air and ventilation in buildings cannot eliminate exposure of nonsmokers to SHS. Therefore, policies prohibiting smoking in MUH, including living units and indoor sharing areas, represent the most effective way to fully protect residents from involuntary exposure to SHS in this environment. Smoke-free MUH policies are favored by most MUH residents, are legally permissible in both government-subsidized and market-rate housing and can result in savings for MUH operators.

Reference Number: 40637

KESKE, R.R., V.W. REES, I. BEHM, B.M. WADLER, AND A.C. GELLER. "Second-Hand Smoke Exposure and Mitigation Strategies Among Home Visitation Workers." *Tobacco Control*, 22: 250–254, 2013 (2 tables, 1 figure, 33 references)

Abstract: A survey was conducted of 316 visitation outreach health workers in Massachusetts as to their exposure to secondhand smoke in the private homes they visited. Eighty-three percent of the respondents had at least 1 hour per month of SHS exposure and 16% reported at least 1 hour per week. But only 22% of the workers had ever counseled clients on maintaining a smoke-free home. Fewer than 30% had ever voiced concerns to their employment agency and only 14% stated that their agency had rules to protect workers from SHS. In 2006, the US Surgeon General concluded that there is no safe limit of SHS.

Home visiting health workers, who currently number approximately 1.7 million in the United States, are not generally protected by workplace smoking bans when making private home visits. This is a previously undocumented population at risk for SHS exposure, and for whom little or no formal policy for protection from exposure appears to exist. The present data provide the first known insight into SHS exposure encountered in the course of home visiting duties for Early Intervention workers, including frequency and duration of exposure, expressed concerns, and strategies used to reduce exposure. Potential strategies for protecting this sizeable workforce are outlined.

Reference Number: 40638

GARCÍA-ESQUINAS, E., A. NAVAS-ACIEN, AND F. RODRÍGUEZ-ARTALEJO. "Exposure to Secondhand Smoke and the Frailty Syndrome in US

Older Citizens." *GeroScience*, 37: 25 (9 pp), 2015 (4 tables, 1 figure, 48 references)

Abstract: In 2004, approximately 33% of male and 35% of female non-smokers worldwide were exposed to secondhand smoke. The serum cotinine concentration, SHS exposure, and frailty (e.g., weakness, low body weight, low physical activity) was determined in 2,059 non-smoking adults aged 60 years and older. The median concentration of serum cotinine was 0.95 ng/mL. About 10% of the population lived with at least one smoker and 6% were frail. Older non-smokers were 5.4× as likely to be frail if they lived with two or more smokers.

In the US nonsmoking older adult population, exposure to SHS was associated with an increased frequency of frailty. More efforts are needed to protect older adults from SHS, especially at home and in other areas not covered by smoke-free regulations.

Reference Number: 40639

GARCÍA-ESQUINAS, E., A. JIMÉNEZ, R. PASTOR-BARRIUSO, M.R. JONES, B. PEREZ-GOMEZ, A. NAVAS-ACIEN, AND M. TELLEZ-PLAZA. "Impact of Declining Exposure to Secondhand Smoke in Public Places to Decreasing Smoking-Related Cancer Mortality in the US Population." *Environment International*, 117: 260–267, 2018 (4 tables, 1 figure, 34 references)

Abstract: A prospective study was conducted on the tobacco-related mortality of 11,856 non-smoking adults aged >40 years who took part in the 1988–1994 or 1999–2004 US National Heath and Nutrition Examination Survey (NHANES). Serum cotinine concentrations were measured by LC/MS and exposure to secondhand smoke at home and work were assessed. The adjusted smoking-related cancer mortality rate declined from 249.1 to 158.4 deaths per 100,000 person between the two surveys. There was also a 70% decrease in the unadjusted baseline serum cotinine concentration.

Table: Median Serum Cotinine Concentrations, Percent, and Duration of Exposure to Occupational SHS (1988–1994 and 1999–2004)

Survey Period	Median Serum Cotinine Concentration	Percent Exposed to SHS	Exposure Time (hrs/day)
1988–1994	0.13 ng/mL	16.5%	3.0
1999–2004	0.04 ng/mL	7.7%	1.0

Source: Adapted from García-Esquinas et al, 2018

Reference Number: 40639

Our results support the effectiveness of smoking reductions in indoor public places to reduce smoking-related cancer mortality among non-smoking workers. Given these findings and the fact that SHS remains an important indoor and outdoor air pollutant in many US states and countries, smoke-free laws and interventions must be extended to all public places as well as to private areas in order to further protect the population from unnecessary and involuntary exposure to an established carcinogen.

4.07 THIRD- AND FOURTH-HAND SMOKE

"This is the first study to show that children of smokers carry tobacco smoke toxicants on their hands, even when nobody around them is smoking at the time."

—Mahabee-Gittens et al, "Preliminary Evidence that High Levels of Nicotine on Children's Hands May Contribute to Overall Tobacco Smoke Exposure" (2018)

Thirdhand smoke is the residue left behind by secondhand smoke as it mixes with the air and deposits on household dust, rugs, carpets, fabrics, and clothes (40702–3). Thirdhand smoke chemicals persist long after smoking has ceased (40703–9). The persistence of nicotine and other toxic chemicals in the house of smokers probably increases the difficulty of quitting (40706–7).

Thirdhand smoke also has economic repercussions, as cars that are smoked in sell for about 9% less than non-smokers' cars. Contaminated private houses, apartments, and hotels incur more costs and increasingly, hotels are becoming more and more smoke-free in all rooms (40708).

Fourth-hand smoke is mainly the result of toxic cigarette butts and remains of tobacco products that are discarded into the environment by smokers. Metals such as aluminum, barium, cadmium, iron, manganese, and strontium leach out of the cigarette butts and into the environment for many years (40709). Just one smoked cigarette filter in 0.5 liters of water will kill 50% of the exposed topsmelts fish (40710).

Discarded cigarettes and butts have not only poisoned children but also domestic dogs, cats, birds, rodents, and turtles (40711).

The tobacco industry is also a major polluter in the cultivation, curing, and manufacturing of cigarettes and a source of deforestation (40712–13).

Reference Number: 40701

MAHABEE-GITTENS, E.M., A.L. MERIANOS, AND G.E. MATT. "Preliminary Evidence that High Levels of Nicotine on Children's Hands May Contribute to Overall Tobacco Smoke Exposure." *Tobacco Control*, 27: 217–219, 2018 (1 table, 16 references)

Abstract: House dust and surfaces are important sources of environmental toxicants such as pesticides and lead. Persistent residue from secondhand smoke (SHS) accumulates in the dust, objects, home surfaces, and rugs, and on smokers' skin and clothes. This thirdhand smoke may expose non-smokers and children to its effects. Twenty-five children who were brought to the ED in 2016 as a result of breathing difficulties and whose parent(s) smoked had salivary samples and hand wipe samples tested for cotinine and nicotine, respectively. All but one child had a detectable oral fluid cotinine concentration (1.2–28.8 ng/mL). All children had detectable nicotine on their hands (18.26–690.94 ng/wipe).

- This is the first study to show that children of smokers carry tobacco smoke toxicants on their hands, even when nobody around them is smoking at the time.
- Children had high levels of hand nicotine, higher than that of non-smokers in our previous work.
- The significant association between hand nicotine and salivary cotinine suggests that hand nicotine may contribute to overall tobacco smoke exposure, independent of SHS, and can serve as a marker of overall tobacco smoke exposure.

Reference Number: 40702

MOHAMED, N.N., S.L. LOY, C.N. MAN, A. AL-MAMUN, AND H.J.J. MOHAMED. "Higher Hair Nicotine Level in Children Compared to Mother Living with Smoking Father in Malaysia." *Environmental Health and Preventative Medicine*, 21: 572–578, 2016 (6 tables, 1 figure, 37 references)

Abstract: Hair nicotine concentrations were determined in 92 fathers, 124 mothers, and 111 children living in Malaysia in 2013 by GC/MS. Residuals of smoke contaminants left on surfaces and dust after smoking are known as thirdhand smoke. Over 50% of the fathers smoked and none of the mothers. Smokers had significantly fewer years of education than non-smokers, as did the mothers who were living with the smokers. Elimination of nicotine occurs at a greater rate in women than in men. Higher median hair concentrations in children with smoking fathers was 6.08 ng/mL compared to 0.22 ng/mL for children living with non-smoking fathers.

There is a positive relationship between fathers' hair nicotine with mother's and children's hair nicotine. Living with smoking fathers can contribute to higher hair nicotine levels in children but not in mothers.

Reference Number: 40703

MATT, G.E., E. HOH, P.J.E. QUINTANA, J.M. ZAKARIAN, AND J. ARCEO. "Cotton Pillows: A Novel Field Method for Assessment of Thirdhand Smoke Pollution." *Environmental Research*, 168: 206–210, 2019 (2 tables, 1 figure, 23 references)

Abstract: Thirdhand smoke (THS) is the residue left behind by SHS as it mixes with the air, rapidly distributes throughout a room, migrates to other rooms, and deposits on household dust, surfaces, and materials. THS chemicals can persist long after smoking has ceased and can emit semi-volatile compounds. Small organic pillows were placed on couches in the living rooms of eight former smokers and four non-smokers for over a 3-week period. The smokers received the pillows 1 week after smoking ceased. Smoking cessation was verified by breath CO and salivary cotinine levels. After the end of 3 weeks, the nicotine content of the pillowcases, fillings, and fabric were determined by LC/GC/GC. The entire pillow absorbed on average 21.5 µg of nicotine, of which 56% was found in the pillowcase, 17% in the fabrics, and 28% in the filling.

Table: Mean Nicotine Concentration Detected in Pillow Components in the Homes of Former Smokers

Pillow Component	Mean Nicotine Concentration (ng/g)
Case	257.0
Fabric	96.8
Filling	17.1

Source: Adapted from Matt et al, 2019

Home environments typically contain large quantities of cotton, polyester, wool, and other fabrics in the form of clothes, carpeting, upholstery, drapes, blankets, towels, bed sheets, toys, etc. If these materials have been in a home during years of smoking, they would be expected to contain a substantial mass of nicotine and other THS compounds, based on our data. Considering that there are other porous materials in a home (e.g., particle board, gypsum board, ceiling tiles), it is to be expected that indoor smoking creates massive depositions of tobacco smoke toxicants throughout a home that will continue to affect indoor environments and their residents after smoking ends. This research contributes to the important policy debate that smoking should be prohibited in all indoor environments to prevent the build-up and the later release of tobacco smoke toxicants. This study also suggests that pre-screened cotton pillows could provide a simple tool to measure nicotine from THS sources and presumably SHS in home environments. More generally, porous materials with large surface areas that readily absorb volatile SHS and THS compounds may provide a method for removing and cleaning-up THS polluted environments, and for severely contaminated homes, removal of porous objects such as sofas, particle board, etc. may be necessary to remediate the home. These findings also have implications for peer-to-peer internet commerce (e.g., eBay, AirBnB) and secondhand merchandise sales. Findings from this study show that it is likely that products from smokers' homes contain tobacco smoke toxicants.

Reference Number: 40704

SHEU, R., C. STÖNNER, J.C. DITTO, T. KLÜPFEL, J. WILLIAMS, AND D.R. GENTNER. "Human Transport of Thirdhand Tobacco Smoke: A Prominent Source of Hazardous Air Pollutants into Indoor Nonsmoking Environments." *Science Advances*, 6: (14 pp), 2020 (1 table, 4 figures, 78 references)

Abstract: THS originates from the contamination of hazardous organic compounds from tobacco smoking onto the smoker's bodies, clothes, indoor furnishing, and surfaces. THS exposure can occur through the inhalation of evaporated gas, resuspended dust, aerosols, and dermal exposure. Surface deposited THS can be oxidized over time by ozone or nitrous oxide in the air to form carcinogenic TSNAs. Unlike SHS, THS is not a source of $PM_{2.5}$. Real time off-gassing from smokers into a non-smoking movie theater was determined with online and offline high-resolution mass spectrometry over 4 consecutive days. No smoking had been allowed in the movie theater tested for 15 years and it was also supplied with 100% fresh air (non recirculation). The theater's air intake was 20+ metres above the base of the building.

> Prominent emission events of THS tracers (e.g., 2,5-dimethylfuran, 2-methylfuran, and acetonitrile) and other tobacco-related volatile organic compounds (VOCs) coincided with the arrival of certain moviegoers and left residual contamination. These VOC emission events exposed occupants to the equivalent of 1 to 10 cigarettes of secondhand smoke, including multiple hazardous air pollutants (e.g., benzene and formaldehyde) at parts-per-billion concentrations. Nicotine and related intermediate-volatility nitrogen-containing compounds, which vaporized from clothes/bodies and recondensed onto aerosol, comprised 34% of observed functionalized organic aerosol abundance. Exposure to THS VOC emission events will be considerably enhanced in poorly ventilated or smaller spaces in contrast with a large, well-ventilated theater—amplifying concentrations and potential impacts on health and indoor chemistry.

Reference Number: 40705

WHITEHEAD, T., C. METAYER, M.H. WARD, M.G. NISHIOKA, R. GUNIER, J.S. COLT, P. REYNOLDS, S. SELVIN, P. BUFFLER, AND S.M. RAPPAPORT. "Is House-Dust Nicotine a Good Surrogate for Household Smoking?" *American Journal of Epidemiology*, 169: 1113–1123, 2009 (7 tables, 32 references)

Abstract: House dust nicotine concentrations were determined in vacuum cleaner samples and samples collected by high volume surface samples by GC/MS in 233 cases and 236 control households in northern California. The nicotine concentrations ranged from not detected (<20 ng/g) to 35,000 ng/g with a median of 265 ng/g. Only about 4% of the households allowed in-house smoking. House dust nicotine concentrations reflected past smoking habits and were elevated for a least 1 month after smoking ceased and reflected the cumulative smoking habits of residents over a period of several years. House dust nicotine concentrations increased with increasing SES and age.

Table: Self Reported Cigarette Use and House Dust Median Nicotine Concentrations

	Median Nicotine Concentration (ng/g)	
Self-Reported Cigarette Use	Yes	No
Mother ever smoked?	465	203
Father ever smoked?	415	185
Mother smoked during pregnancy?	1,634	245
Father smoked during 3 months before conception?	613	215
Anyone smoked during 1 year before birth?	1,109	252

Source: Adapted from Whitehead et al, 2009

In summary, this study confirmed previous findings that house-dust nicotine concentrations are significantly associated with self-reported household smoking. Results suggest that house-dust nicotine can be used as a long-term surrogate for exposure to cigarette smoke in the home. Moreover, indirect evidence indicates that self-reported smoking may vary by parental SES and age, highlighting the utility of an objective surrogate of exposure to cigarette smoke in the home.

Reference Number: 40706

MATT, G.E., P.J.E. QUINTANA, J.M. ZAKARIAN, E. HOH, M.F. HOVELL, M. MAHABEE-GITTENS, K. WATANABE, K. DATUIN, C. VUE, AND D.A. CHATFIELD. "When Smokers Quit: Exposure to Nicotine and Carcinogens Persists from Thirdhand Smoke Pollution." *Tobacco Control*, 26: 548–556, 2016 (4 tables, 37 references)

Abstract: Tobacco smoke pollutants (thirdhand smoke or THS) were determined over a six-month period in the home of 90 smokers who had quit. The small size and mobility of the particle and gas-phase compounds of tobacco smoke allow SHS and THS to penetrate deep into the carpet padding and upholstery, infiltrate closed closets and drawers, and coat every available micro-surface of an indoor environment. If prolonged smoking occurs in casinos or the home or car of a smoker, or permitted hotel rooms, a large reservoir of THS pollutants occurs. SHS only lasts for 2 to 3 hours, but THS can last for months. After 6 months, 9% of the smokers had maintained successful quitting. The urine cotinine concentration in non-smokers should be < 0.1 ng/mL, but 3–6 months later the concentrations were still about 1.5 ng/mL in smokers who had quit, which indicates nicotine deposits at home as a source. Nicotine derived nitrosamines (NNK and NNAL (4-methylnitrosamino)-1-(3-pyriole)-1-butanone) have a long half-life of 18–45 days. Three to six months after smoking cessation, urine NNAL concentrations were 54–72% that of baseline.

Table: Finger Nicotine and Dust Nicotine Concentrations on Non-smokers and Homes after Cohabitating Smoker Successfully Quit Smoking for up to 6 Months

	Baseline	1 Week	1 Month	3 Months	6 Months
Finger nicotine (ng/wipe) non-smoker	49.4	7.5	11.5	9.1	2.9
Dust nicotine concentration (μg/g) home	4.8	5.2	5.0	3.6	3.8

Source: Adapted from Matt et al, 2016

The persistence of THS pollutants in settled house dust and on surfaces presents a potential risk after smoking cessation, especially for infants and young children in the home, and poses interesting questions for future research. It is well documented that successful smoking cessation is often difficult to achieve even after multiple attempts. Consistent with the relapse observed in the present study, 80–90% of quitters overall relapse and begin smoking again within the first 3 months of cessation. The main reasons for such poor cessation outcomes are the addictive properties of nicotine, strong cravings and withdrawal symptoms and conditioned cues for smoking (eg, odours, paraphernalia, situations). Findings from this study raise the possibility that relapse may be mediated or moderated by the continued low-level exposure to nicotine and odorant compounds of THS in a smoker's home that serve as triggers and cues. Chemical exposure to tobacco constituents found in THS may contribute to relapse due to physiological processes associated with cravings. If this is indeed the case, removing THS pollutants from the home environment after smoking cessation might lead to more prolonged and sustained cessation outcomes.

Reference Number: 40707

MATT, G.E., E.M. MAHABEE-GITTENS, J.M. ZAKARIAN, P.J.E. QUINTANA, E. HOH, AND M. MYERS. "Nicotine in Thirdhand Smoke Residue Predicts Relapse from Smoking Cessation: A Pilot Study." *Addictive Behavior*, 98: 106041 (10 pp), 2019 (2 tables, 21 references)

Abstract: THS residue (including nicotine) can linger in the homes of smokers for months and may play a role in relapse after smoking cessation. Sixty-four cigarette smokers who were enrolled in a smoking cessation program and had a specific date to quit were tested (by breath CO <5 ppm) at 1 week and 1, 3, and 6 months after the quit date, and it was determined that they had maintained abstinence. House dust samples were collected at baseline and analyzed for nicotine concentration (μg/g) and nicotine loading (μg/m²) using LC/MS/MS. Independent of age, general, overall, and indoor smoking rates at baseline, a 10-fold increase in dust nicotine loading consistently lowered the odds of remaining abstinent by 50%.

Limitations notwithstanding, the association of nicotine levels in settled house dust and relapse to smoking suggests that in addition to known cues that contribute to relapse (e.g., visual, environmental), THS pollution in a former smoker's home may adversely affect short- and longer-term cessation outcomes. In addition to the removal of established smoking cues, quitters may also benefit from home cleaning efforts that reduce THS pollutants from dust and other reservoirs (e.g., carpets, upholstery, surfaces).

Reference Number: 40708

HOVELL, M.F., AND S.C. HUGHES. "The Behavioral Ecology of Secondhand Smoke Exposure: A Pathway to Complete Tobacco Control." *Nicotine and Tobacco Research*, 11: 1254–1264, 2009 (1 figure, 84 references)

Abstract: The tobacco industry creates more smokers and disease than doctors can prevent by clinical care alone and can only indirectly contribute to prevention of smoking. A Behavioral Ecological Model was applied to smoking cessation and control. Behavioral change is required to reduce the morbidity of most diseases. The tobacco industry knows how to change behavior and has made tobacco a central feature of Western culture. It has also encouraged smokers to demand a right to smoke which will ensure more nicotine addicts. Widespread public smoke-free environments which prevent secondhand smoke exposure (SHSE) will do much to deformalize and change smoking behavior.

Our research and review suggest that emphasis on treatment of disease or smoking cessation services cannot offset the industry's ability to produce new addicts. Similarly, smokeless tobacco products will likely promote new addicts via modeling and social contingencies. Thus, even if they reduce harm relative to smoke-producing cigarettes, they too may yield more harm than benefit. We believe that interventions can reduce SHSe, lower smoking initiation, and increase long-term cessation rates. Moreover, as more and more people get accustomed to a smoke-free environment, this strengthens an antitobacco culture.

Based on the BEM, we expect that emphasis on SHSe elimination in all micro- and macroenvironments, in both outdoor and indoor areas, if achieved, will eliminate places where smokers can smoke without criticism. As criticism increases, it may generalize to other sanctions. For example, used cars sell for about 9% less if contaminated by an owner who smoked compared with an uncontaminated car owned by a nonsmoker. This suggests that the culture is shifting in subtle ways that generate financial costs for contaminating microenvironments. While the evidence is not yet available, we suspect similar and greater costs for contaminated private homes, apartment owners, and hotels. This is implicitly supported by the trend in hotels restricting all smoking in all rooms.

Reference Number: 40709

MOERMAN, J.W., AND G.E. POTTS. "Analysis of Metals Leached from Smoked Cigarette Litter." *Tobacco Control*, 20: i30–i35, 2011 (3 tables, 3 figures, 28 references)

Abstract: The metals leached from smoked tobacco butts (collected from outdoor covered cigarette receptacles) and unsmoked cigarettes were determined at 1, 7, and 34 days by inductively coupled plasma optical emission. The sampling and leaching procedures were designed to simulate natural conditions as much as possible.

Table: Percentage of Metal Leached from Cigarette Butts, Concentration of Metals in Whole American Cigarettes, and Metals Leached Globally

Metal	Concentration in Whole Cigarette (μg/g)	Percentage Leached	Amount Leached Globally (kg/year)
Aluminum	950.0	1.2	9.25
Barium	48.7	21.0	8.76
Cadmium	1.0	14.0	0.12
Iron	394.0	11.0	38.10

Metal	Concentration in Whole Cigarette (μg/g)	Percentage Leached	Amount Leached Globally (kg/year)
Manganese	178.0	23.0	33.90
Strontium	39.6	59.0	19.70
Titanium	106.0	0.7	0.60

Source: Adapted from Moerman and Potts, 2011

The results of this research suggest that littered cigarette butts are point sources for prolonged metal contamination. Furthermore, the apparent rapid release of multiple metals from littered cigarette butts increases the potential for acute harm to local organisms. Given the varying responses of organisms to metal contamination, knowledge of the leaching behaviour of metals from cigarette butts is necessary to assess the effect littered cigarette butts have on local biota and the environment in general. With a better understanding of the toxicity of cigarette butts and how contaminants are leached from them, the environmental hazards posed by littered cigarette butts may no longer be a subject of debate.

Reference Number: 40710

SLAUGHTER, E., R.M. GERSBERG, K. WATANABE, J. RUDOLPH, C. STRANSKY, AND T.E. NOVOTNY. "Toxicity of Cigarette Butts, and Their Chemical Components, to Marine and Freshwater Fish." *Tobacco Control*, 20: i25–i29, 2011 (1 table, 2 figures, 24 references)

Abstract: An estimated 4.5 trillion cigarette butts are discarded worldwide every year, making cigarette butts the most common form of litter. Cigarette butts were found to comprise about 30% of the total litter on shorelines, waterways, and on land in the United States. Test cigarettes, which were regular filter cigarettes (but not flavored or low tar) were smoked by either a machine or a smoker. The leachate from smoked cigarette butts (1–2 cm remnant tobacco), smoked cigarette filters (but no remnant tobacco), and not smoked filters (no remnant tobacco) were tested on saltwater topsmelt and freshwater fathead minnows. The toxicity of the leachate increased from unsmoked filters, to smoked cigarettes filters (no tobacco) and to the most toxic, smoked cigarette butts with tobacco remnant. Artificially smoked cigarettes were more toxic than naturally smoked filter leachates.

Table: Number of Cigarette Butts Required per Liter of Water to Kill 50% of the Fish

Fish species (LC_{50}, butts/L)	Unsmoked Filter	Smoked Filter (no Tobacco Attached)	Smoked Filter with Tobacco Remnant
Topsmelt	5.1	1.0	1.1
Fathead minnow	13.5	4.3	0.97

Source: Adapted from Slaughter et al, 2011

Chemical additives are often introduced to make tobacco products more attractive to consumers. For example, sugars and humectants make smoke milder and easier to inhale, humectants can prolong shelf life, ammonia may enhance the delivery of nicotine and menthol and eugenol effectively numb the throat. In fact, approximately 600 additives were in use by major American cigarette companies in 1994. Many of these chemicals may be harmful to humans as a result of smoking. The major humectants used for cigarettes are glycerol, diethylene glycol and/or propylene glycol which may be carcinogenic to humans. However, little is known about the fate of such additives in cigarette butt leachates.

There are several chemicals in an unsmoked cigarette filter that may contribute to aquatic toxicity. The filter of a filter-tipped cigarette is composed of cellulose acetate fibres. These fibres, each approximately 20 µm in diameter, are treated with titanium dioxide (a delustrant) and over 15 000 of them are packed tightly together, using triacetin (glycerol triacetate) as a binding agent, to create a single filter. Most cigarette filters are surrounded by two layers of paper and/or rayon wrapping, which contain chemicals, such as glues to hold the paper together, and alkali metal salts of organic acids (eg, sodium acetate) in order to maintain burning while the cigarette is being smoked. It is also possible that cigarette filters attached to tobacco absorb toxicants from the adjacent unsmoked tobacco column; however, this has not been investigated in the literature.

Reference Number: 40711

NOVOTNY, T.E., S.N. HARDIN, L.R. HOVDA, D.J. NOVOTNY, M.K. MCLEAN, AND S. KHAN. "Tobacco and Cigarette Butt Consumption in Humans and Animals." *Tobacco Control* 20: i17–i20, 2011 (2 tables, 32 references)

Abstract: Cigarettes contain at least 4,000 chemicals, about 50 of which are carcinogenic. A typical cigarette contains 9 to 30 mg of nicotine, although smoking delivers only 0.5–2 mg. Discarded cigarette butts present health risks to human infants and small animals. Accidental ingestion of cigarettes (and butts) are not uncommon among children. Cigarette and butt poisonings have been reported in domestic birds, dogs, cats, rodents, and turtles, although only 1 fatality due to ingestion has been reported, which occurred in a dog.

Table: Cigarette/Butt Ingestion in Various Domestic Animals

Species	Number of Reported Cases
Birds	2
Dogs	801
Goats	2
Cats	41
Rodent	1
Turtle	1

Source: Adapted from Novotny et al, 2011

This review suggests that cigarette butt consumption by small children and animals is a frequent source of concern and attention for poison control centres, parents and pet owners. The ubiquity of this waste should thus be a concern for policymakers who would seek to reduce the costs of dealing with tobacco ingestions for both parents and providers. We found that severe poisoning by cigarette butts among young children was rare but possible and that poison centres already emphasise the need to reduce the risk of poisoning by tobacco products. It is clear that smokers must treat cigarette butts as toxic waste products and take more care in discarding them; children will mimic parental behaviour by putting these items into their mouths and sometimes consume them in quantities sufficient to be toxic. It is also clear that new, flavoured tobacco products are a source of concern for poisoning among small children because these products, unlike cigarette butts, are not foul tasting.

Veterinary reports of nicotine poisoning are uncommon, but domestic animals may consume them and show serious gastrointestinal, central nervous system and cardiovascular signs. Veterinary poison centre data may be further studied with regard to follow-up on reported exposures and outcomes of these exposures.

Reference Number 40712

DE GRANDA-ORIVE, J.I., C.A. JIMÉNEZ-RUIZ, AND S. SOLANO-REINA. "World Health Organization Positioning. The Impact of Tobacco in the Environment: Cultivation, Curing, Manufacturing, Transport, and Third and Fourth-Hand Smoking." *Archivos de Bronconeumologia*, 54: 357–358, 2018 (15 references)

Abstract: Tobacco cultivation and curing had a major negative impact on the environment, as it is a crop that is rarely rotated with other crops and so is vulnerable to pests and diseases which require the heavy use of pesticides and other chemicals. Deforestation is caused by clearing the land for the tobacco crop. In addition, an estimated 11.4 metric tons of wood is burned for its curing. Fourth-hand smoke is the two-thirds of cigarette butts that ends up in the environment and pollutes the water with toxic chemicals.

> Both tobacco cultivation and curing, then, are associated with farming methods that destroy the environment in developing countries, and while it can provide profits for farmers and local communities, these societies are impoverished by the loss of resources. Data show that farms dedicated to tobacco cultivation are less profitable. Chronic exposure to pesticides, such as dichlorodiphenyltrichloroethane, and another 11 persistent organic pollutants, banned in developed countries, cause health problems, such as birth defects, tumors, genetic changes, and endocrine, blood, neurological and psychiatric disorders, even in individuals who do not work directly with these substances. Sustained exposure to the tobacco plant itself and nicotine absorbed through the skin causes green tobacco sickness, characterized by nausea, vomiting, headache, muscle weakness, and vertigo.
>
> The manufacture and distribution of tobacco clearly contribute to environmental pollution, both as solid and chemical waste, and tobacco companies have admitted that manufacturing and production are the processes that generate most environmental pollution.

Reference Number 40713

NOVOTNY, T.E., AND F. ZHAO. "Consumption and Production Waste: Another Externality of Tobacco Use." *Tobacco Control*, 8: 75–80, 1999 (3 tables, 3 figures, 20 references)

Abstract: The waste products of cigarette smoking are clearly visible on city sidewalks and public beaches. Cellulose acetate cigarette filter fibers persist for at least 18 months or longer. In addition, the tobacco manufacturing process produces liquid, solid, and airborne wastes.

Table: Releases into the Environment by the Tobacco Manufacturing Industry in the United States, 1996

Chemical	Total Quantity Released (kg)
Ammonia	946,144
Ethylene glycol	8,936
Hydrochloric acid	407,371
Hydrogen fluoride	27,937
Methyl ethyl ketone	340,821
Nicotine and nicotine salts	900,377
Nitric acid	58,970
Sulphuric acid	67,228
Toluene	349,622

Source: Adapted from Novotny and Zhao, 1999

Nicotine is an interesting tobacco production waste product. The tobacco manufacturing process and all activities that use tobacco produce solid or liquid wastes with high concentrations of nicotine. Partly in response to information about the addictive nature of nicotine, the market for low-nicotine cigarettes has been growing. Thus, an ironic outcome of this market growth is the increased need to detoxify nicotine in tobacco production waste. A non-recyclable, powdery, nicotine-containing waste is formed during tobacco production, which has average nicotine content of 18g per kg of dry weight. This waste is classified as "toxic and hazardous" by European Union Regulations when the nicotine content exceeds 500mg per kg dry weight. According to a 1997 study from Italy, an average 3 million kilograms of this waste are reported in 1995 by the tobacco industry to be produced from 55300 million cigarettes manufactured in that country. If manufacturing processes and waste outputs per million cigarettes produced in the United States and globally are similar to those reported in Italy, it is estimated that 38870000kg of nicotine waste was produced in the United States in 1995, and 300274000kg was produced globally in the same year. Denicotinisation of waste requires chemical–physical treatments or biological methods with microorganisms to reduce the nicotine content of released wastes below the threshold of 500mg per kg. In the United States, nicotine has been included on the TRI list since November 1994. Nicotine itself is an anti-parasitic plant pesticide.

4.08 YOUTH

"Smoking is a pediatric disease. No one in their 30s or 50s are going to start smoking. When you see a 79-year-old on oxygen from COPD, it is probably because they saved their milk money to buy cigarettes when they were 13."

—Dr. Ted Moon, as quoted by Vance, "Living Well" (2017)

Almost 90% of smokers began smoking before the age of 21 years (40801). Adolescents are very susceptible to tobacco industry advertising. The cigarette brands smoked by youth are typically those that are most heavily advertised (40802). In the 1950s, tobacco companies aggressively marketed to this valuable market by using candy cigarettes to make smoking attractive and normalized (40801). It is estimated that the Joe Camel campaign resulted in nearly 3.5 million new smokers in the United States (40802).

Due to the ban on billboard advertising as a result of the Master Settlement Agreements in 1998, the tobacco industry switched to advertising to stores selling cigarettes or point of sale marketing (40803). Adolescents who were exposed to tobacco outlets more than twice a week were 1.41× as likely to initiate smoking (40803).

Another tobacco ploy to attract youth is through movies in which smoking causes no harm or disease, often showing mature, attractive, White, male characters smoking (40804). For adolescents, whether they watch smoking in movies was as accurate a predictor of future smoking as having one or more parents who smoked (40804).

Cigarettes such as Camels were redesigned to capture the valuable FUBYAS (First Usual Brand Younger Adult Smoker) market by increasing nicotine impact but making cigarettes smoother and easier to inhale (40805). Internal documents showed that RJ Reynolds determined that the most important feature of the cigarette to attract younger adults were smoothness or mildness (40805).

Smoking adolescents' health problems include upper respiratory infections, immature lung development, reduced vital capacity, and lung cancer (40806). Adults who smoke as adolescents may never develop lungs to full capacity (40806).

One study divided smokers aged 11 to 18 years into seven groups and found that the established smoker group increased over 200-fold, from 0.1% at age 11 years to 20.9% at age 18 years (40807). Preventing youth from becoming addicted to cigarettes would greatly reduce the public health costs from nicotine, as 87% of current adult smokers smoke cigarettes before the age of 18 years (40408). Public health messages that emphasize that cigarettes change your brain have been shown to be the most effective anti-smoking message for youth (40408).

The use of cannabis and cigarettes has increased in youth smokers in recent years (40809).

Reference Number: 40801

APOLLONIO, D.E., AND S.A. GLANTZ. "Minimum Ages of Legal Access for Tobacco in the United States From 1863 to 2015." *American Journal of Public Health Law*, 106: 1200–1207, 2016 (1 table, 1 figure, 89 references)

Abstract: A historical overview of minimum age laws (MLAs) to buy tobacco in the United States between 1863 to 2015 was presented. By the late 1800s there was public awareness that those who used tobacco found it difficult to quit. The Bonsak cigarette-making machine, which was patented in 1881 and was 500× faster than making cigarettes by hand, dramatically reduced the price of cigarettes. In the 1950s, tobacco companies were openly marketing to children and promoted the sale of candy cigarettes to children. Restricting the use of tobacco to those under the age of 21 years was common throughout the late 19th century and early 20th century. It was found that almost 90% of smokers began using tobacco before the age of 21 years and that young adults aged 18–20 years provide cigarettes to younger family and friends. If MLAs were increased to 21 years, adult smoking would decline by 12%.

> Throughout most of the 20th century, the tobacco industry aggressively encouraged and defended lowered MLAs for tobacco. This political advocacy reflects the tobacco industry's assessment that recruiting youth smokers is critical to its economic survival. This assessment and the increasing body of evidence on tobacco addiction among young adults suggest that restoring MLAs to 21 years would reduce smoking initiation and prevalence, particularly in youths younger than 18 years.

Reference Number: 40802

PIERCE, J.P., E.A. GILPIN, AND W.S. CHOI. "Sharing the Blame: Smoking Experimentation and Future Smoking-Attributable Mortality due to Joe Camel and Marlboro Advertising and Promotions." *Tobacco Control*, 8: 37–44, 1999 (4 tables, 39 references)

Abstract: Adolescents are susceptible to tobacco industry advertising and promotions. The cigarette brands smoked by youth are typically the brands most heavily advertised. A study was conducted of 2,659 confirmed never-smoking adolescents (aged 12–17 years) in 1993 and 2,779 in 1996 as to their favorite cigarette advertisements and promotions. Overall, in 1993 about 70% of confirmed never-smokers were determined to be receptive to tobacco ads and promotions. Approximately one-third of new cigarette experimenters (5,300/day) is attributable to tobacco ads and promotions. It was found that if 60% of the new experimenters became established, then there would be 4.7 million newly established smokers between 1988 and 1999. Approximately 50% of ever-smokers would be smoking at age 35 years, and 50% of those would die of smoking-attributable disease.

Table: Estimated Number of New Cigarette Smokers in the United States Attributable to Cigarette Advertising and Promotions

Year	All advertising/ promotions	Camel Campaign	Marlboro Campaign	Other Campaigns
1988–1994	4,644,734	2,005,132	971,678	1,667,924
1995–1998	3,305,148	1,459,884	1,002,782	842,482

Source: Adapted from Pierce et al, 1999

The strategies adopted by RJ Reynolds and Phillip Morris to advertise and promote their cigarette brands appear consistent with the objective to capture a share of the youth market outlined in the confidential RJ Reynolds research planning memorandum. The "Joe Camel" campaign certainly caught the attention of children and young people. Furthermore, the subsequent switch in emphasis away from advertising to promotional items suggests that this strategy was identified as a promising tactic for "influencing pre-smokers to try smoking, learn to smoke and become confirmed smokers."

Reference Number: 40803

JOHNS, M., R. SACKS, M. RANE, AND S.M. KANSAGRA. "Exposure to Tobacco Retail Outlets and Smoking Initiation Among New York City Adolescents." *Journal of Urban Health: Bulletin of the New York Academy of Medicine*, 90: 1091–1101, 2013 (3 tables, 1 figure, 30 references)

Abstract: Tobacco companies shifted most of their attention to Point of Sale (PoS) marketing after the Master Settlement Agreements, in 1998 which banned billboard advertisements. In 2010, the tobacco companies were estimated to have spent about $477 million promoting tobacco products in retail environments or PoS. In New York City, a large number of retailers are located near schools and playgrounds. A survey of 11,570 grade 9 to 12 students in NYC was conducted on the number of times they shopped at pharmacies, delis, or corner stores (75% of tobacco retailers in NYC) and smoking behavior. Banning of PoS tobacco displays in New Zealand in 2003 resulted in a significant decline in youth smoking. Adolescents who were exposed to tobacco outlets more than twice per week were 1.41× as likely to initiate smoking and adolescents with low-risk behavior were 1.78× as likely to initiate smoking.

> Visiting tobacco retailers two or more times per week was significantly associated with smoking initiation. The predicted probability of initiating among students visiting retailers seven times or more a week (16 %) was double that of students who never visit retailers (8 %) in a typical week. This relationship was strongest among students who were lowest in risk-taking tendencies—a group that is otherwise at low risk for smoking experimentation and uptake. The strength of the association between initiation and frequent exposure overall was comparable to that of the association between initiation and living with a smoker.

Reference Number: 40804

DOZIER, D.M., M.M. LAUZEN, C.A. DAY, S.M. PAYNE, AND M.R. TAFOYA. "Leaders in Elites: Portrayals of Smoking in Popular Films." *Tobacco Control*, 14: 7–9, 2005 (1 table, 16 references)

Abstract: An analysis was conducted of 100 top US grossing films in 2002 regarding cigarette smoking. Almost two-thirds of the films showed at least one character tobacco incident. Only one-third showed no tobacco use. Major characters in the film were 4× more likely to smoke than minor characters and male characters were 2× as likely to smoke. Of the 453 incidents involving tobacco shown in the 88 films, only 8% showed any consequences to tobacco use and typically it was a verbal reprimand from another character. Only two incidents showed tobacco causing death (0.4%), one due to the lighting of a cigarette to set off a car bomb, and the other when a cigarette attracted a heat-seeking missile.

> What do these impressions communicate about tobacco? First, smoking has no serious consequences. If any, consequences are most likely verbal reprimands. Although tobacco use is the leading cause of preventable deaths in the USA, and the second leading cause of death globally, only 0.4% of smoking incidents in top grossing films led to death. Diseases caused neither of the two deaths. Coupled with the higher frequency of smoking among characters 60 and older, smokers are wrapped in a cloak of invincibility. Those who smoke are important people (major characters) who play leadership roles. Film characters who smoke tend to be members of privileged elites: white, male, and mature. What are the consequences of such fictional portrayals of tobacco use in films? One cross sectional study of 4919 American adolescents showed that exposure to smoking incidents in films was as at least as powerful in predicting smoking initiation as having parents or siblings who smoked.

Reference Number: 40805

WAYNE, G.F., AND G.N. CONNOLLY. "How Cigarette Design Can Affect Youth Initiation into Smoking: Camel Cigarettes 1983–93." *Tobacco Control*, 11: i32–i39, 2002 (4 tables, 4 figures, 62 references)

Abstract: Internal tobacco industry documents were analyzed to determine how cigarettes were redesigned to promote youth smoking. In addition to the extensive Joe Camel marketing and advertising addressed to youth, the Camel cigarette was redesigned to make inhalation less irritating, increasing smoothness but also increasing the nicotine impact. RJ Reynolds (RJR) focused on the FUBYAS market and noted that more than half of all male smokers started before the age of 18 years and 95% before 25 years. RJR determined that the most important feature of the cigarette to attract younger adults were smoothness or mildness.

Table: Design Feature Changes to Cigarettes to Make Them More Appealing and Addictive to Youth

Design Feature	Approach	Description	Results
Tobacco blend	Two-stage blend	Separate blends used at front and back of the cigarette	Allowed strength (high nicotine) in initial puffs, smoothness in later puffs
Puffed tobacco (impregnated with ammonia and CO_2 and heated)	G13-23	Puffed tobacco filler	Improved mouth sensation and throat scratch
Humectant	Glystar (hystar/ glycerin)	Replacement for glycerin/propylene glycol	Increased smoothness
Flavorant	"Combination top dressing"	Chocolate, vanillin, licorice, and "tobacco enhancer"	Significant appeal among 18–24-year-old smokers
Filter	Carbowax	Filter additive	Reduced harshness, mouth sensation, throat dryness, bitterness

Source: Adapted from Wayne and Connolly, 2002

Prior analyses have linked the rise in youth smoking prevalence during the early 1990s to increases in tobacco industry advertising and promotional spending. However, no studies have examined the influence of product design changes on youth smoking initiation and uptake.

Internal tobacco industry documents suggest that product changes that make cigarettes more palatable, easier to smoke, and more addictive, can act in coordination with advertising and promotions targeting youth, shaping product perceptions and promoting initiation. Work must be done to monitor product changes as well as assess advertising and promotional influences as part of future efforts to address youth smoking.

Reference Number: 40806

LAMPALO, M., M. MAJER, I. JUKIĆ, H.S. STANIĆ, B. BARIŠIĆ, AND S. POPOVIĆ-GRLE. "Health Sequelae of Tobacco Exposure in Childhood and Adolescence." *Psychiatria Danubina*, 31: 39–43, 2019 (25 references)

Abstract: This study summarizes effect of cigarette smoking on children and adolescents. Tobacco smoking is one of the most common preventable causes of premature death worldwide. A survey in 2015 found that the rate of smoking in the past 30 days among 15- to 16-year-old students ranged from 37% in Italy and 33% in Bulgaria and Croatia to 10% in Norway, 9% in Moldova, and 6% in Iceland.

Adolescents' habitual smoking causes various health problems including upper respiratory infection, immature lung development, reduced maximum vital capacity, and lung cancer. The evidence indicates that it can lead to the increased number and severity of respiratory illnesses and decreased physical fitness. Adults who smoked as adolescents can have lungs that never fulfill to grow to their potential size and never perform at full capacity. This condition is an outcome of pernicious chemicals found in tobacco smoke, which slow down lung growth, and such damage is permanent. Abundant data suggest that cigarette smoking is generally associated with alterations in mechanisms of the host defense system. Ciliary function is flawed and mucous volume is much increased due to mucous gland hypertrophy, leading to obvious respiratory symptoms such as cough, dyspnea and an increased amount of phlegm.

Reference Number: 40807

MOWERY, P.D., M.C. FARELL, M.L. HAVILARD, J.M. GABLE, AND H.E. WELLS. "Progression to Established Smoking Among US Youths." *American Journal of Public Health*, 94: 331–337, 2004 (3 tables, 1 figure, 24 references)

Abstract: The National Youth Tobacco Survey conducted of middle and high school students in 1999–2000 was used to determine the seven stages of smoking in youths ages 11–18 years.

Table: Percent of Youth Ages 11, 13, 15, and 18 Years Divided into 7 Smoking Stages

Smoking Stage	Age 11 years	Age 13 years	Age 15 years	Age 18 years
Never smoker, not open to smoking	72.2%	44.8%	32.3%	23.0%
Never smoker, open to smoking	13.8%	17.0%	10.9%	5.4%
Early experimenter	13.1%	31.6%	34.7%	32.6%
Late experimenter	0.4%	2.9%	6.7%	7.2%

Smoking Stage	Age 11 years	Age 13 years	Age 15 years	Age 18 years
Former smoker	0.03%	0.4%	1.6%	3.7%
Non-daily current smoker	0.30%	1.4%	4.4%	7.3%
Established smoker	0.10%	1.9%	9.4%	20.9%

Source: Adapted from Mowery et al, 2004

Between the ages of 11 and 18 years, the prevalence of established smoking increases at an increasing rate with age. Our results suggest that parental influences and school antismoking classes can slow or prevent progression to established smoking. Our results also suggest that exposure to smoking at home, peer smoking, and tobacco industry marketing are important risk factors for established smoking. An especially promising public health intervention is paid antismoking media campaigns. Results from 2 state campaigns and preliminary results form the American Legacy Foundation's national truth campaign suggest that paid antismoking media campaigns directed at youths and used as 1 component of a comprehensive tobacco control program can be effective in reducing progression.

Reference Number: 40808

RODITIS, M.L., C. JONES, A.P. DINEVA, AND T.N. ALEXANDER. "Lessons on Addiction Messages from the Real Cost Campaign." *American Journal of Preventive Medicine,* 56 (2S1): S24-S30, 2019 (30 references)

Abstract: Preventing youth from becoming addicted to cigarettes can greatly reduce the public health burden of cigarettes as 87% of daily adult smokers smoked their first cigarette by the age of 18 years. Three focus groups and 4 copy testing studies were conducted on participants between the ages of 12 to 17 years, who either experimented with cigarettes or were susceptible to taking up smoking. The participants were assessed in their response to the FDA "The Real Cost" antismoking campaign.

Youth responded favorably to loss of control messages showing real, often short-term, consequences of addiction, such as choosing to spend money on cigarettes instead of going to a movie and depictions of scenarios that were relatable to youth. Youth also responded favorably to messages describing how nicotine changes the brain. A portion of youth remained skeptical, stating they felt the consequences depicted were unreliable. This study shows that by framing addiction as a loss of control and tying that loss of control to short-term health and social consequences, addiction becomes more concrete and understandable, and the consequences feel more relatable and relevant to youth.

Reference Number: 40809

BERG, C.J., J. PAYNE, L. HENRIKSEN, P. CAVAZOS-REHG, B. GETACHEW, G.L. SCHAUER, AND R. HAARDÖRFER. "Reasons for Marijuana and Tobacco Co-Use Among Young Adults: A Mixed Methods Scale Development Study." *Substance Use and Misuse*, 53: 357–369, 2018 (5 tables, 1 figure, 51 references)

Abstract: Cannabis and tobacco co-use has increased recently among young adults as there appears to be a synergistic effect between these two drugs. Of the 3,418 students (aged 18–25 years) from seven colleges and universities in the state of Georgia, USA, in 2015, phone interviews were conducted of 46 current cannabis users (20 were lifetime users).

An interesting distinction regarding item loadings on the subscales occurred in relation to the items "I use tobacco when I can't use marijuana" (Instrumentality) and "I use marijuana when I can't use tobacco" (Displacement). This might reflect that, in environments where marijuana is not allowed (public places) or otherwise inappropriate (e.g., around family), tobacco might be used in place of marijuana in order to stimulate a physiological effect, even though the effects may differ. However, it is likewise possible that when tobacco is not available, tobacco users might use marijuana to simulate the process of smoking. These findings warrant additional examination.

CHAPTER 5

Vaping (E-cigarettes)

> "Who thought sucking on a battery was a good idea?"
>
> —Hall, "Who Thought Sucking on a Battery Was a Good Idea?" (2020)

> "Researchers have disproved claims by e-cigarette manufacturers that e-cigarette aerosols are only water vapor, glycerol and PG and can be used safely in all environments."
>
> —Walley et al, "A Public Health Crisis: Electronic Cigarettes, Vape, and JUUL" (2019)

Most e-cigarettes (ECs) structurally contain five different components:

1) Mouthpiece.
2) Sensor or user-activated button to ignite the heating coil.
3) Power source (usually a rechargeable battery).
4) Heating element or coil (atomizer).
5) Cartridge or reservoir, which holds a liquid solution (e-liquid or e-juice) containing varying amounts of nicotine, flavorings, and other chemicals.

Researchers and academics use the name "electronic nicotine delivery systems" (ENDS), which is not a name commonly recognized by the public. Thus, e-cigarettes (ECs) will be used throughout this book instead. In many ECs, the user's puffing activates the battery powered heating device, which activates the heating element and vaporizes the e-liquid, which is stored in the cartridge.

5.01 HISTORY

> "The technology for the modern e-cigarette did not originate solely in China in 2003; the idea that Hon Lik was the first person to develop a device that aerosolised a nicotine solution is an oversimplification of the history of the e-cigarette. PM began exploring nicotine aerosol generation technology in 1990 to create a product that would serve as an alternative to NRT therapy for health-concerned smokers and deal with smoke-free environments. . . . Researchers and policymakers should recognise that PM developed e-cigarette technology to evade tobacco control regulations."
>
> —Dutra, Grana, and Glantz, "Philip Morris Research on Precursors to the Modern E-Cigarette Since 1990" (2017)

Tobacco companies developed electronic delivery systems that avoided smoke for many years before Hon Lik was touted as the first person to aerosolize a nicotine solution (50101, 50104–5, 50107). In fact, Hon Lik's original device did not use a coil to heat the e-liquid but a piezoelectric element to produce ultrasonic waves that vaporized the liquid (50101).

The evolution of ECs can generally be divided into five generations, each delivering more and more nicotine and volume of vape (50103):

- **1st Generation (Cig-a-likes):** These ECs looked a lot like cigarettes and some even had an LED tip to mimic the lit tip of a cigarette. Some cig-a-likes were disposable and some rechargeable, but the heating power and volume of nicotine liquid vaporized were low.
- **2nd Generation (Clearomizers):** The clearomizers are transparent and have higher volume e-liquid reservoirs than cig-a-likes and have larger variable voltage batteries (pen-style). These vaporizers are refillable.
- **3rd Generation (Mods):** These much larger devices can contain refillable tanks, which can contain a much larger volume of e-liquids. The coils, batteries, and other parts can be customized or modified by the user, hence the name. Almost all third-generation ECs have a fire button and are therefore mainly manual.
- **4th Generation (Sub Ohm Mods):** More powerful mods allow for large puff volumes and are capable of producing vape clouds.
- **5th Generation (Pods):** The most popular vaping devices include JUUL and come with a fixed voltage. The fifth-generation devices typically look like laser pointers or, in the case of JUUL, like a USB flash drive. In fact, a JUUL EC can be recharged by plugging it into

a computer port. Vaping with JUUL and many other pod brands is totally automatic with no buttons to push, and the devices contain small, replaceable, flavored liquid nicotine pods which contain a greater volume of e-liquid than cig-a-likes. JUUL can deliver more nicotine than a conventional cigarette and it is easier to stealth vape in schools or other public institutions than it is to smoke a cigarette.

The trend in ECs has been to have stronger batteries and coils, which vaporize the e-liquid at higher temperatures to provide a greater hit of nicotine (50109). There are now more than 480 brands of ECs and 7,700 types of flavored e-liquids (50404).

ECs were marketed by the tobacco industry as a safer method of obtaining nicotine, which is how low tar filter cigarettes were marketed in the 1950s and 1960s in order to prevent smokers from quitting (50105). Large tobacco companies now own over half of the EC market (50111). ECs, however, require more suction and stronger puffs than cigarettes and allow toxic aerosols (e.g., silver, iron, aluminum, chromium, and nickel from the heating coils) to enter deeper into the lungs (50106).

Philip Morris International has funded an institute ($1 billion over 12 years) with the deceptive name of "Foundation for a Smoke-Free World," which is basically a front group for the promotion of nicotine in all other forms but smoking (50110).

The slow response of regulators to the new ENDS allowed the cigarette companies the déjà vu of advertising to youth just like they did in the 1940s and 1950s, linking vaping with freedom, rebellion, youth, and glamor and using cartoon figures (50108, 50112–15). Some vaping ads have even had endorsement from physicians, just as cigarette ads did (50112).

The tobacco industry has promoted the incorrect factoid that vaping is 95% safer than cigarette smoking (50114).

Reference Number: 50101

DUTRA, L.M., R. GRANA, AND S.A. GLANTZ. "Philip Morris Research on Precursors to the Modern E-Cigarette Since 1990." *Tobacco Control*, 26: e97–e105, 2017 (2 tables, 2 figures, 78 references)

Abstract: Although the invention of the e-cigarette is often credited to the Chinese pharmacist Hon Lik in 2003, researchers at Philip Morris (PM) described a device in 1990 that included a heater, which would be activated when the device was puffed on. A printhead would spray a liquid nicotine solution onto the heater, which would become an aerosol to be inhaled. In 1994, PM developed the capillary aerosol generator (CAG), which generated an aerosol by pumping the liquid into a small, heated capillary tube. This was part of the PM Ideal Smoke program. The Hon Lik device generated an aerosol by pumping the nicotine liquid into a small cavity where it would be heated by ultrasonic waves (a piezoelectric element).

> The technology for the modern e-cigarette did not originate solely in China in 2003; the idea that Hon Lik was the first person to develop a device that aerosolised a nicotine solution is an oversimplification of the history of the e-cigarette. PM began exploring nicotine aerosol generation technology in 1990 to create a product that would serve as an alternative to NRT therapy for health-concerned smokers and deal with smoke-free environments. PM did not market an e-cigarette until 2013 (the MarkTen) because of concern about triggering FDA regulation of conventional cigarettes and the potential impact of the product on their political relationship with farmers. As PM anticipated, many e-cigarette users report using e-cigarettes to try to quit smoking (but smokers who use e-cigarettes are significantly less likely to quit smoking cigarettes than smokers who do not use e-cigarettes) and in place of conventional cigarettes in smoke-free environments. PM developed e-cigarettes to provide a product for health-concerned smokers and smokers facing increased restrictions on smoking in public. Researchers and policymakers should recognise that PM developed e-cigarette technology to evade tobacco control regulations.

Reference Number: 50102

SANTOS, U.P. "Electronic Cigarettes—The New Playbook and Revamping of the Tobacco Industry." *Jornal Brasileiro de Pneumologia*, 44: 345–346, 2018 (20 references)

Abstract: The smoking epidemic began in the late 1800s with the invention of rapid, automatic cigarette-making machines. It was driven in the 20th century by World War I and World War II, the cinema, and the advertising industry. The tobacco industry has devised a new strategy and now seeks to present itself as a defender of public health by marketing tobacco vape products. But by adding flavors (which are banned in cigarettes), sleek designs, and heavy social media marketing, the tobacco industry actively seeks to attract new users who will become dual users.

Why should a doctor prescribe a product made by the same industry that, despite having been aware of the disastrous health impacts that its products have, has always been slow to admit that there is such an impact and that nicotine is in fact addictive, steadfastly refusing to pay reparations to its victims, as demonstrated by the exhaustive collection of documents produced by the tobacco industry itself and released to the public in recent decades? On the basis of the current knowledge, there is no reason.

Reference Number: 50103

MIYASHITA, L., AND G. FOLEY. "E-cigarettes and Respiratory Health: The Latest Evidence." *Journal of Physiology*, 598: 5027–5038, 2020 (1 table, 2 figures, 77 references)

Abstract: The first known EC, described as a tobacco-free nicotine aerosol device, was invented in the United States by Herbert A. Gilbert in 1963 and was made a commercial success by a Chinese pharmacist, Hon Lik, in 2003. EC technology has changed dramatically from the first-generation cig-a-likes to the popular fifth-generation pods such as JUUL. There are 15,000 different e-liquids available to be inhaled. In 2016, the European Union banned diacetyl (which leads to popcorn lung disease) in e-liquids, but this chemical still appears in e-liquids in Canada and the United States. In 2019, an outbreak of EVALI (e-cigarette or vaping product associated lung injury) occurred in the United States, probably due to vitamin E acetate, a thickening agent added to e-liquids. The main ingredients of e-liquids, propylene glycol and vegetable glycerin, are deemed safe for oral consumption but have not been tested for inhalational toxicity.

Table: Five Generations of E-cigarettes

Generation	Common Name	Power (voltage)	Notes
First	Cig-a-likes	Fixed low voltage (3.7 V)	LED light automatically lights up upon inhalation, looks like a cigarette
Second	Clearomizers Vape pens	Higher variable voltage (3–6 V)	Manual fire button to turn on heater coils and initiate inhalation
Third	Mods	Voltage is adjustable to 8 V	User refills e-liquids into a tank or reservoir

Generation	Common Name	Power (voltage)	Notes
Fourth	Sub ohm mods	Control voltage and/or wattage	Large puff volumes can be inhaled and vapes can be produced, LED screens
Fifth	Pods	3.7 V	No buttons or switches, automatic, uses prefilled liquid pods

Source: Adapted from Miyashita and Foley, 2020

The respiratory system must maintain an intricate balance to maintain pulmonary homeostasis, which involves both physiological and immune-mediated processes. Evidence continues to emerge on the potential of EC use and vaping products to disrupt this equilibrium, leading to a cascade of events that result in lung inflammation, injury, reduced host defence and increased infection risk. As the popularity of ECs continues to rise and new cases of EVALI continue to be reported, it is clear, at least in the USA, that there is an urgent need to revisit the current EC regulatory environment. ECs have a vast appeal to non-smoking adolescents and young adults, which is a cause of great concern. The huge variety of flavours and advances in EC technology have only increased the potential deleterious effects to respiratory health and it is vital that research into the long-term implications of vaping continues.

Reference Number: 50104

UNGER, M., AND D.E. UNGER. "E-cigarettes/Electronic Nicotine Delivery Systems: A Word of Caution on Health and New Product Development." *Journal of Thoracic Disease*, 10: S2588–S2592, 2018 (22 references)

Abstract: There are currently over 460 EC brands and over 7,700 flavors. JUULs are a liquified nicotine salt delivery device with rapidly absorbable nicotine delivery much higher than that of a standard cigarette. The US Food and Drug Administration (FDA) was given regulatory powers over ECs on 8 August 2016. Acute vaping of an aerosol of propylene glycol/glycerol at high voltage and in large amounts induces tissue hypoxia, airway epithelial injury, and airway constriction.

Most e-cigarettes structurally contain four different components including: 1) Cartridge or reservoir, which holds a liquid solution (e-liquid or e-juice) containing varying amounts of nicotine, flavorings, and other chemicals; 2) Heating element or coil (atomizer); 3) Power source (usually a rechargeable battery); and 4) Mouthpiece that the person uses to inhale. It is thus more appropriate to address these products as electronic nicotine delivery systems (ENDS). In many ENDS, user's puffing activates the battery powered heating device, which vaporizes the liquid in the cartridge"

Reference Number: 50105

FAIRCHILD, A., AND J. COLGROVE. "Out of the Ashes: The Life, Death, and Rebirth of the 'Safer' Cigarette in the United States." *American Journal of Public Health*, 94: 192–204, 2004 (2 figures, 105 references)

Abstract: In the 1950s and 1960s tobacco companies began to manufacture filtered, low tar cigarettes and market them as safer. Harm reduction has been advocated by some public health groups, while most believe in smoking control, cessation, and the complete elimination of cigarettes from society. In 1988, RJ Reynolds announced a smokeless tobacco cigarette called Premier, which was removed from the market due to poor consumer response and attack from public health groups. In 1996, RJ Reynolds introduced another, safer cigarette (heat not burn) called Eclipse. Public health officials had attempted to collaborate with the tobacco industry to find a safer cigarette but were found to have been betrayed when tobacco industry documents were made public. Certain advocates for safer tobacco were found later to have been cozy with the tobacco industry.

The later history of tobacco industry deception and manipulation was an important factor contributing to the erosion of public health support for harm reduction; indeed, its own labyrinth of lies made the industry a dishonest partner in the pursuit of supposedly safer products, which must give us pause. But while knowledge of this deception and its consequences now makes the reemergence of the "safer" cigarette unacceptable to many, it neither invalidates the earlier public health quest for harm reduction nor dictates the course we must take.

Reference Number: 50106

KAUR, G., R. PINKSTON, B. MCIEMORE, W.C. DORSEY, AND S. BATRA. "Immunological and Toxicological Risk Assessment of e-cigarettes." *European Respiratory Review*, 27: 170119 (14 pp), 2018 (2 tables, 2 figures, 104 references)

Abstract: The concept of a battery powered nicotine delivery device dates back to 1963 but was first successfully introduced by Hon Lik in China in 2003 and introduced to the United States in 2007. In 2013, sales were $1.7 billion in the US market. Seventy-five percent of vapers use conventional cigarettes as well. The basic design remains the same, an atomizer, reservoir containing e-liquids, and a rechargeable lithium battery. ECs require more suction and stronger puffs than cigarettes, which cause aerosols to reach deeper into the lungs. EC aerosols have high concentrations of silver, iron, aluminum, silicate, and nanoparticles of chromium and nickel from the heating coils and can be 2 to 100× that found in cigarettes. EC maintains nicotine dependence, making it harder to quit. Unlike with cigarettes, there is no set dosage of nicotine in ECs. In 2016, the FDA placed ENDS as well as vape pens, hookah tobacco, and pipe tobacco as tobacco products.

> Based on recent evidence, prolonged exposure to some constituents of e-cig aerosols might result in respiratory complications such as asthma, chronic obstructive pulmonary disease and inflammation. Future studies are warranted that focus on establishing correlations between e-cig types, generations and e-liquid flavours and immunological and toxicological profiles to broaden our understanding about the effects of vaping.

Reference Number: 50107

SLADE, J., G.N. CONNOLLY, AND D. LYMPERIS. "Eclipse: Does It Live Up to its Health Claims?" *Tobacco Control*, 11: ii64–ii70, 2002 (8 tables, 1 figure, 20 references)

Abstract: RJR Tobacco introduced several protype vaping devices that delivered an aerosol rather than smoke: Premier in 1988, Eclipse I in 1996, and Eclipse II in 2000. They look like a cigarette but have a carbon fuel element, a vaporization chamber consisting of 50% shredded tobacco paper and 50% glycerin surrounded by aluminum, and a cooling chamber consisting of shredded tobacco paper and finally a hollow mouthpiece. In the cooling chamber the vapors condense to form aerosols. RJR contended that Eclipse presents less risk for disease associated with smoking. The Massachusetts Department of Public Health conducted an independent analysis of the Eclipse aerosol.

Table: Carcinogen Content of Premier, Eclipse I, and Eclipse II Prototype Vaping Devices

Carcinogen	Premier (1988)	Eclipse I (1996)	Eclipse II (2000)
Acetaldehyde (μg)	41	54	84
BaP (ng)	0.08	0.70	1.20
NNN (mg)	8.5	10.0	26.0
NNK (ng)	2.4	12.0	32.0

Source: Adapted from Slade et al, 2002

There have been many attempts by tobacco companies to develop and market low nicotine and ultralight cigarettes. In 1996, the RJ Reynolds Tobacco Company (RJR) began test marketing Eclipse, a novel cigarette-like smoking article that delivers a nicotine containing aerosol to the user. RJR claimed that Eclipse, by heating rather than burning tobacco, reduces the risk of cancer and other diseases compared to conventional cigarettes. Analyses of this claim had not been conducted. This paper analyses this claim by examining several data sources and scientific studies.

We found that Eclipse appears to be at least as toxic as some commercially available cigarette brands and that the data do not support the health claims made about the product. Consumers may be misled by RJR's health claims into believing that Eclipse is a safer alternative to conventional cigarettes, underscoring the need for regulatory intervention. The analyses reported in this paper have prompted the Massachusetts Department of Public Health to request that the Federal Trade Commission carefully review RJR's advertising and marketing practices for Eclipse and/or seek a temporary injunction banning its sale under the FTC Act.

Reference Number: 50108

GOTTLIEB, M.A. "Regulation of E-cigarettes in the United States and Its Role in a Youth Epidemic." *Children*, 6: 40 (6 pp), 2019 (39 references)

Abstract: Chinese pharmacist and inventor Lik Hon patented the electronic atomization cigarette in the early 2000s. It was based on a piezoelectric ultrasound device to vaporize a nicotine solution and not on wicks and heating coils used in current ECs. Few pediatricians had ever heard of an electronic cigarette when the first attempts at regulation were conducted by the FDA in 2008. Initially, the concept of EC or electronic nicotine delivery systems (ENDS) seemed promising as an alternative to high-risk combusted tobacco products for addicted smokers designed to reduce their exposure to tobacco-specific carcinogens. Instead, ENDS used by youth have been declared an epidemic and there are no clear guidelines for treating addicted teens.

The first decade of federal regulation of ENDS has resulted in very limited controls placed on the market and a true epidemic of use of these products by youth. The FDA will soon determine how or whether these products should remain on the market. What had been a very small industry when the agency first sought to regulate ENDS has grown exponentially and is now largely controlled by the parent companies of cigarette manufacturers. These factors combine to create an especially challenging regulatory environment for FDA. Without the fast-tracked implementation of a strong federal regulatory response to the epidemic of youth vaping, we face the prospect of treating an entire generation of youth who are deeply addicted to nicotine.

Reference Number: 50109

DIBJI, S.A.R., S. GUHA, A. ARAB, B.T. MURRAY, AND M.R. MYERS. "Accuracy of Commercial Electronic Nicotine Delivery Systems (ENDS) Temperature Control Technology." *Public Library of Science One*, 13: (13 pp), 2018 (1 table, 4 figures, 23 references)

Abstract: Fourth generation vaping devices have batteries that are capable of supplying 300 watts of electrical power to the heating coil and have advanced temperature control compared to the earlier generation ENDS, which had only 1–15 watts of electrical power and no temperature control. At higher temperatures, increasing concentrations of formaldehyde, acrolein, and acetaldehyde form. Three different atomizers were tested using thermocouples to measure the temperature at different locations of the vaping device. An e-liquid mixture of 65% VG/35% PG and 3 mg/mL nicotine was employed. At more than 200 watts of power applied to the 4th generation ENDS, coil temperatures can reach temperatures near that of combustion in cigarettes.

The user-controlled temperature control feature of the fourth-generation of ENDS can result in the coil temperature exceeding the e-liquid boiling point very rapidly, within a fraction of the puff duration. Also, the resistance-based temperature measurements of the devices themselves correspond to an average value for the temperature along the coil. The temperature can vary significantly along the coil due to the nature of the heat transfer and vaporization of e-liquid. The temperature measured by the device can substantially underestimate the temperature in the center of the coil.

A wide variation in measured coil temperatures between atomizer samples was observed, with some samples of a given brand manifesting thermocouple temperatures 40 °C below the 300 °C set point, and other samples of the same brand generating more than 100 °C in excess of the set point. This shows that not only are there concerns with the accuracy of the device's own temperature reading but also that variability within a brand should be accounted for in risk assessments. At the 300 °C set point, the measured aerosol temperature by the thermocouple at the mouthpiece was found to be as high as 160 °C.

Reference Number: 50110

LEGG, T., S. PEETERS, P. CHAMBERLAIN, AND A.B. GILMORE. "The Philip Morris–Founded Foundation for a Smoke-Free World: Tax Returns Sheds Light on Funding Activities." *Lancet*, 393: 2487–2488, 2019 (19 references)

Abstract: Philip Morris International (PMI), one of the world's largest tobacco companies, set up the Foundation for a Smoke-Free World with almost $1 billion USD in funding over 12 years. Although the Foundation claims to be an independent and scientific body, hundreds of public health agencies and the World Health Organization (WHO) have taken a strong stance against collaborating with the Foundation. In 2018, the Foundation spent $5.22 million with Ogilvy Public Relations Worldwide, which has worked with Big Tobacco for decades.

Given the tobacco industry's long history of using public relations firms and external scientists in its manipulation of research, the connections between the Foundation and these bodies raise serious concerns about the legitimacy of scientific messages and research originating from the Foundation.

Overall, our findings suggest that as it reaches its second anniversary, the Foundation might be failing. It is seemingly struggling both to recruit independent researchers and to spend its resources on much other than public relations. In line with public rejections from Bloomberg Philanthropies and the Bill & Melinda Gates Foundation, no further funders have materialised. Furthermore, several key staff members left the Foundation in its first 21 months of operation. Now, more than ever, we must reinforce the hitherto successful calls from WHO and the public health community to reject collaboration with the Foundation.

Reference Number: 50111

DE ALMEIDA, L.M., R.P. DA SILVA, A.T.C. DOS SANTOS, J.D. DE ANDRADE, AND M.C. SUAREZ. "Mists, Vapors and Other Illusory Volatilities of Electronic Cigarettes." *Cadernos de saúde pública,* 33: (13 pp), 2017 (4 tables, 22 references)

Abstract: This study discusses the marketing of ECs between 2011 and 2013. Lorillard Tobacco Company acquired Blu in 2012 and Imperial Tobacco in 2013. RJ Reynolds launched VUSE in 2013. The Altria Group (Philip Morris) bought the EC-Green Smoke and a large stake in JUUL. Large tobacco companies now own over half of the EC market. Marketing is based on presenting ECs as a safe, practical, and healthy alternative to cigarettes.

For over 100 years, the tobacco industry's economic survival was guaranteed by the commercialization of a non-essential product that is harmful to human life. With the worldwide organized mobilization against tobacco consumption, the industry realized its profit margins were at risk. Currently, the invention of electronic cigarettes, which did not happen by chance, provides indications that this same industry is watching the entry of possible, seductive products into the market, as an opportunity to eternalize its profits. These are products for which very little is known regarding their possible health impacts and which must be more rigorously evaluated.

In this scenario, it is suspected that a very large risk is on the horizon and threatens to undermine the global effort of the past 20 years to inform the public of the harms associated with smoking. This effort concerns not only the inherent risks of smoking itself, but also the changes in attitude in terms of making smoking in closed spaces socially unacceptable, something that may be lost with the re-introduction of new products in these environments.

The social conscience regarding the risks of smoking may be clouded by the fragrant, flavorized fogs, mists and vapors of e-cigs, since the return to the (already surpassed) behavior of smoking in social environments and the reassuring belief in their harmlessness may lead an old evil to reappear. In the medium and long term, they may even reintroduce smokers of conventional tobacco products into collective closed environments. But, unfortunately, this risk cannot yet be measured, only felt.

This study sought to describe and analyze the marketing positioning discourse used by electronic cigarette sellers to convince potential users to buy and use them.

Reference Number: 50112

DINARDO, P., AND E.S. ROME. "Vaping: The New Wave of Nicotine Addiction." *Cleveland Clinic Journal of Medicine*, 86: 789–798, 2019 (2 tables, 1 figure, 64 references)

Abstract: The first US patent of a device resembling the modern EC was in 1963, but the product never made it to market. The first commercially successful EC was made in Beijing in 2003 and ECs were introduced into the US market in 2007. The JUUL device was developed in 2015 by two Stanford University graduates, and by 2017 it had captured 72% of the EC market in the United States. Israel and Europe have forced the company to replace the 5% nicotine pods to pods with 1.7% nicotine. In December 2016, the US Surgeon General warned that EC use among youths was a major public health concern. Advertising for vaping was not initially banned, as it has been for cigarettes since 1971. Marketing themes for vaping have mimicked themes used successfully by the tobacco industry for decades such as freedom, rebellion, youth, and glamor. Initial vaping websites even contained endorsements by physicians, as occurred in the 1940s and 1950s with cigarettes.

Table: Tobacco Use Among US High School Students in 2018

Tobacco Product	Percent Use in High School Students
E-cigarettes	20.8%
Cigarettes	8.1%
Cigars	7.6%
Smokeless tobacco	5.9%
Hookah	4.1%
Pipe tobacco	1.1%

Source: Adapted from Dinardo and Rome, 2019

- Vaping is a common gateway to tobacco and marijuana use for adolescents and adults.

- The Juul vaping device delivers high nicotine concentrations that may pose a higher risk of nicotine addiction.
- Vaping has had unintended consequences that include poisoning of children who swallowed liquid nicotine, fires and explosions from defective batteries in the devices, and effects on the developing brain.
- Vaping is associated with respiratory illness and, in rare cases, death, likely due to vaporized agents introduced into the lungs. Small amounts of heavy metals, acetone, and other carcinogenic compounds in the vaping aerosol may cause lung damage.

Reference Number: 50113

NOWAK, D., R.A. JÖRRES, AND T. RÜTHER. "E-cigarettes—Prevention, Pulmonary Health, and Addiction." *Deutsches Ärzteblatt International*, 111: 349–355, 2014 (2 tables, 2 figures, 40 references)

Abstract: ECs have been on sale without restrictions in Germany since 2008.

The market share of e-cigarettes will increase. The cigarette industry will enter this market, probably also targeting groups who are not yet consumers—in other words, the young in particular—with cigarettes that are ostensibly "healthier" because they are smoke-free. This would thwart medical professionals' efforts to prevent tobacco use among young people. The cigarette industry may try to accelerate nicotine release and increase the quantity of absorbed nicotine in order to attain the "kick" that is well known in cigarette smoke and to support addiction behavior among consumers in the long term. A further aim of the cigarette industry may be to force users to switch to conventional tobacco products later and thereby win back the market share they have been losing. It is impossible to foretell whether such efforts will be successful. In addition, disposable e-cigarettes might cause an electrical waste problem.

Reference Number: 50114

EISSENBERG, T., A. BHATNAGAR, S. CHAPMAN, S-E. JORDT, A. SHIHADEH, AND E.K. SOULE. "Invalidity of an Oft-Cited Estimate of the Relative Harms of Electronic Cigarettes [Editorial]." *American Journal of Public Health*, 110: 161–162, 2020 (7 references)

Abstract: The oft-cited factoid that EC use is 95% safer than cigarettes is based on a meeting of 12 experts (many with connections to the vaping and tobacco industries) in July 2013 that estimated the risk of EC compared to cigarettes based on a lack of real data and evidence on most criteria. Now, over 6 years later, the EC has changed greatly with power outputs that exceed 10 to 20× the power outlet of the earlier ECs. This leads to a greater amount of aerosol liquid being inhaled into the lungs at higher temperatures. There are over 7,000 different types of e-liquids, most using chemicals that have not been tested for pulmonary toxicity. EC use is associated with wheezing, which is a potential early sign of lung disease, and increases HR, BP, platelet activation and increases heart rate variability (a potential sign of myocardial infarction and coronary artery disease). EC aerosols containing PG and VG cause inflammation of the lungs.

The "95% safer" estimate is a "factoid": unreliable information repeated so often that it becomes accepted as fact. Public health practitioners, scientists, and physicians should expose the fragile status of the factoid emphatically by highlighting its unreliable provenance and its lack of validity today, noting the many changes in e-cigarette devices and liquids, the accumulation of evidence of potential harm, the increased prevalence of use, and the growing evidence that e-cigarette use is associated with subsequent cigarette smoking.

Reference Number: 50115

HUANG, J., Z. DUAN, J. KWOK, S. BINNS, L.E. VERA, Y. KIM, G. SZCZYPKA, AND S.L. EMERY. "Vaping versus JUULing: How the Extraordinary Growth and Marketing of JUUL Transformed the US Retail E-Cigarette Market." *Tobacco Control*, 28: 146–151, 2018 (5 tables, 2 figures, 37 references)

Abstract: JUULs are a sleek, high tech, portable e-cigarette fashioned to look like a computer flash drive and rechargeable at a USB port. They have two basic components: a battery and temperature regulation system and a prefilled e-liquid cartridge called a JUUL pod which also serves as a mouthpiece. In 2017, JUUL accounted for 40% of the retail e-cigarette market in the United States. A JUUL pod contains high levels of nicotine (59 mg/mL) and use nicotine salts found in leaf tobacco, rather than free-base nicotine (which is bitter tasting). It delivers a nicotine peak in 5 minutes, similar to combustible cigarettes. JUUL employs a variety of innovative, engaging, and wide-reaching campaigns on Twitter, Instagram, and YouTube, which also mislead about addiction. Instagram posts by the official JUUL account increased from 26 (with 4,820 Likes) to 73 (with 59,465 Likes).

Table: Characteristics of Instagram Posts by the Official JUUL Account (As of 15 February 2018)

Category	Number of Posts	Number of Likes
Cartoon images	1	277
Celebrity/model images	5	2,491
Customer feedback/testimony	21	6,541
Fashion images	5	3,721
Flavor images	15	6,226
Lifestyle appeal images	43	27,352
Sponsored events	1	1,105
Product images	38	42,687
Total	129	90,400

Source: Adapted from Huang et al, 2018

However, JUUL was one of the first major retail e-cigarette brands that relied heavily on social media to market and promote its products. In particular, we found the number of JUUL-related tweets was highly correlated with quarterly retail sales of JUUL. In addition to Twitter, JUUL was heavily marketed and promoted on Instagram and YouTube. The official JUUL account on Instagram, for example, used a variety of marketing and promotional schemes to attract, engage with and retain followers. The account used artsy, professional-grade photographs to display its products and evoke lifestyle feelings such as relaxation, freedom and sex appeal. Those posts also heavily emphasised JUUL's variety of flavours. Related accounts heavily marketed and promoted JUUL and associated products including pods, skins and accessories.

Reference Number: 50116

PEACE, M., J. POKLIS, AND J. TURNER. "Chasing the Electronic Cigarette Dragon: Characterizing the Evolution and Impact of Design and Content." *National Institute of Justice*, DUNS Number: 10-530-0446 (19 pp), 2019 (1 table, 12 figures)

Abstract: Electronic cigarettes are known as e-cigs, personal vaporizers (PVs), or electronic nicotine delivery systems (ENDS). There are numerous drug forums, blogs, and YouTube videos on how to convert ECs from vaporizing nicotine into delivering other drugs such as THC, methamphetamine, fentanyl, or synthetic cannabinoids. There are also new devices which can aerosolize dried plant material and powders. The Source Orb EC has six different atomizers: coil-less, single coil, and double coil, made up of quartz, ceramic, or titanium. The Wulf Mod Dome Kit was equipped with a coil-less ceramic chamber, while the Vaped Vubbler has a single ceramic titanium coil. The temperatures of these coils and cups are significantly lower than the Nichrome coils of the traditional second and third generation coils. Coils in the fourth generation have a ceramic or glass rod through the coil which prevents wicking from a tank. Most e-liquids containing nicotine sell for $5 to $10 a bottle, while those containing drugs sell for 5 to 10× that. The heavy molecular weight drugs are not effectively aerosolized. Nicotine e-liquids that contain 100% PG produce larger particle sizes than 100% VG or 50:50 e-liquids and produce a greater throat hit.

> Aerosol particles that are less than 0.1 mm are referred to as "ultra-fine" particles and are known to result in deep lung deposition. However, with smaller particle size, there is a greater chance of the drug being exhaled before it can be absorbed into the bloodstream. To combat this, some e-liquid manufacturers will print instructions on the bottle labels that instruct the e-cig user to inhale and hold the aerosol for a certain amount of time. This method would increase the amount of pulmonary deposition that would lead to more drug absorption into the blood, thereby increasing the bioavailability of the aerosolized drug.

Reference Number: 50117

ALLEM, J-P., AND E. FERRARA. "Could Social Bots Pose a Threat to Public Health?" *American Journal of Public Health,* 108: 1005–1006, 2018 (7 references)

Abstract: A social bot (robot) is an automated account that uses artificial intelligence to steer discussions and promote products (e.g., vaping) or messages (e.g., vaping is harmless, vaccination is dangerous, etc.) on social media such as Twitter or Facebook. A bot is designed to resemble the appearance of a human user. A 2017 study of 2.2 million Twitter posts found that social bots were 2× as likely to suggest that ECs could be used for smoking cessation and 2× as likely to promote new EC devices and accessories.

> As research has shown, social bots employed to discuss a particular topic can number in the hundreds of thousands, with the potential to drown out medically sound social media messages from medical experts or health campaigns. The number of social bots and the frequency at which they produce and disseminate content may give the impression that a particular behavior is more prevalent online than it is offline, normalizing unhealthy or medically unsound decisions. Social bots may inflate the perception of a health issue by creating panic on the part of the public, contributing to the spread of rumors or unverified information as with the recent Ebola and Zika virus pandemics.

5.02 JUUL AND OTHER VAPING DEVICES

> "The e-cigarette is the most common tobacco product used by youth, exposing users to nicotine, toxicants, and ultrafine particles, which have the potential to lead to nicotine addiction and harmful health effects."
>
> —Walley et al, "A Public Health Crisis: Electronic Cigarettes, Vape, and JUUL" (2019)

JUUL e-cigarettes (ECs) have captured nearly two-thirds of the market for vaping devices by appealing to youth, as the device resembles a USB drive and has a relatively small vape so that it avoids detection in schools (50201). JUUL has prefilled pods that contain a high concentration of nicotine and are available in eight different flavors. All JUUL pod solutions are cytotoxic at a ten-fold dilution (50201).

One of the major innovations of JUUL, which made it more addictive, was to reduce the harsh bitter taste of nicotine by adding benzoic acid. JUUL contained a much greater concentration of addictive nicotine than the earlier models of ECs (50202). Although Europe has restricted the nicotine concentration of JUUL pods to 20 mg/mL, in contrast to the 35–57 mg/mL found in Canada and the United States, JUUL has been able to get around the regulations by modifying the wick in its European devices to increase the nicotine found in the vape (50204).

Smokers who switched to filter and low tar cigarettes because they were supposedly safer are now switching to vaping, as it is perceived as safer as well (50204).

New vaping devices have more power and they heat the e-liquids to a higher temperature, which produces more nanoparticles and toxic chemicals than the low temperature cig-a-likes (50205–9). The sensation of sweetness that occurs when sucralose is added to the e-liquid is only significant in cartridge and not tank ECs (50210).

Reference Number: 50201

OMAIYE, E.E., K.J. MCWHIRTER, W. LUO, P.A. TIERNEY, J.F. PANKOW, AND P. TALBOT. "High-Nicotine Electronic Cigarette Products: Toxicity of JUUL Fluids and Aerosols Correlates Strongly with Nicotine and Some Flavor Chemical Concentrations." *Chemical Research in Toxicology*: (12 pp), 2019 (2 tables, 7 figures, 35 references)

Abstract: Although cigarette smoking is declining in many countries, youth and adult use of ECs have increased. In June 2018, its was estimated that about 68% of current EC sales were JUUL products. JUUL ECs resemble USB drives and are appealing to youth in that they can create relatively small clouds of aerosol, which avoids detection in schools. At the time of the study, the JUUL prefilled EC pods offered eight different flavors. The JUUL pods contained high concentrations of nicotine (60.9 mg/mL) and enough acid to protonate most of the nicotine so the vape tasted smoother and not as bitter. Fifty-nine flavor chemicals were identified in JUUL pod fluids and three had a concentration greater than 1 mg/mL. All pod fluids were cytotoxic at 1:10 dilution. The cytotoxicity of the collected aerosol materials was highly correlated with nicotine and ethyl maltol concentrations and, to a lesser extent, to total flavor chemical composition and menthol concentration.

> In summary, the current popularity of JUUL products has raised two major concerns for the FDA. The first is the likelihood that JUUL use, which is widespread among middle school and high school students, will addict a new generation of adolescents to nicotine. The second is that these adolescents will eventually migrate to tobacco products that may be more dangerous, such as conventional cigarettes. Our data clearly identify a third concern related to the high nicotine concentration in JUUL products, i.e., the potential for high levels of nicotine, as well as flavor chemicals such as ethyl maltol, to damage or even kill cells at the concentrations used in JUUL pods. Our exposures were acute and produced a maximal cytotoxic response that was strongly correlated with nicotine and ethyl maltol concentrations.

Reference Number: 50202

WALLEY, S.C., K.M. WILSON, J.P. WINICKOFF, AND J. GRONER. "A Public Health Crisis: Electronic Cigarettes, Vape, and JUUL." *Pediatrics*, 143: (11 pp), 2019 (5 figures, 70 references)

Abstract: E-cigarettes have five common components: (1) a mouthpiece, (2) a sensor or user-activated switch to activate the heating coil, (3) a battery, (4) a heating coil or atomizer, and (5) a reservoir, pod, or tank for the e-liquid. JUUL pods contain 0.7 mL of a liquid containing propylene glycol, glycerol, benzoic acid, additives, and flavorants. The nicotine concentrations in Canada and the United States are 5% (59 mg/mL) or 3% (35 g/mL). Each pod is marketed as containing 200 puffs, equivalent to the number of puffs in a pack of 20 cigarettes. Altria (Philip Morris) bought a 35% stake in JUUL for $13 billion. TV ads for cigarettes have been banned in the United States since 1971, but 78% of middle and high school students reported seeing EC ads in 2016.

> The e-cigarette is the most common tobacco product used by youth, exposing users to nicotine, toxicants, and ultrafine particles, which have the potential to lead to nicotine addiction and harmful health effects. In 2018, e-cigarettes were used by 3.6 million middle and high school students, and the popularity of newer e-cigarette brands, such as JUUL, with sleek designs and higher nicotine concentrations is a public health crisis.

Reference Number: 50203

MALLOCK, N., H.L. TRIEU, M. MACZIOL, S. MALKE, A. KATZ, P. LAUX, F. HENKLER-STEPHANI, J. HAHN, C. HUTZLER, AND A. LUCH. "Trendy E-cigarettes Enter Europe: Chemical Characterization of JUUL Pods and its Aerosols." *Archives of Toxicology*, 95: 1985–1994, 2020 (1 table, 3 figures, 48 references)

Abstract: The nicotine concentration in European e-liquids was restricted to 20 mg/mL, compared to 57 mg/mL found in US and Canadian JUUL pods. But this limit can be circumvented by technological adjustments (such as different wicks) that increase the nicotine vaporization and delivery, thereby making the JUUL more addictive. The JUUL pod in Europe had been been modified and the e-liquid now contained 18 mg/mL or 9 mg/mL nicotine. The measured nicotine per puff, however, was 23 μg/puff, 61 μg/puff, and 30 μg/puff for the three devices respectively, which resulted in a more effective delivery in the modified European JUUL pods and e-cigarettes. The legislation needs to be framed in terms of maximum nicotine allowed per puff rather than the e-liquid nicotine concentration.

Table: Chemical Characterisation of Initial and Modified European JUUL Liquids/Vape (Rich Tobacco flavor)

Chemical Characterization of JUUL Liquid	Initial European Design (20 mg/mL Nicotine)	Modified (18 mg/mL Nicotine)	Modified (9 mg/mL Nicotine)
Measured nicotine concentration (mg/mL)	17.2	17.69	9.03
pH of liquid	5.51	5.42	5.40
Benzoic acid (mg/mL)	9.64	12.67	7.02
Total particulate matter (mg/puff)	1.6	3.7	3.7
Nicotine (µg/puff)	23	61	30

Source: Adapted from Mallock et al, 2020

Our data demonstrate that the initial product on the European market generated a similar amount of vapor when compared to the American version and subsequently only achieved relatively low levels of nicotine delivered into the aerosol. Amidst our investigation, a new product design, referred to as "Turbo", was launched. We could show that the degree of vaporization in the newly designed product increased more than twice and therefore can be considered sufficient to compensate for the lower nicotine contents in the liquid. This observation confirmed our initial expectation that nicotine delivery will be increased with technical adaptions. Modified JUUL was shown to deliver approximately the same amounts of nicotine as the American version in our machine vaping set-up and thus could potentially lead to blood nicotine levels that are comparable to tobacco cigarettes as well.

The increased vaporization of modified JUUL is linked to the use of another wick material. It is not related to a higher power delivery as parameters like the resistance of the coil and the battery voltage did not change. The properties of the wick can also have a substantial influence on vaporization via the liquid supply rate. A wick made of an expansible material can resupply the coil faster and more stably with unvaporized e-liquid. It is visible by the naked eye that the constitution of the wick has changed. The initial version of the product showed very high intra- and inter-device deviations, especially when more than 200 puffs were drawn. Vapor generation by modified JUUL is more stable; however the deviations are still high and do not attest a good consistency of nicotine delivery. Since both versions contain the same amount of liquid, only half the number of puffs can be drawn from the modified version (see Supplementary Material). Depending on changes in consumption behavior, this could have an influence on cost.

Reference Number: 50204

PULVERS, K., J.Y. SUN, Y-L. ZHUANG, G. HOLGUIN, AND S-H. ZHU. "Internet-Based Advertising Claims and Consumer Reasons for Using Electronic Cigarettes by Device Type in the US." *Tobacco Regulatory Science*, 3: 516–524, 2017 (1 table, 2 figures, 45 references)

Abstract: An online survey was conducted of 556 vapers using a closed system such as JUUL or cig-a-like and 304 vapers using an open or tank vaping system in 2014. The internet pays a major role in EC sales and many sites make claims as to ECs being safer and effective as a smoking cessation aid. Closed system devices are more likely to be owned by tobacco companies.

Whereas the top 2 reasons for using e-cigarettes were identical across both product types (to quit smoking and perception as healthier), there was variability in the third reason. Users of closed-systems cited using where smoking is banned as their third most common main reason for using e-cigs. In contrast, the third most common main reason for using e-cigarettes endorsed by users of open-systems was because they taste good. Previous literature has established that newer brands, which tend to be open-system, are more likely to offer many different flavors.

Reference Number: 50205

FLOYD, E.L., L. QUEIMADO, J. WANG, J.L. REGENS, AND D.L. JOHNSON. "Electronic Cigarette Power Affects Count Concentration and Particle Size Distribution of Vaping Aerosol." *Public Library of Science One*, 13: (15 pp), 2018 (4 figures, 37 references)

Abstract: ECs have evolved rapidly towards higher power devices that produce more vaping aerosol and a more satisfying experience. EC aerosol was measured in a single coil atomizer with increasing wattage (3 to 12 W) by a Scanning Mobility Particle Sizer and an Aerodynamic Particle Sizer. As the power of the EC increased, the particle count >600 nm increased and particle count <600 nm decreased. As the power increased more e-juice was used up as more particles are formed in the aerosol. JUUL is a G1 EC-low powered fixed voltage battery with a user replaceable cartridge/mouthpiece and uses a high nicotine salt concentration in the e-juice (5% or 50 mg/mL). Particle size affects respiratory tract disposition. Mouth puffing is a two-step procedure of using the oral cavity to draw a puff into the mouth at a low flow rate and then inhaling the puff from the mouth and blowing it out at a high flow rate. Mouth puffing provides time for the aerosol to grow in size and deposit on the oral cavity.

> Based upon the results of the specifically tested EC device, with a single e-juice, puffed at a single flow rate using a simulated mouth puff; vaping aerosol is dynamic and mouth puffed EC aerosol spans a much wider particle size range than previously reported. Although the major portion of particle mass is still well within the respirable size range, it is near the upper bound which shifts the primary deposition site to the oro-pharyngeal region instead of the alveoli. These results demonstrate the dramatic effect increasing power has on EC aerosol particle concentration and mass distribution across a wide range of particle sizes.

Reference Number: 50206

MULDER, H.A., J.L. PATTERSON, M.S. HALQUIST, L. KOSMIDER, J.B. MCGEE, J.L. POKLIS, A. POKLIS, AND M.R. PEACE. "The Effect of Electronic Cigarette User Modifications and E-Liquid Adulteration on the Particle Size Profile of an Aerosolized Product." *Scientific Reports*, 9: 10221 (8 pp), 2019 (3 figures, 40 references)

Abstract: Advanced ECs allow for the adjustment of the coil (heating element) resistance and the battery power (voltage) to deliver more of the drug in the aerosol. This study determined the effects of variable voltage on the particle size of a 12 mg/mL nicotine e-liquid containing either 100% PG, 100% VGF, or a 50:50 mixture. Typically, aerosol particles with a MMAD (Mean Mass Aerodynamic Diameter) of less than 5 μm are readily absorbed into the bloodstream via the lung. The nicotine containing e-liquid of 100% PG produced the largest aerosol particle size and may result in the improved throat-hit sensation. The 100% VG e-liquid produced smaller particle sizes and the 50:50 solution produced the smallest particle sizes. Other drugs tested, such as methamphetamine and methadone, produced similar aerosol particle sizes as nicotine.

The user modifications of glycol composition had some effect on the particle size distribution, but is likely not practically significant, and the coil resistance and battery voltage manipulations had no significant effect. The presence of particles less than 3 μm in size indicates that the drug can reach deep into the lung tissue and deposit into the bloodstream. The presence of particles of this size for methamphetamine and methadone highlight the likelihood of absorption of drugs other than nicotine using electronic cigarettes.

Reference Number: 50207

PRÉVÔT, N., F. DE OLIVEIRA, S. PERINEL-RAGEY, T. BASSET, J-M. VERGNON, AND J. POURCHEZ. "Nicotine Delivery from the Refill Liquid to the Aerosol via High-Power E-Cigarette Device." *Scientific Reports*, 7: 2592 (7 pp), 2017 (4 figures, 19 references)

Abstract: This study determined the particle size (using a cascade impactor) and nicotine content (by LC/MS/MS) of an aerosol produced by a high-powered ENDS. For all tests, the battery power was fixed at 15 W and the atomizers were changed regularly to prevent a degraded or dirty coil. The device tank was filled with 2 mL of e-liquid rather than the maximum 3 mL. The e-liquid contained 21 mg/mL of nicotine rather than the 18 mg/mL as labelled. The e-liquid did not contain any flavor chemicals. An 80/20 PG/VG mixture delivered about 16 mg/mL nicotine aerosol, and this was reduced to 13 mg/mL using a 20/80 PG/VG mixture.

Thus, new generation high-power ENDS associated with propylene glycol/vegetable glycerin (PG/VG) based solvent were very efficient to generate carrier-droplets containing nicotine molecules with a constant concentration. Findings highlighted that a portion of the nicotine in the refill liquid may not be transferred to the user.

Reference Number: 50208

LECHASSEUR, A., S. ALTMEJD, N. TURGEON, G. BUONANNO, L. MORAWSKA, D. BRUNET, C. DUCHAINE, AND M.C. MORISSETTE. "Variations in Coil Temperature/Power and E-Liquid Constituents Change Size and Lung Deposition of Particles Emitted by an Electronic Cigarette." *Physiological Reports*, 7: (9 pp), 2019 (5 figures, 23 references)

Abstract: Particle size of the aerosol was determined by a Condensation Particle Counter and Screening Mobility Particle Sizer spectrometer using power control or temperature setting in a tank vaper. The composition of PG, VG, and flavors were changed in the EC liquid. An inExpose EC extension was employed. Maintaining a higher coil temperature generated a smaller particle size. While nicotine and VG increased particle size, the flavors menthol and maltol did not affect particle size, but vanillin dramatically increased particle size. Larger particle size leads to a reduction in lung disposition.

Table: Distribution of Particle Size in Vape in a 50% Propylene Glycol/ Vegetable Glycerin E-liquid According to Increasing Temperature of the Coil

Temperature of coil (°C)	20.9–30 nm	30–40 nm	40–57.3 nm	57.3–117.6 nm	117.6–881.7 nm
210°	20%	25%	25%	20%	10%
225°	15%	20%	25%	25%	15%
250°	30%	25%	25%	20%	5%

Source: Adapted from Lechasseur et al, 2019

Overall, this study shows that changing the e-cigarette setting and e-liquid composition can alter e-cigarette particle size distribution, leading to changes in lung deposition. This may affect the amount of nicotine that is absorbed, and how much PG/Gly and flavors interact with the alveoli. It also highlights how flavoring agents can drastically alter the physicochemical nature of e-liquids. The physiological impacts of these changes remain to be investigated.

Reference Number: 50209

HARVANKO, A.M., A.K. MCCUBBIN, K.B. ASHFORD, AND T.H. KELLY. "Electronic Cigarette Liquid and Device Parameters and Aerosol Characteristics: A Survey of Regular Users." *Addictive Behaviors*, 84: 201–206, 2018 (2 tables, 3 figures, 22 references)

Abstract: In the spring of 2016, a survey was conducted of 522 adult regular EC users. Over half of those surveyed (59%) reported using another tobacco product such as cigarettes, snus, or cigars, which was associated with a greater nicotine dependence. The majority (74%) of those surveyed used refillable ECs. The voltage of the heating coil could be adjusted by 60% of the participants and the most common wattage chosen was 28.3 W (Wattage = voltage^2/resistance). The typical nicotine concentration of the e-liquid used was 9.1 mg/mL. The ability to change the device voltage and the level of resistance used were associated with a greater nicotine dependence. Some EC power levels chosen reached 100 W and new ECs could obtain 200 W. Propylene glycol was associated with a dry mouth, sore throat, headache, and dizziness. Vegetable glycerin was associated with a large cloud, good smell, good taste, and a throat hit.

Table: Regular EC User Behavior

EC User Behavior	Percent
Currently use ECs and other tobacco products	59.1%
Portion of other tobacco product users trying to quit	84.7%
Former regular users of other tobacco products	34.4%
Began other tobacco use after ECs	1.3%
Never used other tobacco products beside ECs	6.5%
Variable voltage EC	60.4%

Source: Adapted from Harvanko et al, 2018

> This study addresses critical gaps in our understanding of EC use by providing detailed information on EC usage behaviors, device parameter settings, and liquid variables among regular (i.e. daily) EC users. EC users utilize a wide variety of devices, power levels (i.e. watts, volts, ohms) and nicotine concentrations. Our data suggest that no predominant combinations of user-adjustable products have emerged from the broad range of commercial options. Consistent with previous findings, taste was identified as the most important variable when using ECs or purchasing liquids. Desirable effects were associated with the ingredient VG, and undesirable effects with PG, suggesting that these ingredients might influence the reinforcing effects of ECs.

Reference Number: 50210

ROSBROOK, K., H.C. ERYTHROPEL, T.M. DEWINTER, M. FALINSKI, S. O'MALLEY, S. KRISHNAN-SARIN, P.T. ANASTAS, J.B. ZIMMERMAN, AND B.G GREEN. "The Effect of Sucralose on Flavor Sweetness in Electronic Cigarettes Varies Between Delivery Devices." *Public Library of Science One*, 12: (14 pp), 2017 (1 table, 4 figures, 36 references)

Abstract: Sixteen female and 16 male smokers (aged 18 to 45 years) rated flavor intensity, harshness, and liking/disliking in four commercial flavors with and without sucralose (1%) using either cartridge or tank ECs at 4.2 W. Sweetness and flavor intensity were perceived more strongly when olfaction was allowed. The contribution of sucralose to sweetness was only significant in cartridge ECs.

> The present results demonstrate that the perceived sweetness of flavors derives primarily from volatile components of e-liquids that are sensed via retronasal olfaction rather than by taste. Although it was found that the artificial sweetener sucralose can add to the sweetness of flavors, the amount of sweetness added was small and had little effect on flavor liking. However, the added sweetness depended on the e-cigarette delivery system.

5.03 E-LIQUIDS

> "[T]he producers of propylene glycol and members of *Cefic's PO/PG sector group do not support the use of propylene glycol in electronic cigarettes*, nor in artificial (theatrical) fogs due to possible effects on the eye, nose, throat, and respiratory tract membranes, as well as the absence of information on potential long term effects from prolonged inhalation of (fine) droplets of propylene glycol."
>
> —"Propylene Glycol: Safe Use for Over 50 Years," Propylene Glycol, online: www.propylene-glycol.com/what-is-propylene-glycol/mono-propylene-glycol/safety

> "Flavour chemicals are present in almost all e-cigarette fluids currently on the market in the USA and globally. Concerns are rising among public health professionals that flavoured e-cigarette products may make e-cigarette use attractive to youth. Second, high doses of some flavour chemicals may be safe when ingested, but quite unsafe when inhaled."
>
> —Tierney et al, "Flavour Chemicals in Electronic Fluids" (2016)

E-liquids are also known as vape juice and are inhaled when heated in ECs. E-liquids are oily and highly viscous but can be analyzed by a number of methods (50301). Most e-liquids contain highly addictive nicotine, propylene glycol, vegetable glycerin (glycerol), flavorants, and other chemical additives such as benzoic acid, which reduces the bitter taste

Reference Number: 50210

of the basic nicotine (50302–4). Even the European manufacturers of propylene glycol do not recommend that it be inhaled.

The flavorants added to e-liquids are generally regarded as safe but only for oral ingestion and have not been tested for toxicity after heating and inhaling into the lungs (50303, 50310). It has been said that people smoke for the nicotine but die from the tar. It may also be said that people vape for the nicotine but die from the numerous flavoring chemicals (50304) of which about 246 have been detected in vaping liquids (50305).

Candy and fruit flavoring chemicals have been banned in tobacco cigarettes in many countries, including the US and Canada, because they appeal to youth and increase youth initiation to cigarettes. Now the tobacco companies are adding these flavoring chemicals to ECs for the same reasons (50307). Candy and dessert flavors are particularly attractive to vulnerable groups such as youth or non-smokers and should be banned completely (50308).

Chemical flavorants added to e-liquids have been found to cause oxidative stress and inflammation in lung cells (50306), increase nicotine absorption, cause deeper inhalation of the vape, and increase the toxicity of the vapor (50309–10, 50313–15).

Pyrazines, which are chemicals produced in the curing of tobacco leaves and added to cigarettes because they reduce the harshness of the smoke, are also found in most e-liquids (50316).

Alcohol is commonly found in e-liquids, although not in an amount that would cause a positive blood alcohol concentration after the consumption of one pod (50317). A false positive breath alcohol concentration (BrAC) occurred in alcohol-free vaping subjects who vaped a high alcohol containing e-liquid, but the effect lasted for less than 3 minutes (50318).

One common problem is that e-liquids that are labelled as nicotine-free actually contain this drug and pose a risk for addiction and child poisonings (50319–20). In the US, Canada, Australia, and England, the majority of vapers were White and male. Daily vapers bought most of their e-liquids at vape stores (50321).

Reference Number: 50301

FAMELE, M., C. FERRANTI, C. ABENAVOLI, L. PALLESCHI, R. MANCINELLI, AND R. DRAISCI. "The Chemical Components of Electronic Cigarette Cartridges and Refill Liquids: Review of Analytical Methods." *Nicotine and Tobacco Research*, 17: 271–279, 2015 (1 table, 2 figures, 49 references)

Abstract: Electronic cigarettes are battery powered devices that simulate tobacco cigarettes by converting a liquid into an inhalable aerosol. The chemical components of the aerosol differ from the e-liquids due to the heating of the liquid. Various methods are used to determine the chemical makeup of e-liquids, including HPLC/DAD, LC/MS/MS, and GC/MS. The main difficulty in analysis of e-liquids is that they are oily and highly viscous. Propylene glycol is more susceptible to thermal degradation than vegetable glycerin and so results in higher levels of carbonyls.

> It should be noted that examination of liquid composition does not assure the safety of generated aerosol after their use due to the production of new substances during heating and vaporization processes. Therefore, analysis on generated aerosol could be useful to correlate its composition with the chemical components of liquid.

Reference Number: 50302

JENSEN, R.P., R.M. STRONGIN, AND D.H. PEYTON. "Solvent Chemistry in the Electronic Reaction Vessel." *Scientific Reports,* 7: (11 pp), 2017 (9 figures, 48 references)

Abstract: Since ECs do not combust tobacco but heat and vaporize an e-liquid containing nicotine, flavoring chemicals, and PG and/or VG, the aerosol has a different chemical profile than cigarettes. The ECs used in this study had a variable voltage/variable wattage battery and replaceable bottom heating coils of from 1.8 to 2.5 Ohms resistance. The e-liquids tested were a 1:1 PG/VG mixture. More PG/VG mixture was consumed with increasing device power. The single puff samples were analyzed by NMR spectroscopy and numerous volatile aerobic thermal decomposition products of PG and VG were identified. Dual coil clearomizers produced less degradation products than a single coil one. Oxygen initiates the thermal degradation of PG and VG at significantly lower temperatures than anaerobic (pyrolysis) conditions.

> The chemical literature strongly suggests that there should be more compounds produced by PG and GLY than have been reported in e-cigarette aerosols to date. Herein we report an extensive investigation of the products derived from vaporizing PG and GLY under mild, single puff conditions. This has led to the discovery of several new compounds produced under vaping conditions. Prior reports on e-cigarette toxin production have emphasized temperature as the primary variable in solvent degradation.

In the current study, the molecular pathways leading to enhanced PG/GLY reactivity are described, along with the most impactful chemical conditions promoting byproduct production.

Reference Number: 50303

TIERNEY, P.A., C.D. KARPINSKI, J.E. BROWN, W. LUO, AND J.F. PANKOW. "Flavour Chemicals in Electronic Fluids." *Tobacco Control*, 25: e10–e15, 2016 (1 table, 1 figure, 19 references)

Abstract: Some e-liquid manufacturers stated that the ingredients used, including flavoring chemicals, are all food grade or generally recognized as safe (GRAS), however, the Flavor Extracts Manufacturers Association (FEMA) states that this pertains only to oral ingestion and not inhalation. FEMA also states that e-liquid manufacturers' statement that flavor chemicals are safe are false and misleading. Thirty EC fluids were analyzed by GC/MS. Of the 30 fluids, 13 had total flavor chemicals of 1 to 4% by weight (10–40 mg/mL) and nicotine was in the range of 0.6 to 2.4% (6–24 mg/mL). The main chemical flavoring group was aldehydes, which are known as primary irritants of mucosal tissues of the respiratory tract.

- Flavour chemicals are present in almost all e-cigarette fluids currently on the market in the USA and globally. Concerns are rising among public health professionals that flavoured e-cigarette products may make e-cigarette use attractive to youth. Second, high doses of some flavour chemicals may be safe when ingested, but quite unsafe when inhaled. Third, toxic degradation products may be produced by reaction of the flavour chemicals at the high temperatures present during e-cigarette use (aka "vaping").
- Flavoured e-cigarette products do not typically list the levels of specific flavour chemicals present, and most do not identify the major flavour chemicals present.
- The analyses of 30 products on the US market revealed that 13 were more than 1% by weight flavour chemicals. Chemicals identified included aldehydes (eg, benzaldehyde and vanillin) which could cause respiratory irritation.

Reference Number: 50304

OMAIYE, E.E., K.J. MCWHIRTER, W. LUO, P.A. TIERNEY, J.F. PANKOW, AND P. TALBOT. "High Concentrations of Flavor Chemicals are Present in

Electronic Cigarette Refill Fluids." *Scientific Reports*, 9: 2468 (9 pp), 2019 (4 figures, 52 references)

Abstract: It appears that for cigarettes, people smoke for the nicotine but are killed by the tar; with ECs, people vape for the nicotine but are killed by the flavorings. The flavor chemicals in 277 refill fluids for ECs were determined by GC/MS. EC manufacturers have about 16,000 flavor chemicals to choose from, but this study detected only 155. Many refill liquids have a greater concentration of flavoring agents than nicotine. The cytotoxicity of the flavoring compounds were tested using human lung fibroblasts and epithelial cells. Fifty percent of the e-liquids tests had menthol, triacetine, and cinnamaldehyde, and ethyl maltol was found in 80% of the e-liquids tested. Flavoring agents such as menthol exceed the concentrations found in regular cigarettes. Aerosolization of flavor chemicals increases the aldehyde concentration in the EC aerosols inhaled. Although many flavoring agents are on the Flavor and Extract Manufacturers Association (FEMA) Generally Accepted as Safe (GRAS) list, that only applies to ingestion and not inhalation.

> There are now sufficient data to heighten concern about the unregulated use of flavor chemicals in refill fluids, especially at high concentrations. Given the current data, regulation of flavor chemicals in EC products should be addressed, as we have recommended previously. Regulatory agencies could consider limiting the concentrations of flavor chemicals in EC products, requiring a list of flavor ingredients on product labels, restricting use of flavor chemicals that are cytotoxic at low concentrations, such as cinnamaldehyde, or banning the use of flavor chemicals in tobacco products, as suggested by others.

Reference Number: 50305

VARDAVAS, C., C. GIRVALAKI, A. VARDAVAS, S. PAPADAKIS, M. TZATZARAKIS, P. BEHRAKIS, AND A. TSATSAKIS. "Respiratory Irritants in E-Cigarette Refill Liquids Across Nine European Countries: A Threat to Respiratory Health? [Research Letter]." *European Respiratory Journal*, 50: (3 pp), 2017 (1 table, 13 references)

Abstract: GC/MS and LC/MS analyses were conducted on 122 randomly obtained e-liquids in nine EU countries in 2016. A total of 246 different flavor chemicals and additives were detected among the samples analyzed. The most common flavor chemical detected was menthol. All of the e-liquids analyzed were found to have additives that were classified to have some degree of risk and as respiratory irritants by the Globally Harmonised System of Classification.

Table: Potential Respiratory Irritants Detected in E-liquids

Chemical Detected	Positive Samples
Menthol	41.0%
Ethyl maltol	36.9%
Linalool	31.1%
Methyl cyclopentanolone	26.2%
Beta-damascone	18.9%
Ethyl vanillin	18.0%
Beta-lonone	10.7%
Acetyl pyrazine	9.0%
Alpha-lonone	9.0%
Ethyl hexanoate	8.2%
2,5 dimethylpyrazine	4.1%
Alpha-damascone	2.5%
3,4 dimethyoxy benzaldehyde	1.6%
Limonene	0.8%

Source: Adapted from Vardavas et al, 2017

> Our findings indicate potentially hazardous substances in e-cigarettes, contrary to Article 20 of the TPD, which clearly states that with the exception of nicotine, only ingredients that do not pose a risk to human health in heated or unheated form should be used in the liquid. For this reason and in order to ensure the protection of European consumer's health, monitoring of e-cigarette additives is necessary and should be addressed urgently.

Reference Number: 50306

LERNER, C.A., I.K. SUNDAR, H. YAO, J. GERLOFF, D.L. OSSIP, S. MCINTOSH, R. ROBINSON, AND I. RAHMAN. "Vapors Produced by Electronic Cigarettes and E-Juices with Flavorings Induce Toxicity, Oxidative Stress, and Inflammatory Response in Lung Epithelial Cells and in Mouse Lung." *Public Library of Science One*, 10: (26 pp), 2015 (4 tables, 8 figures, 59 references)

Abstract: Oxidative stress and inflammatory response are key events in the development of chronic airway diseases. The aerosols of ENDS can produce oxidants or reactive oxygen species (OX/ROS). Even unvaporized e-liquids were oxidative, and sweet/fruit flavors were more highly oxidative than tobacco flavors. Two ENDS devices were tested: the refillable pen style (eGo Vision Spinner battery) with 2.2-ohm heating element and Blu e-cigarettes, both with pre-loaded e-liquids. Various e-liquids were employed including propylene glycol alone, glycerol alone, and fruit and tobacco flavors. The effects of these compounds were observed on human tissues and cells of the human lung and on mice cells. Dripping rather than filling the clearomizer with e-liquid (which completely submerges the heating coil) generates larger amounts of OX/ROS. Propylene glycol is a humectant and is used as a skin penetrator enhancer for transdermal drug delivery. Propylene glycol may enhance the delivery of nicotine flavors and potential toxic impurities into the lung.

> In conclusion, we showed that 1) OX/ROS are generated by vaporizing ENDS/e-cig e-liquids/e-juices and are further influenced by the state of the heating element, 2) differences in OX/ROS reactivity in e-liquids prior to vaporization is associated with e-liquid flavor, 3) e-liquids can mediate effects on lung cell morphology and affect viability, 4) e-cig aerosols can modulate levels of oxidative stress and inflammation markers in both lung cells and mouse lungs, and 5) e-cig aerosols affect in vivo in lung glutathione redox physiology implicating oxidative stress. These data clearly demonstrate the lung toxicity and hazards of exposure to ENDS/e-cigarettes.

Reference Number: 50307

RAO, P.D., H. NANDING, A.A. STRASSER, AND P.M. WISE. "Pilot Experiment: The Effect of Added Flavorants on the Taste and Pleasantness of Mixtures of Glycerol and Propylene Glycol." *Chemosensory Perception*, 11: 1–9, 2018 (2 tables, 3 figures, 58 references)

Abstract: Twenty young adults (aged 18 to 25 years) rated the sweetness, bitterness, and pleasantness of puffs of propylene glycol/vegetable glycerin (1:1 mixture, diluted to 5% and 2.5% v/v) containing various flavors compared to a control air puff. The flavors included pineapple, banana, vanilla, caramel/malty, menthol, and a burnt aroma. Menthol had no effect but has been found to reduce the irritation of the airway caused by high concentrations of nicotine. The fruity aromas enhanced the sweetness and the vanilla and caramel flavors reduced bitterness.

On September 22, 2009, the FDA banned flavors in cigarettes under the authority of the Family Smoking Prevention and Tobacco Control Act (FSPTCA, 2009). The ban aimed to eliminate candy and fruit-flavored cigarettes, which appealed to youth and young adults, with the goal of reducing initiation of cigarette smoking. Since then, cigarette companies have increasingly marketed e-cigs, particularly to youth, and an increase in e-cig use has been observed. Given that youth are particularly vulnerable to transitioning from experimentation to regular use, the National Cancer Institute and FDA have prioritized prevention of continued use among this age group. On August 8, 2016, the FDA extended their authority to include all tobacco products, including e-cigarettes. The FDA specifically identified flavorants as components of e-cigarettes it claims authority to regulate. Therefore, if sufficient evidence existed to suggest flavorants in e-cigarettes pose a risk to public health, this new authority would permit limits on the kinds or amounts of flavorants in these products.

Reference Number: 50308

HAVERMANS, A., E.J.Z. KRÜSEMANN, J. PENNINGS, K. DE GRAAF, S. BOESVELDT, AND R. TALHOUT. "Nearly 20,000 E-Liquids and 250 Unique Flavor Descriptions: An Overview of the Dutch Market Based on Information from Manufacturers." *Tobacco Control*, 30: 57–62, 2021 (3 figures, 34 references)

Abstract: E-liquid flavors increase the attractiveness of ECs to non-smoking adolescents. E-liquid manufacturers are required to report product information to EU member states. A total of 19,266 e-liquids were reported to be marketed in the Netherlands in 2017. The e-liquids were divided into 16 main flavor categories. The greatest number of e-liquid flavors were fruit (34%), tobacco (16%), and dessert (10%). Overall, at least 56% of all e-liquid flavors were classified as sweet.

Table: Main Flavor Category and Number of E-liquids With That Flavor on the Dutch Market

Main Flavor Category	Number of E-liquids on Dutch Market
Tobacco	2,667
Fruit (other)	2,506
Fruit (berries)	2,164
Dessert	1,710

Main Flavor Category	Number of E-liquids on Dutch Market
Menthol/mint	1,240
Fruit (tropical)	1,173
Other beverages	915
Candy	771
Other sweets	768
Fruit (citrus)	743
Coffee/tea	467
Alcohol	368
Unflavored	266
Nuts	179
Spices	176

Source: Adapted from Havermans et al, 2021

> The fact that e-liquids are marketed with such a large variety of—especially sweet—flavour descriptions is highly concerning in the light of previous research demonstrating the great appeal of such flavours to youth, and therefore underlines a significant need for regulation. That is, in order to reduce e-cigarette appeal for adolescents and young adults, the abundant landscape of flavours in which e-liquids are promoted should be restricted. This can be achieved, for instance, by only allowing e-liquid flavours to be described as one of the 16 general terms that make up the main categories of the flavour wheel (inner wheel). Regulating e-liquid flavour names to the actual 16 categories also prevents the marketing of products with extraordinary names that may be particularly appealing to youth, such as unicorn-themed names. Taking potential flavour regulation a step further, specific flavour categories that are proven to be particularly attractive to vulnerable user groups (ie, non-smokers and youth), such as candy and dessert flavours, could be banned completely.

Reference Number: 50309

LEIGH, N.J., R.I. LAWTON, P.A. HESHBERGER, AND M.L. GONIEWICZ. "Flavourings Significantly Affect Inhalation Toxicity of Aerosol Generated from Electronic Nicotine Delivery Systems (ENDS)" *Tobacco Control*, 25: ii81–ii87, 2016 (3 tables, 1 figure, 27 references)

Abstract: H292 human bronchial epithelial cells were exposed to 55 puffs of freshly generated ENDS aerosol and assessed as to cell viability, metabolic activity, and release of inflammatory mediators (cytokines).

Six types of ENDS devices were purchased from stores, gas stations and vape shops in Buffalo, NY and Daly City, CA (disposable ECs, rechargeable ECs, tank type eGO devices, personal vaporizer, ePipe, and eCigar). The study tested the toxicity of nicotine, of the nicotine carriers (propylene glycol (PG), vegetable glycerin (VG) and PG/VG mixture), and of the various flavors (tobacco, Pina Colada, menthol, coffee, and strawberry), and the impact of increasing voltage of the tank system to 3.3, 4.0, and 4.8 volts. A significant decrease in cell viability occurred with the eGo tanks system, personal vaporizer, and ePipe, as well as with the various flavors (strawberry was the most toxic). Cell viability was not affected by nicotine but was affected by the nicotine carriers, PG, VG, and PG/VG. A higher voltage caused a greater toxic reaction and less cell viability. The increasing battery output voltage increased the release of carbonyl compounds.

> In summary, our data indicate that combinations of product, voltages and flavorings exist that are cytotoxic to airway epithelial cells. ENDS users should use the products with caution until more comprehensive studies are performed.

Reference Number: 50310

HUA, M., E.E. OMAIYE, W. LUO, K.J. MCWHIRTER, J.F. PANKOW, AND P. TALBOT. "Identification of Cytotoxic Flavor Chemicals in Top-Selling Electronic Cigarette Refill Fluids." *Scientific Reports*, 9: 2782 (15 pp), 2019 (1 table, 10 figures, 40 references)

Abstract: In 2014, smokers could choose from 400 models of ECs and about 8,000 different refill fluid flavor names. The cytotoxicity of the flavoring chemicals was tested with embryonic and adult cells. The top 20 EC refill fluids were identified (most had berries/fruit/citrus or candy flavors, including breakfast cereals) and 99 flavor chemicals were identified. Bird Brain had the fewest flavor chemicals (22), while Dewberry Cream had the most chemicals (47). The greater the number of flavor chemicals, the greater the cytotoxic effect. Nicotine, propyl glycol and glycerin were not cytotoxic.

> Maltol and ethyl maltol, which were highly toxic, were present in 19 and 13 of the 20 refill fluids, respectively. The cytotoxicity of refill fluids was directly correlated with ethyl maltol concentrations in the fluids. These data raise concerns about the safety of popular EC refill fluids as those tested all contained concentrations of flavor chemicals that would be cytotoxic

at the concentration in the undiluted fluids. Although flavor chemicals have been used for many years in foods, their introduction into products that are heated and inhaled presents new potential health concerns. Our data may facilitate establishing concentration limits of the dominant flavor chemicals used in EC refill fluids and requirements for labeling the flavor chemicals included in each product.

Reference Number: 50311

HARVANKO, A., R. KRYSCIO, C. MARTIN, AND T. KELLY. "Stimulus Effects of Propylene Glycol and Vegetable Glycerin in Electronic Cigarette Liquids." *Drug and Alcohol Dependence*, 194: 326–329, 2019 (1 table, 2 figures, 14 references)

Abstract: Twelve male and 4 female experienced EC users were tested on an EC with a refillable tank containing 12 mg/mL nicotine, no flavoring chemicals, and varying mixtures of propylene glycol and vegetable glycerin (100/0, 75/25, 50/50, 25/75, and 0/100, respectively). The EC was set at 10 watts. VG produced a greater visible cloud or exhalant.

Liquids containing greater concentrations of VG are more capable of producing visible exhalant and mixtures of PG and VG engender greater airway sensory effects than either ingredient alone.

Reference Number: 50312

KOSMIDER, L., T.R. SPINDLE, M. GAWRON, A. SOBCZAK, AND M.L. GONIEWICZ. "Nicotine Emissions from Electronic Cigarettes: Individual and Interactive Effects of Propylene Glycol to Vegetable Glycerin Composition and Device Power Output." *Food and Chemical Toxicology*, 115: 302–305, 2018 (1 table, 1 figure, 18 references)

Abstract: The nicotine present in the aerosol produced from ECs (known as nicotine yield) is mainly influenced by design features and e-liquid compositions. This study used a smoking machine to test the nicotine vapor concentration in an eG-3 Twist EC at 4.3, 6.7, and 9.6W, using e-liquid mixtures containing 18 mg/mL nicotine and 100% PG, 50% PG/VG, and 100%VG. At 4.3 watts, the nicotine yield was much greater with the PG alone than either VG or PG/VG mixtures. But at 9.6 watts, all three e-liquid mixtures resulted in the same yield.

The differing rate of change in volatility in response to temperature increases may provide an explanation for PG/VG composition having less of an influence on nicotine yield as device wattage was increased. That is, PG has a lower threshold for evaporation, and thus is more volatile, relative to VG at normal temperatures. Temperature increases, such as those induced by increasing the power output of an e-cigarette, can cause solvents such as PG and VG to increase exponentially in their volatility. Importantly, substances such as VG that are ordinarily less volatile can increase to a greater extent in their evaporation rate in response to temperature increases relative to substances such as PG that are already volatile at relatively low temperatures. Thus, high temperatures, such as those produced with the highest power output in the present study, may result in comparable volatility rates for PG and VG and consequently comparable nicotine yields for e-cigarette liquids of all PG/VG compositions. Thus higher nicotine yields may be connected with higher aerosol yield vaporization.

Reference Number: 50313

BEHAR, R.Z., W. LUO, S.C. LIN, Y. WANG, J. VALLE, J.F. PANKOW, AND P. TALBOT. "Distribution, Quantification and Toxicity of Cinnamaldehyde in Electronic Cigarette Refill Fluids and Aerosols." *Tobacco Control*, 25: ii94–ii102, 2016 (4 figures, 36 references)

Abstract: In 2014, over 8,000 refill fluid products were commercially available. Most EC refill liquids contain flavor chemicals, some of which are toxic. Diacetyl, which has a buttery flavor, is associated with bronchiolitis, and irreversible thickening of the lung tissues. Benzaldehyde, which has a fruity taste, was present in three-quarters of the refills containing a cherry flavor and is cytotoxic and genotoxic. The concentration of flavor chemicals in refills have been found to range from 10 to 50 mg/mL (1–5%). Cinnamaldehyde was found in Cinnamon Ceylon flavorings and is highly cytotoxic. In this study, the distribution and concentration of cinnamaldehyde were determined in 39 EC refill fluids by GC/MS. Twenty of the 39 refill fluids contained cinnamaldehyde. The products containing cinnamaldehyde were tobacco, fruit, sweet, cinnamon, and flavored tobacco (including fruit or coffee).

Cinnamaldehyde-containing refill fluids and aerosols are cytotoxic, genotoxic and low concentrations adversely affect cell processes and survival.

These data indicate that cinnamaldehyde in e-cigarette refill fluids/aerosols may impair homeostasis in the respiratory system.

Reference Number: 50314

LISKO, J.G., H. TRAN, S.B. STANFILL, B.C. BLOUNT, AND C.H. WATSON. "Chemical Composition and Evaluation of Nicotine, Tobacco Alkaloids, pH, and Selected Flavors in E-Cigarette Cartridges and Refill Solutions." *Nicotine and Tobacco Research*, 17: 1270–1278, 2015 (3 tables, 2 figures, 46 references)

Abstract: The nicotine alkaloids, flavor chemicals, and pH were measured in 136 e-liquid brands from four manufacturers. ECs are rapidly developing to increase the delivery of nicotine in the vapor and now offer replaceable heating coils with two or more wicks for better vaporization, multiple chamber atomizers, and adjustable voltage. The nicotine used in ECs is extracted from tobacco, and other tobacco constituents are co-extracted. When nicotine is exposed to air, oxidation can occur, which can result in minor alkaloids being produced. A number of the e-liquids tested had impurities exceeding USP specifications. Measurable levels of eucalyptol and pulegone were found in menthol flavored e-liquids. Higher total nicotine content of the e-liquid caused a higher pH due to the inherent alkalinity of nicotine. The free or unprotonated form of nicotine is more readily absorbed and has higher uptake than the protonated form. The pH of the liquids ranged from 5.1 to 9.1. The free-base level of all e-liquids were 60 to 90%. The greater the pH, the greater the free base nicotine concentration.

Table: The Nicotine, pH, Free Nicotine and Total Minor Alkaloids Found in Various E-liquids and Flavors

E-liquid/flavor	Nicotine Concentration	pH	Free Nicotine	Total minor Alkaloids
South beach smoke/peach	12.2 mg/g	8.8	86.3%	70.0 µg/g
V2/peppermint	9.6 mg/g	8.2	62.5%	63.5 µg/g
Premium/pear	10.1 mg/g	8.2	59.6%	95.1 µg/g
Premium/tobacco	20.5 mg/g	8.9	89.3%	154.9 µg/g
eSmoke/green apple	6.2 mg/g	8.6	79.8%	102.9 µg/g
eSmoke/caribbean coconut	11.1 mg/g	8.8	86.9%	262.5 µg/g
eSmoke/morning coffee	16.5 mg/g	8.9	87.4%	419.4 µg/g

Source: Adapted from Lisko et al, 2015

Free nicotine levels calculated from the measurement of pH correlated with total nicotine content. The direct correlation between the total nicotine concentration and pH suggests that the alkalinity of nicotine drives the pH of e-cigarette solutions. A higher percentage of nicotine exists in the more absorbable free form as total nicotine concentration increases. A number of products contained tobacco alkaloids at concentrations that exceed U.S. Pharmacopeia limits for impurities in nicotine used in pharmaceutical and food products.

Reference Number: 50315

ERYTHROPEL, H.C., S.V. JABBA, T.M. DEWINTER, M. MENDIZABAL, P.T. ANASTAS, S.E. JORDT, AND J.B. ZIMMERMAN. "Formation of Flavorant-Propylene Glycol Adducts with Novel Toxicological Properties in Chemically Unstable E-Cigarette Liquids." *Nicotine and Tobacco Research*, 21: 1248–1258, 2019 (1 table, 4 figures, 50 references)

Abstract: Cherry and vanilla flavored e-liquids at PG/VG ratios of 100/0, 70/30, 50/50, and 0/100 were analyzed by GC, LC-R, and 1H-NMR spectroscopy. Flavor aldehydes including benzaldehyde, cinnamaldehyde, citral, ethylvanillin, and vanilla reacted with propylene glycol after mixing and up to 40% of the flavor aldehydes were converted into flavor PG acetals (which could also be detected in commercial e-liquids). Vaping experiments at 4.2V and 4 second puffs showed that 50–80% of the acetals carried over to the vapor.

To fully assess the risk potential of e-liquid use for regulatory purposes, it is imperative that the compounds that a user will actually be exposed to are reported and evaluated, and not only the initial ingredients combined during manufacturing. When selecting compounds to screen for potential toxicity concerns, the chemical instability of the given e-liquid, which will depend on the reactivity of the individual constituents, needs to be taken into account. These can undergo a multitude of reactions given the complex mixture environment of e-liquids, making the prediction of equilibrium concentrations of reaction products, including PG or VG acetals, extremely challenging. In addition, a comprehensive risk assessment should consider the potential reaction products that can result from heating and oxidation in the e-cigarette, requiring characterization and quantification of the e-vapor constituents for toxicity concerns. Finally, the biological stability of compounds present in the e-vapor should be considered, including interactions with pulmonary surfactants, as reactions may occur that change the composition of the e-vapor and could result in new chemical products with unknown toxicities. This suggests that ingredient disclosure of e-liquids by suppliers may be insufficient to inform a comprehensive risk assessment of e-liquids and vaping more generally.

Reference Number: 50316

EL-HAGE, R., A. EL-HELLANI, R. SALMAN, S. TALIH, A. SHIHADEH, AND N.A. SALIBA. "Fate of Pyrazines in the Flavored Liquids of E-cigarettes." *Aerosol Science and Technology*, 52: 377–384, 2018 (3 tables, 2 figures, 46 references)

Abstract: Pyrazines are among the most frequently used additives in the aroma and flavoring of tobacco cigarettes. Pyrazines can also be formed during the curing process of tobacco leaves. The addition of pyrazines reduces the side-stream odor and masks the noxious irritating effects of tobacco smoke to both the user and bystander. Pyrazine reduces the harshness of nicotine so that the user experiences a smoother smoke. Pyrazine has a synergistic effect with nicotine to increase its appeal. A sample of 8 commercial e-liquids with 14 different flavors were tested in the e-liquid and the aerosol by GC/MS. Even though pyrazine was not listed on the label of EC liquids, it was detected in most of the samples. High concentration of three different pyrazine derivatives were detected in tobacco and licorice flavors.

Previous studies that reported synergistic effects of pyrazines on nicotine addiction and their role in "normalizing" smoking among youth suggested that pyrazine emissions in ECIG aerosols may contribute to initiation and adherence of ECIG use. Therefore, pyrazine content should be taken into consideration in any future effort for regulating this category of tobacco products. Moreover, the reproductive toxicity of pyrazine derivatives was recently questioned and other studies showed that their presence in tobacco smoke significantly inhibited oviductal functioning in animals. This raises questions about their safety when inhaled by smokers and, similarly, by vapers.

Reference Number: 50317

POKLIS, J.L., C.E. WOLF II, AND M.R. PEACE. "Ethanol Concentration in 56 Refillable Electronic Cigarettes Liquid Formulations Determined by

Headspace Gas Chromatography with Flame Ionization Detector (HS-GC-FID)." *Drug Testing and Analysis*, 9: 1637–1640, 2017 (2 tables, 11 references)

Abstract: The alcohol concentration of 56 commercially available e-liquids was determined by headspace GC. E-liquids typically contain propylene glycol and/or glycerin, nicotine, and flavoring agents. Only one e-liquid listed alcohol as a component. A positive alcohol concentration was found in 53 of the 56 e-liquids tested (95%) and ranged between 0.07 and 206 mg/mL (16% alcohol v/v).

Table: The Highest Ethanol Content of E-liquids Which Contain Nicotine

Manufacturer	Brand Name	Label	Ethanol Concentration
Top Vapor	Honeydew	30 mL, 6 mg, ingredients: VG, PG organic and/or natural flavors	206.0 mg/mL
Top Vapor	Apple Cider	30 mL. 6 mg, ingredients: VG, PG, organic and/or natural flavors	139.0 mg/mL
S&S MODS	Grumpy's Hooch	Mg: 12 mL: 30, PG50, VG 50, BOD: 1/14: expires 1/15	61.0 mg/mL
Vape Well	Cheery	High 18 mg, 30 mL, ingredients: nicotine, propylene glycol, vegetable glycerin, ethyl alcohol, artificial and natural flavoring	56.0 mg/mL
Grand Traverse Vapor	Blueberry Lemonade	10 mg/12 mL, 50/50 PG/VG	48.0 mg/mL
AVAIL Vapor	Continental Breakfast	10 mL, 2.4% nicotine, PG, VG, water, flavoring	37.2 mg/mL

Source: Adapted from Poklis et al, 2017

Ethanol is a common component found in e-liquid regardless of the labeled active ingredient(s). The ethanol detected in these products may be used in flavorant or a solvent, however, the reason for inclusion cannot be fully ascertained. The implications of vaporizing ethanol as an e-liquid component are largely uninvestigated.

Reference Number: 50318

MINH, A.E. "Effects of Vaping E-Juices with and without Alcohol on the Accuracy of the Alco-Sensor FST Approved Screening Device."

Canadian Society of Forensic Science Journal, 54: 77–85, 2021 (5 tables, 13 references)

Abstract: Two female and three male alcohol-free smokers took 10 puffs from an EC and then blew into an Alco-Sensor FST, a fuel cell breath alcohol screener used by the police in the detection of drinking drivers. Two e-liquids were employed, Lighting Loops (0% alcohol v/v) and Honeydew (21.9% alcohol v/v). No false positive breath alcohol results occurred with the inhalation of alcohol-free Lightning Loops. Immediately after inhalation of the alcohol containing e-liquid, the BrACs were 17, 22, and 38 mg/L, and 2 WARNS occurred in the other subjects (i.e., BrAC 50 mg/dL or greater). Two and one-half minutes after vaping the BrACs were zero. For drinking subjects, the BrAC results were not substantially elevated after 1 minute.

Alcohol is a common component found in commercial E-juices but is often not marked on the product label owing to a lack of regulation. The Mad Vapor Honeydew E-juice was found to contain the highest alcohol concentration of E-juices available and was used in this study to elicit the maximum potential mouth alcohol effect. Our study demonstrated that vaping alcohol-containing E-juices can falsely increase an individual's existing BAC or provide a false positive reading in an alcohol-free individual. It was determined that the effects were short lived and dissipated in less than 3 min after vaping ceased. The dissipation time for mouth alcohol with vaping was substantially shorter than that associated with alcohol consumption.

Reference Number: 50319

OMAIYE, E.E., I. CORDOVA., B. DAVIS, AND P. TALBOT. "Counterfeit Electronic Cigarette Products with Mislabeled Nicotine Concentrations." *Tobacco Regulatory Science*, 3: 347–357, 2017 (1 table, 1 figure, 21 references)

Abstract: This study tested the nicotine concentration of 125 EC refills that were labeled as nicotine-free from China, Nigeria, England, and the United States. Nicotine was detected in 52% of the e-liquids from Nigeria, 4% of those from China, 2% from the United States, and 0% from England. The highest nicotine concentration was 20.4 mg/mL.

We introduce 2 novel issues in the e-cigarette industry, the production of counterfeit refill fluids under a brandjacked label and inclusion of nicotine in 81.3% of the counterfeit products labeled 0 mg/mL. This study emphasizes the need for better control and monitoring of nicotine containing products and sales outlets.

Reference Number: 50320

NIKTIN, D., D.S. TIMBERLAKE, AND R.S. WILLIAMS. "Is the E-Liquid Industry Regulating Itself? A Look at E-Liquid Internet Vendors in the United States." *Nicotine and Tobacco Research*, 18: 1967–1972, 2016 (1 table, 1 figure, 20 references)

Abstract: A random sample of 120 online e-liquid vendors in the United States out of a total of 1,107 was collected in July 2015 and analyzed. Only four vendors successfully prevented the sale of e-liquids to minors, although 87.5% of the e-liquids sold were in child resistant packaging and 53.9% had health warnings.

Irrespective of trade association membership or vendor popularity, online vendors of e-liquids are not taking the proper precautions in preventing sales to minors. The FDA's upcoming deeming rules on e-cigarette products should include explicit requirements for offline and online e-liquid vendors, particularly the use of effective age verification, warning labels, and child-resistant packaging.... Further regulation of the industry may increase the already prevalent use of child-resistant packaging, leading to fewer cases of accidental nicotine poisoning.

Reference Number: 50321

BRAAK, D.C., K.M. CUMMINGS, G.J. NAHHAS, B.W. HECKMAN, R. BORLAND, G.T. FONG, D. HAMMOND, C. BOUDREAU, A. MCNEILL, D.T. LEVY, AND C. SHANG. "Where Do Vapers Buy Their Vaping Supplies? Findings from the International Tobacco Control (ITC) 4 Country Smoking and Vaping Survey." *International Journal of Environmental Research and Public Health*, 16: 388 (13 pp), 2019 (2 tables, 29 references)

Abstract: A survey was conducted of 1,899 vapers about where they purchased their e-liquids in Canada, US, England, and Australia in 2016. Overall, 41.4% of current vapers bought their vaping liquids from vape shops, 27.5% online and 31.1% from other retail locations. In the four countries, the majority of vapers were White and male, and most were characterized as daily vapers.

Table: Locations Where Vapers Buy Their Vaping Supplies in Canada

Smoking/Vaping	Online	Vape Shop	Other Retail
Daily smoking	13.8%	53.6%	32.6%
Non-daily smoking	10.6%	67.2%	22.2%
No smoking at all	11.4%	78.5%	10.2%
Daily vaping	10.9%	72.8%	16.3%
Non-daily vaping	14.4%	54.4%	31.2%

Source: Adapted from Braak et al, 2019

In summary, the findings from this study show that government regulations restricting where NVPs can be legally sold, and how the regulations are enforced, influence both the location where current NVP users report purchasing their NVPs and the types of product they report purchasing. The AU retail sales ban of NVPs appears to have led vapers to rely on online purchasing sources. In contrast, in CA, the lack of enforcement of this law has allowed vape shops to flourish limiting access to NVPs sold alongside cigarettes. In the US and EN, NVP users reported a greater diversity of purchase locations.

5.04 HEALTH RISKS

"This is a chemistry experiment that you're inhaling."

—Butzer, "Inhaling a 'Chemistry Experiment': Colorado Confirms Second Case of Vaping-Related Lung Illness" (2019), quoting Dr. Robin Deterding

"Studies with tobacco and EC funding are 67× more likely to find no harm with ECs compared to studies without a conflict of interest."

—Gotts et al, "What are the Respiratory Effects of E-cigarettes?" (2019)

It may surprise you that this section on the health risks of vaping is as long as the section on the health risks of cigarettes. That is because the tobacco/vaping industry has ignored the health risks of e-cigarettes and has sponsored studies that show vaping is basically safe (50401, 50443) and has encouraged the view that EC vape is mainly harmless water vapor.

There are now over 1,000 brands of e-cigarettes and 7,700 different e-liquids, made in different factories with different quality control and regulations, which will affect health safety (50402). For example, flame retardants have been detected in the urine of EC users due to contamination during the manufacturing of ECs (50403).

There are three main health risks with ECs (50402):

- Nicotine, which is a highly addictive neurotoxin, is associated with neurodegenerative diseases such as Alzheimer's and rheumatoid arthritis. Nicotine is known as a cancer promoter, although it may not be carcinogenic by itself. Nicotine also affects brain development in teenagers, with lifelong impacts.
- Thermal degradation of the solvents used in ECs, such as vegetable glycerin, which produce acrolein, which is associated with chronic obstructive pulmonary disease (COPD), when heated in ECs. Propylene glycol (not found in cigarettes) produces propylene oxide. The solvents also can contain toxic metals.
- Numerous flavoring chemicals which are cytotoxic.

Other toxic compounds found in vape include nickel and other toxic metals (50404, 50406), endotoxins (50405), benzene (50407), formaldehyde (50408, 50410), ultrafine particles (50409), acetaldehyde (50410), and acetone (50410). As the power (voltage) of the ECs increase and the temperature of the coils increase, there is an increasing amount of toxic compounds in the vape (50410).

EC use has been associated with the risk of bladder cancer (50411), cardiac arrhythmias and hypertension (50412), and aortic stiffness (50413). But the greatest health risks occur with numerous lung diseases associated with the inhalation of heated vape containing numerous chemicals deep into the lungs.

The solvents alone, PG and VG, are hydroscopic and disrupt the mucociliary clearance of the airways as it dries out the epithelial cells causing obstruction and inflammation (50416). It is no wonder, then, that one of the most common complaints about vaping is that it makes the throat dry (50424).

EC use increases the risk of numerous lung diseases, including.

- Bronchiolitis and severe coughing (50417)
- Hypersensitive pneumonitis (50418)
- Pneumothoraces (50419)
- COPD (50420, 50426, 50428)
- Diffuse alveolar lung damage (50421, 50429)
- Acute eosinophilic pneumonia (50422)

- Asthma (50423, 50425, 50427)
- Partial respiratory insufficiency (50430)
- Pulmonary inflammation (50431–3)
- Myocardial infarction (50434–5)
- Cancer (50404, 50407, 50411, 50436)
- EVALI (50437–44)

EVALI (E-cigarettes and Vaping Associated Lung Injury) was initially known as acute lipoid pneumonia (50442). EVALI cases peaked in September 2019 and have gradually declined, but have resulted in at least 2,711 hospitalizations and 60 deaths in the United States. EVALI occurred mainly in users of illegal unregulated THC vaping products, but some cases have been linked to nicotine containing products as well.

The statement that vaping is 95% safe, produced by a panel of experts with conflict of interests and funding from the tobacco industry, rings hollow in light of the adverse health risks found in studies without conflict of interest from the tobacco/vaping industries (50445).

Reference Number: 50401

PISINGER, C., N. GODTFREDSEN, AND A.M. BENDER. "A Conflict of Interest is Strongly Associated with Tobacco Industry-Favourable Results, Indicating no Harm of E-cigarettes." *Preventative Medicine*, 119: 124–131, 2019 (2 tables, 3 figures, 32 references)

Abstract: An analysis was conducted of the results/conclusions of EC health risks found in 94 vaping studies published between 2009 and 2015, and of associated conflict of interest. One-third of the studies had a tobacco/vaping industry related conflict of interest. Ninety-five percent of the independent studies showed potentially harmful substances/effects compared to only 39% of the industry-sponsored studies. Nearly 61% of the independent studies expressed concern about smokers' health or negative effects on public health compared to only 15% of the industry conflict of interest studies.

Table: Odds Ratio of No Harm Due to ECs According to Conflict of Interest Studies Compared to Independent Studies

	OR of no potentially harmful substances/ effects of ECs found	OR of no concern about negative health effects of ECs
Industry related sponsor	66.9×	23.3×
Strong or moderate conflict of interest	91.5×	24.9×
Industry benefits if no harm is found	34.2×	32.7×

Source: Adapted from Pisinger et al, 2019

After decades of casting doubt on the health hazards of smoking, the tobacco industry now publicly states that chemicals in the combustion emissions of burned tobacco are toxic and carcinogenic. They recommend "harm reduced" products such as e-cigarettes, snus/snuff or heated tobacco cigarettes as safe alternatives to smoking. However, harm-reducing inventions (i.e., filters and low tar, light, no additive cigarettes) have existed for decades without improving the health of smokers. Internal documents show that the tobacco industry has deliberately promoted low-tar cigarettes knowing that they would offer false reassurance without the health benefits. Furthermore, they failed to implement an effective method of removing the radioactivity in tobacco leaves because it affected the absorption of nicotine (nicotine kick), and, in an internal note they wrote that a "harm reduced" product "could produce extra business from the smokers who would otherwise quit". This indicates that the harm reduction discourse of tobacco companies primarily is a tactical adaptation.

Reference Number: 50402

CAI, H., AND C. WANG. "Graphical Review: The Redox Dark Side of E-cigarettes; Exposure to Oxidants and Public Health Concerns." *Redox Biology*, 13: 402–406, 2017 (3 figures, 38 references)

Abstract: There are now more than 1,000 brands of ECs and 7,764 flavors of e-liquids. There are three main health problems associated with ECs. The first is due to the highly addictive nicotine, which is known to contribute to cardiopulmonary disease, neurodegenerative diseases (e.g., Alzheimer's), and cancer. Nicotine in teenagers affects brain development, with lifelong impacts. The second is the thermal degradation of the solvents used in ECs, such as propylene glycol (not found in cigarettes) and vegetable glycerin. Acrolein forms from VG and induces COPD. PG produces propylene oxide. Toxic metals are also released from the EC solvents. Finally, numerous flavoring chemicals are cytotoxic (e.g., vanillin, ethyl maltol, ethyl vanillin, and menthol). The use of ECs is associated with increased oxidative stress in human bronchial and lung epithelial cells and results in inflammation, cytotoxicity, and increased endothelial cell permeability.

> Furthermore, it is probably not too early to start advising the tobacco users in the community not to rush into the harmful adaptation of e-cigarettes, and for physicians to guide populations at risk of relevant diseases, including cardiopulmonary diseases, neurodegenerative disorders and cancer, to avoid them. These recommended strategies to better control negative impacts of e-cigarettes are summarized in Fig. 4. Increasing the awareness of the oxidative stress aspects of the e-cigarette use may prove beneficial in promoting basic and clinical investigations to further our understandings of the adverse effects of e-cigarettes to better protect against these detrimental consequences.

Reference Number: 50403

WEI, B., M.L. GONIEWICZ, R.J. O'CONNOR, M.J. TRAVERS, AND A.J. HYLAND. "Urinary Metabolite Levels of Flame Retardants in Electronic Cigarette Users: A Study Using the Data from NHANES 2013–2014." *International Journal of Environmental Research and Public Health*, 15: 201, 2018 (4 tables, 19 references)

Abstract: ECs are highly engineered products containing plastic, glass, and metal parts. E-liquids are also highly variable, and nicotine is extracted from tobacco using organic solvents and then concentrated. Thus, users can be exposed to toxic compounds not present in conventional cigarettes. Urinary metabolites of organophosphate flame retardants were determined in 1,201 non-users of tobacco, 298 exclusive cigarette smokers, 22 exclusive cigar smokers, 14 exclusive EC users, and 15 users of smokeless tobacco products. The mean urine concentration of BCEP (a metabolite of TCEP flame retardant) was 0.37, 0.39, 0.38, 0.67, and 0.54 ng/mL for non-users, cigarette users, cigar users, EC users, and smokeless tobacco users, respectively.

> Evaluating the safety of e-cigarettes requires sufficient scientific evidence, including solid measurements from comprehensive laboratory tests.

Considering that e-cigarettes are highly engineered products containing plastic, glass and metal parts, and e-liquids are largely different matrices, many toxic compounds, which are not typical hazards for the users of combustible tobacco products (e.g., cigarettes and cigars), could exist in e-cigarettes and/or e-liquids and pose potential health risk to e-cigarette users. One group of such harmful chemicals is flame retardants (FRs). FRs are widely used in plastics, electronics, and other household and consumer products to reduce product flammability to meet safety standards.

Reference Number: 50404

THIRIÓN-ROMERO, I., R. PÉREZ-PADILLA, G. ZABERT, AND I. BARRIENTIOS-GUTIÉRREZ. "Respiratory Impact of Electronic Cigarettes and 'Low Risk' Tobacco." *Revista De Investigacion Clinica*, 71: 17–27, 2019 (1 table, 2 figures, 86 references)

Abstract: The first commercially available EC device was developed in China in 2003 as an alternative to smoking in places where regular cigarette smoking was banned. It was called RUYAN, which means resembling smoking. It was not designed as a pharmacological tool for smoking cessation, but to deliver enjoyable nicotine and overcome regulations. Nickel in EC aerosol is 2–100× higher than in tobacco smoke due to heating coils (nichrome wire, 80% nickel, 20% chromium) and batteries.

Table: Components of ECs and Potential Health Risks

Components	Chemicals	Potential Risk
Metal material: batteries and heating elements, nichrome wire	Nickel, chromium, iron, aluminum, silica, ceramic	Carcinogens, respiratory disease, autoimmune dysfunction, reproductive diseases
E-liquids	Nicotine, propylene glycol, vegetable glycerin, flavoring chemicals (tobacco, menthol, candy)	Impair immune system, alter macrophage activation, chronic autoimmune degenerative diseases such as Alzheimer's, inflammation, cytotoxicity, respiratory diseases, cytotoxicity, primary irritants
Aerosol	TSNAs, NNN, acrolein, glycidol, formaldehyde, VOCs, PAHS	Carcinogens, lung cancer, asthma, COPD, irritation, headaches, organ damage

Source: Adapted from Thirión-Romero et al, 2019

ENDS, and particularly e-cigarettes, are devices that effectively deposit nicotine in the brain and generate nicotine addiction. These e-cigarette aerosols contain fewer toxins than tobacco smoke, but a safety comparator must be clean air breathing. There is growing evidence of the presence of a variety of toxic products in vaporizing liquids in e-cigarettes that result in chemical, morphologic, and functional deleterious effects in *in vitro* and *in vivo* models. Evidence for acute respiratory damage and toxicity is evolving, but data on mid- and long-term effects are still lacking as well as standardization models to compare different devices.

Reference Number: 50405

LEE, M-S., J.G. ALLEN, AND D.C. CHRISTIANI. "Endotoxin and (1→3)-β-D-Glucan Contamination in Electronic Cigarette Products Sold in the United States." *Environmental Health Perspectives*, 127: 047008 (6 pp), 2019 (3 tables, 1 figure, 57 references)

Abstract: Endotoxins (or lipopolysaccharide (LPS)) are parts of the outer membrane of Gram-negative bacteria and are present in higher concentrations in tobacco smoke. Endotoxins are found in cotton-textile workplaces or agricultural environments such as livestock or dairy. (1→3)-β-glucan is a fungal wall constituent and is used to estimate human exposure to fungi in indoor environments. Both microbial agents were analyzed in 37 EC cartridges and 38 e-liquids sold in the United States. EC cartridges contain fibrous pads (such as cotton) that absorb e-liquid and act as a wick to deliver liquid to the heated coil to be heated into an aerosol. Endotoxins were detected in 17 (23%) of the samples and glucan in 61 (81%) of the samples. Mean glucan concentrations were 3.2× higher in the cartridges compared to e-liquids. The mean glucan concentrations in tobacco and menthol flavored ECs were 10.4× and 3.5× higher, respectively, than the fruit flavored brands.

In conclusion, our findings indicate that some popular EC brands and flavors may be contaminated with microbial toxins. Additional research is needed to confirm our findings and assess potential exposures and health effects in EC users.

Reference Number: 50406

MIKHEEV, V.B., M.C. BRINKMAN, C.A. GRANVILLE, S.M. GORDON, AND P.I. CLARK. "Real-Time Measurement of Electronic Cigarette Aerosol Size

Distribution and Metals Content Analysis." *Nicotine and Tobacco Research*, 18: 1895–1902, 2016 (1 table, 3 figures, 36 references)

Abstract: The aerosol size distribution of vaping puffs was determined using a real-time high resolution aerosol differential mobility spectrometer on blu Classic tobacco, blu Magnificent menthol, V2 RED, Joytech Carolina Crush, and Joytech Hypermist vaping devices. EC aerosols show a bimodal distribution around 11–25 nm diameter (nanoparticle) and 96–175 nm diameter (submicron particles). The highest concentration of nanoparticles appears at the beginning of the puff and consists of less volatile particles such as metals. During a dry puff (with no e-liquid), only nanoparticles were detected.

The presence of high concentrations of nanoparticles in e-cigarette aerosol substantially differentiates it from combustible tobacco smoke. While vaping the e-cigarette, along with submicron size particles that mostly should consist of the main components of e-liquid, the user is also inhaling nano-aerosol that consists of nanoparticles with attached chemicals that has not been fully investigated. Based on the results of this study and limited information available from the literature, we hypothesize that nanoparticles contain metals (and/or metal oxides) and other unknown low volatile chemicals. Our data indicate that at the beginning of each puff, the user inhales nanoparticles only. The same situation may apply if the e-liquid is exhausted from the cartridge or from the tank. If there is not enough liquid in the immediate vicinity of the heated wire, then only nanoparticles can be generated. The toxicological impact of inhaled nanoparticles could be significant, though not necessarily similar to the biomarkers typical of combustible tobacco smoke. Inhaled nanoparticles may deposit efficiently by diffusion in all regions of the respiratory tract. Their small size facilitates uptake into cells and transcytosis across epithelial cells into the blood and lymph circulation systems to reach potentially sensitive target sites such as bone marrow, lymph nodes, spleen, and heart. Access to the central nervous system and ganglia via translocation along axons and dendrites of neurons has also been observed.

Reference Number: 50407

PANKOW, J.F., K. KIM, K.J. MCWHIRTER, W. LUO, J.O. ESCOBEDO, R.M. STRONGIN, A.K. DUELL, AND D.H. PEYTON. "Benzene Formation in Electronic Cigarettes." *Public Library of Science One,* 12: (10 pp), 2017 (1 table, 2 figures, 24 references)

Abstract: ECs use an electrical resistance coil to vaporize mixtures of propylene glycol (PG), glycerol (GL), nicotine, and flavor chemicals. Benzene formation was determined in a JUUL device using refill pods, in an EVOD tank-type atomizer, and in a Subtank Nano with a single coil and cotton wicking material. There are no user options with JUUL, but EVOD and Subtank Nano can vary from 6W to 26W. Benzene is produced from a reaction of glycerol and propylene glycol; benzoic acid; or benzaldehyde. The concentration of benzoic acid in JUUL pod fluids was 44.8 mg/mL and 61.6 mg/mL for nicotine. No benzene was detected in the JUUL device, but high concentrations of 13–25W were found in the other two devices.

> For tobacco smoke from "regular" cigarettes, the 1999 Massachusetts Benchmark Study reported an average benzene delivery of 86 μg/cigarette. Assuming a total puff volume of ~400 mL/cigarette for the smoking protocol used, such deliveries correspond to a smoke benzene concentration of ~200,000 μg/m^3, and so a much higher risk from benzene for chronic use of tobacco cigarettes as compared to e-cigarettes. However, median ambient air concentrations of benzene in locations in the U.S. in 2013 were ~1 μg/m^3. Levels such as these resulted in benzene being named by the 2002 National-Scale Air Toxics Assessment (NATA) as the largest single known cancer-risk air toxic in the U.S. It can therefore be concluded for non-smokers that chronically repeated exposure to benzene from e-cigarettes at levels such as 100 μg/m^3 will not be of negligible risk.

Reference Number: 50408

LESTARI, K.S., M.V. HUMAIRO, AND U. AGUSTINA. "Formaldehyde Vapor Concentration in Electronic Cigarettes and Health Complaints of Electronic Cigarette Smokers in Indonesia." *Journal of Environmental and Public Health*, 2018: (6 pp) (2018) (8 tables, 25 references)

Abstract: In the last 4 years ECs have been widely distributed in Indonesia. Formaldehyde is a degradation product of propylene glycol and glycerin, the result of evaporation and incomplete oxidation. When heated and oxidized, propylene glycol is converted to propylene oxide, which is known as a carcinogen. This study determined the characteristics and urine cotinine concentrations of EC smokers between 2015 and 2016. Formaldehyde vapor concentrations were determined in the EC vapor of the users. The main complaint of the users was dry throat (65%), cough

(25%), sneezing (20%), and sore throat (10%). At 150°C, formaldehyde and acetaldehyde vapor concentrations are 20× that at ambient temperature.

Table: Reasons Given Why E-cigarettes are Used

Reason to use e-cigarettes	Percentage
Harmless	4%
Liking	4%
To quit smoking cigarettes	36%
To reducing smoking cigarettes	12%
To taste the flavor	20%
To try	24%

Source: Adapted from Lestari et al, 2018

In conclusion, mostly the first time users smoked electronic cigarette was at 2015, smoking frequency of electronic cigarettes was mostly twice a day and three times a day. Health complaints were mostly upper airway irritation with acute effect and mostly cotinine urine was positive. The recommendations are to educate users about hazard of electronic cigarettes due to the regulation of electronic cigarettes not being set yet and to conduct further research to identify chemical agents in electronic cigarettes.

Reference Number: 50409

QASIM, H., Z.A. KARIM, J.O. RIVERA, F.T. KHASAWNEH, AND F.Z. ALSHBOOL. "Impact of Electronic Cigarettes on the Cardiovascular System." *Journal of the American Heart Association*, 117: (14 pp), 2017 (3 tables, 3 figures, 179 references)

Abstract: The fourth generation of vaping devices includes sub-ohm tank devices, whose heating coils have a resistance less than one ohm and have temperature controls that allow the vaper to inhale high puff volumes with extremely high e-liquid consumption per puff. Smokers consider ECs to be safer because they emit lower levels of some harmful constituents; however, because of the high sensitivity of the cardiovascular system and the non-linear dose-response curve, even low levels of toxicants could have a pronounced effect and not result in a proportional harm reduction. There is a strong association between human exposure to fine particulate matter (PM2.5) and risk of cardiovascular disease development. These ultrafine airborne particles are absorbed into the bloodstream and target multiple organs. PM2.5 is responsible for more than 90% of the predicted harm in thirdhand smoke.

Table: Potential Effects of E-cigarettes on Biological Systems

System	Effects of E-cigarettes
Pulmonary system	• Upper and lower respiratory tract irritation • Bronchitis • Cough • Emphysema
Immune system	• Inflammation induction • Reduced immune efficiency
Central nervous system	• Behavioral changes • Tremor and muscle spasms
Miscellaneous	• Eye irritation • Contact dermatitis and burns • Nausea and vomiting • Throat and mouth irritation

Source: Adapted from Qasim et al, 2017

Although it was originally argued that e-cigarettes are "harm free," the present prevailing belief is that they are "reduced harm" alternatives to conventional cigarettes. This latter notion is still debatable and not supported by conclusive evidence, especially considering the wide variation between e-cigarette products. Even if that were the case, their harm can still extend to innocent/bystander nonsmokers through secondhand and thirdhand vaping, including children, pregnant women, casino/housekeeping workers, and people with preexisting cardiovascular and other diseases.

Reference Number: 50410

KOSMIDER, L., A. SOBCZAK, M. FIK, J. KNYSAK, M. ZACIERA, J. KUREK, AND M.L. GONIEWICZ. "Carbonyl Compounds in Electronic Cigarette Vapors: Effects of Nicotine Solvent and Battery Output Voltage." *Nicotine and Tobacco Research*, 16: 1319–1326, 2014 (2 tables, 1 figure, 25 references)

Abstract: Propylene glycol and vegetable glycerin are the most common nicotine solvents used in ECs. At high temperatures, both PG and VG can undergo decomposition to low molecular weight carbonyl compounds including carcinogenic acetaldehyde and formaldehyde. Twelve carbonyl chemicals were measured by HPLC in the vape from 10 commercially available e-liquids and 3 control liquids containing pure VG, pure PG and a 50/50 mixture. The voltage of the tested vaping device was increased from 3.2V to 4.0V, to 4.8V. Vape levels of formaldehyde, acetaldehyde and acetone increased (4–200×) with increasing voltage. In general, PG based e-liquids generated higher levels of carbonyl chemicals than VG.

> The striking finding of our study is that levels of carbonyls rapidly increase with increased battery output voltage. Increasing battery output voltage leads to higher temperature of the heating element inside EC. In addition, the increased battery output voltage results in more e-liquid consumed per puff. Our findings show that increasing voltage from 3.2 to 4.8V resulted in 4 to over 200 times increase in formaldehyde, acetaldehyde, and acetone levels. The levels of formaldehyde in vapors from high-voltage devices were in the range of levels reported in tobacco smoke (1.6–52 μg/cigarette). This finding suggests that in certain conditions ECs might expose their users to the same or even higher levels of carcinogenic formaldehyde than tobacco smoke. This finding is essential for the product safety and in the light of forthcoming regulation of the devices.

Reference Number: 50411

FULLER, T.W., A.P. ACHARYA, T. MEYYAPPAN, M. YU, G. BHASKAR, S.R. LITTLE, AND T.V. TARIN. "Comparison of Bladder Carcinogens in the Urine of E-Cigarette Users Versus Non-E-Cigarette Using Controls." *Scientific Reports*, 8: 507 (6 pp), 2018 (2 tables, 2 figures, 20 references)

Abstract: Conventional cigarette smoking is a well-established cause of bladder cancer, and about 50% of bladder cancers are caused by tobacco smoke. Known carcinogens in cigarettes include arsenic, aromatic amines, polyaromatic hydrocarbons (PAH), and aldehydes. Urine samples were collected from 13 non-smoking EC using subjects and 10 non-smoking non-EC using controls and were analyzed for various carcinogens using LC-MS. The mean duration of EC use was 26 months. Two urinary carcinogens were detected, o-toluidine (2.3× higher than controls) and 2-naphthylamine (1.3× higher).

> E-cigarettes are historically unregulated with a wide variety of formulations. Previous studies have identified bladder carcinogens including benz(a)anthracene and benzo(a)pyrene, aromatic amines, and aldehydes in e-cigarette liquids, vapor, or urine. In our comparison of non-smoking non-e-cigarette using controls versus non-smoking e-cigarette users there are

higher levels of bladder carcinogenic 2-naphthylamine and o-toluidine in the urine of e-cigarette users. Most of these subjects last used traditional cigarettes greater than 12 months prior to providing samples for this study. Though evidence supports e-cigarettes are safer overall than traditional cigarettes, the current study suggests e-cigarette use with current variable formulations of liquids and vapor control may not be entirely risk free from bladder cancer perspective.

Reference Number: 50412

LIPPI, G., E.J. FAVALORO, T. MESCHI, C. MATTIUZZI, L. BORGHI, AND G. CERVELLIN. "E-cigarettes and Cardiovascular Risk: Beyond Science and Mysticism." *Seminars in Thrombosis and Hemostasis*, 40: 60–65, 2014 (3 figures, 31 references)

Abstract: Cigarette smoking is the most important cause of premature death and is a major risk factor for cardiovascular disease. Smokeless tobacco is a vague term that includes snuff and chewing tobacco, which often contain flavorings (nuts, vanilla, etc.), ammonium, and sodium carbonate to increase pH and increase the concentration of unprotonated nicotine. ECs were originally designed to look and feel like real cigarettes but do not burn tobacco. The cartridge contained 0 to 20 mg of nicotine and a refill kit can contain 1 gram of nicotine. ECs increase heart rate after the first puff from 73 to 78 ppm and blood pressure (BP) is increased as well. The risk of cardiac arrythmias also increase. ECs also require a stronger vacuum (suction) than cigarettes, allowing for a deeper inhalation of chemicals and particles into the lungs.

Although the regular use of these devices appears less hazardous than traditional cigarettes or other forms of smokeless tobacco, recent studies have shown that various potentially harmful substances, especially nicotine, ultraparticles, and volatile organic compounds, may be effectively inhaled or liberated in exhaled air during repeated e-cigarette puffing. This would enhance the risk of cardiac arrhythmias and hypertension, which may predispose some users to increased risk of cardiovascular events, which may be further magnified by other potential adverse effects such as arrhythmias, increased respiratory, and flow respiratory resistance. Some cases of intoxication have also been described, wherein large amounts of nicotine and other harmful compounds may be effectively absorbed. As the use of e-cigarettes is continuously rising, and it is also considered a potentially effective method for smoking cessation, more focused research is urgently needed to definitely establish the cardiovascular safeness of these devices.

Reference Number: 50413

VLACHOPOULOS, C., N. IOAKEIMIDIS, M. ABDELRASOUL, D. TERENTES-PRINTZIOS, C. GEORGAKOPOULOS, P. PIETRI, C. STEFANADIS, AND D. TOUSOULIS. "Electronic Cigarette Smoking Increases Aortic Stiffness and Blood Pressure in Young Smokers [Letter]." *Journal of the American College of Cardiology*, 67: 2802–2803, 2016 (1 figure, 5 references)

Abstract: Cigarette smoking increases aortic stiffness and BP in users, which are important predictors of cardiovascular risk and of all-cause mortality. Aortic stiffness and BP were determined in 24 smokers (mean age 30 years) after use of cigarettes and ECs.

> To the best of our knowledge, this is the first study dealing with various patterns of EC smoking on aortic stiffness and BP demonstrating that it clearly exerts an unfavorable effect. EC over 30 min induces an unfavorable effect on aortic stiffness similar to TC [tobacco cigarette] smoking. The influence of EC smoking over 5 min on aortic stiffness is not as prompt (peak effect at 15 min) and is less potent compared with the effect of TC smoking.
>
> Given the prognostic role of aortic stiffness and increased BP for future cardiovascular events and mortality, as well as the prolonged exposure to EC smoking throughout the day matched with the strong tendency of this form of smoking to spread worldwide, especially within younger ages, our findings have important implications that could aid recommendations regarding the use of EC smoking.

Reference Number: 50414

GOTTS, J.E., S-E. JORDT, R. MCCONNELL, AND R. TARRAN. "What are the Respiratory Effects of E-cigarettes?" *British Medical Journal*, 366: (16 pp), 2019 (2 tables, 3 figures, 193 references)

Abstract: The lungs are a physiologic marvel, with 2,000 kilometres of capillaries, 300 million alveoli, and a surface area of 70 square metres. With every breath, this highly adapted and delicate organ is exposed to infectious and inflammatory environmental toxins. Globally, more than 1 billion people smoked in 2016. In the United States, 14% of adults were current cigarette users (34.3 million) and 2.8% (6.9 million) were current EC users. JUUL liquid cartridges (pods) contain 5% nicotine (59 mg/mL), much higher than the typical e-liquid (6 to 18 mg/mL). Studies with tobacco and EC funding are 67× more likely to find no harm with ECs compared to studies without a conflict of interest.

> Electronic cigarettes (e-cigarettes) are alternative, non-combustible tobacco products that generate an inhalable aerosol containing nicotine, flavors, propylene glycol, and vegetable glycerin. Vaping is now a multi-billion dollar industry that appeals to current smokers, former smokers, and young people who have never smoked. E-cigarettes reached the market without either extensive preclinical toxicology testing or long term safety trials that would be required of conventional therapeutics or medical devices. Their effectiveness as a smoking cessation intervention, their impact at a population level, and whether they are less harmful than combustible tobacco products are highly controversial. Here, we review the evidence on the effects of e-cigarettes on respiratory health. Studies show measurable adverse biologic effects on organ and cellular health in humans, in animals, and in vitro. The effects of e-cigarettes have similarities to and important differences from those of cigarettes. Decades of chronic smoking are needed for development of lung diseases such as lung cancer or chronic obstructive pulmonary disease, so the population effects of e-cigarette use may not be apparent until the middle of this century. We conclude that current knowledge of these effects is insufficient to determine whether the respiratory health effects of e-cigarettes are less than those of combustible tobacco products.

Reference Number: 50415

STAUDT, M.R., J. SALIT, R.J. KANER, C. HOLLMANN, AND R.G. CRYSTAL. "Altered Lung Biology of Healthy Never Smokers Following Acute Inhalation of E-cigarettes." *Respiratory Research*, 19: 78 (10 pp), 2018 (1 table, 3 figures, 45 references)

Abstract: Five male and five female non-smokers (mean age 40 years) had numerous tests conducted before and 2 hours after having 10 puffs from a Blu-brand EC, followed by another 10 puffs, 10 minutes later. Seven subjects used ECs with nicotine and 3 subjects used ECs without nicotine. Various changes were found, including elevated plasma endothelial microparticles (EMP) levels. EMPs are small (0.2–1.5 μm) vesicles composed of plasma membrane and cytosol released by activated or injured endothelial cells. Elevated EMPs also occurred after cigarette smoking. ECs with nicotine resulted in higher EMP levels than without nicotine.

> The data in the present study suggests that even limited, acute exposure to EC aerosols dysregulates biology of the human lung in vivo. Whether or not chronic exposure to EC will result in lung disease is unknown and can only be evaluated by large scale, long-term studies of individuals who are not ex- or current cigarette smokers who have used only e-cigarettes, a study that would be challenging to carry out at present, as most e-cigarette users have had prior or current cigarette smoke exposure. However, the observed changes in the biology of the small airway epithelium, alveolar macrophages and (indirectly) lung capillary endothelium, may signal that EC use may not be as safe as has been assumed. Therefore, recommending EC as less dangerous than cigarette smoking should be carefully considered until additional studies have been completed to determine which components of EC aerosol and patterns of use are responsible for the damage to airway biology.

Reference Number: 50416

CHAUMONT, M., P. VAN DE BORNE, A. BERNARD, A. VAN MUYLEM, G. DEPREZ, J. ULLMO, E. STARCZEWSKA, R. BRIKI, Q. DE HEMPTINNE, W. ZAHER, AND N. DEBBAS. "Fourth Generation E-Cigarette Vaping Induces Transient Lung Inflammation and Gas Exchange Disturbances: Results from Two Randomized Clinical Trials." *American Journal of Physiology. Lung Cellular and Molecular Physiology*, 316: L705–L719, 2019 (3 tables, 7 figures, 69 references)

Abstract: EC liquids mainly consist of propylene glycol (PG), vegetable glycerin (VG), nicotine, and flavorings. PG is an aliphatic alcohol commonly used as a humectant because of its hydroscopic properties. VG is an oily hydroscopic liquid also used as a humectant. PG and VG can dehydrate airway surface liquids due to their hydroscopic properties, disrupt the mucociliary clearance mechanism, and lead to airway obstruction and inflammation. Hydroscopic and hyperosmotic PG/VG aerosols can deposit deep within the lungs. Twenty-five tobacco smokers (mean age 23 years) were tested with a fourth generation tank vaporizer set at 60 W with dual coils. The e-liquid used was 50% PG/VG with or without 3 mg/mL nicotine. Vaping with nicotine increased the mean serum nicotine concentration from 0.2 ng/mL to 11.4 ng/mL at 30 minutes, which decreased to 5.4 ng/mL 150 minutes post vaping.

> Acute vaping of propylene glycol/glycerol aerosol at high wattage with or without nicotine induces airway epithelial injury and sustained decrement in transcutaneous oxygen tension in young tobacco smokers. Intense vaping conditions also transiently impair arterial oxygen tension in heavy smokers.

Reference Number: 50417

LANDMAN, S.T., I. DHALIWAL, C.A. MACKENZIE, T. MARTINU, A. STEELE, AND K.J. BOSMA. "Life-Threatening Bronchiolitis Related to Electronic Cigarette Use in a Canadian Youth." *Canadian Medical Association Journal*, 191: E1321–1331, 2019 (2 tables, 6 figures, 79 references)

Abstract: A previously healthy 17-year-old male youth presented to a hospital with a 1-week history of severe, intractable cough, shortness of breath, and fever. He was diagnosed with pneumonia, treated with antibiotics, and released. Five days later he returned with more severe symptoms and required supplemental oxygen. He had vaped daily for the last 5 months and alternated between mountain dew, green apple, and cotton candy e-liquids, to which he added THC. He required intubation, invasive mechanical ventilation, and high dose corticosteroids, and was eventually discharged from hospital after 47 days of treatment.

Table: Some Potential Causes of Lung Injury from Vaping

E-liquid Component	Chemical
Carrier solution	• Propylene glycol • Vegetable glycerin
Flavoring chemicals	• Diacetyl • 2,3-Pentanedione • Acetoin
Additives	• Nicotine • THC • CBD • BHO • Other oil-rich additives
Adulterants	• Vitamin E acetate (used mainly with THC)

E-liquid Component	Chemical
Aerosol emissions	• Carbonyls form heating PG and VG • Particulates • Trace metal elements • VOCs
Contaminants	• Bacterial endotoxins • Fungal glucans

Source: Adapted from Landman et al, 2019

This case of acute, life-threatening bronchiolitis resulting in fixed, chronic airflow obstruction in a previously healthy youth highlights the need for further research on the epidemiology of e-cigarette use, its addictive potential, and the short- and long-term risks and mechanisms of injury associated with vaping. Furthermore, there is urgent need for tighter regulation of the vaping industry, including marketing, ingredient disclosure, sale to youth, and testing of e-liquid components that may be "generally regarded as safe" for ingestion but may pose risk of serious harm when inhaled.

Reference Number: 50418

SOMMERFELD, C.G., D.J. WEINER, A. NOWALK, AND A. LARKIN. "Hypersensitivity Pneumonitis and Acute Respiratory Distress Syndrome from E-Cigarette Use." *Pediatrics*, 141: (4 pp), 2018 (12 references)

Abstract: An 18-year-old woman attended ER due to 2 days of increasing dyspnea (labored breathing), coughing, and chest pain in the lungs. She had within the previous 2 weeks started using ECs. She developed respiratory failure with hypoxia and was intubated. She was diagnosed with hypersensitivity pneumonitis and treated with methylprednisolone IV.

With this case, we highlight hypersensitivity pneumonitis as a life-threatening health risk of e-cigarette use in an adolescent patient. Although little is known of e-cigarette health risks, especially in children and adolescents, their use in the pediatric population is growing rapidly. This should prompt pediatricians to discuss the potential harms of e-cigarette use with their patients. Hypersensitivity pneumonitis, lipid pneumonia, and eosinophilic pneumonia should be included in the differential diagnosis of patients who exhibit respiratory symptoms after the use of e-cigarettes.

Reference Number: 50419

BONILLA, A., A.J. BLAIR, S.M. ALAMRO, R.A. WARD, M.B. FELDMAN, R.A. DUTKO, T.K. KARAGOUNIS, A.L. JOHNSON, E.E FOLCH, AND J.M. VYAS. "Recurrent Spontaneous Pneumothoraces and Vaping in an 18-Year-Old Man: A Case Report and Review of the Literature." *Journal of Medical Case Reports*, 13: 283 (6 pp), 2019 (1 table, 2 figures, 40 references)

Abstract: A pneumothorax is an accumulation of air in the pleural space resulting in pain, a shortness of breath, and a deflated lung. Risk factors include tobacco smoking, age (adolescents and young adults), thin body build, and male gender. Approximately 20.8% of high school students and 4.8% of adults (ages 18 to 34 years) use ECs. An 18-year-old male developed two spontaneous pneumothoraces, 2 weeks apart, after smoking mint flavor JUULs.

> Here, we present a case of an 18-year-old patient who presented to our hospital for recurrent right-sided spontaneous pneumothoraces in which each episode occurred within 2 weeks. He reported a history of vaping prior to the onset of each pneumothorax. The pneumothorax was resolved following pigtail chest tube placement. To the best of our knowledge, this is the first report to correlate recurrent pneumothoraces and vaping.
>
> The potential adverse health effects of vaping have not been established, despite their growing use. Previous case reports have suggested that vaping is associated with multiple pulmonary processes including exogenous lipoid pneumonia, bilateral pleural effusion, bronchiolitis, acute eosinophilic pneumonia, and acute hypersensitivity pneumonitis.

Reference Number: 50420

HIGHAM, A., D. BOSTOCK, G. BOOTH, J.V. DUNGWA, AND D. SINGH. "The Effect of Electronic Cigarette and Tobacco Smoke Exposure on COPD Bronchial Epithelial Cell Inflammatory Responses." *International Journal of COPD*, 13: 989–1000, 2018 (1 table, 9 figures, 45 references)

Abstract: The respiratory system is in a delicate balance maintained through the interaction of the airway epithelium, immune cells, and the external environment. The chronic inhalation of particulate matter such as tobacco smoke can interfere with this balance and cause chronic inflammation and tissue damage, leading to airflow reduction and COPD. COPD patients are more susceptible to viral and bacterial infection, which can cause hospitalization and mortality. Cigarette smoke reduces the anti-microbial response of the lungs. Bronchial epithelial cells (BEC) were collected from five patients with COPD and three non-smokers and were then exposed to an EC vapor extract (ECVE).

> This is the first study to use BECs from COPD patients and compare the effect of e-cigs and tobacco cigarettes on these cells. There is evidence that e-cig use is associated with increased rates of chronic bronchitis symptoms in adolescents. Moreover, a recent study has shown that e-cig use is associated with an increase in aberrant neutrophil responses and altered mucin secretion. We now demonstrate a reduced antiviral response following exposure to ECVE. These findings point to a potential negative impact on COPD patients who use e-cigs.

Reference Number: 50421

AGUSTIN, M., C.M. YAMAMOTO, F. CABRERA, AND R. EUSEBIO. "Diffuse Alveolar Hemorrhage Induced by Vaping." *Case Reports in Pulmonology*, 2018: (3 pp), 2018 (3 figures, 18 references).

Abstract: Vaping involves inhaling an aerosol produced by heating a liquid in a portable battery powered device. A 33-year-old man with diabetes and seizure disorder presented to the ER due to labored breathing and coughing up blood. He had taken up vaping using a personalized vaporizer 2 months earlier and had increased exposure time and experimented with new flavors. The CT scan of the chest showed a diffuse ground glass background.

> We report a case of young male presenting with subacute respiratory failure. He was later found to have diffuse alveolar hemorrhage syndrome that is likely induced by aggressive vaping. This adds up to the rising concern on the possible serious complications of this innovative technology designed as a safer alternative to traditional cigarettes.

Reference Number: 50422

ARTER, Z.L., A. WIGGINS, C. HUDSPATH, A. KISLING, D.C. HOSTLER, AND J.M. HOSTLER. "Acute Eosinophilic Pneumonia Following Electronic Cigarette Use." *Respiratory Medicine Case Reports*, 27: (3 pp), 2019 (2 figures, 17 references)

Abstract: A previously healthy 18-year-old woman presented to hospital after one day of fever, difficulty breathing, and chest pains. Two months prior, she had started vaping using a Baby Smoke Beast Mod device with 6% nicotine fluid five times a day for 30 minutes. Her condition deteriorated and she was admitted to ICU. She was diagnosed with acute eosinophilic pneumonia (AEP) and recovered after 6 days of steroid treatment.

> E-cigarettes are gaining widespread popularity over the past few years with 10.8 million adult users in the United States as of 2016. Recent studies have shown that the mechanism of inflammation and cytokine stimulation in e-cigarette users is similar to cigarette smokers, including elevation of IL-6 and IL-8. This raises concerns that e-cigarettes may induce AEP and other lung diseases, as their use promotes pulmonary inflammation by a mechanism similar to traditional cigarette smoking. Due to the wide variety of vaping devices and fluid brands (and constituent chemical compounds), more research is required to determine the exact cause of e-cigarette induced AEP.

Reference Number: 50423

WILLS, T.A., I. PAGANO, R.J. WILLIAMS, AND E.K. TAM. "E-Cigarette Use and Respiratory Disorder in an Adult Sample." *Drug and Alcohol Dependence*, 194: 363–370, 2019 (5 tables, 1 figure, 52 references)

Abstract: A telephone survey was conducted of 8,097 participants (mean age 55 years) in Hawaii in 2016. Approximately 17% had ever had asthma and 4% had ever had COPD. Twenty-two percent had ever used an EC, and the prevalence of cigarette smoking (>100 cigarettes) was 38%. Controlling for covariates and smoking, the OR of EC use and having a pulmonary disorder was 2.58×.

> In summary, this research found an association of e-cigarette use with respiratory disorder in a diverse sample of adults with analyses including demographic and psychosocial controls. Though temporal aspects of the findings need to be clarified, the results are consistent with other studies suggesting a linkage of e-cigarette use with respiratory symptoms at other ages. The fact that findings for respiratory symptoms occurred primarily for nonsmokers argues against several alternative interpretations of the results. Further research is needed to obtain more evidence on whether e-cigarette use may be adding to respiratory disorder in the population among nonsmokers.

Reference Number: 50424

KAISAR, M.A., S. PRASAD, T. LILES, AND L. CUCULLO. "A Decade of E-cigarettes: Limited Research and Unresolved Safety Concerns." *Toxicology*, 365: 67–75, 2016 (2 tables, 2 figures, 88 references)

Abstract: A detailed review was conducted on the health risks of ENDS. An EC has three basic parts: a rechargeable lithium battery, an atomizer, and a reservoir for the e-liquids. Activation of the heating coil within the atomizer generates the nicotine aerosol or mist. Automatic models have an airflow sensor that activates the heating coil when a user starts inhaling. But most models require the user to manually trigger the atomizer. The amount of heat generated by the coil can be increased by adjusting resistance (ohms) and voltage (V), and thus producing a denser vapor with more intense taste and feeling of a throat hit. The lung function of 101 employees in 19 theaters exposed to glycol fog revealed impaired lung function associated with dry throat and cough and upper respiratory tract irritation. A 3% reduction in FEV1/FVC reduction occurs after EC exposure. An elevated number of suspended particles were detected during use of ECs containing nicotine, PG, and 1,2 propanediol, which after evaporation caused residual nicotine to be deposited on solid surfaces. They could then react with atmospheric nitrous oxide to form carcinogenic tobacco-specific nitrosamines.

Table: Some Adverse Effects Found in Toxicological Studies of E-cigarettes

Adverse Effect
• Cytotoxicity, excluding hPF cells
• Oxidative stress and inflammatory response
• Reduction of fraction of exhaled nitric oxide (FENO) and enhancement of total peripheral airway resistance
• 3% reduction in FEV1/FVC
• Irritation of mouth and throat

Source: Adapted from Kaisar et al, 2016

The common belief that "e-cigarettes release merely water-based vapors" is incorrect. In addition to nicotine, e-cigarette vapors contain potentially toxic substances which are solvent byproducts (generated by heat) released in the vapor and/or trace constituents of the flavoring additives. Many of the toxic substances present in TC's sidestream and mainstream smoke are absent or negligible although e-cigarette vapors have been shown to contain traces of heavy metals as well as carcinogenic and teratogenic agents derived from flavoring additives.

Reference Number: 50425

CLAPP, P.W., AND I. JASPERS. "Electronic Cigarettes: Their Constituents and Potential Links to Asthma." *Current Allergy Asthma Reports*, 17: 79 (22 pp), 2017 (1 table, 1 figure, 130 references)

Abstract: Adolescents with asthma were found to be more likely to vape ECs (12.4%) than non-asthmatic adolescents (10.2%). A review was conducted of the components of ECs and their effects on the lungs. Many EC users preferred to vape at high temperatures, as that generates more aerosol per puff. Newer EC devices allow users to heat the e-liquid to excess of 300°C, which can pyrolyze the components of e-liquids.

Table: Biological Effects of Acute and Chronic Exposure to Common E-cigarette Chemicals

E-cigarette Constituent	Acute Exposure	Chronic Exposure
Propylene glycol and vegetable glycerin	• Increased cough • Dry throat • Chest tightness • Increased mucosal secretions • Slight reductions in lung function • Acrolein, acetaldehyde, and formaldehyde generated by thermal degradation are respiratory toxicants	• Workplace exposure associated with reduced FEV1 and FVC • No data for chronic inhalation due to e-cigarettes
Nicotine	• Increased release and metabolism of neurotransmitters • Increased pulse rate and blood pressure • Mobilization of sugar • Increased mucous production and viscosity • Reduced cell proliferation and suppression of apoptosis • Reduced allergen-induced inflammation	• Impaired fetal lung development and possible trans-generational asthma • Reduced inflammatory responses are associated with increased respiratory viral infections

Source: Adapted from Clapp and Jaspers, 2018

Based on the limited data for the health effects of e-cig use in the context of asthma, we identify several key points of concern:

- Many adolescents with asthma hold a positive view of e-cigs, and the prevalence of e-cig usage is greater in adolescent asthmatics as compared to non-asthmatics.
- Current e-cig devices are capable of pyrolyzing e-liquids resulting in thermal degradation on constituents and the generation of known respiratory toxicants such as acrolein, formaldehyde, and acetaldehyde.
- Nicotine exposure during pregnancy may induce epigenetic alterations, disrupt developmental signaling pathways necessary for normal fetal lung development, and play a role in multigenerational transmission of asthma.
- Common flavoring agents, which are often present at high concentrations in e-liquids and e-cig aerosols, are chemically similar to known airway irritants and sensitizers, and have been reported to cause occupational asthma. Moreover, e-cig exposures of some of these chemicals may exceed workplace exposure standards.
- There is no data on the potential long-term effects of e-cig use and incidence or exacerbation of asthma.

Reference Number: 50426

PEREZ, M.F., N.C. ATUEGWU, E.L. MEAD, C. ONCKEN, AND E.M. MORTENSEN. "Adult E-Cigarette Use Associated with a Self-Reported Diagnosis of COPD." *International Journal of Environmental Research and Public Health*, 16: 3938 (10 pp), 2019 (2 tables, 2 figures, 49 references)

Abstract: A comparison was conducted between 2,727 EC users and a matched control group of non-e-cigarette users between 2013 and 2014, using data from 32,320 adults in the PATH survey. COPD is one of the leading causes of morbidity and mortality in the United States and worldwide. The prevalence of current EC use was 5.5% of the total survey and the self-reported COPD diagnosis was 5.9%. The prevalence for COPD among non-smokers was 3.0%. EC users were 2.94× as likely to report COPD as non-smokers.

Table: Age Distribution of E-cigarette Users

Age Group (Years)	Percent
18–24	20.9%
25–34	26.5%
35–44	18.8%
45–54	16.6%
55–64	12.4%
65+	4.9%

Source: Adapted from Perez et al, 2019

Our study adds to the literature by examining the association between e-cigarette use and reported COPD in a large population that is representative of the adults in the US. This association is supported by studies that have shown a decline in lung function with exposure to aerosol from e-cigarettes and similarities in the biological response to conventional tobacco cigarette smoke. In humans, at least one study has reported an association between e-cigarette use and symptoms of chronic bronchitis, and another study using the COPDGene and SPIROMICS cohorts found an association between e-cigarette use and progression of COPD among subjects at risk or in whom the disease had already been established.

Reference Number: 50427

OSEI, A.D., M. MIRBOLOUK, O.A. ORIMOLOYE, O. DZAYE, S.M.I. UDDIN, Z.A. DARDARI, A.P. DEFILIPPIS, A. BHATNAGAR, AND M.J. BLAHA. "The Association Between E-Cigarette Use and Asthma Among Never Combustible Cigarette Smokers: Behavioral Risk Factor Surveillance System (BRFSS) 2016 and 2017." *BMC Pulmonary Medicine*, 19: 180 (6 pp), 2019 (1 table, 1 figure, 32 references)

Abstract: Of the 402,822 never cigarette smokers surveyed in the 2016 and 2017 BRFSS, there were 3,103 (0.8%) current e-cigarette users and 34,074 (8.5%) with self-reported asthma. Current EC users without prior cigarette use were 1.39× as likely to report asthma compared to never EC users and the risk increased to 1.73× in daily EC users. The inhalation of PG/VG mixture for vaping causes respiratory irritation resulting in wheeze and cough.

In a nationally-representative sample of never smokers, we report significantly higher odds of asthma among current e-cigarette users compared to never e-cigarette users. E-cigarettes have been promoted as a less harmful alternative to combustible cigarettes and may play a role in smoking cessation. Study of the health effects of e-cigarettes has been limited by the relatively short time e-cigarettes have been on the market. Some studies have reported e-cigarette related acute toxicity reflected by increased airway resistance, oxidative stress and inflammatory responses, with reported cases of acute respiratory distress after e-cigarette use.

Reference Number: 50428

BOWLER, R.P., N.M. HANSEL, S. JACOBSON, R.G. BARR, B.J. MAKE, M.K. HAN, W.K. O'NEAL, E.C. OELSNER, R. CASABURI, I. BARJAKTAREVIC, C. COOPER, M. FOREMAN, R.A. WISE, D.L. DEMEO, E.K. SILVERMAN, W. BAILEY, K.F. HARRINGTON, P.G. WOODRUFF, AND M.B. DRUMMOND. "Electronic Cigarette Use in US Adults at Risk for or with COPD: Analysis from Two Observational Cohorts." *Journal of General Internal Medicine*, 32: 1315–1322, 2017 (1 table, 4 figures, 28 references)

Abstract: Two long-term observational studies were conducted between 2010 and 2016 on 3,536 (COPDGene) and 1,060 (SPIROMICS) subjects (aged 45–80 years) who were at risk for or had COPD. In that time period, participants in the studies who had ever used ECs increased from 12% to 16%, but current use was stable at about 5%. African Americans were 1.8–2.9× as likely to use menthol flavored ECs.

Current e-cigarette and conventional cigarette users had higher nicotine dependence and consumed more nicotine than those who smoked only conventional cigarettes. E-cigarette users had a heavier conventional cigarette smoking history and worse respiratory health, were less likely to reduce or quit conventional cigarette smoking, had higher nicotine dependence, and were more likely to report chronic bronchitis and exacerbations. Ever e-cigarette users had more rapid decline in lung function, but this trend did not persist after adjustment for persistent conventional cigarette smoking.

Conclusions: E-cigarette use, which is common in adults with or at risk for COPD, was associated with worse pulmonary-related health outcomes, but not with cessation of smoking conventional cigarettes. Although this was an observational study, we find no evidence supporting the use of e-cigarettes as a harm reduction strategy among current smokers with or at risk for COPD.

Reference Number: 50429

BAKRE, S.A., T.S. AL-FARRA, AND S. AL-FARRA. "Diffuse Alveolar Damage and E-cigarettes: Case Report and Review of Literature." *Respiratory Medicine Case Reports*, 28: 100935 (3 pp), 2019 (3 figures, 13 references)

Abstract: Diffuse Alveolar Damage is caused by inhalational injuries, respiratory infections, connective tissue disorders and medication and is commonly found in patients with acute respiratory distress syndrome. A 47-year-old woman who used e-cigarettes for 3 years presented at hospital with a 4-day history of progressively worsening shortness of breath, nasal congestion, fevers, chills, and cough. She required intubation and was treated with antibiotics and steroids. She was discharged 20 days after admission to a long-term care center.

Diffuse alveolar damage does not occur commonly, but it is a life-threatening condition with a high mortality rate of 43–50%. Given the patient's history of chronic e-cigarette use and the evidence of associated adverse effects as demonstrated by the in-vitro studies described above, there may be an association between e-cigarette use and diffuse alveolar damage. Tobacco smoking was not regarded as harmful when it was first introduced, but studies were able to reveal the causal relationships between tobacco smoking and many diseases. This case report suggests a new avenue of investigation for studies to assess the long-term effects of e-cigarettes on humans.

Reference Number: 50430

SUHLING, H., T. WELTE, AND T. FUEHNER. "Three Patients with Acute Pulmonary Damage Following Use of E-cigarettes. A Case Series." *Deutsches Ärzteblatt International*, 117: 177–182, 2020 (1 table, 37 references)

Abstract: Three men in Germany (ages 22, 34, and 48 years) who used ordinary commercially available e-cigarettes daily for at least 3 months were hospitalized due to shortness of breath. Two patients had partial respiratory insufficiency and bilateral pulmonary infiltrates. All recovered after high dose systematic corticosteroid treatment and were discharged from the hospital in 2 to 12 days.

It has been assumed that e-cigarettes are less harmful than tobacco smoke (the "harm reduction" hypothesis) and are being advertised as a smoking cessation aid by the manufacturers (industry association of the e-cigarette industry in Germany, VdeH). The importance of e-cigarettes is being discussed in smoking cessation programs, such as the UK Stop Smoking Service (SSS). In one study of the SSS, the proportion of nicotine abstaining patients was 8% after one year. A randomized controlled trial of smoking cessation using e-cigarettes compared with traditional nicotine substitutes showed that the cessation rate after one year was higher, at 18% vs 9.9%, if tobacco cigarettes were swapped for e-cigarettes. Overall, the assumption that e-cigarettes provide a good option for smoking cessation seems more than questionable though.

Reference Number: 50431

HADDAD, C., R. SALMAN, A. EL-HELLANI, S. TALIH, A. SHIHADEH, AND N.A. SALIBA. "Reactive Oxygen Species Emissions from Supra- and Sub-Ohm Electronic Cigarettes." *Journal of Analytical Toxicology*, 43: 45–50, 2019 (1 table, 4 figures, 44 references)

Abstract: The chemicals in e-liquids when heated by a coil may generate reactive oxygen species (ROS), a class of chemicals which induce oxidative stress in the cell and can lead to pulmonary diseases. The ROS emission rate in vapor were determined in a conventional tank and sub-Ohm EC devices (SODs) such as the SmokTFV8 device (eight coils at 50, 75, 100, 150, and 200 watts of power). A supra-ohm EC (VaporFi with a single coil at 5 and 11 watts and a tobacco cigarette were also tested. Nicotine did not have any effect on ROS, but increasing VG in the e-liquid increased ROS. As the amount of power per unit coil surface area (P/SA) increased, the ROS emissions increased.

Table: Reactive Oxygen Species (ROS) Emission According to Vaping Device and Power Compared to a Tobacco Cigarette

Device	Coil Head	Power (W)	ROS flux (nmole/s)
VaporFi	Single coil	5	0.238
		11	0.696
SmokTFV8 SOD	V8-T8	50	0.114
		75	0.109
		100	0.167
		150	0.241
		200	1.143
Tobacco cigarette	–	–	1.240

Source: Adapted from Haddad et al, 2019

Our results showed that ECIGs intrinsically emit ROS even in the absence of flavorants. ROS levels from conventional tank ECIGs and SODs at high powers could reach tobacco cigarette-like levels. P/SA is a better predictor of ROS emissions than power. In addition, ROS emission is affected by the chemical constituents of the vaped liquid (PG/VG ratio). Toxicant flux is an easy tool to compare the results across different puffing regimens and among studies.

Reference Number: 50432

HIGHAM, A., N.J.W. RATTRAY, J.A. DEWHURST, D.K. TRIVEDI, S.J. FOWLER, R. GOODACRE, AND D. SINGH. "Electronic Cigarette Exposure Triggers Neutrophil Inflammatory Responses." *Respiratory Research*, 17: 56 (11 pp), 2016 (3 tables, 7 figures, 55 references)

Abstract: Neutrophils were isolated and cultured from the peripheral blood from 10 healthy never-smokers and were exposed to an EC vapor extract. Acrolein was found in EC vape but not in the e-liquid itself and can be formed from the heating and vaporization of vegetable glycerin.

We have shown that ECVE causes pro-inflammatory responses from human neutrophils. E-cigs are often used to avoid the unwanted effects of conventional cigarettes, such as causing pulmonary inflammation. Our results raise concerns over the safety of e-cig use.

Reference Number: 50433

SINGH, K.P., G. LAWYER, T. MUTHUMALAGE, K.P. MAREMANDA, N.A. KHAN, S.R. MCDONOUGH, D. YE, S. MCINTOSH, AND I. RAHMAN. "Systematic Biomarkers in Electronic Cigarette Users: Implications for Noninvasive Assessment of Vaping-Associated Pulmonary Injuries." *European Respiratory Journal Open Research*, 5: 00182-2019 (12 pp), 2019 (1 table, 4 figures, 70 references)

Abstract: Venous blood, urine, saliva, and exhaled breath condensate were collected from 26 non-tobacco using subjects (mean age 33.9 years) and 22 EC users (mean age 35.4 years). The EC users had vaped for an average of 2 years, for an average of 7.8 vaping sessions a day. The plasma cotinine concentration was significantly higher in EC users (mean 164.7 ng/mL) compared to non-users (3.9 ng/mL). Various systematic biomarkers of inflammation, oxidative stress, growth factor, vascular mediators, lipid mediators, and extracellular breakdown were determined.

E-cig users have increased levels of biomarkers of inflammation and oxidative stress, reduced pro-resolving anti-inflammatory mediators, and endothelial dysfunction, which may act as risk factors for increasing susceptibility to systemic diseases. The identified noninvasive biomarkers can be used for determining e-cig vaping-associated lung injuries, and for regulatory and diagnostic aspects of vaping in humans.

Reference Number: 50434

ALZAHRANI, T., I. PENA, N. TEMESGEN, AND S.A. GLANTZ. "Association Between Electronic Cigarette Use and Myocardial Infarction." *American Journal of Preventive Medicine*, 55: 455–461, 2018 (1 figure, 2 tables, 28 references)

Abstract: The association between cigarette smoking, EC use, and myocardial infarction (MI) was examined in 36,697 and 33,028 adults from the US National Health Interview Survey in 2014 and 2016, respectively. Both cigarettes and ECs cause an increase in oxidative stress, which is associated with cardiac risk. Both cigarettes and ECs deliver ultrafine particles into the lungs, which also increases the risk of MI. E-cigarette aerosol also induce platelet activation, aggregation, and adhesion. Former and same day EC use was not associated with an increase in risk of MI, whereas former and same day cigarette smoking was associated with an increase in risk of MI. Women had lower odds of MI.

Table: Types of Use of E-cigarettes and Cigarettes and Adjusted Odds Ratio of Myocardial Infarction

Type of Use	E-cigarettes Adjusted OR of MI	Smoking Adjusted OR of MI
Never	1.00×	1.00×
Former	1.06×	1.70×
Some days	1.16×	2.36×
Daily	1.79×	2.72×

Source: Adapted from Alzahrani et al, 2018

Daily e-cigarette use is associated with increased odds of MI independent of and in addition to the risks associated with smoking and other risk factors. Dual use of e-cigarettes and conventional cigarettes—the most common use pattern among e-cigarette users—is more dangerous than using either product alone. From these findings, recreational use of e-cigarettes or use of e-cigarettes for smoking cessation should not be recommended.

Reference Number: 50435

KIVRAK, T., M. SUNBUL, E. DURMUS, R. DERVISOVA, I. SARI, AND O. YESILDAG. "Acute Myocardial Infarction due to Liquid Nicotine in a Young Man." *Therapeutic Advances in Cardiovascular Disease,* 8: 32–34, 2014 (2 figures, 6 references)

Abstract: A 24-year-old man presented at hospital as a result of chest pain due to a myocardial infarction. He had been smoking one pack of cigarettes daily for 4 years. He stopped smoking 1 month prior to admission but had been vaping 16 mg nicotine e-liquid daily instead.

Our patient had no risk factors other than smoking. He had 4 pack-years history of smoking. He had a heart attack while he was using liquid nicotine. Nicotine has toxic effects on endothelium and thus may play a key role in impaired nitric oxide synthase (NOS)-dependent vasodilatation. Nicotine produces morphologic abnormalities on the endothelium causing direct alteration of the vascular reactivity. It effects α-2 receptors by the release of catecholamine and leads to severe vasoconstriction and might cause a myocardial infarction. Nicotine, directly or indirectly through the release of endogenous epinephrine, has been shown to increase platelet aggregation.

Reference Number: 50436

MRAVEC, B., M. TIBENSKY, L. HORVATHOVA, AND P. BABAL. "E-cigarettes and Cancer Risk." *Cancer Prevention and Research*, 13: 137–143, 2020 (1 figure, 97 references)

Abstract: ECs generally contain lower amounts of carcinogens such as TSNAs than combustible cigarettes, although the levels in TSNAs are significantly lower in non-smokers than EC users. The nicotine levels, especially of later generation ECs, are comparable and some provide an even higher nicotine dose than cigarettes. Nicotinic receptors occur on the surface of tumor and immune cells. By binding to these receptors, nicotine can potentiate cancer cell survival and tumor cell proliferation. Nicotine also causes the release of norepinephrine and epinephrine and activates the sympathoadrenal system which may stimulate carcinogenesis and tumor progression, according to recent understanding of the role of the nervous system on cancer.

> Although e-cigarettes seem to be less harmful to users and therefore may appear to be a less hazardous alternative to combustible cigarettes, the above-mentioned facts indicating an interconnection between e-cigarettes containing nicotine and cancer support that much effort should be made to strictly regulate the e-cigarette market. This assumption is also supported by the fact that clinical trials investigating the long-term effects of e-cigarettes on cancer incidence and progression are still absent. Therefore, further preclinical and clinical studies focusing on elucidating the potential mechanisms and pathways interconnecting e-cigarettes and cancer are necessary.

Reference Number: 50437

GHINAI, I., I.W. PRAY, L. NAVON, K. O'LAUGHLIN, L. SAATHOFF-HUBER, B. HOOTS, A. KIMBALL, M.W. TENFORDE, J.R. CHEVINSKY, M. LAYER, N. EZIKE, J. MEIMAN, AND J.E. LAYDEN. "E-Cigarette Product Use, or Vaping, Among Persons with Associated Lung Injury: Illinois and Wisconsin, April–September 2019." *Morbidity and Mortality Weekly Report*, 68: 865–869, 2019 (2 tables, 1 figure, 9 references)

Abstract: Detailed interviews were conducted of 86 patients who attended hospital as a result of vaping-associated lung injury between April and September 2019 in Illinois and Wisconsin. Of the 86 patients, 87% reported using ECs containing THC, 71% reported using ECs containing nicotine, 58% reported using both, and 13% reported exclusively using ECs containing only nicotine.

Table: Characteristics of Patients' EC Use as a Percentage of EVALI Patients

Product Use and Behavior	Percent of Patients
Any use of THC vapes	87%
Any use of nicotine vapes	61%
Exclusive use of THC vapes	29%
Exclusive use of Nicotine vapes	13%
Dank vapes use (THC)	66%

Product Use and Behavior	Percent of Patients
EC with prefilled cartridge	92%
EC with refillable tank	21%
Dab rig or dab pen	16%

Source: Adapted from Ghinai et al, 2019

The findings document that many, but not all, patients with lung injury associated with use of an e-cigarette product reported using THC-containing products. Similar findings have been noted in the national data, which include some of the data presented here. These data also reveal a predominant use of prefilled THC cartridges sold through informal and unregulated markets, although the origin of these products further back in the production and distribution chain is unknown. In addition, these data do not elucidate whether the causative exposure is THC itself or a substance associated with prefilled THC cartridges, such as a cutting agent or adulterant.

Reference Number: 50438

SIEGEL, D.A., T.C. JATLAOUI, E.H. KOUMANS, E.A. KIERNAN, M. LAYER, J.E. CATES, A. KIMBALL, D.N. WEISSMAN, E.E. PETERSEN, S. REAGAN-STEINER, S. GODFRED-CATO, D. MOULIA, E. MORITZ, J.D. LEHNERT, J. MITCHKO, J. LONDON, S.R. ZAKI, B.A. KING, C.M. JONES, A. PATEL, D.M. DELMAN, AND R. KOPPAKA. "Update: Interim Guidance for Health Care Providers Evaluating and Caring for Patients with Suspected E-cigarette, or Vaping, Product Use Associated Lung Injury: United States, October 2019." *Morbidity and Mortality Weekly Report*, 68: 919–927, 2019 (3 tables, 1 figure, 12 references)

Abstract: The Centers for Disease Control and Prevention (CDC) described the multistate outbreak of lung injury in the United States due to vaping as E-cigarettes and Vaping product Associated Lung Injury (EVALI). As of 8 October 2019, 49 US states and the District of Columbia have reported 1,299 cases of EVALI and 26 deaths (ages 17 to 75 years). No single compound or product has emerged as a cause of EVALI.

Therefore, at present, CDC continues to recommend that persons consider refraining from using e-cigarette, or vaping, products that contain nicotine. If adults are using e-cigarette, or vaping, products to quit cigarette smoking, they should not return to smoking cigarettes; they should use evidence-based treatments, including health care provider counseling and FDA-approved medications. If persons continue to use these products, they should carefully monitor themselves for symptoms and see a health care provider immediately if symptoms develop. Irrespective of the ongoing investigation, e-cigarette, or vaping, products should never be used by youths, young adults, or women who are pregnant. There is no safe tobacco product, and the use of any tobacco products, including e-cigarettes, carries a risk. Therefore, persons who do not currently use tobacco products should not start using e-cigarette, or vaping, products.

Reference Number: 50439

LEWIS, N., K. MCCAFFREY, K. SAGE, C-J. CHENG, J. GREEN, L. GOLDSTEIN, H. CAMPBELL, D. FERRELL, N. MALAN, N. LACROSS, A. MALDONADO, A. BOARD, A. HANCHEY, D. HARRIS, S. CALLAHAN, S. ABEREGG, I. RISK, S. WILLARDSON, A. CARTER, A. NAKASHIMA, J. DUNCAN, C. BURNETT, R. ATKINSON-DUNN, AND A. DUNN. "E-Cigarette Use, or Vaping, Practices and Characteristics Among Persons with Associated Lung Injury: Utah, April–October 2019." *Morbidity and Mortality Weekly Report* 68: 953–956, 2019 (3 tables, 10 references)

Abstract: Between 6 August and 15 October 2019, there were 83 confirmed cases of EVALI in Utah, of whom 70 were hospitalized including, 35 who received ICU admission. Most of the patients were male (83%) with a median age of 26 years (range 14 to 66 years). Thirty-four percent used THC more than six times a day and 72% of those used nicotine more than six times a day.

Table: Method of Acquisition of EVALI Patients' THC or Nicotine Containing Products in the 3 Months Prior to Their Illness

Method of Acquisition	THC-containing products	Nicotine-containing products
Friend	44%	19%
Dealer	25%	0%
Online dealer	24%	15%
Out-of-state dispensary	6%	2%
In-state vape shop	1%	49%
Convenience store/gas station	0%	18%
Products found to contain vitamin E acetate	89%	0%

Source: Adapted from Lewis et al, 2019

In this study of 83 Utah residents with EVALI during August–October 2019, approximately 90% of patients were hospitalized, approximately half in ICUs, and more than half of hospitalized patients required some form of respiratory support. Three quarters were treated with steroids. It is not known why some patients have more severe illness; preexisting behaviors and conditions might play a role in injury exposure, onset, and injury progression. Whereas some patients reported preexisting respiratory problems, most were previously in good physical health, although many reported that they self-identified as current or former smokers of combustible marijuana or tobacco. Many patients reported histories of anxiety or depression, which might influence the use or patterns of e-cigarette, or vaping, product use, particularly products containing THC.

Reference Number: 50440

BUTT, Y.M., M.L. SMITH, H.D. TAZELAAR, L.T. VASZAR, K.L. SWANSON, M.J. CECCHINI, J.M. BOLAND, M.C. BOIS, J.H. BOYUM, A.T. FROEMMING, A. KHOOR, I. MIRA-AVENDANO, A. PATEL, AND B.T. LARSEN. "Pathology of Vaping-Associated Lung Injury." *New England Journal of Medicine*: (3 pp), 2019 (1 figure)

Abstract: Lung biopsies from 13 male and 4 female patients (median age 35 years) with a vaping-associated lung injury were examined in detail.

Much recent attention has been given to the possibility that vaping-associated lung injury may represent exogenous lipoid pneumonia. However, none of our cases showed histologic evidence of exogenous lipoid pneumonia and no radiologic evidence thereof has been found; this calls into question the diagnostic utility of identifying lipid-laden macrophages or performing oil red O staining on bronchioloalveolar lavage fluid as a marker of vaping-associated lung injury, as has been proposed.

Reference Number: 50441

NAVON, L., C.M. JONES, I. GHINAI, B.A. KING, P.A. BRISS, K.A. HACKER, AND J.E. LAYDEN. "Risk Factors for E-Cigarette or Vaping, Product Use–Associated Lung Injury (EVALI) Among Adults Who Use E-Cigarette, or Vaping, Products: Illinois, July–October 2019." *Morbidity and Mortality Weekly Report*, 68: 1034–1039, 2019 (3 tables, 7 references)

Abstract: The characteristics of 4,631 e-cigarette or vaping product users without EVALI were compared to 137 patients with EVALI in Illinois between July and October 2019. EVALI patients were 2× as likely to report exclusive use of THC products, 3.1× as likely to use the products more frequently than the comparison group, and 9.2× as likely to have obtained these products from informal sources.

> The findings in this report reinforce current recommendations that persons should not use e-cigarette, or vaping, products that contain THC, or any e-cigarette, or vaping, products obtained from informal sources such as off the street, from a dealer, or from a friend. In addition, because the specific compound or ingredient causing lung injury is not yet known, CDC continues to recommend that persons consider refraining from use of all e-cigarette, or vaping, products while the outbreak investigation continues.

Reference Number: 50442

FUENTES, X.F., R. KASHYAP, T. HAYS, S. CHALMERS, C.L. VON BUCHWALD, O. GAJIC, AND A.G. DE MORAES. "VpALI-Vaping-Related Acute Lung Injury: A New Killer Around the Block." *Mayo Clinic Proceedings*, 94: 2534–2545, 2019 (3 tables, 1 figure, 100 references)

Abstract: The term "VpALI" refers to Acute Lung Injury related to vaping. ECs are battery-powered devices that rapidly heat e-liquids when exposed to heating coils made from metal alloys such as iron, chromium, or nickel. The EC produces a suspension containing fine particles of e-liquids or denominated aerosol, which has been inaccurately called a vapor. ECs are associated with the use of other tobacco products, alcohol, and cannabis.

Table: Pulmonary Symptoms Associated with EC Use

EC Associated Lung Injuries	
• Inhalation injury • Exogenous liquid pneumonia • Hypersensitivity pneumonitis • Acute eosinophilic pneumonia • Diffuse alveolar hemorrhage • Pneumothorax/pneumomediastinum • Acute respiratory distress syndrome	• Respiratory bronchiolitis-interstitial lung disease • Bronchiolitis obliterans • Acute fibrous pneumonitis • Organizing pneumonia • Granulomatous pneumonitis

Source: Adapted from Fuentes et al, 2019

Exposure to EC aerosol can be fatally harmful. Health care providers should encourage discontinuation, and regulatory policies are needed to protect vulnerable groups and individuals promoting population-based interventions. There is a potential risk for addiction, and an association with previous use of illicit drugs has been reported. Regarding using ECs for smoking cessation, it should be analyzed more deeply because there are different compounds in EC aerosols that can have adverse health effects, which, in contrast to tobacco smoke, can cause acute and life-threatening lung injury.

Vaping should be considered as a risk for acute lung injury, and health care providers need to be familiar with this new modality of illicit drug consumption.

Reference Number: 50443

DAVIDSON, K., A. BRANCATO, P. HEERDERKS, W. MANSOUR, E. MATHEIS, M. NARIO, S. RAJAGOPALAN, B. UNDERHILL, J. WININGER, AND D. FOX. "Outbreak of Electronic-Cigarette-Associated Acute Lipoid Pneumonia: North Carolina, July–August 2019." *Morbidity and Mortality Weekly Report*, 68: 784–786, 2019 (2 figures, 6 references)

Abstract: In North Carolina, five patients were hospitalized for hypoxemic respiratory failure between July and August 2019. Three required intensive care of acute respiratory distress syndrome and one required intubation and mechanical ventilation. All of the patients survived. The patients were diagnosed with acute exogenous lipoid pneumonia. All patients had vaped THC concentrates or oils which were purchased on the street.

One potential explanation for acute lipoid pneumonia among these patients is that aerosolized oils inhaled from e-cigarettes deposited within their distal airways and alveoli, inciting a local inflammatory response that impaired vital gas exchange. Lipoid pneumonia has long been described from aspiration of oil into the lungs and has been associated with e-cigarette use in some case reports. Symptoms of lipoid pneumonia are often nonspecific with variable chest imaging, which can lead to delayed or missed diagnosis.

Reference Number: 50444

SCHIER, J.G., J.G. MEIMAN, J. LAYDEN, C.A. MIKOSZ, B. VANFRANK, B.A. KING, P.P. SALVATORE, D.N. WEISSMAN, J. THOMAS, P.C. MELSTROM, G.T. BALDWIN,

E.M. PARKER, E.A. COURTNEY-LONG, V.P. KRISHNASAMY, C.M. PICKENS, M.E. EVANS, S.V. TSAY, K.M. POWELL, E.A. KIERNAN, K.L. MARYNAK, J. ADJEMIAN, K. HOLTON, B.S. ARMOUR, L.J. ENGLAND, P.A. BRISS, D. HOURY, K.A. HACKER, S. REAGAN-STEINER, S. ZAKI, AND D. MEANEY-DELMAN. "Severe Pulmonary Disease Associated with Electronic-Cigarette-Product Use: Interim Guidance." *Morbidity and Mortality Weekly Report*, 68: 787–790, 2019 (1 table, 8 references)

Abstract: As of 27 August 2019, a summary of 215 case of severe pulmonary disease associated with EC use have been reported to CDC by 25 states health departments. All patients had a reported history of vaping and no infectious agents had been found.

> While this investigation is ongoing and the definitive cause of reported illnesses remains uncertain, persons should consider not using e-cigarette products. Those who do use e-cigarette products should monitor themselves for symptoms (e.g., cough, shortness of breath, chest pain, nausea, vomiting, or other symptoms) and seek medical attention for any health concerns. Regardless of the ongoing investigation, persons who use e-cigarette products should not buy these products off the street and should not modify e-cigarette products or add any substances that are not intended by the manufacturer.
>
> E-cigarette products should never be used by youths, young adults, pregnant women, or by adults who do not currently use tobacco products. Adult smokers who are attempting to quit should use evidence-based smoking cessation treatments, including counseling and FDA-approved medications; those who need help quitting tobacco products, including e-cigarettes, should contact their medical provider.

Reference Number: 50445

ANONYMOUS. "E-CIGARETTES: PUBLIC HEALTH ENGLAND'S EVIDENCE-BASED CONFUSION [EDITORIAL]." *Lancet*, 386: 829, 2015

Abstract: This editorial criticizes the Public Health England's promotion that ECs are 95% safer than smoking. It was based on a paper by Nutt et al (2014) that was the result of a 2-day workshop sponsored in part by the Vaping Industry and involving members with clear conflict of interest to both the vaping and the tobacco industry. The Nutt et al paper actually emphasizes that there was a lack of hard evidence regarding the harms of most of the tobacco products evaluated.

> Tobacco is the largest single cause of preventable deaths in England—e-cigarettes may have a part to play to curb tobacco use. But the reliance by PHE on work that the authors themselves accept is methodologically weak, and which is made all the more perilous by the declared conflicts of interest surrounding its funding, raises serious questions not only about the conclusions of the PHE report, but also about the quality of the agency's peer review process. PHE claims that it protects and improves the nation's health and wellbeing. To do so, it needs to rely on the highest quality evidence. On this occasion, it has fallen short of its mission.

5.05 SECOND-, THIRD-, AND FOURTH-HAND VAPE

> "Our results show that bystanders may experience irritation of the respiratory tract as a result of exposure to propylene glycol and glycerol. Systemic effects of nicotine should also be expected if nicotine-containing e-liquid is used, including palpitations, and an increase of the systolic blood pressure."
>
> —Visser et al, "The Health Risks of Electronic Cigarette Use to Bystanders" (2019)

Contrary to the vaping industry's claim that vaping is just harmless water vapor, it contains numerous harmful chemicals other then H_20 (50512); see Appendix 3. ECs produce second-, third-, and fourth-hand toxic vape, similar to second-, third-, and fourth-hand toxic cigarette smoke. Secondhand vape is produced mainly from mainstream vape and contains little sidestream vape, as the ECs do not usually smolder like cigarettes. Secondhand vape also contains little CO, as vaping is typically conducted at lower temperatures that do not result in extensive incomplete combustion.

However, secondhand vape produces other toxic chemicals such as nicotine, acrolein, acetaldehyde, carbonyls, aerosol particulates, and ultrafine particles that non-vapers will be exposed to (50501–4, 50514). COPD patients exposed to secondhand aerosol had inflammatory responses in the lungs and blood (50504). Bystanders also have lung irritation by the large amounts of propylene glycol and VG contained in secondhand vape (50502). The increase in nicotine absorption of bystanders was greater when exposed to vapers who use second generation ECs (tanks) than when cig-a-likes were vaped (50512).

Thirdhand vape, which is secondhand vape deposited on the surfaces, contains nicotine and can persist for months and pose a greater risk than secondhand vape for children and pregnant women (50510–1).

Initial perceptions and misinformation about the safety of vaping has allowed vapers to think they may vape indoors at home, in cars, and in public no smoking areas (50505–8, 50513). However, more and more jurisdictions are banning vaping in no smoking areas (50514).

Discarded cigarette and heat-not-burn (HnB) butts are more toxic to the environment than e-liquids, but batteries and electronic wastes from ECs remain a problem (50515).

Reference Number: 50501

GEISS, O., I. BIANCHI, E. BARAHONA, AND J. BARRERO-MORENO. "Characterisation of Mainstream and Passive Vapours Emitted by Selected Electronic Cigarettes." *International Journal of Hygiene and Environmental Health*, 218: 169–180, 2015 (5 tables, 4 figures, 45 references)

Abstract: The effects of vaping ECs in an indoor environment were determined in a 30 m^3 emission chamber using a smoking machine. Two models of second generation, refillable EC devices were employed. Two main types of e-liquids were tested at three different nicotine concentrations. The traditional e-liquid was 46% and 45% for PG and VG, whereas the velvet refill was 83% VG. The vapor was tested using GC, GC/MS, and HPLC. Using one type of EC, the velvet refill results in 4.4× the emission of acrolein and 1.6× the emission of acetaldehyde compared to the traditional refill, and in the other type of EC the increases were 27× and 4.9×, respectively. Refill e-liquids at a nicotine concentration of 9 mg/mL led to vapor condensate nicotine amounts similar to low nicotine cigarettes. Leakage from the cartridge reservoirs were observed in some ECs.

> Electronic cigarettes tested in this study proved to be sources of propylene glycol, glycerol, nicotine, carbonyls and aerosol particulates. The extent to which people could be passively exposed to these depends on the ventilation rate, room size, indoor climate, room equipment and number of e-cigarettes in use. In addition to exposure to toxicants, consideration must also be given to the generally perceived air quality in microenvironments where vaping is permitted (independently of its toxicity). Sensory assessment of the acceptability of air quality or odour intensity by a human panel could answer this question and should be further explored.
>
> In addition to considering exposure to second-hand vapour, this study shows that active vapers inhale relatively high concentrations of propylene

glycol, glycerol, aerosol particulates and certain carbonyls. This exposure might require further toxicological evaluation.

Reference Number: 50502

VISSER, W.E., W.N. KLERX, H.W.J.M. CREMERS, R. RAMLAL, P.L. SCHWILLENS, AND R. TALHOUT. "The Health Risks of Electronic Cigarette Use to Bystanders." *International Journal of Environmental Research and Public Health,* 16: 1525 (14 pp), 2019 (6 tables, 1 figure, 24 references)

Abstract: The exhaled breath concentrations of nicotine, propylene glycol, VG, formaldehyde, acetaldehyde, acrolein, and tobacco-specific nitrosamines (TSNAs), and heavy metals were determined in 17 volunteers. The ECs used were cig-a-likes and two refillable devices. The e-liquids contained between 11 to 18 mg/mL nicotine, tobacco flavors, and a PG/VG ratio of 50/50 and 80/20 respectively. Nicotine, PG, and TSNAs were detected in the exhaled breath but not the other compounds and no heavy metals. Using this data, bystander exposure was estimated in a hypothetical non-ventilated car with two EC users and in a ventilated office with one EC user.

> Our results show that bystanders may experience irritation of the respiratory tract as a result of exposure to propylene glycol and glycerol. Systemic effects of nicotine should also be expected if nicotine-containing e-liquid is used, including palpitations, and an increase of the systolic blood pressure. Furthermore, due to the presence of TSNAs in some e-liquids, an increased risk of tumors could not be excluded for the "car" scenario. While e-cigarette use can clearly have effects on the health of bystanders, the risks depend on the rate of ventilation, dimensions of the room, and vaping behavior of the e-cigarette user. The presence of TSNAs in e-liquids can be avoided, which will prevent the most serious effect identified (increased risk of tumors).

Reference Number: 50503

GENTZKE, A.S., T.W. WANG, K.L. MARYNAK, K.F. TRIVERS, AND B.A. KING. "Exposure to Secondhand Smoke and Secondhand E-Cigarette Aerosol Among Middle and High School Students." *Preventing Chronic Disease,* 16: 180531 (12 pp), 2019 (3 tables, 1 figure, 29 references)

Abstract: Youth exposure to secondhand smoke (SHS) and secondhand aerosol (SHA) was determined in a survey of US students (grades 6 to 12) in 2015 and 2017 using the National Youth Tobacco Survey (NYTS). Exposure to SHS was 52.6% in 2015 and 50.5% in 2017. Exposure to SHA was 56.7% in 2015 and 55.1% in 2017.

Table: Percent Exposed to SHS or SHA in 2017

Characteristic	SHS Exposure	SHA Exposure
Male	43.2%	22.6%
Female	58.1%	28.5%
Grades 6–8	45.2%	20.0%
Grades 9–12	54.6%	29.9%
Current EC user	71.2%	76.5%

Source: Adapted from Gentzke et al, 2019

Strongly enforced and comprehensive smoke-free policies that include e-cigarettes have several important benefits such as the potential to reduce the social acceptability of tobacco product use, promote smoking cessation, and support efforts to decrease smoking initiation among youth. Some e-cigarettes resemble various combustible tobacco products, and exhaled aerosol can look like tobacco smoke, particularly to youth. In addition to complicating enforcement of existing comprehensive smoke-free policies, use of e-cigarettes in indoor or outdoor places where smoking is otherwise prohibited may contribute to the renormalization of behaviors related to use of tobacco products, because adolescents are particularly vulnerable to visual cues and social norms.

Reference Number: 50504

LAURSEN, K.R., J.H. BØNLØKKE, E. BENDSTRUP, M. BILDE, M. GLASIUS, V.H. GUTZKE, S.P. MOOSAKUTTY, A-C. OLIN, P. RAVN, K. ØSTERGAARD, AND T. SIGSGAARD. "An RCT of Acute Health Effects in COPD-Patients After Passive Vape Exposure from E-cigarettes." *European Clinical Respiratory Journal*, 8: (11 pp), 2020 (4 tables, 3 figures, 41 references)

Abstract: Six female and 10 male non-smoking COPD patients (mean age 68 years) were exposed for 4 hours to passive vape or clean air. The second-hand vape was generated by 2 to 3 vapers who used the most popular ECs and e-liquids sold in Denmark. Particle size in the exposure chamber air was measured, and the median $PM_{2.5}$ was 18 μg/m³ (range 8–333) for the secondhand vaping exposure and less than 6 μg/m³ for clean air. Various health parameters were measured in the COPD patients before, immediately after, and 24 hours after exposure. Eye and throat irritation occurred more in the SHV sessions and FVC and FEV1 decreased slightly as well.

> In conclusion, the results of this study indicate that passive vape is capable of exerting acute small inflammatory responses in lungs and blood as well as throat irritation. Despite the study being modest in its size with a narrow scope, it offers new findings on the potential harm of e-cigarettes. Although more research is required, it is clear that e-cigarette emissions are not merely harmless aerosol.

Reference Number: 50505

MAJMUNDAR, A., J-P. ALLEM, T.B. CRUZ, AND J.B. UNGER. "Where Do People Vape? Insights from Twitter Data." *International Journal of Environmental Research and Public Health*, 16: 3056 (5 pp), 2019 (1 table, 36 references)

Abstract: An analysis was conducted of 290,816 Twitter posts in the United States containing vaping related terms (such as JUUL, e-cigarettes) and the location where the person vaped.

Table: Locations Where Vaping Occurred

Location	Percent
Social venues (party, park, beach)	17.9%
Living space (home, hotel)	16.7%
Stores	15.9%
Transportation (cars, train, bus)	15.5%
School	14.9%
Workplace	11.9%
Healthcare offices (doctor, dentist)	2.0%
Restaurants	1.2%
Prisons, jails	0.7%
Religious institutions (church, chapel, temple)	0.4%
Miscellaneous	13.5%

Source: Adapted from Majmundar et al, 2019

Similar to prior research, this study found that vaping was taking place on school grounds during school hours. Many schools in the United States are in the process of responding to the epidemic of vaping among their students through awareness initiatives and school-level bans of vape products. Findings from this study suggest that vaping on school premises is still a common occurrence and in need of regulation.

References to vaping while in cars and other modes of transportation were common in the data and raises concern over road safety and secondhand and thirdhand aerosol exposure in confined spaces. Educational campaigns need to raise these concerns to motorists to help protect passengers. Future communication strategies could utilize the amber alert systems to curb vaping in public and private modes of transportation similar to messages that raise the issue of driving under the influence. Future research may investigate ways in which vaping in restricted-use locations may influence the perception of already existing anti-smoking regulations.

Reference Number: 50506

SHI, Y., S.E. CUMMINS, AND S-H. ZHU. "Use of Electronic Cigarettes in Smoke-Free Environments." *Tobacco Control,* 26: e19–e22, 2017 (2 tables, 23 references)

Abstract: In 2014, a study was conducted of 952 current EC users in the United States. Overall, nearly 60% of current EC users had vaped in places where smoking was not allowed. Young adults (18–29 years of age) were even more likely to vape where smoking was not allowed (74%). Daily users of ECs were 2× as likely to vape in no smoking areas compared to non-daily users. The most common places where vaping was conducted was in bars and restaurants (31%) and workplaces (24%).

Table: Adjusted Odds Ratio of EC use in Smoke Free Environments

Characteristic	Adjusted OR
Male	1.09×
Daily EC Users	2.08×
Current Smokers	1.78×

Source: Adapted from Shi et al, 2017

There is concern that cigarette smokers may use e-cigarettes as a means to circumvent smoke-free laws. A larger proportion of smokers had used e-cigarettes in smoke-free places than had non-smokers, although the results were not significant. E-cigarettes have been promoted as an alternative to smoking; one that can be used anywhere. The "freedom" appeals and the absence of an explicit e-cigarette ban may have encouraged smokers to vape in places where smoking is banned.

Reference Number: 50507

DREHMER, J.E., E. NABI-BURZA, B.H. WALTERS, D.J. OSSIP, D.E. LEVY, N.A. RIGOTTI, J.D. KLEIN, AND J.P. WINICKOFF. "Parental Smoking and E-Cigarette Use in Homes and Cars." *Pediatrics*, 143: (9 pp), 2019 (4 tables, 43 references)

Abstract: A survey was conducted of 761 parents who used either cigarettes or e-cigarettes, of whom 646 (85%) used cigarettes only, 34 (5%) used ECs only, and 81 (11%) were dual users. Most dual users had strict smoke-free policies (64%) but much fewer had a vape-free home policy (26%). Only 20% of parents who used ECs had strictly enforced vape-free policies for the home and car. Having a youngest child >10 years of age was associated with a reduced likelihood of vape-free policies for home and car.

The findings presented in this study could be used to broaden tobacco control recommendations to include guidance for addressing e-cigarettes in the pediatric health care setting. Parents who were dual users of cigarettes and e-cigarettes were more likely to have strictly enforced smoke-free policies than vape-free policies for the home, suggesting that some may believe exposure to vaping inside the home is safe for children. Additionally, parents who dual use cigarettes and e-cigarettes were less likely to have strictly enforced smoke-free policies for the car and vape-free policies in the home and car than parents who only use traditional cigarettes. The large percentages of parents who have not adopted smoke-free and vape-free homes and cars and the low percentages of parents who are asked or advised about keeping homes and cars smoke free when attending pediatric practices reveals the continued unmet need to implement interventions that will promote safe homes and cars for children.

Reference Number: 50508

NGUYEN, K.H., V.T. TONG, K. MARYNAK, AND B.A. KING. "Perceptions of Harm to Children Exposed to Secondhand Aerosol from Electronic Vapor

Products, Styles Survey, 2015." *Preventing Chronic Disease*, 14, E41 (11 pp), 2017 (2 tables, 20 references)

Abstract: Contrary to vaping marketing claims that electronic vapor products (EVP) aerosol is just harmless water vapor, the aerosols contain nicotine, heavy metals, VOCs, and ultrafine particulates. In 2016, the US Surgeon General's report on EC use among youth and young adults concluded that EC aerosol is not harmless and should be banned, just like cigarettes, for indoor use in smoke-free areas. A survey was conducted of 4,127 adults in the 2015 Styles consumer survey. Overall, 5.3% responded that children's exposure to vaping aerosol caused no harm, 39.9% responded little harm or some harm, 21.5% replied a lot of harm, 33.3% replied that they didn't know. Approximately 14% were current cigarette smokers and 5% were EC users.

> In conclusion, current and former cigarette smokers and EVP users had greater odds than nonusers of perceiving no harm toward children exposed to EVP aerosol. Variations in harm perceptions were also observed across sexes and race/ethnicities. EVP use potentially exposes nonusers, including children, to aerosolized nicotine and other harmful substances. Therefore, clean air—free of both smoke and EVP aerosol—remains the standard to protect health. In coordination with a comprehensive approach to prevent and reduce secondhand smoke exposure and tobacco use by young people, efforts are warranted to educate the public, particularly current and former cigarette smokers and EVP users, about the potential health risks of secondhand EVP aerosol exposure among children.

Reference Number: 50509

PALMISANI, J., A. DI GILIO, L. PALMIERI, C. ABENAVOLI, M. FAMELE, R. DRAISCI, AND G. DE GENNARO. "Evaluation of Second-Hand Exposure to Electronic Cigarette Vaping Under a Real Scenario: Measurements of Ultrafine Particle Number Concentration and Size Distribution and Comparison with Traditional Tobacco Smoke." *Toxics*, 7: 59 (16 pp), 2019 (3 tables, 5 figures, 40 references)

Abstract: Two female subjects (aged 41 and 38 years) vaped 15 different types of e-liquids inside a 49 m^3 room. The number of puffs per vaping session was 40 over a 20-minute period with a puff duration of 3 seconds and a puff interval of 30 seconds. A puff duration of 3 seconds is considered necessary for the EC to reach a stable temperature. The EC employed was second generation with a refillable cartomizer of 1.6 mL volume, set at a voltage of 4.2V and push-button activated heater. The ultra fine particle (UFP) and concentration were determined with a Fast Mobility Particle Size using an electrical mobility technique. The subjects also smoked one cigarette over 20 minutes as a comparison. In total, 44 vaping and smoking sessions were conducted and measured. The particle size distribution of secondhand vape was bimodal with a particle size diameter (Dp) of 10.8 nm and 29.4 nm. Secondhand smoke Dp was unimodal at a particle size diameter of 100 nm. The concentration of UFP from ECs ranged from 6,560 to 40,100 per cm^3 for one vaper and 112,00 to 146,000 particles per cm^3 for the other. This high concentration of nano-sized particles in secondhand vape allows penetration of these particles into the deepest regions of the lung and is of concern, especially for children and elderly people.

> Although the comparison in terms of UFPs number concentration indicates a lower impact indoors of e-cig use with respect to tobacco cigarette consumption, the analysis of the particles size distributions reveals that e-cig second-hand aerosol, exhaled by the user and potentially inhaled by the passive bystanders, is mainly composed of nano-sized particles within the size range of 6–26 nm (particles concentration from 2- at Dp = 25.5 nm to 3800-fold at Dp = 9.31 nm higher than in tobacco environmental smoke). The inhalation exposure to high level concentrations of nano-sized particles may represent a high level risk factor for passive vapers due to particles high penetration capacity and delivering action of chemicals/contaminants in the deepest regions of the human respiratory apparatus. Therefore, this study underlines that the particle size has to be considered a key factor to take into account by experts in indoor air and risk assessment fields, when passive exposure risk evaluation due to both e-cig and tobacco cigarette use is required.

Reference Number: 50510

DAVIS, E.S., M.F. SASSANO, H. GOODELL, AND R. TARRAN. "E-Liquid Autofluorescence Can be Used as a Marker of Vaping Deposition and Third-Hand Vape Exposure." *Scientific Reports*, 7: 7459 (10 pp), 2017 (7 figures, 34 references)

Abstract: Two hundred and sixty-six commercially available e-liquids were tested for autofluorescence, and most were found to autofluorescence at

300 and 405 nm wavelength of light. Thirdhand exposure occurs when tobacco products are smoked or vaped in ambient air and the particles from the smoke or vape are deposited on surfaces where they can be taken up dermally. The e-liquid flavor Grape! was vaped 10 times (70 mL puffs at 100W) onto a sheet of aluminum foil, which was then tested for residue by autofluorescence. Thirdhand vape exposure (THV) can be detected by GC-MS devices but they are expensive and more complicated to operate. Every 70 mL vape puff deposited 0.019% of the vape liquid on the surface.

The ability to detect third-hand exposure may be useful for monitoring exposure in public places such as restaurants and movie theaters where the general public including children and pregnant mothers could be unintentionally exposed to third-hand vaping. However, as a caveat, since tobacco smoke is also highly autofluorescent, care would be needed to ensure that cross-contamination did not occur.

Reference Number: 50511

ENGLAND, L.J., R.E. BUNNELL, T.F. PECHACEK, V.T. TONG, AND T.A. MCAFEE. "Nicotine and the Developing Human: A Neglected Element in the Electronic Cigarette Debate." *American Journal of Preventive Medicine*, 49: 286–293, 2015 (104 references)

Abstract: Nicotine is readily absorbed through the skin and lungs and exerts its effects by binding to the nicotinic acetyl choline receptors (nAChRs). These receptors in the CNS are associated with cognitive functions including memory, selective attention, and emotional processing. Receptors in the respiratory system and the immune system also occur. An estimated 10% of pregnant women in the United States smoke. Nicotine crosses the placenta and adversely affects fetal development. Approximately one-third of SIDS deaths are due to maternal smoking. The act of using an EC is indistinguishable from cigarette smoking to children and thus smoking could be renormalized after 40 years of banning advertisements. The use of ECs indoor could cause a nicotine reservoir indoors that children will absorb through their skin and which will affect their development.

If electronic cigarettes are to play a beneficial role in strategies to end the tobacco epidemic without causing unintended harm, steps to protect the health of millions of pregnant women, infants, children, and adolescents are needed. Because nicotine exposure during vulnerable periods of brain and lung development can have lasting detrimental effects, regulatory and policy measures for electronic cigarettes or nicotine-containing products are imperative. Among these are (1) strong prohibitions on marketing that result in youth uptake; (2) enforceable youth access laws identical or similar to those in effect for other tobacco products; (3) appropriate health warnings for all vulnerable populations; (4) packaging to prevent accidental poisonings; and (5) protection from exposure to secondhand electronic cigarette aerosol. Measures should also include consideration of the impact of pricing on youth initiation and use, and product addiction potential and youth appeal, including youth-oriented flavorings, accessibility of products through placement in retail venues, and social networking potential. Product regulations should address nicotine concentration and total amount of nicotine per vial, as well as quality control. The age of legal sale of electronic cigarettes and other nicotine-containing products should reflect the developmental stages during which adolescents/young adults are most vulnerable to the adverse effects of nicotine. Because the brain does not reach full maturity until the 20s, restricting sales of electronic cigarettes and all tobacco products to individuals aged at least 21 years and older could provide positive health benefits to both youth and young adults.

Reference Number: 50512

MELSTROM, P., C. SOSNOFF, B. KOSZOWSKI, B.A. KING, R. BUNNELL, G. LE, L. WANG, M.H. THANNER, B. KENEMER, S. COX, B.R. DECASTRO, AND T. MCAFEE. "Systematic Absorption of Nicotine Following Acute Secondhand Exposure to Electronic Cigarette Aerosol in a Realistic Social Setting." *International Journal of Hygiene and Environmental Health*, 221: 816–822, 2018 (2 tables, 1 figure, 33 references)

Abstract: Five male and two female never users of nicotine products (ages 28 to 54 years) were exposed to vape over 2 hours from 3 e-cigarette users on 2 occasions using either a disposable e-cigarette or a second generation tank EC. Blood, urine, and saliva samples were collected before and throughout exposure. Serum and saliva cotinine concentrations were measured by LC/MS.

Table: Median Increase in Cotinine Concentration (ng/mL) After 2 Hour Exposure to Vaping Users

	Serum	Saliva	Urine
Cig-a-like	+ 0.007	+ 0.033	+ 0.316
Tank, 2nd generation EC	+ 0.041	+ 0.060	+ 0.948

Source: Adapted from Melstrom et al, 2018

Nonusers of e-cigarettes can experience systemic absorption of nicotine from acute exposure to secondhand e-cigarette aerosol. Although less than levels produced by conventional cigarettes, the exposure is greater than in clean indoor air—the standard. Also, the exposure may vary considerably depending upon solution, device, baseline nicotine levels and the use behavior of e-cigarette users. In addition, aerosols produced from the use of e-cigarettes may contain other harmful constituents that were not measured in this study.

Reference Number: 50513

HAGGART, K., L. ROBERTSON, M-L. BLANK, L. POPOVA, AND J. HOEK. "It's Just Steam: A Qualitative Analysis of New Zealand ENDS Users' Perceptions of Secondhand Aerosol." *Tobacco Control*, 30: 30–35, 2020 (1 table, 43 references)

Abstract: A detailed interview was conducted of 23 male and 16 female current ENDS users (aged 19–65 years) of whom 32 were daily users. Many participants perceived the exhaled clouds to be steam rather than an aerosol containing nicotine, chemical flavors, propylene glycol, vegetable glycerin, and heavy metals. The speed at which secondhand aerosol (SHA) dispersed reinforced the perception that it was simply steam.

Participants had limited understanding of SHA, its constituents or its possible effects on others. They drew on the absence of harm information, and their sensory experiences and perceptions of others' views of vaping, to support the conclusion that SHA posed few, if any, risks to bystanders. Yet despite this perception, some felt they should recognise others' rights to clean air and most would not vape around children to avoid setting an example.

Reference Number: 50514

FIALA, S.C., L.L. DRACH, S.W. PARKMAN, S.L. HARGAND, AND K.E. GIRARD. "Including a Novel Tobacco Product in Oregon's Indoor Clean Air Act: Lessons Learned for Clean Air Policy, 2015–2016." *Public Health Reports*, 133: 230–233, 2018 (16 references)

Abstract: As of 2017, only eight US states had passed comprehensive smoke-free indoor laws that included ECs.

Establishing and maintaining smoke-free air laws is a best practice of comprehensive tobacco prevention programs. It is well accepted that smoke-free laws reduce exposure to secondhand smoke, increase cessation attempts, increase perceptions of harm among young people, and decrease tobacco initiation among young people. However, the innovation and introduction of new tobacco products by the tobacco industry may jeopardize state smoke-free air laws. Electronic cigarettes (e-cigarettes), a relatively new form of tobacco, are now the most widely used product among young people nationally, offsetting declines in tobacco cigarette smoking. Despite the acknowledgment that e-cigarettes have the potential to benefit adult smokers if used as a complete substitute for smoked tobacco products, a 2016 surgeon general report concluded that e-cigarette aerosol was not harmless, citing several experimental studies finding the presence of potentially harmful constituents such as nicotine, heavy metals, ultra-fine particulate matter, and volatile organic compounds.

Reference Number: 50515

BARAN, W., D. MADEJ-KNYSAK, A. SOBCZAK, AND E. ADAMEK. "The Influence of Waste from Electronic Cigarettes, Conventional Cigarettes and Heat-not-Burn Tobacco Products on Microorganisms." *Journal of Hazardous Materials*, 385: (7 pp), 2019 (5 figures, 39 references)

Abstract: The ecotoxicity of leachate obtained from e-liquids, used electronic cigarette cartridges, smoked cigarettes, and heat-not-burn butts (HnBBs) were tested using a multispecies commercial MARA test which includes 11 different microorganisms.

The high ecotoxicity of SCBs in combination with high supply is worrisome and may require intervention, protecting the environment. Their penetration and retention, e.g. in stagnant waters, may adversely affect the activity of microorganisms living therein. What is more, it can also have a negative impact on the efficiency of small biological wastewater treatment plants exposed to significant amounts of SCBs. In contrast, no significant threat to the microorganisms was found on the ECC leachate.

5.06 YOUTH

"With this study, we add longitudinal, causal evidence that the transition from e-cigarettes to combustible products, at least for adolescents, is a 1-way street."

—Klein, "E-cigarettes: A 1-Way Street to Traditional Smoking and Nicotine Addiction for Youth" (2018)

Cigarette use has declined in youth in most countries, but another form of nicotine addiction, e-cigarettes, has increased dramatically (50601–3). Youth are now more likely to obtain nicotine from vaping devices than cigarettes, due to ECs like JUUL (which contain 2–10× the nicotine concentration of earlier ECs), youth friendly flavors, and savvy advertising (50603, 50605–7).

JUUL was able to allow for the inhalation of more nicotine by adding benzoic acid to the highly bitter basic nicotine to create a nicotine salt (50606). One JUUL pod contains 59 mg/mL of nicotine, which is the same amount of nicotine as in 20 regular cigarettes. This was also 2–10× the amount of nicotine found in earlier, less addictive ECs (50606). More problematic is that nearly one-third of high school EC users did not know whether or not their e-liquid contained nicotine (50604).

Nicotine dependency is also higher in EC pod users, as 29% reported needing to vape when they awoke in the morning (50607–8). The inclusion of menthol and other flavors increases the appeal of ECs to youth (50609–14).

EC use in youth was associated with an up to 5× greater risk of cigarette smoking and many studies have shown a progression from EC use to smoking (50616–23). This is due in part to a desensitization of the risk of cigarette smoking, as vapers were 4.7× more likely to no longer perceive cigarettes as posing a great risk or harm. Only 22% of vapers believed that ECs were harmful (50624).

EC use, however, was associated with illicit drug use, mental health problems, and impulsivity in university students (50625). Saliva and urine samples from adolescent EC users detected carcinogenic organic compounds and underscored that EC use was not harmless (50626).

Children (aged 11–16 years) exposed to EC advertisements in the United Kingdom were less likely to perceive smoking 1–2 cigarettes as harmful (50627–28). In the United States and other countries, ECs are only required to carry only a nicotine addiction warning, and text warnings have limited impact with youth when placed next to a colorful EC ad (50629).

In spite of the massive effort by the tobacco industry to encourage EC use, the majority (61%) of students in 10 California schools still had a negative opinion of adolescent peer EC users (50630).

Reference Number: 50601

BUSH, A., J. BHATT, AND J. GRIGG. "E Cigarettes: Tar Wars: The (Tobacco) Empire Strikes Back." *Archives Disease Children*, 104: 1027–1029, 2019 (38 references)

Abstract: A short review of the effects and possible risks of e-cigarettes is presented.

> In summary, the position of the e-cigarette lobby is that because it is known that e-cigarettes do not produce two known harmful substances (tar and carbon monoxide), and because no safety issues have been discovered, e-cigarettes are safer than tobacco. On the contrary, we believe we should align with the large international respiratory societies to say that, unless and until e-cigarettes are proved to be safe, they should be marketed and regulated as cigarettes, because they contain hundreds of unknown and unregulated compounds not found in tobacco. Indeed, we know that there are already substantial safety concerns documented (above). It is solely the responsibility of the industry to prove safety. The regulations that apply to tobacco should be applied to e-cigarettes, including those relating to advertising, packaging, taxation and where they can be used. The only justification for their current use is aligned with nicotine patches and gum for smoking cessation. If we persist in the approach of Public Health England with its disregard for protecting the respiratory health of adolescents and young children, we are risking a further epidemic of devastating lung disease for today's children. Non-evidence-based "consensus" should not be allowed to determine public policy. We call on the Royal College of Paediatrics and Child Health, and the other Royal Colleges, to join the international voices who are deeply concerned about the vaping epidemic, and do all that can be done to counter it.

Reference Number: 50602

HAMMOND, D., J.L. REID, V.L. RYNARD, G.T. FONG, K.M. CUMMINGS, A. MCNEILL, S. HITCHMAN, J.F. THRASHER, M.L. GONIEWICZ, M. BANSAL-TRAVERS, R. O'CONNOR, D. LEVY, R. BORLAND, AND C.M. WHITE. "Prevalence of Vaping and Smoking Among Adolescents in Canada, England, and the United States: Repeat National Cross Sectional Surveys." *British Medical Journal*, 365: 12219 (9 pp), 2019 (5 tables, 31 references)

Abstract: Surveys were conducted of 7,891 teenagers (aged 16–19 years) in Canada, 7,879 in England, and 8,140 in the United States in 2017 and 2018 regarding vaping and smoking. In May 2018, Canada implemented more permissive vaping policies that allowed advertising and point of sale displays. JUUL uses benzoic acid and nicotine salt technology and high nicotine concentration (>50 mg/mL compared to 3–24 mg/mL nicotine of other ECs) to deliver more nicotine and increase its addictiveness. The prevalence of the subjects having vaped within the past 30 days increased by 5% in the United States and 6% in Canada between 2017 and 2018. In England, which has more restrictive vaping polices and limits the nicotine concentration of ECs, there was no increase in teenage vaping.

> Before 2018, there was relatively little evidence of regular vaping among adolescents that might be indicative of nicotine addiction; however, the emergence of JUUL and nicotine salt based products might signal a change. Given how rapidly the vaping market is evolving, it is critically important to continue monitoring vaping behavior, including specific products and more frequent use, among adolescents. Rather than fuel the debate over trade-offs between less harmful nicotine alternatives for adult smokers versus increased initiation among adolescents, the findings should direct attention to regulatory measures that more selectively target these vaping products to adult smokers.

Reference Number: 50603

GLAUSSER, W. "New Vaping Products with Techy Allure Exploding in Popularity Among Youth." *Canadian Medical Association Journal*, 191: E172–E173, 2019 (1 figure)

Abstract: About 80% of adult smokers are addicted before age 21. JUULs became popular among youth because they are small and easy to conceal and look like USB sticks. JUUL's marketing campaigns appeal to youth with attractive young models, social media sponsorship, and flavors like mango and vanilla.

> In September, Juul hit the Canadian market, and other Juul-like products have also begun appearing. In Toronto, for example, colourful billboards advertise the "big hit" of the Vype ePod, owned by Imperial Tobacco. Ontario Premier Doug Ford halted legislation proposed by the previous government to ban vaping wherever smoking is illegal and to limit marketing. Only Ontario, Alberta and Saskatchewan allow e-cigarette advertising in retail stores. A coalition of health organizations, including the Heart & Stroke Foundation and the Canadian Cancer Society, have called on Ford to introduce stronger legislation.

Reference Number: 50604

MOREAN, M.E., G. KONG, D.A. CAVALLO, D.R. CARMENGA, AND S. KRISHNAN-SARIN. "Nicotine Concentration of E-cigarettes Used by Adolescents." *Drug and Alcohol Dependence*, 167: 224–227, 2016 (2 tables, 13 references)

Abstract: A survey was conducted of 513 past-month EC users in grades 8, 10, and 12 in Connecticut schools in 2014. Adolescents reported using a nicotine-free e-liquid (28.5%), a nicotine e-liquid (37.4%), or an e-liquid which may or may not have contained nicotine—they did not know (34.1%). Nicotine users were 10.8× more likely to be cigarette smokers and to use EC 1.04× as frequently. Among nicotine users, cigarette smoking, male gender, and purchasing cigarettes from tobacco shops predicted using higher nicotine concentration e-liquid.

> Despite these limitations, our findings highlight that both youth who are smokers and nonsmokers are obtaining e-cigarettes from a wide range of social and commercial sources and are using e-cigarettes containing e-liquids with varying nicotine concentrations. In our sample, males, smokers, and those who were purchasing their own e-cigarettes generally were using higher nicotine concentrations. Of concern, many adolescents did not know the nicotine concentration of the e-liquids contained in the e-cigarettes they use. Considering the variability of the products offered in the market, regulatory authorities should consider standardizing e-cigarette product labeling or developing other strategies to prevent adolescents from being exposed to nicotine inadvertently.

Reference Number: 50605

KLEIN, J.D. "E-cigarettes: A 1-Way Street to Traditional Smoking and Nicotine Addiction for Youth." *Pediatrics*, 141: (2 pp), 2018 (13 references)

Abstract: Heat-not-burn tobacco products are the latest nicotine-containing innovation offered to smokers by the tobacco industry. ECs in general produce less toxins but yield similar concentrations of volatile organic compounds and higher concentrations of carcinogens than cigarettes. Flavoring increases adolescent use and removal of flavoring decreases it.

With this study, we add longitudinal, causal evidence that the transition from e-cigarettes to combustible products, at least for adolescents, is a 1-way street.

Reference Number: 50606

BARRINGTON-TRIMIS, J.L., AND A.M. LEVENTHAL. "Adolescents' Use of 'Pod Mod' E-cigarettes – Urgent Concerns." *New England Journal of Medicine*, 379: 1099–1102, 2018 (2 figures, 5 references)

Abstract: Pod mods are rechargeable devices that aerosolize a liquid solution containing nicotine, flavorings, VG, and/or PG that is encapsulated in a cartridge. The most recognized and popular pod mod is JUUL, which contains a nicotine salt found in loose-leaf tobacco rather than free-base nicotine. JUULs contain 2–10× the amount of nicotine (59 mg/mL) found in other ECs. Pod mods are easier to conceal from authority figures. In 2001, only 2% of high school students used ECs, but the number increased to 16% by 2015. Ensuing public education campaigns brought the EC use down to 11% in 2016. JUUL has 50% of the EC market. Traditional ECs used solutions with free-base nicotine formulation and could cause adverse user experience. Pod mods are indistinguishable from USB drives. Schools should have a zero tolerance for any tobacco products on school property.

This innovation in nicotine chemistry may be critical with regard to the addictiveness of pod mods. Combustible cigarettes deliver high doses of nicotine as well, but the noxious taste and sensations of the initial smoking experience discourage some young people from continuing to smoke. Pod mods may deliver an addictive dose of nicotine without an aversive user experience or other tobacco-related deterrents—which may be one reason why 80% of 15-to-24-year-olds who try Juul continue using the product and why social media posts saying "addicted to my Juul" are common.

Reference Number: 50607

VOGEL, E.A., J. CHO, R.S. MCCONNELL, J.L. BARRINGTON-TRIMIS, AND A.M. LEVENTHAL. "Prevalence of Electronic Cigarette Dependence Among Youth and Its Association with Future Use." *JAMA Network Open*, 3: (13 pp), 2020 (3 tables, 2 figures, 25 references)

Abstract: A tobacco use survey was conducted of 3,168 grade 12 students (aged 16–18 years) in Southern California from 2016 to 2017. At baseline, 444 youth reported past-year EC use and 52 reported at least one EC dependence symptoms. The most common symptoms were craving, urge, need to use ECs, and abstinence-related concentration and emotional problems. Reporting one or more EC dependence symptoms was associated with a 2.56× greater risk of vaping 6 months later.

In clinical settings, e-cigarette dependence symptom screening questions may identify youth at risk for vaping progression who may benefit from intervention. In regulatory decision-making, dependence is a potential health consequence of e-cigarette use that should be considered. Adolescents are particularly vulnerable to nicotine exposure, and our findings suggest that dependence symptoms associated with nicotine exposure via e-cigarettes are associated with greater risk for escalation of vaping behavior. The development of dependence in youth is an important public health consequence that should not be overlooked.

Reference Number: 50608

BOYKAN, R., M.L. GONIEWICZ, AND C.R. MESSINA. "Evidence of Nicotine Dependence in Adolescents Who Use Juul and Similar Pod Devices." *International Journal of Environmental Research and Public Health*, 16: 2135 (6 pp), 2019 (1 table, 19 references)

Abstract: A urine sample was collected and a survey regarding nicotine dependency was conducted on 517 adolescents (aged 12–21 years). Overall, 14.3% had used ECs and 2.9% had smoked cigarettes in the past week. Twenty-one of the EC vapers were pod users and 27 were non-pod users. Pod users were younger than non-pod users (56% were 12–14 years of age versus 44%). More pod users stated they were daily users compared to non-pod users (63.0% vs 11.0%). The urine cotinine concentrations were 921.2 ng/mL for pod users compared to 148.2 ng/mL for non-pod users.

Table: Percentage of Pod Users and Non-pod Users who Agree with Various Statements About Vaping

	Agree with the Statement	
Statement	Pod Users	Non-Pod Users
If I go too long without vaping, the desire to vape interrupts my thinking	15%	0%
If I go too long without vaping, the desire to vape is so great that I need to vape again	10%	0%
If I go too long without vaping, I get angry or irritable	20%	5%
If I go too long without vaping, I get stressed	20%	9%
I need to vape when I awaken in the morning	29%	0%

Source: Adapted from Boykan et al, 2019

Adolescents who used pod products showed more signs of nicotine dependence than non-pod users. Pediatricians should be vigilant in identifying dependence symptoms in their patients who use e-cigarettes, particularly in those using pod devices.

Reference Number: 50609

KRISHNAN-SARIN, S., B.G. GREEN, G. KONG, D.A. CAVALLO, P. JATLOW, R. GUEORGUIEVA, E. BUTA, AND S.S. O'MALLEY. "Studying the Interactive Effects of Menthol and Nicotine Among Youth: An Examination Using E-cigarettes." *Drug and Alcohol Dependence*, 180: 193–199, 2017 (1 table, 5 figures, 49 references)

Abstract: Menthol-containing tobacco products are popular, especially among youth, and menthol is a common flavor in ECs. It produces analgesia and cooling sensations and has been shown to increase self-administration of nicotine. Menthol may slow nicotine metabolism. The 2011 FDA report that menthol smoking was a risk factor for the development of nicotine dependence was challenged by the tobacco companies. Thirty-one female and 20 male EC users (aged 16–20 years), but never-users of cigarettes, were assigned a 0.6 or 12 mg/mL nicotine EC with three menthol concentrations (0, 0.5, and 3.5%). Both the 0.5 and 3.5% menthol concentrations improved taste and increased coolness.

In summary, we provide human behavioral evidence that inclusion of menthol, even at very low concentrations, can increase the appeal of e-cigarettes among youth. Increased appeal could promote continued use, and transitions to higher, more addicting levels of nicotine, particularly in combination with higher concentrations of menthol comparable to those found in commercial e-liquids. Thus, inclusion of menthol alters appeal and reward from e-cigarettes among youth. Regulations to reduce appeal of e-cigarettes by removing menthol should be strongly considered.

Reference Number: 50610

SCHNELLER, L.M., M. BANSAL-TRAVERS, M.L. GONIEWICZ, S. MCINTOSH, D. OSSIP, AND R.J. O'CONNOR. "Use of Flavored E-cigarettes and the Type of E-cigarette Devices Used Among Adults and Youth in the US: Results from Wave 3 of the Population Assessment of Tobacco and Health Study (2015–2016)." *International Journal of Environmental Research and Public Health*, 16: 2991 (12 pp), 2019 (4 tables, 1 figure, 37 references)

Abstract: Between 2015 and 2016, a PATH study was conducted of 415 youth (aged 12–17 years) and 2,123 adults (18+ years) who used ECs as to what e-liquid flavors they had used in the past 30 days. Youth had 21× greater odds of using non-tobacco flavors in their ECs, and vapers who used a refillable device were 5.42× more likely to use non-tobacco flavors than those who used a closed, non-refillable device.

Table: Percentage of Adult and Youth Who Used E-liquid Flavors in Past 30 Days

E-liquid Flavor	Adult Users	Youth Users
Fruit	30.8%	52.8%
Candy/desserts/sweets	18.2%	24.4%
Menthol/mint	17.9%	10.8%
Tobacco	24.5%	5.1%
Chocolate	0.9%	3.0%
Alcoholic beverage	1.5%	1.1%
Other flavor	3.1%	1.0%

Source: Adapted from Schneller et al, 2019

These findings build upon previous research, and reinforce the FDA concerns about the role of flavors in youth vaping appeal, as many youth are unaware of the addictiveness of e-cigarettes. The addition of flavorings to a variety of tobacco products have been shown to reduce the harshness of the nicotine taste and promotes continued use. Previous research has shown that sweet flavors such as fruit or dessert are more satisfying and act as a positive reinforcer among the young adult population (18–30 years). With a multitude of flavors and flavor combinations available on the market, consumers of all ages have many options that enable them to tailor their choices to their individual preferences and thus may encourage initial and continued use. Although most past 30-day e-cigarette users reported using only one flavor category, non-tobacco flavors are far more common among youth than adults. Results from this study identified differences in flavor preferences among adult versus youth vapers that may imply roles for flavors in both the initiation of youth vaping and in adult vaping, including for smoking cessation.

Reference Number: 50611

PEPPER, J.K., K.M. RIBISL, AND N.T. BREWER. "Adolescents' Interest in Trying Flavoured E-cigarettes." *Tobacco Control*, 25: ii62–ii66, 2016 (2 tables, 3 figures, 24 references)

Abstract: A phone survey was conducted of 1,125 adolescents (aged 13–17 years) in the United States in 2014–2015. The participants were randomly assigned 1 of 5 EC flavors (tobacco, alcohol, menthol, candy, or fruit) and questioned about their interest in trying it, if a friend were to offer, to examine the impact of flavors on interest in trying ECs and on harm beliefs. Nearly 15% believed that ECs do not contain nicotine. Adolescents also believed that fruit flavored ECs were less harmful than tobacco flavored ECs.

Table: The Odds Ratio in Adolescent's Interest in Trying an EC According to Flavor

Flavor	OR in interest in trying EC if offered by a friend
Menthol	4.00×
Candy	4.53×
Fruit	6.49×
Tobacco/alcohol	1.00×

Source: Adapted from Pepper et al, 2016

Candy, fruit, and menthol-flavored e-cigarettes appeal to youth more than tobacco or alcohol-flavored, but the appeal is only partially explained by beliefs about reduced harm. Given adolescents' interest in trying e-cigarettes with certain flavors, policymakers should consider restricting advertisements promoting flavored products in channels that reach large numbers of young people. Future research should examine other reasons for the appeal of individual flavors, such as novelty and perceived prestige.

Reference Number: 50612

MOREAN, M.E., E.R. BUTLER, K.W. BOLD, G. KONG, D.R. CAMENGA, D.A. CAVALLO, P. SIMON, S.S. O'MALLEY, AND S. KRISHNAN-SARIN. "Preferring More E-Cigarette Flavors is Associated with E-Cigarette Use Frequency Among Adolescents but Not Adults." *Public Library of Science One*, 13: (10 pp), 2018 (2 tables, 23 references)

Abstract: Because flavored tobacco encourages youth cigarette smoking, the sale of flavored cigarettes except for menthol in the United States was banned in 2009. No such ban exists for ECs and currently there are 1,000 EC liquid flavors available. A study was conducted of 396 adolescent high school past month EC users and 590 adult past month EC users. Fifteen percent of the adults and 42% of the adolescent EC users were cigarette smokers. Nicotine e-liquids were used by 83% of the adult and 68% of the adolescent EC users. Adolescent EC users preferred fruit, alcohol, and "other"-flavored liquids, while adults preferred tobacco, menthol, mint, coffee, and spice. The flavors that were associated with more days of EC use in adolescents were fruit, dessert, and alcohol.

Despite the limitations, the current study provides preliminary evidence that specific flavor preferences and the use of multiple flavors is associated with more frequent e-cigarette use among youth and that a similar pattern of results was not observed within an adult sample. The findings may be used to spur additional research that contributes to regulatory action. The availability of a wide variety of e-liquid flavors may ease the transition to e-cigarette use and contribute to trajectories of continued e-cigarette use (and potential nicotine exposure) given that youth may be especially attracted to certain flavors and the novelty of trying new flavors.

Reference Number: 50613

MCKELVEY, K., M. BAIOCCHI, AND B. HALPERN-FELSHER. "Adolescents' and Young Adults' Use and Perceptions of Pod-Based Electronic Cigarettes." *JAMA Network Open*, 1: e183535 (13 pp), 2018 (5 tables, 47 references)

Abstract: A survey was conducted on 445 high school students in California between April and June 2018 regarding EC and cigarette use. A total of 229 students (51.5%) had ever heard of JUUL. The percentage of students who had, in the past 30 days, used a pod EC, other EC, or cigarette were 10.0%, 7.4%, and 7.6% respectively. On average, 40% of the participants recognized short- or long-term health risks from using pod ECs.

Table: First Flavor Used in Pod-based and Other ECs by Youth

First Flavor Used	Pod Based ECs	Other ECs
First product used was favored (yes)	73.8%	81.2%
Tobacco	1.6%	0.9%
Menthol or mint	28.6%	12.1%
Fruit	30.2%	46.7%
Dessert or sweets	4.8%	17.8%
Coffee or tea	0.0%	0.9%
Don't know or other	34.9%	21.5%

Source: Adapted from McKelvey, Baiocchi, and Halpern-Felsher, 2018

Although pod-based e-cigarettes are the focus of this study, a broader concern is the emerging dynamic between novel tobacco products such as e-cigarettes and public health prevention efforts. For example, consider how, for years, traditional cigarettes had a clear, unified form for their delivery system and public health messaging was able to target and affect all brands of cigarettes simultaneously. In contrast, e-cigarettes take many forms, with variations seen in devices, flavors, e-liquids, power or voltage, and nicotine content. Hence, amid rapid innovations in e-cigarettes (eg, newer pod-based styles) and in the absence of clear, consistent public health prevention messaging that targets relevant aspects common to all forms of e-cigarettes, further misperception of health risks, including nicotine addiction, could result. Public health prevention efforts are challenged to develop multifaceted and adaptive messaging that is necessary to reduce uptake and consumption of pod-based e-cigarettes and other emergent products among adolescents and young adults who otherwise would likely remain nicotine naive.

Reference Number: 50614

CHEN-SANKEY, J.C., G. KONG, AND K. CHOI. "Perceived Ease of Flavored E-Cigarette Use and E-Cigarette Use Progression Among Youth Never Tobacco Users." *Public Library of Science One*, 14: (11 pp), 2019 (2 tables, 47 references)

Abstract: A US national sales restriction of flavored cigarettes was associated with a 17% reduction in cigarette smoking prevalence and a 58% decrease in cigarette consumption among youth. In contrast, EC use among middle school students has increased from 1.5% in 2011 to 11.3% in 2016. A survey of 6,983 youth never users (aged 12–17 years) was conducted in 2013–2014 and a follow-up survey was conducted in 2014–2015. Overall, 21.2% of the sample perceived flavored EC as easier to use than unflavored. Perceiving flavored EC as easier to use was associated with a 1.32× greater risk of EC initiation.

> Given the alarming consequence of perceiving flavored e-cigarettes easy to use among youth never tobacco users, it is important to prevent the development of such perceptions in this group. Thus, it is critical to understand the factors that contribute to the perception. First, youth have strong preferences for sugar and sweet tastes because sugar releases opioids and dopamine and thus might have addictive potential through providing pleasurable sensations to users. Thus, the sensory similarities between sweet foods and e-cigarette flavors (especially sweet flavors) could reduce psychological barriers to using flavored e-cigarettes. Certain e-cigarette flavors (e.g., tobacco, fruit, candy, desserts, etc.) that are particularly appealing to youth may also play a significant role in influencing the ease of use perceptions than other flavors. Second, the wide availability of flavored e-cigarettes may increase youth's perceived access to the product through a variety of channels including social connections, non-compliant tobacco vendors, and online stores that sell e-cigarettes. Although there has been a recent trend of restricting flavored e-cigarettes among U.S. local jurisdictions, these restrictions were found to lack strictness in constraining the availability of e-cigarette flavors (such as leaving menthol flavors exempted) therefore may still leave youth with abundant access to flavored e-cigarettes.

Additionally, e-cigarette advertisements may portray flavored e-cigarettes as less harmful than unflavored e-cigarettes, thereby fostering the impression that the product is easier to use.

Reference Number: 50615

BOLD, K.W., G. KONG, D.R. CAMENGA, P. SIMON, D.A. CAVALLO, M.E. MOREAN, AND S. KRISHNAN-SARIN. "Trajectories of E-Cigarette and Conventional Cigarette Use Among Youth." *Pediatrics*, 141: (7 pp), 2018 (2 tables, 1 figure, 31 references)

Abstract: Surveys were conducted of 808 high school students at three public schools in Connecticut in 2013, 2014, and 2015. Past month EC users in 2013 were 7.1× as likely to use cigarettes in 2014 than non-EC users, and EC users in 2014 were 3.9× as likely to use cigarettes in 2015. EC use increased from 8.9% of the students in 2013 to 14.5% in 2015.

E-cigarette use was associated with future cigarette use across 3 longitudinal waves, yet cigarette use was not associated with future e-cigarette use. Future research needs to examine mechanisms through which e-cigarette use leads to cigarette use. E-cigarette regulation and prevention programs may help prevent future use of cigarettes among youth.

Reference Number: 50616

MIECH, R., M.E. PATRICK, P.M. O'MALLEY, AND L.D. JOHNSTON. "E-Cigarette Use as a Predictor of Cigarette Smoking: Results from a 1-Year Follow-Up of a National Sample of 12th Grade Students." *Tobacco Control*, 26: e106–e111, 2017 (2 tables, 14 references)

Abstract: An analysis was conducted of 347 students in Grade 12 in 2014 and who were resurveyed the next year regarding EC and cigarette use. Among the 12th grade youth who had never smoked, recent vapers were 4.78× as likely to report past-year smoking at follow-up. First use of cigarette at follow-up was reported by 31% of those who recently vaped, compared to only 7% of those who didn't vape. Vaping did not predict smoking cessation among recent smokers at baseline. Vapers were 4.73× as likely to no longer perceive cigarettes as posing as great a risk of harm.

Desensitization to the dangers of smoking may play a role in explaining how vaping can progress to smoking among youth who have no history of cigarette use. Youth who begin to vape primarily to experiment and because they taste good (the most common reasons for vaping may detect no immediate health consequences and conclude that the dangers of smoking are exaggerated). Empirical support for a desensitization process comes from this study's finding that youth who vaped were significantly more likely to change their perception of the dangers of smoking away from "great risk," among those who had never smoked at baseline.

Reference Number: 50617

ALEYAN, S., A. COLE, W. QIAN, AND S.T. LEATHERDALE. "Risky Business: A Longitudinal Study Examining Cigarette Smoking Initiation Among Susceptible and Non-Susceptible E-Cigarette Users in Canada." *BMJ Open*, 8: (7 pp), 2018 (1 table, 2 figures, 37 references)

Abstract: Susceptibility to future smoking is defined as the lack of a firm commitment not to smoke among never-smokers and is a reliable predictor of cigarette smoking initiation among youth. A survey was conducted of 9,501 high school students from 81 Alberta and Ontario secondary schools between 2013 and 2014 (baseline) and 2015 and 2016 (2 years later). The adjusted OR of a susceptible never-smoker EC user to use cigarettes was 2.78×. For a non-susceptible never-smoker, the adjusted OR was 5.28×.

> Among non-smoking youth who were current e-cigarette users, our findings showed an increased risk of progressing to cigarette use after 2 years. Of concern, low-risk youth at baseline were at an even greater risk of cigarette smoking initiation at follow-up. These results suggest that e-cigarettes are expanding the tobacco cigarette market by attracting low-risk youth who would not have otherwise tried using cigarettes. These findings reinforce the need to adopt regulations aimed at reducing the appeal of e-cigarettes and deterring use among youth populations. Additionally, our results point towards the need for sustained efforts focused on deterring the use of all forms of tobacco products, including e-cigarettes.

Reference Number: 50618

KRISHNAN-SARIN, S., A. JACKSON, M. MOREAN, G. KONG, K.W. BOLD, D.R. CAMENGA, D.A. CAVALLO, P. SIMON, AND R. WU. "E-Cigarette Devices Used by High-School Youth." *Drug and Alcohol Dependence*, 194: 395–400, 2019 (2 tables, 3 figures, 27 references)

Abstract: A study was conducted of 2,945 high school students in Connecticut (mean age 16.3 years) as to EC use. There were 1,047 ever-users of ECs (35.6%), of which 875 (82.2%) had used one of the four devices surveyed. Only 7.3% of EC users currently used cig-a-like, 18.7% users vape/hookah pens, 47.1% used JUULs, and 33.1% used mods.

Table: Use of Vaping Devices in High School Students and OR of Other Tobacco Use

Device	Percent Use	OR of other tobacco use
Disposable, cig-a-like	7.3%	3.43×
Vape/Hookah pen	18.7%	2.49×
JUUL	47.1%	1.29×
Mods/Advanced personal vaporizers	33.1%	2.64×

Source: Adapted from Krishnan-Sarin et al, 2019

> The current study is the first to demonstrate that youth who use e-cigarettes are using a wide variety of devices and in many cases multiple devices. We observed that rates of using mods/advanced personal vaporizers, vape/hookah pens, and the JUUL were much higher than those of cig-a-like products. These results are consistent with prior research suggesting that youth are more likely to use later generation devices compared to cig-a-likes but provide more detailed information about which devices are used. Further, approximately two-thirds of youth reported using multiple device types, and these adolescents were also more likely to use other tobacco products. Among youth reporting use of single device, rates of JUUL use were the highest, which may be related to the phenomenal growth in the sales and availability of this product from 2015 to the time of this survey in June 2017.

Reference Number: 50619

FATAAR, F., AND D. HAMMOND. "The Prevalence of Vaping and Smoking as Modes of Delivery for Nicotine and Cannabis among Youth in Canada, England and the United States." *International Journal of Environmental Research and Public Health*, 16: 411 (13 pp), 2019 (4 tables, 35 references)

Abstract: In 2018, an online survey was conducted of youth (aged 16–19 years) in Canada (n=3,757), England (n=3,819), and the United States (n=3,961) regarding their past 30-day vaping. Overall, 1.7% of youth in

England, 4.7% of youth in Canada, and 5.1% of youth in the United States reported vaping cannabis. The odds of vaping cannabis in the US and Canada were 3× higher than in England. Approximately 20% of youth in all countries reported vaping tobacco and/or cannabis. The prevalence of smoking tobacco in youth was highest in England (11%) followed by Canada (7%), and the United States (7.0%). Less than 2% of youth in all countries reported vaping cannabis only.

Table: Prevalence of Past 30-Day Tobacco and Cannabis Use Among Youth

	England	Canada	US
Vape e-cigarettes only	8.0%	11.6%	12.2%
Vape cannabis only	0.9%	1.7%	1.2%
Vape both e-cigarettes and cannabis	0.7%	3.0%	3.9%
Smoke tobacco only	11.0%	7.8%	7.0%
Smoke cannabis only	2.9%	7.3%	7.3%
Smoke both tobacco and cannabis	5.5%	7.3%	4.9%

Source: Adapted from Fataar and Hammond, 2019

Overall, e-cigarettes have emerged as the most common mode of nicotine delivery among youth across Canada and the U.S., whereas smoking remains the dominant form of delivery for cannabis. However, there are increasing signs that legalization of medical and non-medical cannabis in North America are associated with greater adoption of cannabis extracts, including cannabis oil for vaping, as well as edibles and other smoked extracts. Collectively, these forms are typically far more potent with higher THC levels than dried herb.

Reference Number: 50620

CHAFFEE, B.W., S.L. WATKINS, AND S.A. GLANTZ. "Electronic Cigarette Use and Progression from Experimentation to Established Smoking." *Pediatrics*, 141: (6 pp), 2018 (1 table, 25 references)

Abstract: A survey was conducted about EC and cigarette use in 13,651 US adolescents (ages 12 to 17 years) at baseline and 1 year later (88% retention rate). Ever use of ECs by adolescents was positively associated with progression to established smoking with an OR of approximately 2×.

Table: E-cigarette Use and the Adjusted Odds Ratio of Smoking 100 Cigarettes or Past 30 day Use of Cigarettes

E-cigarette Use	Adjusted OR of Having Smoked 100 Cigarettes or of Having Used Cigarettes in Past 30 Days
Never	1.00×
Ever	1.80×
Former	1.85×
Past 30-day use	1.76×

Source: Adapted from Chaffee et al, 2018

Among adolescent cigarette experimenters, using e-cigarettes was positively and independently associated with progression to current established smoking, suggesting that e-cigarettes do not divert from, and may encourage, cigarette smoking in this population.

Reference Number: 50621

COLEMAN, B.N., B.J. APELBERG, B.K. AMBROSE, K.M. GREEN, C.J. CHOINIERE, R. BUNNELL, AND B.A. KING. "Association Between Electronic Cigarette Use and Openness to Cigarette Smoking Among US Young Adults." *Nicotine and Tobacco Research*, 17: 212–218, 2015 (2 tables, 31 references)

Abstract: An analysis was conducted based on data from the 2012–2013 US National Adult Tobacco Survey of 4,310 adults (aged 18–29 years), who had never smoked cigarettes. Approximately 8% reported the ever use of an EC. Those who used ECs were 2.4× more likely to be open to cigarette smoking. Other risk factors included being male, aged 18–24 years, and less educated, and having ever used a hookah.

Ever use of e-cigarettes and other tobacco products was associated with being open to cigarette smoking. This study does not allow us to assess the directionality of this association, so future longitudinal research is needed to illuminate tobacco use behaviors over time as well as provide additional insight on the relationship between ENDS use and conventional cigarette use among young adult populations.

Reference Number: 50622

BARRINGTON-TRIMIS, J.L., G. KONG, A.M. LEVENTHAL, F. LIU, M. MAYER, T.B. CRUZ, S. KRISHNAN-SARIN, AND R. MCCONNELL. "E-Cigarette Use and

Subsequent Smoking Frequency Among Adolescents." *Pediatrics*, 142: (18 pp), 2018 (3 tables, 2 figures, 33 references)

Abstract: Three surveys in California and Connecticut (baseline 2013–2014; follow-up 2014–2016) from 6,258 adolescents involving cigarette and EC use were evaluated. At follow-up, 9.2% of never smoking youths had used cigarettes (6.2% experimental, 1.9% infrequent smokers, and 1.1% frequent smokers). Among baseline never-smokers, EC users had greater odds of being subsequent experimental (4.58×), infrequent (4.27×), or frequent (3.51×) cigarette smokers.

> With our findings, we suggest that smoking uptake and progression is an adverse public health consequence of high rates of e-cigarette use among youth and young adults. The findings also did not reveal strong evidence of transitioning away from combustible cigarette use as a potential public health benefit of e-cigarette use in young people. Together, these findings reveal that adolescent e-cigarette use may result in an overall adverse impact on the public health of youth and young adults. Additional follow-up in large cohort studies along with data on factors that promote transitioning from e-cigarettes to cigarettes and that inhibit transitioning from cigarettes to e-cigarettes or nonuse are needed to develop targeted interventions to minimize the adverse public health impact of e-cigarettes in the adolescent population.

Reference Number: 50623

SONEJI, S., J.L. BARRINGTON-TRIMIS, T.A. WILLS, A.M. LEVENTHAL, J.B. UNGER, L.A. GIBSON, J. YANG, B.A. PRIMACK, J.A. ANDREWS, R.A. MIECH, T.R. SPINDLE, D.M. DICK, T. EISSENBERG, R.C. HORNIK, R. DANG, AND J.D. SARGENT. "Association Between Initial Use of E-cigarettes and Subsequent Cigarette Smoking Among Adolescents and Young Adults: A Systematic Review and Meta-analysis." *JAMA Pediatrics*, 171: 788–797, 2017 (2 tables, 1 figure, 44 references)

Abstract: A systematic review and meta-analysis were conducted of nine studies consisting of 17,389 adolescents on whether the use of ECs increase the risk of cigarette smoking. The pooled ORs for subsequent cigarette initiation was 3.5× for ever versus never EC users. The OR increased to 4.28× for past 30-day cigarette smoking.

Our systematic review and meta-analysis of 9 longitudinal studies found consistent and strong evidence that e-cigarette use is associated with increased odds of subsequent cigarette smoking initiation and current cigarette smoking among adolescents and young adults after adjusting for known demographic, psychosocial, and behavioral risk factors. To minimize the potential public health harm from e-cigarette use, the US Food and Drug Administration, as well as state and local agencies, will need to engage in regulatory actions to discourage youth use of e-cigarettes and prevent the transition from e-cigarettes to other combustible tobacco products. In addition to the currently enacted age limitations on in-store sales, regulatory actions could include restrictions on advertising campaigns that may be viewed by adolescents, limits to characterizing flavors (eg, fruit- and candy-flavored e-cigarettes), strict standards for reporting actual nicotine content in e-liquid, and requirements for age verification for online and retail sales of these products. Such strong regulation of e-cigarettes could curb use among youth and limit the future population-level burden of tobacco.

Reference Number: 50624

AGAKU, I., S. ODANI, C. VARDAVAS, AND L. NEFF. "Self-Identified Tobacco Use and Harm Perceptions Among US Youth." *Pediatrics*, 141: (9 pp), 2018 (2 tables, 1 figure, 31 references)

Abstract: This study analyzed the data of 20,675 students (grades 6–12) from the 2016 US National Youth Tobacco Survey. Past 30-day use of any tobacco product was reported by 2,350 students (11%), of whom nearly half (47.5%) reported using two or more different tobacco products. Among EC only users, 22% believed ECs were harmful, 71% believed hookahs were harmful, 86% believed smokeless tobacco was harmful, 90% believed cigarettes were harmful, and 80% believed all tobacco products were harmful.

Although noncombustible products, such as e-cigarettes, generally contain fewer toxicants than combustible tobacco products, no tobacco product is harmless. While moderation is encouraged for certain ingested substances that are potentially harmful in excess doses (eg, fat, salt, and sugar), youth use of tobacco products in any amount is unsafe. Multifaceted interventions that address protobacco social and contextual factors (such as youth-oriented tobacco advertising and marketing, product design, and packaging as well as social norms and access) can help reduce tobacco use.

Reference Number: 50625

GRANT, J.E., K. LUST, D.J. FRIDBERG, A.C. KING, AND S.R. CHAMBERLAIN. "E-Cigarette Use (Vaping) is Associated with Illicit Drug Use, Mental Health Problems, and Impulsivity in University Students." *Annals of Clinical Psychiatry*, 31: 27–35, 2019 (3 tables, 42 references)

Abstract: A survey was conducted of 3,572 university students (57% female) in a large midwestern US university in 2016. The prevalence of first 12-month use of ECs was 9.2% and 17.7% reported ever use of ECs. Those who reported use of ECs were more likely to be male, undergraduates, and have lower GPAs. EC use was associated with the use of multiple other drugs (e.g., alcohol, opiates) and vapers were more likely to have mental health histories of ADHD, PTSD, gambling disorder, and anxiety, and to report low self-esteem.

Table: Percentage of University Students (E-cigarette Users vs Non-users) Who Used Illicit Drugs

	University Students	
Drug Use in past 12 months	Used e-cigarettes Within Past 12 Months	Never Used e-cigarettes
Cannabis	72.0%	20.9%
Amphetamines	5.7%	0.3%
Cocaine	14.9%	1.6%
Heroin	1.9%	0.1%
Prescription Opioid pain medication	9.7%	0.9%

Source: Adapted from Grant et al, 2019

Use of e-cigarettes is common in university students and appears to be associated with a variety of mental health and drug use problems. Clinicians should be aware that certain mental health conditions are more common in e-cigarette users. This study indicates the need for longitudinal research into the effects of chronic nicotine consumption on brain function and mental health, especially in young people, since such effects would be common to conventional tobacco smoking and vaping.

Reference Number: 50626

RUBENSTEIN, M.L., K. DELUCCHI, N.L. BENOWITZ, AND D.E. RAMO. "Adolescent Exposure to Toxic Volatile Organic Chemicals from E-cigarettes." *Pediatrics*, 141: (9 pp), 2018 (2 tables, 1 figure, 37 references)

Abstract: Saliva and urine samples were collected from 67 EC only users, 16 dual EC and cigarette users, and 20 control adolescents (mean age 16.4 years). The saliva was analyzed for cotinine by LC/MS/MS and the urine for NNAL and 8 different volatile organic solvents (VOCs). Urine excretion of benzene, ethylene oxide, acrylonitrile, acrolein, and acrylamide was higher in dual users versus EC users only compared with controls. The levels of three toxic VOCs were as high in EC users who did or did not use nicotine. The use of fruit flavored EC liquids was associated with a greater amount of acrylonitrile metabolites.

Although e-cigarette vapor may be less hazardous than tobacco smoke, our findings can be used to challenge the idea that e-cigarette vapor is safe, because many of the volatile organic compounds we identified are carcinogenic. Messaging to teenagers should include warnings about the potential risk from toxic exposure to carcinogenic compounds generated by these products.

Reference Number: 50627

PETRESCU, D.C., M. VASILJEVIC, J.K. PEPPER, K.M. RIBISL, AND T.M. MARTEAU. "What is the Impact of E-Cigarette Adverts on Children's Perceptions of Tobacco Smoking? An Experimental Study." *Tobacco Control*, 26: 421–427, 2017 (2 tables, 1 figure, 56 references)

Abstract: Five hundred and sixty children (aged 11–16 years) in the United Kingdom watched healthy, cool (glamorous), or no advertisements about EC and were tested as to their perceptions of cigarette smoking. Of the 560 children, 153 (27.1%) were eliminated from the study as they had already smoked cigarettes or used ECs. Those children exposed to either type of EC advertisements perceived the harms of smoking one to two cigarettes as less than those who did not see the advertisements.

This study provides the first evidence that exposure to e-cigarette adverts might influence children's perceptions of smoking tobacco cigarettes, reducing their perceived harm of occasional smoking. These results suggest the potential for e-cigarette adverts to undermine tobacco control efforts by reducing a potential barrier (ie, beliefs about harm) to occasional smoking.

Reference Number: 50628

VASILJEVIC, M., A.S.J. WALLIS, S. CODLING, D-L. COUTURIER, S. SUTTON, AND T.M. MARTEAU. "E-Cigarette Adverts and Children's Perceptions of Tobacco Smoking Harms: An Experimental Study and Meta-Analysis." *BMJ Open*, 8: e020247 (8 pp), 2018 (2 tables, 1 figure, 41 references)

Abstract: The perceived harm of occasional tobacco smoking was determined in 1,057 English school children (aged 11–16 years) who were shown 10 glamor ads for e-cigarettes or 10 ads for pens (similar shape to ECs). The children exposed to EC ads had lower perceived harms of smoking one or two cigarettes occasionally than those not seeing EC ads, but it had no effect on perceived harms of smoking 10 cigarettes a day.

Interestingly, children perceived that the harm of occasional tobacco smoking was lower when they were exposed to e-cigarette adverts, even though they rated the e-cigarette adverts as significantly less appealing and professed a lower interest in buying and trying the e-cigarettes when compared with the pens shown in the control condition. These findings may have important ramifications for future research and policy, since they suggest that the cross-product impact of e-cigarette adverts may largely work via an unconscious, implicit route that may not necessarily affect self-reported explicit appeal, but may change perceptions of harm (risk) which feed into children's behavioural decisions.

Reference Number: 50629

WACKOWSKI, O.A., J.M. SONTAG, D. HAMMOND, R.J. O'CONNOR, P.A. OHMAN-STRICKLAND, A.A. STRASSER, A.C. VILLANTI, AND C.D. DELNEVO. "The Impact of E-Cigarette Warnings, Warning Themes and Inclusion of Relative Harm Statements on Young Adult's E-Cigarette Perceptions and Use Intentions." *International Journal of Environmental Research and Public Health*, 16: 184 (14 pp), 2019 (4 tables, 44 references)

Abstract: In the United States and other countries, ECs, unlike cigarettes, are required to only carry one nicotine addiction warning. An online study was conducted of 876 adults (aged 18–29 years) on the effectiveness of different health warnings about ECs. The warnings were sized at approximately 20% of the ECs ad size for JUUL, Vuse, Blu, and MarkTen vaping devices. One-third of the participants were smokers and 54.5% were current EC users. Text-only warnings have limited impact with young adults, especially when placed in a colorful EC ad. Relative harm statements in the warnings, such as warning that e-cigarettes may cause harm to health but are less harmful than cigarettes, were less effective as well.

> Findings from this study suggest that warning labels with messages about harms beyond nicotine's addictiveness are perceived as believable, informative, understandable and credible among young adults, may provide novel information, and may discourage e-cigarette use among young people. While inclusion of a relative harm statement in e-cigarette warnings may not unintentionally increase use by non-smokers, it may reduce recall of warning messages in young adults. Further research should continue to identify warning themes, types and execution formats that accurately inform audiences about e-cigarettes.

Reference Number: 50630

MCKELVEY, K., L. POPOVA, J.K. PEPPER, N.T. BREWER, AND B. HALPERN-FELSHER. "Adolescents Have Unfavorable Opinions of Adolescents Who Use E-cigarettes." *Public Library of Science One*, 13: (10 pp), 2018 (2 tables, 42 references)

Abstract: A survey was conducted of 578 adolescents (aged 14–20 years) from 10 California schools between 2015 and 2016 regarding ECs. The majority of those surveyed (61%) had overall negative opinions about adolescent EC use.

Table: Positive and Negative Traits Ascribed to Adolescent E-cigarette Users

Positive Traits	Negative Traits
• Sexy	• Unattractive
• Cool	• Trashy
• Clean	• Immature
• Smart	• Disgusting
• Healthy	• Inconsiderate

Source: Adapted from McKelvey et al, 2018

> Evidence shows pervasive marketing of e-cigarettes targets adolescents by including attractive models, appealing flavors, and claims of lower harm,

which can lead to formation of positive images of people who use e-cigarettes. However, our findings suggest these messages have not translated to positive images of e-cigarette users among this group of adolescents. It is important to ensure that adolescents do not formulate "cool" or otherwise desirable images of e-cigarette users, as these images could become established, as occurred with the positive cigarette-smoking prototypes, which took decades of concentrated efforts to dispel.

5.07 DUAL USE (CIGARETTES AND E-CIGARETTES)

"Use of e-cigarettes was most common among smokers, and dual users had the highest prevalence of respiratory symptoms. On a population level, this study indicates that the present use of e-cigarettes does not adequately serve as a smoking cessation tool."

—Hedman et al, "Association of Electronic Cigarette Use with Smoking Habits, Demographic Factors, and Respiratory Symptoms" (2018)

Dual use is the current use of cigarettes and ECs is a dangerous practice, as various biomarkers of chemical exposure only decrease for those who vape exclusively and not for dual users (50701). Addiction to nicotine was also greater for dual users than exclusive users of ECs (50702). Only one-quarter of dual users wanted to quit and 35.7% of dual users have a strong urge to use ECs, compared to only 5.6% of exclusive users (50703).

A high proportion of smokers who are dual users had no intention to quit, which shows the marked limitations of ECs as a nicotine cessation tool (50704). Dual users prefer using cigarettes in stressful situations as the nicotine uptake into the brain is faster than ECs (50705).

Dual users were 2× as likely to report adverse events than EC only users and dual use was strongly associated with alcohol problems in students (50706–7). Adverse respiratory symptoms were also more common in dual users (50711). Dual users were exposed to more toxic chemicals and had a higher concentration of biomarkers of exposure to carcinogens than even cigarette smokers (50708, 50710).

Switching from cigarettes to ECs did not confer a significantly lesser stroke risk (50709) or cardiovascular risk (50712). Smokers who reduced the number of cigarettes per day to only 1–10 still had a hazard ratio of 9–12× greater for lung cancer and 6× greater for lung disease than never smokers (50705). Complete cigarette cessation, not reduction, is required to reduce health risk to that of never-smokers.

Reference Number: 50701

CZOLI, C.D., G.T. FONG, M.L. GONIEWICZ, AND D. HAMMOND. "Biomarkers of Exposure Among 'Dual Users' of Tobacco Cigarettes and Electronic Cigarettes in Canada." *Nicotine and Tobacco Research*, 21: 1259–1266, 2019 (2 tables, 45 references)

Abstract: Dual use is the current use of both cigarettes and ECs. Over a period of 3 weeks, 48 dual users (mean age 35.9 years) either used no tobacco product, exclusively used ECs, or continued dual use of cigarettes and ECs. Various urinary biomarkers of tobacco exposure such as 1-HOP, NNAL, and breath CO were determined. Levels of biomarkers decreased significantly when the subjects exclusively vaped but not when they had dual use. The biomarkers were at the highest levels when the subjects exclusively smoked.

Table: Percent Change in Biomarkers After 7 Days of Abstaining, Vaping Exclusively, or Smoking Exclusively (Compared to Dual Use)

Biomarker	Abstained	Vaped Exclusively	Smoked Cigarettes Exclusively
CO	–26%	–41%	+21%
1-HOP	–14%	–31%	+23%
NNAL	–35%	–30%	+8%

Source: Adapted from Czoli et al, 2019

Despite slight reductions in exposure associated with dual use, the findings demonstrate that abstaining from tobacco cigarettes is the most important factor in reducing exposure to toxic smoke constituents. Research evidence indicates that smokers who quit tobacco cigarettes completely reduce their risk of premature death to levels comparable to nonsmokers. However, the potential benefits of smoking reduction, as may be the case of dual use, are less clear. To date, significant health benefits from reducing the amount of tobacco cigarettes smoked have not been demonstrated with respect to various disease outcomes. Although it is plausible that dual use could reduce individual risk if it results in substantial reductions in smoking, the threshold for meaningful reductions is unclear, particularly given that smokers may compensate for reductions in the number of cigarettes they smoke by smoking each cigarette more intensely. This is generally supported by the current findings, in which the differences between dual use and exclusive smoking were modest. Therefore, dual use is likely to have public health benefit to the extent that it leads to complete smoking cessation.

Reference Number: 50702

JANKOWSKI, M., M. KRZYSTANEK, J.E. ZEJDA, P. MAJEK, J. LUBANSKI, J.A. LAWSON, AND G. BROZEK. "E-cigarettes are More Addictive than Traditional Cigarettes—A Study in Highly Educated Young People." *International Journal of Environmental Research and Public Health*, 16: 2279 (10 pp), 2019 (2 tables, 38 references)

Abstract: A study was conducted of 30 cigarette smokers, 30 exclusive EC users, and 30 dual users who were university students (mean age 22.4 years). Nicotine dependence was determined by the Fagerstrom Test for Nicotine Dependence (FTND). About 19% of the exclusive EC users initiated nicotine use by ECs. Dual users stated that cigarette smoking provides more satisfaction than EC use (48.2%). In addition, smoking a cigarette with friends, with coffee, and after a meal or sex was more satisfying than EC use. The mean FTND was 1.6 for cigarette smokers, 3.5 for exclusive EC users, and 4.7 for dual users.

Table. Some Characteristics of Exclusive EC and Dual Users

Characteristic	Exclusive EC users	Dual Users (cigarettes and ECs)
Mean FTND	3.5	4.7
Mean nicotine concentration in e-liquid	6.8 mg/mL	8.3 mg/mL
Millilitres of e-liquid used per day	4.1 mL	4.3 mL
Mean number of different models of EC used	3.6	4.3

Adapted from Jankowski et al, 2019

Several hypotheses should be considered to explain why we found higher levels of addiction among e-cigarette users compared to smokers in contrast to findings in studies by González et al. (2017), Etter et al. (2015), and Liu et al. (2017). First, our study was performed in early 2018, and it is possible that our participants were using more effective e-cigarettes with higher addictive potential. The e-cigarette market is quickly changing with newer models and generations of e-cigarettes available. Advanced e-cigarettes, especially those with a high-capacity battery or accumulator, are characterized by the production of a larger volume of aerosol and delivery of significantly higher doses of nicotine compared to older models (first or second generation) of e-cigarettes. In our study, participants, on average, used four different e-cigarette devices, and more than half of the participants used technically advanced e-cigarettes. Second, our study was performed among young adult e-cigarette users, aged 22.4 ± 2.2 years old, a younger study population compared to those of earlier studies.

Reference Number: 50703

CASE, K.R., D.S. MANTEY, M.R. CREAMER, M.B. HARRELL, S.H. KELDER, AND C.L. PERRY. "E-Cigarette Specific Symptoms of Nicotine Dependence Among Texas Adolescents." *Addictive Behaviors*, 84: 57–61, 2018 (2 tables, 28 references)

Abstract: The symptoms of EC nicotine dependence were determined on 2,891 adolescent high school students (mean age 16 years) in Texas in 2016, using the Hooked on Nicotine Checklist (HONC). Overall, 132 students reported using ECs in the past 30 days, of which 91 were exclusive EC users and 41 were dual users. The most frequently reported symptom of EC dependence was the strong urge to use an EC.

Table: Symptoms of Nicotine Dependence in Dual Users and Exclusive Users of ECs

EC User Group	Want to Quit	Quit Attempt	Really Need	Use <30 minutes after wakening	Strong Urge
Dual user	24.2%	22.9%	32.7%	16.4%	35.7%
Exclusive user	53.3%	45.7%	5.0%	5.7%	5.6%

Source: Adapted from Case et al, 2018

This study provides preliminary support that adolescent e-cigarette users experience symptoms of nicotine dependence specific to e-cigarettes. Importantly, our results show that both exclusive users and dual users of e-cigarettes and combustible tobacco products experience e-cigarette-specific symptoms of dependence suggesting that e-cigarettes may have unique characteristics that influence dependence. Furthermore, as experiencing more symptoms of dependence was inversely associated with wanting to quit e-cigarettes and past-year e-cigarette quit attempts, dependence may be an important barrier to e-cigarette cessation and interventions aimed at reducing dependence may be warranted.

Reference Number: 50704

OWUSU, D., J. HUANG, S.R. WEAVER, T.F. PECHACEK, D.L. ASHLEY, P. NAYAK, AND M.P. ERIKSEN. "Patterns and Trends of Dual Use of E-cigarettes and Cigarettes Among U.S. Adults, 2015–2018." *Preventive Medicine Reports*, 16: (7 pp), 2019 (4 figures, 51 references)

Abstract: Smoking and EC use among US adults was studied in 2015 (n=6051), 2016 (n=6014), 2017 (n=5992), and 2018 (n=5989). If dual use of cigarettes and ECs is a temporary condition, it may indicate a transition state to tobacco and nicotine cessation. However, if there is long-term dual use, there may be increased tobacco harm. Between 2015 and 2018, there was no change in the prevalence of EC use among current, light, moderate, or heavy smokers. Among never or former smokers, the prevalence of EC use increased. The high proportion of dual use among heavy smokers and smokers with no intention to quit smoking is a concern.

> The continued high prevalence of dual use and increased prevalence of current e-cigarette use among never smokers highlight the need for better communication about the risks of prolonged dual use for e-cigarette users, and the risks of nicotine initiation and addiction for nonusers.

Reference Number: 50705

ADRIAENS, K., D. VAN GUCHT, AND F. BAEYENS. "Differences Between Dual Users and Switchers Center Around Vaping Behavior and Its Experiences Rather than Beliefs and Attitudes." *International Journal of Environmental Research and Public Health*, 15: 12 (15 pp), 2018 (4 tables, 1 figure, 22 references)

Abstract: Smokers who consistently smoked less than one or 1–10 CPD still had a hazard ratio of 9–12× greater for lung cancer and 6× greater for respiratory disease than never-users. An online survey was conducted in 2016 regarding EC switchers (from cigarettes) and dual users (EC and cigarettes) on a Dutch website. Fifty-eight male and 43 female dual users and 79 male and 21 female switchers completed the survey. The mean age both groups started smoking was 15 years. Before EC, upon initiation both groups smoked on average 22 CPD. Dual users had decreased tobacco consumption by 82%. Both groups had been vaping for 22 months on average, were highly EC dependent, and used advanced ECs with EC liquids containing 4–8 mg/mL and usually flavors other than tobacco. Dual users preferred cigarettes in stressful situations and when rapid nicotine uptake was required.

> It remains equally important, however, to recognize and respect that not all dual users express the desire to quit smoking completely (in this sample only one third expressed the clear intention to quit smoking within the next month). Relatedly, some smokers who have tried vaping report that e-cigs are lacking some of the unique sensory-hedonic characteristics of cigarette smoking that they highly value and enjoy (e.g., the smell and flavor of burned tobacco), whereas at the same time vaping has some intrinsic characteristics that they dislike (e.g., the more "artificial" vaping experience, the complex technology, the "chemical nature" of e-liquids). To the extent that similar likes and dislikes also play a role in dual users' choice to continue smoking, one cannot but acknowledge that e-cigs have their limits as a tool for THR, and welcome present or future other low-risk nicotine products that better address the hedonics of smoking.

Reference Number: 50706

ABAFALVI, L., M. PÉNZES, R. URBÁN, K.L. FOLEY, R. KAÁN, B. KISPÉLYI, AND P. HERMANN. "Perceived Health Effects of Vaping Among Hungarian Adult E-Cigarette–Only and Dual Users: A Cross-Sectional Internet Survey." *BMC Public Health*, 19: 302 (10 pp), 2019 (4 tables, 1 figure, 42 references)

Abstract: A web-based survey was conducted on 859 EC-only users and 183 dual EC/cigarette users in Hungary in 2015. Dual users were over 2× as likely to report adverse events than EC-only users. The majority of EC-only users reported vaping more than 20 times a day, whereas dual users reported less frequent use. Over 60% of the entire sample were heavy smokers and about 10% were light smokers. The most common adverse events reported were cough, sore/dry throat, and mouth.

Table: The Percentage of E-cigarettes and Dual Users Who Report Adverse Events

Adverse Effect	E-Cigarette–only users	Dual users
Sore/dry mouth and throat	3.5%	4.9%
Cough	2.1%	13.1%
Mouth or tongue sores/inflammation	2.2%	6.0%

Adverse Effect	E-Cigarette–only users	Dual users
Headache	1.5%	2.7%
Heart palpitations	0.9%	3.3%
Any adverse event	11.8%	26.2%

Source: Adapted from Abafalvi et al, 2019

In conclusion, our study supports that the majority of e-cigarette-only users report more perceived beneficial changes in physiological functions while less AEs than dual users. However, perceived short-term benefits of e-cigarette use may reinforce users despite the uncertainty of long-term health consequences.

Reference Number: 50707

OH, K.H., C.M. LEE, B. OH, S-W. OH, H-K. JOH, H.C. CHOI, S.J. KIM, AND S.R. LEA. "The Relationship Between Electronic Cigarette Use with or Without Cigarette Smoking and Alcohol Use Among Adolescents: Finding From the 11th Korea Youth Risk Behavior Web-Based Survey." *Korean Journal of Family Medicine*, 40: 241–247, 2019 (4 tables, 19 references)

Abstract: A web-based survey was conducted of 35,024 male and 32,839 female middle and high school students in Korea in 2015. The rate of EC-only users was 1.2% in male and 0.3% in female students. Dual users of ECs and cigarettes comprised of 5.1% of the male and 1.2% of the female students. The adjusted OR of being a problem alcohol drinker was 7.4× for vaping only, 15.5× in cigarette smokers, and 29.6× in dual users for male students.

Vaping is independently associated with alcohol use problems in Korean students, even those not currently smoking. Moreover, dual use of cigarettes and ECs is strongly associated with alcohol use problems. Therefore, vaping students should be concerned about their hidden alcohol use problems.

Reference Number: 50708

GONIEWICZ, M.L., D.M. SMITH, K.C. EDWARDS, B.C. BLOUNT, K.L. CALDWELL, J. FENG, L. WANG, C. CHRISTENSEN, B. AMBROSE, N. BOREK, D. VAN BEMMEL, K. KONKEL, G. ERIVES, C.A. STANTON, E. LAMBERT, H.L. KIMMEL, D. HATSUKAMI, S.S. HECHT, R.S. NIAURA, M. TRAVERS, C. LAWRENCE, AND A.J. HYLAND. "Comparison of Nicotine and Toxicant Exposure in Users of Electronic and Combustible Cigarettes." *JAMA Network Open*, 1: (16 pp), 2018 (2 tables, 3 figures, 51 references)

Abstract: Fifty biomarkers associated with exposure to tobacco were determined in the urine samples of 2,411 cigarette-only smokers, 247 EC-only users, 792 dual users, and 1,655 never-users (aged mainly between 35 and 54 years). The five major classes of biomarkers tested were nicotine, tobacco-specific nitrosamines (TSNAs), metals, polycyclic aromatic hydrocarbons (PAHs), and volatile organic compounds (VOCs). Compared with exclusive EC users, non-users had a 19–81% lower concentration of nicotine metabolites, TSNA, some metals, and some VOCs. Exclusive EC users showed a 10–98% lower concentration of TSNAs, PAHs, and some VOCs compared to cigarette smokers and similar concentrations of metals and three VOCs. The frequency of EC use was associated with urinary tobacco markers concentrations.

Table: Classes of Biomarkers of Tobacco Exposure and Suspected Diseases or Disorders

Biomarker Class	Suspected Diseases or Disorders
Nicotine/metabolite	Fetal brain and lung development in the fetus, brain development in adolescents, chronic degenerative diseases in the adult, and dependence, addiction, and withdrawal disorders
TSNA	Known carcinogen
Metals	Brain damage, behavioral problems, and hearing and vision problems are related to lead Cadmium is a known carcinogen with adverse effects to the kidneys and bone
PAHs	Carcinogen, respiratory irritation due to naphthalene, and acute myocardial infarction with pyrene
VOCs	Carcinogens and eye, nasal, and respiratory tract irritations

Source: Adapted from Goniewicz et al, 2018

Findings suggest exclusive e-cigarette use results in measurable exposure to tobacco-related constituents; however, compared with cigarette smoking, biomarker concentrations of nicotine and toxicants among e-cigarette–only users were much lower. Dual users exhibited higher concentrations of exposure to nearly all biomarkers compared with cigarette-only smokers. Further investigation can identify contributing factors resulting in higher toxicant concentrations among dual users. Continued daily consumption of tobacco cigarettes among dual users may play a role, as 82% of dual users reported daily cigarette smoking. Biomarker concentrations were positively correlated with cigarette smoking frequency.

Reference Number: 50709

PAREKH, T., S. PEMMASANI, AND R. DESAI. "Risk of Stroke with E-Cigarette and Combustible Cigarette Use in Young Adults." *American Journal of Preventive Medicine*, 53: 446–452, 2019 (1 table, 1 figure, 27 references)

Abstract: Data from the 2016–2017 Behavior Risk Factor Surveillance System and a telephone survey in 2019 were combined to determine the association between EC and risk of stroke in 161,529 participants (aged 18–44 years). There were 133,077 non-smokers, 13,318 current cigarette-only users (7.8%), 3,437 current EC-only users (2.7%), and 4,204 former smokers who were current EC users (2.6%). Most of the current EC-only users were 18–24 years of age and most cigarette-only smokers were 35–44 years of age.

Table: Adjusted Odds Ratio of Stroke According to Smoking Status/ EC Use

Smoking Status/EC Use	Adjusted OR of Stroke
Cigarette only	1.59×
EC only	(not statistically different from non-smokers)
Current EC user, former smoker	2.54×
Current dual user	2.91×

Source: Adapted from Parekh et al, 2019

Sole e-cigarette use is not associated with greater odds of stroke in young adults. However, if young adults have former or current combustible cigarette use, odds of stroke are significantly increased even compared with current sole combustible cigarette use. Switching from combustible cigarettes to e-cigarettes does not confer stroke benefits.

Reference Number: 50710

WANG, J.B., J.E. OLGIN, G. NAH, E. VITTINGHOFF, J.K. CATALDO, M.J. PLETCHER, AND G.M. MARCUS. "Cigarette and E-Cigarette Dual Use and Risk of Cardiopulmonary Symptoms in the Health eHeart Study." *Public Library of Science One*, 13: (14 pp), 2018 (2 tables, 4 figures, 49 references)

Abstract: Among the 39,747 participants of the US Health e-Heart Study, 573 (1.4%) reported EC use only, 1,693 (4.3%) reported cigarette use only and 514 (1.3%) reported dual use. Twenty-two percent of EC users were former smokers and 16% were current smokers. Among EC users, only 21% reported never smoking cigarettes ever. Dual users did not smoke fewer cigarettes than those who smoked cigarettes alone. Dual users were exposed to more toxicants on average.

Table: Percentage of Tobacco Product Users with Various Medical Conditions and Symptoms

Medical Condition	Neither Smoker nor EC User	EC Use Only	Cigarette Use Only	Dual
Breathing difficulty (typically)	1.8%	2.3%	2.5%	2.5%
Chest pains	32.9%	47.3%	52.0%	51.7%
Lost consciousness/syncope	14.1%	15.7%	15.0%	18.7%
COPD	3.7%	6.7%	11.0%	12.2%
Asthma	10.3%	16.5%	14.7%	17.9%
Heart attack	5.9%	5.7%	9.9%	11.2%

Source: Adapted from Wang et al, 2018

E-cigarette only use, compared to no product use, was associated with lower general health scores, higher breathing difficulty scores (typically and past month), and greater proportions of those who responded "yes" to having chest pain, palpitations, coronary heart disease, arrhythmia, COPD, and asthma. These data suggest the added use of e-cigarettes alone may have contributed to cardiopulmonary health risks particularly respiratory health risks.

Reference Number: 50711

HEDMAN, L., H. BACKMAN, C. STRIDSMAN, J.A. BOSSON, M. LUNDBÄCK, A. LINDBERG, E. RÖNMARK, AND L. EKERLJUNG. "Association of Electronic Cigarette Use with Smoking Habits, Demographic Factors, and Respiratory Symptoms." *JAMA Network Open*, 1: (12 pp), 2018 (3 tables, 2 figures, 39 references)

Abstract: Detailed surveys were conducted of 16,325 women and 23,947 men in northern and western Sweden regarding smoking and EC use. Overall, 2% used ECs, 12.3% were current cigarette smokers, and 24.4%

were former smokers. Among current smokers, 9.8% also used ECs, compared to 1.1% of former smokers. ECs use was 1.35× more frequent in men than women and current smokers were 18× as likely to use ECs. Adverse respiratory symptoms were most common in dual users.

Table: Adjusted OR of Respiratory Symptoms According to EC/ Cigarette Use

Smoking Status and E-cigarette Use	Adjusted OR of Reporting Respiratory Symptoms
Non-smoker (no EC use)	1.00×
Non-smoker (EC use)	1.46×
Former smoker (no EC use)	1.27×
Former smoker (EC use)	1.47×
Smoker (no EC use)	2.55×
Smoker (EC use)	4.03×

Source: Adapted from Hedman et al, 2018

Use of e-cigarettes was most common among smokers, and dual users had the highest prevalence of respiratory symptoms. On a population level, this study indicates that the present use of e-cigarettes does not adequately serve as a smoking cessation tool.

Reference Number: 50712

FETTERMAN, J.L., R.J. KEITH, J.N. PALMISANO, K.L. MCGLASSON, R.M. WEISBROD, S. MAJID, R. BASTIN, M.M. STATHOS, A.C. STOKES, R.M. ROBERTSON, A. BHATNAGAR, AND N.M. HAMBURG. "Alterations in Vascular Function Associated with the Use of Combustible and Electronic Cigarettes." *Journal of American Heart Association*, 9: e014570: (9 pp), 2020 (4 tables, 2 figures, 32 references)

Abstract: Cardiovascular disease (CVD) is the primary cause of morbidity and mortality in cigarette smokers. Non-invasive vascular function testing was conducted on 94 non-smokers, 285 cigarette smokers, 36 EC users, and 52 dual users. All EC users reported using second and third generation vaping devices.

Our study provides new insights into the effects of tobacco product use on measures of vascular function in young, healthy adult individuals without CVD risk factors. The most recent nationally representative sample shows that the majority of adult e-cigarette users are either former (36.5%) or current (22.1%) combustible cigarette smokers emphasizing the importance of understanding cardiovascular health measures in these groups. Significantly, we found that e-cigarette use either with or without combustible cigarette use was associated with a similar elevation of AIx, a measure of vascular stiffness, as compared with users of combustible cigarettes alone. Several measures of vascular function did not distinguish between tobacco product users and nonusers in young healthy adults. These data suggest that the abnormalities in vascular stiffness persist in e-cigarette users and that, at least within the limitations of our cohort and measurement approaches, there was no evidence that the use of e-cigarettes reduces cardiovascular injury, dysfunction, or harm associated with the use of combustible tobacco products.

5.08 MISCELLANEOUS

> "The temperatures obtainable within vaping devices places them in the category of a small-scale laboratory pyrolysis apparatus which if not used with precision and care can have unforeseen outcomes."
>
> —Wu and O'Shea, "Potential for Release of Pulmonary Toxic Ketene from Vaping Pyrolysis of Vitamin E Acetate" (2016)

A survey of web posts showed that 92% of people who discussed electronic cigarettes on a smoking cessation website used the term "e-cigarette" or derivations thereof, rather than the term "ENDS" or brand names such as "NJoy" and "blu" (50801). Rather than completely stop smoking, more and more cancer patients are beginning to use e-cigarettes, which they view as safer (50802). Instagram and other social media present positive images of vaping (50803).

ECs can be viewed as crude pyrolysis (heating) devices which, depending on the conditions used by the vaper, can produce extremely toxic ketene gas from vitamin E acetate (50804).

The inhalation patterns of vapers are different from those of cigarette smokers. Vapers inhaled for a longer period of time and tended to cluster their puffs together. After an *ad libitum* session, vapers were able to obtain as much nicotine as they would have obtained from three to four cigarettes (50805).

A review of e-cigarette use through the lens of traditional Jewish law, philosophy, and ethics determined that advertisements of JUULs were misleading and should be banned (50806).

Reference Number: 50801

PEARSON, J.L., M.S. AMATO, X. WANG, K. ZHAO, S. CHA, A.M. COHN, G.D. PAPANDONATOS, AND A.L. GRAHAM. "How US Smokers Refer to E-cigarettes: An Examination of User-Generated Posts from a Web-Based Smoking Cessation Intervention, 2008–2015." *Nicotine and Tobacco Research*, 19: 253–257, 2017 (1 table, 1 figure, 27 references)

Abstract: The terms used by US smokers to refer to electronic cigarette use was determined in a survey of online posts made by 1,023 participants (79% female, median age 47 years) on a smoke cessation website between 2008 to 2015. Ninety-two percent of posters used the term "e-cigarette" or a derivative thereof, while 23.2% used the term "vape." Terms such as "cig-a-like," "e-vape," and "electronic nicotine delivery systems (ENDS)" were not used at all. The term "vape" increased in popularity between 2013 and 2015.

> Researchers may have confidence that "e-cigarette," and, to a lesser extent, "vape" are recognizable terms among US adult smokers referring to ENDS (including accessories, brand names, and actions). Conversely, terms such as "electronic nicotine delivery systems," commonly employed by researchers and public health advocates, are not used by US smokers and may be an impediment to tobacco control research and practice.

Reference Number: 50802

BUCZEK, E.J., K.F. HARRINGTON, P.S. HENDRICKS, AND C.E. SCHMALBACH. "Electronic Cigarette Awareness, Use, and Perceptions Among Cancer Patients." *American Journal of Otolaryngology Open*, 2: (6 pp), 2018 (2 tables, 1 figure, 32 references)

Abstract: A detailed survey was conducted of 39 smoking cancer patients as to EC awareness, use, and knowledge. Tobacco cessation is strongly advocated, as continued smoking can lead to the recurrence of the cancer, poorer quality of life, wound-healing complications, and mortality. The most common source of EC initial exposure was from TV (77%), stores (49%), and friends (36%). Nearly half of the smoking cancer patients reported previous EC use. Compared to cigarette use, ECs were viewed as less likely to satisfy nicotine cravings, to relieve boredom, to have a calming effect, and to taste pleasant. ECs were also viewed as being safer than cigarettes and having less negative social impressions.

Cancer patients are a high-risk population portending worse health outcomes with continued cigarette use. E-cig advertising is pervasive. With the increasing popularity of e-cigs, a high proportion of cancer patients who smoke have used e-cigs because they are perceived as a less unhealthy alternative and more socially acceptable as compared with cigarettes. Additional research and cancer patient education are required because the short- and long-term health risks and benefits of e-cigarettes remain unknown.

Reference Number: 50803

CHU, K-H., J-P. ALLEM, T.B. CRUZ, AND J.B. UNGER. "Vaping on Instagram: Cloud Chasing, Hand Checks and Product Placement." *Tobacco Control*, 26: 575–578, 2016 (1 table, 1 figure, 31 references)

Abstract: A study was conducted of images of ECs on Instagram over a period of 3 months. Of the 2,208 images analyzed, most were advertisements (29%), followed by product images (28%) and activity (18%). Only 12% of other Instagram images were advertisements. Instagram provides marketers with more access to teens than other social media platforms, and this vulnerable population is highly exposed to EC advertising.

Table: Theme of 2,208 Instagram Images: Number, Likes, and Comments

Theme	Number of images (% of total)	Likes	Comments
Activity	18%	7%	0.170%
Product	28%	8%	0.200%
Advertisement	29%	3%	0.075%
Text	2%	4%	0.045%
Other	22%	4%	0.090%

Likes are defined as, for a single image, the number of likes divided by the number of followers of the poster. The percentage reported is the median for each given category. Comments are similarly defined.

Source: Adapted from Chu et al, 2016

Instagram offers a platform for posting pictures of vaping activities and custom vapourisers that can be seen by others online. The e-cigarette-related themes identified in this study could inform the design of communication campaigns that aim to counter social norms about e-cigarettes, and the development of future tobacco regulatory policies.

Reference Number: 50804

WU, D., AND D.F. O'SHEA. "Potential for Release of Pulmonary Toxic Ketene from Vaping Pyrolysis of Vitamin E Acetate." *Proceedings of the National Academy of Sciences of the United States of America*, 117: 6349–6355, 2020 (11 figures, 27 references)

Abstract: According to these authors, a vaping device should be considered as a crude uncontrollable minipyrolysis apparatus capable of causing change to the chemical composition of a vaping mixture in an unpredictable manner. The temperature of heating coils has been measured to be between 110–1,008°C and it can vary depending on the heating coil, resistance, voltage, composition of the vaping mixture, and mode of usage. Vitamin E acetate is a viscous lipid oil which has been added to e-liquids, usually those containing THC or CBD, as it mimics the appearance of high-quality cannabis oil. In experimental studies, the pyrolysis of vitamin E acetate can produce the extremely toxic ketene gas, which may be a contributing factor to the upsurge in pulmonary injuries such as EVALI. In addition, carcinogenic alkenes and benzenes were produced.

> The current worrying trend of increasing vaping-associated lung injuries is due to complex and multifaceted issues encompassing social, physical, biological, and medical sciences. When viewed from a physical science standpoint, medical complications from vaping a diverse set of substrates is perhaps not unexpected as the pyrolytic chemistry of single pure compounds is complex, so what occurs within ill-defined mixtures is a risky venture into the unknown. The temperatures obtainable within vaping devices places them in the category of a small-scale laboratory pyrolysis apparatus which if not used with precision and care can have unforeseen outcomes.

Reference Number: 50805

ST. HELEN, G., K.C. ROSS, D.A. DEMPSEY, C.M. HAVEL, P. JACOB III, AND N.L. BENOWITZ. "Nicotine Delivery and Vaping Behavior During *Ad Libitum* E-Cigarette Access." *Tobacco Regulatory Science*, 2: 363–376, 2016 (4 tables, 2 figures, 42 references)

Abstract: Six female and seven male EC users (aged 19–69 years) were videotaped for 90 minutes while vaping their usual EC. Plasma nicotine concentrations were determined every 15 minutes by GC/MS/MS. Two subjects used cig-a-likes, eight used tanks, and three used RBAs (rebuildable atomizer modules). Subjective effects were also measured. The mean nicotine concentration of the e-liquids used was 9.4 μg/mg. The average number of puffs taken was 64. The average puff duration was 3.5 seconds and the average interval between puffs was 118 seconds. On average, the subjects took a clustered puff 5.2 times. The short, medium, and long clustered puffs were taken 8.6, 2.4, and 0.8× respectively. The average puff duration was longer than for conventional cigarettes. On average, 4 mg of nicotine was delivered to the subjects (similar to a typical cigarette) and the average plasma nicotine concentration was 12.8 ng/mL. There was a gradual increase in plasma nicotine concentration, which peaked at the end of the session, unlike the rapid (almost bolus) peak concentration obtained by smoking cigarettes.

Table: Severity of Withdrawal Symptom Pre- and Post-E-cigarette Use

Withdrawal Symptoms (0, none to 4, severe)	Pre-EC Use	Post-EC Use
Angry, irritable, frustrated	1.2	0.1
Depressed mood, sad	0.0	0.1
Desire or craving to smoke	2.4	0.6
Impatient	1.4	0.8
Dizziness	0.1	0.2
Sore throat	0.1	0.2

Source: Adapted from St. Helen et al, 2016

In conclusion, e-cigarette vaping patterns differed from typical patterns of smoking tobacco cigarettes; most puffs were clustered in groups of 2 to 5 puffs. Patterns of use and vaping behavior variables versus nicotine concentration-time curves also differed from those described with conventional tobacco cigarette smoking. During 90 minutes of ad libitum access, experienced e-cigarette users took in, on average, as much nicotine as 3–4 typical tobacco cigarettes. The average plasma nicotine levels were similar to reported nicotine boosts from smoking one tobacco cigarette, consistent with the intermittent dosing of nicotine from e-cigarettes compared to the more nearly bolus dosing from a tobacco cigarette, which may have implications for e-cigarette abuse liability.

Reference Number: 50806

GROSSMAN, S.G. "Vape, Gods and Judaism—E-cigarettes and Jewish Law." *Rambam Maimonides Medical Journal*, 10: (12 pp), 2019 (86 references)

Abstract: The Halakha, or Jewish law, is the traditional Jewish ethical code. An analysis was conducted as to whether vaping nicotine would be prohibited under Jewish law. Maimonides, one of the most influential Torah scholars of the Middle Ages, states that it is forbidden to mislead people in business, even with words only. A JUUL advertisement that appeared throughout the city of Ra'annana, Israel, stated that JUULs do not actually contain tobacco leaves, misleading consumers into thinking that JUUL is nicotine-free. This is a deception with words, as prohibited by Maimonides.

> E-cigarettes are a relatively new product whose long-term risks we do not yet fully understand. Nevertheless, their popularity, particularly among youth, is undeniable. Although they might be less harmful than combustible cigarettes, e-cigarettes are certainly not harmless, especially for pregnant women as nicotine exposure can harm the fetus, and youth, who are at high risk for nicotine addiction and for whom they all too easily lead to the use of combustible cigarettes. Few modern poskim have weighed in on the permissibility or halakhic issues surrounding e-cigarette use. However, extrapolations from halakhic discussions regarding the use of combustible cigarettes suggest that e-cigarettes should be forbidden, in particular for pregnant women and young people, populations for whom the dangers are amplified and to whom governments have issued warnings against cigarette use. We are obligated as a society to educate the public regarding the dangers of e-cigarettes, to prevent children from obtaining e-cigarettes, and to halt the false advertising and perhaps even the sale of e-cigarettes, especially to youth.

CHAPTER 6

Other Nicotine Delivery Systems

> "The tobacco industry's allegations that approval of HNB products for marketing would improve public health by reducing tobacco-associated deaths and diseases is at best an untested and self-serving hypothesis."
>
> —Paumgartten, "Heat-Not-Burn and Electronic Cigarettes: Truths and Untruths About Harm Reduction" (2018)

Like that mythical beast, the phoenix, the tobacco industry seems to arise again and again from the ashes to promote its nicotine drug delivery systems in newer and different products, all of which promise (as did filter and low tar cigarettes) a lower health risk. The underlying message is that you don't have to quit smoking to reduce your health risks, just switch to another tobacco product.

6.01 HEATED TOBACCO PRODUCTS (HEAT NOT BURN (IQOS))

> "Like cigarettes (and e-cigarettes), IQOS uses an aerosol of ultrafine particles to deliver the nicotine. These ultrafine particles cause heart and lung disease."
>
> —Glantz, "PMI's Own in Vivo Clinical Data on Biomarkers of Potential Harm in Americans Show that IQOS Is not Detectably Different from Conventional Cigarettes" (2018)

Heated tobacco products (HTP) are also known by the tobacco industry as Heat not Burn (HnB) products, but I would prefer the more appropriate name "Char not Burn," as the most popular HTP, IQOS (I Quit Ordinary Smoking), produces a large, charred butt after use (60113). The origin of heated tobacco products can be traced back to the 1960s with project Ariel (60101), and in the 1990s to Accord and Oasis (60102).

IQOS consists of a disposable reconstituted tobacco stick (Heet stick) soaked in propylene glycol and VG and inserted into a computerized heating coil that heats it to 350°C, a higher temperature than ECs, but a lower temperature than combustion cigarettes (60103, 60108). The battery only allows for a 6-minute operation time which means that a puff must be taken every 25 seconds (60112).

The used Heet stick leaves a large amount of charred debris, fluid, and fragments behind, which is required to be cleaned. Increased charring occurs if the IQOS holder is not cleaned, and more toxic chemicals are released as a result (60112).

IQOS was introduced to Japan in 2014 and its use increased nearly fourfold after being shown in a popular entertainment Japanese TV show in 2016 (60106, 60108). As usual, the tobacco industry sponsored studies of which 70% showed alleged health benefits of IQOS (60103).

IQOS was found to emit the same toxic chemicals as ordinary cigarettes, such as tobacco-specific nitrosamines, formaldehyde, acetaldehyde, acrolein, CO, and nicotine (60105, 60108, 60110–12). Philip Morris International submitted a Modified Risk Tobacco Product application to the FDA to show the IQOS had less health risk than cigarettes, but it was rejected (60113–16).

It should also be noted that toxicant reduction does not necessarily reduce health risk. Although the levels of tar have decreased by two-thirds since the 1950s, there has not been a corresponding reduction in lung cancer (60104).

Reference Number: 60101

RISI, S. "On the Origins of the Electronic Cigarette: British American Tobacco's Project Ariel (1962–1967)." *American Journal of Public Health*, 107: 1060–1067, 2017 (1 figure, 71 references)

Abstract: The origins of heated tobacco products can be traced to the BAT project Ariel (1962–1967) lead by Sir Charles Ellis, which tried to reduce the harm of cigarettes by heating and not combusting tobacco. Before the project, it was found that filters (some made of asbestos) were ineffectual in a burning cigarette. Cigarettes usually burn at 700–800°C, which is much hotter than the 250–300°C required to release nicotine from tobacco. The Ariel project developed a cigarette within a cigarette device in which the inner tobacco tube was wrapped in aluminum foil surrounded by an outer cylinder wrapped with tobacco on the outside. The burning of tobacco on the outside would release nicotine from the inside and allow the inhalation of a nicotine vapor without the other combustion chemicals, such as tar. The initial pH of the aerosol of the original Ariel device was 7.8, which was highly irritating compared to the 5.0–6.0 of regular cigarettes, and so citric acid was added to the tobacco to lower the pH to 4.4.

> Ariel marks a turning point for the tobacco industry. Launched by the discovery that nicotine is addictive, the Ariel device is the physical embodiment of the moment when a major tobacco company first grappled with the notion that it might, after all, be in the business of selling nicotine rather than tobacco or cigarettes.

Reference Number: 60102

ELIAS, J., L.M. DUTRA, G. ST. HELEN, AND P.M. LING. "Revolution or Redux? Assessing IQOS Through a Precursor Product." *Tobacco Control*, 27: s102–s110, 2018 (3 tables, 2 figures, 81 references)

Abstract: The Accord, a clean smoke battery-operated HnB tobacco product introduced between 1998 and 2006, was a precursor to IQOS. An analysis was conducted of previously secret Philip Morris (PM) and Philip Morris International (PMI) internal documents to compare the IQOS to the Accord. In October 1998, PM first marketed the Accord in about 20 stores near its main manufacturing/research center in Virginia. It released the same HTP in Osaka, Japan under the name of Oasis. The nicotine uptake for the Accord was slower than for cigarettes, so smokers inhaled more deeply to compensate, which negated potential health benefits. After spending over $400 million, the Accord and Oasis were withdrawn in 2006, because smokers did not like its taste, harshness, and inconvenience, or the difficulty in drawing in the smoke. In 2014, PMI introduced the IQOS to Italy and Japan. In January 2018, the FDA's Tobacco Products Scientific Advisory Committee concluded that PMI had not proven IQOS to be safer than cigarettes.

Table: Comparison of Accord and IQOS

	Accord	IQOS
Tobacco material	Tobacco, glycerol, water, ammonium magnesium phosphate, pectin, betacyclodextrin	Tobacco, glycerol, water, guar gum, cellulose, propylene glycol
Tar (mg/cigarette)	3.0	19.4
Nicotine (mg/cigarette)	0.2	1.29
Tube	Hollow acetate tube	Hollow acetate tube
Peak temperature (°F)	500	310–350

Source: Adapted from Elias et al, 2018

While Accord was a commercial failure, IQOS is situated in a distinct social and regulatory landscape, which may increase its chances of success. The decline of cigarette consumption in developed markets, the increased popularity of e-cigarettes, select public health endorsements of e-cigarettes and the creation of a legal mechanism in the USA through which manufacturers can now make reduced risk claims may all have increased both consumers' willingness to try new tobacco products, and PMI's willingness to designate new products as safer than conventional cigarettes. In 2017, PMI announced the establishment of a US$1billion foundation dedicated to partnering with public health and promoting PMI's portfolio of reduced harm products, chiefly IQOS.

Despite PMI's claims of IQOS's novelty, the product appears to be a successor of Accord on the basis of product design, marketing and aerosol chemistry. Both products work by inserting a modified cigarette into a holder that heats, rather than burns tobacco. Both products' marketing implies reduced harm relative to cigarettes. While PMI claims that IQOS is reduced risk largely because it reduces users' exposure to harmful constituents, IQOS does not drastically improve users' exposure to toxic compounds than relative to Accord. PMI's claims for IQOS's comparative safety on the basis of reduced exposure are further undermined by internal documents from PMI's parent company: as late as 2007, PM executives and scientists claimed that reductions in exposure did not render Accord safer than conventional cigarettes.

Reference Number: 60103

JENSSEN, B.P., S.C. WALLEY, AND S.A. MCGRATH-MORROW. "Heat-not-Burn Tobacco Products: Tobacco Industry Claims No Substitute for Science." *Pediatrics*, 141: (3 pp), 2018 (11 references)

Abstract: Philip Morris International (PMI) created and heavily marketed a Heat-not-Burn tobacco product called IQOS, which consists of a disposable tobacco stick soaked in propylene glycol and inserted into a holder with a computerized heating coil that heats the tobacco to approximately 350°C but does not burn it, as cigarettes do. HnB products were introduced to Japan in 2014 and Italy and Switzerland in 2015. The tobacco industry published 19 out of 27 articles about the health benefits of HnB. There was also a seven-part series in Regulatory Toxicology and Pharmacology, a journal financed in part by the tobacco industry with a concealed proindustry bias. The editor-in-chief of the journal was paid $30,000 by the tobacco industry to publish an article dismissing the health risks of secondhand smoke.

The authors of independent studies have identified cancer-causing chemicals in the smoke emitted by HNB tobacco products. In 1 study, the authors compared the contents of the IQOS with the contents of conventional cigarettes by using a smoking device designed to capture the mainstream aerosol and developed to meet the standards for common cigarettes. The smoke released by the IQOS contains the same harmful constituents of cigarette smoke. These elements include volatile organic compounds near the levels found in cigarette smoke, polycyclic aromatic hydrocarbons at wide comparative ranges, and carbon monoxide. All of these elements, on the basis of extensive research of cigarette smoke, have been shown to cause serious harms to human health. Additionally, the IQOS smoke had nicotine levels similar to levels in conventional cigarette smoke. Nicotine, the highly addictive substance found in tobacco, is unsafe to youth and harms infant and adolescent brain development, as has been summarized in a recent report from the US Surgeon General.

Reference Number: 60104

MALLOCK, N., E. PIEPER, C. HUTZLER, F. HENKLER-STEPHANI, AND A. LUCH. "Heated Tobacco Products: A Review of Current Knowledge and Initial Assessments." *Frontiers in Public Health*, 7: (8 pp), 2019 (2 figures, 90 references)

Abstract: Tobacco is burned at the tip of a cigarette at a temperature between 700 and 950°C, which decreases along the length of the cigarette in the distillation zone at between 200–600°C and to a condensation and filtration zone. Three heated tobacco products (HTP) heat the tobacco to temperatures between 250 and 350°C. Aerosols of HTPs contain a high proportion of water. Toxicant reduction is not necessarily linked to reduced health risk. Even though the levels of tar have decreased by two-thirds in cigarettes since the 1950s, there has been no corresponding decrease in lung cancers.

Although a 99% reduction of some major carcinogens is expected to affect health risks, the magnitude or relevance of such putative reduction is not yet clear. A benefit is likely seen for especially the subset of long-term smokers that are unable to quit or to switch to another nicotine source with less HPHC exposure. However, referring back to the tumor potency models, it should be kept in mind that substantial and relevant health risks are still present. Consequently, HTPs should not be the first option to decrease smoking-associated harm.

Reference Number: 60105

LEIGH, N.J., M.N. PALUMBO, A.M. MARINO, R.J. O'CONNOR, AND M.L. GONIEWICZ. "Tobacco-Specific Nitrosamines (TSNA) in Heated Tobacco Product IQOS." *Tobacco Control*, 27: s37–s38, 2018 (1 figure)

Abstract: HTPs heat processed tobacco to 350°C to release volatile chemicals that are not detectable in typical ECs. Tobacco-specific nitrosamines (TSNAs) are not generated by burning but are formed in the curing of tobacco. Four different TSNAs were detected in aerosols generated by IQOS, Mark Ten ECs, and Marlboro Red cigarettes by LC/MS/MS.

Like combustible products, HTPs emit substantial levels of carcinogenic TSNA. Although HTP emits lower amounts of TSNA than combustible cigarettes, the amounts are significantly higher than from e-cigarettes. Our findings are consistent with prior reports. One limitation of this study is that one puffing protocol was used for all devices. While this was helpful in comparing TSNA and nicotine delivery between devices, machine-based measurements are not represented of human smoking patterns or constitute intake. The tested HTP does not reduce emissions of an important class of tobacco carcinogens to the same degree as other commercially available technologies.

Reference Number: 60106

TABUCHI, T., S. GALLUS, T. SHINOZAKI, T. NAKAYA, N. KUNUGITA, AND B. COLWELL. "Heat-not-Burn Tobacco Product Use in Japan: Its Prevalence, Predictors and Perceived Symptoms from Exposure to Secondhand Heat-not-Burn Tobacco Aerosol." *Tobacco Control*, 27: e25–e33, 2018 (4 tables, 2 figures, 34 references)

Abstract: The increased use of IQOS in Japan occurred as a result of the HnB device being demonstrated on a popular Japanese entertainment TV show (Ame-talk) in April 2016. IQOS use was nearly 4× higher in viewers who saw the show compared than in non-viewers (10.3% vs 2.7%). In a survey, approximately 977 persons were exposed to secondhand smoke from IQOS. Approximately 37% experienced symptoms in reaction to SHS, including sore throat, eye pain, and feeling ill.

Table: Symptoms Experienced by Those Exposed to SHS of HnB Products

Symptom	Percentage
Sore throat	20.6%
Eye pain	22.3%
Feeling ill	25.1%
Other injury or symptom	13.4%
Any symptom	37.0%

Source: Adapted from Tabuchi et al, 2018

A 1-hour entertainment TV programme triggered IQOS diffusion in Japan. In 2017, it was estimated that 3.6% of people currently used IQOS, while 4.7% currently used any HNB or e-cigarette product and 3.4% were dual users. If these estimates are extrapolated straightforwardly to the 86 million Japanese adults aged 17–71 years, approximately 3.1 million currently use IQOS, 4.0 million currently use any HNB or e-cigarette product and 2.9 million are dual users. Tobacco control organisations and governments should continue to monitor HNB tobacco and consider how to regulate it, given its impending, likely rapid global diffusion.

Reference Number: 60107

PAUMGARTTEN, F. "Heat-not-Burn and Electronic Cigarettes: Truths and Untruths About Harm Reduction [Letter]." *Revista da Associação Médica Brasileira*, 64: 104–105, 2018 (6 references)

Abstract: HnB tobacco products heat the tobacco to about 350°C instead of burning it at about 800°C. HnB devices aerosolize nicotine and deliver it efficiently to the bloodstream but do not produce as much CO or generate as many carcinogens. But BNH products are not safe. Nicotine addiction is a chronic and relapsing illness and is a major health problem.

> A possible health benefit for a particular group of smokers, however, does not justify approval of HNB tobacco products for (unrestricted) sales. The tobacco industry's allegations that approval of HNB products for marketing would improve public health by reducing tobacco-associated deaths and diseases is at best an untested and self-serving hypothesis.

Reference Number: 60108

AUER, R., N. CONCHA-LOZANO, I. JACOT-SADOWSKI, J. CORNUZ, AND A. BERTHET. "Heat-not-Burn Tobacco Cigarettes: Smoke by Any Other Name." *JAMA Internal Medicine*, 177: 1050-1052, 2017 (1 table, 6 references)

Abstract: The tobacco industry's response to calls to reduce the harms of cigarette smoking was to launch new HnB tobacco cigarettes. PMI created IQOS, which consisted of disposable tobacco sticks soaked in propylene glycol, inserted into a holder, and pierced with an electric blade that is heated to 350°C, compared to 600–700°C for cigarettes and 200–250°C for vaping. Pilot programs in 2014 in Japan and in 2015 in Switzerland suggest that persons aged 15–39 years are more likely to use IQOS. The harmful ingredients of cigarette smoke are due to incomplete combustion (pyrolysis) and thermogenic degradation. Complete combustion only occurs at temperatures of >1,300°C. The smoke from IQOS was compared to that from a cigarette. The measured temperature was 330°C for IQOS and 684°C for cigarettes. IQOS smoke delivers 84% of the nicotine as cigarettes. Carbon monoxide, formaldehyde, acetaldehyde, and acrolein were detected in IQOS smoke.

> In contrast, our analyses reveal that advertising slogans such as "heat-not-burn" are no substitute for science. Dancing around the definition of smoke to avoid indoor-smoking bans is unethical.

Reference Number: 60109

LEE, A., S.Y. LEE, AND K-S. LEE. "The Use of Heated Tobacco Products is Associated with Asthma, Allergic Rhinitis, and Atopic Dermatitis in

Korean Adolescents." *Scientific Reports*, 9: 17699 (8 pp), 2019 (5 tables, 1 figure, 40 references)

Abstract: Allergic diseases such as asthma, allergic rhinitis (AR), and atopic dermatitis (AD) are serious public health concerns, and between 2009 and 2018 they increased in Korean high school students from 7.4 to 8.9%; 29.6 to 38.7%; and 18.2 to 26.1%, respectively. Heated tobacco products heat a processed tobacco leaf, not a liquid, at a temperature of 350°C. The adjusted OR for users of HTP compared to non-users was 1.78× for asthma, 1.21× for AR, and 1.58× for AD.

> In conclusion, our study findings indicated that the use of multiple tobacco products, including HTPs, might be a risk factor for asthma, allergic rhinitis, and atopic dermatitis. Additionally, the consistent use of each tobacco product we examined was linked to multi-morbidity of asthma, allergic rhinitis, and atopic dermatitis. An initial diagnosis of atopic dermatitis can lead to a subsequent systemic immune response. Healthcare professionals should pay close attention to smoking in adolescents because it is a potential contributor to the development of allergic diseases.

Reference Number: 60110

LEIGH, N.J., P.L. TRAN, R.J. O'CONNOR, AND M.L. GONIEWICZ. "Cytotoxic Effects of Heated Tobacco Products (HTP) on Human Bronchial Epithelial Cells." *Tobacco Control*, 27: s26–s29, 2018 (1 figure, 19 references)

Abstract: Early models of the HTP were developed in the 1980s but were withdrawn as they were not popular with smokers. A newer version of HTP—the IQOS—was launched into the international market in mid-2014. The inhalational toxicity of HTP (IQOS, tobacco flavor), EC (Mark Ten tobacco flavor), and cigarettes (Marlboro Red) were determined on human bronchial epithelial cells (H292). The cells were exposed to 55 puffs from the EC, 12 puffs from HTP, and 8 puffs from cigarettes to equalize nicotine delivery.

Table: The Effect of Exposure to Various Tobacco Products on the Percentage of Metabolically Active Human Bronchial Cells

Tobacco Product	Percent Metabolically Active Cells After Exposure
Air (Control)	100%
Mark Ten E-Cigarette	90%

Tobacco Product	Percent Metabolically Active Cells After Exposure
IQOS	75%
Marlboro Red Cigarette	60%

Source: Adapted from Leigh et al, 2018

Using limited cytotoxic measures, the HTP showed reduced cytotoxicity relative to a combustible cigarette but higher toxicity than an e-cigarette. More comprehensive testing is needed to determine long-term effects of inhaling emissions from HTP.

Reference Number: 60111

SALMAN, R., S. TALIH, R. EL-HAGE, C. HADDAD, N. KAROGHLANIAN, A. EL HELLANI, N.A. SALIBA, AND A. SHIHADEH. "Free-Base and Total Nicotine, Reactive Oxygen Species, and Carbonyl Emissions from IQOS, A Heated Tobacco Product." *Nicotine and Tobacco Research*, 21: 1285–1288, 2019 (2 tables, 22 references)

Abstract: The chemical compounds from the aerosol from a heated tobacco product, I Quit Ordinary Smoking (IQOS), was determined using a custom-made puffing machine. IQOS uses small cigarettes called Heets (reconstituted tobacco, paper, filter, propylene glycol, and vegetable glycerin) which is inserted into a protruding heating blade that heats the tobacco to 330°C. The IQOS heater remains active for 6 minutes after which the battery needs to be recharged. After each sampling session, the IQOS heater was cleaned using the cleaning sticks provided. The PG/VG aerosol was 8/92. The chemical constituents of the IQOS aerosol as determined by HPLC, GC, and fluorescence were compared to that from a Marlboro Red cigarette.

Table: Mean Aerosol Toxicant Content of IQOS compared to Marlboro Red

Aerosol Toxicant	IQOS	Marlboro Red
Total particulate matter (mg/cigarette)	12.93	10.03
Nicotine (mg/cigarette)	0.77	0.80
% Free Base	13.6%	14.5%
pH	6.66	6.11
Formaldehyde (µg/cigarette)	0.85	3.17
Acetaldehyde (µg/cigarette)	301.46	1,059

Aerosol Toxicant	IQOS	Marlboro Red
Valeraldehyde (μg/cigarette)	20.11	0
Glyoxal (μg/cigarette)	3.11	0
Methyl glyoxal (μg/cigarette)	33.51	0

Source: Adapted from Salman et al, 2019

Daily inhalation exposure was calculated assuming a mean daily inhalation volume of 16 m^3/day. We found that pack-a-day equivalent IQOS users would roughly double their daily formaldehyde and total ROS inhalation exposures, respectively, relative to breathing urban air alone. Exposure to acetaldehyde would be more than 100 times greater than from breathing urban air. On the other hand, obtaining a pack-a-day nicotine intake equivalent from IQOS rather than combustible cigarettes under the conditions studied would result in a reduction in daily intake of formaldehyde and acetaldehyde of 70% and 65%, respectively, and an 85% reduction in ROS. How differential exposure of these toxicants translates to differential population morbidity and mortality risk is difficult to estimate; indeed risk assessment for novel modified risk tobacco products is an active area of research and concern of regulatory bodies such as the US Food and Drug Administration.

Reference Number: 60112

DAVIS, B., M. WILLIAMS, AND P. TALBOT. "IQOS: Evidence of Pyrolysis and Release of a Toxicant from Plastic." *Tobacco Control*, 28: 34–41, 2019 (4 figures, 31 references)

Abstract: IQOS operates at a temperature below 350°C, but the pyrolysis of tobacco occurs at temperatures between 200 and 600°C. A tobacco plug (or heet stick) is inserted into the battery powered heater. The IQOS only operates for 6 minutes before it shuts off. Since a maximum of 14 puffs can be taken with each IQOS heet stick, a puff must be taken every 25 seconds. Heet sticks that have not been fully smoked cannot be reinserted back into the IQOS, because they crumble. It is recommended that the cleaning function of the charger be used, followed by cleaning with brushes after use of 20 heet sticks and that large remaining fragments of the tobacco plug remaining be removed with the Hook. In this study, use of the heet sticks left a significant number of debris, fluid, and fragments of cast leaf behind. Although there were no flames, the heet sticks are always charred after use. Charring increased if the holder was not cleaned. The polymer film filter is intended to cool the aerosol, and although there was no direct contact with the heater, the filter melted during use, releasing very toxic formaldehyde cyanohydrin. Using the typical ISO 3308 machine smoking protocols, IQOS could not be tested for more than 6 puffs.

> In conclusion, the iQOS appears to be well manufactured, and performance data were consistent between heatsticks. However, the product has limitations that will affect user topography and the application of standard smoking protocols, such as the ISO 3308, which could not be used for more than six puffs with this product. Users may be forced to smoke at a rapid pace in order to fully maximize heatsticks. Decreasing the interpuff interval could lead to an increase in intake of nicotine 30 and carbonyls 31. This study also showed that the iQOS is not strictly a "heat-not-burn" tobacco product. The iQOS tobacco appeared to char without ignition, and charring increased when cleaning was not done after each use. This study also showed the potential dangers that the polymer-film filter poses. This thin plastic sheet readily melts during iQOS use and releases formaldehyde cyanohydrin, a dangerous toxicant. This study has shown that the iQOS system may not be as harm-free as claimed and also emphasizes the urgent need for further safety testing as the popularity and user base of this product is growing rapidly.

Reference Number: 60113

ST. HELEN, G., P. JACOB III, N. NARDONE, AND N.L. BENOWITZ. "IQOS: Examination of Philip Morris International's Claim of Reduced Exposure." *Tobacco Control*, 27: s30–s36, 2018 (1 table, 45 references)

Abstract: A critical review was conducted of Philip Morris International's (PMI) Modified Risk Tobacco Product (MRTP) application to the FDA showing that IQOS had lower exposures and health risks than cigarettes. PMI reported levels for only 40 of the 93 Harmful and Potentially Harmful Constituents (HPHCs), and all had lower levels than the mainstream smoke of a standard cigarette. However, 22 other HPHCs not included by PMI were >200% higher and 7 were >1,000% higher in IQOS than cigarettes.

> PMI's MRTP application fails to address the important question of whether the aerosol generation process for IQOS produces toxic substances not found in the smoke of combustible cigarettes, which could have been

answered through non-targeted chemical analysis. Combustible tobacco cigarettes reach about 900°C during a puff and smoulder at about 400°C between puffs. The burning process, substances emitted and their levels vary at different temperatures. Distillation, the process during which nicotine and aromas are transferred from tobacco to smoke, occurs below 300°C; pyrolysis occurs at about 300°C–700°C, entails the decomposition of biopolymers, proteins, and other organic materials and generates the majority of substances emitted in smoke; and combustion occurs above 750°C and results in the generation of carbon dioxide, CO and water. HeatSticks are heated to a maximum of 350°C, a temperature sufficient to enable pyrolytic decomposition of some organic materials. Formation of toxic volatile organic compounds, including formaldehyde, acetaldehyde and acrolein, via dehydration and oxidation of the humectants, propylene glycol and glycerin, have been reported in e-cigarette aerosols at similar temperatures as IQOS. In addition, flavouring chemicals in e-cigarettes undergo thermal degradation and contribute significantly to levels of toxic aldehydes emitted in e-cigarette aerosol. Since the constituents of HeatSticks may be different from that of combustible cigarettes, including flavourants and additives, it is plausible that the IQOS aerosol may contain substances not present in tobacco smoke.

Reference Number: 60114

GLANTZ, S.A. "PMO's Own In Vivo Clinical Data on Biomarkers of Potential Harm in Americans Show that IQOS is not Detectably Different from Conventional Cigarettes." *Tobacco Control*, 27: s9–s12, 2018 (1 table, 21 references)

Abstract: All of the major tobacco companies have developed HTPs that generate nicotine aerosols by heating a stick of ground tobacco with added chemicals (such as propylene glycol) without setting the tobacco on fire. PMI is promoting the HTP IQOS as less dangerous cigarettes and applied to the FDA for it to be classified as a MRTP. PMI submitted a 3-month study on 24 biomarkers of potential harm in 47 IQOS users, 32 cigarette smokers, and 9 abstainers. Some of the biomarkers were blood pressure, lung function, and blood total cholesterol and fibrinogen levels. There were no statistically detectable differences between the IQOS and cigarette smokers for 23 of the 24 biomarkers of potential harm. The FDA determined that PMI had not shown a reduced risk of harm for IQOS.

Like cigarettes (and e-cigarettes), IQOS uses an aerosol of ultrafine particles to deliver the nicotine. These ultrafine particles cause heart and lung disease. The adverse health effects of these particles and many of the other toxins do not drop in proportion to reducing the dose, so even low levels of exposure can be dangerous. This effect is why smoke-free environment laws are followed by big drops in heart attacks and other diseases despite the fact that secondhand smokers breathe in much less smoke than the smokers. In addition, while the IQOS does not set the tobacco stick on fire, it heats it to 350°C (660°F), which is still hot enough to cause pyrolysis. There is already independent evidence that IQOS compromises functioning of arteries, a key risk factor for heart disease and heart attacks, as badly as a cigarette.

The clinical studies that PMI reported appropriately did not include cancer because carcinogenic effects take much longer to be manifest than cardiovascular and pulmonary effects. Even if the levels of carcinogens delivered by IQOS are lower than conventional cigarettes on a per-puff basis, these lower exposure levels may not yield proportionately lower cancer risks because both the intensity and duration of exposure impact cancer risk.

Reference Number: 60115

MCKELVEY, K., L. POPOVA, M. KIM, L.K. LEMPERT, B.W. CHAFFEE, M. VIJAYARAGHAVAN, P. LING, AND B. HALPERN-FELSHER. "IQOS Labelling Will Mislead Consumers." *Tobacco Control*, 27: s48–s54, 2018 (6 tables, 31 references)

Abstract: The authors conducted a critical review of Philip Morris International's 2016 application to the US Food and Drug Agency for IQOS as an MRPT, with particular focus on the claim that switching completely to IQOS reduces risk. The review concluded that the evidence was deficient, the measurement instruments employed by PMI suffered from design flaws, and that the reporting of associated findings was misleading.

The claim that switching completely to IQOS could reduce harm and tobacco-related diseases assumes that people who attempt to switch will be successful at cigarette smoking cessation. However, available data suggest that cigarette smokers who try to switch from cigarettes to other tobacco products or who use other tobacco products for smoking cessation are more likely to be nicotine-dependent and experience difficulty with smoking cessation compared with people who do not use these alternative tobacco products. Further, PMI ignores evidence that smokers who use novel tobacco products such as e-cigarettes often use two or more tobacco products in combination instead of switching entirely. Indeed, PMI's own data on IQOS show substantial levels of combined use in their test populations, and epidemiological evidence demonstrates that for other non-cigarette tobacco products, switching completely is not the most common outcome. Among US adults who use electronic cigarettes, 75% to 82% use e-cigarettes in combination with at least one other form of combustible tobacco, and only 20% of e-cigarette users report switching completely from combustible cigarettes. Finally, PMI's proposed warnings do not specifically inform consumers that continuing to smoke while using IQOS could reduce the likelihood of quitting smoking. As such, PMI's data do not support their MRTP claim, and instead both data presented by PMI and in the literature base support the idea that introducing any HTP product (including IQOS) will likely be harmful to population health.

Reference Number: 60116

PAUMGARTTEN, F.J.R. "A Critical Appraisal of the Harm Reduction Argument for Heat-not-Burn Tobacco Products." *Pan American Journal of Public Health*, 42: e161 (6 pp), 2018 (23 references)

Abstract: A critical appraisal was conducted of the tobacco industry's argument that HnB tobacco products such as IQOS reduce the harm and assist in the cessation of cigarette smoking. Smoking cessation is the best way to prevent or reduce the risk of disease, but smokers attempting to entirely quit tobacco may give up and instead switch to HnB or EC products. In addition, HnBs do char and charring increases if the devices are not cleaned after heating the tobacco sticks. PMI recommends cleaning after use of 20 tobacco sticks, but it has been shown that IQOS should be cleaned after each tobacco stick. Some pyrolysis of tobacco constituents may occur at temperatures less than 350°C. The melting of the polymer film filter (at 90°C) also releases formaldehyde cyanohydrin.

Some of the industry arguments for the health benefits of alternative nicotine delivery products may be valid, while others are misleading. Even if HNBs were in fact less harmful than conventional cigarettes, they would still have the potential to cause nicotine dependence and other adverse effects such as damaging the cardiovascular system and the unborn child. The tobacco industry claims that HNBs are safer than conventional cigarettes, but long-term RCTs and/or observational studies are lacking to corroborate this assertion. It is plausible to think that alternative nicotine delivery products such as ECs and HNBs enhance smoking appeal and initiation among young persons. Moreover, HNBs are also likely to discourage quit attempts. It is reasonable to expect that if HNBs and ECs break into the market they will increase the prevalence of smoking. Owing to this possibility and health risks posed by nicotine and other vapor constituents, HNBs should not be made freely available on the market. HNBs, however, might be a harm reduction approach for a group of smokers who are unable to quit or are unwilling to do so. The sale of HNBs for use under medical supervision (prescription only), with strict restrictions on product advertising, is a balanced regulatory alternative that would reconcile the harm reduction needs of nicotine-addicted patients with the collective health goal of achieving the first smoking-free generation in the near future.

6.02 CIGARS/CIGARILLOS/LITTLE CIGARS

"Therefore, since cigar smoke contains higher concentrations of toxins and cancerous substances than does cigarette smoke, it also contributes to increasing the risk of lung cancer and other smoking-related diseases in non-smokers (passive smokers)."

—De Assis Viegas, "Noncigarette Forms of Tobacco Use" (2008)

Cigars and pipes were smoked well before cigarettes were invented (60204). A cigar is defined as any roll of tobacco wrapped in leaf tobacco that does not contain a filter (60201). Little cigar sales soared after the 1964 Surgeon General's report that addressed the health risks of cigarettes and appeared to suggest that cigar and little cigar smoking was safer (60201). One of the conclusions of the 1964 Surgeon General's Report was "The death rates for men smoking less than 5 cigars a day are about the same as for non-smokers."

Traditional smokers of large cigars typically did not inhale the smoke as deeply as cigarette smokers, as cigar smoke was more bitter (i.e., higher pH) than cigarette smoke (lower pH) and delivered more nicotine by the oral route, as cigars did not have a plastic mouthpiece or filter (60201–6). Thus, the lungs were not exposed to as much toxic smoke as with cigarettes, because there was no or little inhalation by traditional cigar smokers. However, cigar smoking is not a safe alternative to cigarettes as its use will still cause an increased risk of nicotine addiction, heart disease, stroke, COPD, and various cancers (60203–6). In addition, cigarette smokers who switch to cigars or pipes inhale more than smokers who only smoked cigars or pipes, and thus do not decrease health risks significantly (60307–8).

Cigars are also more of a secondhand smoke problem than cigarettes, as their sidestream smoke contains 20 times the CO, 5× the respirable particles, and other toxic chemicals in greater quantities and for a longer period of time than cigarettes (60201, 60206).

The tobacco industry bent the rules and skirted regulations by adding flavors not allowed in large cigars, adding sugar, and making little cigars a similar size to cigarettes, and promoted little cigars and cigarillos (LCC) as a cheaper and safer alternative to cigarettes (60209–16).

Another lucrative market for LCCs occurs when smokers replace the tobacco in cheap cigarillos with cannabis to make blunts, and thus incur the potential risk of addiction to both THC and nicotine (60217).

A cheroot is a low-cost cylindrical cigar (not tapered), clipped at both ends during manufacturing and common in Burma and India.

Reference Number: 60201

DE ASSIS VIEGAS, C.A. "Noncigarette Forms of Tobacco Use." *Jornal brasileiro de pneumologia*, 34: 1069–1073, 2008 (25 references)

Abstract: A cigar is defined as any roll of tobacco wrapped in leaf tobacco or any substance containing tobacco that does not have a filter. The prevalence of cigar smoking has decreased in the English-speaking world from 30% to 5%. Cigars contain 100 to 200× more nicotine than cigarettes (100–400 mg vs 1–2 mg). Cigar smoke is more alkaline than cigarette smoke and so a greater absorption of nicotine can occur through the oral cavity, meaning that it doesn't need to be inhaled. Cigar smoke contains highly carcinogenic nitrosamines, hydrocarbons, and aromatic amines, at higher levels than cigarette smoke.

> Finally, it is of note that the secondary smoke from cigars contributes more to environmental pollution than does cigarette smoke, assuming an equal quantity of tobacco burned, as well as that cigars contain a greater quantity of tobacco and burn longer (1-2 h vs. 10 min). Therefore, since cigar smoke contains higher concentrations of toxins and cancerous substances than does cigarette smoke, it also contributes to increasing the risk of lung cancer and other smoking-related diseases in nonsmokers (passive smokers).

Reference Number: 60202

HENNINGFIELD, J.E., R.V. FANT, A. RADZIUS, AND S. FROST. "Nicotine Concentration, Smoke pH and Whole Tobacco Aqueous pH of Some Cigar Brands and Types Popular in the United States." *Nicotine and Tobacco Research*, 1: 163–168, 1999 (3 tables, 1 figure, 21 references)

Abstract: The tobacco pH, smoke pH, and nicotine concentrations were determined in 16 cigar brands including smoke cigars, cigarillos, and large premium brands. The tobacco weight per cigar varied from 0.53 to 21.5 grams and the total nicotine concentration varied from 5.9 to 335.2 mg per cigar. The pH of the tobacco used in cigars varied from 5.7 to 7.8. The pH for large cigars was generally higher than for smaller cigars (which were similar to cigarettes).

Also of importance is whether or not the cigar is held in the mouth between puffs. Cigars, especially those with whole tobacco pH values of 7.0 or greater, could be capable of nicotine delivery via the buccal mucosa particularly with prolonged manipulation of the cigar in the mouth whether or not they have been lighted. For example, an earlier study showed that the pH of several popular brands of smokeless tobacco was within the range of pH values observed in the present study, namely 6.9–8.6. In this range, approximately 7–79% of the nicotine would be in the free base form and therefore readily absorbed through the lips and buccal mucosa. Therefore, these cigars would be capable of delivering substantial amounts of nicotine even if not lit or smoked. The amount of nicotine absorbed in this way, and the relative contribution to total nicotine delivery, remains to be studied in humans.

Cigars contain sufficient amounts of nicotine to cause and substain dependence, although how many needed to be consumed per day would be expected to vary widely because of the enormous variability in nicotine delivery between brands. In addition, the patterns of consumption and pH of the tobacco and tobacco smoke would also determine the level of nicotine exposure. For example, holding a high pH cigar in the mouth may provide a sufficient buccal infusion of nicotine to sustain dependence regardless of whether or not it is smoked.

Reference Number: 60203

ROSENBERRY, Z.R., W.B. PICKWORTH, AND B. KOSZOWSKI. "Large Cigars: Smoking Topography and Toxicant Exposure." *Nicotine and Tobacco Research*, 183–191, 2018 (3 tables, 1 figure, 56 references)

Abstract: Before the advent of mass produced cigarettes in 1875, about two-thirds of the tobacco produced was for cigars. Due in part to less taxation and availability of flavors, the number of large cigars sold in the US rose from 4.7 billion in 2008 to 9.5 billion in 2012. Sixteen male and one female dual user (mean age 36 years) smoked 1 large unflavored cigar (Phillies Blunt) or their usual cigarette ad libitum, and various smoking parameters and toxicant exposures were measured. All participants had initiated smoking cigarettes and then started cigars. Most smoked menthol cigarettes and the mean CPD was 20.5. The mean plasma nicotine concentration at the end of smoking was 38.8 ng/mL for cigarettes and 36.3 ng/mL for cigars. The mean exhaled CO was 30 ppm for cigarettes and 47 ppm for cigars. The mean duration of smoking and the total puff volume was 4.2 minutes and 658 mL for cigarettes and 6.1 minutes and 1660 mL for the large cigar, respectively. Cigarettes delivered more nicotine than cigars when adjusted to the number of grams of tobacco smoked and the total puff volume.

> Despite these limitations, large cigars rapidly delivered significant amounts of nicotine, similar to cigarettes. There was also evidence of inhalation during large cigar smoking as shown by the rapid and significant increase in COex. Large cigars decreased the urge to smoke and had similar satisfaction and liking ratings. Despite differences in puff topography, cigarettes and large cigars delivered similar amounts of nicotine and large cigars delivered significantly more CO. The results support the recent "Deeming Rule" decision that extends FDA authorization to regulate all tobacco products including large cigars.

Reference Number: 60204

ROSTRON, B.L., C.G. COREY, AND R.M. GINDI. "Cigar Smoking Prevalence and Morbidity Among US Adults 2000–2015." *Preventive Medicine Reports*, 14: (4 pp), 2019 (1 table, 14 references)

Abstract: Current adult cigar smoking prevalence in the US was relatively stable between 2000-2015 at about 2.3%. Cigar smoking was associated with the prevalence of heart condition, stroke, and cancer, although in former cigar smokers only, which may be due in part to the fact that the cigar smokers stopped using cigars following disease onset. The type of cigar may also influence the frequency, use, and depth of inhalation.

Table: Risk Ratios of Former Exclusive Cigar Smokers

Disease	Adjusted Prevalence Ratio
Heart conditions	1.33×
Stroke	2.41×
All cancers	1.44×

Source: Adapted from Rostron et al, 2019

This analysis has shown that nearly 200,000 cardiovascular conditions and cancer cases were attributable to exclusive cigar smoking among US adults aged ≥35 years in 2015. These results are consistent with previous findings that cigar smoking is associated with preventable diseases.

Reference Number: 60205

MEDABALA, T., B.N. RAO, M.I. GLAD MOHESH, AND M. PRAVEEN KUMAR. "Effect of Cigarette and Cigar Smoking on Peak Expiratory Flow Rate." *Journal of Clinical and Diagnostic Research*, 7: 1886–1889, 2013 (4 tables, 2 figures, 37 references)

Abstract: Smoking is a known risk factor for chronic obstructive pulmonary disease (COPD) and lung cancer. The peak expiratory flow rate (PEFR) was determined in 49 cigarette smokers (mean age 27.9 years), 10 cigar smokers (mean age 35.3 years), and 64 non-smokers (mean age 26.4 years). The mean PEFR in liters per minute (lpm) was 513 for non-smokers, 409 for cigarette smokers, and 288 for cigar smokers.

Table: Smoking Behavior and Mean Peak Expiratory Flow Rate (PEFR)

Smoking Behavior	Mean PEFR (liters per minute)
Non-smokers	513.4
1–50 cigarette years	494.7
51–100 cigarette years	443.3
101–150 cigarette years	350.9
151–230 cigarette years	300.0
Cigar Smokers	288.0

Source: Adapted from Medabala et al, 2013

In the present study, we found that PEFR was decreased significantly in both cigarette and cigar smokers compared to non-smokers, and our findings are in agreement with the findings of others. One possible reason for the decrease in PEFR could be inflammation which is a common and constant pathological finding in cigarette smokers. Earlier studies have reported that airway flow limitation occurs due to bronchial constriction caused by mediators of inflammation. Inflammation either directly or by increasing smooth muscle tone, indirectly, may cause airway fibrosis. All these changes promote wall thickness leading to airway narrowing and flow limitation. In addition, inflammation causes destruction of the alveolar walls attached to the airway contributing further to airflow limitation by deforming and narrowing the airway lumen.

Pearson correlation analysis shows that there exists a strong negative correlation between intensity of cigarette/cigar smoking and PEFR, i.e. the greater the intensity of cigarette/cigar smoking, the lesser the PEFR value. However, negative correlation was highest in cigarette smoking compared to cigar smoking and this difference may be attributed to small sample size for the cigar smoking group. In spite of difference in the magnitude of decline in PEFR, the negative correlation suggests that both types of tobacco smoking adversely affect the lung function.

Reference Number: 60206

BAKER, F., S.R. AINSWORTH, J.T. DYE, C. CRAMMER, M.J. THUN, D. HOFFMANN, J.L. REPACE, J.E. HENNINGFIELD, J. SLADE, J. PINNEY, T. SHANKS, D.M. BURNS, G.N. CONNOLLY, AND D.R. SHOPLAND. "Health Risks Associated with Cigar Smoking." *Journal of the American Medical Association*, 284: 735–740, 2000 (1 table, 35 references)

Abstract: Cigars consist of a filler (the inner part of the cigar), a binder, and a wrapper, all of which are made with air-cured and fermented tobacco. Cigarettes use a blend of heat-cured and air-cured tobacco but no fermentation. Fermentation adds flavor and aroma to the tobacco but reduces the sugar level from over 5% in cigarettes to about 1% in cigars. The pH of cigar smoke is higher than that of cigarette smoke, which allows for more oral absorption of nicotine. Cigar smoke contains a lower oxygen concentration, due to the poor porosity of the tobacco wrappers compared to the paper wrappers used in cigarettes. TNSA is significantly higher in cigar smoke. Smoking cigars is not a safe alternative to cigarette smoking. Cigar smokers are perceived as relatively well-to-do, well-educated older managers or executives.

Sidestream smoke (the aerosol emitted from the burning cone of a cigar, cigarette, or pipe during the interval between puffs and the portion of the inhaled smoke that is not retained and is exhaled) contributes significant pollutants to the environment in the form of carbon monoxide, nitrogen oxides, respirable suspended particulate matter, nicotine, polycyclic aromatic hydrocarbons, and other compounds, and sidestream smoke from cigars does so to a greater degree than the sidestream smoke of cigarettes, when equal amounts of tobacco are burned. Compared to a single cigarette (0.55g) smoked to 70% of its mass, a large cigar smoked 70% emits about 20 times the carbon monoxide, 5 times the respirable particles, and twice the amount of polycyclic aromatic hydrocarbons.

One study of environmental pollutants from tobacco smoke found the levels of carbon monoxide at cigar banquets and in some smokers' homes equal to carbon monoxide concentrations on crowded California freeways.

Reference Number: 60207

TURNER, J.A., M.W. MCNICOL, AND R.W. SILLETT. "Distribution of Carboxyhaemoglobin Concentrations in Smokers and Non-Smokers." *Thorax*, 41: 25–27, 1986 (2 tables, 13 references)

Abstract: The blood carboxyhemoglobin concentration is an accurate marker of tobacco smoke inhalation. Venous blood samples were collected from 1,255 non-smokers, 1,933 cigarette smokers, 193 cigar smokers, and 106 pipe smokers over a period of 5 years and blood COHb was measured using a CO Oximeter. The mean blood COHb concentration was 0.91% in non-smokers, 4.78% in cigarette smokers, 0.93% in cigar only smokers, and 1.36% in pipe only smokers. Previous cigarette smokers who switched to cigars had a mean COHb of 6.80% and those who switched to pipes had a mean COHb of 3.39%.

A change from cigarettes to other smoking materials appears to confer no benefit from reduced inhalation. The highest mean carboxyhaemoglobin concentrations were found in secondary cigar smokers and the proportion of secondary pipe and cigar smokers with raised concentrations was similar to that of cigarette smokers. These results confirm the importance of cigarette smoking in setting the pattern of inhalation. Most primary pipe and cigar smokers inhale minimally, but most of those who have regularly smoked cigarettes inhale and continue to do so whatever smoking material they use.

Reference Number: 60208

KAUFMAN, D.W., J.R. PALMER, L. ROSENBERG, AND S. SHAPIRO. "Cigar and Pipe Smoking and Myocardial Infarction in Young Men." *British Medical Journal*, 294: 1315–1316, 1987 (1 table, 5 references)

Abstract: Many cigarette smokers switch to cigars or pipes in order to reduce health risks. The effect of cigar and pipe smoking was determined in interviews of 572 men with non-fatal myocardial infarction and 934 hospital controls, who have not smoked cigarettes for at least 2 years. The study was conducted in Massachusetts between 1980 and 1983. The risk of a non-fatal MI among ex-cigarette smokers who smoked at least 5 cigars a day was 4.5× higher than ex-cigarette smokers who did not smoke cigars. This is perhaps due to ex-cigarette smokers continuing to inhale the smoke after they switch to cigars.

> Our results suggest that cigars are not a good alternative to cigarettes in terms of the risk of coronary heart disease; cigarette smokers would better be advised to give up smoking altogether.

Reference Number: 60209

ROSTRON, B.L., C.G. COREY, E. HOLDER-HAYES, AND B.K. AMBROSE. "Estimating the Potential Health Impact of Prohibiting Characterizing Flavors in Cigars Throughout the US." *International Journal of Environmental Research and Public Health*, 16: 3234 (7 pp), 2019 (1 table, 1 figure, 22 references)

Abstract: Between 2008 and 2015, sales of flavored cigars increased by almost 50%. Flavorings are added to cigars and other tobacco products to increase their appeal and reduce the harshness and bitterness of the smoke. Flavored cigar use is common among youth (aged 12–17 years) and in 2014, 65% of ever cigar users reported that their first cigar was flavored. About 64% of middle and high school cigar smokers (about 910,000 youth) smoked flavored ones. About 92% of cigars sold in the US are machine made. Handmade cigars are generally not flavored. A ban on flavored cigars in New York City resulted in a 15–20% reduction in sales. Cigars caused serious health problems, including heart disease and lung cancer, and regular cigar use causes approximately 9,000 premature deaths a year.

> The estimates presented in this study have demonstrated the substantial possible public health benefits of measures prohibiting the sale of cigars

with characterizing flavors other than tobacco in states and localities throughout the US. These benefits would result from reduced mortality among current cigar smokers through increased cessation and from reduced cigar use among young people because of decreased initiation and progression to regular use due to decreased appeal of these products. We find that extending policies that have been implemented in certain US localities on a nationwide basis could result in the prevention of approximately 800 premature cigar-attributable deaths in the US each year with a 90% prediction interval of approximately 400 to 1200 premature deaths avoided. We also find that such policies could result in a reduction of approximately 112,000 current cigar smokers in each cohort of 18 year olds in the US, with a 90% prediction interval of 76,000 to 139,000 users prevented.

Reference Number: 60210

DELNEVO, C.D., AND M. HRYWNA. "'A Whole 'Nother Smoke' or a Cigarette in Disguise: How RJ Reynolds Reframed the Image of Little Cigars." *American Journal of Public Health*, 97: 1368–1375, 2007 (1 table, 3 figures, 87 references)

Abstract: Little cigar sales soared after the 1964 Surgeon General's Report, as they were perceived as safer than cigarettes. RJ Reynolds (RJR) designed the little cigar to be as close to be a cigarette as possible while avoiding the taxation and advertising restrictions imposed on cigarettes. The consumption of 4 billion cigars in 2006 rivalled that of the 1970s.

Almost 40 years ago, RJR engaged in a calculated effort to blur the line between cigarettes and little cigars with Winchester, a little cigar designed for cigarette smokers that was as close to cigarettes as legally possible. Although RJR was chiefly motivated by the television broadcast ban during Winchester's early development, it was the company's careful consideration of price and tax structure that was the dominant factor in its success. The little cigar boom of the early 1970s was largely attributed to the loophole in the broadcast ban, but tax disparities and price played an equal, if not greater, role.

Today there is an overwhelming sense of déjà vu—little cigar sales reached an all-time high in 2006. As was the case 40 years ago, there is a marked disparity between cigarette and little cigar excise taxes. Cigarette excise tax increases in numerous states resulted in a doubling of the average tax (includes state and federal) levied on a pack of cigarettes in the United States between fiscal years 2000 and 2006, from $0.65 to $1.31. In states with high cigarette excise taxes, a pack of little cigars costs less than half as much as a pack of cigarettes. Lower cigar prices are associated with higher rates of cigar use, and price disparities may also encourage product switching. Not surprisingly, little cigar marketing continues to capitalize on tax disparities and cigarette-like characteristics. For example, PrimeTime Cigars's Web site once advertised "so much like cigarettes, it's hard to believe they are cigars!"

In addition to having a lower tax rate, little cigars are free from the costs and restrictions imposed on cigarettes by the Master Settlement Agreement (MSA) and related legislation (e.g., the Comprehensive Smoking Education Act), not the least of which is that little cigar manufacturers are not required to make MSA or escrow payments to states (estimated at $4.30/carton) as cigarette manufacturers must.

Reference Number: 60211

STERLING, K.L., B.A. MAJEED, A. NYMAN, AND M. ERIKSEN. "Risk Perceptions of Little Cigar and Cigarillo Smoking Among Adult Current Cigarette Smokers." *Nicotine and Tobacco Research*, 19: 1351–1358, 2017 (3 tables, 37 references)

Abstract: Adults may endorse the safety of little cigars and cigarillos (LCCs) because these tobacco products are not addressed in anti-smoking campaigns. This study determined the risk perceptions of 1,191 current cigarette smokers regarding LCCs. The current cigarette smokers were divided into three groups: dual cigarette smokers who used LCCs; smokers who were susceptible to LCC smoking; and cigarette smokers who were not susceptible to LCCs. Overall, 47.2% of the participants were dual smokers; 12.7% were susceptible to LCC smoking, and 40.1% were not susceptible. Most of the respondents were White, male, and aged 45 to 59 years. The majority of smokers perceived that LCCs were addictive and that smoking LCCs carried the same level of harm as cigarettes.

Our findings suggest that perceptions of risk were not merely subjective beliefs but were also important determinants of LCC smoking behavior. Compared to cigarette smokers who were not susceptible to LCC use, dual smokers who believed that daily use carried minimal risk and believed that LCCs were not addictive were more likely to intend to smoke the product again in the future. As documented in prior studies, some cigarette smokers presume that all cigars (including LCCs) are less dangerous than and are safer alternatives to cigarettes. That these misperceptions are associated with continued LCC smoking among dual smokers is troubling; dual cigarette and LCC use may increase the likelihood of nicotine dependence, cancer, and other chronic health conditions.

Reference Number: 60212

ANTOGNOLI, E., S.K. GONZALEZ, E. TRAPL, D. CAVALLO, B. LAVANTY, R. LIM, AND S. FLOCKE. "Cigarettes, Little Cigars, and Cigarillos: Initiation, Motivation, and Decision-Making." *Nicotine and Tobacco Research*, 20: S5–S11, 2018 (1 table, 45 references)

Abstract: Cigarette use in the past 50 years has declined in the US and other developed countries, but the use of other tobacco products such as little cigars and cigarillos (LCC) has increased. More than 2× as many youths use two or more tobacco products than smoke cigarettes alone. A study was taken of 28 female and 32 male LCC smokers (mean age 19.7 years) between 2015 and 2016 in Cleveland, Ohio. Thirty participants smoked LCCs only and 30 were dual users (LCCs and cigarettes). Among dual users, 40% smoked LCCs before or at the same time as starting cigarette smoking and 60% started smoking cigarettes first. LCC only smokers reported not smoking cigarettes due to their expense, unpleasant taste/smell, and fear of becoming addicted. Cigarette smokers started used LCCs due to be able to afford one LCC but not a pack of cigarettes.

In this sample of current LCC users, half also used cigarettes. Product initiation order was almost evenly split, but reasons for initiating the second product differed substantially, with immediacy of reducing cravings as a key reason for LCC users to smoke a cigarette and social and financial reasons for cigarette users to smoke an LCC. Our findings support tobacco regulation of LCCs that focuses on packaging and pricing to have equity with cigarettes.

Reference Number: 60213

CHATTOPADHYAY, S., E.M. SMYTH, P. KULKARNI, K.R. BABIK, M. REID, L.E. HITTLE, P.I. CLARK, E.F. MONGODIN, AND A.R. SAPKOTA. "Little Cigars and Cigarillos Harbor Diverse Bacterial Communities that Differ Between the

Tobacco and the Wrapper." *Public Library of Science One*, 14: e0211705 (15 pp), 2019 (5 figures, 58 references)

Abstract: Little cigars are similar in size to cigarettes but are wrapped in brown paper that typically contains some tobacco leaf and provides a similar smoke as cigarettes. Cigarillos (7-minute cigars) are more like a traditional cigar but longer than a cigarette and are inhaled more deeply than a cigarette. Little cigars and cigarillos are taxed at a lower rate than cigarettes and are sold as single or loose without health warnings. Between 1995 and 2008, there was a 316% rise in the sales of little cigars and a 255% rise in the sales of cigarillos. One hundred and thirty-two tobacco and wrapper samples were subjected to DNA extraction and PCR amplification to determine the types of bacteria present. Bacillus and Lactobacillus strains dominated the wrapper and Staphylococcus, and Pseudomonas dominated the tobacco rod.

> This study provides novel evidence that the bacterial communities of little cigars and cigarillos are not only diverse, but also differ significantly between the tobacco and the wrapper, and across brands. During the manufacturing process of a general cigar, different types of leaves are used for making the wrapper, filler and binder. These leaves may be grown in multiple regions of the world, under differing soil and climatic conditions. This manufacturing process, along with differences in the composition of the materials within the tobacco versus the wrapper, likely contributed to the observed variability between these sample types. Tobacco samples also differed in bacterial community composition across brands, which is not unexpected considering that brands are manufactured under different industrial conditions with variable tobacco blends. This can also have an effect on the individual lots that are manufactured, resulting in the variable bacterial community composition that we observed in wrapper samples across lots in this study.

Reference Number: 60214

REILLY, S.M., R. GOEL, Z. BLITZER, R.J. ELIAS, J. FOULDS, J. MUSCAT, AND J.P. RITCHIE JR, "Little Cigars, Filtered Cigars, and their Carbonyl Delivery Relative to Cigarettes." *Nicotine and Tobacco Research*, 20: S99–S106, 2018 (4 figures, 50 references)

Abstract: Little cigars are growing in popularity due to their lower costs relative to cigarettes, their wide variety of flavors, and their similar appearance to cigarettes. The carbonyl (e.g., acetaldehyde, formaldehyde, and acrolein) concentrations were determined in smoke from Winchester, Captain Black Sweets, and two research cigarettes by high-performance liquid chromatography with ultraviolet detection. The smoke was produced by a smoking machine using the International Organization for Standardization (ISO) and Health Canada Intense (HCI) protocols. On a per puff basis, five of the seven carbonyls were higher in the little cigars compared to filtered cigars and cigarettes. On a per unit basis, the carbonyl concentration was greater in the little cigars and filtered cigars than in the cigarettes.

> Altogether, our findings suggest that filtered cigars and little cigars may present a greater risk than cigarettes based upon carbonyl delivery, which supports the FDA's recent regulation of these products and warning letters sent to four of the manufacturers of these products. Multiple studies have shown that little cigar perception is mixed among smokers currently with one study reporting a minority (29.9%) of youth cigar users believing that cigars might be less harmful than cigarettes and another reporting comments about little cigars ranging from "little cigars are much more harmful than cigarettes" to "little cigars do not cause cancer." One last study found that, in general, little cigars are perceived as more favorable than cigarettes regarding potential harm to health. Our findings can be used to combat this mixed perception by informing the public about the levels of some of the carcinogens and toxicants produced by little and filtered cigars. This information can help guide FDA regulation of these combustible tobacco products, specifically in terms of setting appropriate product standards and informing the public of potential toxicant exposures from these products. This paper demonstrates that filtered cigars and little cigars can deliver a higher dose of harmful carbonyls per unit than cigarettes and should be regulated just as, if not even more, rigorously.

Reference Number: 60215

KOSZOWSKI, B., Z.R. ROSENBERRY, A. KANU, L.C. VIRAY, J.L. POTTS, AND W.B. PICKWORTH. "Nicotine and Carbon Monoxide Exposure from Inhalation of Cigarillo Smoke." *Pharmacology, Biochemistry, and Behavior*, 139: 7–14, 2015 (4 tables, 1 figure, 64 references)

Abstract: Cigarillos are typically between the weight of a little cigar and a large cigar. They have increased in popularity due to the existing availability of flavors and a lower tax rate. Twenty-one male and two female cigarette/cigarillos smokers (mean age 39 years) were tested on two occasions separated by at least 24 hours, smoking either a cigarette or a cigarillo. Breath CO and plasma nicotine concentrations were measured as well as puff parameters. Most smokers in this study smoked cigarettes first before trying cigarillos. The average weight of the cigarette used in this study (1.075 g) was about one-third the average weight of a cigarillo (3.097 g). Breath CO was significantly greater in subjects after smoking cigarillos than cigarettes. The subjects smoked both cigarettes and cigarillos in a similar manner.

Table: A Comparison of Smoking Cigarettes and Cigarillos

Parameter	Cigarette Smoking	Cigarillo Smoking
Number of puffs	15	24
Total puff volume	698 mL	1,267 mL
Time to smoke	5.25 minutes	10.33 minutes
Puff volume	50.3 mL	56.2 mL
Increase in plasma nicotine after smoking ended	24.5 ng/mL	21.1 ng/mL
Increase in breath CO after smoking ended	9 ppm	25 ppm

Source: Adapted from Koszowski et al, 2015

The epidemiology of cigar smoking has changed dramatically over the past few decades. Formerly cigar smokers were typically older, affluent males, who smoked large cigars and did not smoke cigarettes. Most such smokers did not inhale the cigar smoke into their lungs unless they were former cigarette smokers. Current use pattern indicates the popularity of cigar use among young, urban African American men and women who typically use both cigars and cigarettes. Furthermore, the characteristics of the cigars themselves have changed. In contrast to the large, unflavored, unfiltered cigar sold singly, little cigars and cigarillos are sold in packages of 5–20, they are often tipped, sometimes filtered, and available in a wide variety of flavors.

Reference Number: 60216

STERLING, K.L., C.S. FRYER, I. PAGANO, AND P. FAGAN. "Little Cigars and Cigarillos Use Among Young Adult Cigarette Smokers in the United

States: Understanding Risks of Concomitant Use Subtypes." *Nicotine and Tobacco Research*, 18: 2234–2242, 2016 (4 tables, 35 references)

Abstract: A web-based survey was conducted on 1,018 cigarette smokers (aged 18–44 years) in 2015 to determine LCC use. Sixty-three percent of the cigarette smokers did not smoke LCCs; 15.1% were LCC tobacco users; 11.1% were LCC blunt users; and 10.5% were LCC-poly users. The age of onset of cigarette smoking was less than 15 years of age for 38.4%; 31.9% were 15 to 17 years old; and 29.7% were 18 years and older. Only 9% had a bachelor's degree or higher for education. Young adult cigarette smokers are at a greater risk of concomitant LCC use. The FDA announced in 2016 that LCCs would finally be regulated and that, as such, flavorings would be restricted.

> Other tobacco product use, marijuana and alcohol use is associated with dual use behaviors among some groups. Cigarette smokers who used other tobacco products (large cigars, hookah and e-cigarettes) had greater odds of LCC-tobacco use. Like LCCs, large cigars, hookah tobacco, and e-cigarettes are often flavored. Though the FDA has regulatory authority over these products, they are still available in characterizing flavors. Flavors mask the harshness, bitterness, and irritation of nicotine and may appeal to younger smokers; of whom 53.3% of our sample smoked a mentholated brand. Lack of regulation has provided the tobacco industry an unprecedented, unobstructed opportunity to market these flavored tobacco products to vulnerable populations.

Reference Number: 60217

GIOVENCO, D.P., E.J. MILLER LO, M.J. LEWIS, AND C.D. DELNEVO. "They're Pretty Much Made for Blunts: Product Features That Facilitate Marijuana Use Among Young Adult Cigarillo Users in the United States." *Nicotine and Tobacco Research*, 19: 1359–1364, 2017 (1 table, 1 figure, 27 references)

Abstract: A survey was conducted of 24 male and 16 female cigarillo users (aged 18–32 years) in the United States in 2013. Approximately one-half of cigarillo smokers used it to exclusively make blunts (tobacco is replaced by cannabis). Many of the former cigarette smokers switched to cigars so they could better manage their tobacco consumption as cigars are larger, burn slower, and last longer. They also felt that there was less stigma against cigar users than cigarette users.

Cigarillo use is prevalent among young adults in the United States, many of whom are using the products as blunts. This study found that product features such as brand, flavor, packaging, and price influence the selection of cigarillos used for this purpose. There is also a strong perception among young adult cigarillo users that cigarillo companies design their products and packaging to make the blunt-making process simple and enjoyable. Better surveillance measures are needed to capture the extent to which cigarillos are used as blunts and which product features are driving category growth.

6.03 PIPES/CHILLUMS

"Comprehensively documenting the deleterious health effects of pipe smoking is important in countering efforts by the tobacco industry to promote pipes as a desirable alternative to cigarettes or cigars. The tobacco industry has repeatedly demonstrated its ability to create new markets by reviving interest in tobacco products that had appeared to have become obsolete, especially among youth and young adults."

—Henley et al, "Association Between Exclusive Pipe Smoking and Mortality From Cancer and Other Diseases" (2004)

Tobacco used to be smoked almost exclusively using pipes before the invention and marketing of cigarettes, but pipes are now one of the least common ways of smoking. In the United States, pipe smokers are now less than 1% of the adult population (60301). Pipe smoking, like cigar smoking, was considered for many years to be safer than smoking cigarettes, but this was mainly due to the fact that these smokers did not usually inhale or inhaled lightly and used less tobacco than cigarette smokers (60306).

Pipe smokers still have elevated risks of larynx cancer, lung cancer, COPD, coronary heart disease, and diabetes compared to non-smokers (60302–4).

Chillums are small, inexpensive conical pipes with an end-to-end channel. They were invented in India, used mainly in rural areas, and their use spread globally in the 1960s (60308). Chillums may be used to smoke tobacco, cannabis, or a tobacco/cannabis mixture (60307). A wet cloth may be placed in the mouth to act as a filter or cool the smoke inhaled from the chillum.

The high exhaled carbon monoxide (eCO) concentration indicates that chillum smoking may be one of the most dangerous methods of smoking (60308).

Reference Number: 60301

HENLEY, S.J., M.J. THUN, A. CHAO, AND E.E. CALLE. "Association Between Exclusive Pipe Smoking and Mortality from Cancer and Other Diseases." *Journal of the National Cancer Institute*, 96: 853–861, 2004 (6 tables, 1 figure, 50 references)

Abstract: Pipes are among the least commonly used tobacco product in the United States. Only 2% of adult men and virtually no women smoked pipes in 1991. This represents a decline from 14% in 1965. The tobacco industry has repeatedly demonstrated its ability to revive interests in obsolete tobacco products such as pipes, snuff, premium cigars, bidis, and hookah, especially in youth, by presenting them as less hazardous than cigarettes and creating a positive association with celebrities and success. This study was based on 15,263 men who were former or current exclusive users of pipes, 123,044 men who never used tobacco products in 1982, and 23,589 men who died during the 18-year follow-up.

Smokers who smoked 1–3 pipes per day had a lower risk of lung cancer than those who smoked more than 11 pipes per day (1.99× vs 7.67×). Those who did not inhale or inhaled only slightly had a lower risk of lung cancer than those who inhaled moderately or deeply (2.85× vs 11.06×).

Table: Multivariable Adjusted Hazard Ratio of Various Diseases in Current Pipe Smokers

Disease	Hazard Ratio
Larynx cancer	13.1×
Lung cancer	5.0×
Oropharynx cancer	3.9×
COPD	3.0×
Coronary heart disease	1.3×

Adapted from Henley et al, 2004

Results from this large, prospective study strongly support a causal relationship between pipe smoking and mortality from cancers of the lung, larynx, esophagus, and oropharynx, and chronic obstructive pulmonary disease. Although the risk of dying from tobacco-associated diseases is lower for pipe smokers than for cigarette smokers, pipe smoking is as harmful as, and perhaps more harmful than, cigar smoking. All tobacco products cause excessive morbidity and mortality. Our findings may help deter efforts by the tobacco industry to imply, directly or indirectly, as it has with smokeless tobacco and low-yield cigarettes, that any tobacco product has negligible adverse health effects.

Reference Number: 60302

CHRISTENSEN, C.H., B. ROSTROM, C. COSGROVE, S.F. ALTEKRUSE, A.M. HARTMAN, J.T. GIBSON, B. APELBERG, M. INOUE-CHOI, AND N.D. FREEDMAN. "Association of Cigarette, Cigar, and Pipe Use with Mortality Risk in the US Population." *JAMA Internal Medicine*, 178: 469–476, 2018 (3 tables, 16 references)

Abstract: A study was conducted of 357,420 exclusive users of pipes, cigars, or cigarettes, and never-users of any type of tobacco product in the National Longitudinal Mortality Study between 1985 and 2011. The majority (79.3–98.0%) of the current and former cigar and pipe smokers were male, compared to 46% of current cigarette smokers.

Table: Multivariable Adjusted Odds Ratio of Various Causes of Death in Current Exclusive Pipe, Cigar, and Cigarette Smokers

Outcome	Current Exclusive Pipe Smoker	Current Exclusive Cigar Smoker	Current Exclusive Cigarette Smoker
Lung cancer	1.51×	3.26×	11.82×
Cardiovascular death	0.96×	1.14×	1.44×
Cerebrovascular death	0.24×	0.50×	1.24×
COPD death	1.12×	2.44×	11.01×
Diabetes death	1.79×	1.18×	1.18×

Source: Adapted from Christensen et al, 2018

Our study provides further evidence that cigar, pipe, and cigarette use confers mortality risks, even among nondaily current cigarette users, with lower risks observed among former users than current users. These data underscore the importance of cessation to reduce mortality and morbidity from combustible tobacco use. These risk estimates are consistent with published estimates from other studies, thus demonstrating the reliability of NLMS risk estimates. Given rapid changes in tobacco products available in the marketplace, our results provide information on the contemporary mortality risks of 2 noncigarette combustible tobacco products.

Reference Number: 60303

CARSTENSEN, J.M., G. PERSHAGEN, AND G. EKLUND. "Mortality in Relation to Cigarette and Pipe Smoking: 16 Years' Observation of 25,000 Swedish Men." *Journal of Epidemiology and Community Health*, 41: 166–172, 1987 (4 tables, 3 figures, 32 references)

Abstract: The mortality rate and cause of death of a random sample of 25,000 Swedish men was determined between 1963 and 1979. The relative lung cancer rate increased according to the amount of tobacco used, no matter if it was cigarettes, pipes, or cigars. There appeared to be no threshold effect at low levels of tobacco use.

The relative risks for several smoking related diseases, including lung cancer, have been reported to be lower in cigar and pipe smokers than in cigarette smokers. These results were mostly obtained in studies performed in the United Kingdom and United States, where the proportion of inhalers is substantially higher among cigarette smokers than among smokers of cigars or pipes. In the present study we found similar risks of lung cancer in cigarette, pipe, and cigar smokers, controlling for amount of tobacco consumed, age when started smoking, and urban/rural residence. The similar proportion of inhalers among Swedish cigarette and pipe smokers may partly explain our results. No information was available on inhalation patterns in Swedish cigar smokers.

Reference Number: 60304

BOFFETTA, P., G. PERSHAGEN, K-H. JÖCKEL, F. FORASTIERE, V. GABORIEAU, J. HEINRICH, I. JAHN, M. KREUZER, F. MERLETTI, F. NYBERG, F. RÖSCH, AND L. SIMONATO. "Cigar and Pipe Smoking and Lung Cancer Risk: A Multicenter Study from Europe." *Journal of the National Cancer Institute*, 91: 697–701, 1999 (4 tables, 25 references)

Abstract: The lung cancer risk among smokers of cigarettes, cigars, cigarillos, and pipes were determined in 5,621 male cases and 7,255 male controls in Germany, Italy, and Sweden between 1988 and 1994.

Table: Risk of Lung Cancer Among Cigar, Pipe, Cigarillo, and Cigarette Users

Type of Smoking	Adjusted OR for Lung Cancer
Cigars only	5.6
Pipes only	7.9
Cigarillos only	12.7
Cigarettes	14.9

Source: Adapted from Boffetta et al, 1999

> A dose-response relationship was present for duration of use and cumulative consumption both for cigars and cigarillos and for pipe tobacco. An effect was also suggested for inhalation of cigar and cigarillo smoke. The dose-response relationships between lung cancer risk and either duration of smoking or average and cumulative consumption were similar for cigar and cigarillo smoking, pipe smoking, and cigarette smoking.... Our results suggest that smoking of European cigars, cigarillos, and pipe tobacco might exert a carcinogenic effect on the lung comparable to that of cigarettes.

Reference Number: 60305

RANDI, G., L. SCOTTI, C. BOSETTI, R. TALAMINI, E. NEGRI, F. LEVI, S. FRANCESCHI, AND C. LA VECCHIA. "Pipe Smoking and Cancers of the Upper Digestive Tract." *International Journal of Cancer*, 121: 2049–2051, 2007 (2 tables, 19 references)

Abstract: The association between exclusive pipe smoking and cancers of the upper digestive tract was determined in 93 male cases and 1,032 male controls in Italy and Switzerland between 1984 and 1999. When age, heavy alcohol consumption, BMI, and education were taken into account, the RR of upper digestive tract cancers for exclusive pipe smoking was 8.7×, and with heavy alcohol consumption it increased to 38.8×.

> Pipe smoke is inhaled less than cigarette one, and hence is associated with lower excess risk of lung cancer, as well as of pancreatic cancer, bladder cancer and of most other tobacco-related neoplasms. However, the epithelia of the oral cavity and oesophagus are in direct contact with pipe smoke, and this explains the substantial excess cancer risk for these sites. Direct contact with tobacco carcinogens may also explain the greater RR for oral and pharyngeal than for oesophageal cancer observed in this as well as in another study. Similar line of reasoning applies to cigar smoking,

which has been associated to a 7-fold increased risk of upper digestive tract cancer in our data.

Reference Number: 60306

PITARD, A., P. BRENNAN, J. CLAVEL, E. GREISER, G. LOPEZ-ABENTE, J. CHANG-CLAUDE, J. WAHRENDORF, C. SERRA, M. KOGEVINAS, AND P. BOFFETTA. "Cigar, Pipe, and Cigarette Smoking and Bladder Cancer Risk in European Men." *Cancer Causes and Control*, 12: 551–556, 2001 (4 tables, 28 references)

Abstract: The risk of bladder cancer was determined in pooled data of 2,279 cases and 5,268 controls in Denmark, France, Germany, and Spain. A total of 154 cases and 1,109 controls were classified as never-smokers. The smokers consisted of 57.2% cigarettes only, 2.3% cigars only, 1.5% pipes only, and 22.3% mixed smokers.

> Despite these limitations, our results might have important public health implications. Even though the consumption of cigar and pipe tobacco declined in most European countries and the USA until the 1980s, cigar smoking has since become increasingly popular again in the USA, especially among teenagers. A similar increase in cigar smoking can be expected to occur in Europe also. Pipe smoking is not on the increase. Thus, our results serve as a warning that, if the prevalence of cigar smoking continues to increase, it is likely to result in an increase in the number of cases of bladder cancer, in addition to cases of lung cancer.
>
> In our study, cigar and pipe smoking was clearly associated with an increased risk of bladder cancer. The precise magnitude of the risk depends on inhalation patterns and average daily consumption, factors that we could not fully address. Control of cigar and pipe smoking is therefore an important component of strategies for the prevention of bladder cancer.

Reference Number: 60307

DUGAD, S., P. BHANDARE, G. KULKARNI, AND S. PATIL. "Sputum Characteristics Among 'Chillum' Smoking 'Sadhus' of Kumbh Mela." *MVP Journal of Medical Sciences*, 5: 96–100, 2018 (4 tables, 4 figures, 9 references)

Abstract: Chillum is a straight conical pipe made of clay, with an end-to-end channel, invented in India but spread globally in the 1960s. The size of the chillum has an effect on smoking, as a short chillum will bring hot smoke directly into the mouth. Chillum can be used to smoke tobacco, cannabis, or a cannabis/tobacco mixture. This study analyzed 50 sputum samples of chillum smoking Sadhus of India.

> Characteristic of sputum of chillum smoking sadhus was almost whitish, mucoid (73%), yellow (20%), followed by greenish (7%).
>
> Chillum smoking was found to be associated with raised differential cell count viz., Neutrophils, Lymphocytes, and Macrophages in sputum which indicates inflammatory changes in the airways and lung parenchyma.

Reference Number: 60308

SINGH, S., M. SOUMYA, A. SAINI, V. MITTAL, U.V. SINGH, AND V. SINGH. "Breath Carbon Monoxide Levels in Different Forms of Smoking." *Indian Journal of Chest Diseases and Allied Sciences*, 53: 25–28, 2011 (1 table, 1 figure, 17 references)

Abstract: In India, many people consider hookah and chillum smoking to be less toxic than cigarettes as the smoke is filtered through water or a wet cloth, respectively. Cigarette smoke contains 2–4% carbon monoxide. The eCO of 18 smokers (mean age 48 years) was determined after smoking a bidi, cigarette, hookah, and chillum.

Table: Mean Exhaled Carbon Monoxide Increase (ppm) at the End of Smoking Various Tobacco Products

Type of Smoking	Mean eCO Increase after Smoking (ppm)
Bidi	3.00
Cigarette	4.94
Hookah (5 min)	25.27
Hookah (1 serve, 15 min)	41.57
Chillum (5 min)	5.55
Chillum (1 serve, 15 min)	45.00

Source: Adapted from Singh et al, 2011

> *Bidi*, the most commonly used mode of smoking, is as toxic as a regular cigarette in terms of end-tidal CO levels. *Hookah* and *chillum* use commonly considered harmless were associated with almost eight-fold higher rise in breath CO levels in comparison to cigarette smoking, indicating substantially higher toxicity. One serve of chillum smoking lead to maximal increase in eCO levels indicating the possibility of chillum being the most dangerous mode of smoking.

6.04 HOOKAHS/WATER PIPES

> "In the past decade, waterpipe smoking has become a global phenomenon among youth, many of whom are unaware of the health risks associated with this form of tobacco consumption. There are currently no evidence based warning label requirements for waterpipe products, whereas warnings on cigarette packaging is a major form of communicating tobacco-related health risks to users and others exposed to their smoke. As a result, waterpipe smokers may be misled to believe that the product is less harmful than other tobacco products that carry warning labels."
>
> —Islam et al, "Effectiveness of Health Warnings for Waterpipe Tobacco Smoking Among College Students" (2016)

> "Many waterpipe smokers will point to labeling on waterpipe tobacco that indicates '0% tar'... This labeling is misleading.... Because 'tar' is a *smoke* constituent, any labeling on a container of non-burning tobacco is technically accurate that there is no 'tar' in the tobacco, just as an unlit cigarette also contains no tar."
>
> —Soule et al, 2015

Hookahs are also known as:

- Waterpipes
- Narghile
- Argileh
- Shiska
- Shui jun dai
- Hubble-bubble
- Gozza
- Borry
- Bora goza
- Qaylan
- Chia
- Mada

A waterpipe typically consists of a lower compartment made up of a water jar covered by a gasket with a hose to connect to the user's mouth and a release valve. The upper compartment consists of a bowl where the tobacco or molasses is placed, then covered by perforated aluminum foil. Burning charcoal is placed on top of the foil (60401, 60413). The smoke is then inhaled through the water, through the tube with a release valve, which cools the smoke and gives the false impression that the water filters the smoke and makes it safer (60403).

The waterpipe has existed for over a millennium, originating in the northwestern provinces of India and spreading to Iran, the Arab world, and Turkey. Hookah use has become increasingly popular in Western countries, especially among youth, in part due to the misconception that hookahs are safer and less addictive than cigarettes (60402, 60404, 60407–8). This misconception is further fueled by misleading waterpipe labelling stating there is no tar or low nicotine concentrations in the tobacco used (60405–6, 60409).

A single puff from a cigarette has a volume of 50 to 100 mL, compared to 511 to 900 mL for a single puff from the waterpipe. In addition, a waterpipe smoking session can last for 30 minutes or more, and involve between 60 and 160 puffs (60404, 60419), which causes substantial secondhand smoke containing high CO, nicotine, and carcinogen levels in waterpipe cafés (60410–3).

Waterpipe use has been associated with similar diseases caused by cigarettes, including nicotine addiction, ischemia heart disease, COPD, emphysema, lung, and other cancers (60414, 60416–7, 60419).

Between 2001 and 2019, there were approximately 1,371 ER visits to hospital for hookah-related injuries including burns, fainting, and dizziness (60421). Due to the charcoal used to heat the tobacco, a high concentration of CO can occur in the hookah café. This resulted in four persons being hospitalized for CO poisoning which required oxygen therapy (60422).

The transmission of infectious microbial agents can occur as most hookah cafés do not clean the waterpipe hoses between smoking sessions. Infections of hepatitis C, tuberculosis, herpes simplex virus, and other respiratory diseases have occurred after waterpipe smoking (60423).

Reference Number: 60401

ST. HELEN, G., N.L. BENOWITZ, K.M. DAINS, C. HAVEL, M. PENG, AND P. JACOB III. "Nicotine and Carcinogen Exposure after Water Pipe Smoking in Hookah Bars." *Cancer Epidemiology Biomarkers and Prevention*, 23: 1055–1066, 2014 (4 tables, 57 references)

Abstract: A water pipe consists of a head, which is connected to a water jar, and one or more hoses with a mouthpiece. A tobacco and moist fruit preparation is placed on the head and covered with perforated aluminum foil and burning charcoal is placed on top. The smoke is inhaled through the water, which cools it. A typical smoking session lasts 45 to 60 minutes. Hookah smoking, due to the burning charcoal, has higher CO, benzene, and carcinogenic polycyclic aromatic hydrocarbons (PAH). Twenty-four female and 31 male hookah smokers (mean age 24.5 years) refrained from smoking for a week and then went to a hookah bar where they smoked for an average of 74 minutes. Pre-exposure, post-exposure, and next day urine samples were analyzed for nicotine, cotinine, and various volatile organic compounds (VOC), including TSNA carcinogens by LC/MS/MS. Urine nicotine concentrations increased by 73×, cotinine by 4×, and NNAL by 2× after hookah use. Eight subjects had pre-exposure urine cotinine levels that indicated smoking. Forty percent of the subjects reported smoking marijuana in the past 30 days.

> Water pipe smoking in a hookah bar is associated with significant nicotine intake and carcinogen exposure. . . . Given the significant intake of nicotine and carcinogens, chronic water pipe use could place users at increased risk of cancer and other chronic diseases.

Reference Number: 60402

SOULE, E.K., T. LIPATO, AND T. EISSENBERG. "Waterpipe Tobacco Smoking: A New Smoking Epidemic Among the Young?" *Current Pulmonary Reports*, 4: 163–172, 2015 (3 figures, 119 references)

Abstract: Waterpipe tobacco smoking (WTS) is becoming more prevalent worldwide, especially among college students and youth, as it is incorrectly perceived as being less harmful than smoking cigarettes. But WTS sessions are typically 45 minutes in duration compared to 5 minutes for cigarettes, and as many as 78–269 puffs taken compared to 10 puffs per cigarette. Many waterpipe smokers will point to a label on waterpipe tobacco that indicates 0% tar. However, tar is the dry particulate matter that occurs in smoke when the tobacco is burned and is not found in tobacco itself, so the label is misleading. WTS is associated with many of the same chronic health conditions found with cigarette smoking, such as lung cancer, COPD, CVD, bronchitis, and asthma.

> Waterpipe tobacco smoking is becoming increasingly popular, particularly among youth and young adults. Evidence suggests that WTS is associated with dependence and acute and long-term negative health outcomes

similar to those of cigarette smoking. Pulmonologists and primary care providers should assess for WTS in addition to other forms of tobacco use and explain to patients that there is no indication that WTS is less risky than cigarette smoking. Based on clinical judgment, patients who report any form of tobacco use, including WTS, should be advised that abstaining from all forms of tobacco will limit their risks of negative health outcomes.

Reference Number: 60403

LIPKUS, I.M., T. EISSENBERG, R.D. SCHWARTZ-BLOOM, A.V. PROKHOROV, AND J. LEVY. "Affecting Perceptions of Harm and Addiction Among College Waterpipe Tobacco Smokers." *Nicotine and Tobacco Research*, 13: 599–610, 2011 (4 tables, 1 figure, 37 references)

Abstract: A waterpipe is a nicotine delivery system in which tobacco smoke passes through water before it is inhaled. A large proportion of college students believe WTS is less harmful and addictive than cigarette smoking. Two web-based studies were conducted on the effect of providing information about the harms of daily waterpipe smoking (WPS) to college studies. In Study 1, which provided incremental information about the harm of WPS, 62% of the participants reported having stopped waterpipe use compared to 33% in the control group.

In sum, our results provide preliminary and promising evidence that minimally intensive interventions delivered online that educate college waterpipe smokers of harm, addiction, and toxicant exposure of waterpipe tobacco use can increase factual understanding of the harms of waterpipe use, perceptions of risk, worry, and desire to quit. In addition, there is some evidence that our most promising intervention materials (Study 1 experimental group) may increase cessation.

Reference Number: 60404

COBB, C., K.D. WARD, W. MAZIAK, A.L. SHIHADEH, AND T. EISSENBERG. "Waterpipe Tobacco Smoking: An Emerging Health Crisis in the United States." *American Journal of Health Behavior*, 34: 275–285, 2010 (2 tables, 2 figures, 94 references)

Abstract: WTS is usually associated with southeast Asia and north Africa, but has spread globally and 10–20% of some young adult populations are current waterpipe users. Many WTS users are tobacco naive and thus it may be a gateway to cigarette use. The duration of a single waterpipe use episode lasts about 1 hour rather than 5 minutes for cigarette smoking. On average, a WTS session involves inhaling about 90L of smoke, compared to 500–600 mL of smoke for a single cigarette. It is estimated that for a single WTS session, there is 1.7× the amount of nicotine inhaled, compared to cigarettes, and 36× the amount of tar. Smoke CO from WTS is 18× greater than for cigarettes.

Table: Puff Topography for Waterpipe Users and Cigarette Smokers

Puff Variable	Waterpipe	Cigarette
Puffs (number)	171	12.7
Puff volume (mL)	530	48.6
Puff duration (seconds)	2.6	1.5
Interpuff interval (seconds)	15.5	21.3

Source: Adapted from Cobb et al, 2010

Waterpipe tobacco smoking is a growing health concern globally and especially among young adults in the United States. Many waterpipe tobacco smokers perceive this behavior to be less lethal and addicting than cigarettes. Converging lines of evidence, including waterpipe smoke analysis, user toxicant exposure, and health effects research contradict this perception, though more study is required.

Reference Number: 60405

VANSICKEL, A.R., A. SHIHADEH, AND T. EISSENBERG. "Waterpipe Tobacco Products: Nicotine Labelling Versus Nicotine Delivery." *Tobacco Control*, 21: 377–379, 2012 (2 figures, 18 references)

Abstract: Thirty-two female and 78 male waterpipe smokers (mean age 21 years) engaged in a 45-minute waterpipe smoking session using 3 brands of tobacco: Nakhla (0.5% nicotine), Starbuzz (0.5% nicotine), and Al Fakher (0.05% nicotine). Blood samples were collected, and plasma nicotine concentrations were measured. The mean peak plasma nicotine concentrations were highest for the Al Fakher brand (11.4 ng/mL), followed by Nakhla (9.9 ng/mL) and Starbuzz. The waterpipe tobacco is also labelled as 0.0% tar, which is correct but misleading, as tar is a smoke product not found in tobacco itself.

Labelling that includes wording such as "0.05% nicotine" and "0.0% tar" may lead consumers to believe that waterpipe tobacco smoking exposes them to little or no toxicants. In fact, the WHO has suggested that "0.0% tar" labels on waterpipe products be banned. Furthermore, health warning labels on waterpipe tobacco products are either non-existent or barely visible, covering an average of just 3.5% of the total surface area of packages. Often, the health warning labels on waterpipe tobacco products refer to cigarettes or smoking in general rather than waterpipe smoking in particular. Overall, the current state of waterpipe tobacco labelling may confuse more than inform consumers. These products should be regulated and held to the same product testing and labelling standards as other tobacco products (eg, cigarettes) because consumers have the right to accurate and non-misleading information regarding the products that they are using.

Reference Number: 60406

ISLAM, F., R.G. SALLOUM, R. NAKKASH, W. MAZIAK, AND J.F. THRASHER. "Effectiveness of Health Warnings for Waterpipe Tobacco Smoking Among College Students." *International Journal of Public Health*, 61: 709–715, 2016 (3 tables, 32 references)

Abstract: The effectiveness of text and pictorial health warnings were determined in 262 ever and 105 current waterpipe users from the University of South Carolina in 2014. Text-only messages and pictorial labels warning about harm to children were the most effective warnings at encouraging waterpipe users to quit.

Table: Percentage of Current and Ever Waterpipe Users that Would be Encouraged to Quit According to the Text of the Health Warning Label

Text Warning	Current Waterpipe Users (% Quit)	Ever Waterpipe Users (% Quit)
WARNING: Hookah smoking is addictive	14.3%	17.6%
WARNING: Hookah smoke causes fatal lung disease in non-smokers	26.7%	36.3%
WARNING: Hookah smoking causes cancer	41.9%	48.5%
WARNING: Hookah smoke can harm your children	54.3%	55.3%
WARNING: Hookah smoking during pregnancy can harm your baby	64.8%	68.7%

Source: Adapted from Islam et al, 2016

In the past decade, waterpipe smoking has become a global phenomenon among youth, many of whom are unaware of the health risks associated with this form of tobacco consumption. There are currently no evidence based warning label requirements for waterpipe products, whereas warnings on cigarette packaging is a major form of communicating tobacco-related health risks to users and others exposed to their smoke. As a result, waterpipe smokers may be misled to believe that the product is less harmful than other tobacco products that carry warning labels. Furthermore, existing health warnings on waterpipe tobacco products worldwide are deliberately misleading and do not conform to the guidelines set in Article 11 of the World Health Organization Framework Convention on Tobacco Control (WHO FCTC). For example, one study found that existing health warning labels on waterpipe tobacco packages in a select number of countries covers only an average of 3.5% of the total surface area of packages, which is significantly less than the 30% required under the FCTC. Often, these health-warning labels refer to cigarettes or smoking in general rather than waterpipe smoking in particular.

Reference Number: 60407

LIPKUS, I.M., D. MAYS, AND K.P. TERCYAK. "Characterizing Young Adults' Susceptibility to Waterpipe Use and Their Reactions to Messages About Product Harms and Addictiveness." *Nicotine and Tobacco Research*, 19: 1216–1223, 2017 (4 tables, 35 references)

Abstract: Epidemiological and laboratory data show that waterpipe smoke is no less dangerous than tobacco smoke, but at the time of the study it was estimated that 18.2% of youth (18–24 years of age) had used a waterpipe. An internet survey was conducted of 508 adults (aged 18–30 years) who had never used a hookah. About 20% smoked cigarettes and 12% used another tobacco product. Participants were classified as susceptible or not and were randomized to receive messages about waterpipe smoking harm or no message (control).

In this study, we addressed two issues that have received very little attention in the waterpipe tobacco literature. The first issue is characterizing young adults who are susceptible to WTS outside of college settings—importantly, our results were not moderated by level of education or being in college or not. Second, to the best of our knowledge, we examined for the first time effects of a risk communication intervention to modify risk appraisals, attitudes, and beliefs that may result in WTS among susceptible young adults. Results point to several psychosocial characteristics that place susceptible versus nonsusceptible young adults at higher risk for WTS uptake, including: underappreciating the harms and addictive potential of waterpipe (ie, risk appraisals); less negative evaluations of waterpipe (ie global attitude); and importantly, a stronger willingness/curiosity to try WTS. Viewing WTS as safer than cigarettes, and more positive and less negative attitudes toward WTS have been linked empirically with WTS. These findings point to crucial factors to target with public health efforts to curb WTS uptake among susceptible young adults.

Reference Number: 60408

OWENS, V.L., T. HA, AND N. SOULAKOVA. "Widespread Use of Flavored E-cigarettes and Hookah Tobacco in the United States." *Preventative Medicine Reports*, 14: (5 pp), 2019 (2 tables, 2 figures, 27 figures)

Abstract: Although the FDA banned the sale of flavored cigarettes (except menthol) in the United States in 2009, flavored e-cigarettes (ECs) and flavored hookah tobacco (FHT) are still available. A survey was conducted of 3,691 current EC users and 658 current hookah users between 2014 and 2015. Among the EC users, 65.5% used flavored ECs and its use was greatest among the 18- to 24-year-old, especially compared to the 45+ years of age, and among women compared to men. The use of flavored ECs was also higher among never-smokers than current smokers of cigarettes. About 87% of hookah tobacco users consumed flavored tobacco.

Table: Characteristics of E-cigarette and Hookah Users

Characteristics	E-cigarette Users	Hookah Tobacco Users
18–24 years of age	18.5%	50.4%
25–44 years of age	42.3%	42.9%
45+ years of age	39.2%	6.7%
Never smoker	8.3%	56.3%
Former smoker	26.1%	11.2%
Current smokers	65.6%	32.5%

Source: Adapted from Owens et al, 2019

Availability of emerging tobacco products in many different flavors could be a reason for tobacco use initiation and continued use of tobacco after cessation of regular-cigarette smoking.

Reference Number: 60409

RIJHWANI, K., S. HASHMI, V.R. MOHANTY, A.Y. BALAPPANAVAR, AND S. KAPOOR. "Hookah Products and Online Retail Marketing Strategies in India: A Content Analysis." *Indian Journal of Cancer*, 55: 261–264, 2018 (22 references)

Abstract: Hookah smoking is becoming extremely popular among youth in India. A content analysis of seven popular hookah websites in India was conducted in 2017. A total of 382 hookah products were advertised on the websites and only 7% had health/age warnings. About 20% of the products claimed to be tobacco/nicotine free. Nearly 90% of the websites described hookah smoking in positive phrases such as the "ultimate way to celebrate" or a way to "enjoy the world of smoking without any doubt of harm."

The marketed aesthetic appeal, claims of being healthy alternatives, and heavy discounts offered makes it challenging to persuade people of its potential harm and addictiveness. Hence, public health practitioners and policy-makers need to raise the issue of various hookah products in the same tune as done for other forms of tobacco. There is a need to emphasize that the product used in hookahs could be containing tobacco and that the smoke nevertheless contains combustion products similar to the smoke from cigarettes. We might make use of the same online platform which has "normalized" hookah use at homes and in the society to create awareness.

Reference Number: 60410

SEIDENBERG, A.B., E.N. ORLAN, M.J. TRAVERS, AND E.L. SUTFIN. "Air Quality and Presence of Air Ventilation Systems Inside Waterpipe Cafés in North Carolina." *Tobacco Control*, 28: 356–358, 2019 (1 figure, 20 references)

Abstract: Air quality (carbon monoxide (CO) and particulate matter ($PM_{2.5}$)) were measured inside and outside of six waterpipe cafés (WC) in North Carolina in September 2015 and January 2016. All cafés reported installing air ventilation systems inside. The median indoor

CO concentrations for the two different times were 42 ppm and 65 ppm, respectively, and the median $PM_{2.5}$ were 379.3 and 484.0 μg/m^3. The levels of these pollutants were much lower outside of the hookah cafés.

In conclusion, unsafe levels of CO and $PM_{2.5}$ were measured within a sample of WCs in NC, despite the reported presence of air ventilation. Prohibiting indoor waterpipe smoking may be required to ensure clean air for employees and customers.

Reference Number: 60411

ZHOU, S., M. WEITZMAN, R. VILCASSIM, J. WILSON, N. LEGRAND, E. SAUNDERS, M. TRAVERS, L-C. CHEN, R. PELTIER, AND T. GORDON. "Air Quality in New York City Hookah Bars." *Tobacco Control*, 24: e193–e198, 2015 (3 tables, 1 figure, 55 references)

Abstract: Hookah bars have been allowed in many US cities if tobacco-free, non-nicotine containing herbal hookah products like shisha are used. Air samples were collected surreptitiously in eight NYC hookah bars during the summer and fall of 2013 over one to two sessions. The number of active hookah pipes ranged from 4 to 18 (average 9) per bar. Airborne nicotine was found in all bars despite the ban on tobacco-based shish in water pipes in NYC. Overall, the levels of indoor air pollution increased with the increasing number of active hookahs smoked. The mean real time $PM_{2.5}$ level was 1,179.9 μg/m^3 and the mean CO concentration was 32 ppm.

These results demonstrate that despite the ban on smoking tobacco products, at the very least, some NYC hookah bars are serving tobacco-based hookahs, and have elevated concentrations of indoor air pollutants and toxicants that may present significant health threats to visitors and employees. Therefore, there is an urgent need for better air quality monitoring in such establishments and policies to combat this emerging public health threat.

Reference Number: 60412

JACOB III, P., A.H.A. RADDAHA, D. DEMPSEY, C. HAVEL, M. PENG, L. YU, AND N.L. BENOWITZ. "Nicotine, Carbon Monoxide, and Carcinogen Exposure after a Single Use of a Waterpipe." *Cancer Epidemiology, Biomarkers & Prevention*, 20: 2345–2353, 2011 (2 tables, 4 figures, 23 references)

Abstract: It is a common and incorrect belief that passing tobacco smoke through water makes is much less harmful than cigarette smoke. The waterpipe apparatus consists of a head, which holds 1 to 20 grams of tobacco, over which is placed perforated aluminum foil with burning charcoal on top. The tobacco is typically mixed with honey, molasses, or fruit pulp. Waterpipe smoke is at a temperature of 450°C, compared to 900°C for cigarette smoke. Eight male and 8 female subjects smoked a waterpipe in a clinical research ward. Expired CO and COHb were measured, as well as TSNA, and PAH metabolites in urine and plasma nicotine concentrations.

> Our study confirms the results of previous studies that waterpipe smokers absorb nicotine in amounts comparable to cigarette smokers, and that they absorb substantially more CO. We also measured excretion of carcinogen biomarkers. Following a single waterpipe smoking session, there were increases in urinary excretion of biomarkers for two classes of carcinogens present in tobacco smoke, tobacco-specific nitrosamines (TSNA) and polycyclic aromatic hydrocarbons (PAHs). The maximum boosts were less than those typically found in habitual cigarette smokers. Absorption of nicotine, CO, and carcinogens was generally higher in mixed tobacco users than in waterpipe-only smokers, presumably due to greater depth of inhalation in the subjects who also smoked cigarettes. Additional studies are needed to confirm that mixed tobacco users smoke differently than waterpipe only smokers. Our study shows that waterpipe smoking results in significant amounts of carcinogen absorption, raising concerns of cancer risk.

Reference Number: 60413

QASIM, H., A.B. ALARABI, K.H. ALZOUBI, Z.A. KARIM, F.Z. ALSHBOOL, AND F.T. KHASAWNEH. "The Effects of Hookah/Waterpipe Smoking on General Health and the Cardiovascular System." *Environmental Health and Preventative Medicine*, 24: 58 (17 pp), 2019 (1 table, 6 figures, 213 references)

Abstract: Waterpipes, most commonly known as hookah, consist of upper and lower compartments. The lower compartment consists of a water jar covered by a gasket with a hose to connect to the user's mouth and a release valve. The upper compartment consists of a bowl where the tobacco, or molasses, is placed, then covered by perforated aluminum foil with burning charcoal placed on top. Compounds found in mainstream hookah smoke include nicotine, CO, carcinogenic polycyclic aromatic hydrocarbons (PAHs), aromatic amines, aldehydes, furanic and phenolic compounds, tar, particulate matter, heavy metals, and ammonia.

Table: Hookah Apparatus Components and Their Use

Component	Use/Properties
Water jar (base)	Glass container usually filled with water; milk, fruit juices, or alcohol may also be used
Pipe (body, stem)	Metallic tube extending from the bowl to the base and partially immersed in water
Bowl (head)	Holds 10–20 grams of tobacco
Saucer	Holds charcoal ashes
Filter	Placed on tip of hose and claimed to reduce tar and nicotine
Mouthpiece (extension)	A short plastic tip used to breathe through
Aluminum foil	Separates the tobacco from the burning charcoal; usually perforated
Charcoal	Special type of round pellets that are easy to light
Tobacco	Two types; moassel (fruit flavor, mosit), or ajami (unflavored)

Source: Adapted from Qasim et al, 2019

Based on the scientific evidence of the toxicity associated with tobacco use (passive/active), an increasing number of states has instituted/is instituting regulations to support/expand legislation of clear indoor air quality to include hookah. As mentioned before, hookah smoke may be associated with similar or even greater inhalation of toxicants in comparison to cigarettes. Since 2016, the FDA finalized a rule extending their control of all tobacco products, including hookah tobacco. FDA now regulates the manufacture, import, packaging, labeling, advertising, promotion, and sale, as well as distribution of hookah tobacco and of all hookah apparatus parts (except the accessories; lighters and tongs). However, some US legislations/policies controlling cigarettes do not similarly apply to hookah. For instance, the "Prevent All Cigarette Trafficking Act" prevents US Postal Service to ship cigarettes but does not interfere with hookah shipping. Additionally, nearly 90% of the largest US cities may allow hookah in bars via exemptions (whereas cigarettes are prohibited). Unfortunately, youth represent a large portion of the hookah user population, and their accessibility is facilitated through online ordering. To control such means of access, major credit card companies should ban online payments for hookah, as they did with cigarettes. Clearly, policies and policy improvements are crucial, and research on hookah and its health effects would be helpful/important to health policy officials seeking to update/refine them.

Reference Number: 60414

BAHELAH, R., J.R. DIFRANZA, F.M. FOUAD, K.D. WARD, T. EISSENBERG, AND W. MAZIAK. "Early Symptoms of Nicotine Dependence Among Adolescent Waterpipe Smokers." *Tobacco Control*, 25: e127-e134, 2016 (5 tables, 43 references)

Abstract: Nicotine dependence (ND) was determined in 160 waterpipe smokers and 24 cigarette smokers in grades 8 and 9 in Lebanon (mean age 14 years). The Hooked on Nicotine Checklist (HONC) and other measures of ND were determined. Among waterpipe smokers 71.3% had at least one HONC symptom and 38.1% had the full syndrome of ND (3+ items). The early symptoms of ND among waterpipe smokers were cravings (25%), feeling addicted (22.5%), and failed quitting attempts (14.5%). The median use of 7.5 waterpipes a month was associated with the first HONC item. At a median of 15 waterpipes a month, the full syndrome of nicotine addiction occurred.

Symptoms of ND develop among adolescent waterpipe smokers at low levels of consumption and frequency of use. Craving for nicotine triggered by waterpipe-specific cues is reported even at this young age. Waterpipe-specific ND prevention and intervention programs for youth are needed.

Reference Number: 60415

JACOB III, P., A.H.A. RADDAHA, D. DEMPSEY, C. HAVEL, M. PENG, L. YU, AND N.L. BENOWITZ. "Comparison of Nicotine and Carcinogen Exposure with Water Pipe and Cigarette Smoking." *Cancer Epidemiology, Biomarkers & Prevention*, 22: 765-772, 2013 (1 table, 2 figures, 28 references)

Abstract: Smoking tobacco by using a waterpipe has gained popularity, as it is perceived as relatively safe compared to cigarettes. Eight men and five women (mean age 24 years) smoked either 3 waterpipes or 11 cigarettes for 4 consecutive days and had various biomarkers measured in the urine, blood, and breath. The temperature of the tobacco is 450°C in the waterpipe and about 600–900°C in cigarettes. CO levels were higher in subjects after smoking a waterpipe than smoking cigarettes, due to the burning charcoal. Waterpipe smokers had a greater exposure to benzene and high molecular weight PAH, but less exposure to TSN, acrolein, and ethylene oxide, and low molecular weight PAH.

> A different pattern of carcinogen exposure might result in a different cancer risk profile between cigarette and water pipe smoking. Of particular concern is the risk of leukemia related to high levels of benzene exposure with water pipe use.

Reference Number: 60416

EL-ZAATARI, Z.M., H.A. CHAMI, AND G.S. ZAATARI. "Health Effects Associated with Waterpipe Smoking." *Tobacco Control*, 24: i31-i43, 2015 (5 tables, 152 references)

Abstract: The worldwide prevalence of WPS is estimated to be 100 million. WPS is becoming increasingly popular among youth, in part because it is wrongly considered to be safer and less addictive than cigarettes. A review was conducted of the short- and long-term risks associated with WPS. Some of the acute effects of WPS are increased heartrate, increased blood pressure, carbon monoxide intoxication, impaired pulmonary function, decreased exercise capacity, and larynx and voice changes.

Table: Adverse Long-Term Health Effects Associated with Waterpipe Smoking

Adverse Long-Term Health Effects
• Ischemic heart disease • Impaired pulmonary function • Chronic obstructive lung disease • Chronic bronchitis • Emphysema • Lung cancer • Oesophageal cancer • Gastric cancer • Low birthweight • Pulmonary problems at birth • Periodontal disease • Larynx and voice changes • Lower bone density and increased fracture risk

Source: Adapted from El-Zaatari et al, 2015

Contrary to the widely held misconception, WPS is associated with a variety of adverse short-term and long-term health effects that should reinforce the need for stronger regulation. In addition, this review highlights the limitations of the published work, which is mostly cross-sectional or retrospective. Prospective studies should be undertaken to assess the full spectrum of health effects of WPS, particularly in view of its growing popularity and attractiveness to youth.

Reference Number: 60417

WARD, K.D., S. AHN, F. MZAYEK, R.A. ALI, S. RASTAM, T. ASFAR, F. FOUAD, AND W. MAZIAK. "The Relationship Between Waterpipe Smoking and Body Weight: Population-Based Findings from Syria." *Nicotine and Tobacco Research*, 17: 34–40, 2015 (3 tables, 53 references)

Abstract: The body mass index (BMI) of 2,134 subjects who had never smoked waterpipes (84.1%), 116 former waterpipe smokers (4.6%), 251 current non-daily smokers (9.9%), and 35 current daily smokers (1.4%) in Aleppo, Syria were measured.

In conclusion, this is the first study to document an association between waterpipe smoking and BMI. Daily waterpipe users, compared to nonusers, have higher BMI, translating into six extra kg (13 pounds) of weight on average, and are three times as likely to be obese. This extra body weight may increase their risk of several health conditions, including metabolic syndrome, diabetes, and coronary heart disease.

Reference Number: 60418

KHABOUR, O.F., K.H. ALZOUBI, N. AL-SHEYAB, M.A. AZAB, A.M. MASSADEH, A.A. ALOMARY, AND T.E. EISSENBERG. "Plasma and Saliva Levels of Three Metals in Waterpipe Smokers: A Case Control Study." *Inhalation Toxicology*, 30: 224–228, 2018 (1 table, 3 figures, 32 references)

Abstract: Blood and oral fluid samples were collected from 88 Jordanian exclusive waterpipe smokers and 40 Jordanian never tobacco users. Plasma and oral fluid concentrations of three heavy metals (cadmium, lead, and zinc) were measured by spectrometry. The concentration of metals was higher in waterpipe smokers than non-smokers.

Table: Plasma and Oral Fluid Cadmium, Lead, and Zinc Concentrations in Waterpipe Smokers and Non-Smokers

Heavy Metal	Waterpipe Smokers	Non-Smokers
Plasma cadmium (µg/dL)	3.3	0.82
Plasma lead (µg/L)	5.2	3.4
Plasma zinc (µg/L)	2.0	1.5
OF cadmium (µg/dL)	5.1	0.64
OF lead (µg/L)	4.8	2.8
OF zinc (µg/L)	0.94	0.45

Source: Adapted from Khabour et al, 2018.

In conclusion, waterpipe tobacco smoking is associated with elevated levels of metals in body fluids. These results provide another demonstration of how waterpipe tobacco smoking exposes smokers to harmful toxicants.

Reference Number: 60419

ABOAZZIA, E., AND T. EISSENBERG. "Waterpipe Tobacco Smoking: What is the Evidence That it Supports Nicotine/Tobacco Dependence?" *Tobacco Control,* 24: i44–i53, 2015 (2 tables, 2 figures, 131 references)

Abstract: The mean nicotine concentration for flavored waterpipe tobacco is 3.4 mg/g, compared to a mean of 13.8 mg/g for cigarette tobacco. However, a single puff from a cigarette is about 50–100 mL in volume, compared to a single waterpipe puff of 511–906 mL. In addition, WPS can take 30 minutes or longer and involve 60–169 puffs.

One of the most compelling indicators of dependence may be continued use despite known adverse health effects . Unfortunately, this compelling indicator is also the most difficult to address empirically today, primarily because the health effects of WTS have been understudied for decades and are only beginning to be revealed. Too often, interviews and surveys reveal that WTS is considered to be a low risk activity or, when health risks are acknowledged, they may be downplayed because they are reported inconsistently and/or perceived to be less than those associated with cigarette smoking. This perception likely is strengthened by labelling of waterpipe tobacco that demonstrably is misleading. In this context of poorly communicated health effects and misleading labelling, the well-documented global spread of WTS is unsurprising and the likelihood of documenting continued use despite known adverse health effects is unlikely.

Reference Number: 60420

KASSEM, N.O.F., N.O. KASSEM, S. LILES, A.T. ZARTH, S.R. JACKSON, R.M. DAFFA, D.A. CHATFIELD, S.G. CARMELLA, S.S. HECHT, AND M.F. HOVELL. "Acrolein Exposure in Hookah Smokers and Non-Smokers Exposed to Hookah Tobacco Secondhand Smoke: Implications for Regulating Hookah Tobacco Products." *Nicotine and Tobacco Research*, 20: 492–501, 2018 (4 tables, 68 references)

Abstract: Urine samples were collected from 105 hookah smokers and 103 non-smokers the morning before and the morning after attending a hookah smoking event indoors. The urine was analyzed by LC/MS/MS for 3-hydroxypropylmercapturic acid (3-HPMA), a metabolite of acrolein. Acrolein is a highly ciliatoxic and toxic respiratory irritant present in tobacco smoke. Urinary cotinine concentrations were also determined. When burned, the sugars that are added to tobacco increase acrolein, formaldehyde, acetone, acetaldehyde, and 2-furfural concentrations in tobacco smoke.

Table: Mean Urinary 3-HPMA Concentrations (pmol/mg creatinine) Before and After Attending an Indoor Hookah Smoking Event

	Pre-Event	Post-Event
Daily hookah smokers	1,991	3,348
Non-daily hookah smokers	1,754	2,352
Non-smokers	1,791	2,497

Source: Adapted from Kassem et al, 2017

Hookah tobacco smoke is a source of acrolein exposure. Findings support regulating hookah tobacco products including reducing humectants and sugar additives, which are precursors of acrolein under certain pyrolysis conditions. We suggest posting health warning signs for indoor smoking in hookah lounges, and encouraging voluntary bans of smoking hookah tobacco in private homes.

Reference Number: 60421

INYANG, N.A., J.T. CHANG, AND B. WANG. "National Estimates of Hospital Emergency Department Visits due to Acute Injuries Associated with Hookah Smoking, United States, 2011–2019." *Injury Epidemiology*, 7: 41 (6 pp), 2020 (2 Tables, 25 references)

Abstract: Between 2011 and 2019 there were an estimated 1,371 ER visits to US hospitals due to hookah-related injuries. Two-thirds of the hookah users were between 18 and 24 years of age. The most frequent acute injuries (AI) were burns (41.5%), followed by syncope (fainting), falls (36.5%), and dizziness or light-headedness (18.3%).

> One of the most serious AIs of hookah smoking is acute CO poisoning. More than half of the cases in our study reported symptoms that are consistent with CO poisoning. This is likely an underestimate due to lack of awareness or misperception of symptoms associated with CO poisoning. Some individuals with CO poisoning symptoms may not relate lightheadedness and dizziness to CO poisoning and therefore not present to EDs. Additionally, some CO symptoms could have been considered intended effects of hookah smoking. A previous study among college students found that a commonly reported reason for hookah smoking is relaxation; hookah smokers associated feelings of relaxation with lightheadedness, dizziness, loss of balance, or a so-called "buzz."

Reference Number: 60422

VON RAPPARD, J., M. SCHÖNENBERGER, AND L. BÄRLOCHER. "Carbon Monoxide Poisoning Following Use of a Water Pipe/Hookah." *Deutsches Arzteblatt International*, 111, 674–679, 2014 (2 tables, 1 figure, 36 references)

Abstract: This study presents three case reports of 4 young adults (aged 16–21 years) who attended an ER as a result of CO poisoning due to waterpipe use in Germany. Hookah uses burning charcoal, which is a potent source of CO emissions. The COHb ranged between 16.7 to 29.6%. The patients were administered high pressure oxygen, and all recovered.

> This case series reveals that CO intoxication due to water-pipe smoking is probably more common than is generally realized. Emergency room staff should be aware of this problem and inquire specifically about water-pipe smoking in patients with nonspecific neurological manifestations.

Reference Number: 60423

ALAGAILI, A.N., T. BRIESE, N.M.S. AMOR, O.B. MOHAMMED, AND W.I. LIPKIN. "Waterpipe Smoking as a Public Health Risk: Potential Risk for Transmission of MERS-CoV." *Saudi Journal of Biological Sciences*, 26: 938–941, 2019 (1 table, 40 references)

Abstract: The Middle East Respiratory Syndrome (MERS-CoV) was first detected in Saudi Arabia in 2012. At the time of the study, there had been a total of 2,147 cases worldwide with 750 fatalities (fatality rate of 35%). Primary infection of MERS-CoV was due to direct contact with camels or their products (e.g., milk). In Saudi Arabia, there were cafés providing waterpipe smoking in areas outside of the cities that were frequently in close proximity to camel markets. A total of 2,489 waterpipe samples were collected but no MERS-CoV RNA was detected.

The risk of transmission of infectious microbial agents through smoking waterpipes is high because smokers cough into hoses and moisture in tobacco smoke promotes the survival of microorganisms inside waterpipe hoses. Moreover, most cafés tend not to clean the waterpipes after each smoking session because washing and cleaning waterpipe parts is labor intensive and time consuming. It is therefore, not surprising, that waterpipe smokers are exposed to microorganisms that may be harmful to health.

Among the important pathogens that are transmitted through waterpipe hoses is Mycobacterium tuberculosis. Waterpipes and mouthpieces have been implicated in an outbreak of pulmonary tuberculosis in Queensland, Australia related infection with Helicobacter pylori to waterpipes smoking in Egypt. The risk of transmission of hepatitis C virus through waterpipes smoking was also demonstrated by Habib et al. (2001). Other viruses that can be transmitted are: Epstein-Barr virus (EBV), herpes simplex virus and respiratory virus. Fungal infections have also been reported to be waterpipes transmitted diseases, when a patient with acute myeloid leukemia showed invasive infection with Aspergillus sp.

6.05 BIDIS/KRETEKS

"Bidi cigarettes are smoked differently than commercial cigarettes. For example, because bidi cigarettes must be re-lit several times because they self-extinguish if they are not puffed at least two times a minute. The time to smoke and number of puffs were significantly greater while smoking a Sher bidi than commercial cigarette smoking in a clinical study. Finally, there is less air dilution through the tendu leaf than conventional cigarettes. Low combustibility forces a smoker to inhale more deeply resulting in greater delivery of carbon monoxide, nicotine, and other components of tobacco smoke. All of these factors may exaggerate the health risks associated with nicotine and other components of bidi smoke."

—Malson, Sims, Murty et al, "Comparison of the Nicotine Content of Tobacco Used in Bidis and Conventional Cigarettes" (2001)

Bidis are hand-rolled tobacco in betel or tendu (Indian Ebony) leaves, which are popular in India (over 50% of the tobacco consumed are bidis) and are much cheaper and less regulated than manufactured cigarettes (60501–4). It is estimated that 100 million people in India smoke bidis compared to 25 million who smoke cigarettes (60505). Bidis are smaller than conventional cigarettes but require a more intensive inhalation and so deliver a similar amount of nicotine (60502, 60504). As with other nicotine delivery systems, there is a misperception that bidis are safer than cigarettes because they only use natural ingredients. However, bidi smoke produces numerous toxic compounds such as styrene, furfural, phenol, CO, acrolein, and acetaldehyde (60503, 60505). Bidis are at least as harmful as cigarettes in regard to effect on cardiopulmonary health (60506). Even so, the popularity of bidis has increased in Canada and other countries (60507). Bidi users are over 10× as likely to use cannabis as well (60510).

Bidi manufacturing is the second largest industry in India, employing at least 4.4 million bidi rollers who are mainly women and children. These bidi rollers suffer from various adverse health effects due to exposure to nicotine dust and toxic chemicals while hand-rolling bidis (60508–9)

Kreteks are also referred to as clove cigarettes, as they are made with a combination of tobacco, ground cloves, and special flavor sauces, and are the most popular form of smoking in Indonesia (60511). Kreteks are also gaining popularity in the United States due to the typical widespread misperception that they are safer than cigarettes (60512).

Reference Number: 60501

WU, W., S. SONG, D.L. ASHLEY, AND C.H. WATSON. "Assessment of Tobacco-Specific Nitrosamines in the Tobacco and Mainstream Smoke of Bidi Cigarettes." *Carcinogenesis*, 25: 283–287, 2004 (2 tables, 48 references)

Abstract: Bidi cigarettes, or bidis, are a handmade tobacco product from India that are sun-cured oriental tobacco flakes rolled in a leaf from the tendu plant (Indian Ebony) or dried betel leaves. It is considered a poor man's cigarette and at the time of the study, over 550 billion bidis were made every year, employing nearly 10 million people. Bidis are slimmer than conventional cigarettes and were sold to adolescents in the US in vanilla, chocolate, grape, and root beer flavors as a safer tobacco product. The formation of TSNAs occurs mainly after the harvesting of the Nicotiana tabacum types and includes all flue-cured, sun-cured, and air-cured tobaccos. Burley tobacco has a higher TNSA and nicotine content than oriental tobacco due to its high nitrate content. TSNAs were measured in the tobacco and the mainstream smoke of 14 brands of bidis. The TSNA levels found in bidis were similar to conventional cigarettes but had a wider variation due to the hand-rolled nature of bidis, with a less homogenous tobacco amount.

> Quitting smoking and avoiding exposure to environmental tobacco smoke are the only established means to significantly reduce the risk associated with exposure to TSNAs and other harmful chemicals in tobacco smoke. The Surgeon General's health warnings should be included on the bidi packaging to help dissuade the misconception that bidi cigarettes are somehow more "natural" products and therefore less hazardous. Additional action is needed to counter the popularity of bidi cigarette smoking, especially among young people and others who may experiment with bidis. Additional research, consumer education and public advocacy are critically needed to help reduce the health consequences of bidi smoking.
>
> In summary, our data clearly show that TSNA levels in mainstream bidi cigarette smoke are comparable with those from conventional cigarettes. The analytical findings agree with other studies that tobacco products, including bidis, contain carcinogenic compounds, which are of concern in the formation of cancer in users of these products.

Reference Number: 60502

MALSON, J.L., E.M. LEE, E.T. MOOLCHAN, AND W.B. PICKWORTH. "Nicotine Delivery from Smoking Bidis and an Additive-Free Cigarette." *Nicotine and Tobacco Research*, 4: 485–490, 2002 (1 table, 1 figure, 31 references)

Abstract: Bidis are hand-rolled cigarettes (tobacco rolled in dried betel leaves) from India, which contain less tobacco (215.3 mg) than conventional cigarettes (738.6 mg). However, the nicotine content of nicotine is greater in bidis (21.2 mg/g) than unfiltered cigarettes (12.0 ng/mg) or filtered conventional cigarettes (16.3 mg/g). Recently, additive-free cigarettes made with natural leaf tobacco containing no chemical additives or reconstituted tobacco have been marketed as healthier and safer. Natural American Spirit non-filter, a popular natural cigarette, delivers substantial amounts of nicotine (2.90 mg) and tar (29.0 mg). One female and nine male cigarette smokers (mean age 24.5 years) who also used bidis smoked ad libitum their own brand of cigarette, Irie Bidi, Sher Bidi, and American Spirit. Numerous blood samples were collected via an indwelling catheter and plasma nicotine concentrations were measured by HPLC. Exhaled CO concentrations were measured as well. The plasma nicotine concentration was lower in the conventional cigarettes than in bidis or natural cigarettes.

Table: The Effects of Smoking American Spirit Natural Cigarettes, Irie Bidi, Sher Bidi, and Own Brand on Various Measurements

Parameter	American Spirit	Irie Bidi	Sher Bidi	Own Brand
Mean peak plasma nicotine concentration (ng/mL)	32.1	26.0	21.4	18.5
Mean exhaled CO concentration (ppm)	3.0	5.0	3.4	4.6
Mean number of puffs per cigarette	14.0	13.6	14.1	10.4
Mean increase in heart rate (bpm)	8.5	6.7	7.1	2.5

Source: Adapted from Malson et al, 2002

In spite of differences in nicotine delivery, participants rated all cigarettes as similar in nicotine content. Overall, the results indicate that bidis and the additive-free cigarette delivered nicotine, CO and (presumably) other toxic components of tobacco smoke in equal or greater amounts than conventional cigarettes. These results do not support an emerging belief that bidi cigarettes are safer than conventional brands.

Reference Number: 60503

OLADIPUPO, O.A., D. DUTTA, AND N.S. CHONG. "Analysis of Chemical Constituents in Mainstream Bidi Smoke." *BMC Chemistry*, 13: 93 (11 pp), 2019 (6 tables, 2 figures, 40 references)

Abstract: A pack of 20 bidis costs $4–$6 in the United States and 8–17 cents in India. A chemical analysis using FTIR spectrometer and GC/MS was conducted on the mainstream bidi smoke from three flavored and one unflavored bidi. The low combustibility of the tendu leaves, as opposed to cigarette paper, causes the smokers to smoke bidis with more intense puffs and a higher puff frequency. The non-porous and higher moisture content of the tendu leaves also results in a higher level of tar and CO in the bidi smoke compared to tobacco smoke.

Table: Some Chemical Compounds Detected in Mainstream Bidi Smoke by GC/MS

Chemical Compounds
• 3-Methylpyridine • Styrene • Cotinine • Nicotine • Furfural • 2-Furanmethoanol • Phenol • 2-Methylphenol • 4-Methylphenol • 2-Methoxyphenol • 2-Methoxy-4-methylphenol

Adapted from Oladipupo et al, 2019

Many chemical constituents detected in the mainstream bidi smoke are known toxins. For example, hydrogen cyanide is a ciliatoxin; both benzene and 1,3-butadiene have been classified as human carcinogens by EPA and International Agency for Research in Cancer (IARC). Damage of respiratory tract cilia by hydrogen cyanide can severely impair mucociliary clearance in smokers and compromise their ability for clearing tar and other environmental pollutants in the respiratory tract. These pollutants deposit and then get absorbed in the body. In addition, exposures to high levels of these compounds are associated with eye irritation, giddiness, nausea, and may lead to cardiovascular and respiratory tract diseases. Nicotine, the major component of bidi smoke, was present at a significantly higher level in the particulate size fraction of <0.25 μm. This implies that nicotine can be readily deposited and absorbed through human lung tissues.

Reference Number: 60504

MALSON, J.L., K. SIMS, R. MURTY, AND W.B. PICKWORTH. "Comparison of the Nicotine Content of Tobacco Used in Bidis and Conventional Cigarettes." *Tobacco Control*, 10: 181–183, 2001 (2 tables, 25 references)

Abstract: The nicotine content of 12 unfiltered brands of bidi cigarettes (hand-rolled cigarettes from India) and 8 popular brands of filtered and unfiltered cigarettes were determined. The bidis tested contain less tobacco than conventional cigarettes but had a higher nicotine concentration.

Table: Characteristics of Bidis and Unfiltered and Filtered Conventional Cigarettes

Cigarette Type	Average Cigarette Weight (mg)	Average Tobacco Weight (mg)	Average Nicotine per Cigarette (mg)
Bidis	506.2	215.3	21.2
Unfiltered cigarettes	916.9	860.6	13.5
Filtered cigarettes	939.0	738.6	16.3

Source: Adapted from Malson et al, 2001

In addition to the well documented accounts of the toxic chemicals in bidi smoke, their delivery of nicotine poses a risk for the development of dependence.

Reference Number: 60505

RAY, C.S., AND P.C. GUPTA. "Bidis and Smokeless Tobacco." *Current Science*, 96: 1324–1334, 2009 (5 tables, 3 figures, 65 references)

Abstract: At the time of the study, over half of the tobacco consumed in India was in the form of bidis and 27% was smokeless tobacco. Over 100 million people in India smoked bidis compared to 25 million who smoked cigarettes. Bidi smoking generates similar or higher blood nicotine concentrations than smoking cigarettes even though bidis contain less tobacco than cigarettes. This is due to the non-porous leaf wrapper which does not permit much less air dilution of the smoke during puffing.

Table: Some Harmful Constituents of Bidi Smoke and Associated Risks

Toxic Compounds	Harm Caused
Nicotine	Addiction, atherogenesis, raised blood pressure
Carbon monoxide	Inhibition of oxygen exchange, stress on the heart
Ammonia	Irritation to the respiratory tract
Hydrogen cyanide	Toxicity to cilia inhibits lung clearance

Toxic Compounds	Harm Caused
Phenol	Tumor promotion
Nitrogen oxides	Inflammation of the lung
Tobacco-specific nitrosamines	Cancer
Acrolein	Toxicity to cilia inhibits lung clearance
Acetaldehyde	DNA damage
Isoprene	DNA damage

Source: Adapted from Ray and Gupta, 2009

Bidi smoking and smokeless tobacco use continue to be practised by a large percentage of the population in India. Smokeless tobacco use is twice as high as bidi smoking among adolescents. Initiation into use of these products among youth leads to lifelong adult use. Bidis are responsible for most of the one million or more yearly smoking related deaths in the country. Political will has been lacking for the implementation of the tobacco control policies already adopted in the law. Lack of political will is also responsible for low taxes on bidis and smokeless tobacco products. To protect health and promote healthy economic development in the country, the lobbying power of the industry needs to be countered. To accomplish this, public support is a must

Reference Number: 60506

DUONG, M., S. RANGARAJAN, X. ZHANG, K. KILLIAN, P. MONY, S. SWAMINATHAN, A.V. BHARATHI, S. NAIR, K. VIJAYAKUMAR, I. MOHAN, R. GUPTA, D. MOHAN, S. RANI, V. MOHAN, R. IQBAL, K. KAZMI, O. RAHMAN, R. YUSUF, L.V.M. PINNAKA, R. KUMAR, P. O'BYRNE, AND S. YUSUF. "Effects of Bidi Smoking on All-Cause Mortality and Cardiorespiratory Outcomes in Men from South Asia: An Observational Community-Based Substudy of the Prospective Urban Rural Epidemiology Study (PURE)." *Lancet Global Health*, 5: e168–e176, 2017 (4 tables, 4 figures, 31 references)

Abstract: Bidis are an inexpensive tobacco hand rolled in dry leaves of the Indian Ebony plant that are smaller than cigarettes and marketed as natural and safe. A 5.6-year follow-up survey was conducted on the respiratory, all-cause mortality, and cardiorespiratory outcomes in 8,438 non-smokers, 3,321 light smokers, 959 heavy cigarette smokers, and 2201 heavy bidi smokers.

Table: Adverse Health Effects of Smoking Cigarettes or Bidis

Adverse Health Effects	Non-smokers	Light smokers	Heavy cigarette smokers	Heavy bidi smokers
Wheeze	1.0	1.20	1.45	1.95
Cough or sputum	1.0	1.46	1.83	2.45
Chest pain	1.0	1.31	1.33	1.63
Deaths	1.0	1.28	1.59	1.56

Source: Adapted from Duong et al, 2017

Our findings have important public health implications. South Asia is the second largest consumer of tobacco in the world, with more than 130 million tobacco smokers. More than half these smokers use bidis, particularly among the poorest and most vulnerable sectors of the population. Therefore the health impact of bidis is highly relevant in this region. Furthermore, as the current trend of bidi exportation continues, the global impact of bidis will rise, particularly among young adults. Our findings suggest that bidis are at least as harmful as cigarettes on cardiorespiratory health and mortality, despite having substantially less tobacco content than cigarettes. Control of bidis should be an integral part of any anti-tobacco framework, both regionally and globally, with the greatest potential effects to be seen in poor and young people.

Reference Number: 60507

CZOLI, C.D., S.T. LEATHERDALE, AND V. RYNARD. "Bidi and Hookah Use Among Canadian Youth: Findings From the 2010 Canadian Youth Smoking Survey." *Preventing Chronic Disease*, 10: (10 pp), 2013 (3 tables, 24 references)

Abstract: A smoking survey was conducted on 28.416 high school students in 2006–2007 and 31,393 students in 2010–2011 (grades 9–12). Ever use of bidis decreased from 2.2% to 1.8% of students. Ever use of hookahs increased from 9.5% to 10.1%. Marijuana use is a consistent predictor of bidi and hookah use. Current cigarette smokers were 6.7× more likely to use bidis and 3.0× more likely to use hookahs than non-smokers. Occasional and frequent alcohol drinkers were 2.8× and 3.6× as likely to be current hookah users, respectively.

The rise in hookah use warrants attention, given its increasing popularity. Efforts aimed at eliminating false perceptions of reduced harm associated with hookah use may show promise in reducing its social acceptability, especially as it becomes more prevalent. Factors associated with bidi and hookah use can help developers of prevention programs, and co-occurrence of substance use supports the development of an integrated approach to prevention that addresses multiple risk behaviors.

Reference Number: 60508

PRAKASH, B., AND U. VYAS. "Association Between Occupational Tobacco Exposure of Health Hazards in Women Laborers of Bidi-Industry of Ajmer." *International Journal of Pharma Medicine and Biological Sciences*, 2: 58-61, 2013 (2 tables, 1 figure, 9 references)

Abstract: Bidi manufacturing is the second largest industry in India and employs mainly women and children from poor socio-economic levels. One hundred non-tobacco-using and 100 tobacco-using women actively involved in bidi rolling were surveyed for health complaints. The high nicotine content and other chemicals in bidi tobacco poses a high risk of systematic illness. The tobacco dust contains TSNAs which can be readily absorbed through the skin, mouth, nose, and lungs.

Table: Percentage of Non-Tobacco Using Female Bidi Workers Who Experience Adverse Health Effects

Health Symptoms	Percentage
Eye irritation	15%
Nail discoloration	13%
Irritation and dryness of throat	12%
Difficulty in breathing	9%
Loss of appetite	8%

Source: Adapted from Prakash and Vyas, 2013

Our conclusion is based on health hazards, so the subjects were aware about the harmful effects of tobacco use. There is a need to improve knowledge of the dangers of smoking among the disadvantaged segments of the population. Results of the present study prove that bidi tobacco is very harmful for the workers who actively or passively inhale tobacco during bidi rolling. The health impact on bidi workers is visible on all age groups. Continuous bidi rolling leads to nicotine directly through skin.

Reference Number: 60509

KHANA, A., D.S. GAUTAM, M. GOKHALE, AND S.K. JAIN. "Tobacco Dust Induced Genotoxicity as an Occupational Hazard in Workers of Bidi Making Cottage Industry of Central India." *Toxicology International*, 21: 18–23, 2014 (5 tables, 6 figures, 21 references)

Abstract: Tobacco is a major cash crop grown widely in India and, at the time of the study, provided the employment for 6 million tobacco farmers, 4.4 million bidi rollers, and 200 thousand tendu leaf pluckers. A bidi roller on average made 500–1000 bidis per day and handled about 225 kg of tobacco, and they are exposed to a large amount of tobacco dust. Peripheral blood samples were collected from 31 bidi rollers and 30 controls (aged 25–68 years) and analyzed for the percentage of chromosome aberrations (CA). None of the subjects smoked or chewed tobacco. The frequency of CA was found to be 3.1% in bidi workers compared to 1.5% in controls.

> The results obtained in this investigation indicate that bidi rollers seem to be facing the occupational hazard of genotoxicity due to handling bidi tobacco and inhalation of tobacco dust. They should be advised to work under well-ventilated conditions.

Reference Number: 60510

DELNEVO, C.D., AND M. HRYWNA. "The Relationship of Cigars, Marijuana, and Blunts to Adolescent Bidi Use." *Public Health Reports*, 121: 603–608, 2006 (4 tables, 27 references)

Abstract: This study explores the relationship between the use of bidis, cigarettes, marijuana, and blunts in 17,429 youths, using data from the 2001 National Household Survey. Overall, 1.1% of youth reported past month bidi use. The Odds Ratio (OR) of past month bidi use was 9.9× for cigarette use, 10.2× for cigar use, 10.5× for marijuana, and 11.5× for blunt past month use.

> Tobacco control efforts that target minority youth must address other tobacco products as well as other substances, like marijuana. Such efforts must dispel the dangerous myth that products such as bidis, cigars, and marijuana are safer than conventional cigarettes. The findings of this study raise the importance of having two public health fields, tobacco control and substance abuse, work together to address youth's use of these

harmful products and tailor efforts appropriately and effectively, with consideration of racial, ethnic, and cultural variations.

Reference Number: 60511

PALIPUDI, K., L. MBULO, S. KOSEN, A. TJANDRA, K. KADARMANTO, F. QURESHI, L. ANDES, D.N. SINHA, AND S. ASMA. "A Cross Sectional Study of Kretek Smoking in Indonesia as a Major Risk to Public Health." *Asian Pacific Journal of Cancer Prevention*, 16: 6883–6888, 2015 (2 tables, 36 references)

Abstract: Kreteks are cigarettes that are made with a combination of tobacco and ground cloves mixed with special flavors called sauces and at the time of the study they were the most popular tobacco product in Indonesia. Kreteks are sometimes referred to as clove cigarettes. Smoking rates in Indonesia increased from 27.0% in 1995 to 34.4% in 2004 due to low tobacco prices and intensive tobacco marketing and advertising. Of those who smoked, 90% smoked kreteks. The predictors of kretek use are being male, having less education, and being less wealthy.

Given the health effects associated with kreteks and the large number of Indonesian adults who smoke them, kretek smoking carries a significant health burden in Indonesia. In 2008, it was estimated that overall smoking prevalence was 34%, and that 63% of males were smokers. Of the 57 million smokers in Indonesia, about half were projected to die of tobacco-related illnesses.

Reference Number: 60512

SOLDZ, S., D.J. HUYSER, AND E. DORSEY. "Characteristics of Users of Cigars, Bidis and Kreteks and the Relationship to Cigarette Use." *Preventive Medicine*, 37: 250–258, 2003 (7 tables, 65 references)

Abstract: In the 1990s, youth use of alternative tobacco products such as cigars, bidis, and kreteks increased. A survey was conducted of 5,016 grade 7–12 students in Massachusetts regarding tobacco use. The mean age of initiation of tobacco use was 12.6 years for cigarettes, 13.5 years for cigars, 13.9 years for bidis, and 14.1 years for kreteks.

Unlike the pattern for cigars, the overwhelming majority of bidi and kretek users were currently smoking cigarettes.

6.06 SMOKELESS TOBACCO (SNUFF/CHEWING TOBACCO)

"There is no safe form of tobacco, and both smoked and smokeless tobacco are carcinogenic to humans."

—Warnakulasuriya and Straif, "Carcinogenicity of Smokeless Tobacco: Evidence from Studies in Humans & Experimental Animals" (2018)

The two main types of smokeless tobacco (SLT) are chewing or spit tobacco, which is held in the mouth or chewed, and snuff, which is ground tobacco leaves inhaled or snuffed into the nose. Currently there are approximately 357 million SLT users worldwide. SLT products are mainly unregulated and underreported (60601).

SLTs have different names and additives in different countries but the SLT health burden is borne mainly by the poorest of the poor (60602–4).

In order to increase the speed of nicotine absorption through the oral cavity or nasal cavities, calcium carbonate or another buffer is added to the tobacco to increase the pH, which increases the amount of free base nicotine that passes through the mucous membranes rapidly and increases the risk of addiction (60605–7).

In the late 1880s to early 1900s, anti-spitting laws were enacted in the United States and other countries to combat TB and other diseases. This helped produce a shift to cigarettes (60608).

The tobacco companies as usual encouraged the use of SLTs to youth by having celebrity baseball player endorsement (60609), adding sweeteners and flavors (60610–11), and promoting limited addition packaging (60612). A new, less messy SLT, dissolvable nicotine, has also been introduced and some stores have placed these novel SLT products next to children's candy (60613).

Cigarette smokers who switched to chewing tobacco had a higher rate of lung cancer, coronary heart disease, and COPD than those who quit completely (60614).

Rural, lower income men had the highest rate of SLT use (33.3%). In contrast to cigarette smokers, who start to show a dramatic decrease in use at the age of 20 years, chewing tobacco users showed no decline until well into their 50s (60615). SLT users who drank 1–2 cups of coffee a day were more than twice as likely to quit as non-coffee drinkers (60616).

SLTs contain almost as many toxic carcinogens as cigarettes and have been associated with oral submucous fibrosis, oral cancers, tobacco pouch keratosis, myocardial infarction, and strokes (60617–26).

SLT use causes a lower blood hemoglobin concentration in pregnant women, which may affect their health and the development of the fetus (60627).

The tobacco industry has tried to minimize the finding of carcinogenic nitrosamines in tobacco by stating that nitrosamines are present in bacon and beer as well, and they are not banned. However, one would have to eat over 300 kilograms (660 pounds) of bacon to have the same amount of nitrosamines as in 10 grams of snuff (60628).

Reference Number: 60601

MEHROTRA, R., AND D.N. SINHA. "Global Challenges in Smokeless Tobacco Control." *Indian Journal of Medical Research*, 148: 1–3, 2018 (17 references)

Abstract: At the time of the study, there were globally approximately 357 million SLT users with 83% living in Southeast Asia. SLT products are mainly unregulated and underreported. SLTs have a significant association with all-cause mortality, ischaemic heart disease, stroke, and oral, esophageal, and pancreatic cancers.

> Tobacco manufacturers regularly try to introduce newer SLT products, increasing consumer appeal by adding attractive flavouring, newer delivery methods and brand mimicking. This is complemented by targeted marketing (towards vulnerable population groups such as youth as well as existing consumers with "intention to quit," etc.), producing an impact on tobacco use behaviour. Awareness of the hazards of smoke tobacco use is also exploited by the tobacco industry, proposing SLT use as a harm reduction measure. The industry advertises SLT products as a safer alternative to cigarettes and advocates switching.

Reference Number: 60602

SIDDIQI, K., S. SHAH, S.M. ABBAS, A. VIDYASAGARAN, M. JAWAD, O. DOGAR, AND A. SHEIKH. "Global Burden of Disease Due to Smokeless Tobacco Consumption in Adults: Analysis of Data from 113 Countries." *BMC Medicine*, 13: 194 (22 pp), 2015 (4 tables, 5 figures, 96 references)

Abstract: SLT consists of numerous types of tobacco products which are consumed through the mouth or nose without burning. It is most prevalent in South and Southeast Asia, where at the time of the study, one-third of the tobacco consumed was SLT. In addition to nicotine, SLT contains over 30 carcinogens including TSNAs. The disability-adjusted life-years (DALYs) lost and reported deaths due to SLT was determined by the 2010 Global Burden of Disease Survey. It was estimated that SLT use led to 1.7 million DALYs lost, 62,283 deaths due to cancers of the mouth, pharynx, and oesophagus, and 204,309 deaths from ischemic heart disease. The WHO needs to consider incorporating regulation of SLTs into the Framework Convention for Tobacco Control.

Table: Smokeless Tobacco Products Consumed Most Commonly Throughout the World

SLT Product	Countries with Highest Consumption	Other Ingredients	Preparation and Use
Snus (Swedish)	Denmark, Finland, Iceland, Norway, Sweden	Water, sodium carbonate, sodium chloride, moisturizers, flavoring	A heat treatment process; placed between the gum and upper lip
Plug, snuff, snus (US)	US, Canada, Mexico	Sweeteners, liquorice	Plug: air-cured Dry or most snuff: finely ground and fire cured; kept between lip and gum; dry snuff can also be inhaled Snus: steam cured Snuff
Chimo	Venezuela, Columbia	Sodium bicarbonate, brown sugar, Mamo'n tree ashes	Tobacco paste made from tobacco leaves; placed between the lip or cheek and gum and left there for some time
Nass (Naswar)	Uzbekistan, Kyrgyzstan, Tajikistan, Afghanistan, Pakistan, Iran	Lime, ash, flavorings (cardamom), indigo	Sundried and powdered; placed between the lip or cheek and gum

Reference Number: 60602

SLT Product	Countries with Highest Consumption	Other Ingredients	Preparation and Use
Tambook	Sudan, Chad	Mixed with moist sodium bicarbonate	Fermented and ground; placed and kept in mouth
Snuff (North and West Africa)	Nigeria, Ghana, Algeria, Cameroon, Chad, Senegal	Dried tobacco leaves mixed with ash	Dried snuff finely ground and inhaled as a pinch Most snuff is placed in mouth
Snuff (South Africa)	South Africa	Dried leaves mixed with ash	Dried snuff finely ground and inhaled as a pinch
Khaini	India, Bangladesh, Nepal, Bhutan	Slaked lime, menthol, flavorings, areca nut	Shredded; kept in mouth between lips and gum
Zarda	Bangladesh, India, Pakistan, Myanmar, Thailand, Indonesia, Nepal, Maldives, Sri Lanka, United Kingdom	Served wrapped in a betel leaf with lime, catechu, areca nuts	Shredded tobacco leaves are boiled with lime and saffron; the mixture is dried, then chewed and spat
Gutkha	India, Pakistan, Bangladesh, Nepal, Myanmar, Sri Lanka, United Kingdom	Betel nut, catechu, flavorings, sweeteners	Commercially manufactured; sucked, chewed, and spat

Source: Adapted from Siddiqi et al, 2015

We have found that SLT is consumed worldwide and that its use results in substantial, potentially avoidable, morbidity and mortality. However, owing to marked differences in the types of products available, patterns of consumption, and associated risks, there are substantial differences in the attributable burden between regions and countries. In particular, SLT consumption in South-East Asia leads to a much greater burden of disease than in Sweden, despite its use being equally prevalent. This is due to the much lower levels of TSNA and pH in SLT products in Sweden compared to those found in SLT in South-East Asia. Similarly, SLT products used in the USA have lower risk estimates than for those used in South-East Asia.

Reference Number: 60603

SINHA, D.N., P.C. GUPTA, A. KUMAR, D. BHARTIYA, N. AGARWAL, S. SHARMA, H. SINGH, M. PARASCANDOLA, AND R. MEHROTRA. "The Poorest of the Poor Suffer the Greatest Burden from Smokeless Tobacco Use: A Study from 140 Countries." *Nicotine and Tobacco Research*, 1529–1532, 2018 (1 table, 1 figure, 21 references)

Abstract: At the time of the study, more than 650,000 deaths had been attributed to SLT use in 133 countries. SLT use was on the rise in Southeast Asian countries, where 80% of the global users lived. A survey was conducted of SLT users in 140 countries. The mean prevalence of SLT was 10.4% in men and 5.6% in women. SLT use was 1.2–8× higher among the poorest areas of the countries compared to the richest areas, and 91% of the SLT users resided in the poorest countries of the world. Taxes on SLT products tend to be lower than on cigarettes and SLT affordability has increased due to this inadequate taxation.

The majority of the burden of SLT use is on lower and lower middle income countries with the greatest burden on the poorest segments of the population in these countries.

Reference Number: 60604

HU, S.S., D.M. HOMA, T. WANG, Y. GOMEZ, K. WALTON, H. LU, AND L. NEFF. "State-Specific Patterns of Cigarette Smoking, Smokeless Tobacco Use, and E-Cigarette Use Among Adults—United States, 2016." *Preventing Chronic Disease, Public Health Research, Practice, and Policy*, 16: E17 (15 pp), 2019 (3 tables, 29 references)

Abstract: The data for this study was obtained from 477,665 adults (aged 18+ years) for the Behavioral Risk Surveillance System. The prevalence of current cigarette smoking among US adults ranged from 8.8% in Utah to 24.8% in West Virginia. Current EC use ranged from 2.4% in DC to 6.7% in Oklahoma. Current smokeless tobacco rates ranged from 1.3% in DC to 9.8% in Wyoming. The states with the highest rates of cigarette smoking generally had a high prevalence of EC use.

Table: Percentage of Adults Who Are Current Cigarette Smokers and Cigarette Smokers in Selected US States, Showing the Range of Current Cigarette Smokers from 8.8% to 24.8%

US State	Cigarette Smokers	SLT Users
Arkansas	23.6%	7.8%
California	11.0%	1.7%
Indiana	21.1%	4.1%
Kentucky	24.5%	7.4%
New Jersey	14.0%	2.3%
New York	14.2%	2.2%
Ohio	22.5%	4.7%
Texas	14.3%	4.3%
Utah	8.8%	3.4%
West Virginia	24.8%	8.5%

Source: Adapted from Hu et al, 2019

In contrast to cigarette smoking patterns, smokeless tobacco use has not declined commensurately among US adults. For instance, the range of smokeless tobacco use prevalence presented in this report (1.3%–9.8%) is consistent with the previously published range from 2011 (1.4%–8.9%). Moreover, during 2011 through 2016, most states either experienced no significant change or an increase in the prevalence of smokeless tobacco use among adults. These trends align with the detected increase in smokeless tobacco consumption and may be partly attributed to increased expenditures for advertising and promotion of these products, from $684.9 million in 2015 to $759.3 million in 2016. Given that smokeless tobacco also causes substantial morbidity and premature mortality, these data demonstrate the public health importance of incorporating smokeless tobacco products within the framework of comprehensive tobacco use prevention and control programs.

Reference Number: 60605

SINGH, I., A. SINGH, R. KOUR, A. SINGH, R. SINGH, AND A. BALI. "Is Sodium Carbonate in Snuff a Causative Factor for Oral Mucosal Lesions: A Cross-Sectional Analysis." *Journal of International Society of Preventive and Community Dentistry*, 8: 339–342, 2018 (3 tables, 1 figure, 20 references)

Abstract: Nicotine absorption through the mucous membrane is directly proportional to pH, and so snuff can be buffered to a pH of 8–9 by the addition of sodium carbonate. This increases its absorption rate and its potential for addiction. The composition of snuff includes water, tobacco, moist preservatives, taste enhancers (salt), acidifiers, and aromas. The tobacco itself is composed of over 2,500 chemical compounds, including nicotine, sodium carbonate, sodium chloride, TSNAs, and polycyclic aromatic hydrocarbons. Oral mucosa lesions (also known as snuff dippers lesions, snuff-induced leukoplakia, and snus induced lesions) were determined in 284 chronic snuff users in the Jammu region of India. Only 2 of the 284 snuff users did not use sodium carbonate.

The present study narrates that there seems to be liaison between the use of sodium carbonate in snuff and oral mucosal lesions.

Reference Number: 60606

BENOWITZ, N.L. "Nicotine and Smokeless Tobacco." *CA: A Cancer Journal for Clinicians*, 38: 244–247, 1988 (2 tables, 13 references)

Abstract: Nicotine is the ultimate cause of diseases related to SLT. Chewing tobacco and snuff are buffered to an alkaline pH to increase the absorption of nicotine through the oral cavity. Nicotine is more slowly absorbed from SLT than from cigarettes, but the peak venous concentrations are similar. The amount of nicotine absorbed is greater with chewing tobacco (4.5 mg nicotine) or snuff (3.6 mg nicotine), compared to 1–2 mg of nicotine from each cigarette. Some of the potential health consequences due to nicotine are coronary and peripheral vascular disease, stroke, hypertension (complications), delayed wound healing, and peptic ulcer disease.

The following may be concluded about nicotine and smokeless tobacco use:

- Systemic absorption and levels of nicotine are similar in users of smokeless tobacco and cigarette smokers.
- Data from the few studies performed to date suggest that smokeless tobacco users develop a dependency similar to that for cigarette smokers. Effective treatment strategies to help smokeless tobacco users quit need to be developed. Smokeless tobacco use in young people also poses a concern for later development of dependence on cigarettes.

- The health hazards known to be caused by cigarette smoking and suspected to be related to chronic nicotine exposure are expected to be a hazard of habitual use of smokeless tobacco. A major concern in young males is accelerated coronary artery disease.

Reference Number: 60607

FANT, R.V., J.E. HENNINGFIELD, R.A. NELSON, AND W.B. PICKWORTH. "Pharmacokinetics and Pharmacodynamics of Moist Snuff in Humans." *Tobacco Control*, 8: 397–392, 1999 (2 tables, 1 figure, 17 references)

Abstract: Ten male smokeless tobacco users (aged 26–45 years) placed 2 grams of moist snuff (Copenhagen, Skoal Long Cut Cherry, or Skoal Original Wintergreen) or one Skoal Bandit pouch (containing approximately 0.5 grams of tobacco) between cheek and gum for 30 minutes. Frequent blood samples were collected, and plasma nicotine concentrations were determined by GC/MS. HR, BP, and subjective measures were also determined. Initial plasma nicotine concentrations were 5.3 ng/mL and the mean maximum increase was 19.5 ng/mL for Copenhagen, 14.9 ng/mL for the other two moist snuff, and only 4.2 ng/mL for Skoal Bandits pouch.

Table: The Effect of Three Moist Snuff Tobacco Products on Plasma Nicotine Heart Rate and Blood Pressure

Parameter	Copenhagen	Skoal Original Wintergreen	Skoal Bandits Pouch
Maximum plasma nicotine increase	19.5 ng/mL	14.9 ng/mL	4.2 ng/mL
Maximum increase in Heart Rate	16.1 bpm	14.7 bpm	4.7 bpm
Maximum increase in systolic blood pressure (mm Hg)	16.8 mm Hg	14.9 mm Hg	9.4 mm Hg
Maximum increase in diastolic blood pressure (mm Hg)	14.2 mm Hg	14.0 mm Hg	10.3 mm Hg

Note: Skoal Long Cut Cherry had the same nicotine increase as Skoal Original Wintergreen

Source: Adapted from Fant et al, 1999

Large amounts of nicotine were delivered rapidly to the bloodstream. The amount of nicotine absorbed and the rate of absorption were related to the pH of the snuff product in aqueous suspension. Cardiovascular and subjective effects were related to the amount of nicotine absorbed.

Reference Number: 60608

O'CONNOR, R.J. "Non-Cigarette Tobacco Products: What Have We Learnt and Where Are We Headed?" *Tobacco Control*, 21: 181–190, 2012 (2 tables, 207 references)

Abstract: This paper suggests that in the following 20 years, as the social disapproval and regulations of cigarette smoking increase, the growing use of non-cigarette tobacco will be a public health challenge. There are three main categories of non-cigarette tobacco: other smoked products, smokeless tobacco, and other nicotine products (e.g., vaping). Use of chewing and snuff tobacco was affected by anti-spitting laws in the late 1800s and early 1900s to combat the spread of tuberculosis and other diseases, resulting in the shift to cigarettes.

A wide variety of non-cigarette forms of tobacco and nicotine exists, ranging from smoked forms to smokeless forms to tobacco-free nicotine. Non-cigarette forms of tobacco are widely available, and their use varies regionally and globally. Smoked forms of tobacco such as bidis, kreteks, and waterpipes have high popularity and are often perceived erroneously as less hazardous than cigarettes, when in fact their health burden is similar. Smokeless tobacco products vary widely around the world in both form and health hazards, with some clearly toxic forms (e.g., South Asia), and some forms with far fewer hazards (e.g., Sweden). Also burgeoning is a market for nicotine delivery systems not directly reliant on tobacco (e.g., ENDS).

Reference Number: 60609

CHAFFEE, B.W., E.T. COUCH, AND S.A. GANSKY. "Adolescents' Smokeless Tobacco Susceptibility by Perceived Professional Baseball Players' Use." *Journal of Public Health Dentistry*, 78: 5–8, 2018 (1 table, 1 figure, 10 references)

Abstract: Despite the decline in cigarette smoking by youth, SLT use among US adolescents at the time of the study had not changed for over a decade. In 2015, SLT use in male high school students equaled that for cigarette smoking. A study was conducted of 360 male baseball players in 25 rural California high schools. Overall, 31% of the adolescent baseball players had tried SLT at least once and 13% had used SLT in the past 30 days. High school SLT never users who perceived that their favorite major league baseball player used it were 2.5–3× more likely to be susceptible to

SLT initiation. In 2012, Major League Baseball (MLB) began prohibiting players from carrying SLT products on their uniforms but did not ban on-field use, which remained prominent during game telecasts. In 2017, the MLB banned all new players from using SLT during the game, but not those who had joined before 2017.

> In this population, perceived smokeless tobacco use by admired MLB players was strongly associated with high school baseball players' current ST [smokeless tobacco] use and with susceptibility to ST initiation among ST never-users. This finding is consistent with studies reporting that emulation of respected male figures, including athletes and during sports participation, contributes to normalizing ST and encouraging initiation. Tobacco related-norms and acceptability are fodder for effective anti-tobacco counseling from oral health professionals. Meanwhile, the potential influence of high-profile figures on youth tobacco use provides reason for further limits on tobacco in professional sports.

Reference Number: 60610

MIAO, S., E.S. BEACH, T.J. SOMMER, J.B. ZIMMERMAN, AND S-E. JORDT. "High-Intensity Sweeteners in Alternative Tobacco Products." *Nicotine and Tobacco Research*, 18: 2169–2173, 2016 (1 table, 1 figure, 33 references)

Abstract: Sweetness in tobacco products influences use initiation and reinforcement and has a special appeal to adolescents. This study determined sweetener content in four snus products, four moist snuffs, five electronic cigarette cartridges, three electronic cigarette refill liquids, and one dissolvable tobacco product by LC/ESI/MS and compared it to four sugar-free candies and two sugar-free drinks. All tested products had very low or undetectable levels of sugars (glucose or sucrose). All SLT products contained synthetic high intensity sweeteners. Snus and dissolvable tobacco exceeded the levels of sugar-free candies by as much as 25×. All snus samples contained sucralose and aspartame, while most snuff samples contained saccharin. The dissolvable tobacco product contained sucralose and sorbitol. Ethyl maltol was the most common sweetener in EC liquids.

> Dissolvables and snus have been promoted as risk-mitigation products due to their relatively low content of nitrosamines and other tobacco toxicants. While not as popular, these products may increase the risk of polytobacco use and their intense sweetness is of concern since it may

appeal especially to adolescents who initiate tobacco product use. In addition to E-cigarettes, other sweet flavored tobacco products such as small cigars have made rapid inroads among adolescent populations and remain unregulated. Among the wide variety of flavors offered intense sweetness appears to be a common denominator in the majority of the newly introduced products. Thus, the regulation of sweetener content may represent an efficient means to control palatability of a wide range of products and to reduce tobacco product use initiation.

Reference Number: 60611

KOSTYGINA, G., AND P.M. LING. "Tobacco Industry Use of Flavourings to Promote Smokeless Tobacco Products." *Tobacco Control*, 25: ii40–ii49, 2016 (2 tables, 2 figures, 139 references)

Abstract: Internal tobacco industry documents were searched to determine the use of flavors in smokeless tobacco products. Fruit, candy, and alcohol flavors are not allowed in cigarettes, but they are still allowed in SLT products, where the sales increased by 65.6% between 2005 and 2011 in the United States, while cigarette sales declined. In 1934, the first wintergreen (methyl salicylate) flavor was introduced to SLT. Methyl salicylate has anti-bacterial properties and has been used to prevent the fermentation process or moist snuff products, and it also inhibits the formation of nitrosonornicotine (a powerful carcinogen). Acetaldehyde (which enhances the addictive effects of nicotine) is produced by the burning of sugars in cigarettes and has been added to SLT because there is no combustion. Among SLT, wintergreen was the most popular flavor (39%), followed by spearmint (12%) and fruit flavor (5%).

Our findings confirm that mint-flavoured and wintergreen-flavoured products may be especially effective in making tobacco more palatable to novices, possibly contributing to the initiation and maintenance of ST use. We found that mint flavouring plays a particularly important role in initiation of smokeless use and subsequent dependence; in a recent study of adult ST users, a majority of participants' first and current choice of product was flavoured, specifically mint or wintergreen, and a significant number of smokeless users switched to a flavoured brand after already initiating ST use with a regular "natural" product. One analysis found the average level of mint in the top five ST products was 50% higher than that of the top five brands of candy, and levels of wintergreen additive were eight times higher. Our study adds to this literature by demonstrating that mint and wintergreen flavour (methyl salicylate) appeal to starters as well as among experienced ST users. This suggests that the mint and wintergreen flavours may also reinforce use and potentially enhance addiction. Furthermore, methyl salicylate may serve other functions unrelated to flavouring, due to its antibacterial properties and thus potential effect on nitrosamines, pH and levels of free nicotine. As an analgesic, it may facilitate continued use of ST despite adverse effects on oral health by reducing pain sensations and masking irritation from oral tobacco.

Reference Number: 60612

COUCH, E.T., J. URATA, AND B.W. CHAFFEE. "Limited-Edition Smokeless Tobacco Packaging: Behind the Camouflage." *Tobacco Induced Diseases*, 17: 58 (3 pp), 2019 (1 figure, 15 references)

Abstract: Moist snuff was the top-selling smokeless tobacco product in the United States and quite popular among young males participating in outdoor sports. This paper argues that limited-edition packaging helps to perpetuate that popularity.

Limited-edition ST packaging is used to communicate masculine stereotypes, introduce games and novelty items, promote CSR campaigns, and reinforce the relationship between ST and sport. Limited-edition packaging plausibly contributes to misconceptions about the environmental impact of tobacco production and increases youth appeal. Restrictions on advertising and promotion through online and direct mail may reduce the appeal and urgency to purchase limited-edition ST products. Moreover, mandatory plain packaging for all products, including ST, would eliminate the use of limited-edition packaging to communicate with the public, thereby reducing tobacco use in the US.

Reference Number: 60613

ROMITO, L.M., M.K. SAXTON, L.L. COAN, AND A.G. CHRISTEN. "Retail Promotions and Perceptions of R.J. Reynolds' Novel Dissolvable Tobacco in a US Test Market." *Harm Reduction Journal*, 8: 10 (10 pp), 2011 (3 tables, 1 figure, 41 references)

Abstract: With declining sales, cigarette manufacturers have been developing and marketing new smokeless tobacco products such as RJ

Reynolds's dissolvable tobacco: Camel Sticks, Strips, and Orbs. The popularity of traditional SLT is limited, as smokers are displeased by the litter and unpleasantness of having to spit frequently. Snus require no spitting, but the used pouch has to be removed from the mouth. Dissolvable tobacco overcomes these objections and delivers 0.6–3.1 mg nicotine per piece. Tobacco promotions are typically more effective in attracting new users than existing smokers.

> For decades, cigarette companies have spent millions of dollars per annum on ST research, consumer profiling, product development, and marketing. In the face of greater smoking restrictions and declining cigarette sales, they have developed new cigarette-branded smokeless, spitless products in an effort to satisfy consumer preferences whilst attempting to expand their consumer base. RJR's latest contribution to the smokeless tobacco market is a new line of dissolvable tobacco products: Camel Orbs, Strips and Sticks. As with other forms of tobacco which are usually purchased on an as-needed basis, this study found that the Camel Dissolvable tobacco products were most commonly sold at so-called "on-the-go" retailers - gas stations and convenience stores. These products were often co-located with Camel cigarettes and other forms of tobacco, such as Camel Snus, thereby emphasizing recognition of their popular brand.

Reference Number: 60614

HENLEY, S.J., C.J. CONNELL, P. RICHTER, C. HUSTEN, T. PECHACEK, E.E. CALLE, AND M.J. THUN. "Tobacco-Related Disease Mortality Among Men Who Switched from Cigarettes to Spit Tobacco." *Tobacco Control*, 16: 22–28, 2007 (3 tables, 1 figure, 34 references)

Abstract: Tobacco-related disease mortality was determined in 4,443 switchers (switched from cigarettes to smokeless tobacco) and 111,952 cigarette smokers who quit using tobacco entirely between 1982 and 2002 using data from the US American Cancer Society Prevention Study II.

Table: Age Adjusted Odds Ratio of Mortality from Various Tobacco-Related Diseases in Cigarette Smokers who Quit Entirely or Switched to SLT

Cause of Death	Quit Entirely	Switcher	Switched to Chew Only	Switched to Snuff Only	Switched to Chew and Snuff
All Causes	1.0	1.19	1.17	1.25	1.27
Lung Cancer	1.0	1.92	1.79	2.28	2.19
Coronary Heart Disease	1.0	1.19	1.25	1.16	0.90
COPD	1.0	1.53	1.53	1.94	0.89

Source: Adapted from Henley et al, 2007

In summary, we found that men who switched from smoking cigarettes to using spit tobacco had a higher rate of death from all causes, lung cancer, coronary heart disease and stroke than those who had never used tobacco or those who were former cigarette smokers and quit using tobacco entirely. Our results support the stand that smokers who want to quit should be offered safe, clinically proved treatments for smoking cessation, including pharmacotherapies such as medicinal nicotine or antidepressants, behavioural counselling and telephone quit lines.

Reference Number: 60615

HOWARD-PITNEY, B., AND M.A. WINKLEBY. "Chewing Tobacco: Who Uses and Who Quits? Findings from NHANES III, 1988–1994." *American Journal of Public Health*, 92: 250–256, 2002 (2 tables, 3 figures, 27 references)

Abstract: A US national sample of 1,340 non-Hispanic Black men, 1,358 Mexican American men, and 1,673 non-Hispanic White men (ages 25–64 years) who participated in the National Health and Nutrition Examination Survey (NHANES III) between 1988–1994 were assessed as to their tobacco product use (SLT and cigarettes). Rural lower income Black and White men had the highest regular use of chewing tobacco (33.3%). In sharp contrast to cigarette smoking, which showed a dramatic decrease in use around the age of 20 years, chewing tobacco use continued throughout adulthood into the 50s.

Survival analyses indicated that the risk for onset of regular chewing tobacco use continued well into middle age and beyond, whereas the risk for onset of regular cigarette smoking became negligible by middle age. If chewing onset occurs across the life span, there are many implications for interventions. For example, prevention messages must address middle-aged and older men, in addition to adolescents and young adults. Because smoking prevention programs focus primarily on adolescents, chewing tobacco programs that are offered only in tandem with school- or community-based smoking programs will likely miss a substantial number of men who initiate use during adulthood. Chewing tobacco prevention and cessation programs are needed beyond high school, vocational school, and college and in settings outside of school. The workplace is a viable setting for delivering programs to middle-aged and older men. In addition, some research suggests that cessation interventions located at the site of health care services, such as dentist offices, can be successful with chewing tobacco users.

Reference Number: 60616

KAUFFMAN, R.M., A.K. FERKETICH, A.G. WEE, J.M. SHULTZ, P. KUUN, AND M.E. WEWERS. "Factors Associated with Smokeless Tobacco Cessation in an Appalachian Population." *Addictive Behavior*, 33: 821–830, 2008 (3 tables, 40 references)

Abstract: In 2006 it was estimated that 6.1% of American men and 0.4% of American women used SLT and that its use was highest in rural areas. This study determined the 12-month quit rate and characteristics of 116 men (ages 19–70 years) enrolled in a cessation program in Appalachia. Quit rates increased with increasing age. Men with a high school education or greater were 7× more likely to quit. Alcohol was not associated with quit rates. However, 44.0% of the subjects who drank 1–2 servings of coffee per day quit compared to 19.7% of those who did not drink coffee.

Table: Demographic Variables and Quit Rates of SLT Users

Demographic Variable	Quit Rate
<30 years of age	3.5%
30–39 years of age	26.0%
40–49 years of age	27.3%
50 years of age or older	33.3%
Started at <10 years of age	11.5%
Started at 11–17 years of age	21.9%
Started at 18 years of age or older	30.8%

Source: Adapted from Kauffman et al, 2008

Despite these limitations, this study identifies several factors which may be important predictors of ST cessation. The finding that coffee consumption is associated with successful quit attempts reemphasizes the need for interventions that take into account other health behaviors when attempting to motivate change. Due to its widespread use, caffeine could play an important role in determining the success of intervention programs at the population level. Further research is needed to refine our understanding of caffeine's role in nicotine addiction. The high degree of agreement between factors associated with ST cessation observed in this sample and previously reported predictors of smoking cessation lends support to the belief that nicotine addiction may be regarded as a unified construct regardless of the nicotine delivery system, however further work is needed to clarify the degree to which findings for smokers may be applied to those using ST products.

Reference Number: 60617

BEGUM, S.F., G. NAGAJOTHI, K.S. LATHA, G. SANDEEP, B. SREEKANTH, C.S. KUMAR, W. RAJENDRA, AND N. MADDU. "Possible Role of Nicotine and Cotinine on Nitroxidative Stress and Antioxidant Content in Saliva of Smokeless Tobacco Consumers." *Practical Laboratory Medicine*, 12: (10 pp), 2018 (4 tables, 8 figures, 43 references)

Abstract: Oral smokeless tobacco products are placed in the mouth or lip and sucked (dipped) or chewed. At the time of the study, India had the second highest number of tobacco users in the world (229 million compared to 311 million in China). Two major types of smokeless tobacco are consumed in India, gutkha (containing tobacco, areca nut, catechu, slaked lime, and flavors) and khaini (containing tobacco and limestone paste). Absorption of nicotine from the SLT occurs through the mucous membrane and accelerates quickly, reaching maximum at 5 minutes and declining over the next 30 minutes. The saliva concentration of nicotine, cotinine, and antioxidant enzymes, and levels of lipid peroxidation and nitric oxide, were determined in 30 controls, 30 gutkha users, and 30 khaini users by spectrophotometric methods. Tobacco consumption is the major factor for the generation of free radicals in cells and tissues. Smokeless tobacco in India is from the Nicotiana rustica tobacco species and thus contains a higher level of nicotine.

Decreased antioxidant status, uric acid levels, up regulation of lipid peroxidation and nitric oxide levels in the saliva of smokeless tobacco users when compared to controls to suggest that evaluation of these parameters may perhaps be used as markers of tobacco induced nitroxidative stress. One of the important implications for nicotine exposure is the production of reactive oxygen species in smokeless tobacco consumers. Chewing of smokeless tobacco is a very dangerous and easily addictable product, awareness has been increased about the harmful effects of smokeless tobacco products and reduces the risk of exposure and consumption. Further studies are needed to correlate the toxic effects of prolonged use of smokeless tobacco on human health.

Reference Number: 60618

NIAZ, K., F. MAQBOOL, F. KHAN, H. BAHADAR, F.L. HASSAN, AND M. ABDOLLAHI. "Smokeless Tobacco (*Paan* and *Gutkha*) Consumption, Prevalence, and Contribution to Oral Cancer." *Epidemiology and Health*, 39: (11 pp), 2017 (3 tables, 4 figures, 100 references)

Abstract: This study summarizes the use of SLT products in Taiwan, India, Pakistan, Tanzania, Cambodia, Indonesia, and Malaysia. SLT consumption is widespread globally and is a major cause of oral submucous fibrosis (OSMF). Gutkha consists of slaked lime, areca nut, chewing tobacco, spices, and catechu packed in a tin or pouches. Paan (which means leaf) consists of tobacco, seeds, quenched lime, spices, and areca nut enfolded in a betel quid or leaf. Gutkha is sweet and children believe it is a form of candy.

In the present study, it was concluded that the extensive use of smokeless tobacco in different forms leads to OSMF, which potentially transforms into a malignant condition in all age groups. The genotoxic and carcinogenic effects of smokeless tobacco in the oral cavity are due to the production of ROS and free radicals. These free radicals and ROS damage the normal DNA and RNA, leading to genotoxicity and, eventually, oral cancer. Previously described mechanisms explain the induction, maintenance, and progression of OSMF due to *paan* and *gutkha*. Strict cessation of smokeless tobacco use and follow-up should be implemented to reduce the incidence of oral cancer. Widespread use of these and other products by children, as well as adolescents, is mostly due to their pleasant taste, low cost, and easy availability. Oral cancer rates are increasing due to

> use of these smokeless tobacco products, in particular among the lower socioeconomic levels that constitute the large majority of the population.

Reference Number: 60619

GUVEN, A., AND F. TOLUN. "Effects of Smokeless Tobacco 'Maras Powder' Use on Nitric Oxide and Cardiovascular Risk Parameters." *International Journal of Medical Sciences*, 9: 786–792, 2012 (2 tables, 41 references)

Abstract: Smokeless tobacco products vary worldwide, from chimo in Venezuela (ground tobacco/spices/sugar) to toombak in Sudan (ground tobacco/sodium bicarbonate) to mawa/gutka in India (ground tobacco area/betel nuts). This study examines maras powder (ground tobacco-Nicotiana rustica/oak or walnut ash), an SLT product popular in south-east Turkey. Nicotiana rustica has 6–10× the nicotine content of the usual Nicotiana tabacum and the wood ash allows for greater absorption of nicotine through the oral cavity. About 1 gram of maras powder is typically placed in the lower lip for 5–10 minutes and then spat out. Blood samples were taken at a clinic from 48 maras users, 50 cigarette smokers, and 45 never-users (mean ages 42, 39, and 39 years, respectively) and were analyzed for cholesterol, LDL-cholesterol, triglycerides, and nitrous oxide concentrations. Plasma NO levels were lower in maras powder and cigarette users than in non-users. Plasma total cholesterol, LDL-cholesterol, and triglyceride levels were higher in maras powder and cigarette users than non-users.

> We conclude that Maras powder has similar adverse effects on nitric oxide level and cardiovascular risk parameters and thereby it appears to be harmful as cigarette smoking.

Reference Number: 60620

WARNAKULASURIYA, S., AND K. STRAIF. "Carcinogenicity of Smokeless Tobacco: Evidence from Studies in Humans & Experimental Animals." *Indian Journal of Medical Research*, 148: 681–686, 2018 (3 tables, 29 references)

Abstract: A short review was conducted as to the carcinogenicity of smokeless tobacco (SLT) in regard to the WHO Collaborating Centre for Oral Cancer. The majority of SLT products are from both Nicotiana tabacum and Nicotiana rustica species of tobacco. After a single oral use of 7.9 grams of SLT, the maximum plasma nicotine concentrations reached were equivalent to those of a single cigarette, but the nicotine levels are sustained for a prolonged period and so overall the amount of nicotine absorbed is 2× that as the amount absorbed from smoking a single cigarette. The metabolism of tobacco is such that the blood and urine cotinine levels are much higher in SLT users than smokers. The main and most potent carcinogens in SLT are TSNAs from the tobacco alkaloids formed during the curing and fermentation and ageing of tobacco leaves. Other carcinogens such as heavy metals, cadmium, uranium, and polonium have been found as well.

Table: Mean Nicotine and TSNAs in Indian and Swedish SLT Products

Substance	Indian SLT Product	Swedish SLT Product
Total nicotine (mg/g wet wt.)	10.0	8.340
Unprotonated nicotine (mg/g wet wt.)	mm Hg9.5	0.750
Total tobacco-specific nitrosamines	37.6	0.723

Source: Adapted from Warnakulasuriya and Straif, 2018

There is no safe form of tobacco, and both smoked and smokeless tobacco are carcinogenic to humans. As summarized here, the evidence for increased risk for oral cancer among people consuming ST products in South Asia is based on case-control and cohort studies mostly conducted in India and Pakistan. Dose-response data demonstrate a higher risk with increased frequency and duration of ST use and further support causal inference. This evidence is further corroborated by studies conducted in the USA reporting increased risks of mouth cancer.

Reference Number: 60621

MAHMOODABAD, S.S.M., M.S. JADGAL, I. ZAREBAN, AND H.F. ZADEH. "The Determinants of Salivary Cotinine Concentration in Smokeless Tobacco Users." *Macedonian Journal of Medical Sciences*, 7: 810–815, 2019 (5 tables, 22 references)

Abstract: At the time of the study, about 20% of the tobacco used globally was smokeless, which is highly addictive due to its high nicotine content. Smokeless tobacco also contains carcinogenic chemicals such as TSNAs, which lead to an increased risk of oral cavity, laryngeal, and oesophageal cancers and cardiovascular diseases. A survey was conducted, and the salivary cotinine concentrations were determined in 150 tradespeople in Chabahar, Iran who used smokeless tobacco in 2018. Most of the SLT users were men (60%) and two-thirds were in the 20–30 years age group. The mean salivary cotinine concentration in men was 887 ng/mL and 611 ng/mL in women. The highest mean salivary concentration (665 ng/mL) was found in the 30–40 years age group. Naswar was the type of SLT most commonly used by women (84%). Men typically used two or more different types (such as pan parag, gutka, or mawa). The mean salivary cotinine concentration was higher in those who used two or more smokeless tobacco products or started using it before the age of 10 years.

Table: Type of Smokeless Tobacco Used and Mean Salivary Cotinine Concentration

Type of Smokeless Tobacco	Salivary Cotinine Concentration (ng/mL)
Pan parag	703.5
Gutka	501.0
Naswar	627.0
Mawar	4.0
Combination of two or more	1068.5

Source: Adapted from Mahmoodabad et al, 2019

Based on the results of the regression analysis, the type of smokeless tobacco used was a strong predictor of the concentration of salivary cotinine in the participants. It is, therefore, necessary for the government to adopt appropriate policies and take educational measures to reduce the vending and use of these substances.

Reference Number: 60622

CHANDRA, P., AND P. GOVINDRAJU. "Prevalence of Oral Mucosal Lesions Among Tobacco Users." *Oral Health and Preventative Dentistry*, 10: 149–153, 2012 (3 tables, 37 references)

Abstract: A study was conducted of 359 tobacco-using patients attending the Department of Oral Medicine and Radiology in India, of whom 265 (74%) had oral mucosa lesions. There were 182 smokers, 98 chewers, and 36 who both chewed and smoked. Bidi smokers had a greater number of oral lesions than cigarette smokers. Quid is an SLT made up of a mixture of areca nut and tobacco.

Table: Tobacco-Associated Lesions in Smokers, Tobacco Chewers, and Dual Users

Tobacco-Associated Lesions	Smokers (%)	Tobacco Chewers (%)	Both Smokers and Tobacco Chewers (%)
Tobacco-associated melanosis	4.0	0.8	0.60
Leukoedema	7.0	0.3	0.30
Smoker's palate	4.6	0	0.50
Tobacco pouch keratosis	0	1.4	0.06
Chewer's mucosa	0	2.0	0.50
Malignancy	0	0.1	0

Source: Adapted from Chandra and Govindaraju, 2012

If tobacco were to be introduced today, it would not be approved for human consumption anywhere in the world. Thousands of compounds are found in both forms of tobacco. Many of these compounds are not only irritants and toxins, they are also carcinogens. The smoking form of tobacco contains over 4000 different chemicals and at least 50 are known carcinogens. Smokeless tobacco contains 28 carcinogens. Tobacco-specific nitrosamines are present in higher concentrations while nicotine content is lower in smokeless tobacco than in the smoking form....

Tobacco smoking affects multiple organ systems resulting in numerous tobacco-related diseases.

Reference Number: 60623

HECHT, S.S., S.G. CARMELLA, A. EDMONDS, S.E. MURPHY, I. STEPANOV, X. LUO, AND D.K. HATSUKAMI. "Exposure to Nicotine and a Tobacco Specific Carcinogen Increase with Duration of Use of Smokeless Tobacco." *Tobacco Control*, 17: 128–131, 2008 (2 tables, 1 figure, 20 references)

Abstract: Urine samples were collected from 212 male American smokeless tobacco users (mean age 33.8 years) and analyzed for biomarkers of nicotine uptake (cotinine) and the tobacco-specific carcinogen (NNK) by examining the biomarker total (NNAL). There was a significant increase in urinary cotinine and total NNAL with increasing duration of daily SLT use, which indicates a growing dependence on SLT with duration of use. The duration of use was significantly greater for Copenhagen long cuts than for Skoal straight long cuts or Kodiak premium wintergreen. The time to the first dip (from the SLT tin) related to an increased nicotine dependence among those who used smokeless tobacco for longer periods of time.

Table: Mean Urinary Cotinine and Total NNAL Concentrations in 212 SLT Users by Years of Daily Use

Years of Daily SLT Use	Mean Urinary Cotinine Concentration (nmol/mL)	Mean Urinary Total NNAL Concentration (pmol/mL)
0–5	23.6	2.47
6–10	37.3	3.21
11–15	43.9	4.26
16–20	44.3	4.07
21+	55.9	5.21

Source: Adapted from Hecht et al, 2008

In summary, the results of this study clearly demonstrate that levels of urinary cotinine and total NNAL in smokeless tobacco users increase with duration of use. Smokeless tobacco users need to be educated about the effects of prolonged use. These findings should be considered when recommending smokeless tobacco use as a harm reduction strategy as increased duration of use may have unintended consequences due to higher than expected toxicant exposure.

Reference Number: 60624

BOFFETTA, P., AND K. STRAIF. "Use of Smokeless Tobacco and Risk of Myocardial Infarction and Stroke: Systematic Review with Meta-Analysis." *British Medical Journal*, 339: (6 pp), 2009 (3 tables, 3 figures, 28 references)

Abstract: Smokeless tobacco is a recognized carcinogen in humans. Eight studies from Sweden and three from the United States were systematically reviewed for analysis of the effect of smokeless tobacco use on the risk of myocardial infarction (MI) and stroke. The relative risk of ever use of SLT and fatal MI was 1.13× and the excess risk was restricted to the current user. The relative risk of a fatal stroke was 1.40× with current SLT use. An association between the use of SLT and the risk of fatal MI and stroke was therefore established.

Animal experiments and studies in humans indicate that smokeless tobacco has mainly short term effects resulting in increased blood pressure and heart rate that are presumably due to nicotine, whereas data on long term use and hypertension are inconclusive. Several studies suggest that increased levels of triglycerides, diabetes, obesity, and metabolic syndrome might be associated with use of smokeless tobacco. Experimental findings of increased damage to the cardiac muscle and poorer myocardial healing in dogs exposed to nicotine may provide a pathophysiological rationale for the increased mortality from myocardial infarction. Furthermore, nicotine induced cardiac arrhythmias in dogs and increased the sensitivity towards arrhythmias and induced ventricular fibrillation in canine hearts that survived myocardial infarction.

Reference Number: 60625

ETEMADI, A., H. KHADEMI, F. KAMANGAR, N.D. FREEDMAN, C.C. ABNET, P. BRENNAN, AND R. MALEKZADEH. "Hazards of Cigarettes, Smokeless Tobacco and Waterpipe in a Middle Eastern Population: A Cohort Study of 50,000 Individuals from Iran." *Tobacco Control*, 26: 674–682, 2017 (4 tables, 2 figures, 50 references)

Abstract: This study examined the effect of cigarette smoking, chewing tobacco (Nass), and smoking waterpipe on overall and cancer-related mortality in 50,045 adults (aged 40–75 years) in northeastern Iran over a median 8-year follow-up period. Cigarette smoking was reported by 17%, chewing tobacco by 7.5%, and waterpipe smoking by 1.1% of the participants. Current use of all forms of tobacco products caused an increase in the risk of cancer mortality.

Table: Adjusted Hazard Ratios (HR) for Overall Mortality and Cancer Mortality According to Cigarette and SLT Use

Tobacco Use	HR for Overall Mortality	HR for Cancer Mortality
Former Cigarette Only	1.30×	1.15×
Current Cigarette Smoker	1.45×	1.75×
Current Nass Only	1.17×	1.40×
Former Cigarette and Current Nass	1.33×	1.65×
Current Cigarette and Current Nass	1.28×	1.67×

Source: Adapted from Etemadi et al, 2017

All types of tobacco increased the risk of earlier death (particularly from cancer) in our study. We also showed increased risk of earlier death, compared to never tobacco users, among people who switched to smokeless tobacco after quitting cigarettes, and those who have regularly smoked the waterpipe for a long time.

These results emphasize the need for global tobacco control and show that the focus of anti-tobacco efforts must not be confined to cigarette smoking.

Reference Number: 60626

DONALD, P.M., G. RENJITH, AND A. ARORA. "Tobacco Pouch Keratosis in a Young Individual: A Brief Description." *Journal of Indian Society of Periodontology*, 21: 249–251, 2017 (2 tables, 1 figure, 8 references)

Abstract: Tobacco pouch keratosis or SLT induced keratosis is a precancerous white mucosal lesion which develops in the area of oral tobacco contact. A 22-year-old male SLT user (tobacco, areca nut, betel leaf) had discomfort while eating, drinking, and speaking. He had a lesion in the upper lip from his left canine to his upper first molar. He was advised to quit using SLT. The oral mucosa of about 98% of the patients return to normal after 2–6 weeks of cessation.

Dental professionals play a vital role in early identification of tobacco-related conditions. As this condition is more prone for malignant transformation, such conditions should not be neglected and the patient has to be educated about the harmful effects of habitual chewing of smokeless tobacco, and preventive measures can be taught for a better quality of life.

Reference Number: 60627

SUBRAMONEY, S., AND P.C. GUPTA. "Anemia in Pregnant Women who Use Smokeless Tobacco." *Nicotine and Tobacco Research*, 10: 917–920, 2008 (1 table, 1 figure, 29 references)

Abstract: Smoking affects blood hemoglobin levels. Adult women who smoke have higher levels but pregnant smokers have lower levels. This study determined the hemoglobin level of 121 pregnant users of SLT and 605 pregnant non-users in Mumbai, India in 2002. The mean hemoglobin level was significantly lower in users (10.00 g/dL) compared to non-users (10.46 g/dL). SLT users were 1.7× as likely to have low hemoglobin levels than non-users.

Thus, based on the present study, the relationship between anemia and smokeless tobacco use during pregnancy appears to be real. Although high hemoglobin levels in midpregnancy are associated with fetal growth restriction and preeclampsia, achieving normal hemoglobin levels during pregnancy will be beneficial for the fetus. Smokeless tobacco use is widely prevalent among women in Southeast Asia and is slowly gaining acceptance across the world as a safe alternative to cigarette smoking.

Reference Number: 60628

HECHT, S.S., AND D. HOFFMANN. "Tobacco-Specific Nitrosamines, An Important Group of Carcinogens in Tobacco and Tobacco Smoke." *Carcinogenesis*, 9: 875–884, 1988 (2 tables, 2 figures, 95 references)

Abstract: The presence of a tobacco-specific nitrosamine (NNN) was reported in tobacco smoke and unburned tobacco in 1974. The first report of another tobacco-specific nitrosamine (NNK) followed in 1978. TSNAs can be formed from nicotine and other alkaloids present in tobacco under mild conditions. Since nitrite is present in saliva and nitrogen oxides are present in mainstream tobacco smoke, TSNAs can be formed as well. Nitrosamines in bacon and beer cannot exceed 0.005 ppm, but 1 gram of snuff contains 30,000× that level. A person would have to consume 300,000 grams of bacon to have the same estimated nitrosamine exposure they would have from 10 grams of snuff.

Table: The Concentration of NNN and NNK in Smokeless Tobacco and Cigarette Smoke

Smokeless Tobacco/Cigarette Smoke	NNN Concentration	NNK Concentration
Chewing tobacco (ppm)	0.67–4.1	0.03–0.38
Moist snuff (ppm)	5.8–64	0.1–3.1
Dry snuff (ppm)	9.4–55	1.9–14
Tobacco (used in betel quid) (ppm)	0.47–0.85	0.13–0.23
Mainstream cigarette smoke (μg/cigarette)	0.066–1.01	0.17–0.43
Sidestream cigarette smoke (μg/cigarette)	0.19–0.86	0.39–1.44

Source: Adapted from Hecht and Hoffmann, 1988

Tobacco-specific nitrosamines are a group of carcinogens that are present in tobacco and tobacco smoke. They are formed from nicotine and related tobacco alkaloids. Two of the nicotine-derived nitrosamines, NNK and NNN, are strong carcinogens in laboratory animals. They can induce tumors both locally and systemically. The induction of oral cavity tumors by a mixture of NNK and NNN, and the organospecificity of NNK for the lung are particularly noteworthy. The amounts of NNK and NNN in tobacco and tobacco smoke are high enough that their total estimated doses to long-term snuff-dippers or smokers are similar in magnitude to the total doses required to produce cancer in laboratory animals. These exposures thus represent an unacceptable risk to tobacco consumers, and possibly to non-smokers exposed for years to environmental tobacco smoke. The permission of such high levels of carcinogens in consumer products used by millions of people represents a major legislative failure. Indeed, the levels of tobacco-specific nitrosamines in tobacco are thousands of times higher than the amounts of other nitrosamines in consumer products that are regulated by government authorities.

6.07 SNUS

"These findings suggest that snus performs no better than nicotine gum as a cigarette substitute, has less appeal, is more toxic and is associated with higher rates of prolonged use."

—Hatsukami, Severson, Anderson et al, "Randomized Clinical Trial of Snus Versus Medicinal Nicotine Among Smokers Interested in Product Switching" (2016)

Snus consists of finely ground moistened tobacco with salt, alkaline agents (to increase the pH of the tobacco to allow for more rapid oral absorption), humectants (such as glycerol), and chemical flavorings. It is marketed in Sweden either loose or portion packed and is consumed by placing it between the gum and upper lip. Snus is banned in all EU countries except Sweden, where it has been manufactured since 1630. It is also allowed in Norway, which is not a member of the EU (60701, 60711).

Traditional moist snuff tobacco (MST) is a shredded, flavored, chemically treated fermented tobacco that is normally sold in a round tin (60704). The use of MST is messy and unsightly, but is now sold in small pouches and has gained popularity in the United States and several other countries (60702–3, 60716–17).

As typical with the tobacco industry and with any tobacco product, snus has been redesigned with new packaging, larger pouch sizes, small and non-graphic warning labels, more nicotine, and new flavors to attract young users (60703–7). Snus has been promoted to prevent cigarette smoking, but young snus users are 2× as likely to initiate cigarette smoking. Nicotine gum is a better smoking cessation tool than snus, as snus has less appeal, is more toxic, and is associated with longer term use and addiction (60706–7, 60713)

Snus use has been associated with cancers, cardiovascular diseases, snuff dippers lesions, heart rate variability, and higher blood pressure in children exposed to their mother's use of snus while in the womb (60708, 60711–12).

Reference Number: 60701

LAWLER, T.S., S.B. STANFILL, H.T. TRAN, G.E. LEE, P.X. CHEN, J.B. KIMBRELL, J.G. LISKO, C. FERNANDEZ, S.P. CAUDILL, B.R. DECASTRO, AND C.H. WATSON. "Chemical Analysis of Snus Products from the United States and Northern Europe." *Public Library of Science One* 15: (17 pp), 2020 (2 tables, 4 figures, 32 references)

Abstract: Snus is a finely ground oral tobacco product that originally consisted of air-dried tobacco, salt, and water and was manufactured in Sweden beginning in the 1630s. Snus now consists of fire-cured and fermented tobacco, humectants, alkaline agents, and flavorings. In the early 1990s, the sale of snus was banned in all EU countries except Sweden. In this study, 56 Northern European and eight US snus products were purchased from stores in the state of Georgia in 2013 and 2014 and were analyzed by GC/MS, GC/MS/MS, and other methods. Menthol was the most common flavor detected. Snus products are generally kept refrigerated at the point of sale to hinder microbial growth and the generation of TSNAs.

Table: Chemical Characteristics of Northern Europe Compared to US Snus

Chemical Component	Northern Europe Snus	US Snus
Mean pH	8.08	6.62
Mean total nicotine (mg/g wet)	14.1	10.8
Mean unprotonated nicotine (%)	55.0%	32.7%
Mean total TSNAs (ng/g wet)	1,430	2,220

Source: Adapted from Lawler et al, 2020

Our results identify potential key differences in the physical and chemical characteristics of snus products from the US and Northern Europe. One manufacturer consistently had products with reduced TSNA levels, likely achieved by reducing or eliminating select microorganisms, to minimize TSNA formation. The lower nitrosamine levels in select snus products clearly demonstrates that tobacco products can be manufactured with reduced carcinogens. The higher levels of unprotonated nicotine in some Swedish products, especially those designated as "strong," was an important finding in this product survey as higher percentages of unprotonated nicotine could increase addictiveness. The higher unprotonated nicotine levels associated with certain product descriptors may be achieved through inclusion of higher nicotine tobacco sources and buffering agents that increase alkalinity. Higher unprotonated nicotine levels, in turn, enhance nicotine adsorption and may increase product addictiveness and contribute to adverse health outcomes linked with sustained and repetitive tobacco exposure.

Reference Number: 60702

SEIDENBERG, A.B., O.A. AYO-YUSUF, AND V.W. REES, "Characteristics of 'American Snus' and Swedish Snus Products for Sale in Massachusetts, USA." *Nicotine and Tobacco Research*, 20: 262–266, 2018 (1 figure, 34 references)

Abstract: Snus use may present lower health risks than cigarettes, but its harm reduction potential can be undermined if the user continues smoking cigarettes as well. Total nicotine, free nicotine, pH, and moisture content were determined in 14 US snus and one Swedish snus products. US snus products contain significantly lower unionized nicotine, pH, and moisture than Swedish snus products. They also have a lower addiction potential than the Swedish products and may be used dually with cigarettes.

Tobacco industry documents reveal that smokeless tobacco product design has been used to facilitate initiation and progression towards nicotine dependence. US Smokeless Tobacco Company utilized a "graduation" strategy whereby new users initiate with low-nicotine products and "graduate" to higher nicotine products. Lower nicotine products are thought to create a less aversive initial sensory experience for novice nicotine users. Thus, the introduction of snus products with lower nicotine delivery potential could conceivably assist in recruiting new tobacco users.

The lower moisture content reported for American snus is consistent with suggestions that snus in the United States is targeted at smokers who would not quit smoking, but would use snus in places where they cannot smoke. In particular, the findings of a market survey reported in a previously secret industry document suggest smokers prefer lower moisture smokeless products, which is similar to the mouth feel from smoking.

Reference Number: 60703

SCHEFFELS, J., AND I. LUND. "Cute as Candy: A Qualitative Study of Perceptions of Snus Branding and Package Design Among Youth in Norway." *BMJ Open*, 7: e012837 (7 pp), 2017 (2 tables, 36 references)

Abstract: Snus sales in Norway were low until the 1990s, when traditional moist snus was reintroduced as a newer, drier tobacco product in small sachets. In 2016, 25% of youth (aged 16–24 years) were daily or occasional users of snus, compared to 16% who smoked. The majority of young users in Norway started without prior smoking. The sale of snus is banned in all EU countries except Sweden, where it is manufactured. Norway is not a member of the EU. Focus groups were conducted on 25 tobacco users and 13 non-users (aged 15–17 years) on snus packaging. In 2008, there were 180 different snus brands sold in Norway. The health warnings on snus packages are non-graphic and smaller than cigarette warnings.

> Our analyses indicate also that snus package design was seen as communicating information about product strength, and that packages with light colours were perceived as less harmful than darker packages. Flavoured products in candy-like boxes were perceived positively, in particular among non-users, and flavour additives (eg, mint or fruit) seemed also to contribute to ideas of product strength and harm. Similarly, product qualities and appeal were attributed to the colour and shape of snus sachets. This is similar to studies of combustible tobacco, where flavour additives have been shown to influence cigarette appeal and make the initial experience of smoking less aversive and more appealing to youth. Experimental studies of cigarette packaging have shown how standardising colours can reduce false beliefs about some cigarette products being less harmful to health than others, and that removing flavour descriptors from cigarette packages can make them less appealing to young women.

Reference Number: 60704

HENDLIN, Y.H., J.R. VEFFER, M.J. LEWIS, AND P.M. LING. "Beyond the Brotherhood: Skoal Bandits' Role in the Evolution of Marketing Moist Smokeless Tobacco Pouches." *Tobacco Induced Diseases*, 15: 46 (11 pp), 2017 (1 table, 3 figures, 91 references)

Abstract: Traditional moist snuff tobacco (MST) is a shredded, flavored, chemically treated fermented tobacco that is normally sold in a round tin. The use of MST is messy and unsightly, and it is now sold in small pouches such as snus in Sweden. MST manufacturers introduced 10 new pouch tobacco products between 2006 and 2017. A search of the previously secret tobacco industry archives showed the strategies to recruit new users of MST. The "brotherhood" refers to the traditional rural, White, and lower socioeconomic males who were the typical users of MST.

> Historical analysis of the U.S. introduction of pouched smokeless products provides insight into how pouched products became the current fastest growing segment of the U.S. smokeless tobacco market. Initially the product's physical ease of use, low nicotine levels and appealing flavors allowed it to be mass marketed to beginners and obviate painstaking one-on-one sampling. The introduction of multiple variants of pouch products with an increased range of nicotine strength, coupled with the claims that the products are "new," led to increases in both pouched MST and sales of snus products in the U.S. A major barrier to widespread adoption has been the difficulty of convincing the public that MST pouches were actually "new" and different, as well as dissonance between MST's traditional masculine rugged advertising and the attempts to position the product for minorities or women. These factors are likely to impact the introduction and promotion of novel tobacco products in future years. Due to similar effects on youth initiation, regulatory policies to decrease youth tobacco initiation (such as restricting characterizing flavors) could be fruitfully applied to pouch products.

Reference Number: 60705

SCHNELLER, L.M., B.R. LINDGREN, P.G. SHIELDS, D.K. HATSUKAMI, AND R.J. O'CONNOR. "Strong Preference for Mint Snus Flavor Among Research Participants." *Addictive Behavior Reports*, 6: 51–55, 2017 (3 tables, 1 figure, 30 references)

Abstract: A study was conducted on the snus preference of 94 male and 57 female cigarette smokers (mean age 37–40 years). The subjects could decide which snus they preferred, either Camel Snus Winterchill (menthol 8.9 mg/g nicotine) or Camel Snus Robust (no menthol, tobacco flavor, 9.5 mg/g nicotine), after smelling and using them for 5 minutes. The subjects could then take the snus home and use it for 7 days. One hundred and ten subjects chose Winterchill and only 41 chose Robust. Nicotine dependence was also higher in those who chose Winterchill (Fagerstrom Test for Nicotine Dependence (FTND) score of 4 vs 3, respectively). Winterchill was found to be more satisfying, less aversive, and having a more intense, pleasant odor compared to Robust.

Table: Fagerstrom Test for Nicotine Dependence (FTND) Percentages in Users who Selected Menthol vs No Menthol Snus

FTND (0 = low; 9 = high)	Selected Winterchill (menthol)	Selected Robust (no menthol)
0	2.7%	0.0%
1	1.8%	4.9%
2	10.9%	19.5%
3	31.8%	43.9%
4	26.4%	19.5%
5	18.2%	12.2%
6	7.3%	0.0%
9	0.9%	0.0%

Source: Adapted from Schneller et al, 2017

The finding that mint flavored oral products appear to be preferred by smokers in a clinical trial context represents a conundrum given other evidence suggesting flavorings make tobacco products more attractive to youth. Flavors may be necessary to interest cigarette smokers in novel nicotine products and draw them away from smoking. Yet, the availability of flavored nicotine products may also function to draw youth into tobacco product use. A potential regulatory approach may be to consider a broad prohibition of characterizing flavors in tobacco products. Regulators could then consider allowing flavorings in designated MRTPs if premarket evidence were provided on population impact of potentially shifting smokers to flavored product vs. potential uptake in youth, taking into account the relative toxicity of a product.

Reference Number: 60705

Reference Number: 60706

DANIELSSON, M., A. LAMMI, S. SIITONEN, J. OLLGREN, L. PYLKKÄNEN, AND T. VASANKARI. "Alarming Development of Dual Snus and Cigarette Usage Among Young Finnish Males." *BMC Public Health*, 19: 1249 (7 pp), 2019 (3 tables, 1 figure, 50 references)

Abstract: A survey was conducted of 1,916 Finnish male conscripts (mean age 19.4 years). Almost 40% of the conscripts were either daily or occasionally smoked. Over one-third of the conscripts used snus and 21% reported dual snus and cigarette use. Over 40% of former snus users smoked daily. The transition from cigarettes to snus use among males was incomplete resulting in dual use. SLT may actually initiate smoking.

Table: Percentage of Daily Smokers, Daily Snus Users, and Daily Users According to Education

Education	Daily Smokers	Daily Snus Users	Daily Dual Users
Basic education	56.8%	13.6%	5.0%
Vocational school	38.8%	15.0%	3.0%
Upper secondary school	9.9%	14.2%	1.2%

Source: Adapted from Danielsson et al, 2019

This study provides a better understanding of the complex tobacco habits of young adult males, who are an important target population for health interventions. Young adults are at a transitional and vulnerable age when the usages of tobacco products, as well as other health habits, are yet to be established. The simultaneous usage of multiple tobacco products as well as the high tendency of transition from one tobacco product to another, should be taken into consideration when planning cessation interventions in health care settings and tobacco control policies at societal levels.

Reference Number: 60707

GROTVEDT, L., L. FORSEN, I. ARIANSEN, S. GRAFF-IVERSEN, AND T.L. HOLMEN. "Impact of Snus Use in Teenage Boys on Tobacco Use in Young Adulthood; a Cohort from the HUNT Study Norway." *BMC Public Health*, 19: 1265 (10 pp), 2019 (3 tables, 2 figures, 45 references)

Abstract: In 2017, 25% of young men and 14% of young women (aged 16–24 years) in Norway used snus. Among young Norwegians, daily snus use (19%) was more common than daily smoking (3%). Snus users and cigarette smokers shared many characteristics and much the same susceptibility to tobacco use. Nicotine dependence may explain the common propensity for future snus use and smoking as the quantity of nicotine delivered from each tobacco product is similar. A longitudinal health study was conducted in Norway of 1,345 male adolescents in 1995–1997, and again in 2006–2008. The use of other tobacco products such as hookahs and cigarettes was almost non-existent in Norway and ECs had not yet entered the market there.

> In this study, adolescent snus only users conferred a doubled risk of smoking, and almost a threefold risk to continue with snus as young adults. Adolescent dual users conferred a threefold risk to still be smokers in adulthood. The transition from smoking to snus only use was less common. Any adolescent tobacco use was associated with increased risk of smoking, including dual use, 11 years later.

Reference Number: 60708

NORDENSTAM, F., M. NORMAN, AND R. WICKSTRÖM. "Blood Pressure and Heart Variability in Preschool Children Exposed to Smokeless Tobacco in Fetal Life." *Journal of the American Heart Association*, 8: e012629 (8 pp), 2019 (3 tables, 3 figures, 49 references)

Abstract: Cigarette smoking during pregnancy is associated with an increased risk of preterm birth, stillbirth, low birthrate, and SIDS. Swedish snus is moist SLT packed in pouches and placed under the lip. If the mother uses snus, the fetus is therefore exposed mainly to nicotine and none of the combustible chemicals found in tobacco smoke. Oscillometric blood pressure was measured in 21 children exposed to snus during fetal life and 19 tobacco free children, all ages 5 to 6 years. Heart rate variability over 24 hours was also determined in 14 control and 16 snus exposed children. Taking the child's age and height into account, the snus exposed children had a mean systolic BP of 4.2 mm Hg higher than controls and had a greater spectral heart rate variability.

> In conclusion, our findings of increased SBP and altered HRV in preschool children, associated with exposure to snus in fetal life, may indicate adverse prenatal programming of nicotine, but the implications for cardiovascular health in later life remain to be studied. Meanwhile, women

should be recommended to abstain from all types of tobacco and nicotine products during pregnancy.

Reference Number: 60709

CRUZ-CANO, R., M. RANGEL-GOMEZ, C. VAN WAGONER, A. KIDANU, M.C. BRINKMANN, AND P.I. CLARK. "The Acceptability of Smokeless Tobacco Product Depends on Nicotine Levels." *Addictive Behavior Reports*, 10: (8 pp), 2019 (3 tables, 3 figures, 39 references)

Abstract: Since nicotine is a weak base, the fraction of either monoprotonated or unprotonated (free-base) nicotine in products can be manipulated by using additives to adjust the pH. The free-base form of nicotine is more easily absorbed through the oral cavity, which is how nicotine is mainly absorbed by smokeless tobacco users. TSNAs are among the most carcinogenic compounds in smokeless tobacco. The Swedish National Food Agency Directive has regulated the concentration of TSNAs and other harmful chemicals in snus since 2016, which may explain in part the lack of association between snus use and oral cancer in Scandinavian countries. In this study, 1 female and 27 male SLT users (median age 21.2 years) consumed various types of smokeless tobacco products and Nicorette gum after 12 hours of nicotine abstinence. Their response to the SLT products was determined by various subjective measures and EEG (electroencephalogram).

Table: The Amount of Free Base Nicotine and Total Nicotine in Various Smokeless Tobacco Products

SLT Product	Free-Base Nicotine (mg)	Total Nicotine (mg)
Red Man Long Cut Wintergreen	3.53	13.5
Copenhagen Long Cut Regular	3.78	11.2
General Swedish snus Classic Blend (Swedish Match)	4.29	7.19
Hawken Rough Wintergreen	0.01	3.99
Nicorette Nicotine Lozenge	1.41	1.43

Source: Adapted from Cruz-Cano et al, 2019

These products have distinct consumer acceptability levels and patterns of use characteristics ranging from starter products to those used primarily by established users. Subjective results showed that smokeless tobacco products with higher levels of nicotine were more successful in reducing craving and more reinforcing than those with lower levels. These results were concordant with the activity present in the EEG recordings where products with high nicotine levels produced larger changes in the amplitude of the event-related signal than those with low levels. This study is fundamental in understanding the relationship between subjective and objective smokeless tobacco acceptability measurements, as mediated by the different levels of nicotine in each product.

Reference Number: 60710

STEPANOV, I., J. JENSEN, L. BIENER, R.L. BLISS, S.S. HECHT, AND D.K. HATSUKAMI. "Increased Pouch Sizes and Resulting Changes in the Amounts of Nicotine and Tobacco-Specific N-Nitrosamines in Single Pouches of Camel Snus and Marlboro Snus." *Nicotine and Tobacco Research*, 14: 1241–1245, 2012 (1 table, 23 references)

Abstract: The nicotine and NNK and NNN (TSNs) concentrations were determined in 60 samples of Camel Snus and 87 samples of Marlboro Snus sold in the United States between 2006 and 2010. The weight of the Camel snus pouch increased from 370 mg to 970 mg and the weight of the Marlboro snus pouch increased from 240 mg to 410 mg.

> We observed an increase in nicotine content in single portions of Camel Snus and Marlboro Snus, and an increase in tobacco-specific N-nitrosamine content in single portions of Camel Snus, due to the increases in pouch size that occurred between 2006 and 2010. This finding stresses the importance of tobacco product regulation and ingredient disclosures.

Reference Number: 60711

KHARAZMI, M., A-P. CARLSSON, P. HALBERG, M. MODIG, L. BJÖRNSTAD, AND J-M. HIRSCH. "Surgical Approach to Snus-Induced Injury of the Oral Mucosa." *Journal of Oral Science*, 56: 91–94, 2014 (9 figures, 10 references)

Abstract: Snus consists of finely ground moistened tobacco with salt, sodium bicarbonate, humectants (such as glycerol), and chemical flavorings. It is marketed in Sweden loose or portion-packed and is consumed by placing it between the gum and upper lip. A mucosal lesion known as snuff dipper's lesion always appears at the site where the snus is usually placed. Lesions and irreversible snus induced mucosal injuries (ISIOMI) are generally less pronounced among users of the portion-packed snus than loose. Local gingival retraction is the most common ISIOMI and occurs in about 20% of users. Two male long-term snus users (ages 44 and 46 years) had successful oral surgery to repair the snus-related damage to the oral mucosa, including teeth root exposure.

> Since snus is the cause of mucosal injury, ongoing use with pinch placement at the location of the ISIOMI should be considered a contraindiction for surgery. Continuous use may jeopardize the outcome of surgery or lead to future recurrence of ISIOMI. Patients should quit using snus six to twelve months prior to surgery for ISIOMI. This period should allow significant recovery of the mucosa and also to some extent confirm that the patient is able to overcome the addiction. Nicotine replacement therapy can be recommended to aid cessation of snus use, but should the patient be unable to quit, referral for professional counselling should be considered. For patients who fail to overcome their addiction before surgery, changing to portion-packed snus if using loose snus, and switching the location of pinch placement, is required.

Reference Number: 60712

BOLINDER, G., L. ALFREDSSON, A. ENGLUND, AND U. DE FAIRE. "Smokeless Tobacco Use and Increased Cardiovascular Mortality Among Swedish Construction Workers." *American Journal of Public Health*, 84: 399–404, 1994 (3 tables, 1 figure, 27 references)

Abstract: Health examinations were conducted on 135,036 male Swedish construction workers between 1971 and 1974, and cause-specific mortality was determined over a 12-year period. The study consisted of 6,297 SLT users, 14,983 smokers (<15 CPD), 13,518 smokers (>15 CPD), 17,437 ex-smokers, 50,225 other tobacco users, and 32,546 non-users. The relative risk of death from cerebrovascular and cardiovascular death was 40% to 60% higher among smokeless tobacco users than among non-users. Smokers therefore face even greater RRs for death due to these diseases.

Table: Adjusted Relative Risk of Death in Smokeless Tobacco and Cigarette Users in Sweden

Cause of Death	SLT Users	Smokers (>15 CPD)	Ex-Smokers (>5 years abstinent)
All CV Disease	1.4×	1.9×	1.1×
All Cancer	1.1×	2.5×	1.3×
All Causes	1.4×	2.2×	1.1×

Source: Adapted from Bolinder et al, 1994

Both smokeless tobacco users and smokers face a higher risk of dying from cardiovascular disease than nonusers. Although the risk is lower for smokeless tobacco users than for smokers, the excess risk gives cause for preventive actions.

Reference Number: 60713

HATSUKAMI, D.K., H. SEVERSON, A. ANDERSON, R.I. VOGEL, J. JENSEN, B. BROADBENT, S.E. MURPHY, S. CARMELLA, AND S.S. HECHT. "Randomized Clinical Trial of Snus Versus Medicinal Nicotine Among Smokers Interested in Product Switching." *Tobacco Control*, 25: 267–274, 2016 (3 tables, 2 figures, 40 references)

Abstract: One hundred and eighty-four female and 207 male smokers (10 CPD for at least 1 year) who wanted to switch completely from cigarettes used either nicotine gum (Nicorette 4 mg) or Camel snus (Winterchill or Robust, 2.5 and 2.6 mg nicotine respectively) for 12 weeks. Breath CO was determined and urine samples were collected to determine TSNAs metabolites, cotinine, and nicotine. The subjects returned to the clinic every 2 weeks and were counseled for 10 minutes as to smoking cessation. The mean FTND was 5.1. The end of treatment 7-day avoidance of cigarettes (breath CO <6 ppm) was achieved for 21.9% of the sinus group and 24.6% of the nicotine gum group.

The results showed no significant differences between those assigned to medicinal nicotine vs snus in amount of product use, levels of cotinine attained, the extent to which the product substituted for smoking and rates of avoidance of cigarettes or any nicotine containing products. Furthermore, there were no differences in suppression of withdrawal from cigarettes. However, the nicotine gum users reported more satisfaction and psychological reward from the product, experienced less carcinogen exposure during the time of peak use, and had fewer participants who continued product use compared to the snus users. These findings suggest that snus performs no better than nicotine gum as a cigarette substitute, has less appeal, is more toxic and is associated with higher rates of prolonged use.

6.08 POLYTOBACCO USE

> "Polytobacco users are significantly more likely to report dependence symptoms than sole-product tobacco users for each category of current tobacco user."
>
> —Sung et al, "Polytobacco Use and Nicotine Dependence Symptoms Among US Adults, 2012–2014" (2018)

Polytobacco use is usually defined as cigarette use in combination with other tobacco product use (60801). Polytobacco use occurs mainly in younger male smokers, where the prevalence is 26% compared to 4% in female smokers (60801, 60807).

Although cigarette smoking has declined in schools, polytobacco use, especially involving e-cigarettes, is increasing (60801–5). This is an increasing problem as youth are more prone to nicotine dependence and life-long use of nicotine, and polytobacco users are more likely to binge drink alcohol and to use cannabis and illicit drugs (60802, 60805, 60809)

The reasons for the increase in polytobacco use are due in part to the tobacco industry bypassing the banning of flavors in cigarettes by flavoring other tobacco products instead (60804), and advertising and giving away free samples, which is also banned for cigarettes (60807–8).

The risk of all-cause mortality or cancer mortality was not significantly affected for an exclusive cigarette smoker compared to polytobacco users. There was no significant decrease in health risk if cigarette smokers also used SLT to cut down on cigarette use (60810).

Reference Number: 60801

BOMBARD, J.M., L.L PEDERSON, D.E. NELSON, AND A.M. MALARCHER. "Are Smokers Only Using Cigarettes? Exploring Current Polytobacco Use Among an Adult Population." *Addictive Behaviors*, 32: 2411–2419, 2007 (2 tables, 40 references)

Abstract: Polytobacco use is defined as cigarette use in combination with other tobacco product use. A 2003 survey of approximately 50,000 adults in 10 US states examined the prevalence of polytobacco use among adult smokers and the associated characteristics. The overall adult prevalence for each tobacco product was 22.4% for cigarettes, 5.7% for cigars, 3.5% for smokeless tobacco, 0.9% for pipes, and 0.3% for bidis. Among all adults, the prevalence of polytobacco use was 3.4%. Among male smokers, the prevalence was 26.0% compared to 4.4% for female cigarette smokers. Polytobacco use was also more likely to occur in the 18- to 29-year-old groups.

> Among men who smoked cigarettes, one out of four used at least one other tobacco product. The high prevalence of current polytobacco use among male cigarette smokers should be a focus for at least two reasons. First, most tobacco control interventions are primarily designed to address cigarette smoking and do not address combined use of additional tobacco products. Second, due to legislative, environmental, and marketing factors, the use of other noncigarette tobacco products (i.e. smokeless tobacco) could increase. Possible reasons why this could occur include higher excise taxes on cigarettes, the increasing number of smoke-free policies being instituted in local communities and at the state level, and industry marketing and advertising for other tobacco products. For example, smokeless tobacco company advertisements have encouraged the use of smokeless tobacco in situations when people cannot smoke cigarettes. A recent study found cigarette smokers with a home ban on smoking were more likely to use smokeless tobacco to maintain their nicotine addiction. The industry also proposes smokeless tobacco as a healthier alternative to smoking, this is important to monitor because some researchers have found that using smokeless tobacco may actually lead to initiation of smoking.

Reference Number: 60802

LEE, Y.O., J.K. PEPPER, A.J. MACMONEGLE, J.M. NONNEMAKER, J.C. DUKE, AND L. PORTER. "Examining Youth Dual and Polytobacco Use with E-cigarettes." *International Journal of Environmental Research and Public Health*, 15: 699 (12 pp), 2018 (3 tables, 22 references)

Abstract: A youth tobacco survey was conducted of 36,933 grade 6–8 students and 32,390 grade 9–12 students in Florida in 2014. It was found that 20.5% of high school students and 8.5% of middle school students used ECs. The vast majority (81.2%) of EC using youth used other tobacco products as well, and a significant proportion used more than four products on average. Youth who perceived that cigarettes were easy to quit were 2.5× as likely to use ECs with another tobacco product.

> Results showed that a large majority of e-cigarette-using youth in Florida also use tobacco products and that a significant proportion used more than four products on average. This suggests that any effort to prevent or address e-cigarette use among youth should be done with consideration for concurrent use with multiple tobacco products. In addition, nearly a fifth of e-cigarette using youth did not use tobacco products. This suggests that some youth are being exposed to nicotine who might not otherwise use tobacco products. This exposure is concerning due to the harms nicotine can pose to brain development during adolescence, the potential for youth to become nicotine dependent and the potential for increased susceptibility for using tobacco products, which contain many deadly carcinogens.

Reference Number: 60803

HUH, J., AND A.M. LEVENTHAL. "Progression of Poly-Tobacco Product Use Patterns in Adolescents." *American Journal of Preventative Medicine*, 51: 513–517, 2016 (4 tables, 12 references)

Abstract: Tobacco product use was determined in 3,396 grade 9 students at 10 public high schools in Southern California in the fall of 2013 (baseline) and 6 months later in the spring of 2014. E-cigarette use was the tobacco product most frequently used, followed by hookahs. The most typical transition was from non-user to EC/hookah user to polytobacco user. Transition from tobacco use to non-use was rare.

Table: The Prevalence of Tobacco Use by Grade 9 Students at Baseline and 6 Months Later

Number of Tobacco Products Used	Baseline	6-Month Follow-up
0	80.5%	70.7%
1	11.4%	14.2%
2	4.1%	8.1%
3	2.7%	4.4%
4	0.8%	2.0%
5	0.5%	0.7%

Source: Adapted from Huh and Leventhal, 2016

Of youths who are using a tobacco product as they enter high school, they are more likely to have done so with use of e-cigarettes or hookah and are at a sizeable risk of progressing to a more diverse pattern of poly-tobacco product use in as short as 6 months. Based on these results, continued surveillance of the growing population of e-cigarette- and hookah-using adolescents in the U.S. may be important to determine whether this group ultimately progresses to more diverse patterns of poly-tobacco product use in future years. Furthermore, strategies to limit access to and demand of e-cigarettes and hookah—the most common entry points for tobacco use—may reduce risk of progression to poly-tobacco product use.

Reference Number: 60804

CULLEN, K.A., S.T. LIU, J.K. BERNAT, W.I. SLAVIT, M.A. TYNAN, B.A. KING, AND L.J. NEFF. "Flavored Tobacco Product Use Among Middle and High School Students—United States, 2014–2018." *Morbidity and Mortality Weekly Report*, 68: 839–844, 2019 (1 table, 1 figure, 10 references)

Abstract: Flavoring chemicals other than tobacco and menthol are not allowed in cigarettes under the 2009 Family Smoking Prevention and Tobacco Control Act. However, flavorings are not prohibited in other tobacco products such as ECs, pipes, and bidis. The 2014–2018 National Youth Tobacco Surveys (NYTS) were analyzed to determine the flavored tobacco products use among US middle school (grades 6–8) and high school (grades 9–12) students. In 2018, 64.1% of youth tobacco product users currently used one or more flavored tobacco product, a decrease from 70.0% in 2014.

Table: Percent Change in Flavored Tobacco Product Use in High School Students between 2014 and 2018

Tobacco Product	Percent Change 2014 to 2018
Electronic cigarettes	+2.7%
Smokeless tobacco	−24.6%
Menthol cigarettes	−8.4%
Hookah tobacco	−26.0%
Cigars	−20.2%
Pipe tobacco	−16.7%

Adapted from Cullen et al, 2019

The high prevalence of flavored tobacco product use among middle and high school students is a concern because flavors can increase the appeal of tobacco products to youths, promote youth initiation of tobacco products, and result in lifelong tobacco product use.

A recent examination of online tobacco retailers found that a sizable proportion of noncigarette tobacco products for sale in the United States are flavored. The recent increase in flavored e-cigarette use among youths might be due, in part, to the recent popularity and increased market share of e-cigarettes shaped like a USB flash drive, such as JUUL; these products can be used discreetly, have a higher nicotine content than earlier generation e-cigarettes, and are available in flavors that appeal to youths.

Reference Number: 60805

WANG, T.W., A. GENTZKE, S. SHARAPOVA, K.A. CULLEN, B.K. AMBROSE, AND A. JAMAL. "Tobacco Product Use Among Middle and High School Students—United States, 2011–2017." *Mortality and Morbidity Weekly Report,* 67: 629–633, 2018 (1 table, 2 figures, 9 references)

Abstract: A National Youth Tobacco Survey was conducted between 2011 and 2017 of the past 30-day use of seven different tobacco products in middle school (grades 6–8), and high school (grades 9–12) students. Current tobacco use decreased over that time. In 2017, 19.6% of high school students reported current use of any tobacco product, including 9.2% who used two or more products. The most common tobacco product was e-cigarettes, used by 11.7% of the high school students and 3.3% of middle school students. Among youth, the symptoms of nicotine dependence were higher for polytobacco product users.

The sustained implementation of population-based strategies, in coordination with the regulation of tobacco products by FDA, are critical to reducing all forms of tobacco product use and initiation among U.S. youths. Strategies to reduce youth tobacco product use include increasing the price of tobacco products, implementing comprehensive smoke-free policies, implementing advertising and promotion restrictions and national public education media campaigns, and raising the minimum age of purchase for tobacco products to 21 years.

Reference Number: 60806

CREAMER, M.R., G.V. PORTILLO, S.L. CLENDENNEN, AND C.L. PERRY. "Is Adolescent Poly-Tobacco Use Associated with Alcohol and Other Drug Use?" *American Journal of Health Behavior*, 40: 117–122, 2016 (3 tables, 19 references)

Abstract: Nearly all tobacco use begins in adolescence and although the use of cigarettes has declined among high school students, the rates of dual and other tobacco product use have increased. This study determined that approximately three-quarters (77%) of high school students did not use any tobacco products, 13% used one tobacco product, and 10% used more than one tobacco product.

Table: Prevalence of Alcohol and Other Drug Use by Number of Tobacco Products Used

Drug Use	0 Tobacco Product Use	1 Tobacco Product Use	More than 1 Tobacco Product Use
Current drinking	22.9%	70.4%	85.8%
Binge drinking	10.4%	47.6%	70.2%
Current cannabis use	13.0%	53.6%	64.5%
Lifetime sniffed glue	5.6%	15.1%	26.0%
Lifetime heroin	0.6%	2.9%	12.8%

Source: Adapted from Creamer et al, 2016

This study provides important information regarding tobacco use, specifically multiple tobacco product use, and its association with alcohol and other drug use. This study confirms that behaviors rarely occur in isolation and that health-compromising behaviors, including tobacco and other drug use, tend to cluster. Intervention programs should target multiple risk behaviors, for example, by including skills training for adolescents around multiple tobacco products, alcohol, and other drug use. Study results echo previous studies that have demonstrated important relationships between dual tobacco use and alcohol, illicit drugs, and increased risk for other health conditions. Therefore, identifying these dual and multiple product users for intervention of other risk behaviors is essential, because these high school students appear to be at particularly high risk. Additionally, multi-component prevention programs not only should focus on cigarette smoking, but also on the use of multiple tobacco products, particularly in light of this increased risk. Furthermore, with the increasing diversity of the tobacco and nicotine product landscape, there is a need to study new and emerging products.

Reference Number: 60807

HAJJ, D.E., P.F. COOK, K.A. JAMES, C. BATTAGLIA, AND A.V. PROCHAZCA. "Newer Forms of Tobacco Products: Characteristics of Poly Users Among Adults Living in Colorado—A Secondary Data Analysis of the Attitudes and Behaviors Survey on Health 2015." *Tobacco Use Insights*, 12: 1–6, 2019 (3 tables, 30 references)

Abstract: Health data was collected from 8,618 adults (18+ years) in Colorado between 2015 and 2016 as to their tobacco and cannabis use. The lifetime prevalence of marijuana was 43.0% while it was 25.8% for cigarettes, 10.6% for hookahs, and 12.6% for other tobacco products. The most frequent cigarette smokers were single/divorced women of 65+ years, of lower socioeconomic status (SES), whereas the typical hookah use was a single/living alone young man of moderate SES.

In Colorado, in 2015, cigarette use was still highest among all forms of tobacco, but the use of other tobacco products such as vaping and hookah is on the rise, especially among young adults. Marijuana and hookah users were demographically similar to each other, and different from the typical cigarette user. These results indicate the need for further study of alternative tobacco product use, especially among young adults.

Reference Number: 60808

LITTLE, M.A., Z. BURSAC, K.J. DEREFINKO, J.O. EBBERT, G.W. TALCOTT, A. HRYSHKO-MULLEN, AND R.C. KLESGES. "Types of Dual and Poly-Tobacco Users in the US Military." *American Journal of Epidemiology*, 184: 211–218, 2016 (3 tables, 51 references)

Abstract: This study examined tobacco use among 13,873 Air Force technical trainees between 2013 and 2014. In the US Air Force, all trainees are tobacco free for the 8 1/2 weeks of basic training and the first 2 weeks of technical training. There were 72.9% non-users, 12.7% mono-users, 7.3% dual users, and 7.1% polytobacco users. In general, mono, dual, and polytobacco users were more likely to be male, under 21 years of age, White, and a high school graduate.

Table: Distribution of Tobacco Product in Air Force Technical Trainees in Mono and Polytobacco Users (2013-2014)

Tobacco Product	Mono Users (%)	Poly Users (%)
Cigarettes	48.5	71.1
Hookahs	36.1	60.6
Cigarillos	33.7	67.6
Smokeless tobacco	30.9	46.4
E-cigarettes	21.8	50.7
Cigars	16.2	44.5
Snus	7.1	21.0
Pipes	2.5	13.4
Roll your own cigarettes	4.0	11.5

Source: Adapted from Little et al, 2016

An additional factor associated with all 5 poly-tobacco use categories (as compared with mono-product use) was receiving tobacco products free at bars or social events. This finding suggests that tobacco-company influences may be a factor in the initiation or maintenance of poly-tobacco use. Because this information was gathered using a question designed to exclude receipt of tobacco products from friends, Airmen may be receiving tobacco products either from strangers at bars and social events or through promotional campaigns by tobacco companies. While the distribution of free cigarettes is now prohibited with passage of the 2009 Family Smoking Prevention and Tobacco Control Act, smokeless tobacco can still be distributed for free in adult venues, and there are currently no restrictions on free distribution of other tobacco products such as e-cigarettes. Future studies should closely evaluate tobacco-company marketing and its relationship to dual and poly-tobacco use, because it is clear from tobacco-company documents that young adults in general, and those in the military in particular, are targeted.

Reference Number: 60809

SUNG, H-W., Y. WANG, T. YAO, J. LIGHTWOOD, AND W. MAX. "Polytobacco Use and Nicotine Dependence Symptoms Among US Adults, 2012–2014." *Nicotine and Tobacco Research*, 20: S88–S98, 2018 (4 tables, 45 references)

Abstract: Two US National Adult Tobacco Surveys (NATS) of 135, 425 adults between 2012 and 2014 were analyzed as to tobacco use and nicotine dependency. Between 2012 and 2014, 25.1% of adults were current users of any tobacco product. Of these current tobacco users, 32.5% were polyusers of tobacco products. The largest group of polyusers used cigarettes and ECs (30.2%). The polyuser prevalence ranged from 38.7% of current cigarette users to 69.3% of cigar users and 86.2% of pipe users. Very few EC users used ECs daily and nearly 90% of current EC users smoked cigarettes or used other tobacco products.

> In conclusion, our results indicate that between 52% and 86% of non-cigarette tobacco users and nearly 40% of cigarette smokers engaged in polytobacco use. Polytobacco users are significantly more likely to report dependence symptoms than sole-product tobacco users for each category of current tobacco user. This study provides baseline insight into the complex and multifaceted matrix of polytobacco use patterns and its relationship to nicotine dependence symptoms shortly before the US Food and Drug Administration (FDA) issued the "Deeming Rule" that extends its tobacco regulatory authority over tobacco products that were not previously regulated, such as e-cigarettes and hookah. Consideration of polytobacco use should be part of FDA tobacco regulatory activities.

Reference Number: 60810

CHOI, K., M. INOUE-CHOI, T.S. MCNEEL, AND N.D. FREEDMAN. "Mortality Risk Associated with Dual- and Poly-Tobacco Product Use in the United States." *American Journal of Epidemiology*, 191: 397–401, 2019 (3 tables, 10 references)

Abstract: Data from the 1991, 1992, 1998, 2005, and 2010 National Health Interview Surveys of 118,144 adults were linked with the National Death Index to determine the mortality rates in polytobacco users in the United States. Overall, 71.6% of those surveyed had never used any tobacco, 26.5% used cigarettes, 1.1% used other combustible tobacco products as well as cigarettes, 0.6% used SLT and cigarettes, and 0.2% were polyusers.

Table: Hazard Ratio for All-Cause and Cancer Mortalities According to Tobacco Product Use

Tobacco Use Behavior	All Cause Mortality HR	Cancer Mortality HR
Exclusive current cigarette use	2.26×	2.75×
Current dual use of cigarettes and cigars or pipes	2.36×	2.71×
Current dual use of cigarettes and smokeless tobacco	2.28×	2.91×
Current poly use of cigarettes and both cigars/pipes and smokeless tobacco	2.49×	1.69×

Source: Adapted from Choi et al, 2019

Cigarette smokers who additionally used other tobacco products smoked as many if not more cigarettes per day than exclusive cigarette smokers. Furthermore, cigarette smokers who additionally used other tobacco products had mortality risks that were as high as and sometimes higher than those of exclusive cigarette smokers. As tobacco use patterns continue to change and diversify, investigators in future studies need to carefully assess the impact of noncigarette tobacco products on cigarette use and determine associated disease risks.

6.09 NICOTINE/DRUG USE

"Marijuana use was associated with greater reported nicotine addiction among adolescents. The findings suggest a role of marijuana in potentiating nicotine addiction and underscore the need for treatments that address use of both smoked substances."

—Rubinstein, Rait, and Prochaska, "Frequent Marijuana Use is Associated with Greater Nicotine Addiction in Adolescent Smokers" (2014)

Teenagers are more vulnerable to nicotine dependence than adults and addiction in youth can occur with minimal exposure, with as few as seven cigarettes a month. Nicotine can be considered as a gateway drug, in that smokers are more likely to use other drugs such as alcohol and cannabis (60901–14).

The nicotine and cannabinoid receptors in the body interact with each other, such that between 25 and 52% of adult current tobacco smokers use cannabis as well (60903). Regular smokers were 2× as likely to develop cannabis use disorder (60904). Cannabis use was also associated with greater nicotine addiction among adolescents (60905).

Adolescents who use ECs were two to three times as likely to have also vaped cannabis (60906–7).

Both drugs, THC and nicotine, increase blood pressure and heart rate independently, which suggests a more increased cardiovascular risk (60908).

Alcohol use in adolescents is associated with the initiation and higher rate of nicotine use (60911). And the opposite is also true: cigarette or EC use was associated with a 3 to 5× increase in the risk of alcohol use disorders (AUDs) (60913). Approximately 50 to 80% of adults with AUDs are also chronic cigarette users (60912, 60915), such that the two founders of Alcoholics Anonymous both died of tobacco-related diseases. Neither quitting smoking nor smoking cessation treatment resulted in worsening alcohol use outcomes (60913).

Smoking cessation rates are low in cocaine and other drug users, but contrary to popular views, smoking cessation does not appear to increase the rate of drug relapse (60916). EC users have been found to vape other drugs such as crack cocaine, fentanyl, ketamine, and opioids in addition to nicotine (60916–18).

Reference Number: 60901

REN, M., AND S. LOTFIPOUR. "Nicotine Gateway Effects on Adolescent Substance Use." *Western Journal of Emergency Medicine*, 20: 696–709, 2019 (2 tables, 174 references)

Abstract: A comprehensive review was conducted using human and animal studies on the effect of nicotine as a gateway drug. Teenagers are more vulnerable to nicotine dependence than adults following minimal tobacco exposure (fewer than seven cigarettes in one month). Adolescent nicotine exposure increased alcohol consumption. A survey of high school seniors showed that 88% of smokers were drinkers compared to 55% of non-smokers. Adolescents who began smoking before the age of 17 years are at a higher risk of alcohol abuse and dependence than those who started smoking after the age of 17 years. Nicotine has been found to interact with opioid receptors and naloxone, and naltrexone used for opioid overdoses have been found to reduce tobacco smoking and craving for cigarettes.

Nicotine interacts with other neurotransmitter systems and as a result increases the rewarding effects of other drugs by enhanced activation of reward circuitry. Developing brains are incredibly susceptible to long-lasting changes from perturbations during maturation, leading to behavioral changes that continue into adulthood. The prevalence of nicotine use among adolescents and the extensive interactions between nicotinic receptors and drugs of abuse highlight the critical need to better understand how nicotine modulates long-term consequences on brain and behavior related to addiction vulnerability.

Reference Number: 60902

VIVEROS, M-P., E.M. MARCO, AND S.E. FILE. "Nicotine and Cannabinoids: Parallels, Contrasts and Interactions." *Neuroscience and Biobehavioral Reviews*, 30: 1161–1181, 2006 (3 tables, 289 references)

Abstract: A review of the nicotinic and cannabinoid systems was conducted, including human and animal studies. There are close to 300 references. Nicotine causes the release of dopamine and is one part of the addiction. Abrupt cessation of tobacco smoking in humans is associated with a depressed mood, anxiety, irritability, and difficulty concentrating, and can result in bradycardia and increased appetite. CB1 receptors are mainly located in the cerebral cortex and sparse in the hypothalamus, midbrain, or spinal medulla. Thus, THC affects pain, motor activity, and short-term memory. CB2 receptors are located peripherally and mainly affect immunological and inflammation functions. Cigarette smoking is a potent gateway to various other drugs, including marijuana, but marijuana may also be a gateway drug to smoking (reverse gateway). Male smokers tend to be more sensitive to the rewarding effects of nicotine, while female smokers tend to be more sensitive to the anxiety relief.

From the evidence reviewed, it seems likely that the joint consumption of nicotine and cannabis will enhance the rewarding effects of low doses, although it is not known whether this would also occur with higher doses, or whether there would be a ceiling effect. The CB_1 receptor antagonist, rimonabant, may be effective at reducing cigarette craving during smoking cessation. However, it might exacerbate the somatic signs of withdrawal since THC has been found to reduce these signs.

Reference Number: 60903

PETERS, E.N., R.P. SCHWARTZ, S. WANG, K.E. O'GRADY, AND C. BLANCO. "Psychiatric, Psychosocial, and Physical Health Correlates of Co-Occurring

Cannabis Use Disorders and Nicotine Dependence." *Drug and Alcohol Dependence*, 134: 228–234, 2014 (3 tables, 83 references)

Abstract: A survey was conducted of 74 adults with current cannabis use disorder and nicotine dependence (CUD + ND), 100 adults with CUD only, and 3,424 with ND only. Concurrent use of cannabis and tobacco occurs in between 41 and 94% of adults who smoke tobacco and half of adults seeking treatment for CUD. Cannabis use is high among tobacco smokers and 25–52% of tobacco smokers use cannabis.

Relative to those with CUD only, respondents with CUD + ND were significantly more likely to meet criteria for bipolar disorder, Clusters A and B personality disorders, and narcissistic personality disorder, and reported engaging in a significantly higher number of antisocial behaviors. Relative to those with ND only, respondents with CUD + ND were significantly more likely to meet criteria for bipolar disorder, anxiety disorders, and paranoid, schizotypal, narcissistic, and borderline personality disorders; were significantly more likely to report driving under the influence of alcohol and being involved in partner violence; and reported engaging in a significantly higher number of antisocial behaviors. CUD + ND was not associated with physical health correlates.

Reference Number: 60904

AGRAWAL, A., M.T. LYNSKEY, P.A.F. MADDEN, M.L. PERGADIA, K.K. BUCHOLZ, AND A.C. HEATH. "Simultaneous Cannabis and Tobacco Use and Cannabis-Related Outcomes in Young Women." *Drug and Alcohol Dependence*, 101: 8–12, 2009 (2 tables, 30 references)

Abstract: Between 2005 and 2007, an analysis was conducted on 3,427 young adult women (aged 21–31 years) regarding co-occurring and simultaneous use of cannabis and tobacco. Women with a history of regular cigarette smoking were 9× more likely to report a lifetime history of cannabis use. Regular smokers were 2× as likely to have CUD and simultaneous users were 1.6× as likely to have CUD.

Simultaneous use was not heritable, and twin similarity was attributable to shared environmental factors (31%). While our study does not determine causality between simultaneous tobacco-cannabis use and cannabis involvement, results indicate that simultaneous use is potentially a marker for more severe psychosocial consequences associated with cannabis use.

Reference Number: 60905

RUBINSTEIN, M.L., M.A. RAIT, AND J.J. PROCHASKA. "Frequent Marijuana Use is Associated with Greater Nicotine Addiction in Adolescent Smokers." *Drug and Alcohol Dependence*, 141: 159–162, 2014 (1 table, 32 references)

Abstract: The degree of nicotine addiction was determined in 165 adolescent smokers (aged 13–17 years, mean age 16.1) using the Hooked on Nicotine Checklist, the Nicotine Dependence Syndrome Scale, and a modified Fagerstrom Tolerance Questionnaire. The participants had smoked on average 3 CPD for an average of 2 years. Most participants reported cannabis use in the past 30 days (80%), 43 (26.1%) reported weekly use, and 62 (37.6%) reported daily use. The frequency of cannabis use was associated with CPD but not alcohol use. Marijuana use was significantly associated with nicotine addiction.

Marijuana use was associated with greater reported nicotine addiction among adolescents. The findings suggest a role of marijuana in potentiating nicotine addiction and underscore the need for treatments that address use of both smoked substances.

Reference Number: 60906

KOWITT, S.D., A. OSMAN, C. MEERNIK, G.A. ZARKIN, L.M. RANNEY, J. MARTIN, C. HECK, AND A.O. GOLDSTEIN. "Vaping Cannabis Among Adolescents: Prevalence and Associations with Tobacco Use from a Cross-Sectional Study in the USA." *BMJ Open*, 9: (6 pp), 2019 (2 tables, 32 references)

Abstract: A study was conducted of 2,835 high school students in North Carolina as to vaping ECs with THC, hash oil, or THC wax. Approximately 10% of the high school students reported ever vaping cannabis. The prevalence of ever-vaping cannabis use was 4.7% in Grade 9 and 15.5% in grade 12 students. The adjusted OR of ever-reporting vaping cannabis was 3.76× for cigar use, 2.32× for waterpipe use, and 3.18× for EC use.

The health effects of vaping cannabis are not yet known. Some research suggests that vaping cannabis may be associated with fewer respiratory symptoms than smoking cannabis. However, this does not mean that adolescents' use of vaporised cannabis is safe—particularly if it leads to earlier initiation of tobacco or cannabis use, concomitant tobacco and cannabis use, increased frequency of use or misuse of tobacco or cannabis, or increased potency of cannabis.

Reference Number: 60907

AUDRAIN-MCGOVERN, J., M.D. STONE, J. BARRINGTON-TRIMIS, J.B. UNGER, AND A.M. LEVENTHAL. "Adolescent E-Cigarette, Hookah, and Conventional Cigarette Use and Subsequent Marijuana Use." *Pediatrics*, 142: (9 pp), 2018 (4 tables, 43 references)

Abstract: Non-cigarette tobacco products may confer a risk of marijuana use similar to combustible cigarettes. This study examined baseline e-cigarettes, hookahs, or combustible cigarette use with ever use marijuana (initiation), current marijuana use (30 days), and current dual use of these products in 2,668 grade 9 students in Los Angeles and again 24 months later, in the fall of 2015. Baseline multiple tobacco product use in grade 9 was associated with a 3.79× OR in combustible marijuana use, 2.90× in vaping cannabis, 3.59× in edible cannabis, and 4.21× in any marijuana product. Baseline e-cigarette use in grade 9 was associated with ORs of 3.15×, 2.08×, 2.99×, and 3.67×, respectively. Nicotine primes the brain's reward system by enhancing the level of pleasure experienced from subsequent exposure to other drugs such as cannabis. Hookah smoking can yield 1.7× as much nicotine. Hookahs are the only tobacco product that combine flavor and combustion.

Table: Odds Ratio of Cigarette, E-cigarettes, or Hookah Users Becoming Current Cannabis Users

Tobacco Product Use	Combustible Cannabis	Vaping Cannabis	Edible Cannabis	Any Cannabis product
Cigarettes	1.39×	1.13×	2.06×	1.43×
E-cigarettes	2.41×	1.30×	2.20×	2.87×
Hookahs	3.04×	3.47×	2.37×	3.07×

Source: Adapted from Audrain-McGovern et al, 2018

The associations between e-cigarette and hookah use and subsequent marijuana use are stronger than those for combustible cigarettes. E-cigarette and hookah use is more prevalent than combustible cigarette smoking among adolescents, yet e-cigarettes and hookah are far less regulated. Federal regulations do not currently restrict youth-targeted advertising and promotion or sales of youth-friendly flavors of e-cigarette and hookah products.

Reference Number: 60908

HINDOCHA, C., T.P. FREEMAN, J.X. XIA, N.D.C. SHABAN, AND H.V. CURRAN. "Acute Memory and Psychotomimetic Effects of Cannabis and Tobacco Both 'Joint' and Individually: A Placebo-Controlled Study." *Psychological Medicine*, 47: 2708–2719, 2017 (1 table, 4 figures, 66 references)

Abstract: Both cannabis receptors (CB1R) and tobacco receptors (nAChR) are densely populated in the hippocampus and amygdala, which suggests a neurobiological overlap. Twelve male and 12 female non-dependent cannabis and tobacco users (mean age 24.5 years) consumed, over the course of four different sessions, each of the following combinations: placebo tobacco and placebo cannabis; both active cannabis (16.1% THC, <1% CBD) and active tobacco (15.48 mg nicotine, 16 mg tar); active cannabis and placebo tobacco; and placebo cannabis and active tobacco. The subjects were tested on verbal and working memory and other measures. Tobacco had no effect on the subjective feeling of being stoned or dizzy. Tobacco offset some of the cannabis impairing effect on memory, but subjects were not aware of this. Cannabis and tobacco increased diastolic BP and independently increased HR, which suggests an increase in CV risk.

> In conclusion, this study found that cannabis impaired working and episodic memory. We found preliminary evidence that tobacco co-administration may offset the effects of cannabis on episodic memory. We characterised the acute subjective and cardiovascular effects of cannabis and tobacco administered together through a shared route of administration (i.e. joints) and found that these effects were similar to cannabis alone. There was no effect of tobacco on cannabis induced psychotomimetic effects.

Reference Number: 60909

MCKEE, S.A., S.S. O'MALLEY, J. SHI, T. MASE, AND S. KRISHNAN-SARIN. "Effects of Transdermal Nicotine Replacement on Alcohol Responses and Alcohol Self-Administration." *Psychopharmacology*, 196: 189–200, 2008 (2 tables, 1 figure, 81 references)

Abstract: Eight female and 11 male non-treatment seeking, non-dependent heavy drinkers (mean age 27 years) who were daily smokers were tested 6 hours after the application of a placebo or nicotine patch (21 mg nicotine). The subjects were given a single alcoholic drink as a primer, and were offered up to eight additional drinks with the caveat that they would receive a monetary reward for each drink not consumed after a 2-hour period. Plasma nicotine and alcohol concentrations were determined. The mean plasma nicotine concentration was 0.5 ng/mL after placebo and 18.6 ng/mL after the nicotine patch. The active patch increased heart rate by 7 bpm and systolic BP by 8 mm Hg. The mean maximum plasma alcohol concentration was 20 to 50 mg/dL and was not dependent on presence or absence of nicotine at most points of time. Nicotine attenuated subjective and physiological alcohol responses and delayed the initiation of drinking.

> Using an established alcohol self-administration paradigm, we found that transdermal nicotine replacement increased the latency to the first drink and consequently appeared to reduce the number of drinks consumed during a 2-h self-administration period, compared to 6 h of nicotine deprivation (i.e., placebo patch), in heavy drinking daily smokers. Although not significantly different, during the active patch condition, a larger proportion of subjects declined all drinks within each hour. When drinks were chosen, however, the number of sips taken per drink, time between sips, and time between drinks were similar for the active and placebo conditions. Thus, we conclude that active nicotine replacement compared to placebo replacement delayed the onset of consumption. Although subjects on nicotine patch compared to placebo patch consumed fewer drinks in the time allotted, it is not possible to determine whether this effect would be maintained, enhanced, or reduced if the time period in which drinks could be chosen was not constrained.

Reference Number: 60910

KING, A., P. MCNAMARA, M. CONRAD, AND D. CAO. "Alcohol-Induced Increases in Smoking Behavior for Nicotinized and Denicotinized Cigarettes in Men and Women." *Psychopharmacology*, 207: 107–117, 2009 (1 table, 4 figures, 64 references)

Abstract: Twenty-two male and 20 female heavy social drinking smokers (aged 21–35 years) consumed either 0.8 g/kg alcohol or placebo alcohol. The subjects then smoked either a nicotinized (0.6 mg nicotine/cigarette) or denicotinized cigarettes (<0.05 mg nicotine/cigarette) over a 3-hour period.

> In summary, the present study showed that alcohol and cigarette co-use may involve more complex factors in women compared with men. For men, consuming an intoxicating dose of alcohol, versus a placebo beverage, produced increases in smoking urge and subsequent measures of

smoking behavior. In women, while alcohol also increased smoking urge, subsequent smoking behaviors were similar regardless of whether the beverage contained an intoxicating dose of alcohol or a placebo. Interestingly, in both sexes, smoking behavior in the laboratory was not specific to nicotine-containing cigarettes.

Reference Number: 60911

DIERKER, L., A. SELVA, J. ROSE, D. HEDEKER, AND R. MERMELSTEIN. "Nicotine Dependence and Alcohol Problems from Adolescence to Young Adulthood." *Dual Diagnosis,* 1: 9 (11 pp), 2016 (2 tables, 25 references)

Abstract: Alcohol use in adolescents is associated with the initiation and escalation of smoking, and vice versa. Individuals with a lifetime diagnosis of alcohol dependence showed higher rates of nicotine dependence. Smoking and drinking behavior were observed in 864 participants annually for 6 years starting in the 9th and 10th grades in schools in Chicago. Treatment of alcohol problems may prevent or reduce the early emergence of nicotine depression syndrome.

Alcohol related problems are a consistent risk factor for nicotine dependence over and above measures of drinking and smoking and this association can be demonstrated from the earliest experiences with smoking in adolescents, through the establishment of more regular smoking patterns across the transition to young adulthood. These findings add to accumulating evidence suggesting that smoking and drinking may be related through a mechanism that cannot be wholly accounted for by exposure to either substance.

Reference Number: 60912

DURAZZO, T.C., V.A. CARDENAS, C. STUDHOLME, M.W. WEINER, AND D.J. MEYERHOFF. "Non-Treatment Seeking Heavy Drinkers: Effects of Chronic Cigarette Smoking on Brain Structure." *Drug and Alcohol Dependence,* 87: 76–82, 2007 (1 table, 4 figures, 60 references)

Abstract: Approximately 50 to 80% of individuals with AUD are also chronic cigarette smokers. High resolution MRIs of the brain were conducted on 17 smoking heavy drinkers (HD), 16 non-smoking HDs, and 20 non-smoking light drinkers. Smoking HDs had smaller volumes in the frontal, parietal, and temporal grey matter than non-smoking light drinkers. Smoking HDs also had smaller temporal and total grey matter volumes than non-smoking HDs. Non-smoking HDs and non-smoking light drinkers did not differ significantly in brain volumes.

> Chronic cigarette smoking is associated with significantly increased risk for development of atherosclerosis, and nicotine has been shown to induce alterations of vascular endothelial function. These processes may impact the functional integrity of the cerebrovasculature and contribute to the decreased regional cerebral blood flow and/or white matter disease reported in chronic smokers. The neocortical GM is particularly vulnerable to the effects of diffuse ischemia. The particulate and gas phases of cigarette smoke contain many toxic compounds (e.g., carbon monoxide, free radicals, nitrosamines, polynuclear aromatic compounds; that may directly or indirectly compromise brain tissue. For example, carbon monoxide (CO) levels are significantly higher in smokers which is associated with lower effective hemoglobin concentrations and, consequently, diminished oxygen carrying capacity of the blood, as well as decreased efficiency of the mitochondrial respiratory chain. Chronic smoking has also been equated to a type of repeated acute (mild) CO poisoning, and is linked to nocturnal hypoxia as well as to respiratory risks such as chronic obstructive pulmonary disease and other conditions that may diminish or compromise lung function. Decreased lung function was associated with poorer neurocognition and increased subcortical atrophy in community-dwelling adults aged 60–64 years.

Reference Number: 60913

MCKELVEY, K., J. THRUL, AND D. RAMO. "Impact of Quitting Smoking and Smoking Cessation Treatment on Substance Use Outcomes: An Updated and Narrative Review." *Addictive Behaviors*, 65: 161–170, 2017 (1 table, 2 figures, 63 references)

Abstract: Smoking rates are 2–4× higher among adolescents and adults with substance use disorders (SUDs) than in the general population, yet smoking cessation treatment is usually not included in most SUD treatment units. Historically, smoking was allowed and even encouraged in addiction treatment programs, as there were concerns that if the patient with SUD ceased smoking, they would relapse or go into depression. Meanwhile, the tobacco industry was providing tax-free cigarettes to many treatment facilities. Alcoholics Anonymous meetings often allowed or encouraged smoking as a form of fellowshipping and often maintained smoking to focus on alcohol addiction. The harm of tobacco use, however, far outweighs any perceived benefit of its use in SUD treatment, as shown in a meta-analysis of 24 studies published between 2006 and 2016. Neither quitting smoking nor smoking cessation treatment resulted in worsening substance use outcomes, including for alcohol.

> Smoking cessation does not appear to have a negative effect, and often has a positive effect on substance use outcomes. Smoking cessation advice should be offered, without hesitation, to smokers who report substance use and those in treatment for substance use disorder.

Reference Number: 60914

ROBERTS, W., K.E. MOORE, M.R. PELTIER, T.L. VERPLAETSE, L. OBERLEITNER, R. HACKER, AND S.A. MCKEE. "Electronic Cigarette Use and Risk of Harmful Alcohol Consumption in the US Population." *Alcoholism, Clinical and Experimental Research*, 42: 2385–2393, 2018 (3 tables, 46 references)

Abstract: A survey of 36,309 adults (mean age 45.6 years) from the National Epidemiologic Survey on Alcohol and Related Conditions—Wave III (NESARC-III) was conducted regarding past year EC and alcohol use. Tobacco cigarette smokers are more likely to have Alcohol Use Disorder and engage in hazardous drinking. Smokers consume more alcohol when they are drinking and smoking at the same time. Of those surveyed who endorsed past-year EC use, 1,347 reported daily use and 401 reported non-daily use of ECs.

Table: Tobacco Cigarette and/or E-cigarette Use and Adjusted Odds Ratio (Compared to Non-Tobacco Product User) of Using Alcohol, Being a Hazardous Drinker, or Having Alcohol Use Disorder

Tobacco and/or E-cigarette Use	Adjusted OR of Using Alcohol	Adjusted OR of Being a Hazardous Drinker	Adjusted OR of AUD
No tobacco cigarette use, non-daily EC use	2.37×	2.16×	2.54×
No tobacco cigarette use, daily EC use	1.21×	1.46×	3.56×
Non-daily tobacco cigarette use, no EC use	3.27×	3.23×	3.18×

Tobacco and/or E-cigarette Use	Adjusted OR of Using Alcohol	Adjusted OR of Being a Hazardous Drinker	Adjusted OR of AUD
Non-daily tobacco cigarette use, non-daily EC use	7.44×	6.99×	5.20×
Non-daily tobacco cigarette use, daily EC use	10.36×	4.67×	2.94×
Daily tobacco cigarette use only	1.75×	2.41×	2.62×

Source: Adapted from Roberts et al, 2018

The results of current study extend our understanding of alcohol use in e-cigarette users in a representative sample of adults in the United States. People who use e-cigarettes, particularly nondaily users, are at increased risk for engaging in harmful alcohol use. Secondary analyses found that among e-cigarette users, those who also endorse nondaily use of tobacco cigarettes show the highest rates of harmful alcohol use outcomes. These findings identify a group of high risk users (i.e., nondaily dual users) that warrant additional research. Although recent data suggest that e-cigarettes may have fewer direct health consequences compared to tobacco cigarettes, these findings highlight the importance of thoroughly examining health correlates of e-cigarette use before drawing conclusions about relative safety.

Reference Number: 60915

KODL, M., S.S. FU, AND A.M. JOSEPH. "Tobacco Cessation Treatment for Alcohol-Dependent Smokers: When Is the Best Time?" *Alcohol Research and Health*, 29: 203–207, 2006 (36 references)

Abstract: Cigarette smoking is very prevalent among people with alcohol use disorders who also tend to have a higher nicotine addiction and lower likelihood of smoking cessation. Persons with AUD prefer sequential treatment for smoking cessation and not simultaneous treatment.

Many patients in substance use treatment or recovery do not receive concurrent or delayed nicotine dependence treatment, which is a pressing issue. This should be addressed by incorporating nicotine dependence treatment into aftercare for patients who are completing substance use treatment. With this program design, all smokers leaving treatment would be urged to participate in combined behavioral and pharmacologic treatment to quit smoking. Sequential treatment delivery has practical limitations, however. For example, if smoking cessation is not offered concurrently with alcohol treatment, it may be difficult to interest individuals in treatment even 4 to 6 weeks later and maintain their participation, as several studies have found. Based on the literature to date, it is difficult to know conclusively that concurrent treatment should be avoided, but this is a possibility. Concurrent tobacco dependence treatment should not be denied, however, if requested by patients.

Reference Number: 60916

WEINBURGER, A.H., AND M. SOFUOGLU. "The Impact of Cigarette Smoking on Stimulant Addiction." *American Journal of Drug and Alcohol Abuse*, 35: 12–17, 2009 (88 references)

Abstract: Cocaine users have much greater rates of cigarette smoking than the general population (70–80% vs 22%), and methamphetamine users have even higher rates of smoking (87–92%). The common pathway to nicotine and stimulant neurons is dopamine release into the nucleus accumbens. Cigarettes usually precede and may be a gateway to stimulant use. Cocaine users who smoke are younger, less educated, and have more legal problems than non-smokers. Smoking cessation rates are very low in smokers with substance use disorders (about 12%), but they can obtain quit rates similar to the general population when enrolled in smoking cessation programs. Smoking cessation does not appear to increase the rate of relapse to other drugs of abuse.

Early exposure to nicotine influences the development of stimulant addiction and smokers with co-morbid drug use have more severe stimulant use and may be more treatment resistant. There are still many unanswered questions regarding the impact of smoking on the course of addiction and treatment of cocaine dependence. Very little research has been conducted on amphetamine use and smoking although the near universality of smoking suggests that this area should receive attention in future research.

Reference Number: 60917

BREITBARTH, A.K., J. MORGAN, AND A.L. JONES. "E-cigarettes—An Unintended Illicit Drug Delivery System." *Drug and Alcohol Dependence*, 192: 98–111, 2018 (1 table, 2 figures, 196 references)

Abstract: A serious concern of EC research is that 34% of the authors of the studies showing the safety of ECs have a conflict of interest and most received funding from EC manufacturers. ECs produce particulate matter (PM_{10}, $PM_{2.5}$, and $PM_{1.0}$). A systematic review of ECs and other vaping devices to vape illicit drugs was conducted on 38 articles that passed the screening procedure for inclusion. Vaping devices can be divided into two broad categories: portable powered by batteries and table-top units.

The most common drug vaped in this study was cannabis (18.0% of vape device users). However, there was evidence of a wide range of drugs being vaped by electronic vaping device users including: MDMA (11.7%); cocaine powder (10.9%); crack cocaine (8.4%); synthetic cathinones, mephedrone (8.5%) and α-PVP (7.1%); synthetic cannabinoids (7.8%); opioids, heroin (7.1%) and fentanyl (7.3%); and other drugs including tryptamines and ketamine.

Reference Number: 60918

THURTLE, N., R. ABOUCHEDID, J.R.H. ARCHER, J. HO, T. YAMAMOTO, P.I. DARGAN, AND D.M. WOOD. "Prevalence of Use of Electronic Nicotine Delivery Systems (ENDS) to Vape Recreational Drugs by Club Patrons in South London." *Journal of Medical Toxicology*, 13: 61–65, 2017 (2 tables, 31 references)

Abstract: One hundred and one patrons (median age 28 years) at a South London nightclub were surveyed as to the use of ECs to vape drugs other than nicotine. Eighty percent of the patrons currently smoked cigarettes and 20% used ECs. Two patrons used ECs to vape liquid cannabis and four other unspecified drugs (5.9%).

5.9% of individuals in this sample reported using ENDS to vape substances other than nicotine. Further work is required in larger populations to determine how common this is, evaluate which agents are being vaped and to inform appropriate public education.

CHAPTER 7

Addiction, Withdrawal, Smoking Cessation, and Public Health Measures

> "For much of the 20th century, smoking was regarded as a socially learned habit and as a personal choice.... It is now recognised that cigarette smoking is primarily a manifestation of nicotine addiction."
>
> —Jarvis, "Why People Smoke" (2004)

Smoking is not a bad habit or personal choice, as the tobacco industry would have smokers believe, but due mainly to the highly addictive neurotoxin and pesticide that is nicotine. Nicotine is far more addictive than either alcohol or cannabis.

7.01 ADDICTION/TOBACCO USE DISORDER

> "To cease smoking is the easiest thing I ever did. I ought to know because I've done it a thousand times."
>
> —attributed to Mark Twain

Among the symptoms listed by the American Psychiatric Association Diagnostic and Statistical Manual of Mental Disorders (DSM-5) are:

- unsuccessful efforts to quit or reduce intake of tobacco;
- inordinate amount of time acquiring or using tobacco products;
- cravings for tobacco;
- failure to attend to responsibilities and obligations due to tobacco use;
- continued use despite adverse social or interpersonal consequences;

- forfeiture of social, occupational, or recreational activities in favor of tobacco use;
- tobacco use in hazardous situations;
- continued use despite awareness of physical or psychological problems directly attributable to tobacco use; and
- need for increasingly larger doses of nicotine in order to obtain the desired effect.

Nicotine addiction is due mainly to it binding with receptors in the brain, which releases numerous neurotransmitters, especially dopamine, resulting in the feeling of pleasure and eventually causing addiction as smokers and other drug users continue to seek that pleasure (70101, 70103–4). As if nicotine is not sufficient to cause and maintain addiction, the tobacco industry has added other flavoring chemicals such as pyrazine which reduce the irritation of tobacco smoke and eases the nicotine uptake into the lungs. Pyrazine also reinforces the learned behavior of smoking and dopamine release (70105).

A good indicator of nicotine dependence is the time to the first use of the tobacco product upon awakening. The shorter the time to the first vape or smoke of the day, the more likely the tobacco product user is addicted to nicotine. The median time of first use was shortest (20 minutes) for polytobacco users of cigarettes, cigars, and e-cigarettes, and users of cigarettes and cigars. The longest median time to first tobacco use was 55 minutes for cigars users (70106).

Stressful situations tended to increase craving in dependent cigarette smokers but did not increase the number of cigarettes smoked (70107).

Slow nicotine metabolizers smoked fewer cigarettes per day, as the nicotine remained in the blood for a longer period of time than for fast metabolizers. Slow metabolizers were not as addicted to nicotine as fast metabolizers and generally were able to be more successful at quitting (70108–9).

Although coffee users were more likely to be smokers, coffee consumption is unlikely to have a major impact on smoking rates (70110).

The use of tobacco products at the workplace varies according to occupation. At some construction sites over 90% of the workers smoke (70111). Other studies regarding nicotine addiction can be found in section 1.02. Written tests to determine the degree of nicotine addiction in smokers can be found in appendices 3 and 4.

Reference Number: 70101

SCHERMAN, A., J.E. TOLOSA, AND C. MCEVOY. "Smoking Cessation in Pregnancy: A Continuing Challenge in the United States." *Therapeutic Advances in Drug Safety*, 9: 457–474, 2018 (3 tables, 1 figure, 154 references)

Abstract: Smoking during pregnancy remains a serious public health issue in the United States, with approximately 360,000 to 500,000 smoke exposed infants born yearly. In 2015, 15.1% of US adults smoked (13% of women, 17% of men), accounting for 480,000 premature deaths yearly. Approximately 9% of women report smoking during pregnancy. When tobacco is smoked, nicotine reaches the brain in 20 seconds and peak concentrations occur in 5 minutes. Nicotine binds to the nicotinic acetylcholine receptors, triggering the release of dopamine, serotonin, and other neurotransmitters. The rapid release of dopamine causes pleasure and addiction. Withdrawal symptoms, such as agitation and depression, negatively reinforce addiction. Withdrawal symptoms occur in about one-half of smokers and start within a few hours after smoking cessation. They peak at the first week and resolve over a month. Nicotine is metabolized by cytochrome CYP2A6. The slower elimination of nicotine may allow for longer periods without withdrawal symptoms and allow more successful quitting. By 1965, over 33% of the women and 62% of non-Hispanic White men smoked. The tobacco industry continues to portray smoking as sexy or edgy behavior, one that the creative or adventurous use. Smoking is a leading cause of preterm births (<37 weeks) and is linked to 5 to 8% of premature births. About 33% of sudden unexpected death in infants is linked to maternal smoking. There are >7,000 chemicals in tobacco smoke. Nicotine and CO readily cross the placenta, and the fetal circulation can exceed the maternal by 15% for CO and by 88% for nicotine. Nicotine may have an impact on brain development. Pregnant smokers receiving counselling have 80% greater cessation rate than those relying on self-help books.

Tobacco addiction is a chronic disease involving a complex interplay between the central nervous system, genetics, the environment, and physical and psychological states. Addiction is positively enforced by activation of the neuronal reward system via rapid release of dopamine in response to nicotine. Without nicotine, withdrawal symptoms arise. This cycle likely plays a critical role in continued smoking despite financial and health costs. Women are more likely than men to report smoking to cope with negative psychological states or emotions including anxiety and depression. Additionally, women who are socioeconomically disadvantaged, who report food insecurity, or report active post-traumatic stress disorder or history of trauma have increased risk of smoking during pregnancy. Prenatal smoking is also more prevalent among younger women aged 20–24 years old.

Reference Number: 70102

COSCI, F., F. PISTELLI, N. LAZZARINI, AND L. CARROZZI. "Nicotine Dependence and Psychological Distress: Outcomes and Clinical Implications in Smoking Cessation." *Psychology Research and Behavior Management*, 4: 119–128, 2011 (1 table, 105 references)

Abstract: The diagnosis of nicotine dependence was included for the first time in the 3rd edition of DSM in 1980. The US Surgeon General in 1988 defined tobacco smoking as an addiction. Nicotine dependence is a chronic and relapsing disease. Although approximately 80% of smokers would like to stop and one-third make at least three serious attempts, less than 50% of smokers succeed in stopping before the age of 60 years. The Clinical Practice Guidelines recommend addressing patients' smoking at each visit using the five A's: Ask about tobacco use; advise smokers to quit; assess smoker's willingness to make a quit attempt; assist with treatment; and arrange follow-up.

Persistent cigarette consumption seems to be strongly influenced by two complex factors: nicotine dependence and psychological distress. While the level of nicotine dependence is commonly measured in smokers attending smoking cessation programs, the assessment of psychological distress is often neglected. However, such an evaluation should be performed, to get a complete picture of the smoker and offer a tailored intervention. Some patients, for instance, might benefit from psychological support aimed at managing psychological distress, a high level of negative affectivity, or a low psychological distress tolerance experienced while quitting.

Reference Number: 70103

BAKER, T.B., N. BRESLAU, L. COVEY, AND S. SHIFFMAN. "DSM Criteria for Tobacco Use Disorder and Tobacco Withdrawal: A Critique and

Proposed Revisions for DSM-5." *Addiction*, 107: 263–275, 2012 (3 tables, 121 references)

Abstract: DSM-4 criteria for tobacco disorder and withdrawal were found to be ambiguous and little used in tobacco research. This paper discusses suggested improvements for inclusion in DSM-5, which would be published in 2013.

> In conclusion, substantive and psychometric considerations strongly support four new criteria for further testing and potential inclusion in the DSM-5. These criteria concern smoking heaviness, latency to smoke upon awakening, craving to smoke, and withdrawal severity. Other criteria were identified for additional research and possible inclusion. Little evidence supported the inclusion of most existing DSM-IV items, but rewording and reformulation might improve their validities.

Reference Number: 70104

ZIEDONIS, D., S. DAS, AND C. LARKIN. "Tobacco Use Disorder and Treatment: New Challenges and Opportunities." *Dialogues in Clinical Neuroscience*, 19: 271–280, 2017 (2 tables, 69 references)

Abstract: Tobacco use is the cause of over 5 million deaths per year globally—more than twice the deaths due to alcohol and illicit drugs combined. The DSM-5 classifies tobacco addiction or nicotine dependence as tobacco use disorder. A typical cigarette has about 6–11 mg of nicotine, resulting in 1–2 mg of nicotine absorbed by the smoker. Cigars can contain 100–444 mg of nicotine, resulting in 9–68 mg absorbed. Nicotine cartridges for EC contain 6–24 mg of nicotine and have 2.5–7.7 mg absorbed for the entire cartridge. Nicotine causes release of dopamine and up-regulation of receptors, resulting in nicotine dependence. Nicotine's short half-life of 1 to 2 hours leads to withdrawal symptoms of irritation, frustration, anxiety, and difficulty concentrating. This study discusses various therapies for the treatment of Tobacco Use Disorder or nicotine dependence. Clinical trials are being conducted on nicotine vaccines, which produce antibodies that bind to nicotine and prevent it from acting on receptors.

Table: The Main Effects of Neurotransmitters Released by Nicotine Binding

Neurotransmitter	Effect
Dopamine	Pleasure, appetite suppression
Norepinephrine	Arousal, appetite suppression
Acetylcholine	Arousal, cognitive enhancement
Glutamate	Learning, memory enhancement
Serotonin	Mood modulation, appetite suppression
Beta-endorphin	Reduction in anxiety and tension
Gamma aminobutyric acid	Reduction in anxiety and tension

Source: Adapted from Ziedonis et al, 2017

Tobacco use continues to increase morbidity and mortality with high rates of tobacco-use disorders among individuals with psychiatric disorders resulting in severe health disparities for that population. New nicotine and tobacco products will continue to be developed, as will new treatments, with increasing person-, home-, and community-facing technologies.

Reference Number: 70105

ALPERT, H.R., I.T. AGAKU, AND G.N. CONNOLLY. "A Study of Pyrazines in Cigarettes and How Additives Might be Used to Enhance Nicotine Addiction." *Tobacco Control*, 25: 444–450, 2016 (3 figures, 109 references)

Abstract: A review of more than 7 million internal tobacco industry documents was conducted to determine how the addictiveness of cigarettes was increased by the addition of pyrazines. Pyrazines are chemosensory agents with the general chemical formula of $C_4H_4N_2$, which are formed at temperatures ≥100°C during tobacco leaf curing and smoking. Pyrazines are among the most important chemicals in tobacco, because they act on chemoreceptors and produce the characteristic aroma and flavor of tobacco and tobacco smoke. Pyrazines reduce the irritation of tobacco smoke in the upper airways and ease the nicotine uptake and absorption of nicotine into the lungs. Pyrazines also reinforce the learned behavior of smoking and enhance the nicotine's release of dopamine.

Nicotine is known as the drug that is responsible for the addicted behaviour of tobacco users, but it has poor reinforcing effects when administered alone. Tobacco product design features enhance abuse liability by

(A) optimising the dynamic delivery of nicotine to central nervous system receptors, and affecting smokers' withdrawal symptoms, mood and behaviour; and (B) effecting conditioned learning, through sensory cues, including aroma, touch and visual stimulation, to create perceptions of pending nicotine reward. This study examines the use of additives called "pyrazines". which may enhance abuse potential, their introduction in "lights" and subsequently in the highly market successful Marlboro Lights (Gold) cigarettes and eventually many major brands.

Reference Number: 70106

ROSTRON, B.L., M.J. SCHROEDER, AND B.K. AMBROSE. "Dependence Symptoms and Cessation Intentions Among US Adult Daily Cigarette, Cigar, and E-Cigarette Users, 2012–2013." *BMC Public Health*, 16: 814 (10 pp), 2016 (3 tables, 2 figures, 36 references)

Abstract: This study determined the dependence symptoms and cessation intentions in users of cigarettes, cigars, and/or ECs, using a US National Adult Tobacco Survey (NATS) conducted of more than 60,000 participants between 2012 and 2013. Of the single tobacco product users, there were 3,963 who used cigarettes daily, 131 who used cigars daily, and 124 who used ECs daily. Of the daily multiple product users, 495 used cigars and cigarettes, 774 used cigarettes and ECs, and 130 used cigars and ECs. The percentage of daily tobacco users who used tobacco within 30 minutes of waking up was approximately 62% (cigarettes), 32% (cigars), 35% (ECs), and 62% for dual users.

Table: Median Time to First Tobacco Product Use After Waking Among Daily Tobacco Users

Tobacco Use	Median Time of First Tobacco Use (approximate)
Cigarettes	25 minutes
Cigars	55 minutes
E-cigarettes	50 minutes
Cigarettes and cigars	20 minutes
Cigarettes and e-cigarettes	26 minutes
Cigarettes and cigars and e-cigarettes	20 minutes

Source: Adapted from Rostron et al, 2016

This study presents important information about dependence symptoms and cessation intentions among US adult daily cigarette, cigar, and e-cigarette users that increases our understanding of dependence among tobacco users. We find that dual users of cigarettes and cigars are more likely to report greater tobacco dependence than exclusive cigarette smokers as measured by cpd and time to first tobacco use and that dual cigarette and cigar users and dual cigarette and e-cigarette users are more likely to report withdrawal and craving symptoms than exclusive cigarette smokers. We also find that many users in all product use categories, including exclusive users of cigars and e-cigarettes, exhibit symptoms of tobacco dependence. These findings have the potential to inform future research and tobacco efforts generally, given that tobacco use and dependence remain significant global public health issues, while being of particular relevance to the US specifically, given that the FDA has extended its authority to regulate all tobacco products including e-cigarettes and cigars.

Reference Number: 70107

CHILDS, E., AND H. DE WIT. "Effects of Acute Psychosocial Stress on Cigarette Craving and Smoking." *Nicotine and Tobacco Research*, 12: 449–453, 2010 (1 figure, 51 references)

Abstract: This study examined cigarette craving and number of cigarettes smoked in 4 female and 13 male daily smokers (mean age 24.2 years), after participation in either the Trier Social Stress Test or a non-stressful control task. Stress significantly increased cigarette craving but did not increase smoking.

In conclusion, the findings from the present study support earlier findings that acute stress increases smoking desire. Acute stress did not significantly increase smoking; however, our findings suggest increased pleasure from smoking and deeper inhalations in more dependent smokers after stress. Our results extend previous findings and suggest relationships between the severity of nicotine dependence and influences of stress on smoking. Future studies should further investigate possible relationships between physiological and subjective responses to stress and stress-induced alterations in smoking.

Reference Number: 70108

DUBROFF, J.G., R.K. DOOT, M. FALCONE, R.A. SCHNOLL, R. RAY, R.F. TYNDALE, A.L. BRODY, C. HOU, A. SCHMITZ, AND C. LERMAN. "Decreased Nicotinic Receptor Availability in Smokers with Slow Rates of Nicotine Metabolism." *Journal of Nuclear Medicine*, 56: 1724 –1729, 2015 (1 table, 2 figures, 48 references)

Abstract: Twelve slow nicotine metabolizing smokers (NMR <0.26) and 12 normal nicotine metabolizing smokers (NMR >0.26) underwent brain imaging after an 18-hour nicotine abstinence. Chronic nicotine exposure increases the nicotinic acetylcholinergic receptor density (i.e., up-regulation) and quitting smoking leads to receptor normalization.

Table: Some Smoking-Related Variables in Normal and Slow Nicotine Metabolizers

	Normal Metabolizers	Slow Metabolizers
CPD (range)	10–40	11–20
Mean Fagerstrom Test for Nicotine Dependence score	5.9	5.4
Mean initial breath CO (PPM)	21.2	25.0
Mean plasm cotinine concentration (ng/mL)	219.7	323.9
Mean 3-HC concentration (ng/mL)	104.6	51.8

Source: Adapted from Dubroff et al, 2015

These data suggest that slow metabolizers of nicotine exhibit reduced nAChR availability in thalamus after 18 hours of abstinence from smoking, compared to normal nicotine metabolizers. Individual variation in hepatic metabolism influences nicotine's half-life in plasma from approximately 2 hours to 4 hours. In addition, nicotine levels in the brain can persist beyond nicotine's plasma half-life. Therefore, differences in thalamic nAChR availability observed between slow and normal metabolizers may reflect differences in nicotine binding to nAChRs due to differences in elimination kinetics.

Reference Number: 70109

CHENOWETH, M.J., R.A. SCHNOLL, M. NOVALEN, L.W. HAWK JR., T.P. GEORGE, P.M. CINCIRIPINI, C. LERMAN, AND R.F. TYNDALE. "The Nicotine Metabolite

Ratio is Associated with Early Smoking Abstinence Even After Controlling for Factors that Influence the Nicotine Metabolite Ratio." *Nicotine and Tobacco Research*, 18: 491–495, 2016 (2 tables, 24 references)

Abstract: The smoking rate for adults in the United States and Canada has plateaued to about 20%. The FDA has approved three medications for smoking cessation; nicotine replacement therapy, bupropion, and varenicline. Nicotine is inactivated primarily by the CYP2A6 enzyme into cotinine, which is further metabolized to 3'-hydroxycotinine (3-HC). The 3-HC/COT ratio is known as the nicotine metabolite ratio (NMR). A lower NMR indicates a slower nicotine clearance and is associated with lower cigarette consumption and nicotine dependence scores, and a higher quit ratio in regard to both placebo and the nicotine patch. Varenicline was more effective in smokers with a higher NMR. This study determined the NMR in 1,246 smokers at intake to a smoking cessation program using LC/MS/MS. A slower rate of nicotine metabolism was set at NMR <0.31. One-week abstinence was determined by an exhaled CO concentration of 8 ppm or less. Lower NMR was associated with a 1.43× greater likelihood of 1-week abstinence. Smokers who were abstinent at 1 week were 9.6× as likely to be abstinent long-term.

Table: Odds Ratio of 1-Week Smoking Abstinence Associated with Various Factors

Predictor of 1-Week Abstinence (Breath CO <8 ppm)	OR of 1-Week Abstinence
Nicotine patch	1.80×
Varenicline	2.58×
Female sex	1.28×
White ethnicity	1.56×
NMR slow metabolism (NMR <0.31)	1.43×

Source: Adapted from Chenoweth et al, 2016

Overall, we showed that NMR, which is associated with treatment outcome at end-of-treatment and 6-month follow-up, was also associated with 1-week abstinence, and factors which account for variability in NMR did not remove the association between NMR and 1-week abstinence. This early period of abstinence represents a period of heightened vulnerability during the process of smoking cessation, and NMR may be useful in informing early efficacy screening approaches for compounds undergoing development. Screening approaches may include NMR-based randomization to treatment, as we have done in this clinical trial, or simply including NMR in analytic models as a covariate known to alter smoking cessation and treatment response.

Reference Number: 70110

WARE, J.J., J-A. TANNER, A.E. TAYLOR, Z. BIN, P. HAYCOCK, J. BOWDEN, P.J. ROGERS, G.D. SMITH, AND R.F. TYNDALE. "Does Coffee Consumption Impact on Heaviness of Smoking?" *Addiction*, 112: 1842–1853, 2017 (3 tables, 4 figures, 52 references)

Abstract: Coffee consumption is associated with smoking. Coffee drinkers are more likely to be smokers than non-coffee drinkers and to smoke more heavily. A Mendelian randomization analysis was conducted of coffee/cigarette smoking of 38,981 persons in various European countries. An in vitro study was also conducted on the effect of the major components of coffee, caffeic acid, quercetin, and p-coumaric acid on the rate of nicotine metabolism.

Taken together, our results suggest that coffee consumption is unlikely to have a major causal impact on cigarette smoking. If it does influence smoking, this is not likely to operate via effects of caffeic acid, quercetin or p-coumaric acid on nicotine metabolism. The observational association between coffee consumption and cigarette smoking may be due to smoking impacting upon coffee consumption, or confounding. It is likely that at least some of the positive observational association between coffee consumption and cigarette smoking is due to the impact of smoking upon coffee intake. There is evidence from Mendelian randomization analysis, using a genetic variant that determines heaviness of smoking, that heavier smoking increases coffee consumption causally.

Reference Number: 70111

PARASHAR, M., R. AGARWALLA, S. DWIVEDI, B. PATVAGEKAR, AND R. PATHAK. "Prevalence and Correlates of Nicotine Dependence Among Construction Site Workers: A Cross-Sectional Study in Delhi." *Lung India*, 33: 496–501, 2016 (3 tables, 28 references)

Abstract: A survey was conducted of 172 male construction workers (mean age 32 years) in Delhi, India. Ninety-one percent used a tobacco product. Of these, 49% used smokeless tobacco, 29% used cigarettes/bidis, and 22% used both.

> The prevalence of tobacco use and dependence on nicotine was very high among construction workers compared to that in the general population. Factors found to be significantly associated include age at initiation, lower income group, smokeless tobacco use, lower education. Sensitization of workers and especially youth through schools and community-based awareness program about the problem, social support for deaddiction is needed. Income generation is strongly recommended as part of cessation strategy for this particular group. Recognition of construction sites as workplaces and proper implementation of the law for both forms of tobacco is needed.

7.02 WITHDRAWAL/SMOKING CESSATION MEASURES

> "Nicotine's short half-life of 1 to 2 hours leads to withdrawal symptoms of irritability, frustration, anxiety, difficulty concentrating, restlessness, depressed mood, insomnia, increased appetite, weight gain, and cravings."
>
> —Ziedonis, Das, and Larkin, "Tobacco Use Disorder and Treatment: New Challenges and Opportunities" (2017)

The American Psychiatric Association Diagnostic and Statistical Manual of Mental Disorder (DSM-5) lists seven common withdrawal symptoms that appear within 24 hours after tobacco use has been reduced or abruptly ceased.

1. Irritability, frustration, or anger.
2. Anxiety.
3. Difficulty concentrating.
4. Increased appetite.
5. Restlessness.
6. Depressed mood.
7. Insomnia.[1]

One of the problems with cigarette addiction is that it is not often considered to be a serious chronic recurring disease that needs to be treated

1 American Psychiatric Association, *Diagnostic and Statistical Manual of Mental Disorders*, 5th ed (Arlington, VA: American Psychiatric Association, 2013) at 575.

over and over again, and not just by prescribing the use of the nicotine patch once, and if that doesn't work, then giving up (70201). Patients with other chronic recurring diseases such as diabetes and asthma who visit a doctor's office are approximately 22× more likely to obtain condition-specific medication than tobacco users (70202). Instead, a wide array of smoking cessation aids should be aggressively employed, including nicotine patches, gums, lozenges, nasal sprays, inhalers, bupropion, and varenicline (70201, 70203, 70208, 70223).

About one-quarter of hospitalized smoking patients relapse after being discharged, most within one hour of release, which shows how smoking cessation aids need to be in place before leaving hospital and not several days or weeks later (70204–5). Physicians who had never tried cigarettes were 2.3× more likely to refer patients to smoking cessation clinics than doctors who had tried at least one cigarette (70206).

Nicotine replacement therapy reliably reduces the withdrawal symptoms of poor concentration, craving for nicotine, irritability, and depression, which can persist for 4 weeks or more (70207–13, 70219).

Normal nicotine metabolizers had greater difficulty in stopping smoking and were more successful with behavioral therapy (70214). Companies that offered employees a smoking cessation program with a redeemable deposit reward had a higher abstinence rate (70215). Smoking cessation programs need to be tailored to women, who may be more susceptible to the effects of cigarettes (70221). Older smokers (>60 years of age) were more successful at smoking cessation with nicotine replacement therapy (NRT) (70222).

It is essential to immediately implement smoking cessation therapy in smoking patients with acute coronary syndrome, COPD, and asthma (70216–17, 70220, 70223, 70226).

ECs are not an effective smoking cessation aid and are not approved as such by the FDA (70218).

The plasma nicotine concentrations obtained by NRT are approximately one-half the level obtained by smoking cigarettes and NRT is not recommended to be used for a long period of time, and so NRT does not appear to cause an increase in cancer or cardiovascular disease risk and is considered safe (70224, 70227-30).

There are numerous similarities between opioid and nicotine dependence, and opioid (and methadone) use can lead to an increase in nicotine self-administration (70231).

Reference Number: 70201

RIGOTTI, N.A. "Strategies to Help a Smoker Who is Struggling to Quit." *Journal of the American Medical Association*, 308: 1573–1580, 2012 (3 tables, 63 references)

Abstract: The author presents a case report of a 50-year-old man with hypertension and a history of depression (treated with fluoxetine) who smokes 15 to 20 cigarettes daily. He has attempted to quit smoking by using a nicotine patch for 5 days and took bupropion for a month, but feels nothing works and is asking about ECs as a smoking cessation tool.

Table: Efficacy of Methods Used to Treat Tobacco Dependency

Intervention	Odds Ratio of Quitting vs No Intervention
Counseling in person (individual	1.39×
Counseling in person (group)	1.98×
Telephone quitline	1.37×
Brief physician advice to quit	1.66×
Bupropion (sustained release)	1.69×
Varenicline	2.27×
Nicotine patch	1.66×
Nicotine gum	1.43×
Nicotine lozenge	2.00×
Nicotine inhaler	1.90×
Nicotine nasal spray	2.02×

Source: Adapted from Rigotti, 2012

The case illustrates a common scenario: a smoker who wants to quit, has little self-confidence that he can do so because past quit attempts failed, and believes that he has "tried everything." In fact, he has not tried everything. He has never used behavioral support, combination pharmacotherapy, or varenicline and his trial of NRT was inadequate. Behavioral support is essential in view of his low self-confidence. It could be encouraged and arranged by referral from your office to a free telephone quitline. I would discourage use of the electronic cigarette because of the absence of scientific data on its safety or efficacy for cessation and the existence of FDA-approved effective options that he has not tried. In place of an unapproved nicotine delivery device, he could combine the nicotine patch with his choice of lozenge, gum, or inhaler. Alternately, he could try varenicline, which is appropriate if his psychiatric status is stable, careful follow-up is insured, and his concerns about using varenicline can be allayed. Combining bupropion and NRT is another option since he had partial success with bupropion in the past and bupropion can be used in individuals taking SSRIs. Whatever his next step, it is important to encourage him to keep trying, assure him that he can succeed, monitor his progress, and continue to offer help at each visit.

Reference Number: 70202

BERNSTEIN, S.L., S. YU, L.A. POST, J. DZIURA, AND N.A. RIGOTTI. "Undertreatment of Tobacco Use Relative to Other Chronic Conditions." *American Journal of Public Health*, 103: e59–e65, 2013 (4 tables, 31 references)

Abstract: Tobacco dependence is a chronic re-occurring condition that increases the risk of diabetes, hypertension, and cardiovascular disease. This study compares the likelihood that a tobacco user will receive medical treatment after a visit to the doctor's office to the likelihood that adults with other chronic medical conditions will receive it. Between 2005 and 2007, a total of 38,004 patient visits were analyzed concerning the rate of medication and behavioral counseling provided to tobacco users versus patients with hypertension, diabetes, hyperlipidemia, asthma, and depression. Patients with hypertension, diabetes, or hyperlipidemia were also more likely to receive behavioral counseling.

Table: Percentage and Adjusted OR of a Patient Receiving Condition-Specific Medication

Condition	Visits Receiving Condition-Specific Medication (%)	Adjusted OR of Receiving Condition-Specific Medication
Tobacco use	4.4	1.0 (Reference)
Hypertension	57.4	32.8×
Diabetes	46.2	20.9×
Hyperlipidemia	47.1	16.5×
Asthma	42.4	22.1×
Depression	53.3	24.0×

Source: Adapted from Bernstein et al, 2014

This study indicates that tobacco use is undertreated by US physicians compared with other common chronic conditions that also cause significant morbidity and mortality and for which evidence-based treatment exists. US clinicians also appear to treat smoking less aggressively than other chronic conditions. The reasons for this are likely multifactorial, but they need to be addressed to improve the health of smokers. Whether the paradigm of disease management can be adapted to the treatment of tobacco dependence is unclear, and we suggest this be considered in future prospective studies.

Reference Number: 70203

POULSEN, P.B., H. SPILLEMOSE, G. NIELSEN, L-L. HERGEL, D. WEDELL-WEDELLSBORG, M. STRAND, AND T. RINGBÆK. "Real-Life Effectiveness of Smoking-Cessation Treatments in General Practice Clinics in Denmark. The Escape Smoke Project." *Respiratory Medicine*, 109: 218–227, 2015 (5 tables, 35 references)

Abstract: The effectiveness of general medical practice smoking cessation treatment on 515 daily smokers (mean age 49 years) was determined in Denmark. The average number of years smoking was 32 and the average number of daily cigarettes was 20. Overall, 23% had mild, 55% had moderate, and 22% had severe nicotine dependency. Smoking cessation methods included NRT, prescription drugs (varenicline, or bupropion), or no medicine.

> In conclusion, the present study has shown that smoking cessation in Danish general practice clinics with both experience and interest in smoking cessation can result in rather high long-term quit rates (36.3%) at a rather low cost. This was especially the case for smokers combining counseling with smoking cessation medicine. The combination of counseling and prescription-based smoking medicine was associated with a 97% higher chance of remaining abstinent after 6 months using prescription-based smoking cessation medicine compared to not using any smoking cessation medicine. General practice, that already has many of the smokers as patients in their clinics for other reasons, is therefore one relevant arena for accelerating effective smoking cessation with the overall goal of decreasing the smoking prevalence and improving the health of smokers.

Reference Number: 70204

MUSSULMAN, L.M., T.S. SCHEUERMANN, B. FASERU, N. NAZIR, AND K.P. RICHTER. "Rapid Relapse to Smoking Following Hospital Discharge." *Preventive Medicine Reports*, 15: (4 pp), 2019 (2 tables, 19 references)

Abstract: As approximately 7 million smokers are hospitalized in the US every year, hospitalization can provide a teachable moment for patients to quit smoking. Although a majority of patients do not smoke during their hospital stay, as most hospitals are smoke-free, most return to smoking after hospitalization. Of the 942 hospitalized smoking patients studied between 2011 and 2013 at two large midwestern hospitals, 25% returned to smoking within 1 day of discharge. Of these, 36.6% relapsed within 1 hour of discharge, 35.3% between 1 and 24 hours after discharge, and 28.1% 1 day after discharge.

Table: Predictors of Relapse within 1 Day of Hospital Discharge

Factors	OR of Rapid Relapse of Cigarette Smoking
Length of stay in hospital	0.72×
Depression	1.42×
Used tobacco during hospitalization	7.37×
Low confidence for quitting	2.07×
Did not set a quit date	1.76×
Heavy smoking index	1.26×

Source: Adapted from Mussulman et al, 2019

Interventions to help patients stay abstinent until they can connect to post-discharge treatment has the potential to reduce the overall health impact caused by tobacco use and reduce health care costs in this vulnerable population. Rapid relapse presents a significant challenge because it allows little time for outpatient cessation supports to connect with smokers. This study provides strong support for the need to enhance hospital-based interventions—prior to and during the process of discharge—to help smokers maintain their intentions to quit after they leave the hospital environment.

Reference Number: 70205

CUMMINS, S.E., A.C. GARNST, K. BRANDSTEIN, G.B. SEYMANN, H. KLONOFF-COHEN, C.A. KIRBY, E.K. TONG, E. CHAPLIN, G.J. TEDESCHI, AND S-H. ZHU. "Hospitalized Smokers: A Factorial RCT of Nicotine Patches and Counseling." *American Journal of Preventative Medicine*, 51: 578–586, 2016 (3 tables, 36 references)

Abstract: Smokers are more likely to be hospitalized than non-smokers and have a period of imposed abstinence while in hospital, but most return to smoking when discharged. A total of 1,270 hospitalized adult smokers from five hospitals between 2011 and 2013 were given a 6-month supply of nicotine patches or telephone counseling upon discharge. The 30-day abstinent rate at 6 months was 22.8% for the nicotine patch and 18.3% for the telephone counseling group.

A second possible explanation for the lack of results lies in the context of the study and the population itself. Hospitals regularly screen for smoking and most smokers stop smoking during their hospital stay. Many also receive pharmacotherapy or counseling while hospitalized. The precipitating health crisis can be a powerful motivator for change even without intervention. The margin of improvement from any additional intervention is likely to be smaller here than in other intervention contexts. The relatively high self-reported abstinence rates overall suggest that the population was motivated to stay quit, making it more difficult to detect an additional intervention effect.

Reference Number: 70206

JANKOWSKI, M., D. KALETA, W.S. ZGLICZYŃSKI, J. GRUDZIĄŻ-SĘKOWSKA, I. WRZEŚNIEWSKA-WAL, M. GUJSKI, W. WIERZBA, AND J. PINKAS. "Cigarette and E-Cigarette Use and Smoking Cessation Practices Among Physicians in Poland." *International Journal of Environmental Research and Public Health*, 16: 3595 (15 pp), 2019 (5 tables, 59 references)

Abstract: A survey was conducted of 423 physicians undergoing mandatory public health training in Poland in 2018, regarding tobacco and EC use and smoking cessation. Current cigarette and EC use was reported by 7.8% and 1.9% of physicians, respectively. Current heated tobacco use was reported by 1.9%. Two-thirds of physicians asked patients about smoking whereas only 21% asked about the use of ECs. Only 5.5% of physicians refer a patient to anti-smoking counseling compared to 3.4% who refer EC users. Physicians who had never tried cigarettes were 2.3× more likely to refer patients to smoking cessation clinics than doctors who had tried at least one cigarette.

An analysis of tobacco use and smoking cessation practices (between 1987 and 2010) among physicians in developing countries shows that smoking prevalence was highest in Central and Eastern Europe (37%) and lowest in Asia (17.5%). However, in some European Union (EU) countries, the prevalence of tobacco use among physicians is equal to or even higher than its prevalence in the general population (28% in Italy; 34% in France; 40% in Greece). The high smoking prevalence among physicians might be explained by occupational stress, which is considered a key factor to addiction. Moreover, this suggests an urgent need to develop smoking cessation promotion program at workplace of physicians. In our study, the prevalence of tobacco use among physicians was 7.8%, which is lower than the prevalence of tobacco use in the general population of Poland (24%).

Reference Number: 70207

JARVIS, M.J. "Why People Smoke." *British Medical Journal*, 328: 277–279, 2004 (4 references)

Abstract: For much of the 20th century and earlier, cigarette smoking was regarded as just a habit. It is now known that chronic smoking is due to nicotine addiction. The terminal half-life of nicotine is 2 hours. Nicotine replacement therapy (gum, patch, or spray) is a reliable way to decrease the severity of withdrawal.

Table: Effects of Nicotine Withdrawal

Symptom	Duration	Incidence
Light-headedness	<48 hours	10%
Sleep disturbance	<1 week	25%
Poor concentration	<2 weeks	60%
Craving for nicotine	<2 weeks	70%
Irritability or aggression	<4 weeks	50%
Depression	<4 weeks	60%
Restlessness	<4 weeks	60%
Increased appetite	<10 weeks	70%

Source: Adapted from Jarvis, 2004

- Smoking usually starts as a symbolic act of rebellion or maturity
- By age 20, 80% of smokers regret having started to smoke
- Nicotine from cigarettes is highly addictive—probably because it is delivered so rapidly to the brain
- Smoking a cigarette, especially the first of the day, feels good mainly because it reverses the symptoms of nicotine withdrawal
- Most smokers who switch to low tar cigarettes or reduce the number of cigarettes they smoke continue to inhale the same amount of nicotine, and hence tar, from the cigarettes they smoke

- Heavy dependence on nicotine is strongly related to socioeconomic disadvantage
- Smoking is a chronic relapsing addictive disease

Reference Number: 70208

LANIADO-LABORÍN, R. "Smoking and Chronic Obstructive Pulmonary Disease (COPD). Parallel Epidemics of the 21st Century." *International Journal of Environmental Research and Public Health*, 6: 209–224, 2009 (1 table, 1 figure, 96 references)

Abstract: According to the World Health Organization (WHO), 100 million deaths were caused by tobacco in the 20th century, and if current trends continue, up to 1 billion deaths will occur in the 21st century. More than 80% of the world's smokers live in low- and middle-income countries. Smoking is the most important causative factor in COPD. WHO estimates that 80 million people have moderate to severe COPD and more than 3 million people died of COPD in 2005. About 50% of smokers eventually develop COPD. Cigarette smoke contains an extremely high concentration of oxidants, which causes an inflammation in the lungs and its airways. Smoking cessation is the single most effective and cost-effective treatment for COPD.

Table: Odds Ratio of Smoking Cessation with Pharmacologic Treatment Compared to Placebo

Treatment	OR of Smoking Cessation (compared to placebo)
Nicotine gum	1.66×
Nicotine patch	1.81×
Nicotine inhaler	2.14×
Bupropion	1.94×
Varenicline	2.66×

Source: Adapted from Laniado-Laborin, 2009

Tobacco use kills more than five million people a year and accounts for 10% of adult deaths worldwide. COPD is a major and growing cause of morbidity and mortality with smoking being recognized as its most important causative factor. Several meta-analyses have shown that all pharmacotherapies for smoking cessation are twice as likely more efficacious than placebo with an abstinence rate in the 25–30% range at one year when pharmacological treatment and behavioral support are combined. Unfortunately, in spite of the initial efficacy of these pharmacotherapies, the number of patients that remain abstinent from smoking at 1 year follow-up is low (range 16–25%).

Reference Number: 70209

GOODFELLOW, L.T., AND J.B. WAUGH. "Tobacco Treatment and Prevention: What Works and Why." *Respiratory Care*, 54: 1082–1090, 2009 (1 table, 44 references)

Abstract: Respiratory therapists frequently have contact with patients with chronic lung diseases and are unable to advise them about smoking cessation. This article summarizes all the current smoking cessation techniques.

Table: Smoking Cessation Medication and Recommended Dosage

Medication	Dosage
Patch	Start at 21 mg unless patient weighs less than 100 lbs or has adverse effects, 14 mg if more than 10 cigarettes per day
Gum	2 mg for less than 25 cigarettes per day 4 mg for more than 25 cigarettes per day
Lozenge	2 mg for those who smoke within 30 minutes of awakening 2 mg for those who smoke more than 30 minutes after awakening Up to 20 lozenges per day; 5 per 6 hour period
Inhaler	Can use 6 to 16 per day Inhale 80 times per cartridge
Nasal spray	Maximum of 40 doses per day (1 dose is 1 spray in each nostril, generally 1–2 doses per hour)
Bupropion SR/XL (aka Zyban, Wellbutrin)	With XL, take one 300 mg pill per day With SR, 150 mg twice daily
Varenicline (aka Chantix)	Start use 1 to 2 weeks prior to quitting smoking Take after eating with full glass of water
Nortriptyline	Start use 1 to 3 weeks prior to quitting Take with a full glass of water
Clonidine	Start use 1 week prior to quitting Take after eating with a full glass of water

Source: Adapted from Goodfellow and Waugh, 2009

Another key element of tobacco control that pertains only to approximately 4% of all health-care providers in the United States is the role of smoker. If you are a smoker, become an ex-smoker. Now is the time to finally stop. By quitting your own addiction, becoming skilled at tobacco abuse counseling, and engaging in social and political action against tobacco, you take on a role that is vital to minimizing and preventing tobacco's terrible toll of death and disability. If advocacy is not for you in a public arena, at least become a role model for respiratory therapy students, who often do not receive adequate tobacco control and prevention counseling or formal training in the treatment of nicotine dependence in their education programs. Share your success story and be an inspiration to someone having a difficult time with their own tobacco control.

Reference Number: 70210

WU, J., AND D.D. SIN. "Improved Patient Outcome with Smoking Cessation: When is it Too Late?" *International Journal of Chronic Obstructive Pulmonary Disease*, 6: 259–267, 2011 (1 table, 3 figures, 69 references)

Abstract: Three conditions are responsible for all tobacco-related morbidity and mortality. Cardiovascular diseases (CVD) account for 40% of all smoking-related deaths. Lung cancer accounts for 20% and COPD accounts for an additional 20% of all tobacco-related deaths. Globally, nearly 3,000 people die each day due to tobacco-related disease and in 2030, it is estimated that tobacco will cause 10 million deaths per year. NRT, bupropion, and varenicline are effective medications for smoking cessation.

The benefits of smoking cessation in CVD are almost immediate. Within 24 hours of smoking cessation, there are significant improvements in blood pressure and heart rate. Within one year of abstinence, the risk of cardiovascular events (eg, myocardial infarction [MI] and stroke) is reduced by half (compared with continued smoking). Between 5 and 15 years post-smoking, the risk of stroke and coronary heart disease is "normalized" to that of never smokers. On the other hand, the benefits of tobacco treatment on the risk of lung cancer take much longer to achieve. The risk of lung cancer declines steadily following smoking cessation such that after about 10 years of abstinence, the risk of lung cancer is between 30% and 50% of that for continued smokers. However, even after decades of smoking cessation, the risk of lung cancer may never reach that of never smokers. The benefits of tobacco treatment to COPD progression are measurable within the first year of abstinence and are particularly striking in female smokers who manage to quit smoking. After several years of smoking cessation, the rate of decline in lung function becomes similar to that of never smokers. However, because the lung function lost during the time the patient smoked is never fully recovered and because lung function declines with age (even in nonsmokers), COPD patients may develop additional symptoms and progress to a higher COPD severity class, even after many years of smoking abstinence.

Reference Number: 70211

PERRET, J.L., B. BONEVSKI, C.F. MCDONALD, AND M.J. ABRAMSON. "Smoking Cessation Strategies for Patients with Asthma: Improving Patient Outcomes." *Journal of Asthma and Allergy*, 9: 117–128, 2016 (4 tables, 103 references)

Abstract: Smoking is as common in people with asthma as it is in the general population, but it is even higher in adolescents with asthma. Smokers with asthma are prone to have worse symptoms and greater deficits in lung function. Smokers with asthma are more likely to smoke regularly and at a younger age.

In general, many smokers make multiple attempts before successfully quitting. In addition to individual vulnerability to tobacco withdrawal symptoms, factors contributing to relapse include social situations when drinking alcohol, being around other smokers, as well as stress related to work, relationships, and finances. A different drug or combination of smoking cessation medications may be trialed for smokers who have been unsuccessful in their quit attempt or, a combination of NRT can be repeated for a subsequent attempt.

Depending on the pharmacotherapy used, 5%–25% of smokers who attempt to quit abstain at 6 months, but around half of those who are abstinent at 1 year, relapse within the next 7 years. Thus, follow-up by a health professional is essential, preferably in the first week and month after quitting and thereafter for at least 5 years.

Reference Number: 70212

JORDAN, C.J., AND Z-X. XI. "Discovery and Development of Varenicline for Smoking Cessation." *Expert Opinion Drug Discovery*, 13: 671–683, 2018 (2 tables, 3 figures, 109 references)

Abstract: Although 70% of cigarette smokers express desire to quit, only about 3% remain abstinent for more than a year. Nicotine acts by binding to the nicotinic acetylcholine receptors (nAChRs) in the central and peripheral nervous system. Nicotine binding stimulates dopamine release. Varenicline tartrate (Chantix) was developed as a smoking cessation agent by Pfizer in 1997 based on the plant-based drug cytosine. It is a highly selective and potent partial agonist of nicotine. Nausea is the most common side-effect of varenicline, followed by insomnia. Varenicline is not recommended for people operating vehicles or heavy equipment, because side effects can include vision and heart rhythm issues and it has been linked to accidents and falls. It is banned for use in pilots.

Varenicline development was based on strong theoretical rationale and preclinical evidence. Clinical studies indicate that varenicline is safe and more effective in sustaining abstinence than placebo, bupropion or nicotine replacement therapies. However, given that continuous abstinence rates across studies remain low (18~30% with varenicline; 4~10% with placebo), novel and more effective medications targeting other nicotinic or glutamate receptors for smoking cessation are required.

Reference Number: 70213

PATEL, K., D. SCHLUNDT, C. LARSON, H. WANG, A. BROWN, AND M. HARGREAVES. "Chronic Illness and Smoking Cessation." *Nicotine and Tobacco Research*, 11: 933–939, 2009 (2 tables, 41 references)

Abstract: A telephone survey was conducted of 3,802 adults (18+ years) from Nashville, Tennessee, in the United States. Seventeen percent of those surveyed were former smokers and 25% were current smokers. Based on self-reported height and weight, 34% were overweight and 28% were obese. A total of 11% reported diabetes, 42% reported hypertension, and 37% had high cholesterol.

Participants with diabetes were more likely to report being former smokers, after adjusting for sociodemographic characteristics, whereas having hypertension or high cholesterol was not associated significantly with smoking status. The likelihood of being a former smoker did not increase as number of diagnosed chronic diseases increased. Participants who were women, older (aged 65+), or single were significantly less likely to be former smokers. Participants with at least a college degree, those with incomes of 50,000+ US dollars, and those who were underweight or obese were more likely to be former smokers.

Reference Number: 70214

SHAHAB, L., E. MORTIMER, L. BAULD, J.A. MCGOWAN, A. MCNEILL, AND R.F. TYNDALE. "Characterising the Nicotine Metabolite Ratio and Its Association with Treatment Choice: A Cross Sectional Analysis of Stop Smoking Services in England." *Scientific Reports*, 7: 17613 (9 pp), 2017 (3 tables, 1 figure, 51 references)

Abstract: Oral fluid samples were collected from 3,045 Stop Smoking participants in England between 2012 and 2013 and were analyzed for trans 3'-hydroxycotinine (3-HC) and cotinine by LC/MS/MS to determine the nicotine metabolite ratio (NMR). Samples with cotinine levels less than 10 ng/mL were excluded, as the persons were light or occasional smokers. Normal metabolizers of nicotine were characterized by an NMR of 0.31+ and slow metabolizers were <0.3. Nicotine dependence and other demographics were determined. Nicotine is metabolized mainly by the liver enzyme CYP2A6 into cotinine and then exclusively by the CYP2A6 into 3-HC. NMR is a biomarker of total nicotine clearance, as cotinine has a long half-life. In total, the NMR of 1,875 participants were measured and the mean NMR was 0.41. Of these, 71.5% were normal (mean 0.53) and 28.5% were slow (mean 0.21). High cigarette dependence was seen in about half of the participants. Normal metabolizers were on average older, White, from lower socioeconomic groups, and in poor physical and psychological health. Greater difficulty in stopping smoking was found in normal metabolizers. Higher NMR occurs in women smokers due to the estrogen induction of the CYP2A6 enzyme. Slower metabolizers benefit more from behavioral counselling alone.

Although significant associations between NMR and ethnicity, physical health and gender were identified, these sociodemographic, smoking, and health-related characteristics did not greatly influence variability in NMR and did not confound allocation to either slow or normal NMR status. This suggests that NMR status is relatively unrelated to these variables, functioning consistently across different populations, thus increasing its potential for use in clinical practice. The unexpected association of NMR status with behavioural support should be explored further in clinical trials and studies of real-world data. Given the association of NMR with pharmacotherapy efficacy, the finding that NMR status is currently not associated with pharmacotherapy choice suggests there is a need for providers to tailor smoking cessation therapies based on rate of nicotine metabolism as patients are not naturally selecting the option that has the highest potential efficacy for them.

Reference Number: 70215

HALPERN, S.D., M.O. HARHAY, K. SAULSGIVER, C. BROPHY, A.B. TROXEL, AND K.G. VOLPP. "A Pragmatic Trial of E-Cigarettes, Incentives, and Drugs for Smoking Cessation." *New England Journal of Medicine*, 378: 2302–2310, 2018 (3 tables, 2 figures, 27 references)

Abstract: Six thousand and six smokers from 54 companies were assigned either to one of four smoking cessation interventions or to usual care. Usual care included access to information regarding the benefits of smoking cessation and a motivational text-messaging service. The other four interventions were free cessation aids; free e-cigarettes; reward incentives plus free cessation aids; or a redeemable $600 deposit plus free cessation aids. The primary outcome was sustained smoking abstinence for 6 months after the quitting date. Sustained rates were 0.5% in the free cessation aids group, 1% in free ECs group, 2% in the rewards group, and 2.9% in the redeemable deposits group.

In summary, this trial showed that among unselected smokers, workplace smoking-cessation programs yielded low rates of smoking abstinence and that offering free cessation aids or free e-cigarettes did not increase abstinence among smokers who were given access to information and motivational text messages. In contrast, programs that offered financial incentives tripled the rates of smoking cessation, reduced employers' costs per successful quit as compared with programs that offered cessation aids alone, and yielded total costs that compared favorably with the costs of employing smokers.

Reference Number: 70216

VOGIATZIS, I., E. TSIKRIKA, V. SACHPEKIDIS, S. PITTAS, AND A. KOTSANI. "Factors Affecting Smoking Resumption After Acute Coronary Syndromes." *Hellenic Journal of Cardiology*, 51: 294–300, 2010 (3 tables, 40 references)

Abstract: A study was conducted of 322 male and 104 female patients (mean age 60 years) who were hospitalized between 2003 and 2006 with an acute coronary syndrome. All were active smokers and smoking was strictly prohibited during hospitalization. They were advised to attend a smoking cessation clinic. The patients were studied for 12 months after discharge. Of the 420 patients, 162 (39%) were abstinent from smoking after 1 year and 258 were relapsed smokers.

> Smokers with acute coronary syndromes should be encouraged to participate in special secondary prevention programmes. Smoking cessation clinics contribute significantly to a reduction in smoking in this particular group of patients.

Reference Number: 70217

VOZORIS, N.T., AND M.B. STANBROOK. "Smoking Prevalence, Behaviours, and Cessation Among Individuals with COPD or Asthma." *Respiratory Medicine*, 105: 477–484, 2011 (4 tables, 1 figure, 28 references)

Abstract: The 2003 Canadian Community Health Survey (CCHS) was used to determine smoking prevalence, nicotine addiction, physician counselling, and smoking cessation in 20,453 people without COPD or asthma, 1,311 people with asthma, and 1,168 with COPD. Women smokers with COPD were 2.5× as likely to have high or very high nicotine addiction and women smokers with asthma were 1.7× as likely.

> The prevalence of smoking in COPD remains substantially higher than in the general smoking population, while the prevalence of smoking in asthma is similar to population levels. Our observation that more women with COPD or asthma smoke, and that such women have higher nicotine addiction levels, may in part explain why the prevalence of COPD and asthma is increasing more among women than men. Greater nicotine addiction may in part explain the high prevalence of continued smoking among individuals with COPD. In contrast, reduced desire to quit smoking and fewer attempts at quitting smoking do not appear to explain the higher prevalence of continued smoking in COPD. These findings suggest that more physician counselling regarding smoking cessation, coupled with more aggressive prescription and increased access to smoking cessation medications, is needed for smokers with COPD and asthma in order to improve cessation rates in these vulnerable populations.

Reference Number: 70218

WU, S.Y-D., M.P. WANG, W.H. LI, A.C. KWONG, V.W. LAI, AND T.H. LAM. "Does Electronic Cigarette Use Predict Abstinence from Conventional Cigarettes Among Smokers in Hong Kong?" *International Journal of Environmental Research and Public Health*, 15: 400 (10 pp), 2018 (4 tables, 49 references)

Abstract: Hong Kong has one of the lowest smoking rates in the developed world (10.7% in 2015) and EC use remains rare (0.7%). The effect of ever-use of EC (many of which do not have nicotine) was studied in 956 daily smokers at baseline and 6 months later.

> To our knowledge, this is the first report showing that ever EC use was not prospectively associated with abstinence from conventional cigarettes in Chinese adult smokers, most of whom were motivated to quit (as they were participants in a quitting contest), in a region where nicotine ECs are not widely available.

Reference Number: 70219

WADGAVE, U., AND L. NAGESH. "Nicotine Replacement Therapy: An Overview." *International Journal of Health Sciences*, 10: 425–435, 2016 (2 tables, 71 references)

Abstract: A review of nicotine replacement therapy (NRT) was conducted. NRT is designed to reduce the motivation to consume tobacco and the physiological and psychomotor withdrawal symptoms.

> Nicotine addiction is the major factor impeding smoking cessation and long-term abstinence. Today, several nicotine medications are available in different forms, doses and flavors and their use has been recommended for all tobacco consumers who do not have medical contraindications. The choice of NRT product should normally be guided by the patient's preference. Current evidence suggests that all of the commercially available forms of NRT (gum, transdermal patch, nasal spray, inhaler and sublingual tablets/lozenges) increase their chances of successfully stopping smoking. NRTs increase the rate of quitting by 50 to 70%.

Reference Number: 70220

AGUIAR, M., F. TODO-BOM, M. FELIZARDO, R. MACEDO, F. CAEIRO, R. SCOTTO-MAYOR, AND A. BUGALHO DE ALMEIDA. "Four Years' Follow Up at a

Smoking Cessation Clinic." *Revista Portuguesa de Pneumologia*, 15: 179–197, 2009 (1 table, 5 figures, 21 references)

Abstract: A retrospective analysis was conducted on 526 patients (mean age 49.5 years) at a smoking cessation clinic between 2003 and 2006. The average smoking burden was 50 pack-years. Respiratory disease was present in 52.1% (COPD 39.9%, other 12.2%) and cardiovascular disease in 14.6%. A psychiatric disorder was diagnosed in 46% of the patients. Most patients reported prior quitting attempts (72.2%). Most patients had undergone NRT (patch 53.3%; gum 1%; both patch and gum 37.3%). Sedative-hypnotics were administered in 86.5% of the patients. Nearly half (42.4%) were successful in smoking cessation.

Table: Reasons for Past Failure to Quit Smoking

Reason for Relapse	Percent
Anxiety	52%
Lack of motivation	25%
Non-supportive environment	20%
Weight gain	3%

Source: Adapted from Aguiar et al, 2009

We found a high percentage of patients with psychiatric pathology and a high degree of nicotine addiction, factors which when allied to stress as a main cause of increased smoking imply paying special attention to cognitive/behavioural support to guarantee better results.

Successful treatment is without a doubt related to compliance with appointments, drug therapy and cognitive/behavioural therapy.

Reference Number: 70221

RAHMANIAN, S.D., P.T. DIAZ, AND M.E. WEWERS. "Tobacco Use and Cessation Among Women: Research and Treatment-Related Issues." *Journal of Women's Health*, 20: 349–357, 2011 (2 figures, 105 references)

Abstract: In the United States, the prevalence of tobacco use among women was 6% in 1924, reached a high of 33% in 1965, and had gradually decreased to 18% by the time of the study. There has been relentless marketing of cigarettes to women by the tobacco industry, including light or low tar cigarettes which has sent the message that these products are safer. Women in general tend to have faster plasma clearance of nicotine and cotinine and may benefit more from smoking cessation than do men. COPD has risen dramatically in women.

> Tobacco use is responsible for many diseases affecting women today, most notably COPD and lung cancer. With the rise in sex-targeted marketing of tobacco products, there has been an alarming increase in the number of tobacco-attributable diseases in women. The literature addressing sex differences in susceptibility to tobacco and its associated diseases as well as sex differences in smoking cessation is at times contradictory. Although this is likely related to study design and selection bias, there is still intriguing evidence that sex differences may exist. Women may be more susceptible to the effects of tobacco and, thus, more likely to succumb to the ramifications of long-term smoking. More importantly, healthcare practitioners may be unaware of the increased susceptibility of women to smoking-related diseases. Even smoking cessation therapies need to be tailored specifically for women, as the reasons why they smoke and the barriers that interfere with successfully quitting are unique to women. Strategies to help alleviate the epidemic of tobacco use in women include a greater emphasis on patient and physician education, devising smoking cessation programs geared specifically toward women, and increasing sex-specific research in tobacco-related diseases.

Reference Number: 70222

SCHOLZ, J., P.C.J.L. SANTOS, C.G. BUZO, N.H.M. LOPES, T.O. ABE, P.V. GAYA, H. PIERRI, C. AMORIM, AND A.C. PEREIRA. "Effects of Aging on the Effectiveness of Smoking Cessation Medication." *Oncotarget*, 7: 30032–30036, 2016 (3 tables, 30 references)

Abstract: Two hundred and thirty-nine smokers over 60 years of age and 701 smokers under 60 years of age attended a smoking cessation clinic and received various treatments, including NRT alone, varenicline, bupropion, and combinations thereof. The smoking cessation rates were determined 1 year later.

> Cessation rates were significantly different among younger and older participants who were using nicotine replacement therapy (NRT) alone. Being over 60 years of age was significantly associated with increased cessation success among those who used NRT alone (OR 2.34, 95% CI: 1.36 to 4.04, p = 0.002). The effectiveness of varenicline and bupropion were not significantly different according to age groups.... Using age as a predictor

for tailoring smoking cessation drugs might potentially lead to a more individualized prescription of smoking cessation therapy. These results should be tested in randomized controlled trials.

Reference Number: 70223

TØNNESEN, P. "Smoking Cessation and COPD." *European Respiratory Review*, 22: 37–43, 2013 (7 tables, 3 figures, 44 references)

Abstract: The main smoking cessation therapies are counselling in combination with varenicline, NRT, or bupropion. The Lung Health Study I examined 5,887 smokers with mild COPD (mean age 48 years) and were involved in repeated smoking cessation attempts over the course of 5 years, with a quit rate of 37%. After 14.5 years, quitters had a higher lung function and higher survival rate.

Table: Odds Ratio of Quitting and Type of Treatment

Treatment	OR of Quitting
Nothing/usual care	1.00×
Counselling alone	1.82×
Counselling + antidepressants	3.32×
Counselling + NRT	5.08×
Counselling + varenicline	4.04×

Source: Adapted from Tønnesen, 2013

An optimal approach to smoking cessation today should contain an adequate support programme, either individual or in groups, in combination with a first-line pharmacological smoking cessation agent, i.e. NRT (two formulations), varenicline or bupropion SR for 3 months. When relapse occurs, re-treatment should be offered. COPD patients need more support than smokers without comorbidities and smoking intervention should have top priority as it is very cost-effective, reduces the decline in lung function and reduces morbidity and mortality.

Reference Number: 70224

TØNNESEN, P. "Which Drug to Be Used in Smoking Cessation?" *Polskie Archiwum Medycyny Wewnętrznej*, 118: 373–375, 2008 (15 references)

Abstract: There are three first-line smoking-cessation medications: nicotine replacement therapy (NRT), varenicline (a partial nicotine receptor agonist), and slow-release bupropion (an anti-depressant which releases dopamine in the central nervous system). The plasma nicotine concentrations subjects reach while using NRT are approximately one-half to two-thirds the levels reached during smoking. All three medications nearly double the quit rate when used for 2–3 months.

> Depression and suicidal attempts have been reported with varenicline use but it is probably not induced by varenicline but by the quitting process per se. It is recommended that the first agent to be used in smoking cessation should be NRT as it is the best documented product with mild side effects. It might be optimal to combine the patch with either gum, inhaler, sublingual tablets or nasal spray. In subjects that have failed with NRT, varinicline should be the choice. Bupropion SR is preferred to subjects with depression or smokers who have failed with the previous two agents, due to the many contra-indications and side effects of bupropion SR. With one of the 3 agents combined with follow-up visits with counselling, one can expect a 1-year quit rate around 20-25%.

Reference Number: 70225

KOTZ, D., W. VIECHTBAUER, C.R. SIMPSON, O.C.P. VAN SCHAYCK, R. WEST, AND A. SHEIKH. "Cardiovascular and Neuropsychiatric Risks of Varenicline and Bupropion in Smokers with Chronic Obstructive Pulmonary Disease." *Thorax*, 905–911, 2017 (5 tables, 1 figure, 30 references)

Abstract: The risk of serious cardiovascular events, self-harm, or depression was determined in 10,426 COPD patients using NRT, 3,574 COPD patients using varenicline, and 350 COPD patients using bupropion for smoking cessation in England between 2007 and 2012.

Table: Hazard Ratios (HRs) of Various Adverse Events Comparing Bupropion or Varenicline and Nicotine Replacement Therapy (NRT)

Adverse Event	HR of Bupropion Compared to NRT	HR of Varenicline Compared to NRT
Ischemic heart disease	1.23×	0.97×
Stroke	0.49×	1.06×
Arrhythmia	3.96×	1.12×
Depression	0.75×	0.68×
Self-harm	0.49×	0.60×

Source: Adapted from Kotz et al, 2017

Very recently, the FDA decided to remove the boxed warning for serious mental health effects from the varenicline and bupropion labels. This decision was based on a large randomised controlled trial that the FDA required the drug companies to conduct, which showed no significant increase in neuropsychiatric events in users of varenicline or bupropion compared with users of nicotine patch or placebo. The trial thereby confirmed evidence from meta-analyses of previous randomised controlled trials and from observational studies indicating that a causal relationship between the use of these drugs and serious adverse cardiovascular and neuropsychiatric events is unlikely.

Reference Number: 70226

JEREMIAS-MARTINS, E., AND J.M. CHATKIN. "Does Everyone Who Quit Smoking Gain Weight? A Real-World Prospective Cohort Study." *Jornal Brasileiro de Pneumologia*, 45: (7 pp), 2019 (3 tables, 2 figures, 31 references)

Abstract: Smokers typically have a lower body weight than non-smokers, in part due to nicotine and additives added by the tobacco industry such as tartaric acid and 2-acetylpyridine, which suppress appetite and increase basal metabolism rate. Approximately 50% of the women and 26% of the men state that their main concern in quitting cigarettes is weight gain. The factors that have been associated with weight gain after smoking cessation include male gender, a higher level of nicotine dependence, and advanced age. A total of 161 patients who attended an outpatient cessation clinic in Brazil who achieved continuous smoke-free status (eCO <10 ppm) for one year were studied as to actual weight gain. Sixty-five percent had no weight gain or less than 5%; 35% had a weight gain of more than 5%; and 11% had a weight gain of more than 10%. The use of bupropion as a smoking cessation aid did not affect weight gain, even though this antidepressant is prescribed for weight loss.

The dissemination of our findings could be useful for encouraging smokers to quit. Patients should be informed that weight gain is not a significant problem in the majority of the cases. The benefits of quitting smoking outweigh any potential risks related to weight gain. Patients should also be informed that systemic arterial hypertension and type 2 diabetes mellitus, as collateral effects of weight gain—often mentioned by patients as major concerns—are uncommon, and that the benefits of quitting by far offset the consequences of weight gain related to smoking cessation. Nevertheless, in our sample, 11% of the patients in whom treatment success was achieved showed weight gains that could be harmful to their health (more than 10% in relation to the baseline weight). Of the patients in whom the treatment failed, only 1.1% showed such weight gains.

Reference Number: 70227

FUJITA, T., A. BABAZONO, Y. HARANO, AND P. JIANG. "Risk of Depressive Disorders After Tobacco Smoking Cessation: A Retrospective Cohort Study in Fukuoka, Japan." *British Medical Journal Open*, 9: e025124 (6 pp), 2019 (3 tables, 2 figures, 37 references)

Abstract: This study compared depressive disorders, according to health checks and insurance claims, in 7,841 smokers in a cessation group and 79,414 smokers who did not quit in Fukuoka, Japan between 2010 and 2014. Overall, 2.3% of the smoking cessation group and 2.3% of the smoking group had a history of health care for depressive disorders.

People who have the intention to stop smoking and who work as health professionals or health-related service providers might have concerns about depressive disorders after smoking cessation. The present results could provide helpful information that could not influence whether prescribed stop-smoking medication is used or not because smoking cessation is not associated with depressive disorders in the long term. Moreover, successful smoking cessation has great benefits for reducing the risk of many diseases, such as lung cancer, chronic obstructive pulmonary disease, cardiovascular disease and other disorders. However, most withdrawal symptoms reach maximum intensity within a few days and some symptoms can persist for more than 45 days. Consequently, health professionals would be required to help people who attempt smoking cessation not to relapse because of withdrawals, especially in the first few months after quitting. In addition, our results could provide evidence that can be used to promote smoking cessation programmes as part of government policies.

Reference Number: 70228

SILVA, A.P., J. SCHOLZ, T.O. ABE, G.G. PINHEIRO, P.V. GAYA, A.C. PERIERA, AND P.C.J.L. SANTOS. "Influence of Smoking Cessation Drugs on Blood Pressure and Heart Rate in Patients with Cardiovascular Disease or High

Risk Score: Real Life Setting." *BMC Cardiovascular Disorders*, 16: (6 pp), 2016 (2 tables, 28 references)

Abstract: The safety of smoking cessation medication on 900 outpatients with cardiovascular disorders (57% had hypertension) was determined by the sequential measurement of diastolic and diastolic blood pressure, heart rate, and exhaled CO over the first 45 days of treatment. A total of 271 patients used NRT alone, 81 used bupropion alone, 154 used varenicline alone, and 283 used NRT and bupropion. For all treatments there was no increase in BP or HR and there was a significant decrease in exhaled CO (on average from 10 ppm to 5 ppm).

> Our data from a real clinical practice suggest that, even in patients with hypertension, CAD, and/or AMI, there are no significant clinical change in blood pressure or heart rate during the use of NRT, varenicline, and/or bupropion.

Reference Number: 70229

LAVELLE, C., C. BIREK, AND D.A. SCOTT. "Are Nicotine Replacement Strategies to Facilitate Smoking Cessation Safe?" *Journal of the Canadian Dental Association*, 69: 592–597, 2003 (2 tables, 59 references)

Abstract: Nicotine replacement therapy (NRT) is the most widely used pharmaceutical aid to assist people to stop smoking. Patches, gums, nasal sprays, and sublingual nicotine replacement products are available.

> However, there is no intention to suggest that NRT is not a generally safe and efficacious aid to smoking cessation. There is no evidence that NRT is either carcinogenic or genotoxic, and its relative safety is reflected in its over-the-counter availability in Canada. The evidence that NRT can double the chances of successful quitting, at least among those motivated to quit, is strong. Except where contraindicated, NRT should be considered by dental health care providers as an appropriate aid to smoking cessation. NRT should preferably be used as part of a broader smoking cessation strategy tailored to the needs of nicotine-dependent tobacco users. It should always be remembered that the nicotine doses delivered by NRT are, generally, lower than those delivered through cigarette smoking and that tobacco smoke contains a large number of toxins and carcinogens in addition to nicotine (e.g., carbon monoxide, cyanide, hydrogen sulfide, arsenic and lead). Furthermore, NRT is designed for short-term use

only (normally up to 12 weeks) and should be discontinued if the smoker continues to relapse over the initial 4-week period. Therefore, even in situations where caution is advised, the risks of NRT administration should be carefully weighed against the continuing detrimental influence of tobacco smoking on multiple facets of health.

Reference Number: 70230

MURRAY, R.P., J.E. CONNETT, AND L.M. ZAPAWA. "Does Nicotine Replacement Therapy Cause Cancer? Evidence from the Lung Health Study." *Nicotine and Tobacco Research*, 11: 1076–1082, 2009 (4 tables, 2 figures, 3 references)

Abstract: A study was conducted of 3,320 participants enrolled in the US Lung Health Study whose replacement nicotine therapy and smoking exposures were recorded for 5 years. Participants were encouraged to use Nicorette gum (2 mg nicotine) for at least 6 months and to cease use after a maximum of 2.5 years. The mean number of pieces of gum used over the course of the 5-year study was 0 (40.1%), 1 (25.5%), 1–3 (16.9%), and >3 (17.5%).

> Our basic question was whether nicotine replacement therapy causes cancer to an extent comparable to that caused by smoking cigarettes. We failed to find evidence of such an effect of nicotine replacement therapy in this study, and the sample size, close monitoring of the use of nicotine replacement therapy and cigarettes, and the well-documented outcomes of this study will be difficult for future studies to match.

Reference Number: 70231

KRAUSE, D., M. WARNECKE, C.S. SCHUETZ, M. SOYKA, K.M. MANZ, L. PROEBSTL, F. KAMP, A.I. CHROBOK, O. POGARELL, AND G. KOLLER. "The Impact of the Opioid Antagonist Naloxone on Experimentally Induced Craving in Nicotine-Dependent Individuals." *European Addiction Research*, 24: 255–265, 2018 (3 tables, 7 figures, 34 references)

Abstract: The first symptoms of nicotine dependence are likely to appear within days or weeks of the first exposure to the occasional cigarette. There are numerous similarities between opioid dependence and nicotine dependence. Opioids like heroin and methadone lead to an increase in nicotine self-administration. Thirteen female and seven male nicotine dependent outpatients (mean age 25 years) were deprived of nicotine for 4 hours before being administered 1.6 mg or 3.2 mg/kg naloxone IV or placebo. Neutral images or smoking-related images were then presented. Nicotine withdrawal was monitored by HR, body temperature, respiration, and other measures.

> Naloxone influenced nicotine withdrawal and strengthened significantly by cue exposure, both on objective measurement and on craving scales. These findings suggest an involvement of the endogenous opioid system in the development and maintenance of nicotine dependence.

7.03 PUBLIC HEALTH MEASURES

> "The irony is that the tobacco industry uses images of health to sell death, while health organizations use images of death to sell health."
>
> —Yossuf Saloojee, World Lung Conference (2007)

Tobacco is the most lethal and dangerous artefact made by humans. It has resulted in more than 100 million deaths in the 20th century alone, more than the total deaths from World War I and World War II combined. Some of the best practices to reduce this epidemic include:

1) Smoke-free public spaces
2) Raising cigarette taxes to at least match the rise in inflation
3) Having a national tobacco smoking quit line and social media presence
4) Prominent health warnings on cigarette packages
5) Mass anti-smoking campaigns
6) Advertising bans, including point of sale
7) Removing all flavors (including menthol) from tobacco products
8) Lowering the nicotine content of tobacco products
9) Promotion of readily available, effective smoking cessation aids
10) Stricter regulation of young smokers
11) Tobacco social and industry denormalization (70301–2, 70306, 70309, 70315, 70316, 70322)

The tobacco epidemic can be divided into four stages, and different countries are in different stages (70303–4). The first recorded prohibition against tobacco use was in Peru in 1586, when a papal decree declared it was a sin for a priest to use tobacco before celebrating mass (70302).

The tobacco industry promoted and provided money and support to smokers' rights groups and compared anti-smoking regulations to Hitler and Nazi regulations; in fact, I myself have been called a tobacco Nazi on social media. In 1992, smokers' rights advocates dressed in Nazi SS uniforms and protested against England's secretary of state (70605).

Canada was the first country to implement color pictorial health warnings, which differ among countries (70307–10). The graphic health warning must now cover 75% of the cigarette package in Canada and has curtailed the tobacco industry from using cigarette packages as an advertising vehicle (70311–12). Internationally, English brand names and English writing on cigarette packages have been used to increase the allure of cigarettes (70313).

Restricting vending machines is an effective but underused method of tobacco control (70314). High school students in US states with minimal tobacco restrictions are more likely to become daily smokers (70315). Lowering the nicotine concentration allowed in tobacco products has been found to lower cravings and withdrawal symptoms and decrease youth smoking initiation (70316).

Although cigarette cessation programs are expensive, with California spending $100 million annually, it has been found that the state saves between $2.2 and $10.7 billion a year in reduced health care costs (70317). The endgame strategy of eventually eliminating all cigarette smoking has received wide public support, even among smokers (70318–20).

Cigarette control needs to start with the cultivation, harvesting, and curing of tobacco to minimize its contamination of nearby water resources and environmental damage. Tobacco leaves accumulate metals, radioactivity, pesticides, insecticides, and fungicide (70321).

Social and tobacco denormalization is an important aspect of tobacco control and smoking cessation. Over 80% of smokers in four countries agree that society disapproves of smoking, and nearly 90% agree that people who are important to the smoker believe that they should not smoke (70322). Disseminating the RICO decision to the general population aids in tobacco industry denormalization (70323).

Reference Number: 70301

SRKALOVIĆ, G. "Lung Cancer: Preventable Disease." *Acta Medica Academica*, 47: 39–49, 2018 (1 table, 2 figures, 40 references)

Abstract: Tobacco use is one of the largest preventable causes of death and disability in developing countries. It has caused 100 million deaths in the 20th century (more than deaths in World War I and World War II combined) and is estimated to cause 1 billion deaths in the 21st century. Cigarette smoking increases the risk of at least 17 types of lung cancer. Tobacco smoking is responsible for 30% of all cancer deaths and 80% of lung cancer deaths. In a developing country like Bosnia and Herzegovina, 47.2% of the men and 30% of the women smoke cigarettes. More than 2,000 children (aged 10–14 years) and 1.05 million adults (aged 15+ years) smoke every day in this country. Some of the best practice measures to reduce the tobacco epidemic are 1) smoke-free public places, 2) raising taxes to 70% of the retail price, 3) offering a national tobacco smoking quit line, 4) prominent warning labels on cigarette packages, 5) mass anti-smoking campaigns, and 6) advertising bans.

> In conclusion lung cancer is very heterogeneous disease with high tumor mutational burden and very hard to treat when metastatic. However, it is preventable disease in very high percentage of cases (85–90%). Goal is to exterminate cigarette smoking. That can be achieved only by concerted effort by members of family, patients themselves, physicians, researchers, non-governmental organizations, political figures and society as whole. Best practices already exist and initial goal should be to start using them.

Reference Number: 70302

CUMMINGS, K.M. "Programs and Policies to Discourage the Use of Tobacco Products." *Oncogene*, 21: 7349–7364, 2002 (5 figures, 106 references)

Abstract: The first recorded prohibition against tobacco use was in Peru in 1586, where a papal decree was issued declaring that it was a sin for any priest to use tobacco before celebrating or administrating mass. By the late 19th century tobacco use was widespread, but most people only used a small amount in the forms of pipes, cigars, chewing tobacco, or a pinch of snuff. Cigarettes replaced cigars and chewing tobacco mainly during World War I, when free cigarettes were widely distributed to the soldiers. Both Thomas Edison and Henry Ford voiced concerns about the detrimental health effects of cigarettes. In the 1930s, efforts to limit smoking were fading, partially as a result of the failure of alcohol prohibition. Cigarette advertisers and marketers were able to associate smoking with images of health, athletic performance, wealth, and social standing, which helped fuel over 30 years of the growth of cigarette smoking. A major milestone in tobacco control was the 1964 US Surgeon General's Report on Smoking and Health. In the United States, due to declines in cigarette smoking, cancers caused by smoking began to decline in the 1980s. Prescription-only nicotine gum for smoking cessation was introduced in 1984 and the nicotine patch in 1992. In 1996, the FDA allowed over-the-counter sales of nicotine gum, patches, and other nicotine replacement therapies.

> The past 50 years has witnessed a dramatic change in attitudes toward and use of tobacco by Americans that has resulted in recent declines in the incidence of lung cancer. Most public health scholars believe that this change has been accelerated by public policy interventions to reduce tobacco use. The research literature suggests that the most potent demand reducing influences on tobacco use have been increasing the financial cost of using tobacco products primarily through taxation, smoke-free policies, comprehensive advertising bans, and paid counter-advertising. Other policies such as the requirement of warning labels on tobacco products, restrictions on tobacco sales to minors and increasing access to stop smoking services appear to have had less direct impact on cigarette consumption, although the potential impact of these policies may not have been fully realized to date.

Reference Number: 70303

EDWARDS, R. "THE PROBLEM OF TOBACCO SMOKING." *British Medical Journal*, 328: 217–219, 2004 (6 figures, 6 references)

Abstract: Typically, a smoking epidemic in a population develops in 4 stages. First a rise, then a decline in smoking prevalence, followed 20 to 30 years later by another rise, then a decline in smoking related diseases.

Table: Stage of the Smoking Epidemic in Various Regions

Stage of Smoking Epidemic	Regions
1	Sub-Saharan Africa
2	China, Japan, Latin America, North Africa
3	Eastern Europe, Southern Europe
4	Western Europe, North America, Australia

Source: Adapted from Edwards, 2004

Cigarette smoking is the single biggest avoidable cause of death and disability in developed countries. Smoking is now increasing rapidly throughout the developing world and is one of the biggest threats to current and future world health. For most smokers, quitting smoking is the single most important thing they can do to improve their health. Encouraging smoking cessation is one of the most effective and cost effective things that doctors and other health professionals can do to improve health and prolong their patients' lives.

Reference Number: 70304

CUMMINGS, K.M., AND R.N. PROCTOR. "The Changing Public Image of Smoking in the United States: 1964–2014." *Cancer Epidemiology, Biomarkers and Prevention*, 23: 32–36, 2014 (52 references)

Abstract: The prevalence of cigarette smoking grew rapidly in the United States (and other developed countries) during the early part of the 20th century due to the development of rapid, automatic cigarette manufacturing machines and mass marketing and media. In 1953, 47% of the US adult population were cigarette smokers (58% of men, 36% of women) and half of all physicians. Due to health concerns, cigarette sales declined and then rebounded as filtered cigarettes were introduced and promoted as safer. The tobacco industry realized in the 1930s that filters were not really filters, as there is no such thing as clean smoke, but smokers were led to believe they were safer. In the years after the 1964 Surgeon General's Report, the tobacco industry increased their PR and disinformation campaigns suggesting that there was no real link between smoking and any disease. The tobacco industry continues to work against efforts by governments to adopt policies that would effectively limit cigarette marketing and protect public health.

Tobacco use behaviors have changed significantly over the past century. After a steep increase in cigarette use rates over the first half of the 20th century, adult smoking prevalence rates started declining from their peak reached in 1964. Improved understanding of the health risks of smoking has been aided by the United States Surgeon General's Reports, issued on a nearly annual basis starting in 1964. Among the many forces driving down smoking prevalence were the recognition of tobacco use as an addiction and cause of cancer, along with concerns about the ill-effects of breathing secondhand smoke. These factors contributed to the declining social acceptance of smoking, especially with the advent of legal restrictions on smoking in public spaces, mass media counter-marketing campaigns, and higher taxes on cigarettes. This paper reviews some of the forces that have helped change the public image of smoking, focusing on the 50 years since the 1964 Surgeon General's report on smoking and health.

Reference Number: 70305

SCHNEIDER, N.K., AND S.A. GLANTZ. "'Nicotine Nazis Strike Again': A Brief Analysis of the Use of Nazi Rhetoric in Attacking Tobacco Control Advocacy." *Tobacco Control*, 17: 291–296, 2008 (4 figures, 97 references)

Abstract: In the 1960s, the Tobacco Institute encouraged articles that compared non-smokers and tobacco control measures to Hitler and Nazi regulations. In the 1980s, the term "health fascism" was employed. In 1992, smokers' rights advocates dressed in Nazi SS uniforms and thanked England's secretary of state for continuing the Nazi policies during a demonstration against the UK National No Smoking Day. The Nazi card was also played to fight against smoke-free environments.

Nazi and health fascism rhetoric has been used and promoted for decades by the tobacco industry around the world. Against the background of Proctor's suggestion that the use of Nazi rhetoric would increase with stronger tobacco control efforts, the current use in Germany is neither new nor a purely German phenomenon, but probably a sign of increasing strength of Germany's tobacco control movement. The use of Nazi and health fascism rhetoric can be regarded as part of an institutionalised practice of the tobacco industry and its front groups to discredit tobacco control activities and prevent the introduction of effective policies. "Playing the Nazi card" is an established strategy developed first in the United States and the United Kingdom, then widely used around the world, so far, predominantly outside countries with a Nazi or fascist history. This imagery is now simply being applied in Germany.

Reference Number: 70306

MARCON, A., G. PESCE, L. CALCIANO, V. BELLISARIO, S.C. DHARMAGE, J. GARCIA-AYMERICH, T. GISLASSON, J. HEINRICH, M. HOLM, C. JANSON, D. JARVIS, B. LEYNAERT, M.C. MATHESON, P. PIRINA, C. SVANES, S. VILLANI, T. ZUBERBIER, C. MINELLI, AND S. ACCORDINI. "Trends in Smoking Initiation in

Europe Over 40 Years: A Retrospective Cohort Study." *Public Library of Science One*, 13: e201881 (14 pp), 2018 (2 tables, 3 figures)

Abstract: The Aging Lungs in Europe Cohorts consortium examined the trends in smoking initiation between 1970 and 2009 in 119,104 subjects (median age 33–52 years) in 17 European countries. In all regions combined, smoking initiation before 1970 peaked at age 18 years in males and 19 years in females and in 2000 peaked at 16 and 15 years, respectively. In young adults (aged 21–35 years) initiation rates were below 2% and were stable or declining over time. Tobacco use in young adolescents is strongly influenced by friends and siblings. Early smoking initiation is often associated with early puberty, perhaps due to a gap between physical and social maturity. The age of puberty, regardless of gender, has been decreasing in Europe.

An increase in tobacco prices may be a successful strategy for young people, who are the most price-sensitive sector of the population. Current policies on prices focus on conventional (boxed) cigarettes, and they seem to induce a shift to cheaper products, including hand-rolled cigarettes. Equalising taxation levels of all tobacco products, as advocated by the Framework Convention on Tobacco Control, may limit smoking initiation especially during early adolescence. Smoking in youth can be reduced by preventing tobacco sales to minors through law enforcement, which may include warnings and fines to non-compliant retailers. Since young adolescents generally obtain cigarettes from their peers, restricting sales to minors can break the chain of tobacco supply to them. Reducing exposure to cigarette advertising (including pack display at the point of sale), and implementing plain packaging in combination with pictorial health warnings, makes youth less likely to try smoking or to smoke again after experimentation. It is still unclear whether smoking bans can reduce the prevalence of smoking, but smoke-free places may produce an environment that is less favourable to experimentation in young adolescents.

Reference Number: 70307

SWAYAMPAKALA, K., J.F. THRASHER, D. HAMMOND, H-H. YONG, M. BANSAL-TRAVERS, D. KRUGMAN, A. BROWN, R. BORLAND, AND J. HARDIN. "Pictorial Health Warning Label Content and Smokers' Understanding of Smoking-Related Risks—A Cross Country Comparison." *Health Education Research*, 30: 35–45, 2015 (2 tables, 2 figures, 30 references)

Abstract: Canada was the first county to implement pictorial health warning labels (HWLs) on cigarettes packaging in 2001. In 2012, Australia, Canada, and Mexico implemented new pictorial HWLs. A longitudinal online interview was conducted of 1,000 adult smokers in each country in 2012–2013 as to their recognition of the health risks associated with smoking cigarettes.

Table: Disease/Toxic Constituent Health Warning Labels in Australia, Canada, and Mexico in 2012–2013

Disease/Toxic Constituent	Australia	Canada	Mexico
Heart disease	+	+	+
Emphysema	+	+	+
Bladder cancer	+	+	–
Blindness	+	+	–
Impotence	–	–	–
Gangrene	+	–	+
Cyanide	+	–	–
Benzene	–	+	–
Nitrosamines	–	–	–
Radioactive polonium 210	–	–	+
Carbon monoxide	+	+	+
Formaldehyde	–	+	+

Source: Adapted from Swayampakala et al, 2014

This longitudinal study provides evidence for the specific influence of HWL content on adult smokers' knowledge of health effects and toxic constituents of tobacco smoke, even as new HWL content is included after populations are exposed to prominent pictorial HWLs for a number of years. Smokers from countries where HWLs contained information about specific tobacco-related diseases or toxins showed higher knowledge about those topics than smokers from countries where HWLs do not include that specific content. This was most evident for blindness in AU, gangrene in AU and MX, benzene in CA and radioactive polonium in MX.

Reference Number: 70308

HALL, M.G., J.R. MENDEL, S.M. NOAR, AND N.T. BREWER. "Why Smokers Avoid Cigarette Pack Risk Messages: Two Randomized Clinical Trials in the

United States." *Social Science and Medicine*, 213: 165–172, 2018 (3 tables, 3 figures, 64 references)

Abstract: A random sample of 2,149 adult smokers were assigned to receive either pictorial warnings (intervention) or test-only (control) health warnings for their cigarette packs for 4 weeks between 2014 and 2015. Another random sample of 719 adult smokers were assigned cigarette pack messages about toxic chemicals in cigarettes (intervention) or the harm of littering cigarette butts (control) for 3 weeks between 2016 and 2017. After the implementation of larger pictorial warnings on cigarette packs in Australia in 2012, it was found that smokers were more likely to place their packs face down and to conceal the new warnings using cases and other means. Smokers avoid cigarette pack risk messages because they evoke aversive types of emotions.

Policymakers should be encouraged by evidence showing that strong cigarette pack risk messages, especially those with images, are likely to encourage smoking cessation and therefore prevent death and disease. However, some smokers will inevitably avoid health warnings and other cigarette pack risk messages. We argue that, in the context of cigarette pack risk messages, avoidance is not maladaptive defensive processing but instead a sign of deeper processing. In other words, smokers avoid warnings precisely because they are hard-hitting and elicit productive types of negative emotions, and avoidance is associated with more quitting-related behaviors.

Reference Number: 70309

CAMPBELL, B.K., T. LE, D. YIP, K.B. GRIFFIN, N.R. GUBNER, AND J.R. GUYDISH. "Sex Differences in Graphic Warning Label Ratings by Addictions Clients." *Tobacco Regulatory Science*, 5: 3–14, 2019 (1 table, 1 figure, 62 references)

Abstract: Four hundred and thirty-two female and 449 male smokers in 24 drug addiction treatment programs in California rated three FDA graphic warning labels (GWL) about smoking risks.

The FDA has sought to strengthen evidence that GWLs will reduce smoking prevalence. This should include investigation of GWLs with specific sub-populations who face tobacco-related disparities, such as women with additional SUDs, a group with high smoking rates resistant to smoking cessation. The current study found that female smokers in SUD treatment did not respond differentially to GWLs of women and/or babies but did rate GWLs more strongly overall than male smokers. Results support the use of graphic images including babies, women, and disease images, when developing warning labels for female smokers, including women with additional SUDs. Our findings contribute to the evidence base for implementation of GWLs on cigarette packaging in the United States as one component of a comprehensive tobacco control policy that may reduce smoking cessation disparities.

Reference Number: 70310

FRANCIS, D.B., N. MASON, J.C. ROSS, AND S.M. NOAR. "Impact of Tobacco-Pack Pictorial Warnings on Youth and Young Adults: A Systematic Review of Experimental Studies." *Tobacco Induced Diseases*, 17: 41 (11 pp), 2019 (2 tables, 2 figures, 64 references)

Abstract: A review was conducted of 28 studies of pictorial warning labels (PWL) on cigarette packs and 2 studies of PWL on smokeless tobacco packs. PWLs were found to increase attention, thinking about harms, fear, disgust, more negative pack attitudes, and intention to quit smoking. PWLs were more effective than text warnings.

Despite evidence to the contrary, young people maintain an optimistic bias towards smoking. Young tobacco users often do not connect tobacco use to long-term health problems. In one study, 60% of adolescents smokers believed they could smoke for a few more years and then quit with no adverse health effects, compared to 48% of adult smokers. Potentially due to its long-term, far-off consequences, young tobacco users do not see the link between tobacco use and many chronic diseases. That said, countries have a limited set of warnings, despite having adolescent, young adult, and adult populations. To maximize the impact of warnings on youth and young adults, we should ensure that we implement content that resonates with younger population groups, in addition to adult smokers. While the main target of PWLs may be adult smokers (and cessation behavior), young people are a critical secondary audience for tobacco warnings.

Reference Number: 70311

WADE, W.S., AND K. WHITE. "The Package as a Weapon of Influence: Changes to Cigarette Packaging Design as a Function of Regulatory Changes in Canada." *Tobacco Prevention and Cessation*, 6: (4 pp), 2020 (1 figure, 13 references)

Abstract: In 2011, the Tobacco Products Labeling Regulation—Cigarettes and Little Cigars (TPLR-CLC) was enacted in Canada. The regulations required that all tobacco packages sold in Canada display a graphic warning covering 75% of the package (up from 50% in 2008). A sample of 1,689 tobacco products from 92 brands sold between 1999 and 2015 were assessed to determine the tobacco companies' response to the legislation. Between 1999 and 2015, the prevalence of cigarette smoking in Canada decreased from 25% to 13%.

Table: Changes in Cigarette Packaging Design Before and After Increased Regulations in Canada (1999–2017)

Cigarette Packaging Characteristics	Pre-Regulation	Post-Regulation
High color saturation	35.8%	91.4%
Flip-top	36.4%	57.1%
Average number of cigarettes in a pack	25.2	22.2
Male figure brands on packs	35.1%	1.4%
Female figure brands on packs	7.8%	11.2%
No logo	18.6%	32.5%

Source: Adapted from Wade and White, 2020

For example, an increase in the use of flip-top packaging can be seen as more modern, elegant, or unique, which serves to increase purchasing preference in younger people, especially younger women, who have been shown to prefer smaller packaging styles. This preference may also explain why we observe that the average number of cigarettes per package has decreased over time. The increased usage of black and yellow as focal colors as well as increased color saturation is likely an effort to attract attention, as research shows that black and yellow are effectively used on tobacco packaging for this purpose. Moreover, these attributes may serve to draw attention away from the unsightly health labels and to differentiate the brand. The increased usage of brand variant labels and female figures on packaging may be an effect of strategies pursued by tobacco companies to influence positive consumer perceptions of their products in light of increasingly restrictive packaging regulations. The decreased usage of males in iconography suggests that manufacturers may be targeting women.

Reference Number: 70312

LEE, J.G.L., A. RICHARDSON, S.D. GOLDEN, AND K.M. RIBISL. "Promotions on Newport and Marlboro Cigarette Packages: A National Study." *Nicotine and Tobacco Research*, 19: 1243–1247, 2017 (1 table, 44 references)

Abstract: The Tobacco Industry marketing has been shown to contribute to death and disability by promoting initiation and continued use of tobacco. The cigarette pack itself is a marketing tool. Promotions on the outside of the pack are used to attract non-smokers or promote brand loyalty. In 2012, cigarettes packs were purchased by auditors in 2,147 stores in the United States. A total of 1,090 Marlboro Red and 1,052 Newport Green cigarette packages were purchased. Marlboro packs were 2× as likely to have promotions as Newport packs (31.7% vs 14.7%). Most of the exterior promotions were for contests (>80%), while almost all interior Marlboro promotions were for a discounted price.

Use of promotions on the interior and exterior of cigarette packs are a mechanism that the tobacco industry uses to sell its products and should be continually assessed for their influence on consumer behavior.

Reference Number: 70313

SMITH, K.C., K. WELDING, C. KLEB, C. WASHINGTON, AND J. COHEN. "English on Cigarette Packs from Six Non-Anglophone Low- and Middle-Income Countries." *International Journal of Public Health*, 63: 1071–1079, 2018 (4 tables, 3 figures, 39 references)

Abstract: Low- and middle-income countries are vital to Big Tobacco. The cigarette pack is a key to cigarette marketing which is on display when cigarettes are purchased and can be seen up to 7,000× a year by pack-a-day smokers. Over 1,000 cigarette packs from Bangladesh, Brazil, China, Egypt, Ukraine, and Vietnam were analyzed. Over two-thirds of the cigarette packages in these non-English-speaking countries studied had some English printed on them. English was used mainly to communicate product features such as size, strength, and flavor, and to convey luxury, quality, and modernity.

Table: Percent of English Brand Names and Any English on Cigarette Packs in Various Low- and Middle-Income Countries

Country	English Brand Names	Any English on Front of Cigarette Packs
Bangladesh	52%	93%
Brazil	77%	82%
China	13%	35%
Egypt	52%	95%
Ukraine	48%	76%
Vietnam	56%	91%
Total	41%	67%

Source: Adapted from Smith et al, 2018

Restricting English use should be incorporated into plain packaging policy to protect populations from deceptive branding practices, specifically presenting cigarettes as an aspirational product.

Reference Number: 70314

VUOLO, M., B.C. KELLY, AND J. KADOWAKI. "Impact of Total Vending Machine Restrictions on US Young Adult Smoking." *Nicotine and Tobacco Research*, 18: 2092–2099, 2016 (2 tables, 1 figure, 48 references)

Abstract: The US Family Smoking Prevention and Tobacco Control Act (TCA) of 2009 banned tobacco vending machine sales except in adult-only facilities. Some US counties/cities have enacted total bans on cigarette vending machines, but currently only 10% of the population is affected by a total ban. The effect of a total cigarette vending machine ban on adult and youth smoking was determined by comparing total cigarette vending machine ban areas to those without total bans.

We find that total vending machine restrictions decrease any recent smoking (OR = 0.451; p <.01), net of other covariates. Though the passage of a restriction does not alter an individual's smoking over time, living longer in an area that has a restriction lowers the propensity that an individual will smoke at all (OR = 0.442; p <.05). We find no effect of total vending machine restrictions on smoking a pack daily.... Total vending machine restrictions appear to be an effective, yet highly underutilized, means of tobacco control.

Reference Number: 70315

BOTELLO-HARBAUM, M.T., D.L. HAYNIE, R.J. IANNOTTI, J. WANG, L. GASE, AND B. SIMONS-MORTON. "Tobacco Control Policy and Adolescent Cigarette Smoking Status in the United States." *Nicotine and Tobacco Research*, 11: 875–885, 2009 (3 tables, 52 references)

Abstract: The effects of state-level tobacco control policies such as smoke-free laws, cigarette pricing, and age limits on purchasing were determined in 13,339 grade 6–10 students in 39 US states between 2001 and 2002.

> Compared with students living in states with strict regulations, those living in states with no or minimal restrictions, particularly high school students, were more likely to be daily smokers. These effects were somewhat reduced when logistic regressions were adjusted for sociodemographic characteristics and cigarette price, suggesting that higher cigarette prices may discourage youth to access and consume cigarettes independent of other tobacco control measures.
>
> ... Strict tobacco control legislation could decrease the potential of youth experimenting with cigarettes or becoming daily smokers. The findings are consistent with the hypothesis that smoking policies, particularly clean indoor air provisions, reduce smoking prevalence among high school students.

Reference Number: 70316

FAULKNER, P., D.G. GHAHREMANI, R. TYNDALE, C.M. COX, A.S. KAZANJIAN, N. PATERSON, S. LOTFIPOUR, G.S. HELLEMANN, N. PETERSEN, C. VIGIL, AND E.D. LONDON. "Reduced-Nicotine Cigarettes in Young Smokers: Impact of Nicotine Metabolism on Nicotine Dose Effects." *Neuropsychopharmacology*, 42: 1610–1618, 2017 (1 table, 4 figures, 47 references)

Abstract: Tobacco smoking remains a leading contributor to preventable disease and death. Smoking cigarettes that deliver less than 1 mg nicotine (less than conventional cigarettes) reduces nicotine dependence and promotes smoking cessation when combined with transdermal nicotine. Forty-six young adult smokers (23 men, mean age 22 years) were tested on cigarettes that delivered 0.027, 0.110, 0.231, or 0.763 mg nicotine and on conventional cigarettes. Plasma nicotine and metabolites (3-hydroxycotinine/cotinine) were determined by LC-MS/MS. Cigarettes with a nicotine of 0.231 mg or less did not alleviate withdrawal deficits in sustained attention.

This study of young smokers provides the first evidence that smoking-induced reductions in craving and withdrawal depend on nicotine dose in normal but not slow metabolizers. Improvement in sustained attention during acute abstinence and ratings of cigarette characteristics both depend on nicotine dose, regardless of metabolizer status. These findings have bearing on the use of reduced-nicotine cigarettes to facilitate smoking cessation and on policy regarding regulation of nicotine content in cigarettes.

Reference Number: 70317

MILLER, L.S., W. MAX, H-Y. SUNG, D. RICE, AND M. ZARETSKY. "Evaluation of the Economic Impact of California's Tobacco Control Program: A Dynamic Model Approach." *Tobacco Control*, 19: i68–i76, 2010 (1 table, 4 figures, 59 references)

Abstract: The long-term net economic impact of the Tobacco Control Program in California was determined using a series of dynamic models of smoking-caused mortality, morbidity, health status, and health care expenditures. The California Tobacco Control Program (CTCP) was established in 1989 with an annual budget of $100 million a year funded mainly by a 25 cent per pack increase in tobacco taxes. It is estimated that CTCP resulted in over 700,000 person-years of life saved and over 150,000 person-years of treatment in 1990. The net health care savings and years of life saved was equivalent to $22 billion, or $107 billion in 1990 dollars. These estimates did not include the impact of CTCP on SHS and its adverse health effects on non-smokers.

Tobacco control programmes are costly. However, the benefits of the programmes are substantial and continue to accrue for many years. Although those who are persuaded not to smoke will live longer, have better health status and require additional healthcare resources during their additional years of life, this impact is outweighed by the value of additional years of life and better health. Public health programmes need to be evaluated with healthcare costs, additional years of life and improved health considered as important outcomes.

Reference Number: 70318

SMITH, E.A., AND R.E. MALONE. "An Argument for Phasing Out Sales of Cigarettes." *Tobacco Control*, 29: 703–708, 2019 (70 references)

Abstract: An endgame plan is discussed about phasing out the sale of cigarettes, initially in areas with few smokers. Combustible tobacco products are the tobacco industry's single largest source of profit and power.

Ending sales deprives the industry of income and represents the most concrete way to denormalise the product. Equally importantly, ending sales could reduce the industry's influence in government and policy-making by challenging its legitimacy. Notably, however, defining the endpoint as ending cigarette sales does not require that no one could or would ever use tobacco (or other nicotine products). Tobacco is a product that has been used in some form for centuries and some use (both ritual and addiction based) is likely to continue; however, it is only since the commercialisation of cigarettes that the problems its use causes have reached epidemic proportions. Not expecting policies to achieve total "prohibition" or zero prevalence recognises this.

Objectors to proposed endgame-advancing policy proposals commonly refer to black markets or the failures and unintended consequences of alcohol prohibition. But these objections, often supported by industry-funded research, typically assume exaggerated proportions of illicit trade. Furthermore, under Prohibition, alcohol use prevalence was high, many users perceived their own use to be unproblematic, and possession and use were criminalised, leading to widespread law-breaking and reduced respect for law enforcement. In contrast, phasing out cigarette sales in a jurisdiction with already-low smoking prevalence (without reference to possession and use) is quite different. Many smokers already perceive their own use to be problematic (eg, they want to quit) and thus might be less likely to seek out illicitly sold cigarettes. Furthermore, eliminating ready access to cigarettes could enhance success in cessation, since smokers experience stronger cravings when they expect to be able to smoke in the near future. While some illegal underground sales (whether home-grown or through neighbouring communities) are likely to occur, these types of activities seem unlikely to represent a black market so substantial that it would become worse for public health than the status quo.

Reference Number: 70319

MYERS, M.L. "The FCTC's Evidence-Based Policies Remain a Key to Ending the Tobacco Epidemic." *Tobacco Control*, 22: i45–i46, 2013 (15 references)

Abstract: Many countries have implemented the Framework Convention on Tobacco Control (FCTC) and have dramatically reduced tobacco consumption. The FCTC measures to control tobacco use include higher tobacco taxes, strong warning labels, smoke-free air laws, aggressive public education, and marketing/advertising restrictions. In the last 7 years, 14 countries in South America alone have adopted 100% smoke-free laws.

> For tobacco control measures to have their desired impact, it is critical to first build a climate that enables government to act as boldly as these new ideas propose. Too many countries have not yet done so. Australia was able to adopt plain packaging for cigarettes in November 2012 precisely because it had built public support for strong tobacco control measures over the years. To succeed, tobacco control efforts must be preceded by a concerted effort to educate the public about the magnitude of the harms caused by tobacco and why the proposed policy is needed. Had Australia acted without first building public support, the tobacco industry's attacks—including the attack on the Australian government as a "national nanny"—could well have been successful and could have set back long-term progress.

Reference Number: 70320

CHUNG-HALL, J., G.T. FONG, P. DRIEZEN, AND L. CRAIG. "Smokers' Support for Tobacco Endgame Measures in Canada: Findings from the 2016 International Tobacco Control Smoking and Vaping Survey." *Canadian Medical Association Journal Open,* 6: E412–E422, 2018 (5 tables, 1 figure, 62 references)

Abstract: The smoking rate in Canada among persons 15 years of age or more declined from 15% in 2013 to an all-time low of 13% in 2015. Canada implemented pictorial health warnings in 2001 and banned all flavors in cigarettes (except menthol, which was banned in 2017). A survey was conducted on 1,390 cigarette smokers and 1,825 dual users (cigarettes and ECs) in 2016 to determine support for tobacco countermeasures. Support was highest for reducing nicotine content (70.2%), raising the legal age for purchase (69.8%), and banning marketing (58.5%) for cigarettes. Regarding EC policies, support was highest for restricting age of vaping (86.1%), restricting nicotine content (64.9%), banning vaping in smoke-free areas (63.4%), and restricting marketing (54.8%) for vaping.

> Our findings show that a majority of Canadian smokers support diverse tobacco endgame measures. Our baseline estimates suggest that

innovative endgame strategies that go beyond current measures are likely to be well received by Canadians, including smokers, who stand to be most affected by new policies. As endgame policies are implemented in Canada, it will be important to study their impact on tobacco use, particularly among vulnerable populations such as Indigenous peoples and groups with low socioeconomic status.

Reference Number: 70321

ARAÚJO, M.C.B., AND M.F. COSTA. "From Plant to Waste: The Long and Diverse Impact Chain Caused by Tobacco Smoking." *International Journal of Environmental Research and Public Health*, 16: 2690 (9 pp), 2019 (1 figure, 66 references)

Abstract: Cigarettes usually have two components: processed tobacco (stick) and a synthetic filter, both of which are wrapped in paper. Cellulose acetate is the main component of the filters. It is a synthetic polymer made from cellulose with the addition of acetic anhydride and acetate, and includes plasticizing compounds. The tobacco plant easily accumulates metals from the soil into the leaves. Residues of pesticides, insecticides, and fungicides also are accumulated in the leaves.

Effective mitigation of problems requires four main lines of action. The first would be the control and supervision of the tobacco production process, especially during planting and handling of the leaves, in order to minimize the contamination of nearby water resources, and protect the health of the farmers by demanding the use of personal protective equipment. The second would be the unrestricted ban on smoking in all public spaces, including beaches, imposing fines in cases of noncompliance with the legislation. This action could be an effective measure for reducing the risk posed to passive smokers and the production of waste, thus minimizing impacts on the biota. The third action would be the promotion of educational campaigns not only addressing the risks for smokers (active and passive) and the environment, especially the aquatic environments and their biota, but also warning about the economic costs incurred with cleaning the environments and treating smoking-related diseases. Finally, it would be necessary to increase taxes on cigarette sales, in order to discourage consumption and combat smuggling.

Reference Number 70322

HAMMOND, D., G.T. FONG, M.P. ZANNA, J.F. TRASHER, AND R. BORLAND. "Tobacco Denormalization and Industry Beliefs Among Smokers from Four Countries." *American Journal of Preventative Medicine*, 31: 225–232, 2006 (3 tables, 45 references)

Abstract: Tobacco social denormalization is a public health strategy that endeavors to change broad social acceptance and normalization of smoking into that of an abnormal practice and not a cool or desirable behavior. Tobacco industry denormalization focuses on the tobacco industry and its conduct and to raise people's awareness of the industry responsibility for disease and its manipulative tactics.

Table: Percentage of Smokers in Canada, the United States, the United Kingdom, and Australia Who Agree with Various Statements About Smoking and the Tobacco Industry

	% Agree			
Smokers' Belief	Canada	United States	United Kingdom	Australia
Society disapproves of smoking	88%	78%	77%	82%
There are fewer and fewer places I feel comfortable smoking	84%	78%	77%	84%
People who are important to me believe I should not be smoking	90%	90%	85%	89%
Tobacco companies can be trusted to tell the truth	19%	19%	25%	15%
Tobacco companies should take responsibility for the harm caused by smoking	51%	45%	38%	45%
Tobacco companies had tried to convince the public that there is little or no health risk from second-hand smoke	45%	52%	42%	51%

Source: Adapted from Hammond et al, 2006

Social denormalization was associated with noticing anti-smoking information, warning labels, and ETS restrictions. Comprehensive ETS restrictions in bars and restaurants may be particularly effective in shaping beliefs about smoking by breaking the association between smoking, drinking, and exciting lifestyles—the very associations portrayed in tobacco marketing. In addition, warning labels and anti-smoking media may influence social norms by communicating the health effects of smoking in a highly visible manner to smokers and those around them. Communicating such information publicly—as opposed to communicating such information in an individual setting such as a doctor's office—may be particularly effective in reducing the perceived acceptability of smoking.

Several interactions were observed between social denormalization and policy exposure. First, noticing warning labels was most strongly associated with social denormalization in Canada. This suggests that the graphic Canadian warnings may be a more powerful denormalizing force than the text warnings present in Australia, the UK, and the United States.

Reference Number 70323

MATHENY, J.D., E.M. STEVENS, S. CHEN, B.A. CHRISTIANSEN, S.D. KOWITT, A. OSMAN, AND D.J. VIDRINE. "The RICO Verdict and Corrective Statements: Catalysts for Policy Change?" *Tobacco Regulatory Science*, 5: 206–228, 2019 (3 tables, 110 references)

Abstract: In 2006, a US federal court found that Philip Morris and other major tobacco companies were in violation of the Racketeer Influenced and Corrupt Organization (RICO) Act. In her 1,682 page ruling, Judge Kessler cited the tobacco companies with 145 distinct acts of racketeering and concluded that fraudulent conduct has permeated all aspects of their operations. The court ordered the tobacco industry to disseminate corrective statements (see Appendix 2) through newspaper, television, package onsets, point of sales, and corporate websites. A website survey was conducted in May 2017 of 1,006 exposed adults and 1,004 unexposed adults regarding the tobacco industry. The exposed group read the RICO statements and conclusion before answering the survey. The unexposed group did not read the RICO information until after filling in the survey. The exposed group was less likely to agree with the statement that lawmakers should trust tobacco companies as much as they do other companies and were more likely to support tobacco control measures. Exposing the general population to the RICO decision should help tobacco industry denormalization.

Table: The Related Federal Court RICO Findings About the Tobacco Companies

United States v. Philip Morris USA Inc.
• Violated civil racketeering law (engaged in an organized conspiracy to commit fraud)
• Committed fraud
• Are likely to continue to commit fraud
• Denied that they control the level of nicotine to create and sustain addiction
• Suppressed and concealed scientific research
• Marketed low tar and light cigarettes as less harmful despite knowing they were not
• Marketed cigarettes to young people to replace smokers who die or quit smoking
• Denied that secondhand smoke harms non-smokers
• Denied the health consequences of smoking

Source: Adapted from Matheny et al, 2019

The scope of tobacco industry influences on public policy has been extensive. Tobacco companies have sought to defeat—separately and in all 50 states—legislation to restrict smoking inside workplaces, raise tobacco taxes, limit tobacco marketing, advance prevention programs or research, and reduce youth access to tobacco. When outright defeat cannot be achieved, the companies work to delay or weaken such measures. Their tactics include contributing to politicians' election campaigns, disseminating public relations campaigns, creating controversy over established facts, using front groups, hiring lobbyists, and "preempting" strong legislation. The companies have been successful at promoting preemption of effective local-level tobacco policies in many states. Most of these preemptive clauses remain in effect.

Actively monitoring and exposing tobacco industry misconduct enables effective tobacco control.

CHAPTER 8

Other Effects of Tobacco Products/Nicotine

"Impaired ability to recover from injuries in smokers may add to the death toll. This includes poor wound healing, delayed bone union, decreased blood flow, diminished tissue oxygenation, and diminished immunity, increased wound infections, and more severe pulmonary complications among injured smokers. All of these factors increased morbidity and make complications more serious in smokers, leading to increased mortality."

—Wen et al, "Excess Injury Mortality Among Smokers: A Neglected Tobacco Hazard" (2005)

Nicotine products cause many other harmful effects to humans, in addition to the well-known cancers, heart, and lung diseases. Their role in impairment of driving, other injuries, fires, psychiatric disorders, sexual dysfunction, and suicides is often not explored. Nicotine is an overlooked drug in postmortem medicolegal cases as a cause of overdose death, suicide, and injuries.

8.01 DEPRESSION AND ANXIETY

"We observed a fourfold higher risk of depression for heavy smokers compared with never smokers. Failure to explain the smoking–depression association with other plausible alternatives might reflect a direct causal influence of smoking on depression."

—Klungsøyr et al, "Cigarette Smoking and Incidence of First Depressive Episode: An 11-Year, Population-Based Follow-Up Study" (2006)

Contrary to common misperceptions, encouraged in part by the tobacco industry, smoking does not relieve depression and anxiety but actually causes it, especially in adolescents (80101–6). It is recommended that smoking cessation should be a focus of depression prevention programs (80103).

Depressed people are more likely to be smokers and more unlikely to stop smoking. Reduced nicotine cigarettes (to a minimally addictive level) did not increase depressive symptoms (80107).

Not only cigarettes are associated with depression, anxiety, and ADHD, but e-cigarettes (ECs), waterpipes, and other tobacco products as well, and they should all should be a focus for the mental health population (80108).

Secondhand smoke exposure at home is associated with depressive symptoms in male students (80109)

Reference Number: 80101

SLOMP, F.M., T.S. BARA, G.L. PICHARSKI, AND M.L. CORDEIRO. "Association of Cigarette Smoking With Anxiety, Depression, And Suicidal Ideation Among Brazilian Adolescents." *Neuropsychiatric Disease and Treatment*, 15: 2799–2808, 2019 (3 tables, 1 figure, 56 references)

Abstract: The association between depression and cigarette smoking was determined in 988 students (aged 11–17 years) in 82 schools in Brazil. There were 240 smokers (24%) who had smoked at least 6 cigarettes per day (CPD) in the past years. Smokers had higher anxiety and depression scores than non-smokers and were 2.05× as likely to have suicidal ideation than non-smokers.

Smoking was found to be associated with anxiety symptoms, depressive symptoms, and poor school performance in Brazilian adolescents; and female smokers reported more suicidal ideation than male smokers.

Reference Number: 80102

CHOI, W.S., C.A. PATTEN, J.C. GILLIN, R.M. KAPLAN, AND J.P. PIERCE. "Cigarette Smoking Predicts Development of Depressive Symptoms Among US Adolescents." *Annals of Behavioral Medicine*, 19: 42–50, 1997 (3 tables, 3 figures, 77 references)

Abstract: A 4-year longitudinal study was conducted examining the relationship between smoking and depression in 6,863 adolescents who had no depressive symptoms at baseline (1989). About 15.3% of the girls developed depressive symptoms compared to 18.1% of the boys. By the end of the study, those who were current established smokers at baseline had a rate of depressive symptoms of 17.3% compared to 10.2% of the non-smokers. Smoking status was the most significant predictor of depression in both girls and boys. Adolescents who stopped smoking between the two interviews were less likely to have depressive symptoms than those who continued to smoke or started smoking.

> Adolescent cigarette smoking may have marked health consequences in terms of depressive symptoms. The reduction of cigarette smoking among adolescents should be a focus of depression prevention interventions. In addition, the development of gender-specific components of prevention interventions may be warranted.

Reference Number: 80103

MORRELL, H.E.R., L.M. COHEN, AND D.E. MCCHARGUE. "Depression Vulnerability Predicts Cigarette Smoking Among College Students: Gender and Negative Reinforcement Expectancies as Contributing Factors." *Addictive Behaviors*, 35: 607–611, 2010 (2 tables, 3 figures, 33 references)

Abstract: The association between vulnerability to depression, smoking behavior, and other factors was studied in 1,214 college students (median age 18 years) in the United States. Approximately 30% of the male students and 29% of the female students reported that they were current smokers. Some college students may be more likely to smoke because they expect smoking and nicotine will relieve a negative mood. A history of depression was associated with a greater risk of smoking in female but not male students.

Table: Incidence of Depression and Anhedonia (inability to feel pleasure) by Number of Cigarettes Per Day in College Students

Cigarettes Per Day	Depression	Anhedonia
0	26.1%	23.9%
<5	29.9%	30.8%
6–10	41.8%	38.2%
11–15	32.3%	38.7%
16–19	60.0%	60.0%
20	40.0%	27.8%
>20	75.0%	50.0%

Source: Adapted from Morrell et al, 2010

Reference Number: 80103

Results from the current study indicate that self-reported depression vulnerability is associated with a greater likelihood of smoking in college students. This relationship was moderated by gender, such that it only remained significant among females. These findings suggest that depression vulnerability may warrant attention as a potential risk factor for smoking among female college students. Female non-smoking college students who are being treated for depression may benefit from receiving information about the dangers of smoking from their health care providers, such as physicians and mental health care professionals, in order to reduce the likelihood of smoking initiation among this potentially vulnerable group. Female college students who smoke and who wish to quit may benefit from cessation programs that emphasize management of depressive symptoms. This strategy is especially important to consider, because the nicotine withdrawal syndrome includes depressed mood as a symptom.

Reference Number: 80104

GOODWIN, R.D., K.L. LAVOIE, A.R. LEMESHOW, E. JENKINS, E.S. BROWN, AND D.A. FEDORONKA. "Depression, Anxiety, and COPD: The Unexamined Role of Nicotine Dependence." *Nicotine and Tobacco Research* 14: 176–183, 2012 (4 tables, 53 references)

Abstract: High rates of mood and anxiety disorders have been found in adults with chronic obstructive pulmonary disease (COPD) and are also found in cigarette smokers and those with nicotine dependence. Approximately half of adults who smoke develop nicotine dependence. A comparison was conducted of 5,572 subjects without COPD and 147 with COPD from the National Comorbidity Survey Replication in the United States between 2001 and 2004. Approximately one-third of those with COPD had lifetime nicotine dependence, 25% had a major depressive disorder, and 18% had a generalized anxiety disorder. Current smokers had a 2.80× adjusted odds ratio (AOR) of having COPD and those with nicotine dependence had an AOR of 6.10×.

These findings provide initial evidence suggesting that the association between anxiety, depression, and COPD may be at least partly attributable to confounding by cigarette smoking and nicotine dependence. Efforts toward prevention of chronic lung disease may be more effective if treatment and prevention efforts aimed at smoking cessation address mental health problems.

Reference Number: 80105

KLUNGSØYR, O., J.F. NYGÅRD, T. SØRENSEN, AND I. SANDANGER. "Cigarette Smoking and Incidence of First Depressive Episode: An 11-Year, Population-Based Follow-Up Study." *American Journal of Epidemiology*, 163: 421–432, 2006 (6 tables, 2 figures, 56 references)

Abstract: Interviews regarding cigarette smoking and depression were conducted on 1,190 Norwegians in 1990 and 2001. In 1990, most light smokers were women and most heavy smokers were men. The prevalence of daily cigarette smoking was 41.4% at baseline. The incidence of a first-time depression between 1990 and 2001 and history of cigarette smoking were determined. The odds ratio/hazard of a first depressive episode was statistically significant for current smokers using 11+ CPD.

Table: Hazard Ratio of First Depressive Episode Between 1990 and 2011 According to Cigarette Smoking History

Smoking Category	Hazard Ratio of Depressive Episode
Past, quit >5 years ago	1.05
Past, quit ≤5 years ago	1.65
Current ≤10 CPDs	1.35
Current 11–20 CPDs	1.87
Current >20 CPDs	3.66

Source: Adapted from Klungsøyr et al, 2006

> We observed a fourfold higher risk of depression for heavy smokers compared with never smokers. Failure to explain the smoking–depression association with other plausible alternatives might reflect a direct causal influence of smoking on depression.

Reference Number: 80106

PIETRA, T., A. WITUSIK, M. PANEK, J. SZEMRAI, AND P. GÓRSKI. "Anxiety, Depression and Methods of Stress Coping in Patients with Nicotine Dependence Syndrome." *Medical Science Monitor*, 17: CR272–CR276, 2011 (1 table, 40 references)

Abstract: A study was conducted of 88 subjects with nicotine dependence and compared to 84 healthy subjects with no nicotine dependence (mean age 41 years). Depression was greater in the nicotine dependent subjects (mean score 12.76) compared to a mean score of 10.76 for non-dependent subjects.

On the basis of the obtained results it can be concluded that:

1) Depression, anxiety as a trait, and anxiety as a state are more severe in subjects with nicotine dependence than in healthy individuals.
2) Smokers demonstrate, in comparison with the control group, higher prevalence of emotion-oriented coping style and distracting themselves with other activities.

Reference Number: 80107

TIDEY, J.W., L.R. PACEK, J.S. KOOPMEINERS, R. VANDREY, N. NARDONE, D.J. DROBES, N.L. BENOWITZ, S.S. DERMODY, A. LEMIEUX, R.L. DENLINGER, R. CASSIDY, M. AL'ABSI, D.K. HATSUKAMI, AND E.C. DONNY. "Effects of 6-Week Use of Reduced-Nicotine Content Cigarettes in Smokers With and Without Elevated Depressive Symptoms." *Nicotine and Tobacco Research*, 19: 59–67, 2017 (3 tables, 49 references)

Abstract: In the United States, approximately 8% of adults have current depression. Depressed people are more likely to be smokers and have a lower likelihood of quitting. A study of 717 smokers (mean age 42 years) was conducted, in which they smoked either a typical cigarette with normal nicotine content (15.8 mg/g) or a reduced nicotine cigarette containing 5.2, 2.4, 1.3, or 0.4 mg/g. One hundred and nine subjects were assessed as being depressed (15.2%). Reduced nicotine content (RNC) had similar effects on both smokers with elevated depressive symptoms and smokers without depression.

Reducing the nicotine content of cigarettes to a minimally addictive level has been proposed as a regulatory approach to reducing tobacco dependence and may improve the responses of smokers to other public health and smoking cessation treatment approaches. Results from our recent clinical trial support this hypothesis, by indicating that participants who were switched to RNC cigarettes for a 6-week period had lower smoking rates, nicotine dependence, abstinence-induced craving, and higher post-intervention quit attempts than those who used NNC cigarettes. As smokers with elevated depressive symptoms have relatively poor responses to smoking cessation treatments, concerns have been raised that these smokers could experience unintended negative consequences if the FDA were to institute a nicotine reduction policy.

Reference Number: 80108

KING, J.L., B.A. BEBOUSSIN, J. SPANGLER, J.C. ROSS, AND E.L. SUTFIN. "Tobacco Product Use and Mental Health Status Among Young Adults." *Addictive Behaviors*, 77: 67–72, 2018 (2 tables, 68 references)

Abstract: In 2013 a study was conducted of 2,370 young adults (mean age 21 years) from 11 colleges in North Carolina and Virginia. Overall, 27.3% reported any past 30-day tobacco use. Cigarettes were the most commonly used tobacco product (15%) followed by waterpipes (9.2%), ECs (5.2%), cigars (5.1%), and smokeless tobacco (SLT) (3.4%). A mental health diagnosis was reported by 249 of the subjects (10.5%). The most frequent diagnoses were depression (5.5%), ADHD (4.5%), and anxiety (0.8%). Cigarettes, ECs, and waterpipes increased the risk of a mental health diagnosis, but cigar and SLT use did not.

Table: Adjusted Odds Ratio of Tobacco Product Use and Mental Health Diagnosis (2+ Diagnosis or Depression Scale Score)

	Adjusted Odds Ratio				
Mental diagnosis	Cigarettes	E-cigarettes	Waterpipe	Cigars	Smokeless Tobacco
Mental health disorder	1.55	1.55	1.30	0.82	1.16
2+ diagnoses	3.16	1.83	1.50	0.59	0.20
Depression scale score	1.03	1.04	1.03	1.01	0.98

Source: Adapted from King et al, 2018

Findings in this study provide further evidence of a potential relationship between non-cigarette tobacco products and mental health status. Tobacco control efforts aimed at reducing tobacco use disparities among mental health populations should focus on non-cigarette tobacco products in addition to cigarettes.

Reference Number: 80109

KIM, N.H., J.H. PARK, D.P. CHOI, J.Y. LEE, AND H.C. KIM. "Secondhand Smoke Exposure and Depressive Symptoms Among Korean Adolescents: JS High

School Study." *Public Library of Science One*, 11: (11 pp), 2016 (4 tables, 40 references)

Abstract: A study was conducted of the effect of secondhand smoke exposure (SHSE) in 495 male and 494 female non-smoking adolescent students from a rural high school in South Korea between 2007 and 2012. Approximately 43% of the male students and 57% of the female students reported SHSE at school or home. SHSE increased the risk of male adolescents of having depressive symptoms.

> We found a significant association between SHSE and depressive symptoms among male adolescents, and in particular, SHSE at home was strongly associated with depressive symptoms. Further studies are needed to clarify the relevant biological or psychological mechanisms, as well as potential reasons for differential associations across sex strata.

8.02 FIRES/BURNS

> "Project Hamlet . . . To burn or not to burn."
>
> —McGuire, 1999

Cigarettes, usually combined with alcohol, have been reported as the leading cause of fire deaths in Canada and the United States (80201–80203, 80205). The risk of a fatal fire was increased 4.1× if there was a smoker in the house and 6.3× if there were two or more smokers (80203). For many years, Philip Morris (PM) opposed self-extinguishing cigarettes that would decrease the fire risk, as the tobacco industry was concerned that it would deter smokers from smoking cigarettes. In 1980, PM researchers began work on developing commercially viable fire-safe cigarettes, titling the research project "Project Hamlet" (80201).

In 2005, Canada became the first country to introduce federal standards for reduced ignition propensity cigarettes, which then spread to all 50 US states by March 2010 (80206). Fatal fires caused by cigarette smoking in the home have declined substantially due to reduced ignition propensity (RIP) cigarettes and a decline in adult smoking, especially inside the home (80204–7). Fire prevention should also focus on the promotion of smoke-free homes (80204–5)

Smoking is so addictive that 52% of patients with COPD continue to smoke at home in spite of the fire risks associated with oxygen tanks and masks for oxygen therapy (80208). Vaping while on oxygen therapy at home has also caused explosions and injuries (80215).

A new risk of burn injuries, especially to the face and hands, are exploding batteries in e-cigarettes (80209–14).

Reference Number: 80201

MCGUIRE, A. "How the Tobacco Industry Continues to Keep the Home Fires Burning." *Tobacco Control,* 8: 67–69, 1999 (2 references)

Abstract: This study provides a short history of the tobacco industry opposition to fire safe cigarettes. Although cigarettes were the leading cause of fire deaths in the United States, associated with 1,000 fire deaths and 3,000 burn injuries a year, fire safe legislation had still not passed Congress by 1999.

> As reported on 60 Minutes, Philip Morris researchers began working on their version of a fire-safe cigarette in 1980, shortly after Moakley introduced his first bill in Congress. The research project was given the code name "Project Hamlet". When asked by the plaintiff's lawyer during a deposition in the case if there was a special meaning in the code name, the researcher replied that it referred to an inside joke at Philip Morris, "To burn or not to burn". Besides discovering that the researchers were literate, the plaintiff's lawyer also revealed documents detailing seven years of scientific work on Project Hamlet. The goal of the project was to create a commercially acceptable fire-safe cigarette. In 1987, Philip Morris researchers sent unmarked packs of "Hamlet" cigarettes to dozens of taste testers who smoked and rated the fire-safe cigarettes. The results, as detailed in a memorandum that went up the chain of command within Philip Morris, indicated that the taste testers could not distinguish any difference between the "Hamlet" cigarettes and Marlboros. Philip Morris had its fire-safe cigarette.

Reference Number: 80202

PATETTA, M.J., AND T.B. COLE. "A Population-Based Descriptive Study of Housefire Deaths in North Carolina." *American Journal of Public Health,* 80: 1116–1117, 1990 (3 tables, 13 references)

Abstract: A study was conducted of 159 house fires resulting in 200 deaths in North Carolina in 1985. Of the 160 victims who were over 15 years of age, blood alcohol concentrations were determined in 130 cases. Fifty-six per cent had a concentration of >0.100 g/100 mL (100 mg/100 mL). The cause of death was carbon monoxide poisoning in 83% of the victims.

Table: Causes of House Fires in North Carolina, 1985

Cause of Fire	Percentage of Deaths
Heating	33%
Smoking	26%
Cooking	9%
Electrical	8%
Arson	6%
Open Flame	5%
Children Playing	5%

Source: Adapted from Patetta and Cole, 1990

We estimate that the lethal combination of alcohol and cigarettes resulted in at least one-fourth of the deaths reported in this study. The proportion of fatally injured persons tested who were impaired with alcohol was greater for housefires (56 percent) than that for any other cause of unintentional injury in North Carolina. The use of fire-safe cigarettes may be the most effective way to prevent deaths associated with drinking and smoking.

Reference Number: 80203

WARDA, L., M. TENEBEIN, AND M.E. MOFFATT. "House Fire Injury Prevention Update: Part I. A Review of Risk Factors for Fatal and Non-Fatal House Fire Injury." *Injury Prevention*, 5: 145–150, 1999 (3 tables, 38 references)

Abstract: A review of risk factors for fatal and non-fatal house fires was conducted using 15 studies. The relative risk for fatal and non-fatal fires was 4.1× if there was a smoker in the household and 6.3× if there were two or more smokers in the household.

Dangerous smoking habits are common. In a survey of American Indian households, 25% had a member who smoked in bed and 38% had a member who drank alcohol and smoked at same time. Interventions attempting to alter smoking and drinking habits are important, however passive interventions are more likely to succeed. The "firesafe" cigarette was designed to reduce fires due to dropped cigarettes on furniture and mattresses, due to its reduced circumference, lower density tobacco, and its less porous, lower citrate paper. The American Fire Safe Cigarette Act of

1990 mandated the development of a standard test method to determine the ignition potential of cigarettes, essential for further legislative efforts. Industry has successfully prevented the application of this legislation.

Reference Number: 80204

KEGLER, S.R., A.M. DELLINGER, M.F. BALLESTEROS, AND J. TSAI. "Decreasing Residential Fire Death Rates and the Association with the Prevalence of Adult Cigarette Smoking – United States, 1999–2015." *Journal of Safety Research*, 67: 197–201, 2018 (2 tables, 2 figures, 20 references)

Abstract: Between 1999 and 2015, residential fires in the United States caused between 2,000 and 3,000 deaths annually. Smoking materials were responsible in 5% of the residential fires but caused 22% of the fire deaths. This study examined national trends in residential fire deaths and adult smoking rates between 1999 and 2015. The annual rates of residential fire deaths according to the National Fire Protection Association (NFPA) decreased from 1.04 per 100,000 in 1999 to 0.80 in 2015.

Table: Residential Fire and Smoking Material Related Fire Death Rates per 100,000 Population

Year	All Residential Fire Death Rate (per 100,000 population)	Smoking Material Fire Death Rate (per 100,000 population)
2000	1.21	0.30
2005	1.03	0.25
2010	0.86	0.17
2015	0.80	0.17

Source: Adapted from Kegler et al, 2018

Decreasing U.S. residential fire death rates, both overall and smoking-related, coincided with a declining prevalence of adult cigarette smoking during 1999–2015. These findings further support tobacco control efforts and fire prevention strategies that include promotion of smoke-free homes. While the general health benefits of refraining from smoking are widely accepted, injury prevention represents a potential benefit that is less recognized.

Reference Number: 80205

RODGERS, K.M., L.R. SWETSCHINSKI, R.E. DODSON, H.R. ALPERT, J.M. FLEMING, AND R.A. RUDEL. "Health Toll from Open Flame and Cigarette-Started

Fires on Flame-Retardant Furniture in Massachusetts, 2003–2016." *American Journal of Public Health*, 109: 1205–1211, 2019 (3 tables, 1 figure, 35 references)

Abstract: The authors evaluated the death and injury data from 34,081 residential fires in the Massachusetts Fire Incident Reporting System between 2003 and 2016. The two major heat sources for fires are smoking material (cigarettes, pipes, etc.) and open flames (candles, lighters, matches). Massachusetts had a 28% reduction of fires due to cigarettes following the state's Fire Safe Cigarette law of 2008, which required all cigarettes sold in the state to self-extinguish without active inhalation. Among furniture fires, the risks of death were 3.4× greater when the fires were ignited by smoking material compared to open flames.

Table: Odds Ratio (OR) of Death, Severe Casualty, and Casualty According to Heat Source

Heat Source	Death OR	Severe Casualty OR	Casualty OR
Open Flame	3.28×	2.11×	1.79×
Smoking Materials	9.15×	4.33×	1.77×

Source: Adapted from Rodgers et al, 2019

> Furniture fires started by smoking materials were associated with more deaths than were furniture fires started by open flames.... Historically, furniture flammability regulations have focused on open flame heat sources, resulting in the addition of toxic flame retardants to furniture. Interventions to reduce deaths should instead focus on smoking materials.

Reference Number: 80206

O'CONNOR, R.J., V.W. REES, K.J. NORTON, K.M. CUMMINGS, G.N. CONNOLLY, H.R. ALPERT, A. SJÖDIN, L. ROMANOFF, Z. LI, K.M. JUNE, AND G.A. GIOVINO. "Does Switching to Reduced Ignition Propensity Cigarettes Alter Smoking Behavior or Exposure to Tobacco Smoke Constituents?" *Nicotine and Tobacco Research*, 12: 1011–1018, 2010 (3 tables, 20 references)

Abstract: In 2005, Canada became the first country to adopt a national standard for reduced ignition propensity (RIP) cigarettes. By March 2010, all 50 US states had adopted similar standards. The effect of RIP cigarettes on smoking behavior and patterns was determined in 170 daily smokers. Eighty-three did not change their cigarette brand (comparison group) and

77 switched to RIP cigarettes (experimental group). Various parameters were measured, such as breath CO, salivary cotinine (by enzyme-linked immunosorbent), and urinary polyaromatic hydrocarbons metabolites, both before and after RIP use for the experimental group and compared with the comparison group.

This study aimed to determine whether short-term switching from conventional to RIP cigarettes is associated with changes in smoking behavior and/or exposure to smoke constituents that suggest increased risk to the smoker. No changes in smoking topography attributable to the RIP design modification were identified in models adjusted for a broad range of potential confounding variables. Switching to RIP-compliant cigarettes did not appear to significantly affect subjects' smoking behaviors. This disconfirms the hypothesis that smokers using RIP cigarettes would take more frequent puffs, or puff more intensively, in response to the self-extinguishing feature.

Reference Number: 80207

BUTRY, D.T., AND D.S. THOMAS. "Cigarette Fires Involving Upholstered Furniture in Residences: The Role That Smokers, Smoker Behavior, and Fire Standard Compliant Cigarettes Play." *Fire Technology*, 53: 1123–146, 2017 (8 tables, 9 figures, 20 references)

Abstract: A major source of residential fires has been the ignition of upholstered furniture by smoldering cigarettes. Fire safe cigarettes (FSC) are made with a wrapping paper that contains regularly spaced bands which increase the likelihood of self-extinguishment and reduced fire risk. It is estimated that between 2002 and 2011, FSCs reduced the average number of residential fires by 45% and fatalities by 23%. In general, FSCs were very effective in reducing fires in large smoking populations that engaged in frequent smoking, but had a limited effect in the lightest smokers.

Differences in the behavior of smokers, along with the disproportional reductions in fires to fatalities and extent of flame spread, demonstrate a non-randomness in the types of fires affected by FSC cigarettes. This can be due to severity of the fire or the presence of occupants, or factors related to their condition. Opportunities exist for new technologies, which can better counter the ignitions from light smokers, to reduce the fire risks from cigarettes. The analysis does not account for changes in the flammability of upholstered furniture over the study period, other than those that may have occurred linearly over time. However, no new regulations occurred over this period. The average life span of upholstered furniture is estimated to be 15 to 17 years, so the mix of furniture found in U.S. greatly varies with respect to age.

Reference Number: 80208

CARLOS, W.G., M.S. BAKER, K.A. MCPHERSON, G.T. BOSSLET, R. SOOD, AND A.M. TORKE. "Smoking-Related Home Oxygen Burn Injuries: Continued Cause for Alarm." *Respiration*, 91: 151–155, 2016 (2 tables, 1 figure, 15 references)

Abstract: Home oxygen therapy is used for the treatment of chronic obstructive lung disease in 800,000 patients in the United States. Up to 52% of persons who use oxygen therapy continue to smoke in spite of the risks. Fifty-five patients were admitted to an Indiana hospital burn unit between 2008 and 2013 for smoking-related home oxygen injuries. Over half (51%) of the patients required intubation for respiratory failure and eight patients (15%) died.

> Smoking remains an addictive habit for many of our patients. In our state, Indiana, USA, the smoking rates are approximately 24% for adults over the age of 18 years. These rates demonstrate the addictive power of cigarettes as many people continue to smoke in spite of the known risks of cardiovascular disease, COPD, and lung cancer. It may be no surprise that many patients who are prescribed oxygen therapy for severe cardiopulmonary diseases continue to smoke in spite of the burn risks as well. Our review of our burn database affirms the high morbidity and mortality that can be seen with these injuries. Similar to other authors, we found these patients to generally be older, have a higher rate of inhalation injury, and a longer hospitalization despite smaller body surface area injuries.

Reference Number: 80209

SEITZ, C.M., AND Z. KABIR. "Burn Injuries Caused by e-Cigarette Explosions: A Systematic Review of Published Cases." *Tobacco Prevention and Cessation*, 4: 32 (8 pp), 2018 (3 tables, 1 figure, 68 references)

Abstract: A systematic review was conducted of 31 articles describing 164 cases of burn injuries from exploding ECs. In addition to burns, serious injuries have been inflicted on the hands, face, eyes, mouth (broken and missing teeth), not to mention broken neck bones and the first ever death reported in 2018. The majority of victims were male (90%) and ranged from 16 to 63 years in age (mean 29 years). The location of the explosion was in the pocket (65%), face (20%), or hands (10%). Common burn areas included the thigh, hand, genitals, and face. Second degree burns occurred in 35% of the victims. In all, 48 victims required skin grafting, with 19 reporting a median hospital stay of 5 days. Several cases involved the EC exploding after touching metal objects (car keys, coins).

Table: The Frequency of Body Areas Burned Due to ECs

Burned Body Area	Frequency
Thighs	76%
Hands	40%
Genitals	13%
Face	13%
Torso	7%
Arms	4%
Buttocks	2%
Feet	2%

Source: Adapted from Seitz and Kabir, 2018

This review has several implications, including the need for regulation of batteries, education regarding battery safety, and leveraging images of the severity of e-cigarette explosion burns to discourage the use of e-cigarettes.

Reference Number: 80210

QUIROGA, L., M. ASIF, T. LAGZIEL, D. BHAT, AND J. CAFFREY. "E-Cigarette Battery Explosions: Review of the Acute Management of the Burns and the Impact on Our Population." *Cureus,* 11: e5355 (9 pp), 2019 (8 figures, 9 references)

Abstract: ECs are lithium-battery powered devices that produce a heated vapor or aerosol and have been available for sale in the United States since 2007. At the time of the study, ECs could cost between $30 and $300. A review was conducted of six male patients (aged 27 to 46 years) who were treated at the Johns Hopkins Bayview Burn Center in Baltimore as a result of injuries related to EC battery explosions. The sequence of events in a typical lithium battery overheating occurs as the electrolyte overheats, which causes increased internal pressure. This is followed by breakage of the sealed end of the battery, causing the electrolyte to ignite, which creates gas expansion and leads to increased internal pressure and combustion of the porous plastic separator film. The internal fire or explosion can be started due to an overcharge, puncture, external heat, or a short circuit. Most burns were caused by a combination of flame and chemical (alkaline lithium salt) burns.

Table: Location of E-cigarette Lithium Batteries Explosions

Where the Battery Exploded	Percent
In pocket	31.3%
Device in use	30.8%
Being charged	24.6%
Storage	9.2%
Unknown	3.5%
Being transported	0.5%

Source: Adapted from Quiroga et al, 2019

In the last few years, the medical community has encountered increasing episodes of burn injuries secondary to e-cigarette battery explosion that has been reflected by the media and news. However, there are no clear published guidelines for managing lithium battery burns. In our series, we were able to manage the majority of our patient's injuries at the bedside, with varying combinations of mechanical/chemical debridement as well as topical wound care. Only one of our six patients required tangential excision and skin grafting. The rest of our patients were treated conservatively with complex wound care with excellent responses.

Reference Number: 80211

SATTESON, E.S., N.J. WALKER, C.J. TUOHY, AND J.A. MOLNAR. "Extensive Hand Thermal and Blast Injury from Electronic Cigarette Explosions: A Case Report." *Hand*, 13: NP1–NP5, 2018 (2 figures, 11 references)

Abstract: Due to the rechargeable lithium battery used in most ECs, there is a risk of injury due to explosion. ECs do not come with product warnings related to the risks of overheating the battery and explosion. A 35-year-old man was using an EC device (Dark Horse atomizer with SIMPL med Mod battery device) when it exploded in his right hand. He reported that when he changed the battery, it began to rapidly heat up until it exploded between his right thumb and index finger. He required multiple surgeries.

> An additional potential harm of the device—blast injuries and burns due to e-cigarette explosion—has also received little attention in the literature. The media, however, has reported several such cases. Multiple local news sources have reported apartment or house fires as a result of e-cigarette explosion while the devices were charging. In some cases, patients sustained burns as a result of these fires.

Reference Number: 80212

MICHAEL, R., N. EBRAHEIM, J. MAIER, M. TANIOS, AND A. KOURI. "Electronic Cigarette Burns: A Case Report and Review of Current Literature." *Case Reports in Orthopedics*, 2019: (5 pp), 2019 (1 table, 4 figures, 12 references)

Abstract: A 40-year-old male EC user presented to the ER with severe burns to his left posterior side due to the spontaneous combustion of an EC in his left pocket. The patient stated that the device was switched off. He required skin grafting and one month later still had intermittent pain, irritation, and numbness.

> Cases of severe burns from spontaneous combustion of ENDS are becoming a concern with their increasing popularity as alternatives to traditional cigarette smoking. A large percentage of these explosions occur in the pants or shirt pockets which makes the severe burns on the thigh and chest or upper extremity a common presentation. The scar contractures around the knee and hip joints pose a significant functional limitation. The psychiatric effects of the incurred trauma and resultant cosmetic deformity are not well documented in the current literature but present a significant issue for these patients. Our patient struggled with the appearance of burns on his legs, yet many patients have been documented to have prominent burns on their faces and hands.

Reference Number: 80213

JIWANI, A.Z., J.F. WILLIAMS, J.A. RIZZO, K.K. CHUNG, B.T. KING, AND L.C. CANCIO. "Thermal Injury Patterns Associated with Electronic Cigarettes." *International Journal of Burns and Trauma*, 7: 1–5, 2017 (1 table, 5 figures, 11 references)

Abstract: Between 2014 and 2016, 10 patients presented at the US Army Institute of Surgical Research Burn Center with burn injuries sustained from e-cigarette explosions. There were nine male and one female victims (aged 18–46 years) with a localized burn in at least one extremity. Seven patients reported the EC battery self-combusting in their pants pocket or lap. One EC ignited during a motorcycle crash.

> To conclude, we report a 10-case series of thermal injuries associated with e-cigarettes over a recent 2-year period. Our cohort was comprised mainly of young adults who sustained mixed partial and full thickness thermal burn injuries as a result of e-cigarette-related explosions. In many documented scenarios, a malfunctioning or over-heated battery is the cause. Our data support the need for increased awareness among healthcare providers and the general public of the potential harms of e-cigarette use, modification, storage, and charging.

Reference Number: 80214

ARCHAMBEAU, B.A., S. YOUNG, C. LEE, T. PENNINGTON, C. VANDERBEEK, D. MIULLI, J. CULHANE, AND M. NEEKI. "E-Cigarette Blast Injury: Complex Facial Fractures and Pneumocephalus." *Western Journal of Emergency Medicine*, 16: 805–807, 2016 (2 figures, 6 references)

Abstract: A 59-year-old man with a history of leukemia, hyperlipidemia, chronic back pain, and right hearing loss was airlifted to a trauma center following an explosion of his EC while he was vaping. The patient had purchased the device 2 days earlier and made no modification. He suffered several maxillofacial fractures, blurred vision, and pneumocephalus (air or gas in the cranial cavity) and required close monitoring in intensive care, neurosurgery, and multiple maxillofacial surgeries to reconstruct his facial bones and soft tissues.

> It is currently thought that EC explosions are caused by the proximity of the heating element to an improperly insulated lithium ion battery and its exposure to volatile liquids. Reported cases of possible EC failures are consistent with lithium ion battery failures observed in other devices. The proposed mechanism of failure has been well documented. The failure is potentially more frequent and more harmful in ECs, due to their configuration and proximity to users' faces.

Reference Number: 80215

LACASSE, Y., M. LÉGARÉ, AND F. MALTAIS. "E-Cigarette Use in Patients Receiving Home Oxygen Therapy." *Canadian Respiratory Journal*, 22: 83–85, 2015 (21 references)

Abstract: A review was conducted on the use of ECs in patients receiving home oxygen therapy. A 72-year-old patient with pulmonary fibrosis suffered facial burns after his nasal prongs caught on fire while he was vaping while receiving oxygen. He was discharged from hospital 5 days later.

Current smokers who are prescribed home oxygen may not benefit from the therapy. In addition to being an obvious fire hazard, there is some evidence that the physiological mechanisms by which home oxygen is believed to operate are inhibited by smoking. Although their effectiveness is yet to be demonstrated, electronic cigarettes (e-cigarettes) are often regarded as an aid to smoking cessation. However, several burn accidents in e-cigarette smokers receiving home oxygen therapy have also been reported, leading Health Canada to release a warning of fire risk to oxygen therapy patients from e-cigarettes. It is the authors' position that patients receiving oxygen should definitely not use e-cigarettes. The authors provide suggestions for addressing the delicate issue of home oxygen therapy in current cigarette and/or e-cigarette smokers.

8.03 INFECTIOUS DISEASES

> "[A]t the beginning of the 21st century, we are facing the convergence of several epidemics including tobacco smoking and influenza, and these epidemics interact by way of increasing disease susceptibility and worsening outcomes."
>
> —Noah, Zhou, and Jaspers, "Alteration of the Nasal Responses to Influenza Virus by Tobacco Smoke" (2012)

Smoking not only causes cancers and other physical damage to the body, but it is also associated with increased risk and severity of infectious pulmonary disease such as pneumonia (80302–5), tuberculosis (80306-7), and influenza (80308). The risk of latent TB infection was nearly 12× greater in smokers compared to non-smokers (80307). Smoking promotes these infections by impairing the mucociliary clearance protective mechanism, weakening immune response, and causing inflammatory responses in the lungs (80301, 80307–8).

Smokers and patients with COPD were more susceptible and suffered more severe outcomes with Middle Eastern Respiratory Syndrome (MERS), a coronavirus similar to COVID-19 (80309–11). Therefore it is no surprise that smokers are more likely to develop COVID-19 and suffer death and severe complications than non-smokers (80312–6).

In fact, one article supposedly showing the benefits of smoking during the COVID-19 pandemic was recently pulled from the European Respiratory Journal due to a conflict of interests by the authors (see Chapter 4, reference 40128).

Although touted as safer, e-cigarettes should be discouraged, as nicotine itself is associated with an increase in the risk of infection from COVID-19 (80316–17).

Reference Number: 80301

XAVIER, R.F., D. RAMOS, J.T. ITO, F.M.M. RODRIGUES, G.N. BERTOLINI, M. MACCHIONE, A.C. DE TOLEDO, AND E.M.C. RAMOS. "Effects of Cigarette Smoking Intensity on the Mucociliary Clearance of Active Smokers." *Respiration*, 86: 479–485, 2013 (2 tables, 3 figures, 52 references)

Abstract: This study determined smoking behavior, nicotine dependence, pulmonary function, CO in exhaled breath, blood COHb, and mucociliary clearance (by a saccharin transit time test) in 14 light (<10 CPD), 34 moderate (11–20 CPD), and 27 heavy (>20 CPD) smokers and 24 controls. A positive correlation was observed between saccharin transit time, exhaled CO, and CPD.

Smoking impairs mucociliary clearance and increases respiratory infection frequency and severity in subjects with and without smoking-related chronic lung diseases.

Reference Number: 80302

NUORTI, J.P., J.C. BUTLER, M.M. FARLEY, L.H. HARRISON, A. MCGEER, M.S. KOLCZAK, AND R.F. BREIMAN. "Cigarette Smoking and Invasive Pneumococcal Disease." *New England Journal of Medicine*, 342: 681–689, 2000 (5 tables, 57 references)

Abstract: The rate of invasive pneumococcal disease is greatest among young children and the elderly. Previous studies had shown that smokers accounted for 50% of otherwise healthy adults with invasive pneumococcal disease. A comparison was conducted of 228 immunocompetent patients (aged 18–64 years) with invasive pneumococcal disease and 301 control subjects in the general population between 1995 and 1996.

Table: Smoking Status and Adjusted Odds Ratio for Invasive Pneumococcal Disease

Smoking Status	Adjusted OR for Invasive Pneumococcal Disease
Current smoker (1–14 CPD)	2.3×
Current smoker (15–24 CPD)	3.7×
Current smoker (25+ CPD)	5.5×
Former smoker (<5 years)	3.5×
Non-smoker (1–4 h/day SHSE)	2.4×
Non-smoker (>4 h/day)	3.9×

Source: Adapted from Nuorti et al, 2000

> Our results indicate that cigarette smoking is the strongest independent risk factor for invasive pneumococcal disease among immunocompetent, nonelderly adults and that 51 percent of the disease burden in this population group can be attributed statistically by this modifiable risk factor. We found that the current number of cigarettes smoked per day, the number of pack-years of smoking, and the time since quitting showed clear dose–response relations with the risk of pneumococcal disease. Increased risk was also independently associated with exposure to environmental tobacco smoke, chronic illness, a low level of education, black race, male sex, and living with young children who were in day care.

Reference Number: 80303

ALMIRALL, J., J. BLANQUER, AND S. BELLO. "Community-Acquired Pneumonia Among Smokers." *Archivos de Bronconeumologia*, 50: 250–254, 2014 (66 references)

Abstract: Smoking increases susceptibility to bacterial lung infection, even in those exposed to secondhand smoke (SHS). The most important risk factor in community acquired pneumonia (CAP) is age and its incidence increases 3–4× starting at age 65 years. Smoking is one of the most important risk factors for CAP, due to Legionella pneumophia, and it increases the risk for legionnaire's disease by 3.48–7.49×. Periodontal disease can contribute to CAP and smoking increases the risk of periodontitis by 6×. It is recommended that for patients hospitalized with CAP, the smoking cessation be combined with vaccination.

> At present there is enough scientific evidence to show that active tobacco smoking is an important risk factor for CAP; it has a direct and independent effect on the risk of CAP, but it may also act indirectly causing chronic bronchitis or COPD which, in turn, are well recognized risk factors for CAP. Therefore, one of the main ways of preventing CAP is to intervene in the smoking habit, advising its cessation, which may reduce the risk of CAP by 50% after five years. Furthermore, a reduction of 14% in IPD has been observed every year and, after 10 years, the risk of IPD can reach the same level as that of non-smokers.

Reference Number: 80304

BROWN, J., AND M. LIPMAN. "Community-Acquired Pneumonia in HIV-Infected Individuals." *Current Infectious Disease Reports*, 16: 397 (9 pp), 2014 (1 figure, 78 references)

Abstract: Community acquired pneumonia is an important cause of morbidity and mortality in HIV-positive individuals. The incidence of CAP has decreased with the use of antiretroviral therapy. Cigarette smoking is a consistent risk factor for pneumonia. Smoking cessation and immunization would help prevent CAP in HIV-infected individuals.

> HIV-infected populations continue to have high rates of exposure to cigarette smoke. Rates of pneumonia are higher in smokers than nonsmokers, and smoking cessation can reduce this. A recent systematic review of the literature by De et al. found that current smokers have an increased risk of bacterial pneumonia compared to nonsmokers (hazard ratio 1.73, 95 % confidence interval 1.44–2.06) and that this risk is reduced by smoking cessation. In a prospective study in France, a hazard ratio of 0.48 was found in former smokers compared to current smokers ($p = 0.02$). The use of medication as an adjunct to smoking cessation has been extensively investigated in HIV-uninfected smokers, but there are few data from HIV-infected people. A small study of the safety and tolerability of

varenicline tartrate in 36 HIV-positive individuals found a high frequency of adverse events, most commonly nausea, but excellent quit rates with a 42 % abstinence rate at 12 weeks. There were no adverse changes in HIV viral loads detected.

Reference Number: 80305

CECERE, L.M., E.C. WILLIAMS, H. SUN, C.L. BRYSON, B.J. CLARK, K.A. BRADLEY, AND D.H. AU. "Smoking Cessation and the Risk of Hospitalization for Pneumonia." *Respiratory Medicine*, 106: 1055–1062, 2012 (3 tables, 44 references)

Abstract: Smoking is a leading cause of COPD and is a strong predictor of pneumonia resulting in hospitalization and death. A study was conducted of 6,720 current, 13,625 former, and 4,890 never smokers, all of whom were US veterans enrolled in the Ambulatory Care Quality Improvement Project between 1997 and 1998. Compared to current smokers, never smokers had a hazard ratio of 0.48 for hospitalization due to pneumonia. Only former smokers who had quit over 10 years earlier had a lower hazard ratio (0.62) than current smokers.

> In a large cohort of more than 25,000 Veterans, we found smoking cessation was an important predictor of the risk of pneumonia, but that risk was dependent on the duration of cessation as well as whether patients had recognized COPD. These findings are important as they suggest that smoking cessation could be beneficial in reducing future hospitalizations for pneumonia, but that in order to be effective, quitting smoking likely needs to occur early in life. Furthermore, these results suggest that as clinicians discuss the importance of smoking cessation that they may include the concept that benefits to lung health may be most substantial prior to developing lung disease, such as COPD.

Reference Number: 80306

WATKINS, R.E., AND A.J. PLANT. "Does Smoking Explain Sex Differences in the Global Tuberculosis Epidemic?" *Epidemiology and Infection*, 134: 333–339, 2006 (2 tables, 1 figure, 31 references)

Abstract: Globally, there are about 70% more male victims of tuberculosis (TB) than female. In 2001, a cross-sectional ecological analysis was conducted of 22 high TB countries to determine if gender differences in TB rates could be related to gender smoking differences. Per capita cigarette consumption was compared with the country's TB rate and explained 33% of the variance in the gender ratio of TB.

> Our findings suggest that significant gains in global TB control would be associated with the modification of smoking behaviour in high-prevalence settings, and support the strengthening of population-level interventions which are aimed at decreasing tobacco consumption. The inclusion of tobacco-control strategies within TB prevention and control programmes should enhance the effectiveness of current initiatives. The association of smoking with an often highly stigmatized disease may also facilitate behaviour change and a reduction in the growing rates of tobacco-related morbidity and mortality in developing countries.

Reference Number: 80307

ALTET, N., I. LATORRE, M.Á. JIMÉNEZ-FUENTES, J. MALDONADO, I. MOLINA, Y. GONZÁLEZ-DÍAZ, C. MILÀ, E. GARCÍA-GARCÍA, B. MURIEL, R. VILLAR-HERNÁNDEZ, M. LAABEI, A-C. GÓMEZ, P. GODOY, M.L. DE SOUZA-GALVÃO, S. SOLANO, C.A. JIMÉNEZ-RUIZ, AND J. DOMÍNGUEZ. "Assessment of the Influence of Direct Tobacco Smoke on Infection and Active TB Management." *Public Library of Science One*, 12: e0182998 (17 pp), 2017 (6 tables, 1 figure, 45 references)

Abstract: Tuberculosis (TB) killed 1.4 million people globally in 2015. Active and passive tobacco smoke exposure are risk factors for latent tuberculosis infection and active TB progression. A study was conducted of 175 participants with active TB and 350 from contact tracing studies, of whom 41 were secondary TB cases. The proportion of tobacco smokers was significantly higher in TB patients than contacts (43%). An association was found between smoking and TB outcome due to the weakened host immune system response caused by tobacco smoke. The risk of latent TB infection was 11.57× that of non-smokers. But only a low number of active TB patients quit smoking.

> Smoking had a negative effect on radiological manifestations, delaying time of sputum conversion. Our data establish a link between tobacco smoking and TB due to a weakened IFN-γ response caused by direct tobacco smoke.

Reference Number: 80308

NOAH, T.L., H. ZHOU, AND I. JASPERS. "Alteration of the Nasal Responses to Influenza Virus by Tobacco Smoke." *Current Opinion in Allergy and Clinical Immunology*, 12: 24–31, 2012 (5 figures, 43 references)

Abstract: A review was conducted on the effect of tobacco smoke on influenza virus infection. Epidemiological studies have shown that smokers are more susceptible to the influenza virus than non-smokers. In a 16-year morbidity study of 300,000 veterans, pneumonia and influenza deaths were higher in cigarette smokers than non-smokers. Experiments using human respiratory cells in vitro tended to indicate that tobacco smoke increased viral replication and suppressed anti-viral pathways. The nasal mucosal response to a live attenuated influenza virus vaccine was determined in young adult smokers, those exposed to SHS, and unexposed controls. Active smoking or SHS had a measurable impact on early innate respiratory mucosal defense responses to the influenza virus.

> [A]t the beginning of the 21st century, we are facing the convergence of several epidemics including tobacco smoking and influenza, and these epidemics interact by way of increasing disease susceptibility and worsening outcomes. Smoking and exposure to second hand tobacco smoke remain extremely prevalent worldwide, and their effects on susceptibility to infection could be shared by other common oxidant particulate pollutants such as biomass burning or diesel exhaust. Exposure to all of these factors is likely to increase with increasing global urbanization. It is therefore important that we begin to understand the mechanisms underlying increased infection risk with tobacco smoke exposure. Data from recent experimental studies point toward suppression by tobacco smoke of several specific antiviral and innate host defense pathways, both epithelial and non-epithelial, at the level of the respiratory mucosa. Studies also suggest a possible role for antioxidant strategies in reducing risk.

Reference Number: 80309

HALIM, A.A., B. ALSAYED, S. EMBARAK, T. YASEEN, AND S. DABBOUS. "Clinical Characteristics and Outcome of ICU Admitted MERS Corona Virus Infected Patients." *Egyptian Journal of Chest Diseases and Tuberculosis*, 65: 81–87, 2016 (3 tables, 30 references)

Abstract: Middle East Respiratory Syndrome (MERS) is caused by a coronavirus from camels and was first reported in Saudi Arabia in September 2012. A study was conducted of 32 laboratory confirmed MERS corona virus ICU patients. There were 20 male (62.5%) and 12 female (37.5%) patients with a mean age of 44 years. The main symptoms were fever (96.9%), cough (93.8%), and dyspnea (90.6%). Eighteen patients survived and 14 patients died.

> Current smoking and smoking severity were statistically significantly (P <0.01) higher in the expired group compared to the survivors. Non smoking and ex smoking were statistically significantly (P <0.01) higher among the survivors in comparison with the expired group. Also, there was statistically (P <0.05) significant positive correlation between mortality and smoking severity (r = 0.640). These results are in agreement with previous data that showed that smoking is a major determinant of morbidity and mortality in respiratory tract infection, especially in populations that smoke heavily. Cigarette smoking is associated with a variety of alterations in cellular and humoral immune system functions. These alterations include a decreased level of circulating immunoglobulins, a depression of antibody responses to certain antigens, a decrease in CD4+ lymphocyte counts, an increase in CD8+ lymphocyte counts, depressed phagocyte activity and a decreased release of pro inflammatory cytokines. Smoking increases CD8+ lymphocytes in the cell mediated system and suppresses the host defense against infections

Reference Number: 80310

SEYS, L.J.M., W. WIDAGDO, F.M. VERHAMME, A. KLEINJAN, W. JANSSENS, G.F. JOOS, K.R. BRACKE, B.L. HAAGMANS, AND G.G. BRUSSELLE. "DPP4, the Middle East Respiratory Syndrome Coronavirus Receptor, is Upregulated in Lungs of Smokers and Chronic Obstructive Pulmonary Disease Patients." *Clinical Infectious Diseases*, 66: 45–53, 2018 (3 tables, 4 figures, 37 references)

Abstract: The alveolar epithelial cells are the main target for MERS-CoV and as such, smoking and chronic lung diseases such as COPD are risk factors for contracting the infection. Smoking is the most common cause of COPD and causes an inflammatory response in the lungs, which results in an excessive release of chemokines and cytokines and a high influx of immune cells. The MERS-CoV receptor dipeptidyl peptidase 4

(DPP4, also known as CD26) was found to be upregulated in lung samples of smokers and persons with COPD.

> In conclusion, because smoking is the most common etiology of COPD, our data highlight the association between chronic exposure to cigarette smoking and DPP4 upregulation in the lungs, as well as partially explain the increased susceptibility of smokers and COPD patients to MERS-CoV infection.

Reference Number: 80311

KOMIYAMA, M., AND K. HASEGAWA. "Smoking Cessation as a Public Health Measure to Limit the Coronavirus Disease 2019 Pandemic." *European Cardiology Review*, 15: e16 (2 pp), 2020 (21 references)

Abstract: Smoking is a risk factor for many lung infections such as influenza and pneumonia, which smokers are more likely to contract and exhibit more severe symptoms than non-smokers. Smoking was reported to be a risk factor for Middle East Respiratory Syndrome Coronavirus (MERS-CoV). Nicotine also affects the immune system by promoting the secretion of catecholamines and corticosteroids and inhibits the production of interleukon-22, which suppresses lung inflammation and repairs damaged cells.

> COVID-19 is primarily a disease of the respiratory tract, and virus entry into cells, viral replication and virion release occur within the respiratory tract. Angiotensin-converting enzyme (ACE) 2 converts the vasoconstrictor angiotensin II to vasoprotective angiotensin. Multiple studies have shown that ACE-2 is a host receptor for SARS-CoV-2. SARS-CoV-2 enters cells through ACE-2 receptors present in mucosal epithelial cells and alveolar tissues in a clathrin-dependent process. Cigarette smoking increases the expression of ACE-2 in pulmonary tissues, which could in part account for the increased risk of infection. Additionally, the WHO has noted that smokers perform repeated hand to face reciprocal movements, which contribute to increased opportunity for virus entry.

Reference Number: 80312

RUSSO, P., S. BONASSI, R. GIACCONI, M. MALAVOLTA, C. TOMINO, AND F. MAGGI. "COVID-19 and Smoking: Is Nicotine the Hidden Link?" *European Respiratory Journal*, 55: 2001116 (3 pp), 2020 (1 figure, 11 references)

Abstract: Smoking is associated with a negative expression of SARS-CoV-2 and a worse progression of this disease. Nicotine upregulates the angiotensin converting enzyme II (ACE-2) and downregulates the compensating receptor axis, which contributes to cardiovascular and pulmonary diseases.

In addition, nicotine increases the expression and/or activity of ACE in the lung, an increase which has been found also in the serum of smokers, and that required at least 20 min to return to control level. ACE-2 serves as a physiologically relevant cellular entry receptor for SARS-CoV, for the human respiratory coronavirus NL63, and probably for SARS-CoV-2. ACE binds the SARS-CoV-2 S protein, and through its tissutal expression mediates the localisation and the efficiency of the infection.

Reference Number: 80313

DAS, G., N. MUKHERJEE, AND S. GHOSH. "Neurological Insights of COVID-19 Pandemic." *ACS Chemical Neuroscience*, 11: 1206–1209, 2020 (2 figures, 10 references)

Abstract: The COVID-19 virus is more virulent and contagious than either SARS-CoV or MERS-CoV. The COVID-19 pandemic owes much of its killer properties to its unique structure, which has a stronger binding affinity with the human angiotensin-converting enzyme 2 (hACE2) protein that the virus uses as the entry point into the human cell. The virus may target the brain as well as the lung and cause neurological manifestations.

Table: Risks of Neurological COVID-19 Infection

Risks
• Smoking (nAChR)
• Age
• Genetic susceptibility of individual
• Mutations in endemic strains
• Individuals with prevailing neural aliment
• Individuals with any existing aliment

Source: Adapted from Das et al, 2020

Smokers Are Living on the Edge

The likeliness of COVID-19 patients to contract neurological infections can be exacerbated by secondary factors like smoking, which according to a pilot study can enhance the chances of contracting COVID-19 based neuroinfections due to the functional interactions between hACE2 and the nicotinic receptor (nAChR). The same study reports that due to the coexpression of hACE2 and nicotinic acetylcholine receptor (nAChR) in many cells, there exists a functional link between them. So, when smokers smoke, it augments the expression of the hACE2 due to the nicotine stimulation of nAChR. Hence, while performing autopsies on brains of COVID-19 patients, it would be wise to conduct smoker versus nonsmoker based analysis as this will help to shed light on smoking being an additional risk factor in COVID-19 patients along with age and already existing ailments.

Reference Number: 80314

LIU, W., Z-W. TAO, L. WANG, M-L. YUAN, K. LIU, L. ZHOU, S. WEI, Y. DENG, J. LIU, H-G. LIU, M. YANG, AND Y. HU. "Analysis of Factors Associated with Disease Outcomes in Hospitalized Patients with 2019 Novel Coronavirus Disease." *Chinese Medical Journal*, 133: 1032–1038, 2020 (4 tables, 21 references)

Abstract: A study was conducted of 78 patients with COVID-19 pneumonia admitted to hospital in Wuhan, China between December 2019 and January 2020 and re-evaluated 2 weeks after admission. Nearly three-quarters of the patients sought treatment due to a fever. Eleven patients (14.1%) had deteriorated, and 67 patients (85.9%) had stabilized or improved over the 2 weeks. Patients in the deteriorated group were older than patients who stabilized or improved (66 years vs 33 years) and 27.3% had a history of smoking compared to 3.0% in the stabilized/improved group.

Table: Multivariate Analysis of Risk Factors for COVID-19 Disease Progression

Variables	OR of Disease Progression (or deterioration)
Age (≥60 years vs <60 years)	8.5×
History of smoking (yes vs no)	14.3×
Maximum body temperature (≥37.3°C vs <37.3°C)	9.0×
Respiratory failure (yes vs no)	8.8×
Severe illness (yes vs no)	2.5×
Albumin (<40 g/L vs ≥40 g/L)	7.4×
C-reactive protein (>8.2 mg/L vs 8.2 mg/L)	10.5×

Source: Adapted from Liu et al, 2020

Several factors that led to the progression of COVID-19 pneumonia were identified, including age, history of smoking, maximum body temperature at admission, respiratory failure, albumin, and C-reactive protein. These results can be used to further enhance the ability of management of COVID-19 pneumonia.

Reference Number: 80315

SANCHEZ-RAMINEZ, D.C., AND D. MACKEY. "Underlying Respiratory Diseases, Specifically COPD, and Smoking are Associated with Severe COVID-19 Outcomes: A Systematic Review and Meta-Analysis." *Respiratory Medicine*, 171: (10 pp), 2020 (1 table, 3 figures, 42 references)

Abstract: A meta-analysis was conducted of 13,184 COVID-19 patients in 22 studies published between January and April 2020 regarding various factors that affect the severity of this disease. Pooled analysis showed the risk of severe COVID-19 outcomes was increased by 4.21× in persons with respiratory diseases, 1.98× for current smokers, and 3.46× for former smokers.

Results suggested that underlying respiratory diseases, specifically COPD, and smoking increase the odds of having severe COVID-19 outcomes. This is an important finding considering the high prevalence of COPD and smoking worldwide and the rapid spread of SARS-CoV-2.

Reference Number: 80316

OLDS, J.L., AND N. KABBANI. "Is Nicotine Exposure Linked to Cardiopulmonary Vulnerability to COVID-19 in the General Population?" *Federation of European Biochemical Societies Journal*, 287: 3651–3655, 2020 (2 figures, 28 references)

Abstract: Nicotine causes an acute increase in systematic and diastolic blood pressure as well as increased ACE activity. Smoking causes cardiovascular and pulmonary illness and is associated with higher mortality rates in various respiratory viruses, including influenza.

Tobacco formulations are not just nicotine and often contain a varied mixture of >5000 chemicals, with potential carcinogenic, cardiovascular, and respiratory properties. In addition to nicotine, cigarettes contain toxins such as carbon monoxide and polycyclic aromatic hydrocarbons, which also perturb the function of the cardiovascular, pulmonary, and immune systems, and at this point, such toxins may also contribute to COVID-19 disease outcome. Chronic nicotine use, systemically through cigarette and other tobacco products or indirectly possibly through secondhand smoke, now presents an important factor in COVID-19 vulnerability in various populations. Alarming increases in nicotine consumption and mortalities throughout Asia and rises in the distribution of electronic cigarette (e-cig) formulations worldwide point to augmenting vulnerability to respiratory infecting viruses such as COVID-19.

Reference Number: 80317

JAVELLE, E. "Electronic Cigarette and Vaping Should be Discouraged During the New Coronavirus SARS-CoV-2 Pandemic." *Archives of Toxicology*, 94: 2261–2262, 2020 (3 references)

Abstract: ECs have been shown to cause lung injury as early as 2012 and were implicated in the EVALI (e-cigarette or vaping product associated lung injury) outbreak in 2019.

The emergent SARS-CoV-2 virus, transmitted most readily by droplets and aerosolization, is highly contagious and manifests with severe or critical bilateral pneumonia in 20% of symptomatic infected cases. The severe coronavirus disease-2019 (COVID-19) resulting from SARS-CoV-2 infection is characterized by respiratory distress and decreased oxygen saturation. However, even in asymptomatic patients, imaging features representing diffuse peripheral bilateral ground-glass opacities have been highlighted, and asymptomatic carrier state is suspected to be contagious.

8.04 INJURIES AND MOTOR VEHICLE ACCIDENT COLLISIONS

"Cigarette smoking has been identified as a leading cause of fire disaster and resulting death, fatal automobile accidents, and injuries at work and other unintentional injuries."

—Merril, "Injury-Related Deaths According to Environmental, Demographic, and Lifestyle Factors" (2019)

A neglected tobacco hazard has been the increased risk of traffic collisions and other injuries seen in smokers. Among the causes for this effect are:

- Driving distraction (impaired attention)

- Carbon monoxide and hypoxia effects
- Smoking induced blurring of vision
- Fatigue
- Cigarette smoke deposited on the inside of the windshield reducing visibility
- Direct toxicity from other smoke chemicals reducing visual acuity
- Increased risk taking by smokers
- Non-use of seatbelt by smokers
- Impairment due to nicotine withdrawal symptoms (80401–2, 80404, 80406–8).

In Japan, the hazard ratio of being killed in a motor vehicle accident was 1.54× for male smokers who consumed more than 20 CPDs (80401). Subjects driving a computerized course were found to be impaired 30 minutes after the use of a hookah or waterpipe (80403). The odds ratio of injury death in traffic collisions in Thailand was as high in current smokers as in persons with a history of drunk driving (80405). Teen drivers who smoked also had an increased collision risk (80406).

In Ontario, the OR of a motor vehicle collision for smokers declined after a law was enacted banning a driver or passenger from smoking in a car with anyone under the age of 16 years (80402).

In addition to traffic collisions, smokers also had a greater risk of unintentional injury from stumbles, falls from heights, burns or scalds, suicides, and job-related injuries (80408, 80413).

Recruits who were smokers suffered more injuries during basic training at the US Army Ordnance Advanced Individual Training (80410) and as such tremendous success has been obtained in smoking cessation programs and smoking has declined from about 50% of military personnel to 24% in 2018 (80410). Smoking was associated with lower levels of physical fitness, less mental sharpness, and increased risk of injury (80410).

Counties in the United States with a greater percentage of adult smokers had greater death rates from accidents and other adverse events (80411).

In addition to an increased risk of injury, smokers who have blunt trauma injuries were more likely to develop acute lung injury as well (80412).

Not only does smoking cause impairment of cognitive functioning, but abrupt cessation of smoking can lead to impairment by pilots and a decrease in flight safety (80414).

Reference Number: 80401

IGARASHI, A., J. AIDA, T. SAIRENCHI, T. TSUBOYA, K. SUGIYAMA, S. KOYAMA, Y. MATSUYAMA, Y. SATO, K. OSAKA, AND H. OTA. "Does Cigarette Smoking Increase Traffic Accident Death During 20 Years Follow-Up in Japan? The Ibaraki Prefectural Health Study." *Journal of Epidemiology*, 29: 192–196, 2019 (2 tables, 20 references)

Abstract: A 20-year follow-up study was conducted on 33,108 men and 63,366 women in Japan regarding smoking habits and traffic deaths. During the period of 1993 to 2013, there were 168 traffic deaths among men and 128 traffic deaths among women. Among men, after adjustment for age and alcohol intake, the HR of a traffic fatality was 1.13× for ex-smokers, 1.32× for light smokers (<20 cigarettes/day), and 1.54× for smokers (>20 cigarettes per day). There were no increases in HR among women smokers.

> There is a possible mechanism that might explain the association between smoking and traffic accident deaths. Distraction or impaired attention due to smoking during driving a car may have contributed to the rate of accidents observed. A series of smoking behaviors, such as searching for cigarettes, lighting a cigarette, dropping ash on an ashtray, and extinguishing a burning cigarette are considered to distract the driver.

Reference Number: 80402

PEDERSON, L.L., J. KOVAL, E. VINGILIS, J. SEELEY, A.R. IALOMITEANU, C.M. WICKENS, R. FERRENCE, AND R.E. MANN. "The Relationship Between Motor Vehicle Collisions and Cigarette Smoking in Ontario: Analysis of CAMH Survey Data from 2002 to 2016." *Preventive Medicine Reports*, 13: 327–331, 2018 (1 table, 1 figure, 34 references)

Abstract: Tobacco smoking in motor vehicles has been found to be a risk factor in motor vehicle collisions (MVCs), at a rate ranging from 1.61 to 1.88×. Cigarette smoking was found to be a contributing factor in 1–5% of distracted motor vehicle collisions. The prevalence of motor vehicle collisions among smokers was determined using the Centre for Addiction and Mental Health survey of Ontario adults between 2002 and 2016. On 21 January 2009, a law was enacted in Ontario to prevent a driver or passenger from smoking in a car if there was anyone under the age of 16 years in the car as well. Overall, 18.2% of adults reported being current smokers. There were more smokers before the law (20.3%) than after the law (15.9%). After adjusting for confounders such as drinking or age, the OR of a MVC was 1.209× before the law and 1.155× after the law.

Table: Comparison MVC Percentage in Smokers and Non-smokers, Before and After the 2009 Law

	2002–2008	2010–2016
Smokers	9.39%	7.01%
Non-smokers	7.08%	6.02%
Difference	2.31%	0.99%

Source: Adapted from Pederson et al, 2019

Overall, current smokers showed a 33% higher prevalence of collision involvement in the period from 2002 to 2016, compared to nonsmokers. This finding adds to the evidence that smoking is still a road safety risk, despite reductions in prevalence of tobacco use. These findings are consistent with the suggestion that road safety campaigns might benefit from addressing these risks in addition to those posed by drugs like alcohol and cannabis. However, the data do not provide insight into how tobacco use is related to what appears to be higher prevalence of collisions than that among nonsmokers. Research on the possible causal pathways between tobacco use and collision involvement is needed. Two explanations include the possibility that smoking increases distractions to the driver or that there are neurological and/or physiological effects of the toxicants in tobacco that interfere with driving.

Reference Number: 80403

ELIAS, W., N. ASSY, I. ELIAS, T. TOLEDO, M. YASSIN, AND A. BOWIRRAT. "The Detrimental Danger of Water-Pipe (Hookah) Transcends the Hazardous Consequences of General Health to the Driving Behavior." *Journal of Translational Medicine*, 10: 126 (14 pp), 2012 (7 tables, 2 figures, 76 references)

Abstract: A single waterpipe session exposes the smoker to 1.7× the nicotine, 3–9× the CO, and 56× the smoking volume compared to a single cigarette. Seventy exclusive waterpipe smokers and 70 controls undertook a computerized driving task of 10 km duration on three occasions. The waterpipe smokers (WPS) were tested before, immediately after, and

30 minutes after smoking a waterpipe. After waterpipe smoking, the heart rate of the waterpipe smokers increased from 80 to 95 ppm and the oxygen saturation decreased significantly from 97.9 to 97.3%. WPS had 1.33× the OR of accident and 1.31× the OR of crossing the lane on the computerized driving task.

> Chronic tobacco consuming appears to be associated with deficiencies in executive functions, general intellectual abilities, risk taking and sensation seeking behaviors, impairs performance of the cognitive and motor tasks necessary for safe driving and for reducing collision risk. In addition, smoke-induced eye blurring and cough and the resultant fatigue and even decreased vision of smokers due to deposited smoke on the automobile windshield. Our results converge with recent data for road crashes that point to the increasing presence of drugs other than alcohol (especially chronic tobacco, cannabis and depressants of the central nervous system) in injured and fatally injured drivers. Studies from Spain and also the United States have shown smokers to have a 50% higher risk of road crashes than nonsmokers. Another study from Canada showed that 30–39-year-old males who had been at-fault in crashes were 1.5 times more likely to be smokers.

Reference Number: 80404

SAADAT, S., AND M. KARBAKHSH, M. "Association of Waterpipe Smoking and Road Traffic Crashes." *BMC Public Health,* 10: 639 (6 pp), 2010 (4 tables, 1 figure, 50 references)

Abstract: Road traffic injuries (RTI) are a major global public health and safety problem and over 50% of RTI deaths occur in young adults (15–44 years). Daily waterpipe smoking is estimated to occur in 100 million users. The waterpipe is also called the hookah, hubble-bubble, boory, goza, shisha (glass in Arabic), narghile (coconut in Persian), and qalian. A telephone survey was conducted of 2,070 motor vehicle owners in Iran regarding MVCs and their smoking habits. About 15% (308) were involved in MVCs. After adjustment for daily drive time, the OR of an MVC was 1.68× for cigarette smoking and 2.41× for waterpipe smoking.

Table: Possible Factors for the Increased Risk of Motor Vehicle Collisions in Smokers

Possible Factors for Increased Risk of MVCs in Smokers
• Distraction (from smoking behaviors while driving)
• Carbon monoxide and Hypoxia effects
• Cigarette smoke deposited on the inside of the windshield reducing visibility, especially at night
• Smoking-induced blurring of eyes
• Fatigue
• Increased risk taking by smokers
• Seatbelt non-use is more common in smokers

Source: Adapted from Saadat and Karbakhsh, 2010

The relationship between smoking and RTC seems to be far from distraction and carbon monoxide effect on the driver as the risk exists regardless of whether drivers refrain from smoking while driving or not. Other mechanisms which have been implicated are as follow: cognitive impairments secondary to chronic nicotine exposure, risk taking and sensation seeking behaviors, smoke-induced eye blurring and cough and the resultant fatigue and even decreased vision of smokers due to deposited smoke on the automobile windshield. Prior injury history and risky behaviors such as seat belt non-use have also been reported to be more common in smokers.

Reference Number: 80405

YIENGPRUGSAWAN, V., J. BERECKI-GISOLF, C. BAIN, R. MCCLURE, S-A. SEUBSMAN, AND A.C. SLEIGH. "Predictors of Injury Mortality: Findings from a Large National Cohort in Thailand." *BMJ Open*, 4: e004668 (8 pp), 2014 (4 tables, 34 references).

Abstract: The rate of injury death over 5 years since 2005 and various exposure variables were determined in 87,037 long distance-learning students in Thailand. The incidence of injury deaths per 10,000 years were 9 for never smokers and 21 for smokers in transport-related deaths and 10 for never smokers and 35 for current smokers in non-transport-related deaths. Statistically significant predictors of injury death were being male (OR 3.897), being a current smoker (OR 1.56), and a history of drunk driving (OR 1.49).

Injury remains a major public health challenge world-wide, causing one-tenth of global mortality with a heavy burden in developing countries.

According to WHO, at least 1.2 million people are killed from road crashes and an estimated 50 million are injured on roads worldwide each year. Violence and non-transport injuries accounted for more than 1.3 million deaths and many suffered from serious physical and mental consequences.

Reference Number: 80406

HUTCHENS, L., T.M. SENSERRICK, P.E. JAMIESON, D. ROMER, AND F.K. WINSTON. "Teen Driver Crash Risk and Associations with Smoking and Drowsy Driving." *Accident Analysis and Prevention*, 40: 869–876, 2008 (4 tables, 49 references)

Abstract: A telephone survey was conducted of 550 drivers aged 14 to 22 years in the United States in 2005. One hundred and twenty-eight (23.3%) were current smokers and 202 (40.7%) had been in a car crash as a driver. The OR of crash involvement was 1.8× for driving drowsy alone, 1.3× for the length of licensure, and 2.1× for being a current smoker.

Possible explanations given for smoking's association with increased odds of injury include distractibility involved in the act of smoking (including one-handed steering), direct toxicity (potentially contributing to reduced night vision, vision performance, and reduced driving performance), and associated risk-taking. Another plausible rationale is that people who risk their health by engaging in smoking may also be more risk-taking drivers; for example, smokers have been found to wear their safety belts less often. Smoking was also significantly associated with sensation seeking and alcohol consumption in our study. However, we found that smoking was associated with higher crash odds despite holding constant general risk-taking tendencies as well as other indicators of risky behavior tendencies, such as drinking alcohol, using marijuana, and driving without a seatbelt. Hence, it appears that the risk conferred by smoking is unique to this form of risky behavior.

Reference Number: 80407

CHOI, K-C., S.A. KIM, N.R. KIM, AND M-H. SHIN. "Association Between Smoking and Unintentional Injuries Among Korean Adults." *Chonnam Medical Journal*, 54: 184–189, 2018 (3 tables, 33 references)

Abstract: The Korean Community Health survey of 214,621 adults in 2009 was used to determine the risk of unintentional injury and smoking. After adjusting for demographic, socioeconomic, health, and lifestyle variables, cigarette smoking was associated with the risk of unintentional injury in a dose-response manner.

Table: The Odds Ratio (OR) of Unintentional Injury and Smoking Status

Smoking Status	OR of Injury
Former Smoker	1.19×
<10 CPD	1.22×
11–20 CPD	1.33×
>20 CPD	1.40×

Source: Adapted from Choi et al, 2018

Although the underlying mechanism by which smoking contributes to unintended injuries is not yet clear, it can be explained by the following mechanisms. First, carelessness caused by smoking can increase the risk of injury. Distraction due to smoking and blurred vision due to smoke can cause injury. In addition, current smokers are more likely to have a car accident than non-smokers. Second, smoking decreases functions such as muscle strength, agility, and walking balance, increasing the risk of injury or reducing the likelihood of survival. Third, smoking is associated with other risk-taking behaviors, and smokers tend to behave more dangerously than non-smokers. Smokers have lower rates of use of driving safety belts, more traffic violations, and more drunk driving instances than non-smokers. In this study, we found that former smokers had a higher risk of unintentional injuries than non-smokers. These results suggest that the relationship between smoking and unintentional injuries may be influenced by the weakening of physiological function due to smoking and personal characteristics related to smoking habits, as well as the carelessness caused by smoking. In this study, the association between smoking and unintentional injuries appeared stronger in younger than in older individuals. This seems to be due to the greater tendency of young adults to engage in risk-taking behaviors as compared to older people.

Reference Number: 80408

HUNG, C-I., C-Y. LIU, AND C-H. YANG. "Unintentional Injuries Among Psychiatric Outpatients with Major Depressive Disorder." *Public Library of Science One*, 11: (10 pp), 2016 (5 tables, 27 references)

Abstract: The rate of unintentional injury (UI) such as burns, traffic accidents, and falls, was determined in 141 outpatients with major depressive disorder (MDD) at baseline and in 119 outpatients a year later. The severity of depression was determined using the Hamilton Depression Rating Scale. About 28% of the outpatients had experienced one or more unintentional injuries over the 1-year period. The OR of predicating unintentional injury both at baseline and follow-up was 11.89× for cigarette smoking, 0.05× for male gender, and 1.12× for age.

> Cigarette smoking was an independent risk factor related to UI. Smokers have significant dose-response excesses of injury death after controlling for other confounding factors. The associations of cigarette smoking and UI might result from several reasons, including direct toxicity, distractibility, smoking-associated medical conditions, psychiatric comorbidities, and behavioral characteristics.

Reference Number: 80409

GRIER, T.L., S. MORRISON, J.J. KNAPIK, M. CANHAM-CHERVAK, AND B.H. JONES. "Risk Factors for Injuries in the U.S. Army Ordnance School." *Military Medicine*, 176: 1292–1299, 2011 (4 tables, 40 references)

Abstract: A health survey and the Army Physical Fitness Test were administered to 3,751 male and 498 female soldiers, mainly between the ages of 17 and 19 years, who attended a US Army Ordnance School Advanced Individual Training. The time loss injury incidence was also determined. Thirty-six percent of the men and 61% of the women experienced at least 1 injury during the advanced training. Smoking cessation and fitness training are recommended to reduce injuries among soldiers in the Ordnance School.

Table: Cigarettes Per Day in the 30 Days Prior to Basic Training and the Hazard Ratio of Injury

CPD in 30 Days Prior to Basic Training	Hazard Ratio
0 (non-smokers)	1.00
<10	1.25
10–20	1.46
>20	1.87

Source: Adapted from Grier et al, 2011

The relationship between tobacco use and injury may also be because of greater risk-taking behaviors. In an Air Force study, recruits who were cigarette smokers had higher scores than nonsmokers on various measures of risk taking. These included an overall measure of risk taking, in addition to greater rebelliousness, less seat belt use, more risky sex, more favorable view of illegal drug use, more alcohol use, more binge drinking, less physical activity, less intake of fruits and vegetables, and greater intake of high fat foods. In civilian studies, smokers had more motor vehicle accidents, more traffic violations, less seat belt use, less physical activity, more alcohol consumption, and lower intake of fruits and vegetables compared to nonsmokers. Another hypothesis to explain the association between injuries and tobacco use may be a decrease in fatigue resistance. In a study investigating muscle skeletal fatigue and smoking history, investigators found that skeletal muscle of smokers was more fatigable than that of age- and physical activity- matched nonsmokers. The authors suggested that smoking may have an acute and reversible effect on skeletal muscle fatigability caused by carbon monoxide in cigarette smoke.

Reference Number: 80410

BROWN, J.M., E.M.A. GOODELL, J. WILLIAMS, AND R.M. BRAY. "Socioecological Risk and Protective Factors for Smoking Among Active Duty U.S. Military Personnel." *Military Medicine*, 183: e231–e239, 2018 (3 tables, 62 references)

Abstract: In 1980, more than 50% of military personnel were smokers. In 1986, substantial smoking prevention and cessation programs were introduced and have resulted in the decline of current smokers in the military to 24% (although higher than 19% of the general population). A survey was conducted of 4,748 military personnel who were current smokers.

At the individual level, we found that male Army personnel had the highest levels of risk for heavy smoking and the greatest likelihood of high nicotine dependence and may potentially be at higher risk for negative performance outcomes. It is well documented that smoking is associated with lower levels of physical fitness, less mental sharpness, and increased risk of injury.

Reference Number: 80411

MERRILL, R.M. "Injury-Related Deaths According to Environmental, Demographic, and Lifestyle Factors." *Journal of Environmental and Public Health*, 2019: (12 pp), 2019 (5 tables, 76 references)

Abstract: Relative mortality rates of unintentional injuries were determined in the United States between 2011 and 2015 according to various factors, including cigarette smoking. Counties with a greater percentage of adult smokers had a greater rate of deaths from accidents and adverse events.

Table: Relative Risk of Mortality from Accidents and Adverse Events According to the Percentage of Adult Smokers in Each US County

% Adult Smokers	Relative Risk of Mortality
9.3–23.6	1.00
23.6–26.1	1.29
26.2–28.5	1.44
28.6–40.3	1.69

Source: Adapted from Merrill, 2019

Cigarette smoking has been identified as a leading cause of fire disaster and resulting death, fatal automobile accidents, and injuries at work and other unintentional injuries. Reasons for the association between smoking and injuries may be because of distraction, direct toxicity, and confounding factors (smoking-related medical conditions, personality, and behavior characteristics).

Reference Number: 80412

CALFEE, C.S., M.A. MATTHAY, M.D. EISNER, N. BENOWITZ, M. CALL, J-F. PITTET, AND M.J. COHEN. "Active and Passive Cigarette Smoking and Acute Lung Injury After Severe Blunt Trauma." *American Journal of Respiratory Critical Care Medicine*, 183: 1660–1665, 2011 (3 tables, 2 figures, 48 references)

Abstract: Plasma cotinine concentrations were determined by LC/MS/MS in 144 adult patients with severe blunt trauma admitted to San Francisco General Hospital between 2005 and 2009. The patients were then followed to test for the development of Acute Lung Injury (ALI). Using a plasma cotinine cut-off of more than 3.08 ng/mL, 44% of the patients were classified as active smokers. Patients with evidence of alcohol abuse had a higher median plasma cotinine concentration (26 ng/mL compared to 0.2 ng/mL). Increasing plasma concentrations were associated with an increased risk of ALI. Moderate to heavy passive cigarette exposure also increased the risk of ALI.

Table: Odds Ratio of Developing Acute Lung Injury in Smoking Blunt Trauma Patients

Predictor	OR for Developing ALI
Moderate to heavy passive smoke exposure (cotinine 0.19–3.08 ng/mL)	2.88
Acute smoking (cotinine >3.08 ng/mL)	2.83
Injury Severity Scale (per one-point increase)	1.04

Source: Adapted from Calfee et al, 2011

In conclusion, both active and moderate to heavy passive smoking are independently associated with the development of acute lung injury after severe blunt trauma. This finding has important implications for public health, because both major trauma and ALI are common causes of death, as well as for our understanding of the pathogenesis of ALI.

Reference Number: 80413

WEN, C.P., S.P. TSAI, T.Y. CHENG, H.T. CHAN, W.S.I. CHUNG, AND C.J. CHEN. "Excess Injury Mortality Among Smokers: A Neglected Tobacco Hazard." *Tobacco Control*, 14, i28–i32, 2005 (3 tables, 1 figure, 48 references)

Abstract: The mortality risks of 25,848 smokers, 6,515 ex-smokers, and 31,956 non-smoking males were determined in Taiwan over a period of 12–18 years. After adjusting for age and alcohol use, the OR of death from MVCs was 1.88× for smokers compared to non-smokers and 1.69× for all fatal injuries. The study found that nicotine withdrawal symptoms such as irritation and listlessness after not smoking for several hours may also play a role in fatal injuries. When smokers quit, the excess risk of accidental death was no longer elevated above that of non-smokers. In addition to second- and thirdhand smoke, non-smokers may be at an increased risk of various injuries such as MVCs caused by smokers.

Table: Relative Risk of Injury According to Cigarettes Per Day (CPD) Smoked

Cause of Death	<10 CPD	11–20 CPD	>20 CPD
Motor vehicle collision	1.54	2.01	1.95
Fall	1.92	1.83	4.20
Job related	1.38	3.36	5.72
Suicide	0.72	1.64	1.77

Source: Adapted from Wen et al, 2005

Smokers may have a higher propensity for risk taking, by not wearing a seatbelt in cars, having more traffic violations, or driving while drunk. In addition, workers who smoke cause more workplace injuries than nonsmokers. Our data also suggest Taiwanese smokers have increased job related injuries.

Reference Number: 80414

GIANNAKOULAS, G., A. KATRAMADOS, N. MELAS, I. DIAMANTOPOULOS, AND E. CHIMONAS. "Acute Effects of Nicotine Withdrawal Syndrome in Pilots During Flight." *Aviation, Space, and Environmental Medicine*, 74: 247–251, 2003 (4 tables, 25 references)

Abstract: Twenty male smoking military pilots (mean age 33.7 years) of the Hellenic Air Force were tested on a computerized mental arithmetic, visual vigilance, and free-recall tasks before and after 12 hours abstinence from smoking.

Table: Reported Symptoms in Military Pilots After 12 Hours Abstinence from Cigarettes

Reported Symptoms	Percent Reporting
Nervousness	45%
Craving for tobacco	40%
Tension-anxiety	35%
Fatigue	25%
Difficulty in concentration	20%
Vigilance decrement	20%
Disorders of fine adjustment	15%

Reported Symptoms	Percent Reporting
Prolonged RT during flight	10%
Anger-irritability	10%
Drowsiness	10%
Increased appetite	10%
Impaired judgement	10%

Source: Adapted from Giannakoulas et al, 2003

The most frequent symptoms reported during nicotine deprivation were nervousness, craving for tobacco, tension-anxiety, fatigue, difficulty in concentration, decrease in alertness, disorders of fine adjustments, prolonged reaction times, anger-irritability, drowsiness, increase in appetite, and impairment of judgement. Systolic BP and heart rate tended to decrease and diastolic BP tended to rise during withdrawal, although the differences were not statistically significant. Finally, all tests recorded an impairment of cognitive functions during abstinence.... Abrupt cessation of smoking may be detrimental to flight safety and the smoking withdrawal syndrome may influence flying parameters.

8.05 MISCELLANEOUS

"The substantial harmful effects of tobacco smoking on fertility and reproduction have become apparent but are not generally appreciated. Tobacco smoke contains more than 4000 kinds of constituents, including nicotine, tar, carbonic monoxide, polycyclic aromatic hydrocarbons, and heavy metals."

—Dai, Wang, and Qiao, "The Hazardous Effects of Tobacco Smoking on Male Fertility" (2015)

Tobacco, mainly cigarette use, also causes a number of other adverse effects, including:

- premenstrual disorders (80501);
- lower sperm quantity and quality in the sons of fathers who smoked (80502);
- sperm motility (80503);
- poor oral health and increased cavities (80504);
- grinding teeth during sleeping (80505);

- decreased breastfeeding duration (80506); and
- substantial increase in smoking rates in Manhattan after 9/11 (80507).

Reference Number: 80501

DEL MAR FERNÁNDEZ, M., A. MONTES-MARTÍNEZ, M. PIÑEIRO-LAMAS, C. REGUEIRA-MÉNDEZ, AND B. TAKKOUCHE. "Tobacco Consumption and Premenstrual Syndrome: A Case-Control Study." *Public Library of Science One* 14: (12 pp), 2019 (4 tables, 1 figure, 33 references)

Abstract: Premenstrual disorders are defined as recurrent physical and psychic symptoms (such as malaise, anxiety) in the luteal phase of the menstrual cycle and are divided into two main categories, premenstrual syndrome (PMS) and the more severe premenstrual dysphoric disorder (PMDD). Tobacco affects the regulation of sex hormones such as estrogen. A study was conducted of 285 cases of PMS and controls and 88 cases of PMDD and controls at three hospitals in Spain. The OR of PMS was 1.78× and of PMDD 2.92× for current smokers compared to never smokers.

This study shows that smokers are more likely to develop PMS and PMDD.

Reference Number: 80502

AXELSSON, J., S. SABRA, L. RYLANDER, A. RIGNELL-HYDBOM, C.H. LINDH, AND A. GIWERCMAN. "Association Between Paternal Smoking at the Time of Pregnancy and the Semen Quality in Sons." *Public Library of Science One*, 13: (12 pp), 2018 (4 tables, 2 figures, 36 references)

Abstract: The semen quality (volume, sperm count) was determined in 104 male Swedes (aged 17 to 20 years) and information about paternal smoking was also gathered. Serum nicotine concentrations were also determined by LC/MNS in mothers from samples collected between 6 to 35 weeks of pregnancy, initially collected for rubella screening. Among the 104 participants, 24% were smokers and 76% were non-smokers. The smokers had an average of 7 CPDs and had smoked for 2.9 years. Thirty-two percent of the men reported that their fathers had smoked.

Table: Mean Semen Parameters in Young Men of Smoking or Non-Smoking Fathers

Semen parameter	Smoking father	Non-smoking father
Concentration (million/mL)	29	67
Total sperm count (millions)	61	170
Semen volume (mL)	2.3	2.7
Normal morphology (%)	7.1	9.3

Source: Adapted from Axelsson et al, 2018

Men of smoking fathers had about 50% lower total sperm count and sperm concentration than men of non-smoking fathers, also after adjustments for the levels of maternal exposure to nicotine during the pregnancy and the socioeconomic status.

Reference Number: 80503

DAI, J-B., Z-X. WANG, AND Z-D. QIAO. "The Hazardous Effects of Tobacco Smoking on Male Fertility." *Asian Journal of Andrology*, 17: 954–960, 2015 (1 table, 1 figure, 110 references)

Abstract: Tobacco smoke contains more than 4,000 constituents including nicotine, tar, carbon monoxide, polyaromatic hydrocarbons, radioactive compounds, and heavy metals. Nicotine is detected in the serum and semen of smokers. Nicotine metabolites (cotinine and 3'-HC) are also found in the seminal fluid and cause a dose-related decrease in sperm motility. Other chemicals in cigarette smoke also affect male fertility and cigarette smoking is a risk factor of erectile impotence.

The substantial harmful effects of tobacco smoking on fertility and reproduction have become apparent but are not generally appreciated.... Because of the complexity of tobacco smoke components, the toxicological mechanism is notably complicated. Most studies have reported reduced semen quality, reproductive hormone system dysfunction and impaired spermatogenesis, sperm maturation, and spermatozoa function in smokers compared with nonsmokers. Underlying these effects, elevated oxidative stress, DNA damage, and cell apoptosis may play important roles collaboratively in the overall effect of tobacco smoking on male fertility. In this review, we strive to focus on both the phenotype of and the molecular mechanism underlying these harmful effects, although current studies regarding the mechanism remain insufficient.

Reference Number: 80504

SHARMA, S., S.K. MISHRA, AND N. MITTAL. "Influence of Tobacco Dependence on Caries Development in Young Adult Males: A Cross-Sectional Study." *Journal of Conservative Dentistry*, 21: 597–601, 2018 (2 tables, 22 references)

Abstract: The nicotine dependence (Fagerstrom test) and caries development (decayed-missing-filled teeth/decayed-missing-filled-surfaces (DMFT/DMFS)) were determined in 100 men who used smokeless tobacco (Group A), 100 men who smoked (Group B), and 100 who did not use tobacco in India. The DMFT/DMFS index was higher for Group A, followed by Group B, and the lowest for Group C. Long-term tobacco use decreases taste receptors and depresses the salivary buffering capacity by 20%, which results in a more acidic pH. Smokeless tobacco also contains sugar (up to 34%) to neutralize the bitter taste of tobacco.

> Both forms of tobacco proved to be a significant risk factor for increased caries development. Thus, tobacco use should be an important factor in caries risk assessment of patients.

Reference Number: 80505

RINTAKOSKI, K., J. AHLBERG, C. HUBLIN, F. LOBBEZOO, R.J. ROSE, H. MURTOMAA, AND J. KAPRIO. "Tobacco Use and Reported Bruxism in Young Adults: A Nationwide Finnish Twin Cohort Study." *Nicotine and Tobacco Research*, 12: 679–683, 2010 (2 tables, 19 references)

Abstract: High levels of smoking lead to increased levels of nicotine and dopamine release and can result in bruxism (grinding or teeth clenching during sleep). A survey was conducted of 3,065 pairs of twins in Finland as to smoking history and bruxism. The OR of bruxism was 2.45× in heavy smokers.

> Given the observed associations with both heavy smoking and smokeless tobacco and a dose–response relationship, the present results support our hypothesis of a link between nicotine intake and bruxism.

Reference Number: 80506

HIGGINS, T.M., S.T. HIGGINS, S.H. HEIL, G.J. BADGER, J.M. SKELLY, I.M. BERNSTEIN, L.J. SOLOMON, Y. WASHIO, AND A.M. PRESTON. "Effects of Cigarette

Smoking Cessation on Breastfeeding Duration." *Nicotine and Tobacco Research* 12: 483–488, 2010 (1 table, 3 figures, 23 references)

Abstract: A study was conducted of 158 smoking pregnant, mainly socio-economically disadvantaged women (most <25 years of age) who were either assigned to an incentive-based smoking cessation treatment or receiving no therapy. Smoking status was verified by an enzymatic immunoassay test for cotinine in urine. Maternal reports of breastfeeding were determined at 2, 4, 8, 12, and 24 weeks postpartum and were compared between the two groups. At 8 weeks postpartum, the smoking cessation mothers were 2.7× as likely to breastfeed than non-smoking cessation mothers.

> The present findings suggest that increased breastfeeding duration may be added to what is already a long list of health benefits of maternal smoking cessation. We previously reported that smoking cessation with this same incentives-based intervention increases fetal growth. Both outcomes have the potential to contribute long-standing infant/child health benefits.

Reference Number: 80507

VLAHOY, D., S. GALEA, H. RESNICK, J. AHERN, J.A. BOSCARINO, M. BUCUVALAS, J. GOLD, AND D. KILPATRICK. "Increased Use of Cigarettes, Alcohol, and Marijuana Among Manhattan, New York, Residents After the September 11th Terrorist Attacks." *American Journal of Epidemiology*, 155: 988–996, 2002 (4 tables, 39 references)

Abstract: The incidence of cigarette, alcohol, and cannabis use among 988 Manhattan residents was determined 5–8 weeks after the 11 September 2001 terrorist attacks on the World Trade Center. Approximately 10% reported an increase in smoking, 25% an increase in alcohol consumption, and 3% an increase in cannabis use after the attack. Those who reported an increase in smoking or cannabis use were more likely to experience post-traumatic stress disorder. Depression was more common in those who reported an increase in consumption of cigarettes, cannabis, or alcohol.

> Our results suggest that there was a substantial increase in substance use (cigarette smoking, alcohol consumption, or marijuana use) in Manhattan after the September 11th, 2001, terrorist attacks. Persons who reported an increase in substance use also had a higher prevalence of current PTSD and of current depression than did persons who did not increase their use of these substances. Overall, 28.8 percent of those in our sampling frame

reported an increase in use of any substance; this corresponds to 265,000 persons (95 percent confidence interval: 236,000, 294,000) living in Manhattan below 110th Street. Persons who reported an increase in cigarette smoking or marijuana use were also more likely to have both PTSD and depression, while those who reported an increase in alcohol consumption were more likely to have depression only. Thus, our results suggest that increased substance use may be a significant problem in the immediate aftermath of a disaster and that it may frequently cooccur with PTSD and depression.

8.06 PAIN

> "In patients who completed evaluation in an outpatient pain clinic, current cigarette smokers reported significantly greater pain intensity and pain interference with functioning. Symptoms were more pronounced in smokers with more severe nicotine dependence."
>
> —Weingarten et al, "An Assessment of the Association Between Smoking Status, Pain Intensity, and Functional Interference in Patients with Chronic Pain" (2008)

Former smokers and current smokers were nearly 2× as likely to suffer back pain than non-smokers (80601). Smokers with the highest nicotine dependence had a greater degree of pain that interfered with daily life (80602–3).

Smokers who discontinued smoking at least 3 weeks before surgery had better outcomes and less severe pain than smokers who quit just before surgery (80604).

Reference Number: 80601

ZIMMERMANN-STENZEL, M., J. MANNUSS, S. SCHNEIDER, AND M. SCHILTENWOLF. "Smoking and Chronic Back Pain. Analyses of the German Telephone Health Survey 2003." *Deutsches Arzteblatt International*, 105: 441–448, 2008 (3 tables, 1 figure, 22 references)

Abstract: In 2003, the effect of smoking on chronic back pain was determined in a telephone survey conducted of 8,318 people in Germany. Daily smokers in general were 1.65× as likely to have chronic back pain as non-smokers, and former smokers were 1.65× as likely. As the duration of tobacco smoking increases the greater the OR of developing back pain increases. The age of initiation of smoking and amount of tobacco consumed had no effect.

Table: Mean OR of Chronic Back Pain with Increasing Years of Smoking (Compared to 10 Years of Smoking)

Years of Smoking	OR of Chronic Back Pain
16 to 20	1.89×
21 to 25	1.84×
26 to 30	3.24×
31 to 35	5.15×
>35	9.35×

Source: Adapted from Zimmermann-Stenzel et al, 2008

Nevertheless, the deleterious effect of nicotine is undisputed. On the one hand, nicotine consumption leads in the short term to vasoconstriction and in the long term to arteriosclerosis, resulting not only in an inadequate supply of blood to the internal organs but also to deficient nutrition of bone and thus to reduced bone formation. Decreased perfusion of structures in and around the spinal column, particularly the intervertebral disks and the muscles of the back, seem to promote degenerative changes. On the other hand, tobacco smoke contains poisons and carcinogenic substances that lead to increased bone resorption.

Reference Number: 80602

WEINGARTEN, T.N., S.M. MOESCHLER, A.E. PTASZYNSKI, W.M. HOOTEN, T.J. BEEBE, AND D.O. WARNER. "An Assessment of the Association Between Smoking Status, Pain Intensity, and Functional Interference in Patients with Chronic Pain." *Pain Physician*, 11: 643–653, 2008 (2 tables, 2 figures, 48 references)

Abstract: A study was conducted of 500 consecutive new patients at an outpatient pain clinic regarding smoking status, a Brief Pain Inventory (BPI), and the Fagerstrom Test for Nicotine Dependence (FTND) of whom 230 completed and returned the survey. Thirty-three respondents (14.7%) were current smokers, 33.8% were former smokers, and 51.1% had never smoked or used tobacco products. Current smokers were more likely to be male, younger, and unemployed compared to former/never smokers. Smokers reported greater pain intensity and smokers with severe nicotine dependence (FTND of 6 or higher) had higher BPI scores.

Table: Impact of Smoking Status on Pain Intensity Interference with Life Activities (Higher the Number=Greater Pain Interference)

BPI Pain Interference	Smokers	Non-smokers
General Activity	6.44	5.00
Mood	6.44	4.56
Walking	5.47	4.59
Normal Work	6.55	5.21
Relationships	4.74	3,59
Sleep	7.19	4.97
Life Enjoyment	6.66	5.27

Source: Adapted from Weingarten et al, 2008

In patients who completed evaluation in an outpatient pain clinic, current cigarette smokers reported significantly greater pain intensity and pain interference with functioning. Symptoms were more pronounced in smokers with more severe nicotine dependence.

Reference Number: 80603

ZVOLENSKY, M.J., K. MCMILLAN, A. GONZALEZ, AND G.J.G. ASMUNDSON. "Chronic Pain and Cigarette Smoking and Nicotine Dependence Among a Representative Sample of Adults." *Nicotine and Tobacco Research*, 11: 1407–1414, 2009 (5 tables, 25 references)

Abstract: A survey (NCS-R) was conducted of 9,282 English speaking adults in the United States between 2001 and 2003. Self-reported cigarette smoking and chronic pain were determined. Chronic pain is defined as pain that persists for longer than 3 months with no known biological reasons.

Table: Odds Ratio of Chronic Neck or Back Pain or Unexplained Chronic Pain and Smoking Status

Smoking Status	Past Year Chronic Neck or Back Pain	Lifetime Medically Unexplained Chronic Pain
Current smoker	1.18×	1.47×
Lifetime nicotine dependence	1.95×	1.68×
Past year nicotine dependence	2.30×	2.08×

Adapted from Zvolensky et al, 2009

Reference Number: 80603

After adjusting for sociodemographic variables and the presence of a lifetime substance use disorder, individuals with a lifetime history of chronic neck or back pain were significantly more likely to be current smokers and to be diagnosed with lifetime as well as current nicotine dependence. Although there was no significant incremental relation between current chronic neck and back pain and being a current smoker, there was a significant association with lifetime and current nicotine dependence. Similar relations were evident among those with and without medically unexplained chronic pain in regard to smoking status and lifetime and current nicotine dependence.

Reference Number: 80604

ZHAO, S., F. CHEN, D. WANG, H. WANG, W. HAN, AND Y. ZHANG. "Effect of Preoperative Smoking Cessation on Postoperative Pain Outcomes in Elderly Patients with High Nicotine Dependence." *Medicine*, 98: (6 pp), 2019 (5 tables, 1 figure, 36 references)

Abstract: Thirty percent of surgical patients in the United States are smokers. Nicotine dependence increases the 30-day mortality and incidence of pulmonary complications after surgery. The Fagerstrom Test of Nicotine Dependence (FTND) was administered to 107 male patients who underwent thoracoscopic radical lung cancer surgery. The patients were divided into three groups; high nicotine dependence and discontinued smoking less than 3 weeks prior to surgery (Group A); high nicotine dependence and discontinued smoking greater than 3 weeks before surgery (Group B); and non-smokers (Group C).

Among cases requiring thoracoscopic radical lung cancer surgery, highly nicotine-dependent patients who are deprived of cigarettes before surgery require a larger quantity of postoperative sufentanil and experience more severe pain than nonsmokers. Furthermore, patients who discontinued smoking 3 weeks or more before surgery had better pain outcomes than those who stopped smoking within 3 weeks. Based on the current findings, preoperative smoking cessation as early as possible may contribute to postoperative pain relief in this patient population.

8.07 PREGNANCY

"Smoking also causes reduced fertility, low birth weight in newborns, and cancer of the cervix and uterus."

—Kollath-Cattano et al, "Adult Smokers' Responses to 'Corrective Statements' Regarding Tobacco Industry Deception" (2014)

Nicotine, CO, and numerous other toxic chemicals pass rapidly from the smoking or SHS-exposed pregnant parent into the developing fetus, causing an increased risk of miscarriage, stillbirths, premature births, low birth weight, sudden infant death, and poorer infant cognition (80701–2).

Nicotine appears to be the primary toxin which affects the development of the fetal lung, and so the use of NRT or ECs are not recommended to wean off cigarette use during pregnancy (80703–4). The use of regular full-flavor cigarettes is associated with nicotine dependence during pregnancy and should be restricted (80705).

Not only is the developing fetus affected by smoking, but the pregnant parent will also suffer increased risks of serious psychological distress, stroke, myocardial infarction, and bronchitis (80706–7). Long-term health problems in the fetus may also be caused by smoking during pregnancy (80708–9).

Secondhand smoke exposure will increase the risk of implantation failure and decrease the OR of a successful live birth for those who undergo in vitro fertilization procedure (80710).

Human milk that is purchased online and supposedly guaranteed to be from a non-smoker has in fact been found to contain nicotine and cotinine (58%), and 10% of the human milk samples contained bovine DNA (80711).

Reference Number: 80701

COOPER, S., S. ORTON, J. LEONARDI-BEE, E. BROTHERTON, L. VANDERBLOEMEN, K. BOWKER, F. NAUGHTON, M. USSHER, K.E. PICKETT, S. SUTTON, AND T. COLEMAN. "Smoking and Quit Attempts During Pregnancy and Postpartum: A Longitudinal UK Cohort." *BMJ Open*, 7: (12 pp), 2017 (3 tables, 2 figures, 52 references)

Abstract: A study was conducted of 850 pregnant women (16 years of age or older) who were current smokers or had smoked within 3 months before pregnancy at two perinatal hospitals in Nottingham, England.

Smoking in pregnancy is associated with increased risk of miscarriage, stillbirths, premature births, low birth weight, sudden infant death, and poorer infant cognition. The motivation to quit appears to be strongest in early pregnancy.

Many pregnant women who smoke attempt quitting throughout their pregnancy, but this makes little difference to overall smoking rates. After giving birth, most smokers seem less inclined to make further quit attempts and many who quit in early pregnancy return to smoking. Women who quit in late pregnancy may be most likely to return to smoking after childbirth, while those who stopped prior to pregnancy may be least likely to relapse. It is therefore imperative to discuss smoking with women, including recent ex-smokers, throughout pregnancy and postpartum, and to continue to offer and provide specialist stop smoking support.

Reference Number: 80702

ZHAO, R., Y. WU, F. ZHAO, Y. LV, D. HUANG, J. WEI, C. RUAN, M. HUANG, J. DENG, D. HUANG, AND X. QIU. "The Risk of Missed Abortion Associated with the Levels of Tobacco, Heavy Metals and Phthalate in Hair of Pregnant Women." *Medicine*, 96: (5 pp), 2017 (3 tables, 33 references)

Abstract: Missed abortion (MA) is also known as silent miscarriage in which the embryo or fetus dies, but the body does not recognize the loss and continues to release hormones. MA occurs in about 15% of pregnancies. Hair nicotine, cotinine, heavy metals, and phthalate concentrations by GC were determined in 42 patients with MA and 57 matched pregnant controls. The odds ratio of hair nicotine and cotinine in those with MA were 3.31× and 2.16× respectively compared to those without MA.

Pregnant women's exposure to tobacco smoke also increases the risk of placental abruption, placenta previa, fetal growth restriction, and other pregnancy complications. Active smokers inhale first hand smoke, while passive smokers inhale side stream smoke which due to the incomplete combustion of tobacco, contains harmful material more dangerous than the mainstream smoke. Tobacco smog effect in pregnant women is mainly due to harmful substances (such as nicotine, cotinine, carbon monoxide, and tar) in the smoke which causes pathological changes leading to the placenta calcification, which in turn includes shorter microvilli, reduced placental blood vessels, and increased collagen content of fuzzy matrix. It can furthermore increase the placenta and uterus contraction, reducing the flow of blood leading to miscarriage and stillbirth. Nicotine and cotinine from the mother's body may enter the placental circulation having direct toxic effect on the fetal cardiovascular system.

Reference Number: 80703

MCEVOY, C.T., AND E.R. SPINDEL. "Pulmonary Effects of Maternal Smoking on the Fetus and Child: Effects on Lung Development, Respiratory Morbidities, and Life Long Lung Health." *Paediatric Respiratory Reviews*, 21: 27–33, 2017 (3 figures, 66 references)

Abstract: A detailed review was conducted on the effect that birthing parents smoking during pregnancy has on a developing fetus. Smoking during pregnancy is the largest preventable cause of abnormal in-utero lung development. Only about 50% of the smokers are able to quit smoking during pregnancy. Approximately 5% of pregnant people smoke in Sweden and Austria, and the percentage increases to 15–20% in the Netherlands, England, and Turkey and to 40% in Greece. Nicotine appears to be the primary chemical in tobacco smoke that affects lung development in the fetus and thus vaping and other nicotine replacement therapy is not advised.

The developing fetal lung is very sensitive to the effects of in-utero tobacco products with altered lung function demonstrated at birth that is long lasting and likely affects subsequent lung trajectory and lung health. These deficits are consistently demonstrated as decreases in forced expiratory flows, among others, and are likely mediated by nicotine crossing the placenta to interact with nicotinic acetylcholine receptors that are expressed in fetal lung. Since in-utero smoke/nicotine exposure may prohibit the individual from attaining their maximum lung growth, they are more predisposed to premature pulmonary aging and the development of chronic obstructive pulmonary disease. This makes continued focus on cessation of all tobacco products prior to and during pregnancy of vital importance. Given the difficulty of fighting nicotine addiction and the continued use of tobacco products by millions during pregnancy, it is also critical to identify primary prevention approaches to block the effects of in-utero nicotine on lung development.

Reference Number: 80704

SHERIF, N.A., S.M. KAMEL, O.S. AL-ASHKAR, O.A. SHARAKI, E.A. SEIF, AND E.A. HEGAZY. "Detection of Cotinine in Neonate Meconium as a Marker for Nicotine Exposure in utero." *East Mediterranean Health Journal*, 10: 96–105, 2004 (5 tables, 5 figures, 20 references)

Abstract: Salivary and urine cotinine concentrations were measured by RIA in 10 active smoking, 10 passive smoking, and 10 non-smoking mothers at a perinatal hospital in Egypt between 1999 and 2000. Meconium samples from the neonate were also analyzed for cotinine. Cotinine is the main metabolite of nicotine (70%) and has a long plasma half-life (30 hours), which makes it an ideal marker for nicotine exposure. The neonates whose mothers smoked had an average of a 490 g lower birth weight than neonates from non-smoking mothers.

Table: Salivary and Neonate Meconium Cotinine Concentrations in Active, Passive, and Non-Smoking Mothers

Smoking group	Maternal salivary cotinine concentration (ng/mL)	Neonate meconium cotinine concentration (ng/mL)
Active smokers	91.8	367.2
Passive smokers	13.7	263.4
Non-smokers	1.1	185.0

Adapted from Sherif et al, 2004

Cotinine was detected in maternal urine and saliva and these two biological fluids are good markers for tobacco smoking. Similarly, its presence in meconium proved that meconium is an ideal biological marker for testing direct fetal exposure to tobacco smoking in the neonatal period. This is important as it may throw light on the cause of a neonate with low birth weight. Added to this, collection of meconium is easy and non-invasive.

Reference Number: 80705

HIGGINS, S.T., R. REDNER, C.A. ARGER, A.N. KURTI, J.S. PRIEST, AND J.Y. BUNN. "Use of Higher-Nicotine/Tar-Yield (Regular Full Flavor) Cigarettes is Associated with Nicotine Dependence and Smoking During Pregnancy Among U.S. Women." *Preventive Medicine* 104: 57–62, 2017 (1 table, 2 figures, 35 references)

Abstract: This study included data from the National Survey on Drug Use and Health (NSDUH) of 199,784 women of reproductive age (15–44 years) between 2005 and 2014, including 43,432 current smokers of whom 1,671 were pregnant. Smoking prevalence was 14.3% in pregnant women and

22.7% in non-pregnant women. But full-flavored cigarettes were used by 54.8% of the pregnant smokers compared to 38.9% of non-pregnant smokers. Smoking prevalence decreased from 19.7% to 12.5% to 11.1% in the first, second, and third trimester, respectively.

In terms of tobacco control and regulatory implications, the present study provides strong evidence from nationally representative samples that use of full-flavor cigarettes is associated with increased odds of nicotine dependence and a lower likelihood of quitting smoking during pregnancy. The literature on use of full-flavor compared to lower estimated yield cigarettes indicates that on average use of the former is associated with approximately 25% greater nicotine exposure. Considering that nicotine is a well-documented neuro-toxin and contributor to other adverse developmental effects of maternal cigarette smoking, there is little question that this association of full-flavor cigarette use and smoking in pregnancy has the potential for serious adverse public health impacts. Efforts to reduce use of full-flavor cigarettes among pregnant women are certainly warranted. Perhaps use of full-flavor cigarettes should be an indicator for providing financial incentives or some other more intensive cessation intervention.

Reference Number: 80706

GOODWIN, R.D., K. CHESLACK-POSTAVA, D.B. NELSON, P.H. SMITH, D.S. HASIN, T. JANEVIC, N. BAKOYIANNIS, AND M.M. WALL. "Serious Psychological Distress and Smoking During Pregnancy in the United States: 2008–2014." *Nicotine and Tobacco Research*, 605–614, 2017 (4 tables, 68 references)

Abstract: The association between acute smoking (within the past 30 days) or recent smoking (within last year but not within the last 30 days) and serious psychological distress (SPD) was determined in 1,020 pregnant people (>18 years of age) in the United States between 2008 and 2014. Overall, 14.1% of pregnant people were current smokers, whereas 40% of those with acute SPD were current smokers. The OR of smoking was 5.05× for past month SPD and 2.37× for those with past year SPD.

SPD and smoking during pregnancy are strongly associated; this relationship is present across all sociodemographic groups and the prevalence of smoking in pregnancy has remained relatively unchanged over the past decade both in the presence and absence of SPD.

Reference Number: 80707

ROELANDS, J., M.G. JAMISON, A.D. LYERLY, AND A.H. JAMES. "Consequences of Smoking During Pregnancy on Maternal Health." *Journal of Women's Health*, 18: 867–872, 2009 (5 tables, 30 references)

Abstract: The prevalence of various conditions and complications were determined in 640,813 smoking and 20,567,169 non-smoking pregnancy-related hospital discharges in the United States between 2000 and 2004.

Table: Odds Ratio of Medical Complications Occurring in Smokers During Pregnancy

Medical Complication	OR of Smokers compared to Non-smokers
Stroke	1.7×
Myocardial Infarction	4.6×
Pulmonary Embolus	2.5×
Deep Vein Thrombosis	1.3×
Pneumonia/influenza	2.9×
Bronchitis	15.2×
Bacterial Infection	1.2×
Pyelonephritis	1.2×

Source: Adapted from Roelands et al, 2009

In the end, maternal health cannot be separated from fetal health. A stroke or an MI during pregnancy will have immediate consequences for maternal and fetal health, and it will most certainly also have a long-term impact on the health of the woman, her children, and her family. Often, the best way to ensure the health of the fetus is to ensure the health of the pregnant woman. Identifying and preventing maternal complications of smoking during pregnancy are important components of improving the health of both women and the fetuses they carry.

Reference Number: 80708

MARITZ, G.S., AND M. MUTEMWA. "Tobacco Smoking: Patterns, Health Consequences for Adults, and the Long-term Health of the Offspring." *Global Journal of Health Science*, 4: 62–75, 2012 (107 references)

Abstract: A general review was conducted of the problems with tobacco smoking globally. In the 1990s, approximately one billion people were daily smokers; of these, 47% were adult men and 12% were adult women. The cost of smoking was estimated to be about $197 billion a year at the time of the study, and smoking reduces the lifespan of smokers by 13 to 14 years. Nicotine activates the same reward pathways in the brain as cocaine or amphetamine, although to a lesser degree, and increases the level of dopamine in the brain, resulting in feelings of pleasure and well-being. Tobacco smoking and secondhand smoke are causative for childhood asthma and COPD, which affected nearly 3 billion people worldwide at the time of the study. COPD is irreversible and difficult to treat. Thirdhand smoke is cigarette smoke deposits on the furniture, clothing, and rugs and is the residue of secondhand smoke. Nicotine is readily absorbed to these surfaces and is of particular risk to toddlers and young children.

> In conclusion, although tobacco use decreases in developed countries, it increases in developing countries. A decrease in tobacco use is not accompanied by an immediate decrease in tobacco related diseases. The prevalence in these diseases only decrease after several decades.

Reference Number: 80709

RANTAKALLIO, P., E. LÄÄRÄ, AND M. KOIRANEN. "A 28 Year Follow-up of Mortality Among Women Who Smoked During Pregnancy." *British Medical Journal*, 311: 477–480, 1995 (5 tables, 21 references)

Abstract: The smoking habits and mortality over a 28-year period was determined in 12,066 pregnant women who gave birth in northern Finland in 1966.

> The mortality ratio adjusted for age, place of residence, years of education and marital status was 2.3 (95% confidence interval 1.8 to 2.8) for the women who smoked during pregnancy and 1.6 (1.1 to 2.2) for those who stopped smoking before the second month of pregnancy, both compared with non-smokers. Among the smokers the relative mortality was higher for typical diseases related to tobacco intake, such as respiratory and oesophageal cancer and diseases of the cardiovascular and digestive organs and also for accidents and suicides. . . . The risk of premature death seems to be higher in women who smoke during pregnancy than in other women who smoke. This may be explained either by the low proportion

of those who stop later and the high proportion of heavy smokers or by other characteristics of these subjects that increase the risk.

Reference Number: 80710

BENEDICT, M.D., S.A. MISSMER, A. VAHRATIAN, K.F. BERRY, A.F. VITONIS, D.W. CRAMER, AND J.D. MEEKER. "Secondhand Tobacco Smoke Exposure is Associated with Increased Risk of Failed Implantation and Reduced IVF Success." *Human Reproduction*, 26: 2524–2531, 2011 (4 tables, 32 references)

Abstract: Infertility and early pregnancy loss are prevalent in people exposed to secondhand smoke. A study was conducted of 1,909 non-smoking women during 3,270 IVF treatments in Boston between 1994 and 2003. The follicular fluid nicotine concentration was measured for cotinine by enzyme-linked immunosorbent assay (ELISA). Most women (69%) had never actively smoked and the remainder were ex-smokers. Male factor and tubal inflammation/occlusion were the most common causes of infertility. The ORs for implantation failure and successful live birth were 1.52× and 0.75× respectively in women exposed to SHS.

> In conclusion, we found a significant increase in the risk of implantation failure following IVF among women exposed to STS compared with those who were unexposed based on cotinine concentrations measured in the FF of the women at oocyte retrieval from repeated treatment cycles. We also observed a significant decrease in the odds of achieving a successful live birth among STS-exposed women. These findings are likely of great public health significance due to continued widespread STS exposure worldwide.

Reference Number: 80711

GERAGHTY, S.R., K. MCNAMARA, J.J. KWIEK, L. ROGERS, M.A. KLEBANOFF, M. AUGUSTINE, AND S.A. KEIM. "Tobacco Metabolites and Caffeine in Human Milk Purchased via the Internet." *Breastfeeding Medicine*, 10: 419–424, 2015 (2 tables, 1 figure, 48 references)

Abstract: One hundred and two human milk samples were purchased over the internet in 2012. Of the samples bought, 45% arrived at temperatures greater than 4°C and 74% contained high counts of pathogenic bacteria. In addition, 10% of the samples had bovine DNA detected. Detectable levels of nicotine were found in 58% of the samples and 4% of the milk samples had a high enough nicotine/cotinine concentration to indicate active smoking. Caffeine was detected in almost all of the human milk samples (97%).

Table: Nicotine and Cotinine Concentrations Detected in Human Milk Samples Bought Over the Internet

Seller #	Nicotine (ng/mL)	Cotinine (ng/mL)	Claims in Ad
1	33.6	133.6	My milk is from a smoke-free house, it is alcohol-free, drug-free
2	23.2	17.8	Smoke-free, drug-free and medication-free home
3	2.4	6.3	No mention of smoking
4	1.0	5.4	Healthy . . . With two healthy babies (both breast fed) smoke- and alcohol-free
5	15.8	1.3	I am a healthy drug-free . . . woman

Source: Adapted from Geraghty et al, 2015

> In 102 milk samples that we bought via the Internet, we detected evidence of active smoking, secondhand smoke exposure, and almost ubiquitous caffeine consumption. Buyers of human milk on the Internet should be aware that advertisements do not always include information as to what substances may be present. Some sellers may misrepresent their health behaviors or be unaware of lifestyle factors that can lead to exposure to nicotine and caffeine. We showed that infants fed human milk bought from unfamiliar sources may unknowingly be exposed to nicotine and caffeine.

8.08 PSYCHIATRIC DISORDERS

> "Overall, there is now substantial evidence that tobacco smoking is associated with changes in brain structure and neural circuitry in brain regions and systems clearly implicated in many psychiatric disorders."
>
> —Boksa, "Smoking, Psychiatric Illness and the Brain" (2017)

Tobacco companies used to supply free cigarettes to psychiatric institutions and hospitals, in order to capture this lucrative market, as the rates of smoking were 2–5× greater for individuals with psychiatric disorders, especially schizophrenia, depression, anxiety, and bipolar disorders (80801–4, 80807–9). As usual, the tobacco industry sponsored research to promote the idea that individuals with schizophrenia need tobacco to self-medicate (80801–2).

This has resulted in many family members of psychiatric patients (and doctors!) believing the following misconceptions, which have prevented patients from being treated for nicotine addiction and smoking cessation (80105):

- People with mental illness must smoke to manage mental health symptoms.
- Quitting smoking would make mental health symptoms worse.
- Smoking helps people with mental illness cope with the stress in their lives.
- Smoking does not cause health problems for loved ones.

However, studies have shown that patients with schizophrenia can quit smoking without threatening their mental health recovery (80806).

The prevalence of current smoking in persons attending ERs was highest for substance use disorder (58%), psychiatric illness (41%), and alcohol use (39%), and lowest for pregnancy (14%) (80809). Adolescents also suffer from increased psychiatric problems from smoking (80810–12).

Cigarettes with reduced nicotine content should be available for the psychiatric population to encourage smoking cessation and decrease adverse health impacts (80813).

Alternative tobacco products such as e-cigarettes, snuff, and smokeless tobacco are also associated with an increased risk for psychiatric problems (80813).

Over 70% of unhoused people are current cigarette smokers and those with PTSD are more likely to smoke heavily and have difficulty in stopping (80818).

Reference Number: 80801

QUIGLEY, H., AND J.H. MACCABE. "The Relationship Between Nicotine and Psychosis." *Therapeutic Advances in Psychopharmacology*, 9: (13 pp), 2019 (130 references)

Abstract: A review was conducted of the association of nicotine and psychosis. Psychosis causes difficulties in establishing what is real and what is not and results in disordered thinking and speech, hallucinations, and delusions. One of the most severe psychiatric disorders is schizophrenia. Rates of smoking in individuals with psychiatric disorders were found to be 2–3× greater than those without such disorders. Tobacco smokers with psychiatric disorders who display patterns of heavy smoking and nicotine dependence are less likely to quit smoking. The tobacco industry has directly funded research promoting the idea that individuals with schizophrenia need tobacco to self-medicate and are less susceptible to the harms of tobacco smoking.

> Cigarette smoking is strongly associated with psychotic disorders such as schizophrenia. For several decades it was assumed that the relationship could be explained by reverse causation; that smoking was secondary to the illness itself, either through self-medication or a process of institutionalization, or was entirely explained by confounding by cannabis use or social factors. However, studies have exposed that such hypotheses cannot fully explain the association, and more recently a bidirectional relationship has been proposed wherein cigarette smoking may be causally related to risk of psychosis, possibly *via* a shared genetic liability to smoking and psychosis.

Reference Number: 80802

BOKSA, P. "Smoking, Psychiatric Illness and the Brain [Editorial]." *Journal of Psychiatry and Neuroscience*, 42: 147–149, 2017 (34 references)

Abstract: Rates of smoking are 2–5× greater in people with several mental disorders, including schizophrenia, mood disorders, anxiety disorders, ADHD, binge eating disorder, bulimia, and substance use disorders (SUDs). Smoking is a factor that should be considered as potentially confounding in the study of psychiatric illnesses as it may be a causal factor or result in worsening severity of the disorder. Smoking is associated with an increased risk for dementia or Alzheimer's disease and smoking during pregnancy is associated with a more severe form of ADHD in children.

> Overall, there is now substantial evidence that tobacco smoking is associated with changes in brain structure and neural circuitry in brain regions and systems clearly implicated in many psychiatric disorders. These findings, together with the high prevalence of smoking in psychiatric populations, highlight the importance of including smoking as a potential confounding variable in studies examining neural mechanisms of psychiatric disorders.

Reference Number: 80803

GÜLSEN, A., AND UYGUR, B. "Psychological Features of Smokers." *Respiratory Care*, 63: 1492–1497, 2018 (4 tables, 2 figures, 29 references)

Abstract: Smoking tobacco is common among patients with psychiatric disorders such as depression, anxiety, and schizophrenia. These patients are highly resistant to smoking cessation. An analysis was conducted in 124 never smokers (mean age 42.2 years) and compared to 127 smokers attending a clinic to quit smoking (mean age 41.5 years) in Turkey in 2015. The subjects filled in psychological questionnaires and the nicotine dependence of the smokers was determined by FTND. Seven smokers (6%) were mild, 39 (32%) were moderate, and 78 (63%) were heavy smokers.

Table: Psychiatric Disorders and Smoking Rate

Psychiatric Disorders	Smoking Rate
Obsessive-compulsive	13–37%
Anxiety	17–47%
Somatoform	48%
Depression	18–60%
Bipolar disorder	52–70%
Psychoticism	57–66%
Schizophrenia	40–90%

Source: Adapted from Gülsen and Uygar, 2018

This study examined the association between psychological symptoms and smokers' daily cigarette consumption and nicotine addiction levels. Psychiatric symptoms (somatization, anxiety, depression, hostility, and paranoia) were observed more frequently in people with high-grade nicotine addiction (FTND score 7–10 points). The psychological symptom questionnaire may be important in identifying patients who need psychiatric counseling.

Reference Number: 80804

MOLLA, Z., L. DUBE, W. KRAHL, AND M. SOBOKA. "Tobacco Dependence Among People with Mental Illness: A Facility-based Cross-Sectional Study from Southwest Ethiopia." *BMC Research Notes*, 10: 289 (7 pp), 2017 (3 tables, 24 references)

Abstract: A cross-sectional study was conducted of 305 male and 117 female mental health service users (mean age 32 years) in southwest Ethiopia in 2014. The prevalence of current tobacco use in this group was overall 31% (2.6% in women and 42% in men). According to FTND, current tobacco dependence occurred in 18.5% of the patients. The most common psychiatric diagnosis was schizophrenia (39.1%), followed by major depressive disorder (33.2%) and bipolar disorder (16.8%). The prevalence of tobacco dependence was 16.9% in bipolar disorders and 29.1% in schizophrenia patients.

Table: Reasons for Smoking and Percentage of People with Mental Disorders

Reason for Tobacco Smoking	Rate of Mental Disorders
Enjoyment	19.4%
Smoking is fashionable	10.0%
Increase self-esteem	7.6%
Self-treatment for illness	2.4%
To relieve long-standing stress	1.4%
Family history of tobacco smoking	33.6%
Having friends who use tobacco	45.7%

Source: Adapted from Molla et al, 2017

A high prevalence of tobacco dependence was found among people with schizophrenia attending mental health services in Ethiopia. Psychotherapy and pharmacotherapy are crucial interventions to reduce tobacco dependence and tobacco related medical illness in this population. Patients with tobacco dependence also tended to use other substances like khat and alcohol which have their independent impact on medication adherence and prognosis of the mental illness. A comprehensive treatment approach is necessary to reduce the risk associated with these substances.

Reference Number: 80805

ASCHBRENNER, K.A., L.B. DIXON, J.A. NASLUND, J.C.M. BIENVENIDA, K.L. MCMANUS, S.J. BARTELS, AND M.F. BRUNETTE. "An Online Survey of Family Members' Beliefs and Attitudes About Smoking and Mental Illness." *Journal of Dual Diagnosis*, 13: 179–183, 2017 (1 table, 17 references)

Abstract: An online survey was conducted on 256 family members of individuals with mental illness. One-quarter of respondents agreed that people with mental illness must smoke to manage mental health symptoms.

Table: National Alliance on Mental Illness Family Members' Views and Beliefs on Smoking and Mental Disorder

Survey Items	Strongly Agree	Agree	Neither Agree nor Disagree	Disagree	Strongly Disagree
People with mental illness must smoke to manage mental health symptoms	10.2%	13.3%	34%	25.8%	16.8%
Quitting smoking would make mental health symptoms worse	6.6%	18.4%	31.3%	28.9%	14.8%
Smoking helps people with mental illness cope with the stress in their lives	21.5%	52.3%	12.5%	8.2%	5.7%
Smoking does not cause health problems for my loved one	1.25%	3.5%	5.5%	26.6%	63.3%

Adapted from Aschbrenner et al, 2017

Misconceptions about smoking and mental illness and uncertainty about the safety of cessation treatment may interfere with family support for quitting smoking among people with mental illness.

Reference Number: 80806

HIGUCHI, Y., M. FUJIWARA, N. NAKAYA, M. FUJIMORI, C. HAYASHIBARA, R. SO, I. SHINKAWA, K. SATO, Y. YADA, M. KODAMA, H. TAKENAKA, Y. KISHI, K. KAKEDA, Y. UCHITOMI, N. YAMADA, AND M. INAGAKI. "Change in Smoking Cessation Stage Over 1 Year in Patients with Schizophrenia: A Follow Up Study in Japan." *BioMed Central Psychiatry*, 19: 367 (9 pp), 2019 (3 tables, 2 figures, 44 references)

Abstract: The smoking quitting rate of 350 schizophrenic patients over 1 year was determined in Japan. Initially there were 113 current smokers (32.2%) and 68 were former smokers (19.4%). One hundred and four schizophrenic smokers were studied for 1 year and only seven had received smoking cessation treatment at baseline, even though 79 (70%)

were interested in smoking cessation. In total, six patients (5.8%) stopped smoking after 1 year.

Older theories have considered smoking in patients with schizophrenia to be inevitable, as follows: "Tobacco is self-medication for schizophrenia patients because nicotine relieves their symptoms. Therefore, they would not quit smoking, and smoking cessation may exacerbate their illness." However, more recent evidence has failed to support these claims, and has shown that patients with schizophrenia can quit smoking without threatening their mental health recovery. Smoking cessation should be an integral part of serious mental illness treatment.

Reference Number: 80807

ELKHOLY, H., N. NAGY, G.R.A. TAHA, M. ELHABIBY, M. YOSEF, AND L. AZZAM. "Stop Turning a Blind Eye: Tobacco Smoking Among Egyptian Patients with Schizophrenia." *Frontiers in Psychiatry*, 9: (8 pp), 2019 (6 tables, 46 references)

Abstract: Patients with schizophrenia have higher reported smoking rates (58 to 90%) than the general population and a higher smoking rate than patients with other psychiatric illness. Schizophrenia is associated with a mortality rate 2 to 3× higher than the general population. A comparison was conducted of 222 male and 114 female patient smokers with schizophrenia and with 93 male and 57 female smokers with no mental illness. Morning smoking preference was positively correlated to conceptual disorganization and excitement in schizophrenia. Over 40% of the patients smoked over 20 cigarettes per day, whereas only 21% of those without schizophrenia smoked that much. Thirty-nine percent of the schizophrenic patients smoked their cigarette within 5 minutes of waking compared to 15% of the non-schizophrenic group. There was a positive association between the number of cigarettes smoked and the severity of delusions, hostility, anxiety, and poor impulse control. Age, gender, and schizophrenia were significant predictors for FTND scores.

Tobacco is one of the major contributors to major health issues in patients with schizophrenia. It seems that there is some sort of a link between some positive symptoms of schizophrenia and severity of tobacco dependence, number of cigarettes, and morning smoking preference. It also appears that patients with schizophrenia who have more severe positive symptoms are more inclined to quit smoking; mainly for health concerns. Thus, it is important for psychiatrists to investigate these issues when assessing patients with schizophrenia especially their readiness to quit. Addressing tobacco use during the course of treatment might have positive outcomes on schizophrenia and health related issues and consequently lead to decreased morbidities and mortalities.

Reference Number: 80808

ŠAGUD, M., B. VUKSAN-ĆUSA, N. JAKŠIĆ, A. MIHALJEVIĆ-PELEŠ, M.R. KUZMAN, AND N. PIVAC. "Smoking in Schizophrenia: An Updated Review." *Psychiatria Danubina*, 30: 216–223, 2018 (1 table, 54 references)

Abstract: Smoking in schizophrenia continues to be a hot topic as 50–60% of the patients smoke and have higher rates of nicotine dependence than the general population. Smoking was found to be associated with 2× the risk of natural cause mortality in patients with schizophrenia and bipolar disorders. Smoking causes proinflammatory effects and is a risk factor for metabolic syndrome. This syndrome is characterized by hypertension, dyslipidemia, insulin resistance, and obesity. Fetal exposure to nicotine due to parental smoking was found to increase the odds of schizophrenia in the offspring. Adolescents who smoked more than 10 CPD or who started smoking before the age of 13 years had an increased rate of psychosis.

Smoking is related to schizophrenia in many ways. Ameliorative effects of smoking in patients with schizophrenia, and the lack of insight to harmful smoking effects, might reinforce heavy smoking. In spite of some observed benefits of smoking in patients with schizophrenia, smoking has a deleterious impact on patient's psychiatric and somatic health. The ultimate goal should be complete smoking cessation, given that smoking only one cigarette daily carries a much greater risk of developing coronary heart disease and stroke compared to never smokers. The delivery of smoking cessation treatments should be a high priority in settings that provide care to patients with schizophrenia. In theory, beneficial effects of brain functioning exerted by nicotine, might be obtained by substances mimicking those effects, which is an important area for future research.

Reference Number: 80809

KANTROW, S.P., S.E. JOLLEY, E.G. PRICE-HAYWOOD, X. WANG, T-S. TSENG, D. ARNOLD, L.F. BROWN, C. LEONARDI, R.A. SCRIBNER, E.J. TRAPIDO, AND

H-YI. LIN. "Using the Emergency Department to Investigate Smoking in Young Adults." *Annuals of Epidemiology*, 30: 44–49, 2019 (5 tables, 1 figure, 34 references)

Abstract: This study determined the smoking status of 55,777 emergency department (ED) attendees (aged 18–30 years) in Louisiana between 2013 and 2016. Sixty-one percent of the subjects were women, 55% were Black, 35% were White, and 8% were Hispanic. The prevalence of current smoking was 21.7% for women and 42.5% for men. Smoking prevalence was concurrent with substance use disorder (58%), psychiatric illness (41%), and alcohol use (39%), and lowest with pregnancy (14%.).

Several diagnoses are strongly associated with smoking, including psychiatric diseases (schizophrenia, bipolar illness) and substance use disorder. When these conditions were the primary ED diagnosis in our study, we observed the expected association with higher smoking prevalence. By contrast, the lowest smoking prevalence by diagnosis was observed for women evaluated for pregnancy and was similar to national rates reported for this population. Further analysis of high-risk groups in our study suggested that among those who reported illicit drug use, women were at higher risk of smoking than men, and younger adults had the highest risk of smoking. For those subjects who reported alcohol use, current smoking varied by race/ethnicity and age, with the highest risk observed among Hispanic and younger adults.

Reference Number: 80810

HALPERIN, A.C., S.S. SMITH, E. HEILIGENSTEIN, D. BROWN, AND M.F. FLEMING. "Cigarette Smoking and Associated Health Risks Among Students at Five Universities." *Nicotine and Tobacco Research*, 12: 96-104, 2010 (4 tables, 47 references)

Abstract: A health screening survey was conducted on 2,091 students at five colleges as to their tobacco use. Nearly 25% of the students reported any current smoking, though 41% of the smokers reported smoking less than 1 cigarette per day. Students who experienced depression had 2.3× the odds of being a dependent smoker and had 2.1× the odds of seeking mental health counselling.

While most college students who use tobacco are light or intermittent smokers (LITS), student health center clinicians need to be made aware that these students are at risk for nicotine dependence as well as more immediate harms due to their smoking and associated behavioral risks. Our analysis leads us to recommend that campus clinic providers systematically identify students who smoke at any level and seize the opportunity to address tobacco use in conjunction with fitness, risky drinking and driving, depression, and other mental health issues to improve health status and decrease morbidity.

Reference Number: 80811

BUCHY, L., K.S. CADENHEAD, T.D. CANNON, B.A. CORNBLATT, T.H. MCGLASHAN, D.O. PERKINS, L.J. SEIDMAN, M.T. TSUANG, E.F. WALKER, S.W. WOODS, R. HEINSSEN, C.E. BEARDEN, D. MATHALON, AND J. ADDINGTON. "Substance Use in Individuals at Clinical High Risk of Psychosis." *Psychological Medicine*, 45: 2275–2284, 2015 (3 tables, 1 figure, 38 references)

Abstract: Tobacco, alcohol, and cannabis use was determined in 735 youth at clinical high risk (CHR) of developing psychosis and in 278 control subjects at baseline, 6 months later, and 12 months later. CHR youth had higher cannabis and tobacco use severity at baseline and 1 year later than controls. Ninety of the CHR subjects developed psychosis.

Table: Substance Use Percentage in Youth at a Clinical High Risk of Developing Psychosis and Controls

Substance Use	CHR	Controls
Alcohol use	37.1%	50.9%
Alcohol abuse	2.2%	0.0%
Tobacco use	22.2%	7.2%
Tobacco abuse	0.7%	0.0%
Cannabis use	19.6%	9.4%
Cannabis abuse	3.5%	0.0%

Source: Adapted from Buchy et al, 2015

In conclusion, CHR participants used more cannabis and tobacco and less alcohol at baseline and over 1 year compared with controls. CHR participants also had higher lifetime prevalence and frequency of cannabis use, began using cannabis at a younger age, and were more likely to use during the day than evening and alone than socially. Substance use was similar across CHR participants who did and did not transition to psychosis.

Finally, after 2 years in the study, CHR participants with a symptomatic, prodromal progression or psychotic clinical outcome reported greater tobacco use at baseline than controls, though only the latter survived when controlling for cannabis use. CHR participants with a psychotic clinical outcome after 2 years had significantly higher baseline cannabis use than controls when adjusting for cannabis use.

Reference Number: 80812

BANYAL, N., D. BHATTACHARYYA, AND P. YADAV. "Study to Determine the Prevalence of Substance Use and Factors Associated with it, In First-Episode of Psychosis." *Industrial Psychiatry Journal*, 27: 264–270, 2018 (6 tables, 33 references)

Abstract: An analysis was conducted of 79 first-episode psychiatric patients in northern India. Nearly three-quarters of the patients (58) were psychoactive substance users. Tobacco was the most commonly used substance in 56% of the patients. Alcohol was reported by 41% and cannabis by 9%. The mean age of the 58 psychoactive substance users was 36.3 years compared to 35.7 years of the non-users. The mean age of the cannabis users was significantly lower at 24.6 years. Substance use was more common in male compared to female patients.

The results of the present study show a high prevalence of psychoactive substance abuse when compared with the prevalence in the general population. This high percentage demands that a thorough history of substance use and its laboratory screening is essential in the cases of treatment-naïve, first-episode psychosis, and further appropriate treatment of comorbid status needs to be addressed, as it affects the course of illness, response, and chances of relapses.

Reference Number: 80813

HIGGINS, S.T., S.H. HEIL, S.C. SIGMON, J.W. TIDEY, D.E. GAALEMA, J.R. HUGHES, M.L. STITZER, H. DURAND, J.Y. BUNN, J.S. PRIEST, C.A. ARGER, M.E. MILLER, C.L. BERGERIA, D.R. DAVIS, J.M. STRECK, D.D. REED, J.M. SKELLY, AND L. TURSI. "Addiction Potential of Cigarettes with Reduced Nicotine Content in Populations with Psychiatric Disorders and Other Vulnerabilities to Tobacco Addiction." *JAMA Psychiatry*, 74: 1056–1064, 2017 (3 tables, 1 figure, 48 references)

Abstract: One hundred and twenty female and 49 male daily smokers (mean age 35.6 years) who were either opioid dependent, with affective disorders, or socioeconomically disadvantaged women smoked their usual brand of cigarettes and cigarettes containing 0.4, 2.3, 5.2, and 15.8 mg/g nicotine. Addiction potential of the cigarettes was determined using the Cigarette Purchase Task (CPT) and other measures such as the Minnesota Nicotine Withdrawal Scale. Smoking is prevalent in 21% of the US adult population, 32% of those with affective disorders, 92% of opioid addicts, and 30% of disadvantaged women of reproductive age.

Reducing the nicotine content of cigarettes may decrease their addiction potential in populations that are highly vulnerable to tobacco addiction. Smokers with psychiatric conditions and socioeconomic disadvantage are more addicted and less likely to quit and experience greater adverse health impacts. Policies to reduce these disparities are needed; reducing the nicotine content in cigarettes should be a policy focus.

Reference Number: 80814

GOLDENSON, N.I., R. KHODDAM, M.D. STONE, AND A.M. LEVENTHAL. "Associations of ADHD Symptoms with Smoking and Alternative Tobacco Product Use Initiation During Adolescence." *Journal of Pediatric Psychology*, 43: 613–624, 2018 (4 tables, 45 references)

Abstract: The use of alternative tobacco products such as ECs and hookahs was determined in 1,921 grade 9 students in California between 2014 and 2015. Overall ADHD symptoms, inattention, and hyperactivity-impulsive symptoms were assessed at baseline. Past 6-month EC, hookah, and cigarette use was determined at three semi-annual follow-ups.

Table: Odds Ratio of Initiation of EC, Hookah, or Cigarette Use with ADHD Symptoms in Grade 9 Students

ADHD Symptoms	OR Initiation of ECs	OR Initiation of Hookahs	OR Initiation of Cigarettes
Overall ADHD symptoms	1.22×	1.08×	1.03×
Inattention	1.13×	1.02×	0.95×
Hyperactivity-impulsivity	1.26×	1.13×	1.12×

Source: Adapted from Goldenson et al, 2018

ADHD symptoms may be a risk factor for the increased likelihood of e-cigarette use initiation during high school, as well as single- and poly-product use initiation outcomes. The trajectory shape of e-cigarette use over time does not appear to differ by ADHD, suggesting that the increased odds of e-cigarette use uptake conferred by ADHD result in a pattern that is similar to the general population of youth who initiate use of e-cigarettes during high school. Targeting ADHD-related behavioral factors, particularly those related to hyperactivity and impulsivity, in pediatric psychology treatment and prevention may be beneficial for preventing e-cigarette use. Data are needed to uncover the mechanisms underlying this risk pathway and advance tobacco use etiologic theory and prevention practice in the current era in which e-cigarettes are popular among youth.

Reference Number: 80815

WILENS, T.E., M. VITULANO, H. UPADHYAYA, J. ADAMSON, R. SAWTELLE, L. UTZINGER, AND J. BIEDERMAN. "Cigarette Smoking Associated with Attention-Deficit Hyperactivity Disorder." *Journal of Pediatrics*, 153: 414–419, 2008 (2 tables, 2 figures, 33 references)

Abstract: This study compared 80 smokers with ADHD to 86 smokers without ADHD (mean age 19.2 years). Subjects with ADHD reported earlier onset of smoking, at 13.9 years, compared to 15.2 years for smokers without ADHD. Smokers with ADHD had greater rates of nicotine dependence according to the modified Fagerstrom Tolerance Questionnaire (mFTQ) compared to smokers without ADHD. Environmental factors such as current parental smoking, peer smoking, and living with smokers increased the risk of smoking in youth with ADHD much more than in control smokers.

Table: Association Between Number of ADHD Symptoms and Percent Who are Smokers

Number of ADHD Symptoms	Percentage Who Are Smokers
0–2	15%
3–5	17%
6–8	25%
9–11	28%
12–14	36%
15–18	51%

Source: Adapted from Wilens et al, 2008

Despite these limitations, our results show that ADHD is associated with earlier onset and more severe cigarette smoking compared with matched non-ADHD controls. A direct relationship between ADHD symptoms and the risk and severity of cigarette smoking exists. Given the significant risk that youth with ADHD have for developing a severe form of nicotine dependence, clinically relevant prevention and early intervention strategies aimed at cigarette smoking in youth with ADHD are necessary.

Reference Number: 80816

LEVENTHAL, A.M., D.R. STRONG, S. SUSSMAN, M.G. KIRKPATRICK, J.B. UNGER, J.L. BARRINGTON-TRIMIS, AND J. AUDRAIN-MCGOVERN. "Psychiatric Comorbidity in Adolescent Electronic and Conventional Cigarette Use." *Journal of Psychiatric Research*, 73: 71–78, 2016 (2 tables, 2 figures, 47 references)

Abstract: A survey was conducted of 2,557 grade 9 students (mean age 14 years) attending high school in Los Angeles regarding conventional cigarette and/or EC use and emotional psychiatric comorbidity. Over 75% did not ever use conventional cigarettes/ECs; 12% reported EC use only; 5% reported conventional cigarette use; and 6% reported dual use. Alcohol problems, cannabis abuse, and drug abuse was greatest in dual users, followed by conventional and EC users. Inability to feel pleasure (anhedonia), depression, and panic disorders was higher in EC-only users than non-users. Emotionally healthy adolescents who would not use conventional cigarettes may be at risk to use ECs.

Adolescent e-cigarette use is characterized by emotional problems midway between non-use and conventional cigarette use. Dual use is associated with pervasive psychiatric comorbidity.

Reference Number: 80817

FU, Q., M.G. VAUGHN, L-T. WU, AND A.C. HEATH. "Psychiatric Correlates of Snuff and Chewing Tobacco Use." *Public Library of Science One*, 9: 113196 (14 pp), 2014 (4 tables, 28 references)

Abstract: The data from the National Epidemiologic Survey on Alcohol and Related Conditions (NESARC) of 43,093 adults in the United States between 2001 and 2002 were used to determine the association between snuff and chewing tobacco use and psychiatric correlates. The prevalence of lifetime exclusive snuff use, exclusive chewing tobacco use, and dual use was 2.16%, 2.52%, and 2.79% respectively. SLT use was more common in White men, those aged 18–34 years, and residents of rural and suburban areas. After controlling for sociodemographic variables and cigarette smoking, the odds of exclusive chewing tobacco use were 1.53× in persons with panic disorders and 1.41× in persons with specific phobias. The odds of an alcohol use disorder were 1.97×, 2.01×, and 2.99× for exclusive snuff, exclusive chewing tobacco, and dual use, respectively.

> Despite those limitations, this is the first epidemiologic study of its kind to systematically examine the associations between each specific type of lifetime smokeless tobacco use and lifetime diagnoses of psychiatric disorders. As such, these findings provide new insights into the link of psychiatric disorders with snuff use and chewing tobacco. These links are understudied yet given the public health significance of smokeless tobacco product and their health consequences. Thus, elucidating the link between psychiatric and behavioral health factors is important. These results can guide treatment efforts on smokeless tobacco by targeting the treatment for alcohol use disorder among all smokeless tobacco users, the treatment for panic disorder, specific phobia, and inhalant use disorder among exclusive chewing tobacco users, as well as the treatment for cannabis use disorder among exclusive snuff users. It is helpful to realize that persons who use both snuff and chewing tobacco are different from exclusive users of snuff or chewing tobacco regarding their comorbid antisocial and schizoid personality disorders, respectively.

Reference Number: 80818

BAGGETT, T.P., E.G. CAMPBELL, Y. CHANG, L.M. MAGID, AND N.A. RIGOTTI. "Posttraumatic Stress Symptoms and Their Association with Smoking Outcome Expectancies Among Homeless Smokers in Boston." *Nicotine and Tobacco Research*, 18: 1526–1532, 2016 (2 tables, 58 references)

Abstract: At the time of the study, it was estimated that 70 to 80% of unhoused adults in Boston were current cigarette smokers. In the unhoused population, people with PTSD have a higher prevalence of smoking, smoke more heavily, and have more difficulty quitting than persons without PTSD. A survey was conducted of 228 unhoused men and 72 women (mean age 48 years) in Boston in 2014. Forty-six percent of the total had symptoms of PTSD, while among those who smoked, over two-thirds screened positive for PTSD.

Table: Characteristics of the Homeless with and Without Post-Traumatic Stress Disorder

Characteristic	PTSD Screen +ve	PTSD Screen –ve
Psychiatric severity (0–1)	0.51	0.23
Hallucinations (past month)	38	5
Age at first homelessness (years)	29.3	34.3
Cigarettes per day (mean)	12.5	11.9
Cigarette dependence (0–10)	4.7	3.9

Symptoms of posttraumatic stress are common among homeless smokers and strongly associated with positive smoking outcome expectancies. These findings suggest that smoking cessation programs for homeless smokers should incorporate PTSD screening, be sensitive to the needs and vulnerabilities of trauma survivors, and incorporate behavioral counseling approaches that attempt to identify and address the potential linkages between posttraumatic stress symptoms and smoking behavior in a fashion that emphasizes autonomy and control.

8.09 SEXUAL DYSFUNCTION

"Erectile function is directly proportional to the degree of exposure to smoking, and quitting smoking improves male sexual function in all age groups between 30-60 years of age regardless of pack-year and education level."

—Sahin et al, "Effect of Smoking Cessation on Sexual Functions in Men Aged 30 to 60 Years" (2020)

Cigarette smoking is a risk factor for female sexual dysfunction (80901). Over 70% of female sex workers in Brazil were current cigarette smokers (80902).

Nicotine gum (6 mg) reduced the physiological and psychological arousal in young non-smoking men (80903). The risk of erectile dysfunction was increased by 55% in male smokers (80905), and smoking cessation improved sexual functioning for male smokers of all ages (80906).

Reference Number: 80818

It is suggested that educating younger men (who are the most resistant to smoking cessation) regarding erectile dysfunction and smoking may further encourage them to quit (80907).

Reference Number: 80901

CHOI, J., D.W. SHIN, S. LEE, M.J. JEON, S.M. KIM, B. CHO, AND S.M. LEE. "Dose-Response Relationship Between Cigarette Smoking and Female Sexual Dysfunction." *Obstetrics and Gynecology Science*, 58: 302–308, 2015 (4 tables, 1 figure, 29 references)

Abstract: Female sexual dysfunction (FSD) is defined as a recurrent or persistent deficiency in sexual desire and arousal, difficulty or absence of reaching orgasm, and genital pain. Female sexual functioning was determined in 838 never/former smokers and 62 current smokers who were sexually active, premenopausal women in South Korea between 2010 and 2011.

> In premenopausal women, current smoking was an independent risk factor for FSD. And cumulative smoking dose and nicotine dependency were associated with higher risk of FSD.

Reference Number: 80902

DEVÓGLIO, L.L., J.E. CORRENTE, M.H. BORGATO, AND I. DE GODOY. "Smoking Among Female Sex Workers: Prevalence and Associated Variables." *Jornal Brasileira de Pneumologia*, 43: 6–13, 2017 (6 tables, 35 references)

Abstract: The prevalence of cigarette smoking and associated variables were determined in 83 female sex workers (FSW) (mean age 26.8 years) in Botucotu, Brazil in 2014. Over 70% were smokers and 6% were former smokers, compared to 8.3% prevalence of smokers in the general population of women in Brazil. The mean age of smoking initiation was 16 years and the mean CPD was 22.3. Nearly 90% of the FSW consumed alcohol regularly and 52% used illegal drugs. The level of nicotine dependence was high, and the majority of the subjects showed no intention of quitting. Smoking was associated with alcohol consumption and illicit drug use.

> Women have experienced several withdrawal symptoms, two of which are irritability and increased appetite; other studies have shown that, in addition to increasing the chances of relapse, these symptoms make smoking cessation difficult in women.

Reference Number: 80903

HARTE, C.B., AND C.M. MESTON. "Acute Effects of Nicotine on Physiological and Subjective Sexual Arousal in Nonsmoking Men: A Randomized, Double-Blind, Placebo-Controlled Trial." *Journal of Sexual Medicine*, 5: 110–121, 2008 (46 references)

Abstract: Twenty-eight sexually functional heterosexual men (mean age 21 years) received either Nicorette polarcrilex gum (6 mg nicotine) or placebo gum 40 minutes prior to viewing an erotic film. Physiological measures of penile circumference and subjective changes in mood and sexual responses were measured. Higher heart rates and BP occurred at 30 minutes after nicotine than placebo. Nicotine caused a 23% reduction in physiological sexual arousal.

> The results of the present investigation provide the first empirical evidence that an intermediate dose of isolated nicotine significantly reduces erectile response in healthy, young, nonsmoking men. That nicotine significantly reduces physiological sexual arousal provides support to the hypothesis that nicotine may be the primary pharmacological agent responsible for genital hemodynamic disruption, by acting on the nervous system centrally or peripherally. The findings of this study have important clinical implications. That nicotine intake significantly impairs erectile function acutely in never smokers could be used as a convincing public health claim to assist in the prevention of cigarette smoking in adolescents and adults.

Reference Number: 80904

SAHIN, M.O., V. SEN, G. GUNDUZ, AND O. UCER. "Effect of Smoking Cessation on Sexual Functions in Men Aged 30 to 60 Years." *International Brazilian Journal of Urology*, 46: 642–648, 2020 (4 tables, 27 references)

Abstract: This study determined the effect of smoking and cessation in erectile dysfunction on 202 smoking cessation male patients (ages 30–60 years) before and after 6 months of cigarette cessation in Turkey between 2017 and 2018. The amount of tobacco exposure was determined in pack-years and the International Index of Erectile Function (IIEF) was filled out by the patients before and 6 months after smoking cessation. Free radicals, aromatic compounds, and superoxide anions in the cigarette smoke can disrupt dilation in the penile artery and arterioles by impairing nitrous oxide synthesis and degradation.

Table: International Index of Erectile Function Measures and Scores at Baseline (Smoking) and After 6 Months Smoking Cessation

IIEF Measures	Baseline (smoking)	Smoking Cessation for 6 months
Erectile function	21.6	24.2
Intercourse satisfaction	10.5	11.7
Orgasmic function	8.6	8.9
Sexual desire	6.8	7.7
Overall satisfaction	7.2	8.0

Adapted from Sahin et al, 2020

Erectile function is directly proportional to the degree of exposure to smoking, and quitting smoking improves male sexual function in all age groups between 30-60 years of age regardless of pack-year and education level.

Reference Number: 80905

CAO, S., X. YIN, Y. WANG, H. ZHOU, F. SONG, AND Z. LU. "Smoking and Risk of Erectile Dysfunction: Systematic Review of Observational Studies with Meta-Analysis." *Public Library of Science One*, 8: e60443, 2013 (1 table, 3 figures, 19 references)

Abstract: The effect of current and ex-smoking on the risks of erectile dysfunction (ED) was determined in four prospective cohort studies and four case-control studies of 28,586 subjects, all published between 2000 and 2010.

In this meta-analysis, the results of the included case-control studies were significantly heterogeneous and may have over-estimated the association between smoking and ED. Therefore, we consider the consistent results of prospective studies are more accurate. The results from prospective cohort studies suggested that the risk of ED was increased by 51% for current smokers and 20% for ex-smokers, as compared with never-smokers. The results also suggested that the increased risk of ED associated with smoking may decrease after stopping smoking.

Reference Number: 80906

MILLETT, C., L.M. WEN, C. RISSEL, A. SMITH, J. RICHTERS, A. GRULICH, AND R. DE VISSER. "Smoking and Erectile Dysfunction: Findings from a

Representative Sample of Australian Men." *Tobacco Control*, 15: 136–139, 2006 (1 table, 23 references)

Abstract: A survey was conducted of 8,367 Australian men (aged 16–59 years) as to smoking status and erectile dysfunction. More than one-quarter (27.2%) of respondents were current smokers, 20.9% smoked less than 21 CPD, and 6.3% smoked more than 20 CPD. Compared with non-smokers, the adjusted OR of erectile dysfunction was 1.24× and 1.39×, respectively. Older age, lower level of education, and taking medication for cardiovascular disease were also associated positively with erectile dysfunction, whereas moderate alcohol consumption (1–4 drinks per day) was associated with a less likelihood of having erectile dysfunction.

Australia has one of the world's most comprehensive tobacco control strategies which is reflected in our relatively low smoking rates. However, to meet the ambitious targets set by some states for further reductions—for example, a 1% annual reduction in prevalence in the most populous state, New South Wales—will require more effective targeting of high-risk groups. Younger men (aged 16–44 years) are more likely to smoke than other groups yet may be least likely to change their behaviour in response to conventional tobacco control approaches. Increasing awareness about the link between smoking and erectile dysfunction in combination with other strategies may increase motivation for more young men to quit smoking. Lower levels of smoking would probably result in a reduction of prevalence of erectile dysfunction which affects about one million men in Australia.

8.10 SPORTS

"Nicotine use, including SLT, fulfils all the WADA criteria for inclusion in the WADA prohibited list as a doping product. Hence chewing and spit tobacco, as well as dry and moist snuff must therefore be discouraged, particularly when sports activities are considered."

—Chagué et al, "Smokeless Tobacco, Sport and the Heart" (2015)

The use of nicotine is not prohibited by the World Anti-Doping Agency (WADA) and is in widespread use (25 to 50%) in sports teams such as football and hockey (81001). WADA has banned THC and alcohol but not caffeine or nicotine, which may be more ergogenic than the previous two drugs (81002).

Reference Number: 80906

Nicotine patches or gum did not improve the performance of a one-hour cycling trial (81003). Over 50% of alpine skiers use the Swedish SLT, snus, and testing showed that it is not ergogenic in athletes, although it did cause throat discomfort, nausea, and dizziness (81004). Athletes use SLT for increased concentration, relaxation, weight control, or ritual, but nicotine does have several negative cardiovascular effects such as increased BP, HR, cardiac arrhythmias, and cardiac vasoconstriction (81005).

In cardiovascular exercises where the heart rate is monitored to determine how heavy the exercise is, the test may have to be adjusted, as nicotine patches can increase the exercise HR by a mean of 6.7 bpm (81006).

Vaping e-cigarettes and smoking cigarettes have additional negative effects on athletic performance compared to nicotine patches or gum (81007–8).

Reference Number: 81001

MÜNDEL, T. "Nicotine: Sporting Friend or Foe? A Review of Athlete Use, Performance Consequences and Other Considerations." *Sports Medicine*, 47: 2497–2506, 2017 (3 tables, 62 references)

Abstract: The use of nicotine is not prohibited by the World Anti-Doping Agency. At the time of the study, nicotine use was widespread in professional team sports such as hockey and football (25 to 50%). Together with caffeine, nicotine was considered to be the most widely consumed psychoactive substance in the world. Nicotine also occurs in small amounts in cauliflowers, potatoes, and tomatoes. Nicotine (3-(1-methyl-2-pyrrolidinyl) pyridine) is the principal alkaloid of the nightshade family of plants and constitutes approximately 3% or less of the dry weight. Nicotine was extensively used as an insecticide. Recreationally, nicotine is consumed as a stimulant and ironically the primary therapeutic use of nicotine is to treat nicotine dependence. Nicotine is structurally similar to acetylcholine, it is an nAChR agonist, and it activates dopamine release. A literature review was conducted of 16 studies on the effects of nicotine on athletic performance. The majority of studies have demonstrated no significant effect of nicotine on athletic performance.

Table: Absorption of Nicotine with Different Forms of Nicotine Delivery System

Nicotine Administration and Dose	Mean Maximum Serum Nicotine Concentration (ng/mL)	Time to Maximum (minutes)	Bioavailability
Smoking (1 cigarette over 5 minutes, 2 mg nicotine/cigarette)	15–30	5–8	80–90%
Smokeless tobacco (Snus, 60 minutes, 11 mg nicotine)	11	60	24–32 %
Gum (4 mg nicotine over 30 minutes)	9	45	63%
Transdermal patch (15 mg nicotine over 16 hours)	11–14	6–9 hours	75–100 %
Inhaler (10 mg cartridge over 20 minutes)	8	30	51–56%
Sublingual tablet (2 mg nicotine over 20 to 30 minutes)	4	60	65%
E-cigarette (18 mg/mL over 65 minutes)	14–16	70–75	–

Adapted from Mündel, 2017

Athletes' own beliefs are that consumption of nicotine/nicotine-containing substances (i.e. ST) proves ergogenic by preventing xerostomia (dry mouth as saliva secretion is augmented), controlling weight as satiety is stimulated, improving reaction time and concentration, and helping relaxation and desirable arousal-attention. These latter reports are substantiated by a meta-analysis determining that the stimulant effects of nicotine enhance aspects of cognition and attention, namely motor abilities, attention and memory.

Reference Number: 81002

PESTA, D.H., S.S. ANGADI, M. BURTSCHER, AND C.K. ROBERTS. "The Effects of Caffeine, Nicotine, Ethanol, and Tetrahydrocannabinol on Exercise Performance." *Nutrition and Metabolism*, 10: 71–85, 2013 (4 tables, 173 references)

Abstract: A general review was conducted on the ergogenic effects of caffeine, nicotine, alcohol, and THC on exercise. There are 173 references. There is only enough scientific evidence that caffeine (and possibly nicotine) are ergogenic aids, although the World Anti-Doping Agency has banned THC and alcohol. Caffeine is both water and fat soluble and has a half-life of 5 to 6 hours. Caffeine reduces fatigue and increases concentration and alertness. Caffeine is more effective as a tablet than in coffee.

The physiological effects of the above mentioned substances are well established. However, the ergogenic effect of some of the discussed drugs may be questioned and one has to consider the cohort tested for every specific substance. However, only caffeine has enough strength of evidence to be considered an ergogenic aid. Cannabis and alcohol are ergolytic for sports performance, and nicotine needs confirmation with further research.

Reference Number: 81003

MÜNDEL, T., S.D. HOULTHAM, M.J. BARNES, AND S.R. STANNARD. "Nicotine Supplementation Does Not Influence Performance of a 1h Cycling Time-Trial in Trained Males." *Frontiers in Physiology*, 10: (8 pp), 2019 (1 table, 3 figures, 28 references)

Abstract: The use of nicotine among professional and elite athletes is high (nicotine metabolites have been detected in 23.4% of urine samples) and its use is not banned by the World Anti-Doping Agency. Ten well-trained male cyclists (mean age 34 years) engaged in three self-paced cycling time trials. Before each trial, the participants were given two of the following: 2 mg nicotine gum for 30 minutes; placebo gum for 30 minutes; 7 mg nicotine patch for 10 hours; or a placebo patch for 10 hours. Blood samples were collected and plasma cotinine concentrations were determined by HPLC. Various physiological measures were determined as well. The highest mean plasma cotinine concentration obtained was approximately 45 ng/mL after the use of the patch.

In conclusion: (i) nicotine administration, whether via gum or transdermal patch, did not exert an ergogenic or ergolytic effect on self-paced cycling performance of ~1 h; (ii) systemic delivery of nicotine was greatest when using a transdermal patch; and (iii) nicotine administration did not alter any of the psycho-physiological measures observed.

Reference Number: 81004

ZANDONAI, T., E. TAM, P. BRUSEGHINI, F. PIZZOLATO, L. FRANCESCHI, M. BARALDO, C. CAPELLI, P. CESARI, AND C. CHIAMULERA. "The Effects of Oral Smokeless Tobacco Administration on Endurance Performance." *Journal of Sport and Health Science*, 7: 465–472, 2018 (1 table, 4 figures, 46 references)

Abstract: Sweden is the only European Union country that has a special exemption to manufacture and sell snus. Snus is a smokeless tobacco product that is placed between the upper lip and gums for approximately 30 minutes and then discarded. Approximately 74% of alpine skiers have used snus and 50% continue to use it. Snus users have expectancies about its positive effect on sports performance. But nicotine also impairs myocardial perfusion and reduces stroke volume. Fourteen healthy male non-smokers and non-snus users (mean age 23 years) were administered 8 mg of nicotine (pH 8.5) in 1 gram of a Catch White Eucalyptus snus package or placebo and performed various physical tasks. Blood samples were collected and plasma nicotine and cotinine concentrations were determined. The mean plasma nicotine and cotinine concentrations were 4.5 ng/mL and 27.8 ng/mL, respectively. Five subjects reported adverse events such as throat discomfort, nausea, and dizziness.

This study showed that muscular and cerebral oxygenation increased significantly with snus administration during an endurance exercise until exhaustion, but this did not affect fatigue perception and TTE. The results showed that snus could not be considered an ergogenic substance in non-tobacco users.

Reference Number: 81005

CHAGUÉ, F., C. GUENANCIA, A. GUDJONCIK, D. MOREAU, Y. COTTIN, AND M. ZELLER. "Smokeless Tobacco, Sport and the Heart." *Archives of Cardiovascular Diseases*, 108: 75–83, 2015 (3 tables, 1 figure, 51 references)

Abstract: One form of smokeless tobacco (SLT) is snuff, which is finely shredded tobacco that is sniffed through the nose or placed between the cheek and gums. Chewing tobacco, another form, is a wad of tobacco placed inside the cheek. Current guidelines recommend that tobacco, including SLT, must be avoided for 2 hours before and after a sports session. A review was conducted of the health effects of SLT. Athletes used SLT for enhanced concentration, relaxation, weight control, or ritual.

There are various negative cardiovascular effects of nicotine such as increased arrhythmias, increased BP, increased HR, and increased coronary vasoconstriction.

Table: Effects of Nicotine on Various Parameters After Submaximal Exercise

Parameter	Nicotine Effect
Heart rate	Increased
Blood pressure	Increased
Stroke volume	Diminished
Cardiac output	Increased
Systematic vascular resistance	Increased
Myocardial oxygen consumption	Increased
Myocardial blood flow	Increased

Adapted from Chagué et al, 2014

Current data in the literature strongly suggest that SLT has deleterious effects on human health, including the cardiovascular system. SLT is used extensively in the sports community worldwide and, more precisely, in snow sports in Europe. Nicotine use is reported in the monitoring program and is under current consideration by the WADA. Nicotine use, including SLT, fulfils all the WADA criteria for inclusion in the WADA prohibited list as a doping product. Hence, chewing and spit tobacco, as well as dry and moist snuff, must therefore be discouraged, particularly when sports activities are considered.

Reference Number: 81006

NAKAGATA, T., K. FUKAO, H. KOBAYASHI, S. KATAMOTO, AND H. NAITO. "The Effects of Transferal Nicotine Patches on the Cardiorespiratory and Lactate Response During Exercise from Light to Moderate Intensity: Implications for Exercise Prescription during Smoking Cessation." *Medicina*, 55: 348 (10 pp), 2019 (1 table, 1 figure, 51 references)

Abstract: Eight male non-smokers and six current smokers (aged 20–26 years) performed a submaximal exercise test using an electromagnetic cycle ergometer at 30–210 W, with or without a nicotine patch. The nicotine patch (17.5 mg nicotine) or placebo patch was applied to the left arm 8 to 10 hours prior to the exercise test. The mean HR was increased by

9–10 bpm at rest and 6.7 bpm during exercise with the TNP. There were no significant differences in the other parameters (such as oxygen consumption) measured.

> When comparing the cardiorespiratory responses and La [lactate] concentration during exercise at light to moderate intensity between TNP and control conditions, the HR during exercise was 6.7 bpm higher on average in the TNP condition than in the control condition. However, there were no significant differences in VO_2, VE, RPE, and La between both conditions. For practical recommendations for exercise prescription during smoking cessation in the clinical setting, clinicians should consider that TNPs increase the HR at rest and during exercise when prescribing exercise intensity and monitoring physiological responses, and the HRR methods may be preferred when prescribing the exercise intensity based on HR. The results of our study are valuable for clinicians and health professionals to guide them when prescribing exercise programs combined with TNPs during smoking cessation in the clinical setting, such as cardiology and respiratory medicine, for healthy young individuals.

Reference Number: 81007

SUMARTININGSIH, S., H-F. LIN, AND J-C. LIN. "Cigarette Smoking Blunts Exercise-Induced Heart Rate Response Among Young Adult Male Smokers." *International Journal of Environmental Research and Public Health*, 16: 1032 (9 pp), 2019 (4 tables, 35 references)

Abstract: The heart rate response (HRR) and heart rate variability (HRV) were determined in 24 young male adult smokers (mean age 23 years) who were administered 0 mg nicotine EC, 3 mg nicotine EC, and two 1.5 mg cigarettes before a standardized exercise test. The time to exhaustion showed no statistically significant differences under the three different conditions. Exercise-induced HRV was decreased after the smoking of cigarettes. The HRV standard deviation of normal to normal intervals (SDNN) during exercise increased under the 3 mg dose of nicotine.

> The results showed that a significant acute autonomic cardiac modulation during exercise is induced by an acute episode of EC and TC smoking.

Reference Number: 81008

PRICE, T.B., S. KRISHNAN-SARIN, AND D.L. ROTHMAN. "Smoking Impairs Muscle Recovery from Exercise." *American Journal of Physiology, Endocrinology and Metabolism*, 285: E116–E122, 2003 (1 table, 3 figures, 48 references)

Abstract: Eight smokers (20+ CPD) and 10 age- and weight-matched non-smoking controls (mean age 23.5 years) performed single leg toe raises for 20–24 minutes to determine the effect of smoking on glycogen recovery. Plasma samples were collected before and during 4 hours of recovery and assayed for glucose and insulin. NMR spectroscopy was used to determine muscle glycogen and glucose-6-phosphate levels. Cigarette smoking impairs the insulin-dependent portion of muscle recovery from glycogen-depleting exercise.

Table: Muscle Glycogen Depletion Data in Smokers and Controls After Exercise

	Smokers	Controls
Baseline glycogen (mmol/L)	69.0	71.8
Exercise time (min)	20	24
Total work (kJ)	29.4	30.6
Total glycogen depleted (mmol/L)	45.6	53.5
Total glycogen resynthesized (mmol/L)	14.3	26.8

Adapted from Price et al, 2003

Habitual nicotine exposure has been linked to insulin resistance and is a major risk factor for coronary and peripheral vascular disease, as well as hypertension. Habitual nicotine exposure also impairs insulin-dependent glucose metabolism. Insulin resistance is known to be a major risk factor in the development of adult-onset diabetes. Furthermore, there may be a dose-response relationship between smoking and the risk of diabetes. Exercise is known to increase skeletal muscle insulin sensitivity and glycogen synthesis in both healthy controls and insulin-resistant subjects; however, the mechanisms by which exercise enhances and/or nicotine impairs muscle insulin sensitivity are not well characterized.

8.11 SUICIDE

"The public health authorities never mention the main reason many Americans have for smoking heavily, which is that smoking is a fairly sure, fairly honorable form of suicide."

— Kurt Vonnegut, "Welcome to the Monkey House" (1968)

Psychiatric patients who smoke suffer more delusion, hallucinations, and thoughts of suicide or suicide attempts than non-smokers (81101).

Nicotine is a water- and slightly fat-soluble alkaloid from the tobacco plant of the nightshade family. The initial symptoms of poisoning can occur within 15 minutes of ingestion. Death is rapid and is the result of seizures, respiratory depression, or cardiac arrhythmias (81102, 81105). Nicotine e-liquids for vaping are a new, readily available source of this toxic poison for use in suicides (81102–5).

One study describes two suicides in which the victims extracted and concentrated nicotine from tobacco according to instructions on the internet (81106).

Reference Number: 81101

DE OLIVEIRA, R.M., J.L.F. SANTOS, AND A.R.F. FUREGATO. "Tobacco Addiction in the Psychiatric Population and in the General Population." *Revista Latino-Americana de Enfermagem*, 25: e2945 (9 pp), 2017 (2 tables, 3 figures, 44 references)

Abstract: Although only 22% of the US adult population suffer from a mental disorder, these patients consumed half of the cigarettes produced in the country. High tobacco consumption has significant financial impacts, as these smokers can spend up to 30% of monthly income on cigarettes. Tobacco dependence was determined in 134 psychiatric patients who smoked in Brazil. About three-quarters of the smokers had severe mental disorders such as schizophrenia. In the psychiatric patients, 27.3% had a very high tobacco addiction compared to 4.2% of the general smoking population. Psychiatric patients who smoked were found to suffer more outbreaks (delusions and hallucinations), thoughts of suicide or suicide attempts, and psychiatric hospitalizations.

Table: Adjusted Odds Ratio of Degree of Tobacco Dependence and Risk of Suicide

Current Risk of Suicide	Adjusted OR of Degree of Tobacco Dependence
No risk	1.00
Low	1.14
Moderate	1.22
High	1.22

Adapted from de Oliveira et al, 2017

Nicotine dependence was greater in the psychiatric population than in the general population. The independent factors associated with severe dependence were sex, age group, diagnosis, and current risk of suicide. These results trigger reflection among nurses on the need to focus more attention on a neglected subject in mental health services.

Reference Number: 81102

LEE, J., Y. YOU, J.S. PARK, J.H. MIN, I. YOO, W. JEONG, Y. CHO, S. RYU, S. KIM, S.U. CHO, S.K. OH, J. LEE, H.J. AHN, AND S.M. JUNG. "Liver Donation After Brain Death Following Intentional Ingestion of 99% E-Cigarette Liquid Nicotine 10 mL." *Experimental and Clinical Transplantation*, 18: 120–122, 2020 (2 figures, 10 references)

Nicotine is a water and lipid soluble liquid alkaloid with a half-life of 1 to 4 hours. Approximately 90% of nicotine is metabolized to cotinine by the cytochrome oxidase in the liver. The initial symptoms of nicotine toxicity can occur 15 minutes after consumption. Death is rapid, within one hour, due to respiratory depression or cardiac arrhythmias. A 26-year-old man with a history of severe depression consumed 10 mL of a 99% liquid nicotine e-cigarette refill (990 mg/mL) in a suicide attempt. He was found unconscious and after 13 days in ICU was declared brain dead. His urine cotinine concentration 12 hours after admission was 1,115 ng/mL.

Our present case suggests that liver from a donor with fatal nicotine intoxication may be successfully used for transplant in the absence of extensive steatosis and other contraindications for donation.

Reference Number: 81103

PARK, E.J., AND Y.G. MIN. "The Emerging Method of Suicide by Electronic Cigarette Liquid: A Case Report." *Journal of Korean Medical Science*, 33: e52, 2018 (1 figure, 9 references)

Abstract: A 27-year-old man was admitted to an ER in a comatose state after suffering a cardiac arrest due to the intentional ingestion of EC liquids containing 16 to 18 mg/mL nicotine. He also had seizures but survived. A 17-year-old woman was also admitted to hospital after a cardiac arrest and tonic clonic movements after consuming Pure Nicotine (210 mg/mL nicotine). She also survived but required rehabilitation.

> With an increase in EC usage and the absence of proper regulation for EC liquid, the concerns about the toxic exposure have also increased. The presented cases caution that the ingestion of EC liquid can be one of the emerging methods for suicide, especially in young adults and adolescents.

Reference Number: 81104

BELKONIENE, M., J. SOCQUET, D. NIEMBA-FREIBURGHAUS, AND C. PELLATON. "Near Fatal Intoxication by Nicotine and Propylene Glycol Injection: A Case Report of an e-Liquid Poisoning." *BMC Pharmacology and Toxicology*, 20: 28 (6 pp), 2019 (1 table, 2 figures, 26 references)

Abstract: A 51-year-old male smoker and EC user with a personality disorder injected himself with 10 mL of an e-liquid, which contained 100 mg/mL nicotine and propylene glycol, in a suicide attempt. Upon admission to hospital 30 minutes later, he had tachycardia, an elevated BP, and psychomotor agitation and mydriatic pupils which were poorly responsive to light. Two hours post injection, the patient became stuporous with bradypnea and fell into a coma. He was transferred to ICU and despite treatment remained in a profound coma for 3 hours. Ten hours post injection, he was alert and responsive. His blood nicotine and cotinine concentrations 18 hours later were 12 ng/mL and 3,210 ng/mL, respectively.

> The relation between the dose of nicotine and its effects is complex. Low doses of nicotine stimulates the nAChRs, while higher doses, as in the case of injection, induce a biphasic clinical reaction. In the early phase, stimulation of the nAChRs occurs in both systems of the ANS like we observed in our patient presenting subsequent hypertension, tachycardia,

abdominal pain, tremor and agitation. In the late phase (2h post-injection in our case), a global desensitisation of all nAChRs appears, causing central nervous system depression, ganglionic blockade and curare-like effects on the NMJ with coma and respiratory failure.

Reference Number: 81105

JUN, Y.J., H. KIM, AND S. PARK. "Suicidal Death by Nicotine Poisoning." *Korean Journal of Legal Medicine*, 38: 167–170, 2014 (1 figure, 10 references)

Abstract: A 56-year-old man was found dead in his office after he had been searching for information about poisons on the internet. Nicotine was detected in the blood and gastric samples by GC-MS, and the peripheral blood nicotine concentration was 58,000 ng/mL. A positive postmortem blood alcohol concentration was 0.192 g/100mL.

Although recipes for extraction of nicotine from tobacco are easily found on the internet and nicotine liquid can be bought easily as refill for e-cigarettes, suicidal nicotine poisoning is rarely reported in our country. We assume that it is because tobacco is such a commonly used substance that it might be regarded as just a harmful chemical related to chronic diseases, but not as a fatal poison leading to sudden death. In addition, it is also reported as a potential chemical weapon because it is highly toxic and can be purchased without restriction. Therefore, careful concern and restriction against uncontrolled administration of high doses of nicotine should be required.

Reference Number: 81106

CORKERY, J.M., J. BUTTON, A. VENTO, AND F. SCHIFANO. "Two UK Suicides Using Nicotine Extracted from Tobacco Employing Instructions Available on the Internet." *Forensic Science International*, 199: e9–e13, 2010 (1 table, 43 references)

Abstract: Fatalities caused by nicotine are rare, especially from the ingestion of a tobacco extract, as the emetic effects of nicotine cause the victim to vomit and expel the nicotine from the stomach before it can be substantially absorbed and reach toxic levels in the blood. This study examines two suicides in the United Kingdom due to the ingestion of nicotine extracted from tobacco using instructions from the internet. The instructions given were to soak 150 grams of tobacco in water for several days and then slowly heat the liquid and let it simmer until the liquid became a gooey residue, which would then be added to a drink. In one case, a 19-year-old male was found dead in the bedroom of his father's home, surrounded by empty tobacco packs and a saucepan with a brown gooey residue. He had vomited and there were large quantities of feces surrounding the body. His postmortem blood nicotine concentration (which was not detected in the initial toxicology screen) was 5,500 ng/mL and cotinine concentration was 2,500 ng/mL. Nicotine poisoning as a cause of death was initially overlooked due to the wide prevalence of smoking. In another case, a 32-year-old man had a postmortem blood nicotine concentration of 1,000 ng/mL.

> This paper has described what are believed to be the first two suicides in the UK by nicotine derived from tobacco using instructions available on the Internet reported in the scientific literature. Although unusual, other such deaths and unsuccessful attempts have been reported using this method, as well as other instances of poisoning by other tobacco products and nicotine pesticide solutions.

CHAPTER 9

Nicotine Metabolites/Chemicals

The most common nicotine metabolite is cotinine, which can be used to measure second- and thirdhand smoke exposure. Anabasine is a chemical that is found in tobacco products such as cigarettes but not in nicotine replacement therapy (NRT) (e.g., patches, gums). As a result, anabasine can be used to distinguish smokers who have relapsed from NRT and are still smoking.

9.01 COTININE

> "Our results indicate that the level of urinary cotinine is independently associated with the degree of nicotine dependence and that cotinine is a valid biomarker for prediction of high nicotine dependence."
>
> —Jung et al, "Can Urinary Cotinine Predict Nicotine Dependence Level in Smokers?" (2012)

The major metabolite of nicotine, cotinine, is a good biomarker of nicotine exposure and can be measured in human blood (serum), saliva, urine, milk, and cervical mucosa (91010).

Reference Number: 90101

JARVIS, M.J., M.A. RUSSELL, N.L. BENOWITZ, AND C. FEYERABEND. "Elimination of Cotinine from Body Fluids: Implications for Noninvasive Measurement of Tobacco Smoke Exposure." *American Journal of Public Health*, 78: 696–698, 1988 (1 table, 1 figure, 18 references)

Abstract: Cotinine can be measured in human blood, saliva, urine, milk, and cervical mucosa. Three non-smokers and two occasional smokers ingested 28 mg of nicotine base every day for 5 days in order to achieve a high cotinine concentration and low nicotine concentration. Blood, urine, and saliva samples were collected daily for 2 weeks and analyzed for cotinine by gas chromatography (GC). The initial mean cotinine concentrations following the period of nicotine ingestion were 249 ng/mL in plasma, 350 ng/mL in saliva, and 1,394 ng/mL in urine. The average ratios of cotinine saliva to plasma were 0.97:1.15. The half-life of cotinine average 16–19 hours in the different body fluids.

> We conclude that cotinine samples from blood, saliva, and urine are equally applicable to the whole range of issues requiring estimates of nicotine exposure from tobacco smoking. After cessation of smoking cotinine concentrations in all body fluids may be expected to decline to nonsmoking levels within four days in the majority of cases, with an upper limit of seven days. Choice of fluid for sampling will depend on practical rather than pharmacokinetic considerations. Marked differences in half-life between smokers and nonsmokers seem unlikely on present evidence, but the precise magnitude of any difference must await larger studies.

Reference Number: 90102

JARVIS, M., H. TUNSTALL-PEDOE, C. FEYERABEND, C. VESEY, AND Y. SALLOOJEE. "Biochemical Markers of Smoke Absorption and Self Reported Exposure to Passive Smoking." *Journal of Epidemiology and Community Health*, 38: 335–339, 1984 (2 tables, 2 figures, 18 references)

Abstract: The self-reported degree of exposure to secondhand smoke (SHS) was determined in 100 non-smoking patients (mean age 56.8 years) in an outpatient clinic. Blood, breath, saliva, and urine samples were also collected. Plasma, saliva, or urine nicotine and cotinine concentrations were determined by GC. Blood carboxyhemoglobin (COHb) and expired breath carbon monoxide (CO) concentrations were also determined.

Table: Some Biochemical Measures (Mean Concentrations) of Non-smokers Compared to Smokers

Biochemical Measure (mean)	Non-Smokers	Smokers
Expired CO (ppm)	5.6	20.8
COHb (%)	0.9	3.9
Plasma nicotine (ng/mL)	0.9	14.8
Saliva nicotine (ng/mL)	4.8	672.5
Urine nicotine (ng/mL)	8.3	1,749.9
Plasma cotinine (ng/mL)	1.5	275.2
Saliva cotinine (ng/mL)	1.7	309.9
Urine cotinine (ng/mL)	5.0	1,391.0

Adapted from Jarvis et al, 1984

This study shows that cotinine can provide a sensitive guide to the extent of recent daily life passive smoker exposure in stringently defined non-smokers. The concentration of this marker was systematically related to self report of the degree of exposure over the past three days. Furthermore, non-invasive samples of saliva and urine provided essentially the same information as did invasive and less easily gathered blood samples and for epidemiological applications may be preferable. Our data showed a high degree of intercorrelation between the concentration of cotinine in urine and saliva and do not permit a recommendation of one sample over the other. However, we note two recent publications which indicate that urinary cotinine may be the marker of choice.

Reference Number: 90103

BERNERT JR., J.T., J.E. MCGUFFEY, M.A. MORRISON, AND J.L. PIRKLE. "Comparison of Serum and Salivary Cotinine Measurements by a Sensitive High-Performance Liquid Chromatography-Tandem Mass Spectrometry Method as an Indicator of Exposure to Tobacco Smoke Among Smokers and Nonsmokers." *Journal of Analytical Toxicology*, 24: 333–339, 2000 (2 tables, 3 figures, 26 references)

Abstract: Salivary cotinine concentrations are a good biomarker of exposure to nicotine and are especially useful as samples can be easily and quickly collected from children. Serum and saliva were collected from 156 non-smokers and 51 smokers and were analyzed for cotinine by LC/MS/MS after a liquid-liquid extraction. The salivary and serum cotinine concentrations correlation was excellent ($r^2 = 0.9968$). Salivary cotinine concentrations were very stable during storage at room temperature. At least 1 mL of saliva is required to ensure full sensitivity of the analysis. The mean saliva/serum cotinine concentration ratio was 1.27:1.

> The results indicate that a simple, linear relationship can be developed to describe serum and saliva cotinine concentrations in an individual, and the expression describing this relationship can be used to estimate with reasonable accuracy (approximately ± 10%) the serum cotinine concentration in an individual .... It was further confirmed that saliva cotinine samples are generally quite stable during storage after collection, even at ambient temperatures, and this sample matrix appears to be well-suited to the requirements of many epidemiologic investigations.

Reference Number: 90104

SHARMA, P., N. SANE, S.D. ANAND, P. MARIMUTTHU, AND V. BENEGAL. "Assessment of Cotinine in Urine and Saliva of Smokers, Passive Smokers, and Nonsmokers: Method Validation Using Liquid Chromatography and Mass Spectrometry." *Indian Journal of Psychiatry*, 61: 270–276, 2019 (2 tables, 3 figures, 16 references)

Abstract: The most commonly used biomarker of tobacco exposure is cotinine, which is the primary metabolite of nicotine and has a half-life of 16–18 hours. Serum and saliva cotinine concentrations were determined by LC/MS/MS in 56 smokers, 15 passive smokers, and 27 non-smokers. The urine cotinine concentration was on average 5× greater than the serum cotinine concentration, which agrees with other studies that found a ratio of 5–6×. There was no correlation between reported Fagerstrom Test for Nicotine Dependence (FTND) score and cotinine concentration.

Table: Mean Urine and Saliva Cotinine Concentrations in Non-smokers, Passive Smokers, and Smokers

Tobacco Exposure	Mean Urine Cotinine Concentration (ng/mL)	Mean Saliva Cotinine Concentration (ng/mL)
Non-smokers	13.6	9.5
Passive smokers	36.6	18.3
Smokers	1,043.6	327.4

Adapted from Sharma et al, 2019

We report an ESI-LC-MS/MS method to identify smoked tobacco exposure (active or passive) in saliva and urine. Results were in correlation with last use or exposure. LOD and LOQ were 0.50 and 1.1 ng/ml, respectively, and recovery was 99.56% showing the sensitivity of the method. We conclude that the current analytical procedure can identify and validate the smoking status and exposure in our population.

Reference Number: 90105

HALEY, N.J., D.W. SEPKOVIC, AND D. HOFFMANN. "Elimination of Cotinine from Body Fluids: Disposition in Smokers and Non-smokers." *American Journal of Public Health*, 79: 1046–1048, 1989 (1 table, 26 references)

Abstract: Urine samples were collected four times daily for 5 days from nine smokers (aged 23 to 38 years) who abstained from smoking and 10 non-smokers (aged 26 to 45 years) who were exposed to environmental tobacco smoke (ETS) for 80 minutes, twice daily for 2 days. Urine cotinine concentrations were determined by radio-immuno-assay (RIA). The ETS was produced from the sidestream smoke of four cigarettes and obtained a chamber air concentration of 25 ppm CO, 0.9 ppm Nox, and 160 μg/m^3 formaldehyde. The mean biological half-life of cotinine was 16.5 hours for smokers and 27.3 hours for non-smokers exposed to ETS.

The slower clearance in nonsmokers of a terminal nicotine metabolite suggests that other tobacco smoke constituents or their metabolites may also remain in the bodies of nonsmokers for longer periods of time. Conceivably, this could increase the potential for endogenous formation of carcinogenic tobacco specific nitrosamines. In considering the toxic burden imposed by exposure to cigarette smoke, it is important to note that sidestream smoke which is a major fraction of ETS contains higher levels of certain toxic cocarcinogenic and cancerogenic compounds than does mainstream smoke. Before dilution with room air, some of these compounds are 50- to 100-fold higher in the sidestream smoke compared to their concentration in mainstream smoke.

Reference Number: 90106

CARABALLO, R.S., D.B. HOLIDAY, S.D. STELLMAN, P.D. MOWERY, G.A. GIOVINO, J.E. MUSCAT, M.P. ERIKSEN, J.T. BERNERT, P.A. RICHTER, AND L.T. KOZLOWSKI. "Comparison of Serum Cotinine Concentration within and across Smokers of Menthol and Nonmenthol Cigarette Brands among Non-Hispanic

Black and Non-Hispanic White U.S. Adult Smokers, 2001–2006." *Cancer Epidemiology, Biomarkers and Prevention*, 20: 1329–1340, 2011 (3 tables, 37 references)

Abstract: About 72% of nicotine is converted to cotinine in the human body. The half-life of cotinine is 16 to 19 hours, much longer than the half-life of nicotine (2 hours). For smokers with a consistent daily use of cigarettes, the serum cotinine concentration varies only about 15–20%. Black smokers usually have higher serum cotinine concentrations than White smokers, which, it was hypothesized, may be due to differences in menthol cigarette use. This study determined the serum cotinine concentration and smoking habits of 1,379 White and 564 Black cigarette smokers between 2001 and 2006. The serum cotinine concentration was determined by LC/MS/MS. Smoking a menthol brand did not result in a higher serum cotinine concentration than smoking a non-menthol brand in either Black or White smokers.

Table: The Mean Serum Cotinine Concentration in Non-Hispanic Black and White Smokers and Number of Cigarettes Per Day

Cigarettes per Day	Mean Serum Cotinine Concentration for Non-Hispanic White Smokers (ng/mL)	Mean Serum Cotinine Concentration for Non-Hispanic Black Smokers (ng/mL)
1–3	32.7	94.4
4–9	100.9	194.6
10–16	195.3	263.9
17+	239.7	292.2

Adapted from Caraballo et al, 2011

The higher levels of cotinine observed in black smokers compared with white smokers are not explained by their higher preference for menthol cigarette brands.

Reference Number: 90107

JUNG, H-S., Y. KIM, J. SON, Y-J. JEON, H-G. SEO, S-H. PARK, AND B.R. HUH. "Can Urinary Cotinine Predict Nicotine Dependence Level in Smokers?" *Asian Pacific Journal of Cancer Prevention*, 13: 5483–5488, 2012 (3 tables, 2 figures, 31 references)

Abstract: The Fagerstrom Test for Nicotine Dependence was used to divide 381 current smokers into low (53.8%), moderate (33.3%), and high dependence (12.9%) groups. Urine samples were collected and the urine cotinine concentrations were measured by an immunoassay method. The mean urine cotinine concentrations were 523.3 ng/mL for the low, 982.6 ng/mL for the moderate, and 1,378.3 ng/mL for the high dependence nicotine groups.

Our results indicate that the level of urinary cotinine is independently associated with the degree of nicotine dependence and that cotinine is a valid biomarker for prediction of high nicotine dependence. To the best of our knowledge, this study is the first to suggest an available cut-off value of urinary cotinine as an indicator of high nicotine dependence. Because cotinine is an alkaloid found in tobacco and a metabolite of nicotine, its levels in the body have predominantly been used to verify smoking status in individuals. However, our data potentially suggest an extended application of urinary cotinine levels, particularly for assessment of nicotine dependence. We suggest that the optimal cut-off value for the urinary cotinine level that can predict high nicotine dependence is 1,000 ng/mL (sensitivity: 71.4%, specificity: 74.4%).

Reference Number: 90108

PATEL, V.D., K.B. JADHAV, V.S. SHAH, AND N.D. GUPTA. "Comparative Assessment of Salivary Cotinine Level and Psychological Dependence Among Tobacco Users." *Dental Research Journal*, 14: 125–130, 2017 (1 table, 2 figures, 19 references)

Abstract: This study determined salivary cotinine concentration (SCC) by NicAlert stripes (ELISA) and nicotine dependence by FTND in 26 smokers and 28 smokeless (chewing) tobacco users. Nicotine has a half-life of 2 hours and is not a valid marker of smoking, whereas cotinine with a half-life of 17 hours is a more stable marker. It takes about 24 hours for saliva cotinine to appear after oral ingestion of nicotine. The SLT group had higher SCCs compared to the smokers. There is only a weak correlation between SCC and nicotine dependence.

Salivary cotinine and psychological dependence through FTND scoring are not strongly correlating with each other. This indicates that dependence over tobacco is a separate phenomenon and cannot be assessed by salivary cotinine level. It is well accepted that salivary cotinine level is influenced by age of individual, duration of habit, and type of tobacco consumption.

Reference Number: 90109

BOFFETTA, P., S. CLARK, M. SHEN, R. GISLEFOSS, R. PETO, AND A. ANDERSEN. "Serum Cotinine Level as Predictor of Lung Cancer Risk." *Cancer Epidemiology, Biomarkers and Prevention*, 15: 1184–1188, 2006 (5 tables, 1 figure, 37 references)

Abstract: The serum concentration of samples of 1,741 Norwegians who developed lung cancer and 1,741 matched controls who were lung cancer-free, stored frozen since 1973 onwards, were analyzed by an immunoassay method. Cotinine is a biomarker of tobacco smoke exposure. Mean serum cotinine concentrations were lower in smokers with lung cancer than controls. The OR of lung cancer increased linearly with the serum cotinine concentration. The OR of lung cancer with serum cotinine concentration was similar in both men and women.

Table: Serum Cotinine Concentrations and Odds Ratio (OR) of Lung Cancer

Serum Cotinine Concentration (ng/mL)	OR of Lung Cancer
5.1–24.7	0.93
24.8–114.7	2.80
114.8–180.2	7.67
180.3–223.8	13.8
223.9–271.9	24.9
272.0–317.7	30.4
317.8–378.8	33.1
>378.8	55.1

Source: Adapted from Boffetta et al, 2006

The results of this study show a strong, linear dose-response relationship between serum cotinine level and lung cancer risk among smokers. Measurement of serum cotinine, in particular if based on repeated samples, would reduce exposure misclassification and improve the estimate of the carcinogenic effect of smoking. In our study, serum cotinine level, when measured in a single sample taken several years before onset of cancer, was a strong predictor of risk, suggesting that this approach is more relevant than traditional (e.g., questionnaire-based methods to measure exposure to tobacco smoke and to investigate its carcinogenic effects).

One important result of our analysis is the lack of a clear plateau in the relative risk of individuals with high serum cotinine levels.

Reference Number: 90110

FIX, B.V., R.J. O'CONNOR, N. BENOWITZ, B.W. HECKMAN, K.M. CUMMINGS, G.T. FONG, AND J.F. THRASER. "Nicotine Metabolite Ratio (NMR) Prospectively Predicts Smoking Relapse: Longitudinal Findings from ITC Surveys in Five Countries." *Nicotine and Tobacco Research*, 19: 1040–1047, 2017 (5 tables, 37 references)

Abstract: The ratio of 3'-hydroxycotinine (3-HC) to cotinine is known as the nicotine metabolite ratio (NMR) and is a biomarker of nicotine metabolism. Higher NMR indicates a faster nicotine metabolism and has been found to be associated with higher daily cigarette consumption and less success at smoking cessation. Saliva samples were collected from 874 participants in the International Tobacco Control Surveys in the United States, the United Kingdom, Mauritius, Mexico, and Thailand and were analyzed by LC/MS/MS to determine 3-HC and cotinine concentrations, and hence the NMR.

Given that high metabolizers are typically more nicotine dependent (based on quantity/time to first cigarette assessments), and experience more withdrawal based cravings, their smoking may be contingent primarily upon blood nicotine levels. Thus, high metabolizers who persist through the 2–4 weeks of withdrawal-based cravings may be at lower risk of relapse in the long term. Indeed, faster metabolizers have been shown to experience greater nicotine withdrawal and higher rates of withdrawal over the first week of abstinence predicted more favorable cessation outcomes in a small pilot study.

Reference Number: 90111

EISNER, M.D., J. BALMES, E.H. YELIN, P.P. KATZ, S.K. HAMMOND, N. BENOWITZ, AND P.D. BLANC. "Directly Measured Secondhand Smoke Exposure and COPD Health Outcomes." *BMC Pulmonary Medicine*, 6: 12 (11 pp), 2006 (5 tables, 1 figure, 43 references)

Abstract: Seventy-seven non-smokers with COPD were measured as to SHS exposure using urine cotinine concentrations, 3-HC (most abundant urinary metabolite of nicotine), and a personal badge which measured environmental exposure. Urinary metabolites were measured by LC/MS/MS and nicotine by GC. The majority of the subjects were White, female (mean age 64.4 years), and although not current smokers, 72% were former smokers. Patients with the highest urine cotinine concentrations were associated with worse COPD severity, disease-specific quality of life, and dyspnea. But personal nicotine badges showed no association.

> There is no question that cigarette smoking is the dominant risk factor for COPD and is the most important factor driving the progression of airflow obstruction. Our results implicate SHS exposure as another important factor influencing disease severity and health status in this health condition. Because SHS is a modifiable risk factor, clinicians should assess SHS exposure in their patients and counsel its avoidance. In public health terms, the effects of SHS exposure on this vulnerable subpopulation provide a further rationale for laws prohibiting public smoking.

9.02 OTHER METABOLITES/CHEMICALS

> "Compared to cigarettes in Australia and Canada, tobacco blends used in U.S. cigarettes have higher levels of tobacco-specific nitrosamines (TSNAs), chemicals known to cause cancer. TSNA levels have increased since 1964."
>
> — A. McIntyre, "Tobacco Companies Have Made Cigarettes Deadlier than Ever" (2014)

The main metabolites of nicotine (90201–3) are:

- Cotinine
- 3-Hydroxycotinine
- Norcotinine
- Nornicotine
- Cotinine Oxide
- Nicotine Oxide

The minor alkaloids found in tobacco other than nicotine are anabasine and anatabine.

Analysis of urine and other biological samples for nicotine metabolites can indicate when smoking ceased and amount of tobacco smoked. It is also useful in determining the smoker's compliance to NRT, nicotine addiction, and uptake (90203). For example, the minor alkaloid anabasine is found in tobacco, but not in NRT such as patches or gums.

Therefore, a urinary anabasine concentration greater than 2 ng/mL indicates that the smoker continued to smoke cigarettes, which is risky if the smoker is using the nicotine patch as well (90203–4).

Breath carbon monoxide concentrations are a good indicator of recent smoking; however, e-cigarettes do not generate as much CO as combustible cigarettes and so this test is of limited value in EC users (90205). Interestingly, due to extensive smoke-free laws, the mean breath CO concentration in non-smokers due to SHS has fallen from 10 ppm to 5 ppm or less (90205).

Another gas found in cigarette smoke is cyanide (HCN), which is found as thiocyanate in saliva and has been found at elevated concentrations in EC users as well (90206).

The levels of potent carcinogens, TSNAs, have not decreased in US cigarettes since 1979, despite tobacco companies' pledge to lower them (90207).

Reference Number: 90201

MCGUFFEY, J.E., B. WEI, J.T. BERNERT, J.C. MORROW, B. XIA, L. WANG, AND B.C. BLOUNT. "Validation of a LC-MS/MS Method for Quantifying Urinary Nicotine, Six Nicotine Metabolites and the Minor Tobacco Alkaloids—Anatabine and Anabasine—in Smokers' Urine." *Public Library of Science One*, 9: e101816 (13 pp) 2014 (7 tables, 2 figures, 47 references)

Abstract: The six primary metabolites of nicotine are cotinine, hydroxycotinine, norcotinine, nornicotine, cotinine oxide, and nicotine oxide. Two minor tobacco alkaloids (nicotine is the major alkaloid), anabasine and anatabine, are found in tobacco smokers but not in people who use NRT. This LC/MS/MS method is used to detect the six nicotine metabolites, anabasine, and anatabine in urine to determine compliance in smokers with NRT. The urine metabolites are reported as free and total. Acetone precipitation is used prior to analysis to clean up the urine sample.

> The optimized assay produces linear responses across the analyte concentrations typically found in urine collected from daily smokers. Because matrix ion suppression may influence accuracy, we include a discussion of conventions employed in this procedure to minimize matrix interferences.

Simplicity, low cost, low maintenance combined with high mean metabolite recovery (76–99%), specificity, accuracy (0–10% bias) and reproducibility (2–9% C.V.) make this method ideal for large high through-put studies.

Reference Number: 90202

JACOB III, P., L. YU, A.T. SHULGIN, AND N.L. BENOWITZ. "Minor Tobacco Alkaloids as Biomarkers for Tobacco Use: Comparison of Users of Cigarettes, Smokeless Tobacco, Cigars, and Pipes." *American Journal of Public Health*, 89: 731–736, 1999 (3 tables, 1 figure, 29 references)

Abstract: Nicotine and its major metabolite cotinine cannot be used as biomarkers of cigarette smokers, because subjects using NRT such as a patch or gum will also test positive. Tobacco products, however, in addition to containing the major alkaloid nicotine (96%), also contain minor alkaloids such as nornicotine (2%), anabasine (0.5%), and anatabine (1.5%), but the actual concentration varies among different tobacco products. Nornicotine is also a nicotine metabolite in the human body, in addition to appearing in tobacco products. The concentrations of nicotine is lowest in SLT and highest in cigarette tobacco. The concentration of nornicotine is lowest in SLT and highest in cigar tobacco. Urine and plasma concentrations of these tobacco alkaloids were determined in 12 cigarette smokers, 9 SLT users, 5 pipe smokers, and 8 cigar smokers by GC/MS.

Table: Mean Urine Tobacco Alkaloid Concentrations (ng/mL) in the 24-hour Urine of Cigarette, Cigar, and Pipe Smokers and SLT Users

Alkaloid	Cigarette Smoking	SLT Use	Cigar Smoking	Pipe Smoking
Anabasine	6.2	6.3	2.8	5.9
Anatabine	11.5	14.9	2.8	4.6
Nornicotine	50.0	30.7	35.9	21.1
Nicotine	1,160	732	289	1,107
Cotinine	1,280	919	670	1,358

Adapted from Jacob III et al, 1999

Of interest is the finding that the elimination half-life of nicotine as based on urine excretion data averaged 11 hours. This is much longer than the half-life of 2 to 3 hours based on plasma concentrations. The likely explanation is the slow release of nicotine from high-affinity tissue-binding sites. The resulting levels of nicotine are too low to be measured in plasma, but they can be measured in urine because concentrations of nicotine are much higher in urine than in plasma. In contrast, for cotinine, which is much less avidly taken up by tissue, the elimination half-lives derived from urine excretion data and plasma level data are similar.

In conclusion, we provide novel data on minor alkaloid levels in tobacco products and in users of these products. Our data suggest that measurement of anabasine and anatabine levels in urine has the potential to be a biomarker of tobacco consumption and exposure to toxic substances in tobacco, even in individuals using nicotine-containing medications, since such medications do not contain these alkaloids.

Reference Number: 90203

XU, X., M.M. IBA, AND C.P. WEISEL. "Simultaneous and Sensitive Measurement of Anabasine, Nicotine, and Nicotine Metabolites in Human Urine by Liquid Chromatography–Tandem Mass Spectrometry." *Clinical Chemistry*, 50: 2323–2330, 2004 (4 tables, 2 figures, 20 references)

Abstract: Determination of nicotine metabolites can provide a useful tool to estimate compliance to NRT, nicotine addiction, and uptake of nicotine. Patients abstaining from cigarettes typically have a urine anabasine concentration less than 2 ng/mL. A sensitive method was developed to determine urinary nicotine, cotinine, nicotine oxide, cotinine oxide, 3-hydroxycotinine, nornicotine, and a minor tobacco alkaloid (anabasine) by LC/MS/MS. This method was employed to detect the nicotine/metabolites and anabasine concentration for eight smokers during NRT (Table). Of the eight patients shown in the table, only patient 2 with a urinary anabasine concentration of 1.0 ng/mL stopped completely using cigarettes. Patients 3 and 6 had minimally used cigarettes.

Table: Urine Nicotine/Metabolite and Anabasine Concentrations (ng/mL) in Eight Patients Undergoing NRT

Patient No.	Nicotine	Cotinine	Nornicotine	Cotinine Oxide	3-HC	Anabasine
1	400	1,241	67.1	455.0	6,820	12.8
2	658	256	19.0	23.0	102	1.0
3	275	970	42.6	223.0	2,970	3.1
4	5,180	2,097	265.0	1,290.0	22,700	27.9
5	1,570	1,267	91.0	468.0	4,280	12.2

Patient No.	Nicotine	Cotinine	Nornicotine	Cotinine Oxide	3-HC	Anabasine
6	599	435	21.9	55.2	398	3.3
7	1,440	959	51.6	136.0	1,330	6.5
8	5,470	1,282	125.0	321.0	2,420	20.3

Source: Adapted from Xu et al, 2004

This sensitive LC/MS/MS procedure can be used to determine nicotine metabolism profiles of smokers, people during nicotine replacement therapy, and passively exposed nonsmokers. This method avoids the need for a time-consuming and labor-intensive sample enrichment step and thus allows for high-throughput sample preparation and automation.

Reference Number: 90204

HOFFNAGLE, A.N., T.J. LAHA, P.M. RAINEY, AND S.M.H. SADRAZADEH. "Specific Detection of Anabasine, Nicotine, and Nicotine Metabolites in Urine by Liquid Chromatography–Tandem Mass Spectrometry." *American Journal of Clinical Pathology*, 126: 880–887, 2006 (4 tables, 2 figures, 26 references)

Abstract: A LC/MS/MS method was developed to determine the concentration of nicotine, cotinine, 3-hydroxycotinine, nornicotine, and the tobacco alkaloid anabasine in urine samples. The method is highly specific and requires only the centrifugation of the urine for sample preparation. Nornicotine is usually less than 1% of the total nicotine metabolites.

For patients undergoing surgery, tobacco and nicotine use have important implications. Several studies have shown delayed healing times and increased infection and thrombosis rates in nicotine users after surgical procedures. Tobacco use in patients receiving organ transplants is associated with elevated rates of graft loss, mortality, and new cancer. For these reasons, assays for nicotine and its metabolites in the urine of patients scheduled for elective surgery or organ transplantation can help in planning surgery and establishing transplant eligibility.

Reference Number: 90205

MARRONE, G.F., D.M. SHAKLEYA, K.B. SCHEIDWEILER, E.G. SINGLETON, M.A. HUESTIS, AND S.J. HEISHMAN. "Relative Performance of Common

Biochemical Indicators in Detecting Cigarette Smoking." *Addiction*, 106: 1325–1334, 2011 (4 tables, 1 figure, 24 references)

Abstract: Breath CO is a good indicator of recent smoking, but its short half-life (2–3 hours) limits its sensitivity. Urine and saliva cotinine have a longer half-life of 10–30 hours and can be measured by immunoassay test strips or GC/MS or LC/MS/MS. Breath CO and urine and saliva cotinine concentrations were determined in 90 non-treatment seeking smokers and 82 non-smokers. Smokers had a mean breath CO concentration of 19.2 ppm (range 2–66 ppm in heavy smokers, 1–43 ppm in light smokers) and 2.1 ppm for non-smokers (range 0–6 ppm for those not exposed to SHS and 1–7 ppm for those exposed to SHS). Saliva NicAlert strip test results for cotinine were variable.

In surveys of smokers and non-smokers in areas with strong smoke-free laws, the breath carbon monoxide cut-off that discriminates most effectively appears to be ≥5 p.p.m. rather than the ≥10 p.p.m. cut-off often used. These findings may not generalize to clinical trials, regions with different carbon monoxide pollution levels or areas with less stringent smoke-free laws.

Reference Number: 90206

FLIEGER, J., J. KAWKA, AND M. TATARCZAK-MICHALEWSKA. "Levels of the Thiocyanate in the Saliva of Tobacco Smokers in Comparison to e-Cigarette Smokers and Nonsmokers Measured by HPLC on a Phosphatidylcholine Column." *Molecules*, 24: 3790 (14 pp), 2019 (5 tables, 4 figures, 49 references)

Abstract: Tobacco smoke is a mixture of gases such as CO and hydrogen cyanide (HCN). A main metabolite of HCN is thiocyanate (SCN^-), which is excreted in the oral fluids and has a long half-life of 10–14 days. The saliva thiocyanate concentrations were determined in eight cigarette smokers, eight EC smokers, and eight non-smokers by high-performance liquid chromatography with ultraviolet detection. The ranges of salivary HCN were 121.25–187.5 ng/L, 121.24–244.1 ng/L, and 33.03–187.54 ng/L in smokers, EC users, and non-smokers, respectively.

The measurement of thiocyanate concentration in saliva is widely recognized as a non-invasive method for assessing exposure to nicotine smoke. Salivary thiocyanate is responsible for neurological and endocrine disorders; therefore, the development of a fast and reliable method for its determination is important from a clinical diagnosis point of view. The study presented is the first to assess and compare SCN^- levels in the saliva of e-cigarette smokers compared to non-smokers and smokers.

Reference Number: 90207

STEPANOV, I., A. KNEZEVICH, L. ZHANG, C.H. WATSON, D.K. HATSUKAMI, AND S.S. HECHT. "Carcinogenic Tobacco–Specific N-Nitrosamines in US Cigarettes: Three Decades of Remarkable Neglect by the Tobacco Industry." *Tobacco Control*, 21: 44–48, 2012 (2 tables, 2 figures, 30 references)

Abstract: Tobacco-specific nitrosamines (TSNAs), such as N-Nitrosonornicotine (NNN) and Nicotine-Derived Nitrosamine Ketone (NNK), are found in high amounts in both unburned tobacco and tobacco smoke and are well-known carcinogens. In 1999, RJ Reynolds and Philip Morris announced that they would reduce TSNAs in their cigarettes. In April 2010, 17 different brands of cigarettes were purchased from retail stores in Minnesota and the smoke and tobacco filler were analyzed for TNSAs by LC/MS/MS. The TSNA concentration averaged 2.54 µg/g tobacco in 2010 compared to 2.1 µg/g for tobacco in 1979. In contrast, the TSNA concentration in cigarettes sold in Canada in 2005 were 0.35 µg/g.

> We found no indication that any meaningful attempt was made to reduce or even control TSNA levels in the new varieties of the popular brands Marlboro and Camel introduced over the last decade. In light of the recently granted regulatory authority to the FDA over tobacco products, regulation of TSNA levels in cigarette tobacco should be strongly considered to reduce the levels of these potent carcinogens in cigarette smoke.

Nicotine Timeline

Pre-1492 Different types of tobacco are widely cultivated and used throughout North and South America, the Middle East, and China for thousands of years.

1492 After first throwing it into the sea, Italian explorer Christopher Columbus eventually brings tobacco back with him from the Americas to Europe where it soon becomes popular.

1604 King James I of England writes "A Counterblaste to Tobacco," one of the first antismoking documents.

1665 Smoking is thought to ward off the Great Plague and was made compulsory at Eton.

1730 First US tobacco factory making snuff opens in Virginia.

1794 US Congress passes the first excise tax on tobacco products.

1828 Hand-rolled cigarettes become popular in Europe.

1843 The French Company SEITA begins large scale manufacture of cigarettes.

1858 Fears about the health risks of smoking are first raised in the British medical journal Lancet.

1862 The US government places the first federal tax on the sale of raw tobacco to help pay for the Civil War.

1913 The first prepackaged, pre-blended, or "modern" cigarette, "Camels," is produced by RJ Reynolds.

1918 Cigarettes are included in rations for US soldiers in World War I, causing many soldiers to become addicted to nicotine.

1929 Public relations campaigns designed by ad man Edward Bernays makes public smoking by women acceptable.

1933 The US Agricultural Adjustment Act provides price support for tobacco farms.

1945 During World War II, tobacco is declared a protected crop. So many cigarettes are shipped to the troops overseas that there is a national shortage in the US.

1950 The first clinical study linking smoking to lung cancer is published in the Journal of the American Medical Association.

1952 Filter cigarettes are developed by the tobacco firm Lorillard, the "Micronite" filter on "Kent" cigarettes. The health risks of smoking are raised in a Reader's Digest article, "Cancer by the Carton."

1954 The Tobacco Industry Research Committee is formed by the US tobacco companies to counteract health risks claims.

1964 Surgeon General Luther Terry issues the landmark report "Smoking and Health," although it does not consider that nicotine was addictive.

1965 The first Surgeon General's health warning appears on US cigarette packs as ordered by Congress.

1971 Cigarette advertising in television and radio is banned in the US.

1977 The American Cancer Society holds its first Great American Smokeout to encourage quitting.

1984 The FDA approves Nicorette (a nicotine gum substitute) to aid smoking cessation.

1986 Surgeon General C. Everett Koop's report "The Health Consequences of Involuntary Smoking" is the catalyst for smoke-free workplaces and public places, due to the dangers of second-hand smoke.

1989 Smoking is banned on all US domestic airline flights.

1992 The transdermal patch NicoDerm is introduced as a smoking cessation aid.

1994 Mississippi is the first US state to sue the tobacco companies for health costs associated with smoking.

1995 The FDA classifies nicotine as a drug.

1998 Forty-six US states settle with the tobacco companies for $206 billion over 25 years in the Master Settlement Agreement.

1999 The US Federal government files a $22 billion civil suit against the tobacco companies under the Racketeer Influences and Corrupt Organization (RICO) Act.

2001 Canada becomes the first country to make mandatory pictorial health warnings on cigarette packages.

2002 Ireland becomes the first European country to have a nationwide smoking ban.

2006 The tobacco companies are found guilty of conspiracy, fraud, and violations of the RICO Act (see Appendix 2).
US Surgeon General concludes that there is no risk-free level of secondhand smoke.

2007 E-cigarettes first sold in the US.

2010 Canada bans all flavorings in cigarettes except menthol, which will be banned in 2017.

2012 In the United States, the percentage of smokers declines to below 20% for the first time since 1900s.

2017 Nicotine-containing vaping devices are first legally sold in Canada and the rate of vaping in youth soars.

2019 A severe pulmonary disease strikes down over 3,000 mainly younger vapers across the United States, killing at least 30; it is called EVALI (e-cigarette or vaping product associated lung injury).

2020 The COVID-19 pandemic spreads globally as the tobacco industry encourages the view that cigarette smokers are more resistant to the disease.

RICO Decision Against the Tobacco Companies

On 17 August 2006, after seven years of trial, Judge Gladys Kessler of the United States District Court of the District of Columbia issued a landmark 1,683-page opinion[1] holding that the tobacco companies were liable for conspiracy and violations of the Racketeer Influenced and Corrupt Organizations Act (RICO) and ordered the tobacco companies to issue the following corrective statements.

TEXT OF COURT-ORDERED CORRECTIVE STATEMENTS: *UNITED STATES V PHILIP MORRIS USA INC.*

Adverse Health Effects of Smoking

A Federal Court has ordered Altria, R.J. Reynolds Tobacco, Lorillard, and Philip Morris USA to make this statement about the health effects of smoking.

- Smoking kills, on average, 1,200 Americans. Every day.
- More people die every year from smoking than from murder, AIDS, suicide, drugs, car crashes, and alcohol, **combined**.
- Smoking causes heart disease, emphysema, acute myeloid leukemia, and cancer of the mouth, esophagus, larynx, lung, stomach, kidney, bladder, and pancreas.
- Smoking also causes reduced fertility, low birth weight in newborns, and cancer of the cervix.

---

1 See *US v Phillip Morris USA, Inc*, 449 F.Supp.2d 1 (2006).

Addictiveness of Smoking and Nicotine

A Federal Court has ordered Altria, R.J. Reynolds Tobacco, Lorillard, and Philip Morris USA to make this statement about the health effects of smoking.

- Smoking is highly addictive. Nicotine is the addictive drug in tobacco.
- Cigarette companies intentionally designed cigarettes with enough nicotine to create and sustain addiction.
- It's not easy to quit.
- When you smoke, the nicotine actually changes the brain—that's why quitting is so hard.

Lack of Significant Health Benefit from Smoking "Low Tar," "Light," "Ultra Light," "Mild," and "Natural" Cigarettes

A Federal Court has ordered Altria, R.J. Reynolds Tobacco, Lorillard, and Philip Morris USA to make this statement about the health effects of smoking.

- Many smokers switch to low tar and light cigarettes rather than quitting because they think low tar and light cigarettes are less harmful. They are not.
- "Low tar" and "light" cigarette smokers inhale essentially the same amount of tar and nicotine as they would from regular cigarettes.
- All cigarettes cause cancer, lung disease, heart attacks, and premature death—lights, low tar, ultra lights, and naturals.
- There is no safe cigarette.

Manipulation of Cigarette Design and Composition to Ensure Optimum Nicotine Delivery

A Federal Court has ordered Altria, R.J. Reynolds Tobacco, Lorillard, and Philip Morris USA to make this statement about designing cigarettes to enhance the delivery of nicotine.

- Altria, R.J. Reynolds Tobacco, Lorillard, and Philip Morris USA intentionally designed cigarettes to make them more addictive.
- When you smoke, the nicotine actually changes the brain—that's why quitting is so hard.

Adverse Health Effects of Exposure to Secondhand Smoke

A Federal Court has ordered Altria, R.J. Reynolds Tobacco, Lorillard, and Philip Morris USA to make this statement about the health effects of secondhand smoke.

- Secondhand smoke kills over 38,000 Americans each year.
- Secondhand smoke causes lung cancer and coronary heart disease in adults who do **not** smoke.
- Children exposed to secondhand smoke are at an increased risk for sudden infant death syndrome (SIDS), acute respiratory infections, ear problems, severe asthma, and reduced lung function.
- There is no safe level of exposure to secondhand smoke.

RELATED FEDERAL FINDINGS (*US V PHILLIP MORRIS USA, INC*)

The court found that Phillip Morris:

- violated civil racketeering laws (engaged in an organized conspiracy to commit fraud);
- committed fraud;
- are likely to continue to commit fraud;
- denied that they control the level of nicotine to create and sustain addiction;
- suppressed and concealed scientific research;
- marketed low tar and light cigarettes as less harmful though they knew they were not;
- marketed cigarettes to young people to replace smokers who die or quit smoking;
- denied that secondhand smoke harms non-smokers;
- denied that smoking is addictive; and
- denied the health consequence of smoking.

APPENDIX 3

Some Chemicals Found in E-cigarettes and Their Health Risks

Chemical name	Health risk
1-methyl phenanthrene	cytotoxic/eye/skin irritation
1,3 butanediol	unknown risk
1,3-propanediol	unknown risk
2-acetylpyrorole	skin irritant
2,3-dimethylpyrazine	cytotoxic
2,3-pentanedione	skin/eye irritation
2,3,5-trimethylpyrazine	cytotoxic
3-hexen-1-ol	eye irritation
3-methyl-1-butanol	cytotoxic
acetic acid	respiratory tract irritation
aromadenendrene	cytotoxic/skin irritation
benzyl acetate	cytotoxic, eye/respiratory tract irritation
benzyl alcohol	cytotoxic
butyl butyrate	eye irritation, behavioral effects
camphor	cytotoxic, systemic organ irritation, behavioral effects
cinnamaldehyde	skin/respiratory tract irritation
cinnamyl alcohol	unknown risks
coumarin	systemic organ irritation/behavioral effects
diacetyl	popcorn lung disease
diethylene glycol	systemic organ irritation
ethylbutyrate	mild effects/behavioral effects
ethylene glycol	harmful effects

Chemical name	Health risk
ethylmaltol	cytotoxic
ethylvanillin	unknown risks
glycerin	eye/skin/respiratory tract irritation
hydroxyacetone	cytotoxic
i-butyric acid	respiratory tract irritation
isoamyl acetate	eye/skin/respiratory tract irritation
isobutyl acetate	eye/skin/respiratory tract irritation
isopentyl isovalerate	unknown risks
l-menthyl acetate	respiratory tract irritation
limonene	unknown risks
maltol	cytotoxic
menthone	unknown risks
methyl cyclopentenolone	unknown risks
methylanthranilate	unknown risks
methylcinnamate	unknown risks
methylsalicylate	neurotoxic/cardiovascular effects
myosamine	carcinogenic
n-hexanol	unknown risks
nicotyrine	unknown risks
o-tolualdehyde	unknown risks
p-cymene	skin irritation
propylene glycol	respiratory tract irritation
safrole	unknown risks
thujone	unknown risks
trans-2-heen-1-ol	unknown risks
vanillin	cytotoxic
β-damascone	skin irritation
γ-decalactone	respiratory tract irritation

Source: adapted from Omare et al, "A Review of Tobacco Abuse and Its Epidemiological Consequences." *Journal of Public Health: From Theory to Practice*, 30: 1485–1500, 2022.

Fagerstrom Test for Nicotine Dependence

Question	Answer (score)
How soon after waking do you smoke your first cigarette?	Within 5 minutes (3) Within 5–30 minutes (2) 31–60 minutes (1)
Do you find it difficult to refrain from smoking in places where it is forbidden (e.g., church, library, movie theaters, etc.)?	Yes (1) No (0)
Which cigarette would you hate to give up?	The first in the morning (1) Any other (0)
How many cigarettes a day do you smoke?	10 or less (0) 11–20 (1) 21–30 (2) 31 or more (3)
Do you smoke frequently in the morning?	Yes (1) No (0)
Do you smoke even if you are sick in bed most of the day?	Yes (1) No (0)

Scoring
1–2 = low dependence
3–4 = low to moderate dependence
5–7 = moderate dependence
8 + = high dependence

The Hooked On Nicotine Checklist (HONC)

	YES	NO
1. Have you ever tried to quit, but couldn't?		
2. Do you smoke NOW because it is really hard to quit?		
3. Have you ever felt like you were addicted to tobacco?		
4. Do you ever have strong cravings to smoke?		
5. Have you ever felt like you really needed a cigarette?		
6. Is it hard to keep from smoking in places where you are not supposed to?		
When you haven't used tobacco for a while OR when you tried to stop smoking:		
7. Did you find it hard to concentrate because you couldn't quit?		
8. Did you feel more irritable because you couldn't smoke?		
9. Did you feel a strong need or urge to smoke?		
10. Did you feel nervous, restless, or anxious because you couldn't smoke?		

The HONC is a measure of severity of diminished autonomy. Individuals who score a zero on the HONC by answering NO to all ten questions enjoy full autonomy over their use of tobacco. In schools and clinics, smokers who have scores above zero can be told that they are already hooked.

Source: adapted from DiFranza et al, "Measuring the Loss of Autonomy Over Nicotine Use in Adolescents: The DANDY (Development and Assessment of Nicotine Dependence in Youths) Study." *Archives of Pediatrics and Adolescent Medicine*, 156: 397–403, 2002.

APPENDIX 6

Abbreviations and Chemical Symbols

1-HOP	1-hydroxypyrene
3-HC	(trans) 3'-hydroxycotinine
3-HC-Glu	(trans) 3'-hydroxycotinine glucuronide
A	Arterial
ACS	Acute Coronary Syndrome
AD	Alzheimer's Disease
ADHD	Attention Deficit/Hyperactivity Disorder
AE	Adverse Events
AEP	Acute Eosinophillic Pneumonia
ALI	Acute Lung Injury
AMA	American Medical Association
AMD	Age Related Macular Degeneration
AMI	Acute Myocardial Infarction
AOR	Adjusted Odds Ratio
APA	American Psychiatric Association
AR	Allergic Rhinitis
ARDS	Acute Respiratory Distress Syndrome
ARED	Age Related Eye Diseases
ARM	Age Related Macular Disease
AU	Australia
AUC	Area Under the Curve (drug concentration vs time of dose)
AUD	Alcohol Use Disorder
AV	Arterial/Venous

BAL	Bronchoalveolar Lavage
BAT	British American Tobacco
BCEP	Bis (2-Chloroethyl) Phosphate
BEC	Bronchial Epithelial Cells
BEM	Behavioral Ecological Model
BLL	Blood Lead Levels
BMI	Body Mass Index
BP	Blood Pressure
BPI	Brief Pain Intensity
BPM	Beats per Minute (heart rate)
BRFSS	Behavioral Risk Factor Surveillance System
CA	Chromosome Aberrations
CAC	Coronary Artery Calcification
CAD	Coronary Artery Disease
CAG	Capillary Aerosol Generator
CAP	Community Acquired Pneumonia
Cd	Cadmium
CDC(P)	Center for Disease Control (and Prevention)
CHD	Coronary Heart Disease
CI	Canadian Intense (puffing parameters)
CIAR	Clean Indoor Air Regulation
CMA	Canadian Medical Association
cm	centimetre
Cmax	Maximum Concentration
CMSS	Combined Mainstream and Sidestream Smoke
CNS	Central Nervous System
CO	Carbon Monoxide
CO_2	Carbon Dioxide
CO_2e	Carbon Dioxide Equivalents
COHb	Carboxyhemoglobin
COPD	Chronic Obstructive Pulmonary Disease
COVID-19	2019 Novel Coronavirus Disease
CPD	Cigarettes Per Day
CPET	Cardiopulmonary Exercise Testing
Cs	Cesium
CSR	Corporate Social Responsibility
CTCP	California Tobacco Control Program

CTR	Council for Tobacco Research
CVD	Cardiovascular Disease
CYP2A6	Cytochrome P450 2A6
DA	Dopamine
DAD	Diffuse Alveolar Damage
DALY	Disability Adjusted Life Years
DAP-BL	Diammonium Phosphate Blended Leaf
D/I	Desmosine/Isodesmosine Ratio
DIET	Dry Ice Expanded Tobacco
DIY	Do It Yourself
DM	Diabetes Mellitus
DMFS	Decayed Missing Field Surfaces
DMFT	Decayed, Missing, Filled Teeth
DNA	Deoxyribonucleic Acid
Dp	Particle Diameter
DW	Dry Weight
EC	Electronic Cigarette
EC50	Effective Concentration 50%
ECA	Electronic Cigarette Aerosol
ECC	Electronic Cigarette Cartridge
ECG	Electrocardiogram
ECIG	Electronic Cigarette
e-cigarettes	Electronic Cigarettes
eCO	Exhaled Carbon Monoxide
ECS	Electronic Cigarette Smoke
ECVC	Electronic Cigarette Vapor Extract
ED	Emergency Department
EDS	Excessive Daytime Sleepiness
EEG	Electroencephalogram
E-liquids	Electronic Liquids for use in ECs, usually containing nicotine, PG, and/or VG
ELISA	Enzyme-Linked Immunoadsorption Assay
ENDS	Electronic Nicotine Delivery System
ER	Emergency Room
et al	and others
ETS	Environmental Tobacco Smoke (Second- and Thirdhand Smoke)
EU	European Union

EVALI	E-cigarette or Vaping product Associated Lung Injury
EVP	Electronic Vaping Products
FAHS	Flight Attendants Health Study
FAMRI	Flight Attendant Medical Research Institute International
FCTC	Framework Convention on Tobacco Control
FDA	Food and Drug Administration (US)
FEMA	Flavor Extracts Manufacturers Association
FEV1	Forced Expired Volume in 1 Second
FFA	Free Fatty Acid
FM	Factory Made (cigarettes)
FR	Flame Retardant
FSC	Fire Safe Cigarette
FSD	Female Sexual Dysfunction
FTND	Fagerstrom Test for Nicotine Dependence
FUBYAS	First Usual Brand Younger Adult Smoker
GABA	Gamma Aminobutyric Acid
GC	Gas Chromatography
GC/MS	Gas Chromatography - Mass Spectrometry
GEP	Good Epidemiology Practice
GHS	Globally Harmonized Classification System
GL	Glycerol
GWL	Graphic Warning Labels
GTS	Green Tobacco Syndrome (or Sickness)
H_2O_2	Hydrogen peroxide
hACE2	Human Angiotensin Converting Enzyme
HCI	Health Canada Intensive
HCN	Hydrogen Cyanide
HDL	High Density Lipoprotein
HnB	Heat not Burn
HnBB	Heat not Burn Butts
HNC	Hair Nicotine Concentration
HONC	Hooked On Nicotine Checklist
HPHC	Harmful and Potentially Harmful Constituents
HPLC	High Pressure Liquid Chromatography
HR	Hazard Ratio
HR	Heart Rate
HRQL	Health Related Quality of Life

HRR	Heart Rate Response
HRV	Heart Rate Variability
HTP	Heated Tobacco Products (another name for HnB)
HWL	Health Warning Labels
IIEF	International Index of Erectile Function
Inter alia	among other things
IOP	Intraocular Pressure
IPD	Invasive Pneumococcal Disease
IQOS	I Quit Ordinary Smoking (a HnB nicotine delivery system)
ISIOMI	Irreversible Sinus-Induced Oral Mucosal Injuries
ISO	International Organization for Standardization
ITL	Imperial Tobacco Limited
IV	Intravenous
IVF	In Vitro Fertilization
JT	Japan Tobacco
JTL	Japan Tobacco Limited
LC50	Lethal Concentration 50%
LCC	Little Cigars or Cigarillos
LDC	Less Developed Countries
LDL	Low Density Lipoprotein
LHS	Lung Health Study
LOD	Limit of Detection
LOQ	Level of Quantitation
lpm	litres (of breath) per minute
LPS	Lipopolysaccharide
LV	Left Ventricle
m3	Cubic metres
MARA	Microbial Assay for Risk Assessment
MDD	Major Depressive Disorder
MERS	Middle Eastern Respiratory Syndrome
mFTQ	Modified Fagerstrom Tolerance Questionnaire
mg	milligram
MI	Myocardial Infarction
mL	millilitre
MLA	Minimum Legal Age
MLB	Major League Baseball
MMAD	Mean Mass Aerodynamic Diameter

MOE	Margin of Error
MP	Maras Powder
MRL	Maximum Residual Level
MRTP	Modified Risk Tobacco Products
MS	Mainstream Smoke
MS	Multiple Sclerosis
MST	Moist Snuff Tobacco
MUH	Multi-Unit Housing
MX	Mexico
MYO	Make Your Own (cigarettes)
n =	number in the group
nAChR	Nicotinic Acetylcholine Receptors
NAFLD	Nonalcoholic Fatty Liver Disease
NAB	N-Nitrosoanabasine
NAT	N-Nitrosoanatabine
NATS	National Adult Tobacco Survey
NC	North Carolina
NCCLS	Northern California Childhood Leukemia Study
ND	Nicotine Dependence
NHANES	National Health and Nutrition Examination Survey
NIC	Nicotine
NMESARC-III	National Epidemiologic Survey on Alcohol and Related Conditions—Wave III
ng	nanogram
nm	nanometre
NMR	Nicotine Metabolite Ratio (3-HC/Cotinine), (the higher the ratio the faster the rate of nicotine metabolism)
NNA	Nicotine-Derived Nitrosamine Aldehyde
NNAL	4-(methyl nitrosamine)-1-(3-pyridyl)-1-butanol
NNC	Normal Nicotine Concentration (cigarette)
NNK	Nicotine-Derived Nitrosamine Ketone
NNN	N-Nitrosonornicotine
NO	Nitrous Oxide
NRT	Nicotine Replacement Therapy
NSDUH	National Survey on Drug Use and Health
NYTS	National Youth Tobacco Survey
OF	Oral Fluid

OR	Odds Ratio
OX/ROS	Oxidant or Reactive Oxygen Species
PAD	Peripheral Artery Disease
PAF	Population Attributable Fraction
PAH	Polyaromatic Hydrocarbons
PATH	Population Assessment of Tobacco and Health
Pb	Lead
PEFR	Peak Expiratory Flow Rate
PET	Positron Emission Tomography
PG	Propylene Glycol
PHE	Public Health England
PLWH	People Living With HIV
PM	Particulate Matter
PM_1	Ultrafine Particulate Matter (1.0 microns or less in diameter)
$PM_{2.5}$	Fine Particulate Matter (2.5 microns or less in diameter)
PM_{10}	Coarse Particulate Matter (10 microns or more in diameter)
PMDD	Premenstrual Dysphoric Disorder
PMI	Philip Morris International
PMM	Personal Machine Made (cigarettes)
PMS	Premenstrual Syndrome
Po-210	Polonium 210 (a chemical used as poison by the KGB)
POAG	Primarily Open-Angle Glaucoma
POS	Point of Sale
ppb	parts per billion
PPE	Personal Protective Equipment
ppm	parts per million
PTSD	Post-Traumatic Stress Disorder
PURE	Prospective Urban Rural Epidemiology Study
PV	Personal Vaporizer
PWL	Pictorial Warning Labels
QoL	Quality of Life
Ra	Radium
RA	Rheumatoid Arthritis
RCT	Randomized Control Trials
R+D	Research and Development
RIA	Radio-Immuno-Assay
RICO	Racketeer Influenced and Corrupt Organizations

RIP	Reduced Ignition Propensity (for cigarettes)
RNC	Reduced Nicotine Content (cigarettes)
RR	Relative Risk
RYO	Roll Your Own (cigarettes)
S	second
SCB	Smoked Cigarette Butts
SCC	Salivary Cotinine Concentration
SCN–	thiocyanate
SES	Socio-Economic Status
SHA	Secondhand Aerosol
SHS	Secondhand Smoke
SHSE	Secondhand Smoke Exposure
SHTS	Secondhand Tobacco Smoke
SHV	Secondhand Vape
SIDS	Sudden Infant Death Syndrome
SL	Smoking Load
SLT	Smokeless Tobacco
SMI	Serious Mental Illness
SOD	Sub Ohm Devices
SPD	Serious Psychological Distress
SPE	Solid Phase Extraction
SR	Sustained Release
SRF	Smoking Research Foundation
SS	Sidestream Smoke
ST	Smokeless Tobacco
SUN	Seguimiento Universidad de Navarra
T2D	Type 2 Diabetes
T2DM	Type 2 Diabetes Mellitus
TABP	Type A Behavior Pattern
TB	Tuberculosis
TC	Traditional (Tobacco) Cigarettes
TCA	Tobacco Control Act
THC	Tetrahydrocannabinol
THR	Tobacco Harm Reduction
THS	Thirdhand Smoke
TN	Tennessee
TNCO	Tar, Nicotine, and Carbon Monoxide

TNP	Transdermal Nicotine Patches
TPD	Tobacco Products Directive
TPM	Total Particulate Matter
TSNA (TSA)	Tobacco-Specific Nitrosamines
TTE	Time To Exhaustion
μg	microgram
UFP	Ultra Fine Particles
UPLC-MS	Ultra High-Pressure Liquid Chromatograph- Mass Spectrograph
UV	Ultraviolet
v	Venous
V	Volts
VaD	Vascular Dementia
VG	Vegetable Glycerin (also known as Glycerol)
VLNC	Very Low Nicotine Concentration (cigarettes)
VLS	Vanishing Lung Syndrome
VOC	Volatile Organic Compounds
vs	versus
W	Watt
WADA	World Anti-Doping Agency
WC	Waterpipe Café
WHO	World Health Organization
WMHI	White Matter Hyperintensities
WPS	Waterpipe Smoking
wt	weight
WTS	Waterpipe Tobacco Smoking
×	times

APPENDIX 7

Abstracted Studies

ALPHABETICAL BY FIRST AUTHOR (**reference number in bold parentheses**)

A

ABAFALVI, L., M. PÉNZES, R. URBÁN, K.L. FOLEY, R. KAÁN, B. KISPÉLYI, AND P. HERMANN. "Perceived Health Effects of Vaping Among Hungarian Adult E-Cigarette-Only and Dual Users: A Cross-Sectional Internet Survey." *BMC Public Health*, 19: 302 (10 pp), 2019 **(50706)**

ABOAZZIA, E., AND T. EISSENBERG. "Waterpipe Tobacco Smoking: What is the Evidence That it Supports Nicotine/Tobacco Dependence?" *Tobacco Control,* 24: i44–i53, 2015 **(60419)**

ADAMS, J.D., K.J. O'MARA-ADAMS, AND D. HOFFMANN. "Toxic and Carcinogenic Agents in Undiluted Mainstream Smoke and Sidestream Smoke of Different Types of Cigarettes." *Carcinogenesis*, 8: 729–731, 1987 **(40602)**

ADRIAENS, K., D. VAN GUCHT, AND F. BAEYENS. "Differences Between Dual Users and Switchers Center Around Vaping Behavior and Its Experiences Rather than Beliefs and Attitudes." *International Journal of Environmental Research and Public Health,* 15: 12 (15 pp), 2018 **(50705)**

AGAKU, I., S. ODANI, C. VARDAVAS, AND L. NEFF. "Self-Identified Tobacco Use and Harm Perceptions Among US Youth." *Pediatrics*, 141: (9 pp), 2018 **(50624)**

AGRAWAL, A., M.T. LYNSKEY, P.A.F. MADDEN, M.L. PERGADIA, K.K. BUCHOLZ, AND A.C. HEATH. "Simultaneous Cannabis and Tobacco Use and Cannabis-Related Outcomes in Young Women." *Drug and Alcohol Dependence*, 101: 8–12, 2009 **(60904)**

AGUIAR, M., F. TODO-BOM, M. FELIZARDO, R. MACEDO, F. CAEIRO, R. SCOTTO-MAYOR, AND A. BUGALHO DE ALMEIDA. "Four Years' Follow Up at a Smoking Cessation Clinic." *Revista Portuguesa de Pneumologia*, 15: 179–197, 2009 **(70220)**

AGUSTIN, M., C.M. YAMAMOTO, F. CABRERA, AND R. EUSEBIO. "Diffuse Alveolar Hemorrhage Induced by Vaping." *Case Reports in Pulmonology*, 2018: (3 pp), 2018 **(50421)**

AKTER, S., A. GOTO, AND T. MIZOUE. "Smoking and the Risk of Type 2 Diabetes in Japan: A Systematic Review and Meta-Analysis." *Journal of Epidemiology*, 27: 553–561, 2017 **(40510)**

ALAGAILI, A.N., T. BRIESE, N.M.S. AMOR, O.B. MOHAMMED, AND W.I. LIPKIN. "Waterpipe Smoking as a Public Health Risk: Potential Risk for Transmission of MERS-CoV." *Saudi Journal of Biological Sciences*, 26: 938–941, 2019 **(60423)**

AL-DELAIMY, W.K., AND A. WOODWARD. "Questionnaire and Hair Measurement of Exposure to Tobacco Smoke." *Journal of Exposure Analysis and Environmental Epidemiology*, 10: 378–384, 2000 **(30503)**

ALEXANDRE, M., S.J. PANDOL, F.S. GORELICK, AND E.C. THROWER. "The Emerging Role of Smoking in the Development of Pancreatitis." *Pancreatology*, 11: 469–474, 2011 **(40516)**

ALEYAN, S., A. COLE, W. QIAN, AND S.T. LEATHERDALE. "Risky Business: A Longitudinal Study Examining Cigarette Smoking Initiation Among Susceptible and Non-Susceptible E-Cigarette Users in Canada." *BMJ Open*, 8: (7 pp), 2018 **(50617)**

ALKAM, T., AND T. NABESHIMA. "Molecular Mechanisms for Nicotine Intoxication." *Neurochemistry International,* 125: 117–126, 2019 **(10305)**

ALLEM, J-P., AND FERRARA, E. "Could Social Bots Pose a Threat to Public Health?" *American Journal of Public Health,* 108: 1005–1006, 2018 **(50117)**

ALMIRALL, J., J. BLANQUER, AND S. BELLO. "Community-Acquired Pneumonia Among Smokers." *Archivos de Bronconeumologia*, 50: 250–254, 2014 **(80303)**

ALPERT, H.R., I.T. AGAKU, AND G.N. CONNOLLY. "A Study of Pyrazines in Cigarettes and How Additives Might be Used to Enhance Nicotine Addiction." *Tobacco Control*, 25: 444–450, 2016 **(70105)**

ALTET, N., I. LATORRE, M.Á. JIMÉNEZ-FUENTES, J. MALDONADO, I. MOLINA, Y. GONZÁLEZ-DÍAZ, C. MILÀ, E. GARCÍA-GARCÍA, B. MURIEL, R. VILLAR-HERNÁNDEZ, M. LAABEI, A-C. GÓMEZ, P. GODOY, M.L. DE SOUZA-GALVÃO, S. SOLANO, C.A. JIMÉNEZ-RUIZ, AND J. DOMÍNGUEZ. "Assessment of the Influence of Direct Tobacco Smoke on Infection and Active TB Management." *Public Library of Science One*, 12: e0182998 (17 pp), 2017 **(80307)**

ALZAHRANI, T., I. PENA, N. TEMESGEN, AND S.A. GLANTZ. "Association Between Electronic Cigarette Use and Myocardial Infarction." *American Journal of Preventive Medicine*, 55: 455–461, 2018 **(50434)**

ANONYMOUS. "E-Cigarettes: Public Health England's Evidence-Based Confusion [Editorial]." *Lancet*, 386: 829, 2015 **(50445)**

ANONYMOUS. "Nicotine Poisoning in Calf." *Canadian Journal of Comparative Medicine and Veterinary Science*, 7: 250, 1943 **(10426)**

ANTOGNOLI, E., S.K. GONZALEZ, E. TRAPL, D. CAVALLO, B. LAVANTY, R. LIM, AND S. FLOCKE. "Cigarettes, Little Cigars, and Cigarillos: Initiation, Motivation, and Decision-Making." *Nicotine and Tobacco Research*, 20: S5–S11, 2018 **(60212)**

ANTONIEWICZ, L., A. BRYNEDAL, L. HEDMAN, M. LUNDBÄCK, AND J.A. BOSSON. "Acute Effects of Electronic Cigarette Inhalation on the Vasculature and the Conducting Airways." *Cardiovascular Toxicology*, 19: 441–450, 2019 **(10316)**

ANTONIO, A.M.M., P. FAGAN, F.D. HAMAMURA, I.J.N. LAGUA, J. LIU, D.J. PARK, P. POKHREL, T.A. HERZOG, I. PAGANO, K. CASSEL, A. SY, D. JORGENSEN, T. LYNCH, C. KAWAMOTO, C.J. BOUSHEY, A. FRANKE, M.S. CLANTON, E.T. MOOLCHAN, AND L.A. ALEXANDER. "Menthol Cigarette Smoking and Obesity in Young Adult Daily Smokers in Hawaii." *Preventive Medicine Reports*, 2: 946–952, 2015 **(40309)**

AOKI, Y., T. IKEDA, N. TANI, A. SHIDA, S. ORITANI, AND T. ISHIKAWA. "Evaluation of the Distribution of Nicotine Intravenous Injection: An Adult Autopsy Case Report with a Review of Literature." *International Journal of Legal Medicine*, 134: 243–249 2019 **(30602)**

APOLLONIO, D.E., AND S.A. GLANTZ. "Minimum Ages of Legal Access for Tobacco in the United States From 1863 to 2015." *American Journal of Public Health Law*, 106: 1200–1207, 2016 **(40801)**

ARAÚJO, M.C.B., AND M.F. COSTA. "From Plant to Waste: The Long and Diverse Impact Chain Caused by Tobacco Smoking." *International Journal of Environmental Research and Public Health*, 16: 2690 (9 pp), 2019 **(70321)**

ARBES JR., S.J., H. AGÚSTSDÓTTIR, AND G.D. SLADE. "Environmental Tobacco Smoke and Periodontal Disease in the United States." *American Journal of Public Health*, 91: 253–257, 2001 **(40613)**

ARCHAMBEAU, B.A., S. YOUNG, C. LEE, T. PENNINGTON, C. VANDERBEEK, D. MIULLI, J. CULHANE, AND M. NEEKI. "E-Cigarette Blast Injury: Complex Facial Fractures and Pneumocephalus." *Western Journal of Emergency Medicine*, 16: 805–807, 2016 **(80214)**

ARMITAGE, A.K., C.T. DOLLERY, C.F. GEORGE, T.H. HOUSEMAN, P.J. LEWIS, AND D.M. TURNER. "Absorption and Metabolism of Nicotine from Cigarettes." *British Medical Journal*, 4: 313–316, 1975 **(20103)**

ARTER, Z.L., A. WIGGINS, C. HUDSPATH, A. KISLING, D.C. HOSTLER, AND J.M. HOSTLER. "Acute Eosinophilic Pneumonia Following Electronic Cigarette Use." *Respiratory Medicine Case Reports*, 27: (3 pp), 2019 **(50422)**

ASCHBRENNER, K.A., L.B. DIXON, J.A. NASLUND, J.C.M. BIENVENIDA, K.L. MCMANUS, S.J. BARTELS, AND M.F. BRUNETTE. "An Online Survey of Family Members' Beliefs and Attitudes About Smoking and Mental Illness." *Journal of Dual Diagnosis*, 13: 179–183, 2017 **(80805)**

AUDRAIN-MCGOVERN, J., M.D. STONE, J. BARRINGTON-TRIMIS, J.B. UNGER, AND A.M. LEVENTHAL. "Adolescent E-Cigarette, Hookah, and Conventional Cigarette Use and Subsequent Marijuana Use." *Pediatrics*, 142: (9 pp), 2018 **(60907)**

AUER, R., N. CONCHA-LOZANO, I. JACOT-SADOWSKI, J. CORNUZ, AND A. BERTHET. "Heat-not-Burn Tobacco Cigarettes: Smoke by Any Other Name." *JAMA Internal Medicine*, 177: 1050-1052, 2017 **(60108)**

AUNE, D., L.J. VATTEN, AND P. BOFFETTA. "Tobacco Smoking and the Risk of Gallbladder Disease." *European Journal of Epidemiology*, 31: 643–653, 2016 **(40509)**

AUNE, D., S. SCHLESINGER, T. NORAT, AND E. RIBOLI. "Tobacco Smoking and the Risk of Abdominal Aortic Aneurysm: A Systematic Review and Meta Analysis of Prospective Studies." *Scientific Reports*, 8: 4786 (9 pp), 2018 **(40536)**

AXELSSON, J., S. SABRA, L. RYLANDER, A. RIGNELL-HYDBOM, C.H. LINDH, AND A. GIWERCMAN. "Association Between Paternal Smoking at the Time of Pregnancy and the Semen Quality in Sons." *Public Library of Science One*, 13: (12 pp), 2018 **(80502)**

B

BACHINGER, E., M. MCKEE, AND A. GILMORE. "Tobacco Policies in Nazi Germany: Not as Simple as it Seems." *Public Health*, 122: 497–505, 2008 **(40110)**

BAGGETT, T.P., E.G. CAMPBELL, Y. CHANG, L.M. MAGID, AND N.A. RIGOTTI. "Posttraumatic Stress Symptoms and Their Association with Smoking Outcome Expectancies Among Homeless Smokers in Boston." *Nicotine and Tobacco Research*, 18: 1526–1532, 2016 **(80818)**

BAHELAH, R., J.R. DIFRANZA, F.M. FOUAD, K.D. WARD, T. EISSENBERG, AND W. MAZIAK. "Early Symptoms of Nicotine Dependence Among Adolescent Waterpipe Smokers." *Tobacco Control*, 25: e127-e134, 2016 **(60414)**

BAKER, F., S.R. AINSWORTH, J.T. DYE, C. CRAMMER, M.J. THUN, D. HOFFMANN, J.L. REPACE, J.E. HENNINGFIELD, J. SLADE, J. PINNEY, T. SHANKS, D.M. BURNS, G.N. CONNOLLY, AND D.R. SHOPLAND. "Health Risks Associated with Cigar Smoking." *Journal of the American Medical Association*, 284: 735–740, 2000 **(60206)**

BAKER, T.B., N. BRESLAU, L. COVEY, AND S. SHIFFMAN. "DSM Criteria for Tobacco Use Disorder and Tobacco Withdrawal: A Critique and Proposed Revisions for DSM-5." *Addiction*, 107: 263–275, 2012 **(70103)**

BAKRE, S.A., T.S. AL-FARRA, AND S. AL-FARRA. "Diffuse Alveolar Damage and E-Cigarettes: Case Report and Review of Literature." *Respiratory Medicine Case Reports*, 28: 100935 (3 pp), 2019 **(50429)**

BANKS, E., G. JOSHY, R.J. KORDA, B. STRAVRESKI, K. SOGA, S. EGGER, C. DAY, N.E. CLARKE, S. LEWINGTON, AND A.D. LOPEZ. "Tobacco Smoking and Risk of 36 Cardiovascular Disease Subtypes: Fatal and Non-Fatal Outcomes in a Large Prospective Australian Study." *BMC Medicine*, 17: 128 (18 pp), 2019 **(40537)**

BANOŽIĆ, M., S. JOKIĆ, Đ. AČKAR, M. BLAŽIĆ, AND D. ŠUBARIĆ. "Carbohydrates–Key Players in Tobacco Aroma Formation and Quality Determination." *Molecules*, 25: 1734 (13 pp), 2020 **(40401)**

BANYAL, N., D. BHATTACHARYYA, AND P. YADAV. "Study to Determine the Prevalence of Substance Use and Factors Associated with it, In First-Episode of Psychosis." *Industrial Psychiatry Journal*, 27: 264–270, 2018 **(80812)**

BARAN, W., D. MADEJ-KNYSAK, A. SOBCZAK, AND E. ADAMEK. "The Influence of Waste from Electronic Cigarettes, Conventional Cigarettes and Heat-not-Burn Tobacco Products on Microorganisms." *Journal of Hazardous Materials*, 385: (7 pp), 2019 **(50515)**

BARRINGTON-TRIMIS, J.L., AND A.M. LEVENTHAL. "Adolescents' Use of 'Pod Mod' E-Cigarettes—Urgent Concerns." *New England Journal of Medicine*, 379: 1099–1102, 2018 **(50606)**

BARRINGTON-TRIMIS, J.L., G. KONG, A.M. LEVENTHAL, F. LIU, M. MAYER, T.B. CRUZ, S. KRISHNAN-SARIN, AND R. MCCONNELL. "E-Cigarette Use and Subsequent Smoking Frequency Among Adolescents." *Pediatrics*, 142: (18 pp), 2018 **(50622)**

BAUMUNG, C., J. REHM, H. FRANKE, AND D.W. LACHENMEIER. "Comparative Risk Assessment of Tobacco Smoke Constituents Using the Margin of Exposure Approach: The Neglected Contribution of Nicotine." *Scientific Reports*, 6: (9 pp), 2016 **(10303)**

BEGUM, S.F., G. NAGAJOTHI, K.S. LATHA, G. SANDEEP, B. SREEKANTH, C.S. KUMAR, W. RAJENDRA, AND N. MADDU. "Possible Role of Nicotine and Cotinine on Nitroxidative Stress and Antioxidant Content in Saliva of Smokeless Tobacco Consumers." *Practical Laboratory Medicine*, 12: (10 pp), 2018 **(60617)**

BEHAR, R.Z., W. LUO, S.C. LIN, Y. WANG, J. VALLE, J.F. PANKOW, AND P. TALBOT. "Distribution, Quantification and Toxicity of Cinnamaldehyde in Electronic Cigarette Refill Fluids and Aerosols." *Tobacco Control*, 25: ii94–ii102, 2016 **(50313)**

BEHERA, D., R. UPPAL, AND S. MAJUMDAR. "Urinary Levels of Nicotine and Cotinine in Tobacco Users." *Indian Journal of Medical Research*, 118: 129–13, 2003 **(30301)**

BELKONIENE, M., J. SOCQUET, D. NIEMBA-FREIBURGHAUS, AND C. PELLATON. "Near Fatal Intoxication by Nicotine and Propylene Glycol Injection: A Case Report of an e-Liquid Poisoning." *BMC Pharmacology and Toxicology*, 20: 28 (6 pp), 2019 **(81104)**

BENDEL, G.S., H.M. HILLER, AND A. RALSTON. "Nicotine Toxicity Secondary to Aftermarket Modifications to a Vaping Device." *Military Medicine*, usab223: (4 pp), 2021 **(10425)**

BENEDICT, M.D., S.A. MISSMER, A. VAHRATIAN, K.F. BERRY, A.F. VITONIS, D.W. CRAMER, AND J.D. MEEKER. "Secondhand Tobacco Smoke Exposure is Associated with Increased Risk of Failed Implantation and Reduced IVF Success." *Human Reproduction*, 26: 2524–2531, 2011 **(80710)**

BENOWITZ, N.L. "Emerging Nicotine Delivery Products. Implications for Public Health." *Annals of the American Thoracic Society*, 11: 231–235, 2014 **(10213)**

BENOWITZ, N.L. "Nicotine Addiction." *New England Journal of Medicine*, 362: 2295–2303, 2010 **(10210)**

BENOWITZ, N.L. "Nicotine and Smokeless Tobacco." *CA: A Cancer Journal for Clinicians*, 38: 244–247, 1988 **(60606)**

BENOWITZ, N.L. "Pharmacology of Nicotine: Addiction, Smoking-Induced Disease, and Therapeutics." *Annual Review of Pharmacology and Toxicology*, 49: 57–71, 2009 **(20402)**

BENOWITZ, N.L., AND A.D. BURBANK. "Cardiovascular Toxicity of Nicotine: Implications for Electronic Cigarette Use." *Trends in Cardiovascular Medicine*, 26: 515–523, 2016 **(10309)**

BENOWITZ, N.L., AND P. JACOB III. "Trans-3-Hydroxycotinine: Disposition Kinetics, Effects and Plasma Levels During Cigarette Smoking." *British Journal of Clinical Pharmacology*, 51: 53–59, 2001 **(30201)**

BENOWITZ, N. L., J. HUKKANEN, AND P. JACOB III. "Nicotine Chemistry, Metabolism, Kinetics and Biomarkers." *Handbook Experimental Pharmacology*, 192: 29–60, 2009 **(20401)**

BENOWITZ, N.L., P. JACOB III, R.T. JONES, AND J. ROSENBERG. "Interindividual Variability in the Metabolism and Cardiovascular Effects of Nicotine in Man." *Journal of Pharmacology and Experimental Therapeutics*, 221: 368–372, 1982 **(20403)**

BERG, C.J., J. PAYNE, L. HENRIKSEN, P. CAVAZOS-REHG, B. GETACHEW, G.L. SCHAUER, AND R. HAARDÖRFER. "Reasons for Marijuana and Tobacco Co-Use Among Young Adults: A Mixed Methods Scale Development Study." *Substance Use and Misuse*, 53: 357–369, 2018 **(40809)**

BERNERT JR., J.T., J.E. MCGUFFEY, M.A. MORRISON, AND J.L. PIRKLE. "Comparison of Serum and Salivary Cotinine Measurements by a Sensitive High-Performance Liquid Chromatography-Tandem Mass Spectrometry Method as an Indicator of Exposure to Tobacco Smoke Among Smokers and Nonsmokers." *Journal of Analytical Toxicology*, 24: 333–339, 2000 **(90103)**

BERNSTEIN, S.L., S. YU, L.A. POST, J. DZIURA, AND N.A. RIGOTTI. "Undertreatment of Tobacco Use Relative to Other Chronic Conditions." *American Journal of Public Health*, 103: e59–e65, 2013 **(70202)**

BERO, L.A. "Tobacco Industry Manipulation of Research." *Public Health Reports*, 120: 200–208, 2005 **(40127)**

BERRIDGE, M.S., S.M. APANA, K.K. NAGANO, C.E. BERRIDGE, G.P. LEISURE, AND M.V. BOSWELL. "Smoking Produces Rapid Rise of [11C] Nicotine in Human Brain." *Psychopharmacology*, 209: 383–394, 2010 **(20204)**

BINSWANGER, I.A., E.A. CARSON, P.M. KRUEGER, S.R. MUELLER, J.F. STEINER, AND W.J. SABOL. "Prison Tobacco Control Policies and Deaths from Smoking in United

States Prisons: Population Based Retrospective Analysis." *British Medical Journal*, 349: g4542 (19 pp), 2014 **(40628)**

BLUM, A. "The Tobacco Reliquary." *Canadian Medical Association Journal*, 187: 133–134, 2015 **(40101)**

BOFFETTA, P., AND K. STRAIF. "Use of Smokeless Tobacco and Risk of Myocardial Infarction and Stroke: Systematic Review with Meta-Analysis." *British Medical Journal*, 339: (6 pp), 2009 **(60624)**

BOFFETTA, P., G. PERSHAGEN, K-H. JÖCKEL, F. FORASTIERE, V. GABORIEAU, J. HEINRICH, I. JAHN, M. KREUZER, F. MERLETTI, F. NYBERG, F. RÖSCH, AND L. SIMONATO. "Cigar and Pipe Smoking and Lung Cancer Risk: A Multicenter Study from Europe." *Journal of the National Cancer Institute*, 91: 697–701, 1999 **(60304)**

BOFFETTA, P., S. CLARK, M. SHEN, R. GISLEFOSS, R. PETO, AND A. ANDERSEN. "Serum Cotinine Level as Predictor of Lung Cancer Risk." *Cancer Epidemiology, Biomarkers and Prevention*, 15: 1184–1188, 2006 **(90109)**

BOKSA, P. "Smoking, Psychiatric Illness and the Brain [Editorial]." *Journal of Psychiatry and Neuroscience*, 42: 147–149, 2017 **(80802)**

BOLD, K.W., G. KONG, D.R. CAMENGA, P. SIMON, D.A. CAVALLO, M.E. MOREAN, AND S. KRISHNAN-SARIN. "Trajectories of E-Cigarette and Conventional Cigarette Use Among Youth." *Pediatrics*, 141: (7 pp), 2018 **(50615)**

BOLINDER, G., L. ALFREDSSON, A. ENGLUND, AND U. DE FAIRE. "Smokeless Tobacco Use and Increased Cardiovascular Mortality Among Swedish Construction Workers." *American Journal of Public Health*, 84: 399–404, 1994 **(60712)**

BOMBARD, J.M., L.L PEDERSON, D.E. NELSON, AND A.M. MALARCHER. "Are Smokers Only Using Cigarettes? Exploring Current Polytobacco Use Among an Adult Population." *Addictive Behaviors*, 32: 2411–2419, 2007 **(60801)**

BONILLA, A., A.J. BLAIR, S.M. ALAMRO, R.A. WARD, M.B. FELDMAN, R.A. DUTKO, T.K. KARAGOUNIS, A.L. JOHNSON, E.E FOLCH, AND J.M. VYAS. "Recurrent Spontaneous Pneumothoraces and Vaping in an 18-Year-Old Man: A Case Report and Review of the Literature." *Journal of Medical Case Reports*, 13: 283 (6 pp), 2019 **(50419)**

BONITA, R., J. DUNCAN, T. TRUELSEN, R.T. JACKSON, AND R. BEAGLEHOLE. "Passive Smoking as Well as Active Smoking Increases the Risk of Acute Stroke." *Tobacco Control*, 8: 156–160, 1999 **(40615)**

BOTELLO-HARBAUM, M.T., D.L. HAYNIE, R.J. IANNOTTI, J. WANG, L. GASE, AND B. SIMONS-MORTON. "Tobacco Control Policy and Adolescent Cigarette Smoking Status in the United States." *Nicotine and Tobacco Research*, 11: 875–885, 2009 **(70315)**

BOULTON, M., M. STANBURY, D. WADE, J. TILDEN, D. BRYAN, J. PAYNE, AND B. EISENGA. "Nicotine Poisoning After Ingestion of Contaminated Ground Beef—Michigan, 2003." *Morbidity and Mortality Weekly Report*, 52: 413–416, 2003 **(10402)**

BOWLER, R.P., N.M. HANSEL, S. JACOBSON, R.G. BARR, B.J. MAKE, M.K. HAN, W.K. O'NEAL, E.C. OELSNER, R. CASABURI, I. BARJAKTAREVIC, C. COOPER, M. FOREMAN, R.A. WISE, D.L. DEMEO, E.K. SILVERMAN, W. BAILEY, K.F. HARRINGTON, P.G. WOODRUFF, AND M.B. DRUMMOND. "Electronic Cigarette Use in US Adults at Risk for or with COPD: Analysis from Two Observational Cohorts." *Journal of General Internal Medicine*, 32: 1315–1322, 2017 **(50428)**

BOYKAN, R., M.L. GONIEWICZ, AND C.R. MESSINA. "Evidence of Nicotine Dependence in Adolescents Who Use Juul and Similar Pod Devices." *International Journal of Environmental Research and Public Health*, 16: 2135 (6 pp), 2019 **(50608)**

BRAAK, D.C., K.M. CUMMINGS, G.J. NAHHAS, B.W. HECKMAN, R. BORLAND, G.T. FONG, D. HAMMOND, C. BOUDREAU, A. MCNEILL, D.T. LEVY, AND C. SHANG. "Where Do Vapers Buy Their Vaping Supplies? Findings from the International Tobacco Control (ITC) 4 Country Smoking and Vaping Survey." *International Journal of Environmental Research and Public Health*, 16: 388 (13 pp), 2019 **(50321)**

BRANDT, A.M. "Inventing Conflicts of Interest: A History of Tobacco Industry Tactics." *American Journal of Public Health*, 102: 63–71, 2012 **(40122)**

BRAUN, M., E-L. FROMM, A. GERBER, D. KLINGELHÖFER, R. MÜLLER, AND D.A. GRONEBERG. "Particulate Matter Emissions of Four Types of One Cigarette Brand with and Without Additives: A Laser Spectrometric Particulate Matter Analysis of Secondhand Smoke." *BMJ Open*, 9: e024400 (7 pp), 2019 **(40406)**

BREITBARTH, A.K., J. MORGAN, AND A.L. JONES. "E-Cigarettes—An Unintended Illicit Drug Delivery System." *Drug and Alcohol Dependence*, 192: 98–111, 2018 **(60917)**

BRODY, A.L., A.G. MUKHIN, J. LA CHARITE, K. TA, J. FARAHI, C.A. SUGAR, M.S. MAMOUN, E. VELLIOS, M. ARCHIE, M. KOZMAN, J. PHUONG, F. ARLONIO, AND M.A. MANDELKERN. "Up-Regulation of Nicotinic Acetylcholine Receptors in Menthol Cigarette Smokers." *International Journal of Neuropsychopharmacology*, 16: 957–966, 2013 **(40308)**

BROWN, J., AND M. LIPMAN. "Community-Acquired Pneumonia in HIV-Infected Individuals." *Current Infectious Disease Reports*, 16: 397 (9 pp), 2014 **(80304)**

BROWN, J.M., E.M.A. GOODELL, J. WILLIAMS, AND R.M. BRAY. "Socioecological Risk and Protective Factors for Smoking Among Active Duty U.S. Military Personnel." *Military Medicine*, 183: e231–e239, 2018 **(80410)**

BUCHY, L., K.S. CADENHEAD, T.D. CANNON, B.A. CORNBLATT, T.H. MCGLASHAN, D.O. PERKINS, L.J. SEIDMAN, M.T. TSUANG, E.F. WALKER, S.W. WOODS, R. HEINSSEN, C.E. BEARDEN, D. MATHALON, AND J. ADDINGTON. "Substance Use in Individuals at Clinical High Risk of Psychosis." *Psychological Medicine*, 45: 2275–2284, 2015 **(80811)**

BUCZEK, E.J., K.F. HARRINGTON, P.S. HENDRICKS, AND C.E. SCHMALBACH. "Electronic Cigarette Awareness, Use, and Perceptions Among Cancer Patients." *American Journal of Otolaryngology Open*, 2: (6 pp), 2018 **(50802)**

BUSH, A., J. BHATT, AND J. GRIGG. "E Cigarettes: Tar Wars: The (Tobacco) Empire Strikes Back." *Archives Disease Children*, 104: 1027–1029, 2019 **(50601)**

BUTRY, D.T., AND D.S. THOMAS. "Cigarette Fires Involving Upholstered Furniture in Residences: The Role That Smokers, Smoker Behavior, and Fire Standard Compliant Cigarettes Play." *Fire Technology*, 53: 1123–146, 2017 **(80207)**

BUTT, Y.M., M.L. SMITH, H.D. TAZELAAR, L.T. VASZAR, K.L. SWANSON, M.J. CECCHINI, J.M. BOLAND, M.C. BOIS, J.H. BOYUM, A.T. FROEMMING, A. KHOOR, I. MIRA-AVENDANO, A. PATEL, AND B.T. LARSEN. "Pathology of Vaping-Associated Lung Injury." *New England Journal of Medicine*: (3 pp), 2019 **(50440)**

C

CAI, H., AND C. WANG. "Graphical Review: The Redox Dark Side of E-Cigarettes; Exposure to Oxidants and Public Health Concerns." *Redox Biology*, 13: 402–406, 2017 **(50402)**

CALFEE, C.S., M.A. MATTHAY, M.D. EISNER, N. BENOWITZ, M. CALL, J-F. PITTET, AND M.J. COHEN. "Active and Passive Cigarette Smoking and Acute Lung Injury After Severe Blunt Trauma." *American Journal of Respiratory Critical Care Medicine,* 183: 1660–1665, 2011 **(80412)**

CAMPBELL, B.K., T. LE, D. YIP, K.B. GRIFFIN, N.R. GUBNER, AND J.R. GUYDISH. "Sex Differences in Graphic Warning Label Ratings by Addictions Clients." *Tobacco Regulatory Science*, 5: 3–14, 2019 **(70309)**

CAO, S., X. YIN, Y. WANG, H. ZHOU, F. SONG, AND Z. LU. "Smoking and Risk of Erectile Dysfunction: Systematic Review of Observational Studies with Meta-Analysis." *Public Library of Science One*, 8: e60443, 2013 **(80905)**

CARABALLO, R.S., AND K. ASMAN. "Epidemiology of Menthol Cigarette Use in the United States." *Tobacco Induced Diseases*, 9: (14 pp), 2011 **(40304)**

CARABALLO, R.S., D.B. HOLIDAY, S.D. STELLMAN, P.D. MOWERY, G.A. GIOVINO, J.E. MUSCAT, M.P. ERIKSEN, J.T. BERNERT, P.A. RICHTER, AND L.T. KOZLOWSKI. "Comparison of Serum Cotinine Concentration within and across Smokers of Menthol and Nonmenthol Cigarette Brands among Non-Hispanic Black and Non-Hispanic White U.S. Adult Smokers, 2001–2006." *Cancer Epidemiology, Biomarkers and Prevention*, 20: 1329–1340, 2011 **(90106)**

CARLOS, W.G., M.S. BAKER, K.A. MCPHERSON, G.T. BOSSLET, R. SOOD, AND A.M. TORKE. "Smoking-Related Home Oxygen Burn Injuries: Continued Cause for Alarm." *Respiration*, 91: 151–155, 2016 **(80208)**

CARSTENSEN, J.M., G. PERSHAGEN, AND G. EKLUND. "Mortality in Relation to Cigarette and Pipe Smoking: 16 Years' Observation of 25,000 Swedish Men." *Journal of Epidemiology and Community Health*, 41: 166–172, 1987 **(60303)**

CARTER, S.M., AND S. CHAPMAN. "Smoking, Disease, and Obdurate Denial: The Australian Tobacco Industry in the 1980s." *Tobacco Control*, 12: iii23–iii30, 2003 **(40106)**

CASALDELLI-MAIA, J.M., A. VENTRIGLIO, AND D. BHUGRA. "Tobacco Smoking: From Glamour to Stigma. A Comprehensive Review." *Psychiatry and Clinical Neurosciences*, 70: 24–33, 2015 **(40107)**

CASE, K.R., D.S. MANTEY, M.R. CREAMER, M.B. HARRELL, S.H. KELDER, AND C.L. PERRY. "E-Cigarette Specific Symptoms of Nicotine Dependence Among Texas Adolescents." *Addictive Behaviors*, 84: 57–61, 2018 **(50703)**

CATALDO, J.K., J.J. PROCHASKA, AND S.A. GLANTZ. "Cigarette Smoking is a Risk Factor for Alzheimer's Disease: An Analysis Controlling for Tobacco Industry Affiliation." *Journal of Alzheimer's Disease*, 19: 465–480, 2010 **(40112)**

CAVALIERI, C., L. BOLZONI, AND M. BANDINI. "Nicotine Determination in Mushrooms by LC-MS/MS with Preliminary Studies on the Impact of Drying on Nicotine Formation." *Food, Additives and Contaminants*, 27: 473–477, 2010 **(10506)**

CECERE, L.M., E.C. WILLIAMS, H. SUN, C.L. BRYSON, B.J. CLARK, K.A. BRADLEY, AND D.H. AU. "Smoking Cessation and the Risk of Hospitalization for Pneumonia." *Respiratory Medicine*, 106: 1055–1062, 2012 **(80305)**

CHAFFEE, B.W., E.T. COUCH, AND S.A. GANSKY. "Adolescents' Smokeless Tobacco Susceptibility by Perceived Professional Baseball Players' Use." *Journal of Public Health Dentistry*, 78: 5–8, 2018 **(60609)**

CHAFFEE, B.W., S.L. WATKINS, AND S.A. GLANTZ. "Electronic Cigarette Use and Progression from Experimentation to Established Smoking." *Pediatrics*, 141: (6 pp), 2018 **(50620)**

CHAGUÉ, F., C. GUENANCIA, A. GUDJONCIK, D. MOREAU, Y. COTTIN, AND M. ZELLER. "Smokeless Tobacco, Sport and the Heart." *Archives of Cardiovascular Diseases*, 108: 75–83, 2015 **(81005)**

CHANDRA, P., AND P. GOVINDRAJU. "Prevalence of Oral Mucosal Lesions Among Tobacco Users." *Oral Health and Preventative Dentistry*, 10: 149–153, 2012 **(60622)**

CHANG, J.T., B. WANG, C.M. CHANG, AND B.K. AMBROSE. "National Estimates of Poisoning Events Related to Liquid Nicotine in Young Children Treated in US Hospital Emergency Departments, 2013–2017." *Injury Epidemiology*, 6: 10 (6 pp), 2019 **(10411)**

CHATTOPADHYAY, S., E.M. SMYTH, P. KULKARNI, K.R. BABIK, M. REID, L.E. HITTLE, P.I. CLARK, E.F. MONGODIN, AND A.R. SAPKOTA. "Little Cigars and Cigarillos Harbor

Diverse Bacterial Communities that Differ Between the Tobacco and the Wrapper." *Public Library of Science One*, 14: e0211705 (15 pp), 2019 **(60213)**

CHAUMONT, M., P. VAN DE BORNE, A. BERNARD, A. VAN MUYLEM, G. DEPREZ, J. ULLMO, E. STARCZEWSKA, R. BRIKI, Q. DE HEMPTINNE, W. ZAHER, AND N. DEBBAS. "Fourth Generation E-Cigarette Vaping Induces Transient Lung Inflammation and Gas Exchange Disturbances: Results from Two Randomized Clinical Trials." *American Journal of Physiology. Lung Cellular and Molecular Physiology*, 316: L705–L719, 2019 **(50416)**

CHEN, A., M. MACHIORLATTI, N.M. KREBS, AND J.E. MUSCAT. "Socioeconomic Differences in Nicotine Exposure and Dependence in Adult Daily Smokers." *BMC Public Health*, 19: 375 (10 pp), 2019 **(10203)**

CHEN, E.Y., A. SUN, C-S. CHEN, A.J. MINTZ, AND W-C. CHIN. "Nicotine Alters Mucin Rheological Properties." *Lung Cellular and Molecular Physiology*, 307: L149–L157, 2014 **(10310)**

CHENOWETH, M.J., M. NOVALEN, L.W. HAWK JR., R.A. SCHNOLL, T.P. GEORGE, P.M. CINCIRIPINI, C. LERMAN, AND R.F. TYNDALE. "Known and Novel Sources of Variability in the Nicotine Metabolite Ratio in a Large Sample of Treatment-Seeking Smokers." *Cancer Epidemiology, Biomarkers and Prevention*, 23: 1773–1782, 2014 **(20307)**

CHENOWETH, M.J., R.A. SCHNOLL, M. NOVALEN, L.W. HAWK JR., T.P. GEORGE, P.M. CINCIRIPINI, C. LERMAN, AND R.F. TYNDALE. "The Nicotine Metabolite Ratio is Associated with Early Smoking Abstinence Even After Controlling for Factors that Influence the Nicotine Metabolite Ratio." *Nicotine and Tobacco Research*, 18: 491–495, 2016 **(70109)**

CHEN-SANKEY, J.C., G. KONG, AND K. CHOI. "Perceived Ease of Flavored E-Cigarette Use and E-Cigarette Use Progression Among Youth Never Tobacco Users." *Public Library of Science One*, 14: (11 pp), 2019 **(50614)**

CHIBA, M., AND R. MASIRONI. "Toxic and Trace Elements in Tobacco and Tobacco Smoke." *Bulletin of the World Health Organization*, 70: 269–275, 1992 **(10107)**

CHILDS, E., AND H. DE WIT. "Effects of Acute Psychosocial Stress on Cigarette Craving and Smoking." *Nicotine and Tobacco Research*, 12: 449–453, 2010 **(70107)**

CHOI, A., M. LE, T. RAHIM, C. ROSE, AND T. KOSATSKY. "Electronic Cigarette Exposures Reported to the British Columbia Drug and Poison Information Centre: An Observational Case Series." *Canadian Medical Association Journal Open*, 7: E462–E471, 2019 **(10409)**

CHOI, J., D.W. SHIN, S. LEE, M.J. JEON, S.M. KIM, B. CHO, AND S.M. LEE. "Dose-Response Relationship Between Cigarette Smoking and Female Sexual Dysfunction." *Obstetrics and Gynecology Science*, 58: 302–308, 2015 **(80901)**

CHOI, K., M. INOUE-CHOI, T.S. MCNEEL, AND N.D. FREEDMAN. "Mortality Risk Associated with Dual- and Poly-Tobacco Product Use in the United States." *American Journal of Epidemiology*, 191: 397–401, 2019 **(60810)**

CHOI, K-C., S.A. KIM, N.R. KIM, AND M-H. SHIN. "Association Between Smoking and Unintentional Injuries Among Korean Adults." *Chonnam Medical Journal*, 54: 184–189, 2018 **(80407)**

CHOI, W.S., C.A. PATTEN, J.C. GILLIN, R.M. KAPLAN, AND J.P. PIERCE. "Cigarette Smoking Predicts Development of Depressive Symptoms Among US Adolescents." *Annals of Behavioral Medicine*, 19: 42–50, 1997 **(80102)**

CHOSI, S.H., J.E. TERRELL, C.R. BRADFORD, T. GHANEM, M.E. SPECTOR, G.T. WOLF, I.M. LIPKUS, AND S.A. DUFFY. "Does Quitting Smoking Make a Difference Among Newly Diagnosed Head and Neck Cancer Patients?" *Nicotine and Tobacco Research*, 18: 2216–2224, 2016 **(40544)**

CHRISTENSEN, C.H., B. ROSTROM, C. COSGROVE, S.F. ALTEKRUSE, A.M. HARTMAN, J.T. GIBSON, B. APELBERG, M. INOUE-CHOI, AND N.D. FREEDMAN. "Association of Cigarette, Cigar, and Pipe Use with Mortality Risk in the US Population." *JAMA Internal Medicine*, 178: 469–476, 2018 **(60302)**

CHU, K-H., J-P. ALLEM, T.B. CRUZ, AND J.B. UNGER. "Vaping on Instagram: Cloud Chasing, Hand Checks and Product Placement." *Tobacco Control*, 26: 575–578, 2016 **(50803)**

CHUNG-HALL, J., G.T. FONG, P. DRIEZEN, AND L. CRAIG. "Smokers' Support for Tobacco Endgame Measures in Canada: Findings from the 2016 International Tobacco Control Smoking and Vaping Survey." *Canadian Medical Association Journal Open*, 6: E412–E422, 2018 **(70320)**

CLAPP, P.W., AND I. JASPERS. "Electronic Cigarettes: Their Constituents and Potential Links to Asthma." *Current Allergy Asthma Reports*, 17: 79 (22 pp), 2017 **(50425)**

CLARK, K.D., N. WARDROBE-WONG, J.J. ELLIOTT, P.T. GILL, N.P. TAIT, AND P.D. SNASHALL. "Cigarette Smoke Inhalation and Lung Damage in Smoking Volunteers." *European Respiratory Journal*, 12: 395–399, 1998 **(40523)**

COBB, C., K.D. WARD, W. MAZIAK, A.L. SHIHADEH, AND T. EISSENBERG. "Waterpipe Tobacco Smoking: An Emerging Health Crisis in the United States." *American Journal of Health Behavior*, 34: 275–285, 2010 **(60404)**

COLEMAN, B.N., B.J. APELBERG, B.K. AMBROSE, K.M. GREEN, C.J. CHOINIERE, R. BUNNELL, AND B.A. KING. "Association Between Electronic Cigarette Use and Openness to Cigarette Smoking Among US Young Adults." *Nicotine and Tobacco Research*, 17: 212–218, 2015 **(50621)**

COLLISHAW, N. "Blowing Smoke: The History of Tobacco-Specific Nitrosamines in Canadian Tobacco." *Tobacco Control*, 26: 365–370, 2017 **(40412)**

CONNOLLY, G.N., G.D. WAYNE, D. LYMPERIS, AND M.C. DOHERTY. "How Cigarette Additives Are Used to Mask Environmental Tobacco Smoke." *Tobacco Control*, 9: 283–291, 2000 **(40408)**

COOKE, J.P. "New Insights into Tobacco-Induced Vascular Disease: Clinical Ramifications." *Methodist DeBakey Cardiovascular Journal*, 11: 156–159, 2015 **(10315)**

COOPER, S., S. ORTON, J. LEONARDI-BEE, E. BROTHERTON, L. VANDERBLOEMEN, K. BOWKER, F. NAUGHTON, M. USSHER, K.E. PICKETT, S. SUTTON, AND T. COLEMAN. "Smoking and Quit Attempts During Pregnancy and Postpartum: A Longitudinal UK Cohort." *BMJ Open*, 7: (12 pp), 2017 **(80701)**

CORKERY, J.M., J. BUTTON, A. VENTO, AND F. SCHIFANO. "Two UK Suicides Using Nicotine Extracted from Tobacco Employing Instructions Available on the Internet." *Forensic Science International*, 199: e9–e13, 2010 **(81106)**

COSCI, F., F. PISTELLI, N. LAZZARINI, AND L. CARROZZI. "Nicotine Dependence and Psychological Distress: Outcomes and Clinical Implications in Smoking Cessation." *Psychology Research and Behavior Management*, 4: 119–128, 2011 **(70102)**

COUCH, E.T., J. URATA, AND B.W. CHAFFEE. "Limited-Edition Smokeless Tobacco Packaging: Behind the Camouflage." *Tobacco Induced Diseases*, 17: 58 (3 pp), 2019 **(60612)**

CREAMER, M.R., G.V. PORTILLO, S.L. CLENDENNEN, AND C.L. PERRY. "Is Adolescent Poly-Tobacco Use Associated with Alcohol and Other Drug Use?" *American Journal of Health Behavior*, 40: 117–122, 2016 **(60806)**

CROGHAN, I.T., J.O. EBBERT, J.T. HAYS, D.R. SCHROEDER, A.M. CHAMBERLAIN, V.L. ROGERS, AND R.D. HURT. "Impact of a Countywide Smoke-Free Workplace Law on Emergency Department Visits for Respiratory Diseases: A Retrospective Cohort Study." *BMC Pulmonary Medicine*, 15: 6 (8 pp), 2015 **(40518)**

CRUZ-CANO, R., M. RANGEL-GOMEZ, C. VAN WAGONER, A. KIDANU, M.C. BRINKMANN, AND P.I. CLARK. "The Acceptability of Smokeless Tobacco Product Depends on Nicotine Levels." *Addictive Behavior Reports*, 10: (8 pp), 2019 **(60709)**

CULLEN, K.A., S.T. LIU, J.K. BERNAT, W.I. SLAVIT, M.A. TYNAN, B.A. KING, AND L.J. NEFF. "Flavored Tobacco Product Use Among Middle and High School Students—United States, 2014–2018." *Morbidity and Mortality Weekly Report*, 68: 839–844, 2019 **(60804)**

CUMMINGS, K.M. "Programs and Policies to Discourage the Use of Tobacco Products." *Oncogene*, 21: 7349–7364, 2002 **(70302)**

CUMMINGS, K.M., AND R.N. PROCTOR. "The Changing Public Image of Smoking in the United States: 1964–2014." *Cancer Epidemiology, Biomarkers and Prevention*, 23: 32–36, 2014 **(70304)**

CUMMINGS, K.M., A. BROWN, AND R. O'CONNOR. "The Cigarette Controversy." *Cancer Epidemiology, Biomarkers & Prevention*, 16: 1070–1076, 2007 **(40104)**

CUMMINS, S.E., A.C. GARNST, K. BRANDSTEIN, G.B. SEYMANN, H. KLONOFF-COHEN, C.A. KIRBY, E.K. TONG, E. CHAPLIN, G.J. TEDESCHI, AND S-H. ZHU. "Hospitalized Smokers: A Factorial RCT of Nicotine Patches and Counseling." *American Journal of Preventative Medicine*, 51: 578–586, 2016 **(70205)**

CZOLI, C.D., G.T. FONG, M.L. GONIEWICZ, AND D. HAMMOND. "Biomarkers of Exposure Among 'Dual Users' of Tobacco Cigarettes and Electronic Cigarettes in Canada." *Nicotine and Tobacco Research*, 21: 1259–1266, 2019 **(50701)**

CZOLI, C.D., S.T. LEATHERDALE, AND V. RYNARD. "Bidi and Hookah Use Among Canadian Youth: Findings From the 2010 Canadian Youth Smoking Survey." *Preventing Chronic Disease*, 10: (10 pp), 2013 **(60507)**

D

DAI, J-B., Z-X. WANG, AND Z-D. QIAO. "The Hazardous Effects of Tobacco Smoking on Male Fertility." *Asian Journal of Andrology*, 17: 954–960, 2015 **(80503)**

DANIELSSON, M., A. LAMMI, S. SIITONEN, J. OLLGREN, L. PYLKKÄNEN, AND T. VASANKARI. "Alarming Development of Dual Snus and Cigarette Usage Among Young Finnish Males." *BMC Public Health*, 19: 1249 (7 pp), 2019 **(60706)**

DAS, G., N. MUKHERJEE, AND S. GHOSH. "Neurological Insights of COVID-19 Pandemic." *ACS Chemical Neuroscience*, 11: 1206–1209, 2020 **(80313)**

DAVIDSON, K., A. BRANCATO, P. HEERDERKS, W. MANSOUR, E. MATHEIS, M. NARIO, S. RAJAGOPALAN, B. UNDERHILL, J. WININGER, AND D. FOX. "Outbreak of Electronic-Cigarette-Associated Acute Lipoid Pneumonia: North Carolina, July–August 2019." *Morbidity and Mortality Weekly Report*, 68: 784–786, 2019 **(50443)**

DAVIES, P., S. LEVY, A. PAHARI, AND D. MARTINEZ. "Acute Nicotine Poisoning Associated with a Traditional Remedy for Eczema." *Archives of Disease in Childhood*, 85: 500–502, 2001 **(10415)**

DAVIS, B., A. RAZO, E. NOTHNAGEL, M. CHEN, AND P. TALBOT. "Unexpected Nicotine in Do-It-Yourself Electronic Cigarette Flavourings." *Tobacco Control*, 25 (E1): e67–e68, 2016 **(10424)**

DAVIS, B., M. WILLIAMS, AND P. TALBOT. "IQOS: Evidence of Pyrolysis and Release of a Toxicant from Plastic." *Tobacco Control*, 28: 34–41, 2019 **(60112)**

DAVIS, E.S., M.F. SASSANO, H. GOODELL, AND R. TARRAN. "E- Liquid Autofluorescence Can be Used as a Marker of Vaping Deposition and Third-Hand Vape Exposure." *Scientific Reports*, 7: 7459 (10 pp), 2017 **(50510)**

DE ALMEIDA, L.M., R.P. DA SILVA, A.T.C. DOS SANTOS, J.D. DE ANDRADE, AND M.C. SUAREZ. "Mists, Vapors and Other Illusory Volatilities of Electronic Cigarettes." *Cadernos de saúde pública*, 33: (13 pp), 2017 **(50111)**

DE ASSIS VIEGAS, C.A. "Noncigarette Forms of Tobacco Use." *Jornal brasileiro de pneumologia*, 34: 1069–1073, 2008 **(60201)**

DE CASTRO, M.R.P., T. MATSUO, AND S.O.V. NUNES. "Clinical Characteristics and Quality of Life of Smokers at a Referral Center for Smoking Cessation." *Jornal brasileiro de pneumologia*, 36: 67–74, 2010 **(40532)**

DE GRANDA-ORIVE, J.I., C.A. JIMÉNEZ-RUIZ, AND S. SOLANO-REINA. "World Health Organization Positioning. The Impact of Tobacco in the Environment: Cultivation, Curing, Manufacturing, Transport, and Third and Fourth-Hand Smoking." *Archivos de Bronconeumologia*, 54: 357–358, 2018 **(40712)**

DE OLIVEIRA, P.P.V., C.B. SIHLER, L. DE MOURA, D.C. MALTA, M.C. DE ALBUQUERQUE TORRES, S.M.C.P. LIMA, A.L.A. DE LIMA, C.E. LEITE, V.L. DA COSTA-E-SILVA, J. SOBEL, AND T.M. LANZIERI. "First Reported Outbreak of Green Tobacco Sickness in Brazil." *Cadernos Saúde Pública*, 26: 2263–2269, 2010 **(10120)**

DE OLIVEIRA, R.M., J.L.F. SANTOS, AND A.R.F. FUREGATO. "Tobacco Addiction in the Psychiatric Population and in the General Population." *Revista Latino-Americana de Enfermagem*, 25: e2945 (9 pp), 2017 **(81101)**

DEL MAR FERNÂNDEZ, M., A. MONTES-MARTÎNEZ, M. PIÑEIRO-LAMAS, C. REGUEIRA-MÉNDEZ, AND B. TAKKOUCHE. "Tobacco Consumption and Premenstrual Syndrome: A Case-Control Study." *Public Library of Science One* 14: (12 pp), 2019 **(80501)**

DELNEVO, C.D., AND M. HRYWNA. "'A Whole 'Nother Smoke' or a Cigarette in Disguise: How RJ Reynolds Reframed the Image of Little Cigars." *American Journal of Public Health*, 97: 1368–1375, 2007 **(60210)**

DELNEVO, C.D., AND M. HRYWNA. "The Relationship of Cigars, Marijuana, and Blunts to Adolescent Bidi Use." *Public Health Reports*, 121: 603–608, 2006 **(60510)**

DERBY, K.S., K. CUTHRELL, C. CABERTO, S.G. CARMELLA, A.A. FRANKE, S.S. HECHT, S.E. MURPHY, AND L. LE MARCHAND. "Nicotine Metabolism in Three Ethnic/Racial Groups with Different Risks of Lung Cancer." *Cancer Epidemiology, Biomarkers and Prevention*, 17: 3526–3535, 2008 **(20305)**

DEVÓGLIO, L.L., J.E. CORRENTE, M.H. BORGATO, AND I. DE GODOY. "Smoking Among Female Sex Workers: Prevalence and Associated Variables." *Jornal Brasileira de Pneumologia*, 43: 6–13, 2017 **(80902)**

DHAR, P. "Measuring Tobacco Smoke Exposure: Quantifying Nicotine/Cotinine Concentration in Biological Samples by Colorimetry, Chromatography and Immunoassay Methods." *Journal of Pharmaceutical and Biomedical Analysis*, 35: 155–168, 2004 **(30104)**

DIBJI, S.A.R., S. GUHA, A. ARAB, B.T. MURRAY, AND M.R. MYERS. "Accuracy of Commercial Electronic Nicotine Delivery Systems (ENDS) Temperature Control Technology." *Public Library of Science One*, 13: (13 pp), 2018 **(50109)**

DIERKER, L., A. SELVA, J. ROSE, D. HEDEKER, AND R. MERMELSTEIN. "Nicotine Dependence and Alcohol Problems from Adolescence to Young Adulthood." *Dual Diagnosis*, 1: 9 (11 pp), 2016 **(60911)**

DIFRANZA, J.R. "Can Tobacco Dependence Provide Insights into Other Drug Addictions?" *BioMed Central Psychiatry*, 16: 365 (11 pp), 2016 **(10209)**

DINARDO, P., AND E.S. ROME. "Vaping: The New Wave of Nicotine Addiction." *Cleveland Clinic Journal of Medicine*, 86: 789–798, 2019 **(50112)**

DONALD, P.M., G. RENJITH, AND A. ARORA. "Tobacco Pouch Keratosis in a Young Individual: A Brief Description." *Journal of Indian Society of Periodontology*, 21: 249–251, 2017 **(60626)**

DOZIER, D.M., M.M. LAUZEN, C.A. DAY, S.M. PAYNE, AND M.R. TAFOYA. "Leaders in Elites: Portrayals of Smoking in Popular Films." *Tobacco Control*, 14: 7–9, 2005 **(40804)**

DRAZEN, J.M., S. MORRISSEY, AND E.W. CAMPION. "The Dangerous Flavors of E-Cigarettes." *The New England Journal of Medicine*, 380: 679–680, 2019 **(10204)**

DREHMER, J.E., E. NABI-BURZA, B.H. WALTERS, D.J. OSSIP, D.E. LEVY, N.A. RIGOTTI, J.D. KLEIN, AND J.P. WINICKOFF. "Parental Smoking and E-Cigarette Use in Homes and Cars." *Pediatrics*, 143: (9 pp), 2019 **(50507)**

DUBROFF, J.G., R.K. DOOT, M. FALCONE, R.A. SCHNOLL, R. RAY, R.F. TYNDALE, A.L. BRODY, C. HOU, A. SCHMITZ, AND C. LERMAN. "Decreased Nicotinic Receptor Availability in Smokers with Slow Rates of Nicotine Metabolism." *Journal of Nuclear Medicine*, 56: 1724 –1729, 2015 **(70108)**

DUELL, A.K., J.F. PANKOW, AND D.H. PEYTON. "Nicotine in Tobacco Product Aerosols: 'It's Déjà Vu All Over Again.'" *Tobacco Control*, 29: 656–662, 2019 **(20111)**

DUGAD, S., P. BHANDARE, G. KULKARNI, AND S. PATIL. "Sputum Characteristics Among 'Chillum' Smoking 'Sadhus' of Kumbh Mela." *MVP Journal of Medical Sciences*, 5: 96–100, 2018 **(60307)**

DUONG, M., S. RANGARAJAN, X. ZHANG, K. KILLIAN, P. MONY, S. SWAMINATHAN, A.V. BHARATHI, S. NAIR, K. VIJAYAKUMAR, I. MOHAN, R. GUPTA, D. MOHAN, S. RANI, V. MOHAN, R. IQBAL, K. KAZMI, O. RAHMAN, R. YUSUF, L.V.M. PINNAKA, R. KUMAR, P. O'BYRNE, AND S. YUSUF. "Effects of Bidi Smoking on All-Cause Mortality and Cardiorespiratory Outcomes in Men from South Asia: An Observational Community-Based Substudy of the Prospective Urban Rural Epidemiology Study (PURE)." *Lancet Global Health*, 5: e168–e176, 2017 **(60506)**

DURAZZO, T.C., V.A. CARDENAS, C. STUDHOLME, M.W. WEINER, AND D.J. MEYERHOFF. "Non-Treatment Seeking Heavy Drinkers: Effects of Chronic Cigarette Smoking on Brain Structure." *Drug and Alcohol Dependence*, 87: 76–82, 2007 **(60912)**

DUTRA, L.M., R. GRANA, AND S.A. GLANTZ. "Philip Morris Research on Precursors to the Modern E-Cigarette Since 1990." *Tobacco Control*, 26: e97–e105, 2017 **(50101)**

E

ECHEVERRÍA, J., AND H.M. NIEMEYER. "Nicotine in the Hair of Mummies from San Pedro de Atacama (Northern Chile)." *Journal of Archeological Science*, 40: 3561–3568, 2013 **(30509)**

EDWARDS, R. "The Problem of Tobacco Smoking." *British Medical Journal*, 328: 217–219, 2004 **(70303)**

EISNER, M.D., C. IRIBARREN, E.H. YELIN, S. SIDNEY, P.P. KATZ, G. SANCHEZ, AND P.D. BLANC. "The Impact of SHS Exposure on Health Status and Exacerbations Among Patients with COPD." *International Journal of COPD*, 4: 169-176, 2009 **(40614)**

EISNER, M.D., J. BALMES, E.H. YELIN, P.P. KATZ, S.K. HAMMOND, N. BENOWITZ, AND P.D. BLANC. "Directly Measured Secondhand Smoke Exposure and COPD Health Outcomes." *BMC Pulmonary Medicine*, 6: 12 (11 pp), 2006 **(90111)**

EISSENBERG, T., A. BHATNAGAR, S. CHAPMAN, S-E. JORDT, A. SHIHADEH, AND E.K. SOULE. "Invalidity of an Oft-Cited Estimate of the Relative Harms of Electronic Cigarettes [Editorial]." *American Journal of Public Health*, 110: 161–162, 2020 **(50114)**

EL-HAGE, R., A. EL-HELLANI, R. SALMAN, S. TALIH, A. SHIHADEH, AND N.A. SALIBA. "Fate of Pyrazines in the Flavored Liquids of E-Cigarettes." *Aerosol Science and Technology*, 52: 377–384, 2018 **(50316)**

ELIAS, J., L.M. DUTRA, G. ST. HELEN, AND P.M. LING. "Revolution or Redux? Assessing IQOS Through a Precursor Product." *Tobacco Control*, 27: s102–s110, 2018 **(60102)**

ELIAS, W., N. ASSY, I. ELIAS, T. TOLEDO, M. YASSIN, AND A. BOWIRRAT. "The Detrimental Danger of Water-Pipe (Hookah) Transcends the Hazardous Consequences of General Health to the Driving Behavior." *Journal of Translational Medicine*, 10: 126 (14 pp), 2012 **(80403)**

ELKHOLY, H., N. NAGY, G.R.A. TAHA, M. ELHABIBY, M. YOSEF, AND L. AZZAM. "Stop Turning a Blind Eye: Tobacco Smoking Among Egyptian Patients with Schizophrenia." *Frontiers in Psychiatry*, 9: (8 pp), 2019 **(80807)**

EL-RAMADY, H., N. ABDALLA, T. ALSHAAL, A. ELHENAWY, M. AMER, A. SHEHATA, J. PAULSEN, M. KLEINWÄCHTER, D. SLEMAR, AND H. TAHA. "Contamination of Plant Foods with Nicotine: An Overview." *Report of the DBG, Kommission VIII*, Munich, (4 pp), 2015 **(10503)**

EL-ZAATARI, Z.M., H.A. CHAMI, AND G.S. ZAATARI. "Health Effects Associated with Waterpipe Smoking." *Tobacco Control*, 24: i31–i43, 2015 **(60416)**

ENGLAND, L.J., R.E. BUNNELL, T.F. PECHACEK, V.T. TONG, AND T.A. MCAFEE. "Nicotine and the Developing Human: A Neglected Element in the Electronic Cigarette Debate." *American Journal of Preventive Medicine*, 49: 286–293, 2015 **(50511)**

EPPERSON, A.E., T.O. JOHNSTON, N.C. SCHLEICHER, AND L. HENRIKSEN. "The Price of Natural American Spirit Relative to Other Cigarette Brands." *Nicotine and Tobacco Research*, 21: 1715–1720, 2019 **(40410)**

ERYTHROPEL, H.C., S.V. JABBA, T.M. DEWINTER, M. MENDIZABAL, P.T. ANASTAS, S.E. JORDT, AND J.B. ZIMMERMAN. "Formation of Flavorant-Propylene Glycol

Adducts with Novel Toxicological Properties in Chemically Unstable E-Cigarette Liquids." *Nicotine and Tobacco Research*, 21: 1248–1258, 2019 **(50315)**

ETEMADI, A., H. KHADEMI, F. KAMANGAR, N.D. FREEDMAN, C.C. ABNET, P. BRENNAN, AND R. MALEKZADEH. "Hazards of Cigarettes, Smokeless Tobacco and Waterpipe in a Middle Eastern Population: A Cohort Study of 50,000 Individuals from Iran." *Tobacco Control*, 26: 674–682, 2017 **(60625)**

EUROPEAN RESPIRATORY JOURNAL EDITORS. "Retraction Notice for: 'Characteristics and Risk Factors for COVID-19 Diagnosis and Adverse Outcomes in Mexico: An Analysis of 89,756 Laboratory-Confirmed COVID-19 Cases.': Theodoros V. Giannouchos, Roberto A. Sussman, José M. Mier, Konstantinos Poulas and Konstantinos Farsalinos. *Eur Respir J* 2020; In Press." *European Respiratory Journal*, 57: (1 p), 2021 **(40128)**

F

FAIRCHILD, A., AND J. COLGROVE. "Out of the Ashes: The Life, Death, and Rebirth of the 'Safer' Cigarette in the United States." *American Journal of Public Health*, 94: 192–204, 2004 **(50105)**

FALLIN, A., AND S.A. GLANTZ. "Tobacco-Control Policies in Tobacco-Growing States: Where Tobacco Was King." *The Milbank Quarterly*, 93: 319–358, 2015 **(40125)**

FAMELE, M., C. FERRANTI, C. ABENAVOLI, L. PALLESCHI, R. MANCINELLI, AND R. DRAISCI. "The Chemical Components of Electronic Cigarette Cartridges and Refill Liquids: Review of Analytical Methods." *Nicotine and Tobacco Research*, 17: 271–279, 2015 **(50301)**

FANT, R.V., J.E. HENNINGFIELD, R.A. NELSON, AND W.B. PICKWORTH. "Pharmacokinetics and Pharmacodynamics of Moist Snuff in Humans." *Tobacco Control*, 8: 397–392, 1999 **(60607)**

FASSA, A.G., R.D. MEUCCI, N.S. FIORI, M.L.V. CARRETT, AND N.M.X. FARIA. "Urinary Cotinine in Tobacco Farmers in Southern Brazil." *Revista de Saúde Pública*, 52: 70 (8 pp), 2018 **(10119)**

FASSA, A.G., N.M.X. FARIA, R.D. MEUCCI, N.S. FIORI, V.I. MIRANDA, AND L.A. FACCHINI. "Green Tobacco Sickness Among Tobacco Farmers in Southern Brazil." *American Journal of Industrial Medicine*, 57: 726–735, 2014 **(10125)**

FATAAR, F., AND D. HAMMOND. "The Prevalence of Vaping and Smoking as Modes of Delivery for Nicotine and Cannabis among Youth in Canada, England and the United States." *International Journal of Environmental Research and Public Health*, 16: 411 (13 pp), 2019 **(50619)**

FAULKNER, P., D.G. GHAHREMANI, R. TYNDALE, C.M. COX, A.S. KAZANJIAN, N. PATERSON, S. LOTFIPOUR, G.S. HELLEMANN, N. PETERSEN, C. VIGIL, AND E.D. LONDON. "Reduced-Nicotine Cigarettes in Young Smokers: Impact of Nicotine

Metabolism on Nicotine Dose Effects." *Neuropsychopharmacology*, 42: 1610–1618, 2017 **(70316)**

FENG, S., O. CUMMINGS, AND G. MCINTIRE. "Nicotine and Cotinine in Oral Fluid: Passive Exposure vs Active Smoking." *Practical Laboratory Medicine*, 12: (8 pp), 2018 **(30402)**

FETTERMAN, J.L., R.J. KEITH, J.N. PALMISANO, K.L. MCGLASSON, R.M. WEISBROD, S. MAJID, R. BASTIN, M.M. STATHOS, A.C. STOKES, R.M. ROBERTSON, A. BHATNAGAR, AND N.M. HAMBURG. "Alterations in Vascular Function Associated with the Use of Combustible and Electronic Cigarettes." *Journal of American Heart Association*, 9: e014570: (9 pp), 2020 **(50712)**

FEYERABEND, C., R.M. INGS, AND M.A. RUSSELL. "Nicotine Pharmacokinetics and Its Application to Intake from Smoking." *British Journal of Clinical Pharmacokinetics*, 19: 239–247, 1985 **(20104)**

FIALA, S.C., L.L. DRACH, S.W. PARKMAN, S.L. HARGAND, AND K.E. GIRARD. "Including a Novel Tobacco Product in Oregon's Indoor Clean Air Act: Lessons Learned for Clean Air Policy, 2015–2016." *Public Health Reports*, 133: 230–233, 2018 **(50514)**

FIORI, N.S., A.G. FASSA, N.M.X. FARIA, R.D. MEUCCI, V.I. MIRANDA, AND D.C. CHRISTIANI. "Wheezing in Tobacco Farm Workers in Southern Brazil." *American Journal of Industrial Medicine*, 58: 1217–1228, 2015 **(10126)**

FIX, B.V., R.J. O'CONNOR, N. BENOWITZ, B.W. HECKMAN, K.M. CUMMINGS, G.T. FONG, AND J.F. THRASER. "Nicotine Metabolite Ratio (NMR) Prospectively Predicts Smoking Relapse: Longitudinal Findings from ITC Surveys in Five Countries." *Nicotine and Tobacco Research*, 19: 1040–1047, 2017 **(90110)**

FLIEGER, J., J. KAWKA, AND M. TATARCZAK-MICHALEWSKA. "Levels of the Thiocyanate in the Saliva of Tobacco Smokers in Comparison to e-Cigarette Smokers and Nonsmokers Measured by HPLC on a Phosphatidylcholine Column." *Molecules*, 24: 3790 (14 pp), 2019 **(90206)**

FLOYD, E.L., L. QUEIMADO, J. WANG, J.L. REGENS, AND D.L. JOHNSON. "Electronic Cigarette Power Affects Count Concentration and Particle Size Distribution of Vaping Aerosol." *Public Library of Science One*, 13: (15 pp), 2018 **(50205)**

FOULDS, J., J.A. STAPLETON, N. BELL, J. SWETTENHAM, M.J. JARVIS, AND M.A. RUSSELL. "Mood and Physiological Effects of Subcutaneous Nicotine in Smokers and Never-Smokers." *Drug and Alcohol Dependence*, 44: 105–115, 1997 **(10323)**

FRACOL, M., R. DORFMAN, L. JANES, S. KULKARNI, K. BETHKE, N. HANSEN, AND J. KIM. "The Surgical Impact of E-Cigarettes: A Case Report and Review of the Current Literature." *Archives of Plastic Surgery*, 44: 477–481, 2017 **(10322)**

FRANCIS, D.B., N. MASON, J.C. ROSS, AND S.M. NOAR. "Impact of Tobacco-Pack Pictorial Warnings on Youth and Young Adults: A Systematic Review of Experimental Studies." *Tobacco Induced Diseases*, 17: 41 (11 pp), 2019 **(70310)**

FRANCIS, J.A., A.K. SHEA, AND J.M. SAMET. "Challenging the Epidemiologic Evidence on Passive Smoking: Tactics of Tobacco Industry Expert Witnesses." *Tobacco Control*, 15: iv68–iv76, 2006 **(40603)**

FRASCH, H.F., AND A.M. BARBERO. "In Vitro Human Epidermal Permeation of Nicotine from Electronic Cigarette Refill Liquids and Implications for Dermal Exposure Assessment." *Journal of Exposure Science & Environmental Epidemiology*, 27: 618–624, 2017 **(20112)**

FROST, K., E. FRANK, AND E. MAIBACH. "Relative Risk in the News Media: A Quantification of Misrepresentation." *American Journal of Public Health*, 87: 842–845, 1997 **(40114)**

FU, Q., M.G. VAUGHN, L-T. WU, AND A.C. HEATH. "Psychiatric Correlates of Snuff and Chewing Tobacco Use." *Public Library of Science One*, 9: 113196 (14 pp), 2014 **(80817)**

FUENTES, X.F., R. KASHYAP, T. HAYS, S. CHALMERS, C.L. VON BUCHWALD, O. GAJIC, AND A.G. DE MORAES. "VpALI-Vaping-Related Acute Lung Injury: A New Killer Around the Block." *Mayo Clinic Proceedings*, 94: 2534–2545, 2019 **(50442)**

FUJITA, T., A. BABAZONO, Y. HARANO, AND P. JIANG. "Risk of Depressive Disorders After Tobacco Smoking Cessation: A Retrospective Cohort Study in Fukuoka, Japan." *British Medical Journal Open*, 9: e025124 (6 pp), 2019 **(70227)**

FULLER, T.W., A.P. ACHARYA, T. MEYYAPPAN, M. YU, G. BHASKAR, S.R. LITTLE, T.V. TARIN. "Comparison of Bladder Carcinogens in the Urine of E-Cigarette Users Versus Non-E-Cigarette Using Controls." *Scientific Reports*, 8: 507 (6 pp), 2018 **(50411)**

FURER, V., M. HERSCH, N. SILVETZKI, G.S. BREUER, AND S. ZEVIN. "Nicotiana Glauca (Tree Tobacco) Intoxication—Two Cases in One Family." *Journal of Medical Toxicology*, 7: 47–51, 2011 **(10507)**

G

GANGULY, K., B. LEVÄNEN, L. PALMBERG, A. ÅKESSON, AND A. LINDÉN. "Cadmium in Tobacco Smokers: A Neglected Link to Lung Disease?" *European Respiratory Review*, 27: (8 pp), 2018 **(10129)**

GAOP, W., M. SANNA, M. HEFLER, AND C.P. WEN. "Air Pollution is not the 'New Smoking': Comparing the Disease Burden of Air Pollution and Smoking Across the Globe, 1990-2017." *Tobacco Control*, 29: (4 pp), 2019 **(40126)**

GARCÍA-ESQUINAS, E., A. JIMÉNEZ, R. PASTOR-BARRIUSO, M.R. JONES, B. PEREZ-GOMEZ, A. NAVAS-ACIEN, AND M. TELLEZ-PLAZA. "Impact of Declining Exposure to Secondhand Smoke in Public Places to Decreasing Smoking-Related Cancer Mortality in the US Population." *Environment International*, 117: 260–267, 2018 **(40639)**

GARCÍA-ESQUINAS, E., A. NAVAS-ACIEN, AND F. RODRÍGUEZ-ARTALEJO. "Exposure to Secondhand Smoke and the Frailty Syndrome in US Older Citizens." *Gero-Science*, 37: 25 (9 pp), 2015 **(40638)**

GEISS, O., I. BIANCHI, E. BARAHONA, AND J. BARRERO-MORENO. "Characterisation of Mainstream and Passive Vapours Emitted by Selected Electronic Cigarettes." *International Journal of Hygiene and Environmental Health*, 218: 169–180, 2015 **(50501)**

GENDREAU, P.L., AND F. VIATRO. "The Unbearable Lightness of 'Light' Cigarettes: a Comparison of Smoke Yields in Six Varieties of Canadian 'Light' Cigarettes." *Canadian Journal of Public Health*, 96: 167–172, 2005 **(40211)**

GENTZKE, A.S., T.W. WANG, K.L. MARYNAK, K.F. TRIVERS, AND B.A. KING. "Exposure to Secondhand Smoke and Secondhand E-Cigarette Aerosol Among Middle and High School Students." *Preventing Chronic Disease*, 16: 180531 (12 pp), 2019 **(50503)**

GERAGHTY, S.R., K. MCNAMARA, J.J. KWIEK, L. ROGERS, M.A. KLEBANOFF, M. AUGUSTINE, AND S.A. KEIM. "Tobacco Metabolites and Caffeine in Human Milk Purchased via the Internet." *Breastfeeding Medicine*, 10: 419–424, 2015 **(80711)**

GHINAI, I., I.W. PRAY, L. NAVON, K. O'LAUGHLIN, L. SAATHOFF-HUBER, B. HOOTS, A. KIMBALL, M.W. TENFORDE, J.R. CHEVINSKY, M. LAYER, N. EZIKE, J. MEIMAN, AND J.E. LAYDEN. "E-Cigarette Product Use, or Vaping, Among Persons with Associated Lung Injury: Illinois and Wisconsin, April–September 2019." *Morbidity and Mortality Weekly Report*, 68: 865–869, 2019 **(50437)**

GIANNAKOULAS, G., A. KATRAMADOS, N. MELAS, I. DIAMANTOPOULOS, AND E. CHIMONAS. "Acute Effects of Nicotine Withdrawal Syndrome in Pilots During Flight." *Aviation, Space, and Environmental Medicine*, 74: 247–251, 2003 **(80414)**

GILL, N., G. SANGHA, N. POONAI, AND R. LIM. "E-Cigarette Liquid Nicotine Ingestion in a Child: Case Report and Discussion." *Canadian Journal of Emergency Medicine*, 17: 699–703, 2015 **(10414)**

GIOVANNUCCI, E., E.B. RIMM, A. ASCHERIO, G.A. COLDITZ, D. SPIEGELMAN, M.J. STAMPFER, AND W.C. WILLETT. "Smoking and Risk of Total and Fatal Prostate Cancer in United States Health Professionals." *Cancer Epidemiology, Biomarkers and Prevention*, 8: 277–282, 1999 **(40541)**

GIOVENCO, D.P., E.J. MILLER LO, M.J. LEWIS, AND C.D. DELNEVO. "They're Pretty Much Made for Blunts: Product Features That Facilitate Marijuana Use Among Young Adult Cigarillo Users in the United States." *Nicotine and Tobacco Research*, 19: 1359–1364, 2017 **(60217)**

GIVELBER, D., AND L. STRICKLER. "Junking Good Science: Undoing Daubert v Merrill Dow Through Cross-Examination and Argument." *American Journal of Public Health*, 96: 33–37, 2007 **(40105)**

GLANTZ, S.A. "PMO's Own In Vivo Clinical Data on Biomarkers of Potential Harm in Americans Show that IQOS is not Detectably Different from Conventional Cigarettes." *Tobacco Control*, 27: s9–s12, 2018 **(60114)**

GLANTZ, S.A., AND E. GIBBS. "Changes in Ambulance Calls After Implementation of a Smoke-Free Law and Its Extension to Casinos." *Circulation*, 128: 811–813, 2013 **(40623)**

GLAUSSER, W. "New Vaping Products with Techy Allure Exploding in Popularity Among Youth." *Canadian Medical Association Journal*, 191: E172–E173, 2019 **(50603)**

GOLDENSON, N.I., R. KHODDAM, M.D. STONE, AND A.M. LEVENTHAL. "Associations of ADHD Symptoms with Smoking and Alternative Tobacco Product Use Initiation During Adolescence." *Journal of Pediatric Psychology*, 43: 613–624, 2018 **(80814)**

GONIEWICZ, M.L., D.M. SMITH, K.C. EDWARDS, B.C. BLOUNT, K.L. CALDWELL, J. FENG, L. WANG, C. CHRISTENSEN, B. AMBROSE, N. BOREK, D. VAN BEMMEL, K. KONKEL, G. ERIVES, C.A. STANTON, E. LAMBERT, H.L. KIMMEL, D. HATSUKAMI, S.S. HECHT, R.S. NIAURA, M. TRAVERS, C. LAWRENCE, AND A.J. HYLAND. "Comparison of Nicotine and Toxicant Exposure in Users of Electronic and Combustible Cigarettes." *JAMA Network Open*, 1: (16 pp), 2018 **(50708)**

GOODFELLOW, L.T., AND J.B. WAUGH. "Tobacco Treatment and Prevention: What Works and Why." *Respiratory Care*, 54: 1082–1090, 2009 **(70209)**

GOODWIN, R.D., K. CHESLACK-POSTAVA, D.B. NELSON, P.H. SMITH, D.S. HASIN, T. JANEVIC, N. BAKOYIANNIS, AND M.M. WALL. "Serious Psychological Distress and Smoking During Pregnancy in the United States: 2008–2014." *Nicotine and Tobacco Research*, 605–614, 2017 **(80706)**

GOODWIN, R.D., K.L. LAVOIE, A.R. LEMESHOW, E. JENKINS, E.S. BROWN, AND D.A. FEDORONKA. "Depression, Anxiety, and COPD: The Unexamined Role of Nicotine Dependence." *Nicotine and Tobacco Research* 14: 176–183, 2012 **(80104)**

GOTTLIEB, M.A. "Regulation of E-Cigarettes in the United States and Its Role in a Youth Epidemic." *Children*, 6: 40 (6 pp), 2019 **(50108)**

GOTTS, J.E., S-E. JORDT, R. MCCONNELL, AND R. TARRAN. "What are the Respiratory Effects of E-Cigarettes?" *British Medical Journal*, 366: (16 pp), 2019 **(50414)**

GOVINDARAJAN, P., H.A. SPILLER, M.J. CASAVANT, T. CHOUNTHIRATH, AND G.A. SMITH. "E-Cigarette and Liquid Nicotine Exposures Among Young Children." *Pediatrics*, 141: (8 pp), 2018 **(10412)**

GRANT, J.E., K. LUST, D.J. FRIDBERG, A.C. KING, AND S.R. CHAMBERLAIN. "E-Cigarette Use (Vaping) is Associated with Illicit Drug Use, Mental Health Problems, and Impulsivity in University Students." *Annals of Clinical Psychiatry*, 31: 27–35, 2019 **(50625)**

GRAY, N. "The Consequences of the Unregulated Cigarette." *Tobacco Control*, 15: 405–408, 2006 **(40207)**

GRIER, T.L., S. MORRISON, J.J. KNAPIK, M. CANHAM-CHERVAK, AND B.H. JONES. "Risk Factors for Injuries in the U.S. Army Ordnance School." *Military Medicine*, 176: 1292–1299, 2011 **(80409)**

GROSSMAN, S.G. "Vape, Gods and Judaism—E-Cigarettes and Jewish Law." *Rambam Maimonides Medical Journal*, 10: (12 pp), 2019 **(50806)**

GROTVEDT, L., L. FORSEN, I. ARIANSEN, S. GRAFF-IVERSEN, AND T.L. HOLMEN. "Impact of Snus Use in Teenage Boys on Tobacco Use in Young Adulthood; a Cohort from the HUNT Study Norway." *BMC Public Health*, 19: 1265 (10 pp), 2019 **(60707)**

GRZYWA-CELIŃSKA, A., I. DROGOŃ, J. EMERYK-MAKSYMIUK, I. CHMIELEWSKA, AND J. MILANOWSKI. "Not Only Cigarettes—Other Culprits of Lung Cancer." *Annals of Agricultural and Environmental Medicine*, 26: 661–664, 2019 **(40501)**

GUBNER, N.R., A. KOZAR-KONIECZNA, I. SZOLTYSEK-BOLDYS, E. SLODCZYK-MANKOWSKI, J. GONIEWICZ, A. SOBCZAK, P. JACOB III, N.L. BENOWITZ, AND M.L. GONIEWICZ. "Cessation of Alcohol Consumption Decreases Rate of Nicotine Metabolism in Male Alcohol-Dependent Smokers." *Drug and Alcohol Dependence*, 163: 157–164, 2016 **(20306)**

GUBNER, N.R., D.D. WILLIAMS, A. PAGANO, B.K. CAMPBELL, AND J. GUYDISH. "Menthol Cigarette Smoking Among Individuals in Treatment for Substance Use Disorder." *Addictive Behaviors*, 80: 135–141, 2018 **(40307)**

GULLU, H., M. CALISKAN, O. CIFTCI, D. ERDOGAN, S. TOPCU, E. YILDIRIM, E. YILDIRIR, AND H. MUDERRISOGLU. "Light Cigarette Smoking Impairs Coronary Microvascular Functions as Severely as Smoking Regular Cigarettes." *Heart*, 93: 1274–1277, 2007 **(40535)**

GÜLSEN, A., AND UYGUR, B. "Psychological Features of Smokers." *Respiratory Care*, 63: 1492–1497, 2018 **(80803)**

GUPTA, R., J. LUO, R.H. ANDERSON, AND A. RAY. "Clean Indoor Air Regulation and Incidence of Hospital Admissions for Acute Coronary Syndrome in Kanawha County, West Virginia." *Preventing Chronic Disease*, 8: A77 (8 pp), 2011 **(40621)**

GUPTA, S.K., N.L. BENOWITZ, P. JACOB III, C.N. ROLF, AND J. GORSLINE. "Bioavailability and Absorption Kinetics of Nicotine Following Application of a Transdermal System." *British Journal of Clinical Pharmacology*, 36: 221–227, 1993 **(20113)**

GUTHRIE, S.K., J.K. ZUBIETA, L. OHL, L. NI, R.A. KOEPPE, S. MINOSHIMA, AND E.F. DOMINO. "Arterial/Venous Plasma Nicotine Concentrations Following Nicotine Nasal Spray." *European Journal of Clinical Pharmacology*, 55: 639–643, 1999 **(20203)**

GUVEN, A., AND F. TOLUN. "Effects of Smokeless Tobacco 'Maras Powder' Use on Nitric Oxide and Cardiovascular Risk Parameters." *International Journal of Medical Sciences*, 9: 786–792, 2012 **(60619)**

H

HADDAD, C., R. SALMAN, A. EL-HELLANI, S. TALIH, A. SHIHADEH, AND N.A. SALIBA. "Reactive Oxygen Species Emissions from Supra- and Sub-Ohm Electronic Cigarettes." *Journal of Analytical Toxicology*, 43: 45–50, 2019 **(50431)**

HAGGART, K., L. ROBERTSON, M-L. BLANK, L. POPOVA, AND J. HOEK. "It's Just Steam: A Qualitative Analysis of New Zealand ENDS Users' Perceptions of Secondhand Aerosol." *Tobacco Control*, 30: 30–35, 2020 **(50513)**

HAHN, E.J., M.K. RAYENS, N. YORK, C.T.C. OKOLI, M. ZHANG, M. DIGNAN, W.K. AL-DELAIMY. "Effects of a Smoke-Free Law on Hair Nicotine and Respiratory Symptoms of Restaurant and Bar Workers." *Journal of Occupational and Environmental Medicine*, 48: 906–913, 2006 **(30506)**

HAJEK, P., M.L. GONIEWICZ, A. PHILLIPS, K.M. SMITH, O. WEST, AND H. MCROBBIE. "Nicotine Intake from Electronic Cigarettes on Initial Use and After 4 Weeks of Regular Use." Nicotine and Tobacco Research, 17: 175–179, 2015 **(20108)**

HAJJ, D.E., P.F. COOK, K.A. JAMES, C. BATTAGLIA, AND A.V. PROCHAZCA. "Newer Forms of Tobacco Products: Characteristics of Poly Users Among Adults Living in Colorado—A Secondary Data Analysis of the Attitudes and Behaviors Survey on Health 2015." *Tobacco Use Insights*, 12: 1–6, 2019 **(60807)**

HALEY, N.J., D.W. SEPKOVIC, AND D. HOFFMANN. "Elimination of Cotinine from Body Fluids: Disposition in Smokers and Non-smokers." *American Journal of Public Health*, 79: 1046–1048, 1989 **(90105)**

HALIM, A.A., B. ALSAYED, S. EMBARAK, T. YASEEN, AND S. DABBOUS. "Clinical Characteristics and Outcome of ICU Admitted MERS Corona Virus Infected Patients." *Egyptian Journal of Chest Diseases and Tuberculosis*, 65: 81–87, 2016 **(80309)**

HALL, M.G., J.R. MENDEL, S.M. NOAR, AND N.T. BREWER. "Why Smokers Avoid Cigarette Pack Risk Messages: Two Randomized Clinical Trials in the United States." *Social Science and Medicine*, 213: 165–172, 2018 **(70308)**

HALPERIN, A.C., S.S. SMITH, E. HEILIGENSTEIN, D. BROWN, AND M.F. FLEMING. "Cigarette Smoking and Associated Health Risks Among Students at Five Universities." *Nicotine and Tobacco Research*, 12: 96-104, 2010 **(80810)**

HALPERN, S.D., M.O. HARHAY, K. SAULSGIVER, C. BROPHY, A.B. TROXEL, AND K.G. VOLPP. "A Pragmatic Trial of E-Cigarettes, Incentives, and Drugs for Smoking Cessation." *New England Journal of Medicine*, 378: 2302–2310, 2018 **(70215)**

HAMILTON, D.A., M.C. MAHONEY, M. NOVALEN, M.J. CHENOWETH, D.F. HEITJAN, C. LERMAN, R.F. TYNDALE, AND L.W. HAWK JR. "Test-Retest Reliability and Stability of the Nicotine Metabolite Ratio Among Treatment-Seeking Smokers." *Nicotine and Tobacco Research*, 17: 1505–1509, 2015 **(30102)**

HAMMOND, D., G.T. FONG, M.P. ZANNA, J.F. TRASHER, AND R. BORLAND. "Tobacco Denormalization and Industry Beliefs Among Smokers from Four Countries." *American Journal of Preventative Medicine*, 31: 225–232, 2006 **(70322)**

HAMMOND, D., J.L. REID, V.L. RYNARD, G.T. FONG, K.M. CUMMINGS, A. MCNEILL, S. HITCHMAN, J.F. THRASHER, M.L. GONIEWICZ, M. BANSAL-TRAVERS, R. O'CONNOR, D. LEVY, R. BORLAND, AND C.M. WHITE. "Prevalence of Vaping and Smoking Among Adolescents in Canada, England, and the United States: Repeat National Cross Sectional Surveys." *British Medical Journal*, 365: 12219 (9 pp), 2019 **(50602)**

HAMMOND, D., M. CHAITON, A. LEE, AND N. COLLISHAW. "Destroyed Documents: Uncovering the Science That Imperial Tobacco Sought to Conceal." *Canadian Medical Association Journal*, 181: 691–698, 2009 **(40113)**

HARRIS, B. "The Intractable Cigarette 'Filter Problem.'" *Tobacco Control*, 20 (suppl 1): i10–i16, 2011 **(40201)**

HARTE, C.B., AND C.M. MESTON. "Acute Effects of Nicotine on Physiological and Subjective Sexual Arousal in Nonsmoking Men: A Randomized, Double-Blind, Placebo-Controlled Trial." *Journal of Sexual Medicine*, 5: 110–121, 2008 **(80903)**

HARVANKO, A., R. KRYSCIO, C. MARTIN, AND T. KELLY. "Stimulus Effects of Propylene Glycol and Vegetable Glycerin in Electronic Cigarette Liquids." *Drug and Alcohol Dependence*, 194: 326–329, 2019 **(50311)**

HARVANKO, A.M., A.K. MCCUBBIN, K.B. ASHFORD, AND T.H. KELLY. "Electronic Cigarette Liquid and Device Parameters and Aerosol Characteristics: A Survey of Regular Users." *Addictive Behaviors*, 84: 201–206, 2018 **(50209)**

HATSUKAMI, D.K., H. SEVERSON, A. ANDERSON, R.I. VOGEL, J. JENSEN, B. BROADBENT, S.E. MURPHY, S. CARMELLA, AND S.S. HECHT. "Randomized Clinical Trial of Snus Versus Medicinal Nicotine Among Smokers Interested in Product Switching." *Tobacco Control*, 25: 267–274, 2016 **(60713)**

HATSUKAMI, D.K., L.F. STEAD, AND P.C. GUPTA. "Tobacco Addiction." *Lancet*, 371: 2027–2038, 2008 **(10205)**

HAVERMANS, A., E.J.Z. KRÜSEMANN, J. PENNINGS, K. DE GRAAF, S. BOESVELDT, AND R. TALHOUT. "Nearly 20,000 E-Liquids and 250 Unique Flavor Descriptions: An Overview of the Dutch Market Based on Information from Manufacturers." *Tobacco Control*, 30: 57–62, 2021 **(50308)**

HECHT, S.S., AND D. HOFFMANN. "Tobacco-Specific Nitrosamines, An Important Group of Carcinogens in Tobacco and Tobacco Smoke." *Carcinogenesis*, 9: 875–884, 1988 **(60628)**

HECHT, S.S., S.G. CARMELLA, A. EDMONDS, S.E. MURPHY, I. STEPANOV, X. LUO, AND D.K. HATSUKAMI. "Exposure to Nicotine and a Tobacco Specific Carcinogen Increase with Duration of Use of Smokeless Tobacco." *Tobacco Control*, 17: 128–131, 2008 **(60623)**

HEDMAN, L., H. BACKMAN, C. STRIDSMAN, J.A. BOSSON, M. LUNDBÄCK, A. LINDBERG, E. RÖNMARK, AND L. EKERLJUNG. "Association of Electronic Cigarette Use with Smoking Habits, Demographic Factors, and Respiratory Symptoms." *JAMA Network Open*, 1: (12 pp), 2018 **(50711)**

HEDSTRÖM, A.K., J. HILLERT, T. OLSSON, AND L. ALFREDSSON. "Smoking and Multiple Sclerosis Susceptibility." *European Journal of Epidemiology*, 28: 867–874, 2013 **(40520)**

HEFLER, M., R. HOPKINS, AND D.P. THOMAS. "Successes and Unintended Consequences of the Northern Territory's Smoke-Free Prisons Policy: Results from a Process Evaluation." *Public Health Research and Practice*, 26: (8 pp), 2016 **(40631)**

HEINRICH, J., B. HÖLSCHER, M. SEIWERT, C.L. CARTY, G. MERKEL, AND C. SCHULZ. "Nicotine and Cotinine in Adults' Urine: The German Environmental Survey 1998." *Journal of Exposure Analysis and Environmental Epidemiology*, 15: 74–80, 2005 **(30302)**

HENDLIN, Y.H., AND S.A. BIALOUS. "The Environmental Externalities of Tobacco Manufacturing: A Review of Tobacco Industry Reporting." *Ambio*, 49: 17–34, 2020 **(10112)**

HENDLIN, Y.H., J.R. VEFFER, M.J. LEWIS, AND P.M. LING. "Beyond the Brotherhood: Skoal Bandits' Role in the Evolution of Marketing Moist Smokeless Tobacco Pouches." *Tobacco Induced Diseases*, 15: 46 (11 pp), 2017 **(60704)**

HENLEY, S.J., C.J. CONNELL, P. RICHTER, C. HUSTEN, T. PECHACEK, E.E. CALLE, AND M.J. THUN. "Tobacco-Related Disease Mortality Among Men Who Switched from Cigarettes to Spit Tobacco." *Tobacco Control*, 16: 22–28, 2007 **(60614)**

HENLEY, S.J., M.J. THUN, A. CHAO, AND E.E. CALLE. "Association Between Exclusive Pipe Smoking and Mortality from Cancer and Other Diseases." *Journal of the National Cancer Institute*, 96: 853–861, 2004 **(60301)**

HENNINGFIELD, J.E., J.M. STAPLETON, N.L. BENOWITZ, R.F. GRAYSON, AND E.D. LONDON. "Higher Levels of Nicotine in Arterial Than in Venous Blood After Cigarette Smoking." *Drug and Alcohol Dependence*, 33: 23–29, 1993 **(20201)**

HENNINGFIELD, J.E., R.V. FANT, A. RADZIUS, AND S. FROST. "Nicotine Concentration, Smoke pH and Whole Tobacco Aqueous pH of Some Cigar Brands and Types Popular in the United States." *Nicotine and Tobacco Research*, 1: 163–168, 1999 **(60202)**

HERGENS, M-P., L. ALFREDSSON, G. BOLINDER, M. LAMBE, G. PERSHAGEN, AND W. YE. "Long-Term Use of Swedish Moist Snuff and the Risk of Myocardial Infarction Amongst Men." *Journal of Internal Medicine*, 262: 351–359, 2007 **(10312)**

HIGGINS, S.T., R. REDNER, C.A. ARGER, A.N. KURTI, J.S. PRIEST, AND J.Y. BUNN. "Use of Higher-Nicotine/Tar-Yield (Regular Full Flavor) Cigarettes is Associated with Nicotine Dependence and Smoking During Pregnancy Among U.S. Women." *Preventive Medicine* 104: 57–62, 2017 **(80705)**

HIGGINS, S.T., S.H. HEIL, S.C. SIGMON, J.W. TIDEY, D.E. GAALEMA, J.R. HUGHES, M.L. STITZER, H. DURAND, J.Y. BUNN, J.S. PRIEST, C.A. ARGER, M.E. MILLER, C.L. BERGERIA, D.R. DAVIS, J.M. STRECK, D.D. REED, J.M. SKELLY, AND L. TURSI. "Addiction Potential of Cigarettes with Reduced Nicotine Content in Populations with Psychiatric Disorders and Other Vulnerabilities to Tobacco Addiction." *JAMA Psychiatry*, 74: 1056–1064, 2017 **(80813)**

HIGGINS, T.M., S.T. HIGGINS, S.H. HEIL, G.J. BADGER, J.M. SKELLY, I.M. BERNSTEIN, L.J. SOLOMON, Y. WASHIO, AND A.M. PRESTON. "Effects of Cigarette Smoking Cessation on Breastfeeding Duration." *Nicotine and Tobacco Research* 12: 483–488, 2010 **(80506)**

HIGHAM, A., D. BOSTOCK, G. BOOTH, J.V. DUNGWA, AND D. SINGH. "The Effect of Electronic Cigarette and Tobacco Smoke Exposure on COPD Bronchial Epithelial Cell Inflammatory Responses." *International Journal of COPD*, 13: 989–1000, 2018 **(50420)**

HIGHAM, A., N.J.W. RATTRAY, J.A. DEWHURST, D.K. TRIVEDI, S.J. FOWLER, R. GOODACRE, AND D. SINGH. "Electronic Cigarette Exposure Triggers Neutrophil Inflammatory Responses." *Respiratory Research*, 17: 56 (11 pp), 2016 **(50432)**

HIGUCHI, Y., M. FUJIWARA, N. NAKAYA, M. FUJIMORI, C. HAYASHIBARA, RYUHEI SO, I. SHINKAWA, K. SATO, Y. YADA, M. KODAMA, H. TAKENAKA, Y. KISHI, K. KAKEDA, Y. UCHITOMI, N. YAMADA, AND M. INAGAKI. "Change in Smoking Cessation Stage Over 1 Year in Patients with Schizophrenia: A Follow Up Study in Japan." *BioMed Central Psychiatry*, 19: 367 (9 pp), 2019 **(80806)**

HINDOCHA, C., T.P. FREEMAN, J.X. XIA, N.D.C. SHABAN, AND H.V. CURRAN. "Acute Memory and Psychotomimetic Effects of Cannabis and Tobacco Both 'Joint' and Individually: A Placebo-Controlled Study." *Psychological Medicine*, 47: 2708–2719, 2017 **(60908)**

HIRSHBEIN, L.D. "Politics, Profit, and Psychiatric Diagnosis. A Case Study of Tobacco Use Disorder." *American Journal of Public Health*, 104: 2076–2084, 2014 **(10211)**

HISCOCK, R., K. SILVER, M. ZATOŃSKI, AND A.B. GILMORE. "Tobacco Industry Tactics to Circumvent and Undermine the Menthol Cigarette Ban in the UK." *Tobacco Control*, 29: e38–e42, 2020 **(40311)**

HOFFNAGLE, A.N., T.J. LAHA, P.M. RAINEY, AND S.M.H. SADRAZADEH. "Specific Detection of Anabasine, Nicotine, and Nicotine Metabolites in Urine by Liquid Chromatography–Tandem Mass Spectrometry." *American Journal of Clinical Pathology*, 126: 880–887, 2006 **(90204)**

HOLM, A.L., AND R.M. DAVIS. "Clearing the Airways: Advocacy and Regulation for Smoke-Free Airlines." *Tobacco Control*, 13: i30–i36, 2004 **(40605)**

HOLOYDA, B., K. FRITZ, AND G.L. XIONG. "A Case of Nicotine Overdose-Induced Seizure on an Inpatient Psychiatric Unit." *Annals of Clinical Psychiatry*, 28: 66–67, 2016 **(10418)**

HOREL, S., AND T. KEYZER. "Covid 19: How Harm Reduction Advocates and the Tobacco Industry Capitalised on the Pandemic to Promote Nicotine." *British Medical Journal*, 373: n1303, 2021 **(40129)**

HOVELL, M.F., AND S.C. HUGHES. "The Behavioral Ecology of Secondhand Smoke Exposure: A Pathway to Complete Tobacco Control." *Nicotine and Tobacco Research*, 11: 1254–1264, 2009 **(40708)**

HOWARD-PITNEY, B., AND M.A. WINKLEBY. "Chewing Tobacco: Who Uses and Who Quits? Findings from NHANES III, 1988–1994." *American Journal of Public Health*, 92: 250–256, 2002 **(60615)**

HU, S.S., D.M. HOMA, T. WANG, Y. GOMEZ, K. WALTON, H. LU, AND L. NEFF. "State-Specific Patterns of Cigarette Smoking, Smokeless Tobacco Use, and E-Cigarette Use Among Adults—United States, 2016." *Preventing Chronic Disease, Public Health Research, Practice, and Policy*, 16: E17 (15 pp), 2019 **(60604)**

HUA, M., E.E. OMAIYE, W. LUO, K.J. MCWHIRTER, J.F. PANKOW, AND P. TALBOT. "Identification of Cytotoxic Flavor Chemicals in Top-Selling Electronic Cigarette Refill Fluids." *Scientific Reports*, 9: 2782 (15 pp), 2019 **(50310)**

HUANG, J., Z. DUAN, J. KWOK, S. BINNS, L.E. VERA, Y. KIM, G. SZCZYPKA, AND S.L. EMERY. "Vaping versus JUULing: How the Extraordinary Growth and Marketing of JUUL Transformed the US Retail E-Cigarette Market." *Tobacco Control*, 28: 146–151, 2018 **(50115)**

HUH, J., AND A.M. LEVENTHAL. "Progression of Poly-Tobacco Product Use Patterns in Adolescents." *American Journal of Preventative Medicine*, 51: 513–517, 2016 **(60803)**

HUNG, C-I., C-Y. LIU, AND C-H. YANG. "Unintentional Injuries Among Psychiatric Outpatients with Major Depressive Disorder." *Public Library of Science One*, 11: (10 pp), 2016 **(80408)**

HURT, R.D., L.C. DALE, K.P. OFFORD, G.G. LAUGER, L.B. BASKIN, G.M. LAWSON, N-S. JIANG, AND P.J. HAURI. "Serum Nicotine and Cotinine Levels During Nicotine-Patch Therapy." *Clinical Pharmacology and Therapeutics*, 54: 98–106, 1993 **(20114)**

HUTCHENS, L., T.M. SENSERRICK, P.E. JAMIESON, D. ROMER, AND F.K. WINSTON. "Teen Driver Crash Risk and Associations with Smoking and Drowsy Driving." *Accident Analysis and Prevention*, 40: 869–876, 2008 **(80406)**

HUTTUNEN, R., T. HEIKKINEN, AND J. SYRJÄNEN. "Smoking and the Outcome of Infection." *Journal of Internal Medicine*, 269: 258–269, 2011 **(40513)**

HYLAND, A., R. GOLDSTEIN, A. BROWN, R. O'CONNOR, AND K.M. CUMMINGS. "Happy Birthday Marlboro: The Cigarette Whose Taste Outlasts Its Customers." *Tobacco Control*, 15: 75–77, 2006 **(40102)**

I

IFFLAND, Y., R. MÜLLER, D. GRONEBERG, AND A. GERBER. "High Particulate Matter Emission from Additive-Free Natural American Spirit Cigarettes." *Springer Plus*, 5: 1958 (7 pp), 2016 **(40409)**

IGARASHI, A., J. AIDA, T. SAIRENCHI, T. TSUBOYA, K. SUGIYAMA, S. KOYAMA, Y. MATSUYAMA, Y. SATO, K. OSAKA, AND H. OTA. "Does Cigarette Smoking Increase Traffic Accident Death During 20 Years Follow-Up in Japan? The Ibaraki Prefectural Health Study." *Journal of Epidemiology*, 29: 192–196, 2019 **(80401)**

IIDA, K., AND R.N. PROCTOR. "'The Industry Must be Inconspicuous': Japan Tobacco's Corruption of Science and Health Policy via the Smoking Research Foundation." *Tobacco Control*, 27: e3–e11, 2018 **(40124)**

IKKA, T., H. YAMASHITA, I. KURITA, Y. TANAKA, F. TANIGUCHI, A. OGINO, K. TAKEDA, N. HORIE, H. HOJO, F. NANJO, AND A. MORITA. "Quantitative Validation of Nicotine Production in Tea (Camellia sinensis L.)." *Public Library of Science One*, 13: (13 pp), 2018 **(10505)**

INTRIAGO, M., G. MALDONADO, J. CÁRDENAS, AND C. RÍOS. "Clinical Characteristics in Patients with Rheumatoid Arthritis; Difference Between Genders." *The Scientific World Journal*: (6 pp), 2019 **(40505)**

INYANG, N.A., J.T. CHANG, AND B. WANG. "National Estimates of Hospital Emergency Department Visits due to Acute Injuries Associated with Hookah Smoking, United States, 2011–2019." *Injury Epidemiology*, 7: 41 (6 pp), 2020 **(60421)**

ISLAM, F., R.G. SALLOUM, R. NAKKASH, W. MAZIAK, AND J.F. THRASHER. "Effectiveness of Health Warnings for Waterpipe Tobacco Smoking Among College Students." *International Journal of Public Health*, 61: 709–715, 2016 **(60406)**

J

JACOB III, P., A.H.A. RADDAHA, D. DEMPSEY, C. HAVEL, M. PENG, L. YU, AND N.L. BENOWITZ. "Comparison of Nicotine and Carcinogen Exposure with Water Pipe and Cigarette Smoking." *Cancer Epidemiology, Biomarkers & Prevention*, 22: 765–772, 2013 **(60415)**

JACOB III, P., A.H.A. RADDAHA, D. DEMPSEY, C. HAVEL, M. PENG, L. YU, AND N.L. BENOWITZ. "Nicotine, Carbon Monoxide, and Carcinogen Exposure after a Single Use of a Waterpipe." *Cancer Epidemiology, Biomarkers & Prevention*, 20: 2345–2353, 2011 **(60412)**

JACOB III, P., L. YU, A.T. SHULGIN, AND N.L. BENOWITZ. "Minor Tobacco Alkaloids as Biomarkers for Tobacco Use: Comparison of Users of Cigarettes, Smokeless Tobacco, Cigars, and Pipes." *American Journal of Public Health*, 89: 731–736, 1999 **(90202)**

JACOB III, P., N.L. BENOWITZ, AND A.T. SHULGIN. "Recent Studies of Nicotine Metabolism in Humans." *Pharmacology, Biochemistry, and Behavior*, 30: 249–253, 1988 **(20202)**

JACOBS, E.J., C.C. NEWTON, B.D. CARTER, D. FESKANICH, N.D. FREEDMAN, R.L. PRENTICE, AND W.D. FLANDERS. "What Proportion of Cancer Deaths in the Contemporary United States is Attributable to Cigarette Smoking?" *Annals of Epidemiology*, 25: 179–182, 2015 **(40539)**

JAMROZIK, K. "Estimate of Deaths Attributable to Passive Smoking Among UK Adults: Database Analysis." *British Medical Journal*, 330: 812 (5 pp), 2005 **(40625)**

JANKOWSKI, M., D. KALETA, W.S. ZGLICZYŃSKI, J. GRUDZIĄŻ-SĘKOWSKA, I. WRZEŚNIEWSKA-WAL, M. GUJSKI, W. WIERZBA, AND J. PINKAS. "Cigarette and E-Cigarette Use and Smoking Cessation Practices Among Physicians in Poland." *International Journal of Environmental Research and Public Health*, 16: 3595 (15 pp), 2019 **(70206)**

JANKOWSKI, M., M. KRZYSTANEK, J.E. ZEJDA, P. MAJEK, J. LUBANSKI, J.A. LAWSON, AND G. BROZEK. "E-Cigarettes are More Addictive than Traditional Cigarettes—A Study in Highly Educated Young People." *International Journal of Environmental Research and Public Health*, 16: 2279 (10 pp), 2019 **(50702)**

JARVIS, M., H. TUNSTALL-PEDOE, C. FEYERABEND, C. VESEY, AND Y. SALLOOJEE. "Biochemical Markers of Smoke Absorption and Self Reported Exposure to Passive Smoking." *Journal of Epidemiology and Community Health*, 38: 335–339, 1984 **(90102)**

JARVIS, M.J. "Dietary Nicotine.... Unless Subjects Eat 90 kg Tomatoes a Day." *British Medical Journal*, 308: 62, 1994 **(10501)**

JARVIS, M.J. "Why People Smoke." *British Medical Journal*, 328: 277–279, 2004 **(70207)**

JARVIS, M.J., M.A. RUSSELL, AND C. FEYERABEND. "Absorption of Nicotine and Carbon Monoxide from Passive Smoking Under Natural Conditions of Exposure." *Thorax*, 38: 829–833, 1983 **(20106)**

JARVIS, M.J., M.A. RUSSELL, N.L. BENOWITZ, AND C. FEYERABEND. "Elimination of Cotinine from Body Fluids: Implications for Noninvasive Measurement of Tobacco Smoke Exposure." *American Journal of Public Health*, 78: 696–698, 1988 **(90101)**

JAVELLE, E. "Electronic Cigarette and Vaping Should be Discouraged During the New Coronavirus SARS-CoV-2 Pandemic." *Archives of Toxicology*, 94: 2261–2262, 2020 **(80317)**

JAYES, L.R., E. RATSCHEN, R.L. MURRAY, S. DYMOND-WHITE, AND J. BRITTON. "Second-Hand Smoke in Four English Prisons: An Air Quality Monitoring Study." *BMC Public Health*, 16: 119 (8 pp), 2016 **(40629)**

JENSEN, K., D. NIZAMUTDINOV, M. GUERRIER, S. AFROZE, D. DOSTAL, AND S. GLASER. "General Mechanisms of Nicotine-Induced Fibrogenesis." *Federation of American Societies for Experimental Biology Journal*, 26: 4778–4787, 2012 **(10318)**

JENSEN, R.P., R.M. STRONGIN, AND D.H. PEYTON. "Solvent Chemistry in the Electronic Reaction Vessel." *Scientific Reports,* 7: (11 pp), 2017 **(50302)**

JENSSEN, B.P., S.C. WALLEY, AND S.A. MCGRATH-MORROW. "Heat-not-Burn Tobacco Products: Tobacco Industry Claims No Substitute for Science." *Pediatrics*, 141: (3 pp), 2018 **(60103)**

JEREMIAS-MARTINS, E., AND J.M. CHATKIN. "Does Everyone Who Quit Smoking Gain Weight? A Real-World Prospective Cohort Study." *Jornal Brasileiro de Pneumologia*, 45: (7 pp), 2019 **(70226)**

JIWANI, A.Z., J.F. WILLIAMS, J.A. RIZZO, K.K. CHUNG, B.T. KING, AND L.C. CANCIO. "Thermal Injury Patterns Associated with Electronic Cigarettes." *International Journal of Burns and Trauma*, 7: 1–5, 2017 **(80213)**

JOHN, U., J. RIEDEL, H-J. RUMPF, U. HAPKE, AND C. MEYER. "Associations of Perceived Work Strain with Nicotine Dependence in a Community Sample." *Occupational and Environmental Medicine*, 63: 207–211, 2006 **(10208)**

JOHNS, M., R. SACKS, M. RANE, AND S.M. KANSAGRA. "Exposure to Tobacco Retail Outlets and Smoking Initiation Among New York City Adolescents." *Journal of Urban Health: Bulletin of the New York Academy of Medicine*, 90: 1091–1101, 2013 **(40803)**

JOHNSON, S.E., B.N. COLEMAN, AND C.L. SCHMIDTT. "It's Complicated: Examining Smokers' Relationships with Their Cigarette Brands." *Psychology of Addictive Behaviors*, 30: 887–894, 2016 **(40414)**

JONES, M.R., H. WIPFLI, S. SHAHRIR, E. AVILA-TANG, J.M. SAMET, P.N. BREYSSE, AND A. NAVAS-ACIEN. "Secondhand Tobacco Smoke: An Occupational Hazard for Smoking and Non-smoking Bar and Nightclub Employees." *Tobacco Control*, 22: 308–314, 2013 **(40626)**

JONES, W.J., AND G.A. SILVESTRI. "The Master Settlement Agreement and Its Impact on Tobacco Use 10 Years Later: Lessons for Physicians about Health Policy Making." *Chest*, 137: 692–700, 2010 **(40117)**

JOO, H., M-H. LIM, M. HA, H-J. KWON, S.J. YOO, K-H. CHOI, AND K-C. PAIK. "Secondhand Smoke Exposure and Low Blood Lead Levels with Attention-Deficit Hyperactivity Disorder and Its Symptom Domain in Children: A Community-Based Case-Control Study." *Nicotine and Tobacco Research*, 94–101, 2017 **(40607)**

JORDAN, C.J., AND Z-X. XI. "Discovery and Development of Varenicline for Smoking Cessation." *Expert Opinion Drug Discovery*, 13: 671–683, 2018 **(70212)**

JUN, Y.J., H. KIM, AND S. PARK. "Suicidal Death by Nicotine Poisoning." *Korean Journal of Legal Medicine*, 38: 167–170, 2014 **(81105)**

JUNG, H-S., Y. KIM, J. SON, Y-J. JEON, H-G. SEO, S-H. PARK, AND B.R. HUH. "Can Urinary Cotinine Predict Nicotine Dependence Level in Smokers?" *Asian Pacific Journal of Cancer Prevention*, 13: 5483–5488, 2012 **(90107)**

K

KAISAR, M.A., S. PRASAD, T. LILES, AND L. CUCULLO, L. "A Decade of E-Cigarettes: Limited Research and Unresolved Safety Concerns." *Toxicology*, 365: 67–75, 2016 **(50424)**

KAISERMAN, M.J., AND W.S. RICKERT. "Handmade Cigarettes: It's the Tube that Counts." *American Journal of Public Health*, 82: 107–109, 1992 **(40212)**

KAMBOJ, A., H.A. SPILLER, M.J. CASAVANT, T. CHOUNTHIRATH, AND G.A. SMITH. "Pediatric Exposure to E-Cigarettes, Nicotine, and Tobacco Products in the United States." *Pediatrics*, 137: (9 pp), 2016 **(10410)**

KANAMORI, Y., AND R.E. MALONE. "Conveying Misinformation: Top-Ranked Japanese Books on Tobacco." *Tobacco Induced Diseases*, 9: 3 (6 pp), 2011 **(40118)**

KANTROW, S.P., S.E. JOLLEY, E.G. PRICE-HAYWOOD, X. WANG, T-S. TSENG, D. ARNOLD, L.F. BROWN, C. LEONARDI, R.A. SCRIBNER, E.J. TRAPIDO, AND H-Y. LIN. "Using the Emergency Department to Investigate Smoking in Young Adults." *Annuals of Epidemiology*, 30: 44–49, 2019 **(80809)**

KAPAYA, M., D.V. D'ANGELO, V.T. TONG, L. ENGLAND, N. RUFFO, S. COX, L. WARNER, J. BOMBARD, T. GUTHRIE, A. LAMPKINS, AND B.A. KING. "Use of Electronic Vapor Products Before, During and After Pregnancy Among Women with a Recent Live Birth—Oklahoma and Texas, 2015." *Morbidity and Mortality Weekly Report*, 68: 189–194, 2019 **(10311)**

KARACONJI, I.B. "Facts About Nicotine Toxicity." *Arhiv za higijenu rada i toksikologiju*, 56: 363–371, 2005 **(20101)**

KARZLOWSKI, L.T., N.A. DRESCHEL, S.D. STELLMAN, J. WILKENFELD, E.B. WEISSS, AND M.E. GOLDBERG. "An Extremely Compensatible Cigarette by Design: Documentary Evidence on Industry Awareness and Reactions to the Barclay Filter Design Cheating the Tar Testing System." *Tobacco Control*, 14: 64–70, 2005 **(40205)**

KASSEM, N.O.F., N.O. KASSEM, S. LILES, A.T. ZARTH, S.R. JACKSON, R.M. DAFFA, D.A. CHATFIELD, S.G. CARMELLA, S.S. HECHT, AND M.F. HOVELL. "Acrolein Exposure in Hookah Smokers and Non-Smokers Exposed to Hookah Tobacco Secondhand Smoke: Implications for Regulating Hookah Tobacco Products." *Nicotine and Tobacco Research*, 20: 492–501, 2018 **(60420)**

KAUFFMAN, R.M., A.K. FERKETICH, A.G. WEE, J.M. SHULTZ, P. KUUN, AND M.E. WEWERS. "Factors Associated with Smokeless Tobacco Cessation in an Appalachian Population." *Addictive Behavior*, 33: 821–830, 2008 **(60616)**

KAUFMAN, D.W., J.R. PALMER, L. ROSENBERG, AND S. SHAPIRO. "Cigar and Pipe Smoking and Myocardial Infarction in Young Men." *British Medical Journal*, 294: 1315–1316, 1987 **(60208)**

KAUR, G., R. PINKSTON, B. MCIEMORE, W. C. DORSEY, AND S. BATRA. "Immunological and Toxicological Risk Assessment of e-Cigarettes." *European Respiratory Review*, 27: 170119 (14 pp), 2018 **(50106)**

KEGLER, S.R., A.M. DELLINGER, M.F. BALLESTEROS, AND J. TSAI. "Decreasing Residential Fire Death Rates and the Association with the Prevalence of Adult Cigarette Smoking—United States, 1999–2015." *Journal of Safety Research*, 67: 197–201, 2018 **(80204)**

KEMP, C.B., C.A. SPEARS, T.F. PECHACEK, AND M.P. ERIKSEN. "Adults' Perceptions of Nicotine Harm to Children." *Pediatrics,* 142: (8 pp), 2018 **(10308)**

KEMP, P.M., G.S. SNEED, C.E. GEORGE, AND R.E. DISTEFANO. "Postmortem Distribution of Nicotine and Cotinine from a Case Involving the Simultaneous Administration of Multiple Nicotine Transdermal Systems." *Journal of Analytical Toxicology*, 21: 310–313, 1997 **(30601)**

KESKE, R.R., V.W. REES, I. BEHM, B.M. WADLER, AND A.C. GELLER. "Second-Hand Smoke Exposure and Mitigation Strategies Among Home Visitation Workers." *Tobacco Control*, 22: 250–254, 2013 **(40637)**

KHABOUR, O.F., K.H. ALZOUBI, N. AL-SHEYAB, M.A. AZAB, A.M. MASSADEH, A.A. ALOMARY, AND T.E. EISSENBERG. "Plasma and Saliva Levels of Three Metals in Waterpipe Smokers: A Case Control Study." *Inhalation Toxicology*, 30: 224–228, 2018 **(60418)**

KHANA, A., D.S. GAUTAM, M. GOKHALE, AND S.K. JAIN. "Tobacco Dust Induced Genotoxicity as an Occupational Hazard in Workers of Bidi Making Cottage Industry of Central India." *Toxicology International*, 21: 18–23, 2014 **(60509)**

KHARAZMI, M., A-P. CARLSSON, P. HALBERG, M. MODIG, L. BJÖRNSTAD, AND J-M. HIRSCH. "Surgical Approach to Snus-Induced Injury of the Oral Mucosa." *Journal of Oral Science*, 56: 91–94, 2014 **(60711)**

KIM, N.H., J.H. PARK, D.P. CHOI, J.Y. LEE, AND H.C. KIM. "Secondhand Smoke Exposure and Depressive Symptoms Among Korean Adolescents: JS High School Study." *Public Library of Science One* 11: (11 pp), 2016 **(80109)**

KIM, S., B.J. APELBERG, E. AVILA-TANG, L. HEPP, D. YUN, J.M. SAMET, P.N. BREYSSE. "Utility and Cutoff Value of Hair Nicotine as a Biomarker of Long-Term Tobacco Smoke Exposure, Compared to Salivary Cotinine." *International Journal of Environmental Research and Public Health*, 11: 8368–8382, 2014 **(30507)**

KIM, W.H., A. NAVAS-ACIEN, F. DOMINICI, E. AVILA-TNAG, G. ONICESCU, P. BREYSSE, AND J.M. SAMET. "Determinants of Hair Nicotine Concentration in Nonsmoking Women and Children: A Multicounty Study of Secondhand Smoke Exposure in Homes." *Cancer Epidemiology, Biomarkers & Prevention*, 18: 3407–3414, 2009 **(30502)**

KING, A., P. MCNAMARA, M. CONRAD, AND D. CAO. "Alcohol-Induced Increases in Smoking Behavior for Nicotinized and Denicotinized Cigarettes in Men and Women." *Psychopharmacology*, 207: 107–117, 2009 **(60910)**

KING, B.A., S.D. BABB, M.A. TYNAN, AND R.B. GERZOFF. "National and State Estimates of Secondhand Smoke Infiltration Among US Multiunit Housing Residents." *Nicotine Tobacco Research*, 15: 1316-1321, 2013 **(40636)**

KING, J.L., B.A. BEBOUSSIN, J. SPANGLER, J.C. ROSS, AND E.L. SUTFIN. "Tobacco Product Use and Mental Health Status Among Young Adults." *Addictive Behaviors*, 77: 67–72, 2018 **(80108)**

KIVRAK, T., M. SUNBUL, E. DURMUS, R. DERVISOVA, I. SARI, AND O. YESILDAG. "Acute Myocardial Infarction due to Liquid Nicotine in a Young Man." *Therapeutic Advances in Cardiovascular Disease,* 8: 32–34, 2014 **(50435)**

KLEIN, J.D. "E-Cigarettes: A 1-Way Street to Traditional Smoking and Nicotine Addiction for Youth." *Pediatrics*, 141: (2 pp), 2018 **(50605)**

KLEMSDAL, T.O., K. GJESDAL, AND K. ZAHLSEN. "Physical Exercise Increases Plasma Concentration of Nicotine During Treatment with a Nicotine Patch." *British Journal of Pharmacology*, 39: 677–679, 1995 **(20115)**

KLUNGSØYR, O., J.F. NYGÅRD, T. SØRENSEN, AND I. SANDANGER. "Cigarette Smoking and Incidence of First Depressive Episode: An 11-Year, Population-Based Follow-Up Study." *American Journal of Epidemiology*, 163: 421–432, 2006 **(80105)**

KNOTTENBELT, C.M., S. BAWAZEER, J. HAMMOND, D. MELLOR, AND D.G. WATSON. "Nicotine Hair Concentration in Dogs Exposed to Environmental Tobacco Smoke: A Pilot Study." *Journal of Small Animal Practice*, 53: 623–626, 2012 **(30508)**

KODL, M., S.S. FU, AND A.M. JOSEPH. "Tobacco Cessation Treatment for Alcohol-Dependent Smokers: When Is the Best Time?" *Alcohol Research and Health*, 29: 203–207, 2006 **(60915)**

KOLAPPAN, C., AND P.G. GOPI. "Tobacco Smoking and Pulmonary Tuberculosis." *Thorax,* 57: 964–966, 2002 **(40511)**

KOMIYAMA, M., AND K. HASEGAWA. "Smoking Cessation as a Public Health Measure to Limit the Coronavirus Disease 2019 Pandemic." *European Cardiology Review*, 15: e16 (2 pp), 2020 **(80311)**

KOSMIDER, L., A. SOBCZAK, M. FIK, J. KNYSAK, M. ZACIERA, J. KUREK, AND M.L. GONIEWICZ. "Carbonyl Compounds in Electronic Cigarette Vapors: Effects of Nicotine Solvent and Battery Output Voltage." *Nicotine and Tobacco Research*, 16: 1319–1326, 2014 **(50410)**

KOSMIDER, L., T.R. SPINDLE, M. GAWRON, A. SOBCZAK, AND M.L. GONIEWICZ. "Nicotine Emissions from Electronic Cigarettes: Individual and Interactive Effects of Propylene Glycol to Vegetable Glycerin Composition and Device Power Output." *Food and Chemical Toxicology*, 115: 302–305, 2018 **(50312)**

KOSTYGINA, G., AND P.M. LING. "Tobacco Industry Use of Flavourings to Promote Smokeless Tobacco Products." *Tobacco Control*, 25: ii40–ii49, 2016 **(60611)**

KOSZOWSKI, B., Z.R. ROSENBERRY, A. KANU, L.C. VIRAY, J.L. POTTS, AND W.B. PICKWORTH. "Nicotine and Carbon Monoxide Exposure from Inhalation of Cigarillo Smoke." *Pharmacology, Biochemistry, and Behavior*, 139: 7–14, 2015 **(60215)**

KOSZOWSKI, B., Z.R. ROSENBERRY, L.C. VIRAY, J.L. POTTS, AND W.B. PICKWORTH. "Make Your Own Cigarettes: Toxicant Exposure, Smoking Topography, and Subjective Effects." *Cancer Epidemiology, Biomarkers and Prevention*, 23: 1793–1803, 2014 **(40411)**

KOTZ, D., W. VIECHTBAUER, C.R. SIMPSON, O.C.P. VAN SCHAYCK, R. WEST, AND A. SHEIKH. "Cardiovascular and Neuropsychiatric Risks of Varenicline and Bupropion in Smokers with Chronic Obstructive Pulmonary Disease." *Thorax*, 905–911, 2017 **(70225)**

KOWITT, S.D., A. OSMAN, C. MEERNIK, G.A. ZARKIN, L.M. RANNEY, J. MARTIN, C. HECK, AND A.O. GOLDSTEIN. "Vaping Cannabis Among Adolescents: Prevalence and Associations with Tobacco Use from a Cross-Sectional Study in the USA." *BMJ Open*, 9: (6 pp), 2019 **(60906)**

KOZLOWSKI, L.T., N.Y. MEHTA, C.T. SWEENEY, S.S. SCHWARTZ, G.P. VOGLER, M.J. JARVIS, AND R.J. WEST. "Filter Ventilation and Nicotine Content of Tobacco in Cigarettes from Canada, the United Kingdom, and the United States." *Tobacco Control*, 7: 369–375, 1998 **(40209)**

KOZLOWSKI, L.T., R.C. FRECKER, V. KHOUW, AND M.A. POPE. "The Misuse of 'Less Hazardous' Cigarettes and Its Detection: Hole-Blocking of Ventilated Filters." *American Journal of Public Health*, 70: 1202–1203, 1980 **(40206)**

KRAUSE, D., M. WARNECKE, C.S. SCHUETZ, M. SOYKA, K.M. MANZ, L. PROEBSTL, F. KAMP, A.I. CHROBOK, O. POGARELL, AND G. KOLLER. "The Impact of the Opioid Antagonist Naloxone on Experimentally Induced Craving in Nicotine-Dependent Individuals." *European Addiction Research*, 24: 255–265, 2018 **(70231)**

KRISHNAN-SARIN, S., A. JACKSON, M. MOREAN, G. KONG, K.W. BOLD, D.R. CAMENGA, D.A. CAVALLO, P. SIMON, AND R. WU. "E-Cigarette Devices Used by High-School Youth." *Drug and Alcohol Dependence*, 194: 395–400, 2019 **(50618)**

KRISHNAN-SARIN, S., B.G. GREEN, G. KONG, D.A. CAVALLO, P. JATLOW, R. GUEORGUIEVA, E. BUTA, AND S.S. O'MALLEY. "Studying the Interactive Effects of Menthol and Nicotine Among Youth: An Examination Using E-Cigarettes." *Drug and Alcohol Dependence*, 180: 193–199, 2017 **(50609)**

KULI, M.C., S.A. BIALOUS, S. MUNTHALI, AND W. MAX. "Tobacco Growing and the Sustainable Development Goals, Malawi." *Bulletin of the World Health Organisation*, 95: 362–367, 2017 **(10113)**

KUNITZ, S.J. "Historical Influences on Contemporary Tobacco Use by Northern Plains and Southwestern American Indians." *American Journal of Public Health*, 106: 246–255, 2016 **(10101)**

L

LACASSE, Y., M. LÉGARÉ, AND F. MALTAIS. "E-Cigarette Use in Patients Receiving Home Oxygen Therapy." *Canadian Respiratory Journal*, 22: 83–85, 2015 **(80215)**

LAJOUS, M., L. TONDEUR, G. FAGHERAZZI, B. DE LAUZON-GUILLAIN, M-C. BOUTRON-RUAUALT, AND F. CLAVEL-CHAPELON. "Childhood and Adult Secondhand Smoke and Type 2 Diabetes in Women." *Diabetes Care*, 36: 2720–2725, 2013 **(40618)**

LAMPALO, M., M. MAJER, I. JUKIĆ, H.S. STANIĆ, B. BARIŠIĆ, AND S. POPOVIĆ-GRLE. "Health Sequelae of Tobacco Exposure in Childhood and Adolescence." *Psychiatria Danubina*, 31: 39–43, 2019 **(40806)**

LANDMAN, S.T., I. DHALIWAL, C.A. MACKENZIE, T. MARTINU, A. STEELE, AND K.J. BOSMA. "Life-Threatening Bronchiolitis Related to Electronic Cigarette Use in a Canadian Youth." *Canadian Medical Association Journal*, 191: E1321–1331, 2019 **(50417)**

LANIADO-LABORÍN, R. "Smoking and Chronic Obstructive Pulmonary Disease (COPD). Parallel Epidemics of the 21st Century." *International Journal of Environmental Research and Public Health*, 6: 209–224, 2009 **(70208)**

LAURIA, V.T., E.F. SPERANDIO, T.L.W. DE SOUSA, W. DE OLIVEIRA VIEIRA, M. ROMITI, A.R. DE TOLEDO GAGLIARDI, R.L. ARANTES, AND V.Z. DOURADO. "Evaluation of Dose-Response Relationship Between Smoking Load and Cardiopulmonary Fitness in Adult Smokers: A Cross-Sectional Study." *Revista Portuguesa de Pneumologia*, 23: 79–84, 2017 **(40507)**

LAURSEN, K.R., J.H. BØNLØKKE, E. BENDSTRUP, M. BILDE, M. GLASIUS, V.H. GUTZKE, S.P. MOOSAKUTTY, A-C. OLIN, P. RAVN, K. ØSTERGAARD, AND T. SIGSGAARD. "An RCT of Acute Health Effects in COPD-Patients After Passive Vape Exposure from E-Cigarettes." *European Clinical Respiratory Journal*, 8: (11 pp), 2020 **(50504)**

LAVELLE, C., C. BIREK, AND D.A. SCOTT. "Are Nicotine Replacement Strategies to Facilitate Smoking Cessation Safe?" *Journal of the Canadian Dental Association*, 69: 592–597, 2003 **(70229)**

LAWLER, T.S., S.B. STANFILL, H.T. TRAN, G.E. LEE, P.X. CHEN, J.B. KIMBRELL, J.G. LISKO, C. FERNANDEZ, S.P. CAUDILL, B.R. DECASTRO, AND C.H. WATSON. "Chemical Analysis of Snus Products from the United States and Northern Europe." *Public Library of Science One* 15: (17 pp), 2020 **(60701)**

LECHASSEUR, A., S. ALTMEJD, N. TURGEON, G. BUONANNO, L. MORAWSKA, D. BRUNET, C. DUCHAINE, AND M.C. MORISSETTE. "Variations in Coil Temperature/Power and E-Liquid Constituents Change Size and Lung Deposition of Particles Emitted by an Electronic Cigarette." *Physiological Reports*, 7: (9 pp), 2019 **(50208)**

LEE, A., S.Y. LEE, AND K-S. LEE. "The Use of Heated Tobacco Products is Associated with Asthma, Allergic Rhinitis, and Atopic Dermatitis in Korean Adolescents." *Scientific Reports*, 9: 17699 (8 pp), 2019 **(60109)**

LEE, J., Y. YOU, J.S. PARK, J.H. MIN, I. YOO, W. JEONG, Y. CHO, S. RYU, S. KIM, S.U. CHO, S.K. OH, J. LEE, H.J. AHN, AND S.M. JUNG. "Liver Donation After Brain Death Following Intentional Ingestion of 99% E-Cigarette Liquid Nicotine 10 mL." *Experimental and Clinical Transplantation*, 18: 120–122, 2020 **(81102)**

LEE, J.G.L., A. RICHARDSON, S.D. GOLDEN, AND K.M. RIBISL. "Promotions on Newport and Marlboro Cigarette Packages: A National Study." *Nicotine and Tobacco Research*, 19: 1243–1247, 2017 **(70312)**

LEE, M-S., J.G. ALLEN, AND D.C. CHRISTIANI. "Endotoxin and (1→3)-β-D-Glucan Contamination in Electronic Cigarette Products Sold in the United States." *Environmental Health Perspectives*, 127: 047008 (6 pp), 2019 **(50405)**

LEE, Y.O., AND S.A. GLANTZ. "Menthol: Putting the Pieces Together." *Tobacco Control*, 20: ii1–ii7, 2011 **(40310)**

LEE, Y.O., J.K. PEPPER, A.J. MACMONEGLE, J.M. NONNEMAKER, J.C. DUKE, AND L. PORTER. "Examining Youth Dual and Polytobacco Use with E-Cigarettes." *International Journal of Environmental Research and Public Health*, 15: 699 (12 pp), 2018 **(60802)**

LEGG, T., S. PEETERS, P. CHAMBERLAIN, AND A.B. GILMORE. "The Philip Morris–Founded Foundation for a Smoke-Free World: Tax Returns Sheds Light on Funding Activities." *Lancet*, 393: 2487–2488, 2019 **(50110)**

LEIGH, N.J., M.N. PALUMBO, A.M. MARINO, R.J. O'CONNOR, AND M.L. GONIEWICZ. "Tobacco-Specific Nitrosamines (TSNA) in Heated Tobacco Product IQOS." *Tobacco Control*, 27: s37–s38, 2018 **(60105)**

LEIGH, N.J., P.L. TRAN, R.J. O'CONNOR, AND M.L. GONIEWICZ. "Cytotoxic Effects of Heated Tobacco Products (HTP) on Human Bronchial Epithelial Cells." *Tobacco Control*, 27: s26–s29, 2018 **(60110)**

LEIGH, N.J., R.I. LAWTON, P.A. HESHBERGER, AND M.L. GONIEWICZ. "Flavourings Significantly Affect Inhalation Toxicity of Aerosol Generated from Electronic Nicotine Delivery Systems (ENDS)" *Tobacco Control*, 25: ii81–ii87, 2016 **(50309)**

LERNER, C.A., I.K. SUNDAR, H. YAO, J. GERLOFF, D.L. OSSIP, S. MCINTOSH, R. ROBINSON, AND I. RAHMAN. "Vapors Produced by Electronic Cigarettes and E-Juices with Flavorings Induce Toxicity, Oxidative Stress, and Inflammatory Response in Lung Epithelial Cells and in Mouse Lung." *Public Library of Science One*, 10: (26 pp), 2015 **(50306)**

LESTARI, K.S., M.V. HUMAIRO, AND U. AGUSTINA. "Formaldehyde Vapor Concentration in Electronic Cigarettes and Health Complaints of Electronic Cigarette Smokers in Indonesia." *Journal of Environmental and Public Health*, 2018: (6 pp) (2018) **(50408)**

LEVENTHAL, A.M., D.R. STRONG, S. SUSSMAN, M.G. KIRKPATRICK, J.B. UNGER, J.L. BARRINGTON-TRIMIS, AND J. AUDRAIN-MCGOVERN. "Psychiatric Comorbidity in Adolescent Electronic and Conventional Cigarette Use." *Journal of Psychiatric Research*, 73: 71–78, 2016 **(80816)**

LEWANDER, W., H. WINE, R. CARNEVALE, J. LINDENMAYER, E. HARVEY, C. HALL-WALKER, L. LAMBRIGHT, AND E. MANZO. "Ingestion of Cigarettes and Cigarette Butts by Children—Rhode Island, January 1994–July 1996." *Morbidity and Mortality Weekly Report*, 46: 125–128, 1997 **(10405)**

LEWIS, N., K. MCCAFFREY, K. SAGE, C-J. CHENG, J. GREEN, L. GOLDSTEIN, H. CAMPBELL, D. FERRELL, N. MALAN, N. LACROSS, A. MALDONADO, A. BOARD, A. HANCHEY, D. HARRIS, S. CALLAHAN, S. ABEREGG, I. RISK, S. WILLARDSON, A. CARTER, A. NAKASHIMA, J. DUNCAN, C. BURNETT, R. ATKINSON-DUNN, AND A. DUNN. "E-Cigarette Use, or Vaping, Practices and Characteristics Among Persons with Associated Lung Injury: Utah, April–October 2019." *Morbidity and Mortality Weekly Report* 68: 953–956, 2019 **(50439)**

LHOTTA, K., H.J. RUMPELT, P. KÖNIG, G. MAYER, AND F. KRONENBERG. "Cigarette Smoking and Vascular Pathology in Renal Biopsies." *Kidney International*, 61: 648–654, 2002 **(40526)**

LI, V.C., AND S. TANG. "China's New Road for Tobacco Control: Tobacco Crop Substitution [Editorial]." *American Journal of Public Health*, 108: 1316–1317, 2018 **(10114)**

LIANG, Y., X. ZHENG, D.D. ZENG, X. ZHOU, S.J. LEISCHOW, AND W. CHUNG. "Exploring How the Tobacco Industry Presents and Promotes Itself in Social Media." *Journal of Medical Internet Research*, 17: e24 (18 pp), 2015 **(40108)**

LIPKUS, I.M., D. MAYS, AND K.P. TERCYAK. "Characterizing Young Adults' Susceptibility to Waterpipe Use and Their Reactions to Messages About Product Harms and Addictiveness." *Nicotine and Tobacco Research*, 19: 1216–1223, 2017 **(60407)**

LIPKUS, I.M., T. EISSENBERG, R.D. SCHWARTZ-BLOOM, A.V. PROKHOROV, AND J. LEVY. "Affecting Perceptions of Harm and Addiction Among College Waterpipe Tobacco Smokers." *Nicotine and Tobacco Research*, 13: 599–610, 2011 **(60403)**

LIPPI, G., E.J. FAVALORO, T. MESCHI, C. MATTIUZZI, L. BORGHI, AND G. CERVELLIN. "E-Cigarettes and Cardiovascular Risk: Beyond Science and Mysticism." *Seminars in Thrombosis and Hemostasis*, 40: 60–65, 2014 **(50412)**

LISKO J.G., H. TRAN, S.B. STANFILL, B.C. BLOUNT, AND C.H. WATSON. "Chemical Composition and Evaluation of Nicotine, Tobacco Alkaloids, pH, and Selected Flavors in E-Cigarette Cartridges and Refill Solutions." *Nicotine and Tobacco Research*, 17: 127 –1278, 2015 **(50314)**

LITTLE, M.A., Z. BURSAC, K.J. DEREFINKO, J.O. EBBERT, G.W. TALCOTT, A. HRYSHKO-MULLEN, AND R.C. KLESGES. "Types of Dual and Poly-Tobacco Users in the US Military." *American Journal of Epidemiology*, 184: 211–218, 2016 **(60808)**

LIU, W., Z-W. TAO, L. WANG, M-L. YUAN, K. LIU, L. ZHOU, S. WEI, Y. DENG, J. LIU, H-G. LIU, M. YANG, AND Y. HU. "Analysis of Factors Associated with Disease Outcomes in Hospitalized Patients with 2019 Novel Coronavirus Disease." *Chinese Medical Journal*, 133: 1032–1038, 2020 **(80314)**

LIU, Y., A.G. WHEATON, L.B. MURPHY, F. XU, J.B. CROFT, AND K.J. GREENLUND. "Chronic Obstructive Pulmonary Disease and Arthritis Among US Adults, 2016." *Preventing Chronic Disease*, 16: E93 (10 pp), 2019 **(40504)**

LUCK, W., AND H. NAU. "Nicotine and Cotinine Concentrations in Serum and Milk of Nursing Smokers." *British Journal of Clinical Pharmacology*, 18: 9–15, 1984 **(30203)**

LUI, Y., M. DAI, Y. BI, M. XU, Y. XU, M. LI, T. WANG, F. HUANG, B. XU, J. ZHANG, X. LI, W. WANG, AND G. NING. "Active Smoking, Passive Smoking, and Risk of Nonalcoholic Fatty Liver Disease (NAFLD): A Population-Based Study in China." *Journal of Epidemiology*, 32: 115–121, 2013 **(40525)**

M

MA, Y., H. TONG, X. ZHANG, M. WANG, J. YANG, M. WU, R. HAN, M. CHEN, X. HU, Y. YUAN, G. PAN, Y. ZOU, S. XU, AND F. PAN. "Chronic Obstructive Pulmonary Disease in Rheumatoid Arthritis: A Systematic Review and Meta-Analysis." *Respiratory Research*, 20: 144 (8 pp), 2019 **(40503)**

MADDATU, J., E. ANDERSON-BAUCUM, AND C. EVANS-MOILINA. "Smoking and Risk of Type 2 Diabetes." *Translational Research : The Journal of Laboratory and Clinical Medicine* 184: 101–107, 2017 **(40508)**

MAESSEN, G.C., A.M. WIJNHOVEN, R.L. NEIJZEN, M.C. PAULUS, D.A.M. VAN HEEL, B.H.A. BOMERS, L.E. BOERSMA, B. KONYA, AND M.A.G. VAN DER HEYDEN. "Nicotine Intoxication by E-Cigarette Liquids: A Study of Case Reports and Pathophysiology." *Clinical Toxicology*, 58: (8 pp), 2019 **(10420)**

MAHABEE-GITTENS, E.M., A.L. MERIANOS, AND G.E. MATT. "Preliminary Evidence that High Levels of Nicotine on Children's Hands May Contribute to Overall Tobacco Smoke Exposure." *Tobacco Control*, 27: 217–219, 2018 **(40701)**

MAHMOODABAD, S.S.M., M.S. JADGAL, I. ZAREBAN, AND H.F. ZADEH. "The Determinants of Salivary Cotinine Concentration in Smokeless Tobacco Users." *Macedonian Journal of Medical Sciences*, 7: 810–815, 2019 **(60621)**

MAJMUNDAR, A., J-P. ALLEM, T.B. CRUZ, AND J.B. UNGER. "Where Do People Vape? Insights from Twitter Data." *International Journal of Environmental Research and Public Health*, 16: 3056 (5 pp), 2019 **(50505)**

MAKADIA, L.D., P.J. ROPER, J.O. ANDREWS, AND M.S. TINGEN. "Tobacco Use and Smoke Exposure in Children: New Trends, Harm, and Strategies to Improve Health Outcomes." *Current Allergy and Asthma Reports*, 17: 55 (15 pp), 2017 **(40519)**

MALLOCK, N., E. PIEPER, C. HUTZLER, F. HENKLER-STEPHANI, AND A. LUCH. "Heated Tobacco Products: A Review of Current Knowledge and Initial Assessments." *Frontiers in Public Health*, 7: (8 pp), 2019 **(60104)**

MALLOCK, N., H.L. TRIEU, M. MACZIOL, S. MALKE, A. KATZ, P. LAUX, F. HENKLER-STEPHANI, J. HAHN, C. HUTZLER, AND A. LUCH. "Trendy E-cigarettes Enter Europe: Chemical Characterization of JUUL Pods and its Aerosols." *Archives of Toxicology*, 95: 1985–1994, 2020 **(50203)**

MALSON, J.L., E.M. LEE, E.T. MOOLCHAN, AND W.B. PICKWORTH. "Nicotine Delivery from Smoking Bidis and an Additive-Free Cigarette." *Nicotine and Tobacco Research*, 4: 485–490, 2002 **(60502)**

MALSON, J.L., K. SIMS, R. MURTY, AND W.B. PICKWORTH. "Comparison of the Nicotine Content of Tobacco Used in Bidis and Conventional Cigarettes." *Tobacco Control*, 10: 181–183, 2001 **(60504)**

MANOUCHEHRINIA, A., C.R. TENCH, J. MAXTED, R.H. BIBANI, J. BRITTON, AND C.S. CONSTANTINESCU, "Tobacco Smoking and Disability Progression in Multiple Sclerosis: United Kingdom Cohort Study." *Brain*, 136: 2298–2304, 2013 **(40521)**

MARCON, A., G. PESCE, L. CALCIANO, V. BELLISARIO, S.C. DHARMAGE, J. GARCIA-AYMERICH, T. GISLASSON, J. HEINRICH, M. HOLM, C. JANSON, D. JARVIS, B. LEYNAERT, M.C. MATHESON, P. PIRINA, C. SVANES, S. VILLANI, T. ZUBERBIER, C. MINELLI, AND S. ACCORDINI. "Trends in Smoking Initiation in Europe Over 40 Years: A Retrospective Cohort Study." *Public Library of Science One*, 13: e201881 (14 pp), 2018 **(70306)**

MARITZ, G.S., AND M. MUTEMWA. "Tobacco Smoking: Patterns, Health Consequences for Adults, and the Long-term Health of the Offspring." *Global Journal of Health Science*, 4: 62–75, 2012 **(80708)**

MARRONE, G.F., D.M. SHAKLEYA, K.B. SCHEIDWEILER, E.G. SINGLETON, M.A. HUESTIS, AND S.J. HEISHMAN. "Relative Performance of Common Biochemical Indicators in Detecting Cigarette Smoking." *Addiction*, 106: 1325–1334, 2011 **(90205)**

MASANOTTI, G.M., E. ABBAFATI, E. PETRELLA, S. VINCIGUERRA, AND F. SRACCI. "Intensive Tobacco Cultivations, A Possible Public Health Risk?" *Environmental Science and Pollution Research*, 26: 12616–12621, 2019 **(10108)**

MATHENY, J.D., E.M. STEVENS, S. CHEN, B.A. CHRISTIANSEN, S.D. KOWITT, A. OSMAN, AND D.J. VIDRINE. "The RICO Verdict and Corrective Statements: Catalysts for Policy Change?" *Tobacco Regulatory Science*, 5: 206–228, 2019 **(70323)**

MATT, G.E., E. HOH, P.J.E. QUINTANA, J.M. ZAKARIAN, AND J. ARCEO. "Cotton Pillows: A Novel Field Method for Assessment of Thirdhand Smoke Pollution." *Environmental Research*, 168: 206–210, 2019 **(40703)**

MATT, G.E., E.M. MAHABEE-GITTENS, J.M. ZAKARIAN, P.J.E. QUINTANA, E. HOH, AND M. MYERS. "Nicotine in Thirdhand Smoke Residue Predicts Relapse from Smoking Cessation: A Pilot Study." *Addictive Behavior*, 98: 106041 (10 pp), 2019 **(40707)**

MATT, G.E., P.J.E. QUINTANA, J.M. ZAKARIAN, E. HOH, M.F. HOVELL, M. MAHABEE-GITTENS, K. WATANABE, K. DATUIN, C. VUE, AND D.A. CHATFIELD. "When Smokers Quit: Exposure to Nicotine and Carcinogens Persists from Thirdhand Smoke Pollution." *Tobacco Control*, 26: 548–556, 2016 **(40706)**

MAX, W., AND T. TSOUKALAS. "Economics on Trial: The Use and Abuse of Economic Methods in Third Party Tobacco Litigation." *Tobacco Control*, 15: iv77–iv83, 2006 **(40116)**

MAYER, B. "How Much Nicotine Kills a Human? Tracing Back the Generally Accepted Lethal Dose to Dubious Self-Experiments in the Nineteenth Century." *Archives of Toxicology* 88: 5–7, 2014 **(10401)**

MCEVOY, C.T., AND E.R. SPINDEL. "Pulmonary Effects of Maternal Smoking on the Fetus and Child: Effects on Lung Development, Respiratory Morbidities, and Life Long Lung Health." *Paediatric Respiratory Reviews*, 21: 27–33, 2017 **(80703)**

MCGRATH, J.J., S. RACICOT, C.T.C. OKOLI, S.K. HAMMOND, AND J. O'LOUGHLIN. "Airborne Nicotine, Secondhand Smoke, and Precursors to Adolescent Smoking." *Pediatrics*, 141: S63-S73, 2018 **(40620)**

MCGUFFEY, J.E., B. WEI, J.T. BERNERT, J.C. MORROW, B. XIA, L. WANG, AND B.C. BLOUNT. "Validation of a LC-MS/MS Method for Quantifying Urinary Nicotine, Six Nicotine Metabolites and the Minor Tobacco Alkaloids—Anatabine and Anabasine—in Smokers' Urine." *Public Library of Science One*, 9: e101816 (13 pp) 2014 **(90201)**

MCGUIRE, A. "How the Tobacco Industry Continues to Keep the Home Fires Burning." *Tobacco Control*, 8: 67–69, 1999 **(80201)**

MCKEE, S.A., S.S. O'MALLEY, J. SHI, T. MASE, AND S. KRISHNAN-SARIN. "Effects of Transdermal Nicotine Replacement on Alcohol Responses and Alcohol Self-Administration." *Psychopharmacology*, 196: 189–200, 2008 **(60909)**

MCKELVEY, K., J. THRUL, AND D. RAMO. "Impact of Quitting Smoking and Smoking Cessation Treatment on Substance Use Outcomes: An Updated and Narrative Review." Addictive Behaviors, 65: 161–170, 2017 **(60913)**

MCKELVEY, K., L. POPOVA, J.K. PEPPER, N.T. BREWER, AND B. HALPERN-FELSHER. "Adolescents Have Unfavorable Opinions of Adolescents Who Use E-Cigarettes." *Public Library of Science One*, 13: (10 pp), 2018 **(50630)**

MCKELVEY, K., L. POPOVA, M. KIM, L.K. LEMPERT, B.W. CHAFFEE, M. VIJAYARAGHAVAN, P. LING, AND B. HALPERN-FELSHER. "IQOS Labelling Will Mislead Consumers." *Tobacco Control*, 27: s48–s54, 2018 **(60115)**

MCKELVEY, K., M. BAIOCCHI, AND B. HALPERN-FELSHER. "Adolescents' and Young Adults' Use and Perceptions of Pod-Based Electronic Cigarettes." *JAMA Network Open*, 1: e183535 (13 pp), 2018 **(50613)**

MCKNIGHT, R.H., AND H.A. SPILLER. "Green Tobacco Sickness in Children and Adolescents." *Public Health Reports*, 120: 602–605, 2005 **(10128)**

MCNEELY, E., I. MORDUKHOVICH, S. STAFFA, S. TIDEMAN, AND B. COULL. "Legacy Health Effects Among Never Smokers Exposed to Occupational Secondhand Smoke." *Public Library of Science One* 14: (12 pp), 2019 **(40604)**

MEDABALA, T., B.N. RAO, M.I. GLAD MOHESH, AND M. PRAVEEN KUMAR. "Effect of Cigarette and Cigar Smoking on Peak Expiratory Flow Rate." *Journal of Clinical and Diagnostic Research*, 7: 1886–1889, 2013 **(60205)**

MEHMETI, E., T. KILIC, C. LAUR, AND S. CARRARA. "Electrochemical Determination of Nicotine in Smokers' Sweat." *Microchemical Journal*, 158: (7 pp), 2020 **(30405)**

MEHROTRA, R., AND D.N. SINHA. "Global Challenges in Smokeless Tobacco Control." *Indian Journal of Medical Research*, 148: 1–3, 2018 **(60601)**

MEIJA, P., L. DOOFMAN, A. CHEYNE, L. NIXON, L. FRIEDMAN, M. GOTTLIEB, AND R. DAYNARD. "The Origins of Personal Responsibility Rhetoric in News Coverage of the Tobacco Industry." *American Journal of Public Health*, 104: 1048–1051, 2014 **(40123)**

MELSTROM, P., C. SOSNOFF, B. KOSZOWSKI, B.A. KING, R. BUNNELL, G. LE, L. WANG, M.H. THANNER, B. KENEMER, S. COX, B.R. DECASTRO, AND T. MCAFEE. "Systematic Absorption of Nicotine Following Acute Secondhand Exposure to Electronic Cigarette Aerosol in a Realistic Social Setting." *International Journal of Hygiene and Environmental Health*, 221: 816–822, 2018 **(50512)**

MENDEL, J.R., S.A. BAIG, M.G. HALL, M. JEONG, M.J. BYRON, J.C. MORGAN, S.M. NOAR, K.M. RIBISL, AND N.T. BREWER. "Brand Switching and Toxic Chemicals in Cigarette Smoke: A National Study." *Public Library of Science One*, 13: (10 pp), 2018 **(40413)**

MERCINCAVAGE, M., K. LOCHBUEHLER, E.P. WILEYTO, N.L. BENOWITZ, R.F. TYNDALE, C. LERMAN, AND A.A. STRASSER. "Association of Reduced Nicotine Content Cigarettes with Smoking Behaviors and Biomarkers of Exposure Among Slow and Fast Nicotine Metabolizers. A Nonrandomized Clinical Trial." *JAMA Network Open*, 1: 1346 (17 pp), 2018 **(20303)**

MERRILL, R.M. "Injury-Related Deaths According to Environmental, Demographic, and Lifestyle Factors." *Journal of Environmental and Public Health*, 2019: (12 pp), 2019 **(80411)**

MIAO, S., E.S. BEACH, T.J. SOMMER, J.B. ZIMMERMAN, AND S-E. JORDT. "High-Intensity Sweeteners in Alternative Tobacco Products." *Nicotine and Tobacco Research*, 18: 2169–2173, 2016 **(60610)**

MICHAEL, R., N. EBRAHEIM, J. MAIER, M. TANIOS, AND A. KOURI. "Electronic Cigarette Burns: A Case Report and Review of Current Literature." *Case Reports in Orthopedics*, 2019: (5 pp), 2019 **(80212)**

MIECH, R., M.E. PATRICK, P.M. O'MALLEY, AND L.D. JOHNSTON. "E-Cigarette Use as a Predictor of Cigarette Smoking: Results from a 1-Year Follow-Up of a National Sample of 12th Grade Students." *Tobacco Control*, 26: e106–e111, 2017 **(50616)**

MÍGUEZ-BURBANO, M.J., M. VARGAS, C. QUIROS, J.E. LEWIS, L. ESPINOZA, AND D. ASTHANA. "Menthol Cigarettes and the Cardiovascular Risks of People Living with HIV." *Journal of the Association of Nurses AIDS Care*, 25: 427–435, 2014 **(40306)**

MIKHEEV, V.B., M.C. BRINKMAN, C.A. GRANVILLE, S.M. GORDON, AND P.I. CLARK. "Real-Time Measurement of Electronic Cigarette Aerosol Size Distribution and Metals Content Analysis." *Nicotine and Tobacco Research*, 18: 1895–1902, 2016 **(50406)**

MILBERGER, S., R.M. DAVIS, C.E. DOUGLAS, J.K. BEASLEY, D. BURNS, T. HOUSTON, AND D. SHOPLAND. "Tobacco Manufacturers' Defence Against Plaintiffs' Claim of Cancer Causation: Throwing Mud at the Wall and Hoping Some of it Will Stick." *Tobacco Control*, 15: iv17–iv26, 2006 **(40119)**

MILLER, E.I., H-R.K. NORRIS, D.E. ROLLINS, S.T. TIFFANY, C.M. MOORE, M.J. VINCENT, A. AGRAWAL, AND D.G. WILKINS. "Identification and Quantification of Nicotine Biomarkers in Human Oral Fluid from Individuals Receiving Low-Dose Transdermal Nicotine: A Preliminary Study." *Journal of Analytical Toxicology*, 34: 357–366, 2010 **(30403)**

MILLER, L.S., W. MAX, H-Y. SUNG, D. RICE, AND M. ZARETSKY. "Evaluation of the Economic Impact of California's Tobacco Control Program: A Dynamic Model Approach." *Tobacco Control*, 19: i68–i76, 2010 **(70317)**

MILLETT, C., L.M. WEN, C. RISSEL, A. SMITH, J. RICHTERS, A. GRULICH, AND R. DE VISSER. "Smoking and Erectile Dysfunction: Findings from a Representative Sample of Australian Men." *Tobacco Control*, 15: 136–139, 2006 **(80906)**

MILLQVIST, E., AND M. BENDE. "Capsaicin Cough Sensitivity is Decreased in Smokers." *Respiratory Medicine*, 95: 19–21, 2001 **(10319)**

MILLS, S.D., R. KURTZMAN, S.D. GOLDEN, A.Y. KONG, AND K.M. RIBISL. "Cultivating New Directions: The Changing Role of Tobacco in North Carolina's Economy." *North Carolina Medical Journal*, 79: 30–33, 2018 **(10111)**

MINH, A.E. "Effects of Vaping E-Juices with and without Alcohol on the Accuracy of the Alco-Sensor FST Approved Screening Device." *Canadian Society of Forensic Science Journal*, 54: 77–85, 2021 **(50318)**

MINH, H.V., K.B. GIANG, N.M. BICH, AND N.T. HUONG. "Tobacco Farming in Rural Vietnam: Questionable Economic Gain but Evident Health Risks." *BMC Public Health*, 9: 24 (10 pp), 2009 **(10110)**

MISHRA, A., P. CHATURVEDI, S. DATTA, S. SINUKUMAR, P. JOSHI, AND A. GARG. "Harmful Effects of Nicotine." *Indian Society of Medical & Paediatric Oncology*, 36: 24–31, 2015 **(10301)**

MITHANI, M.H., S.A. KHAN, S. EL KHALID, R. SIDDIQUE, H. HUMAYUN, AND A.S. AWAN. "Awareness of Risk Factors and Fallacies Associated with Urinary Bladder Cancer in Our Population: A Prospective Survey." *Journal of the Pakistan Medical Association*, 68: 55–58, 2018 **(40540)**

MIYASHITA, L., AND G. FOLEY. "E-Cigarettes and Respiratory Health: The Latest Evidence." *Journal of Physiology*, 598: 5027–5038, 2020 **(50103)**

MIYAUCHI, M., M. HAYASHIDA, K. HIRATA, K. HIRATA, AND H. YOKOTA. "Gastric Lavage Guided by Ultrathin Transnasal Esophagogastroduodenoscopy in a Life-Threatening Case of Tobacco Extract Poisoning: A Case Report." *Journal of Nippon Medical School*, 80: 307–311, 2013 **(30303)**

MOERMAN, J.W., AND G.E. POTTS. "Analysis of Metals Leached from Smoked Cigarette Litter." *Tobacco Control*, 20: i30–i35, 2011 **(40709)**

MOHAMED, N.N., S.L. LOY, C.N. MAN, A. AL-MAMUN, AND H.J.J. MOHAMED. "Higher Hair Nicotine Level in Children Compared to Mother Living with Smoking Father in Malaysia." *Environmental Health and Preventative Medicine*, 21: 572–578, 2016 **(40702)**

MOLLA, Z., L. DUBE, W. KRAHL, AND M. SOBOKA. "Tobacco Dependence Among People with Mental Illness: A Facility-based Cross-Sectional Study from Southwest Ethiopia." *BMC Research Notes*, 10: 289 (7 pp), 2017 **(80804)**

MOREAN, M.E., E.R. BUTLER, K.W. BOLD, G. KONG, D.R. CAMENGA, D.A. CAVALLO, P. SIMON, S.S. O'MALLEY, AND S. KRISHNAN-SARIN. "Preferring More E-Cigarette Flavors is Associated with E-Cigarette Use Frequency Among Adolescents but Not Adults." *Public Library of Science One*, 13: (10 pp), 2018 **(50612)**

MOREAN, M.E., G. KONG, D.A. CAVALLO, D.R. CARMENGA, AND S. KRISHNAN-SARIN. "Nicotine Concentration of E-Cigarettes Used by Adolescents." *Drug and Alcohol Dependence*, 167: 224–227, 2016 **(50604)**

MORRELL, H.E.R., L.M. COHEN, AND D.E. MCCHARGUE. "Depression Vulnerability Predicts Cigarette Smoking Among College Students: Gender and Negative Reinforcement Expectancies as Contributing Factors." *Addictive Behaviors*, 35: 607–611, 2010 **(80103)**

MOWERY, P.D., M.C. FARELL, M.L. HAVILARD, J.M. GABLE, AND H.E. WELLS. "Progression to Established Smoking Among US Youths." *American Journal of Public Health*, 94: 331-337, 2004 **(40807)**

MRAVEC, B., M. TIBENSKY, L. HORVATHOVA, AND P. BABAL. "E-Cigarettes and Cancer Risk." *Cancer Prevention and Research*, 13: 137–143, 2020 **(50436)**

MUGGLI, M.E., J.O. EBBERT, C. ROBERTSON, AND R.D. HURT. "Waking a Sleeping Giant: The Tobacco Industry's Response to the Polonium-210 Issue." *American Journal of Public Health*, 98: 1643–1650, 2008 **(40405)**

MULDER, H.A., J.L. PATTERSON, M.S. HALQUIST, L. KOSMIDER, J.B. MCGEE, J.L. POKLIS, A. POKLIS, AND M.R. PEACE. "The Effect of Electronic Cigarette User Modifications and E-Liquid Adulteration on the Particle Size Profile of an Aerosolized Product." *Scientific Reports*, 9: 10221 (8 pp), 2019 **(50206)**

MÜNDEL, T. "Nicotine: Sporting Friend or Foe? A Review of Athlete Use, Performance Consequences and Other Considerations." *Sports Medicine*, 47: 2497–2506, 2017 **(81001)**

MÜNDEL, T., S.D. HOULTHAM, M.J. BARNES, AND S.R. STANNARD. "Nicotine Supplementation Does Not Influence Performance of a 1h Cycling Time-Trial in Trained Males." *Frontiers in Physiology*, 10: (8 pp), 2019 **(81003)**

MURRAY, R.P., J.E. CONNETT, AND L.M. ZAPAWA. "Does Nicotine Replacement Therapy Cause Cancer? Evidence from the Lung Health Study." *Nicotine and Tobacco Research*, 11: 1076–1082, 2009 **(70230)**

MUSSULMAN, L.M., T.S. SCHEUERMANN, B. FASERU, N. NAZIR, AND K.P. RICHTER. "Rapid Relapse to Smoking Following Hospital Discharge." *Preventive Medicine Reports*, 15: (4 pp), 2019 **(70204)**

MYERS, M.L. "The FCTC's Evidence-Based Policies Remain a Key to Ending the Tobacco Epidemic." *Tobacco Control*, 22: i45–i46, 2013 **(70319)**

N

NAKAGATA, T., K. FUKAO, H. KOBAYASHI, S. KATAMOTO, AND H. NAITO. "The Effects of Transferal Nicotine Patches on the Cardiorespiratory and Lactate Response During Exercise from Light to Moderate Intensity: Implications for Exercise Prescription during Smoking Cessation." *Medicina*, 55: 348 (10 pp), 2019 **(81006)**

NAVON, L., C.M. JONES, I. GHINAI, B.A. KING, P.A. BRISS, K.A. HACKER, AND J.E. LAYDEN. "Risk Factors for E-Cigarette or Vaping, Product Use–Associated Lung Injury (EVALI) Among Adults Who Use E-Cigarette, or Vaping, Products: Illinois, July–October 2019." *Morbidity and Mortality Weekly Report*, 68: 1034–1039, 2019 **(50441)**

NEBOT, M., M.J. LÓPEZ, C. ARIZA, M. PÉREZ-RÍOS, M. FU, A. SCHIAFFINO, G. MUÑOZ, E. SALTÓ, AND E. FERNÁNDEZ. "Impact of the Spanish Smoking Law on Exposure to Secondhand Smoke in Offices and Hospitality Venues: Before-and-After Study." *Environmental Health Perspectives*, 117: 344–347, 2009 **(40627)**

NEGRI, E., M.G. FRANZOSI, C. LA VECCHIA, L. SANTORO, A. NOBILI, AND G. TOGNONI. "Tar Yield of Cigarettes and Risk of Acute Myocardial Infarction." *British Medical Journal*, 306: 1567–1570, 1993 **(40538)**

NGUYEN, K.H., V.T. TONG, K. MARYNAK, AND B.A. KING. "Perceptions of Harm to Children Exposed to Secondhand Aerosol from Electronic Vapor Products, Styles Survey, 2015." *Preventing Chronic Disease*, 14, E41 (11 pp), 2017 **(50508)**

NIAZ, K., F. MAQBOOL, F. KHAN, H. BAHADAR, F.L. HASSAN, AND M. ABDOLLAHI. "Smokeless Tobacco (*Paan* and *Gutkha*) Consumption, Prevalence, and Contribution to Oral Cancer." *Epidemiology and Health*, 39: (11 pp), 2017 **(60618)**

NIKTIN, D., D.S. TIMBERLAKE, AND R.S. WILLIAMS. "Is the E-Liquid Industry Regulating Itself? A Look at E-Liquid Internet Vendors in the United States." *Nicotine and Tobacco Research*, 18: 1967–1972, 2016 **(50320)**

NOAH, T.L., H. ZHOU, AND I. JASPERS. "Alteration of the Nasal Responses to Influenza Virus by Tobacco Smoke." *Current Opinion in Allergy and Clinical Immunology*, 12: 24–31, 2012 **(80308)**

NOBLE, M.J., B. LONGSTREET, R.G. HENDRICKSON, AND R. GERONA. "Unintentional Pediatric Ingestion of Electronic Cigarette Nicotine Refill Liquid Necessitating Intubation." *Annals of Emergency Medicine*, 69: 94–97, 2017 **(10413)**

NORDENSTAM, F., M. NORMAN, AND R. WICKSTRÖM. "Blood Pressure and Heart Variability in Preschool Children Exposed to Smokeless Tobacco in Fetal Life." *Journal of the American Heart Association*, 8: e012629 (8 pp), 2019 **(60708)**

NOVOTNY, T.E., AND F. ZHAO. "Consumption and Production Waste: Another Externality of Tobacco Use." *Tobacco Control*, 8: 75–80, 1999 **(40713)**

NOVOTNY, T.E., S.N. HARDIN, L.R. HOVDA, D.J. NOVOTNY, M.K. MCLEAN, AND S. KHAN. "Tobacco and Cigarette Butt Consumption in Humans and Animals." *Tobacco Control* 20: i17–i20, 2011 **(40711)**

NOWAK, D., R.A. JÖRRES, AND T. RÜTHER. "E-Cigarettes—Prevention, Pulmonary Health, and Addiction." *Deutsches Ärzteblatt International*, 111: 349–355, 2014 **(50113)**

NUORTI, J.P., J.C. BUTLER, M.M. FARLEY, L.H. HARRISON, A. MCGEER, M.S. KOLCZAK, AND R.F. BREIMAN. "Cigarette Smoking and Invasive Pneumococcal Disease." *New England Journal of Medicine*, 342: 681–689, 2000 **(80302)**

O

O'CONNOR, R.J. "Non-Cigarette Tobacco Products: What Have We Learnt and Where Are We Headed?" *Tobacco Control*, 21: 181–190, 2012 **(60608)**

O'CONNOR R.J., V.W. REES, K.J. NORTON, K.M. CUMMINGS, G.N. CONNOLLY, H.R. ALPERT, A. SJÖDIN, L. ROMANOFF, Z. LI, K.M. JUNE, AND G.A. GIOVINO. "Does Switching to Reduced Ignition Propensity Cigarettes Alter Smoking Behavior or Exposure to Tobacco Smoke Constituents?" *Nicotine and Tobacco Research*, 12: 1011–1018, 2010 **(80206)**

O'CONNOR, R.J., L.M. SCHNELLER, R.V. CARUSO, W.E. STEPHENS, Q. LI, J. YUAN, AND G.T. FONG. "Toxic Metal and Nicotine Content of Cigarettes Sold in China, 2009 and 2012." *Tobacco Control*, 24: iv55–iv59, 2015 **(10106)**

O'MALLEY, S.S., R. WU, S.T. MAYNE, AND P.I. JATLOW. "Smoking Cessation is Followed by Increases in Serum Bilirubin, an Endogenous Antioxidant Associated with Lower Risk of Lung Cancer and Cardiovascular Disease." *Nicotine and Tobacco Research*, 16: 1145–1149, 2014 **(40543)**

OH, K.H., C.M. LEE, B. OH, S-W. OH, H-K. JOH, H.C. CHOI, S.J. KIM, AND S.R. LEA. "The Relationship Between Electronic Cigarette Use with or Without Cigarette Smoking and Alcohol Use Among Adolescents: Finding From the 11th Korea Youth Risk Behavior Web-Based Survey." *Korean Journal of Family Medicine*, 40: 241–247, 2019 **(50707)**

OKAMOTO, M., T. MIYAKE, K. KITAI, S. FURUKAWA, S. YAMAMOTO, H. SENBA, S. KANZAKI, A. DEGUCHI, M. KOIZUMI, T. ISHIHARA, H. MIYAOKA, O. YOSHIDA, M. HIROOKA, T. KUMAGI, M. ABE, B. MATSUURA, AND Y. HIASA. "Cigarette Smoking is a Risk Factor for the Onset of Fatty Liver Disease in Nondrinkers: A Longitudinal Cohort Study." *Public Library of Science One*, 13: e0195147 (13 pp), 2018 **(40524)**

OLADIPUPO, O.A., D. DUTTA, AND N.S. CHONG. "Analysis of Chemical Constituents in Mainstream Bidi Smoke." *BMC Chemistry*, 13: 93 (11 pp), 2019 **(60503)**

OLDS, J.L., AND N. KABBANI. "Is Nicotine Exposure Linked to Cardiopulmonary Vulnerability to COVID-19 in the General Population?" *Federation of European Biochemical Societies Journal*, 287: 3651–3655, 2020 **(80316)**

OLSSON, A.R., T. SKOGH, AND G. WINGREN. "Comorbidity and Lifestyle, Reproductive Factors, and Environmental Exposures Associated with Rheumatoid Arthritis." *Ann. Rheum. Dis*, 60: 934-939, 2001 **(40506)**

OMAIYE, E.E., I. CORDOVA, B. DAVIS, AND P. TALBOT. "Counterfeit Electronic Cigarette Products with Mislabeled Nicotine Concentrations." *Tobacco Regulatory Science*, 3: 347–357, 2017 **(50319)**

OMAIYE, E.E., K.J. MCWHIRTER, W. LUO, P.A. TIERNEY, J.F. PANKOW, AND P. TALBOT. "High-Nicotine Electronic Cigarette Products: Toxicity of JUUL Fluids and Aerosols Correlates Strongly with Nicotine and Some Flavor Chemical Concentrations." *Chemical Research in Toxicology*: (12 pp), 2019 **(50201)**

OMAIYE, E.E., K.J. MCWHIRTER, W. LUO, P.A. TIERNEY, J.F. PANKOW, AND P. TALBOT. "High Concentrations of Flavor Chemicals are Present in Electronic Cigarette Refill Fluids." *Scientific Reports*, 9: 2468 (9 pp), 2019 **(50304)**

ONG, E.K., AND S.A. GLANTZ. "Constructing 'Sound Science' and 'Good Epidemiology': Tobacco, Lawyers, and Public Relations Firms." *American Journal of Public Health*, 91: 1749–1757, 2001 **(40115)**

ONUKI, M., K. YOKOYAMA, K. KIMURA, H. SATO, R.B. NORDIN, L. NAING, Y. MORITA, T. SAKAI, Y. KOBAYASHI, AND S. ARAKI. "Assessment of Urinary Cotinine as a Marker of Nicotine Absorption from Tobacco Leaves: A Study on Tobacco Farmers in Malaysia." *Journal of Occupational Health*, 45: 140–145, 2003 **(10121)**

ORDONEZ, J.E., K.C. KLEINSCHMIDT, AND M.B. FORRESTER. "Electronic Cigarette Exposures Reported to Texas Poison Centers." *Nicotine and Tobacco Research*, 25: 209–211, 2014 **(10408)**

OSEI, A.D., M. MIRBOLOUK, O.A. ORIMOLOYE, ET AL. "The Association Between E-Cigarette Use and Asthma Among Never Combustible Cigarette Smokers: Behavioral Risk Factor Surveillance System (BRFSS) 2016 and 2017." *BMC Pulmonary Medicine*, 19: 180 (6 pp), 2019 **(50427)**

OWENS, V.L., T. HA, AND N. SOULAKOVA. "Widespread Use of Flavored E-Cigarettes and Hookah Tobacco in the United States." *Preventative Medicine Reports*, 14: (5 pp), 2019 **(60408)**

OWILI, P.O., M.A. MUGA, AND H-W. KUO. "Gender Differences in the Association Between Environmental Tobacco Smoke and Birth Weight in Africa." *International Journal of Environmental Research and Public Health*, 15: 1409 (10 pp), 2018 **(40608)**

OWUSU, D., J. HUANG, S.R. WEAVER, T.F. PECHACEK, D.L. ASHLEY, P. NAYAK, AND M.P. ERIKSEN. "Patterns and Trends of Dual Use of E-Cigarettes and Cigarettes Among U.S. Adults, 2015–2018." *Preventive Medicine Reports*, 16: (7 pp), 2019 **(50704)**

P

PAIK, J.H., S. KANG, A. DUREY, J.H. KIM, AND A.J. KIM. "Symptomatic Bradycardia due to Nicotine Intoxication." *Revista Brasileira de terapia intensiva*, 30: 121–126, 2018 **(10419)**

PALIPUDI, K., L. MBULO, S. KOSEN, A. TJANDRA, KADARMANTO, F. QURESHI, L. ANDES, D.N. SINHA, AND S. ASMA. "A Cross Sectional Study of Kretek Smoking in Indonesia as a Major Risk to Public Health." *Asian Pacific Journal of Cancer Prevention*, 16: 6883–6888, 2015 **(60511)**

PALMISANI, J., A. DI GILIO, L. PALMIERI, C. ABENAVOLI, M. FAMELE, R. DRAISCI, AND G. DE GENNARO. "Evaluation of Second-Hand Exposure to Electronic Cigarette Vaping Under a Real Scenario: Measurements of Ultrafine Particle Number Concentration and Size Distribution and Comparison with Traditional Tobacco Smoke." *Toxics*, 7: 59 (16 pp), 2019 **(50509)**

PANKOW, J.F., K. KIM, K.J. MCWHIRTER, W. LUO, J.O. ESCOBEDO, R.M. STRONGIN, A.K. DUELL, AND D.H. PEYTON. "Benzene Formation in Electronic Cigarettes." *Public Library of Science One,* 12: (10 pp), 2017 **(50407)**

PAPASTEFANOU, C. "Radioactivity of Tobacco Leaves and Radiation Dose Induced from Smoking." *International Journal of Environmental Research and Public Health*, 6: 558–567, 2009 **(10109)**

PARASHAR, M., R. AGARWALLA, S. DWIVEDI, B. PATVAGEKAR, AND R. PATHAK. "Prevalence and Correlates of Nicotine Dependence Among Construction Site Workers: A Cross-Sectional Study in Delhi." *Lung India*, 33: 496–501, 2016 **(70111)**

PAREKH, T., S. PEMMASANI, AND R. DESAI. "Risk of Stroke with E-Cigarette and Combustible Cigarette Use in Young Adults." *American Journal of Preventive Medicine*, 53: 446–452, 2019 **(50709)**

PARK, E.J., AND Y.G. MIN. "The Emerging Method of Suicide by Electronic Cigarette Liquid: A Case Report." *Journal of Korean Medical Science*, 33: e52, 2018 **(81103)**

PARK, S-J., H-S. LIM, K. LEE, AND S-J. YOO. "Green Tobacco Sickness Among Tobacco Harvesters in a Korean Village." *Safety and Health at Work*, 9: 71–74, 2018 **(10116)**

PATEL, K., D. SCHLUNDT, C. LARSON, H. WANG, A. BROWN, AND M. HARGREAVES. "Chronic Illness and Smoking Cessation." *Nicotine and Tobacco Research*, 11: 933–939, 2009 **(70213)**

PATEL, V.D., K.B. JADHAV, V.S. SHAH, AND N.D. GUPTA. "Comparative Assessment of Salivary Cotinine Level and Psychological Dependence Among Tobacco Users." *Dental Research Journal*, 14: 125–130, 2017 **(90108)**

PATETTA, M.J., AND T.B. COLE. "A Population-Based Descriptive Study of Housefire Deaths in North Carolina." *American Journal of Public Health*, 80: 1116–1117, 1990 **(80202)**

PAULY, J.L., A.B. MEPANI, J.D. LESSES, K.M. CUMMINGS, AND R.J. STRECK. "Cigarettes with Defective Filters Marketed for 40 Years: What Philip Morris Never Told Smokers." *Tobacco Control*, 11; i51–i61, 2002 **(40203)**

PAULY, J.L., AND G. PASZKIEWICZ. "Cigarette Smoke, Bacteria, Mold, Microbial Toxins, and Chronic Lung Inflammation." *Journal of Oncology*, 2011: Article 819129 (13 pp), 2011 **(40404)**

PAULY, J.L., H.J. LEE, E.L. HURLEY, K.M. CUMMINGS, J.D. LESSES, AND R.J. STRECK. "Glass Fiber Contamination of Cigarette Filters: An Additional Health Risk to the Smoker?" *Cancer Epidemiology, Biomarkers, and Prevention*, 7: 967–979, 1998 **(40215)**

PAULY, J.L., S.J. STEGMEIER, A.G. MAYER, J.D. LESSES, AND R.J. STRECK. "Release of Carbon Granules from Cigarettes with Charcoal Filters." *Tobacco Control*, 6: 33–40, 1997 **(40204)**

PAUMGARTTEN, F. "Heat-not-Burn and Electronic Cigarettes: Truths and Untruths About Harm Reduction [Letter]." *Revista da Associação Médica Brasileira*, 64: 104–105, 2018 **(60107)**

PAUMGARTTEN, F.J.R. "A Critical Appraisal of the Harm Reduction Argument for Heat-not-Burn Tobacco Products." *Pan American Journal of Public Health*, 42: e161 (6 pp), 2018 **(60116)**

PAUWELS, C.G.G.M., W.N.M. KLERX, J.L.A. PENNINGS, A.W. BOOTS, F.J. VAN SCHOOTEN, A. OPPERHUIZEN, AND T. TALHOUT. "Cigarette Filter Ventilation and Smoking Protocol Influence Aldehyde Smoke Yields." *Chemical Research in Toxicology*, 31: 462–471, 2018 **(40210)**

PEACE, M., J. POKLIS, AND J. TURNER. "Chasing the Electronic Cigarette Dragon: Characterizing the Evolution and Impact of Design and Content." *National Institute of Justice*, DUNS Number: 10-530-0446 (19 pp), 2019 **(50116)**

PEARSON, J.L., M.S. AMATO, X. WANG, K. ZHAO, S. CHA, A.M. COHN, G.D. PAPANDONATOS, AND A.L. GRAHAM. "How US Smokers Refer to E-cigarettes: An Examination of User-Generated Posts from a Web-Based Smoking Cessation Intervention, 2008–2015." *Nicotine and Tobacco Research*, 19: 253–257, 2017 **(50801)**

PEDERSON, L.L., J. KOVAL, E. VINGILIS, J. SEELEY, A.R. IALOMITEANU, C.M. WICKENS, R. FERRENCE, AND R.E. MANN. "The Relationship Between Motor Vehicle Collisions and Cigarette Smoking in Ontario: Analysis of CAMH Survey Data from 2002 to 2016." *Preventive Medicine Reports*, 13: 327–331, 2018 **(80402)**

PEINEMANN, F., S. MOEBUS, N. DRAGANO, S. MÖHLENKAMP, N. LEHMANN, H. ZEEB, R. ERBEL, K-H. JÖCKEL, AND B. HOFFMANN. "Secondhand Smoke Exposure and Coronary Artery Calcification Among Nonsmoking Participants of a Population-Based Cohort." *Environmental Health Perspectives*, 119: 1556–1561, 2011 **(40617)**

PEPPER, J.K., K.M. RIBISL, AND N.T. BREWER. "Adolescents' Interest in Trying Flavoured E-Cigarettes." *Tobacco Control*, 25: ii62–ii66, 2016 **(50611)**

PEREZ, M.F., N.C. ATUEGWU, E.L. MEAD, C. ONCKEN, AND E.M. MORTENSEN. "Adult E-Cigarette Use Associated with a Self-Reported Diagnosis of COPD." *International Journal of Environmental Research and Public Health*, 16: 3938 (10 pp), 2019 **(50426)**

PÉREZ-DE-ARCELUS, M., E. TOLEDO, M.Á. MARTÍNEZ-GONZÁLEZ, N. MARTÍN-CALVO, A. FERNÁNDEZ-MONTERO, AND J. MORENO-MONTAÑÉS. "Smoking and Incidence of Glaucoma." *Medicine*, 96: (6 pp), 2017 **(40528)**

PEŘINOVÁ, P., E. FEKETEOVÁ, D. KEMLINK, P. KOVALSKÁ, K. CHLEBUŠOVÁ, J. NEPOŽITEK, V. IBARBURU, E. KRÁLÍKOVÁ, S. NEVŠÍMALOVÁ, AND K. ŠONKA. "Smoking Prevalence and Its Clinical Correlations in Patients with Narcolepsy-Cataplexy." *Prague Medical Report*, 117: 81–89, 2016 **(10321)**

PERRET, J.L., B. BONEVSKI, C.F. MCDONALD, AND M.J. ABRAMSON. "Smoking Cessation Strategies for Patients with Asthma: Improving Patient Outcomes." *Journal of Asthma and Allergy*, 9: 117–128, 2016 **(70211)**

PESTA, D.H., S.S. ANGADI, M. BURTSCHER, AND C.K. ROBERTS. "The Effects of Caffeine, Nicotine, Ethanol, and Tetrahydrocannabinol on Exercise Performance." *Nutrition and Metabolism*, 10: 71–85, 2013 **(81002)**

PETERS, E.N., R.P. SCHWARTZ, S. WANG, K.E. O'GRADY, AND C. BLANCO. "Psychiatric, Psychosocial, and Physical Health Correlates of Co-Occurring Cannabis Use Disorders and Nicotine Dependence." *Drug and Alcohol Dependence*, 134: 228–234, 2014 **(60903)**

PETERS, R., R. POULTER, J. WARNER, N. BECKETT, L. LURCH, AND C. BULPITT. "Smoking, Dementia and Cognitive Decline in the Elderly, a Systematic Review." *BMC Geriatrics*, 8: 36 (7 pp), 2008 **(40530)**

PETRESCU, D.C., M. VASILJEVIC, J.K. PEPPER, K.M. RIBISL, AND T.M. MARTEAU. "What is the Impact of E-Cigarette Adverts on Children's Perceptions of Tobacco Smoking? An Experimental Study." *Tobacco Control*, 26: 421–427, 2017 **(50627)**

PETRIDOU, E., N. KOURI, A. POLYCHRONOPOULOU, K. SIAFAS, M. STOIKIDOU, AND D. TRICHOPOULOS. "Risk Factors for Childhood Poisoning: A Case-Control Study in Greece." *Injury Prevention*, 2: 208–211, 1996 **(10403)**

PETTICREW, M.P., K. LEE, AND M. MCKEE. "Type A Behavior Pattern and Coronary Heart Disease: Philip Morris's 'Crown Jewel.'" *American Journal of Public Health*, 102: 2018–2025, 2012 **(40111)**

PICCIOTTO, M.R., AND Y.S. MINEUR. "Molecules and Circuits Involved in Nicotine Addiction: The Many Faces of Smoking." *Neuropharmacology*, 76: (22 pp), 2014 **(10202)**

PICHINI, S., O. GARCIA-ALGAR, L. MUÑOZ, O. VALL, R. PACIFICI, C. FIGUEROA, J.A. PASCUAL, D. DIAZ, AND J. SUNYER. "Assessment of Chronic Exposure to Cigarette Smoke and Its Change During Pregnancy by Segmental Analysis of Maternal Hair Nicotine." *Journal of Exposure Analysis and Environmental Epidemiology*, 13: 144–151, 2003 **(30504)**

PIERCE, J.P., E.A. GILPIN, AND W.S. CHOI. "Sharing the Blame: Smoking Experimentation and Future Smoking-Attributable Mortality due to Joe Camel and Marlboro Advertising and Promotions." *Tobacco Control*, 8: 37–44, 1999 **(40802)**

PIETRA, T., A. WITUSIK, M. PANEK, J. SZEMRAI, AND P. GÓRSKI. "Anxiety, Depression and Methods of Stress Coping in Patients with Nicotine Dependence Syndrome." *Medical Science Monitor*, 17: CR272–CR276, 2011 **(80106)**

PISINGER, C., N. GODTFREDSEN, AND A.M. BENDER. "A Conflict of Interest is Strongly Associated with Tobacco Industry-Favourable Results, Indicating no Harm of E-Cigarettes." *Preventative Medicine*, 119: 124–131, 2019 **(50401)**

PITARD, A., P. BRENNAN, J. CLAVEL, E. GREISER, G. LOPEZ-ABENTE, J. CHANG-CLAUDE, J. WAHRENDORF, C. SERRA, M. KOGEVINAS, AND P. BOFFETTA. "Cigar, Pipe, and Cigarette Smoking and Bladder Cancer Risk in European Men." *Cancer Causes and Control*, 12: 551–556, 2001 **(60306)**

POKLIS, J.L., C.E. WOLF II, AND M.R. PEACE. "Ethanol Concentration in 56 Refillable Electronic Cigarettes Liquid Formulations Determined by Headspace Gas Chromatography with Flame Ionization Detector (HS-GC-FID)." *Drug Testing and Analysis*, 9: 1637–1640, 2017 **(50317)**

POLLAY, R.W., AND T. DEWHIRST. "A Premiere Example of the Illusion of Harm Reduction Cigarettes in the 1990s." *Tobacco Control*, 12: 322–332, 2003 **(40216)**

POMERLEAU, C.S., AND O.F. POMERLEAU. "Euphoriant Effects of Nicotine in Smokers." *Psychopharmacology*, 108: 460–465, 1992 **(10212)**

POMPERLEAU, O.F., M. HARIHARAN, C.S. POMERLEAU, O.G. CAMERON, AND S.K. GUTHRIE. "Differences Between Smokers and Never-Smokers in Sensitivity to Nicotine: A Preliminary Report." *Addiction*, 88: 113–118, 1993 **(20404)**

POPE III, C.A., R.T. BURNETT, M.C. TURNER, A. COHEN, D. KREWSKI, M. JERRETT, S.M. GAPSTUR, AND M.J. THUN. "Lung Cancer and Cardiovascular Disease Mortality Associated with Ambient Air Pollution and Cigarette Smoke: Shape of the Exposure-Response Relationships." *Environmental Health Perspectives*, 119: 1616–1621, 2011 **(40610)**

POULSEN, P.B., H. SPILLEMOSE, G. NIELSEN, L-L. HERGEL, D. WEDELL-WEDELLSBORG, M. STRAND, AND T. RINGBÆK. "Real-Life Effectiveness of Smoking-Cessation Treatments in General Practice Clinics in Denmark. The Escape Smoke Project." *Respiratory Medicine*, 109: 218–227, 2015 **(70203)**

PRAKASH, B., AND U. VYAS. "Association Between Occupational Tobacco Exposure of Health Hazards in Women Laborers of Bidi-Industry of Ajmer." *International Journal of Pharma Medicine and Biological Sciences*, 2: 58-61, 2013 **(60508)**

PRÉVÔT, N., F. DE OLIVEIRA, S. PERINEL-RAGEY, T. BASSET, J-M. VERGNON, AND J. POURCHEZ. "Nicotine Delivery from the Refill Liquid to the Aerosol via High-Power E-Cigarette Device." *Scientific Reports*, 7: 2592 (7 pp), 2017 **(50207)**

PRICE, T.B., S. KRISHNAN-SARIN, AND D.L. ROTHMAN. "Smoking Impairs Muscle Recovery from Exercise." *American Journal of Physiology, Endocrinology and Metabolism*, 285: E116–E122, 2003 **(81008)**

PROCHASKA, J.J., AND N.L. BENOWITZ. "Current Advances in Research in Treatment and Recovery: Nicotine Addiction." *Science Advances*, 5: (23 pp), 2019 **(10207)**

PUBLIC LIBRARY OF SCIENCE ONE. 14: e0220300 (15 pp), 2019 **(20107)**

PULVERS, K., J.Y. SUN, Y-L. ZHUANG, G. HOLGUIN, AND S-H. ZHU. "Internet-Based Advertising Claims and Consumer Reasons for Using Electronic Cigarettes by Device Type in the US." *Tobacco Regulatory Science*, 3: 516–524, 2017 **(50204)**

Q

QASIM, H., A.B. ALARABI, K.H. ALZOUBI, Z.A. KARIM, F.Z. ALSHBOOL, AND F.T. KHASAWNEH. "The Effects of Hookah/Waterpipe Smoking on General Health and the Cardiovascular System." *Environmental Health and Preventative Medicine*, 24: 58 (17 pp), 2019 **(60413)**

QASIM, H., Z.A. KARIM, J.O. RIVERA, F.T. KHASAWNEH, AND F.Z. ALSHBOOL. "Impact of Electronic Cigarettes on the Cardiovascular System." *Journal of the American Heart Association*, 117: (14 pp), 2017 **(50409)**

QUIGLEY, H., AND J.H. MACCABE. "The Relationship Between Nicotine and Psychosis." *Therapeutic Advances in Psychopharmacology*, 9: (13 pp), 2019 **(80801)**

QUIROGA, L., M. ASIF, T. LAGZIEL, D. BHAT, AND J. CAFFREY. "E-Cigarette Battery Explosions: Review of the Acute Management of the Burns and the Impact on Our Population." *Cureus,* 11: e5355 (9 pp), 2019 **(80210)**

R

RABINOFF, M., N. CASKEY, A. RISSLING, AND C. PARK. "Pharmacological and Chemical Effects of Cigarette Additives." *American Journal of Public Health*, 97: 1981–1991, 2007 **(40407)**

RAHMANIAN, S.D., P.T. DIAZ, AND M.E. WEWERS. "Tobacco Use and Cessation Among Women: Research and Treatment-Related Issues." *Journal of Women's Health*, 20: 349–357, 2011 **(70221)**

RANDI, G., L. SCOTTI, C. BOSETTI, R. TALAMINI, E. NEGRI, F. LEVI, S. FRANCESCHI, AND C. LA VECCHIA. "Pipe Smoking and Cancers of the Upper Digestive Tract." *International Journal of Cancer*, 121: 2049–2051, 2007 **(60305)**

RANTAKALLIO, P., E. LÄÄRÄ, AND M. KOIRANEN. "A 28 Year Follow-up of Mortality Among Women Who Smoked During Pregnancy." *British Medical Journal,* 311: 477–480, 1995 **(80709)**

RAO, P.D., H. NANDING, A.A. STRASSER, AND P.M. WISE. "Pilot Experiment: The Effect of Added Flavorants on the Taste and Pleasantness of Mixtures of Glycerol and Propylene Glycol." *Chemosensory Perception*, 11: 1–9, 2018 **(50307)**

RATSCH, A.M., A. MASON, L. RIVE, F.E. BOGOSSIAN, AND K.J. STEADMAN. "The Pituri Learning Circle: Central Australian Aboriginal Women's Knowledge and Practices Around the Use of *Nicotiana* spp., as a Chewing Tobacco." *Rural and Remote Health*, 17: 4044 (15 pp), 2017 **(10104)**

RAY, C.S., AND P.C. GUPTA. "Bidis and Smokeless Tobacco." *Current Science*, 96: 1324–1334, 2009 **(60505)**

REILLY, S.M., R. GOEL, Z. BLITZER, R.J. ELIAS, J. FOULDS, J. MUSCAT, AND J.P. RITCHIE JR. "Little Cigars, Filtered Cigars, and their Carbonyl Delivery Relative to Cigarettes." *Nicotine and Tobacco Research*, 20: S99–S106, 2018 **(60214)**

REN, M., AND S. LOTFIPOUR. "Nicotine Gateway Effects on Adolescent Substance Use." *Western Journal of Emergency Medicine*, 20: 696–709, 2019 **(60901)**

REN, M., M. ZHANG, H. YANG, AND H. SHI. "Reducing the Nicotine Content of Tobacco by Grafting with Eggplant." *BMC Plant Biology*, 20: 285 (10 pp), 2020 **(10115)**

RICHMOND, S.A., I. PIKE, J.L. MAGUIRE, AND A. MACPHERSON. "E-Cigarettes: A New Hazard for Children and Adolescents." *Paediatrics and Child Health*, 23: 255–259, 2018 **(10421)**

RIGOTTI, N.A. "Strategies to Help a Smoker Who is Struggling to Quit." *Journal of the American Medical Association*, 308: 1573–1580, 2012 **(70201)**

RIJHWANI, K., S. HASHMI, V.R. MOHANTY, A.Y. BALAPPANAVAR, AND S. KAPOOR. "Hookah Products and Online Retail Marketing Strategies in India: A Content Analysis." *Indian Journal of Cancer*, 55: 261–264, 2018 **(60409)**

RINTAKOSKI, K., J. AHLBERG, C. HUBLIN, F. LOBBEZOO, R.J. ROSE, H. MURTOMAA, AND J. KAPRIO. "Tobacco Use and Reported Bruxism in Young Adults: A Nationwide Finnish Twin Cohort Study." *Nicotine and Tobacco Research*, 12: 679–683, 2010 **(80505)**

RIORDAN, M., G. RYLANCE, AND K. BERRY. "Poisoning in Children 5: Rare and Dangerous Poisons." *Archives of Disease in Childhood*, 87: 407–410, 2002 **(10404)**

RIQUINHO, D.L., AND E.A. HENNINGTON. "Health, Environment and Working Conditions in Tobacco Cultivation: A Review of the Literature." *Revista Ciência & Saúde Coletiva*, 17: 1587–1600, 2012 **(10123)**

RISI, S. "On the Origins of the Electronic Cigarette: British American Tobacco's Project Ariel (1962–1967)." *American Journal of Public Health*, 107: 1060–1067, 2017 **(60101)**

ROBERTS, W., K.E. MOORE, M.R. PELTIER, T.L. VERPLAETSE, L. OBERLEITNER, R. HACKER, AND S.A. MCKEE. "Electronic Cigarette Use and Risk of Harmful Alcohol Consumption in the US Population." *Alcoholism, Clinical and Experimental Research*, 42: 2385–2393, 2018 **(60914)**

ROBINSON, A., H. SWEETING, AND K. HUNT. "UK News Media Representations of Smoking, Smoking Policies and Tobacco Bans in Prisons." *Tobacco Control*, 27: 622-630, 2018 **(40632)**

RODGERS, K.M., L.R. SWETSCHINSKI, R.E. DODSON, H.R. ALPERT, J.M. FLEMING, AND R.A. RUDEL. "Health Toll from Open Flame and Cigarette-Started Fires on Flame-Retardant Furniture in Massachusetts, 2003–2016." *American Journal of Public Health*, 109: 1205–1211, 2019 **(80205)**

RODITIS, M.L., C. JONES, A.P. DINEVA, AND T.N. ALEXANDER. "Lessons on Addiction Messages from the Real Cost Campaign." *American Journal of Preventive Medicine*, 56 (2S1): S24-S30, 2019 **(40808)**

ROELANDS, J., M.G. JAMISON, A.D. LYERLY, AND A.H. JAMES. "Consequences of Smoking During Pregnancy on Maternal Health." *Journal of Women's Health*, 18: 867–872, 2009 **(80707)**

ROH, S. "Scientific Evidence of the Addictiveness of Tobacco and Smoking Cessation in Tobacco Litigation." *Journal of Preventative Medicine and Public Health*, 51: 1–5, 2018 **(10201)**

ROMITO, L.M., M.K. SAXTON, L.L. COAN, AND A.G. CHRISTEN. "Retail Promotions and Perceptions of R.J. Reynolds' Novel Dissolvable Tobacco in a US Test Market." *Harm Reduction Journal*, 8: 10 (10 pp), 2011 **(60613)**

ROSBROOK, K., H.C. ERYTHROPEL, T.M. DEWINTER, M. FALINSKI, S. O'MALLEY, S. KRISHNAN-SARIN, P.T. ANASTAS, J.B. ZIMMERMAN, AND B.G GREEN. "The Effect of Sucralose on Flavor Sweetness in Electronic Cigarettes Varies Between Delivery Devices." *Public Library of Science One*, 12: (14 pp), 2017 **(50210)**

ROSENBERRY, Z.R., W.B. PICKWORTH, AND B. KOSZOWSKI. "Large Cigars: Smoking Topography and Toxicant Exposure." *Nicotine and Tobacco Research*, 183–191, 2018 **(60203)**

ROSS, K.C., N.R. GUBNER, R.F. TYNDALE, L.W. HAWK JR., C. LERMAN, T.P. GEORGE, P. CINCIRIPINI, R.A. SCHNOLL, AND N. BENOWITZ. "Racial Differences in the Relationship Between the Rate of Nicotine Metabolism and Nicotine Intake from Cigarette Smoking." *Pharmacology, Biochemistry, and Behavior*, 148: 1–7, 2016 **(20304)**

ROSTRON, B.L., C.G. COREY, AND R.M. GINDI. "Cigar Smoking Prevalence and Morbidity Among US Adults 2000–2015." *Preventive Medicine Reports*, 14: (4 pp), 2019 **(60204)**

ROSTRON, B.L., C.G. COREY, E. HOLDER-HAYES, AND B.K. AMBROSE. "Estimating the Potential Health Impact of Prohibiting Characterizing Flavors in Cigars Throughout the US." *International Journal of Environmental Research and Public Health*, 16: 3234 (7 pp), 2019 **(60209)**

ROSTRON, B.L., M.J. SCHROEDER, AND B.K. AMBROSE. "Dependence Symptoms and Cessation Intentions Among US Adult Daily Cigarette, Cigar, and E-Cigarette Users, 2012–2013." *BMC Public Health*, 16: 814 (10 pp), 2016 **(70106)**

RUBENSTEIN, M.L., K. DELUCCHI, N.L. BENOWITZ, AND D.E. RAMO. "Adolescent Exposure to Toxic Volatile Organic Chemicals from E-Cigarettes." *Pediatrics*, 141: (9 pp), 2018 **(50626)**

RUBENSTEIN, M.L., N.L. BENOWITZ, G.M. AUERBACK, AND A-B. MOSCIKI. "Rates of Nicotine Metabolism and Withdrawal Symptoms in Adolescent Light Smokers." *Pediatrics*, 122: e643–647, 2008 **(20302)**

RUBINSTEIN, M.L., M.A. RAIT, AND J.J. PROCHASKA. "Frequent Marijuana Use is Associated with Greater Nicotine Addiction in Adolescent Smokers." *Drug and Alcohol Dependence*, 141: 159–162, 2014 **(60905)**

RUIZ-ESQUIDE, V., AND R. SAMNARTÍ. "Tobacco and Other Environmental Risk Factors in Rheumatoid Arthritis." *Reumatologia Clinica*, 8: 342–350, 2012 **(40502)**

RUSSELL, M.A., M. JARVIS, R. IYER, AND C. FEYERABEND. "Relation of Nicotine Yield of Cigarettes to Blood Nicotine Concentrations in Smokers." *British Medical Journal*, 280: 972–976, 1980 **(20105)**

RUSSELL, M.A., M.J. JARVIS, G. JONES, AND C. FEYERABEND. "Non-Smokers Show Acute Tolerance to Subcutaneous Nicotine." *Psychopharmacology*, 102: 56–58, 1990 **(10206)**

RUSSO, P., S. BONASSI, R. GIACCONI, M. MALAVOLTA, C. TOMINO, AND F. MAGGI. "COVID-19 and Smoking: Is Nicotine the Hidden Link?" *European Respiratory Journal*, 55: 2001116 (3 pp), 2020 **(80312)**

S

SAADAT, S., AND M. KARBAKHSH. "Association of Waterpipe Smoking and Road Traffic Crashes." *BMC Public Health,* 10: 639 (6 pp), 2010 **(80404)**

ŠAGUD, M., B. VUKSAN-ĆUSA, N. JAKŠIĆ, A. MIHALJEVIĆ-PELEŠ, M.R. KUZMAN, AND N. PIVAC. "Smoking in Schizophrenia: An Updated Review." *Psychiatria Danubina*, 30: 216–223, 2018 **(80808)**

SAHIN, M.O., V. SEN, G. GUNDUZ, AND O. UCER. "Effect of Smoking Cessation on Sexual Functions in Men Aged 30 to 60 Years." *International Brazilian Journal of Urology*, 46: 642–648, 2020 **(80904)**

SAJI, S., S.S. PATIL, M. ALLEYN, R. LOCKEY, AND N. KOLLIPUTI. "Nicotine in E-Cigarette Smoke: Cancer Culprit?" *Journal of Cell Communication and Signaling*, 14: 127–128, 2020 **(10307)**

SALEEON, T., W. SIRIWONG, H.L. MALDONADO-PÉREZ, AND M.G. ROBSON. "Green Tobacco Sickness Among Thai Traditional Tobacco Farmers, Thailand." *International Journal of Occupational and Environmental Medicine*, 6: 169–176, 2015 **(10117)**

SALEEON, T., W. SIRIWONG, H.L. MALDONADO-PÉREZ, AND M.G. ROBSON. "Salivary Cotinine Levels as a Biomarker for Green Tobacco Sickness in Dry Tobacco Production Among Thai Traditional Tobacco Farmers." *Roczniki Państwowego Zakładu Higieny*, 67: 121–130, 2016 **(10122)**

SALMAN, R., S. TALIH, R. EL-HAGE, C. HADDAD, N. KAROGHLANIAN, A. EL HELLANI, N.A. SALIBA, AND A. SHIHADEH. "Free-Base and Total Nicotine, Reactive Oxygen Species, and Carbonyl Emissions from IQOS, A Heated Tobacco Product." *Nicotine and Tobacco Research*, 21: 1285–1288, 2019 **(60111)**

SANCHEZ, P., J-L DUCASSÉ, M. LAPEYRE-MESTRE, O. MARTINET, P. ROUGÉ, M.F. JORDA, AND B. CATHALA. "Nicotine Poisoning as a Cause of Cardiac Arrest?" *Clinical Toxicology,* 34: 475–476, 1996 **(10423)**

SANCHEZ-RAMINEZ, D.C., AND D. MACKEY. "Underlying Respiratory Diseases, Specifically COPD, and Smoking are Associated with Severe COVID-19 Outcomes: A Systematic Review and Meta-Analysis." *Respiratory Medicine*, 171: (10 pp), 2020 **(80315)**

SANNER, T., AND T.K. GRIMSRUD. "Nicotine: Carcinogenicity and Effects on Response to Cancer Treatment—A Review." *Frontiers in Oncology*, 5: 196 (10 pp), 2015 **(10302)**

SANTOS, U.P. "Electronic Cigarettes—The New Playbook and Revamping of the Tobacco Industry." *Jornal Brasileiro de Pneumologia*, 44: 345–346, 2018 **(50102)**

SARGENT, R.P., R.M. SHEPARD, AND S.A. GLANTZ. "Reduced Incidence of Admissions for Myocardial Infarction Associated with Public Smoking Ban: Before and After Study." *British Medical Journal*, 328: 977–980, 2004 **(40622)**

SATORA, L., H. GOSZCZ, E. GOMÓŁKA, AND W. BIEDROŃ. "Green Tobacco Sickness in Poland." *Polskie Archiwum Medycyny Wewnetrznej*, 119: 184–185, 2009 **(10118)**

SATTESON, E.S., N.J. WALKER, C.J. TUOHY, AND J.A. MOLNAR. "Extensive Hand Thermal and Blast Injury from Electronic Cigarette Explosions: A Case Report." *Hand*, 13: NP1–NP5, 2018 **(80211)**

SAVANT, S.C., H.G. SHAH, V. SODANI, AND D. BAJAJ. "Nicotine Hair Test- A Potential Biomarker for Smoke Analysis." *Bhavnagar University Journal of Dentistry*, 3: 56-60, 2013 **(30501)**

SCHEFFELS, J., AND I. LUND. "Cute as Candy: A Qualitative Study of Perceptions of Snus Branding and Package Design Among Youth in Norway." *BMJ Open*, 7: e012837 (7 pp), 2017 **(60703)**

SCHERMAN, A., J.E. TOLOSA, AND C. MCEVOY. "Smoking Cessation in Pregnancy: A Continuing Challenge in the United States." *Therapeutic Advances in Drug Safety*, 9: 457–474, 2018 **(70101)**

SCHICK, S., AND S. GLANTZ. "Philip Morris Toxicological Experiments with Fresh Sidestream Smoke: More Toxic Than Mainstream Smoke." *Tobacco Control*, 14: 396–404, 2005 **(40601)**

SCHIER, J.G., J.G. MEIMAN, J. LAYDEN, C.A. MIKOSZ, B. VANFRANK, B.A. KING, P.P. SALVATORE, D.N. WEISSMAN, J. THOMAS, P.C. MELSTROM, G.T. BALDWIN, E.M. PARKER, E.A. COURTNEY-LONG, V.P. KRISHNASAMY, C.M. PICKENS, M.E. EVANS, S.V. TSAY, K.M. POWELL, E.A. KIERNAN, K.L. MARYNAK, J. ADJEMIAN, K. HOLTON, B.S. ARMOUR, L.J. ENGLAND, P.A. BRISS, D. HOURY, K.A. HACKER, S. REAGAN-STEINER, S. ZAKI, AND D. MEANEY-DELMAN. "Severe Pulmonary Disease Associated with Electronic-Cigarette-Product Use: Interim Guidance." *Morbidity and Mortality Weekly Report*, 68: 787–790, 2019 **(50444)**

SCHIPPER, E.M., L.C.G. DE GRAAFF, B.C.P. KOCH, Z. BRKIC, E.B. WILMS, J. ALSMA, AND S.C.E. SCHUIT. "A New Challenge: Suicide Attempt Using Nicotine Fillings for Electronic Cigarettes." *British Journal of Clinical Pharmacology*, 78: 1469–1471, 2014 **(10417)**

SCHNEIDER, N.K., AND S.A. GLANTZ. "'Nicotine Nazis Strike Again': A Brief Analysis of the Use of Nazi Rhetoric in Attacking Tobacco Control Advocacy." *Tobacco Control*, 17: 291–296, 2008 **(70305)**

SCHNELLER, L.M., B.R. LINDGREN, P.G. SHIELDS, D.K. HATSUKAMI, AND R.J. O'CONNOR. "Strong Preference for Mint Snus Flavor Among Research Participants." *Addictive Behavior Reports*, 6: 51–55, 2017 **(60705)**

SCHNELLER, L.M., M. BANSAL-TRAVERS, M.L. GONIEWICZ, S. MCINTOSH, D. OSSIP, AND R.J. O'CONNOR. "Use of Flavored E-Cigarettes and the Type of E-Cigarette Devices Used Among Adults and Youth in the US: Results from Wave 3 of the Population Assessment of Tobacco and Health Study (2015–2016)." *International Journal of Environmental Research and Public Health*, 16: 2991 (12 pp), 2019 **(50610)**

SCHNOLL, R.A., F. PATTERSON, E.P. WILEYTO, R.F. TYNDALE, N. BENOWITZ, AND C. LERMAN. "Nicotine Metabolite Rate Predicts Successful Cessation with Transdermal Nicotine: A Validation Study." *Pharmacology, Biochemitry, and Behavior*, 92: 6–11, 2009 **(20301)**

SCHOLZ, J., P.C.J.L. SANTOS, C.G. BUZO, N.H.M. LOPES, T.O. ABE, P.V. GAYA, H. PIERRI, C. AMORIM, AND A.C. PEREIRA. "Effects of Aging on the Effectiveness of Smoking Cessation Medication." *Oncotarget*, 7: 30032–30036, 2016 **(70222)**

SEIDENBERG, A.B., C.L. JO, AND K.M. RIBISL. "Knowledge and Awareness of Added Sugar in Cigarettes." *Nicotine and Tobacco Research*, 21: 1689–1694, 2019 **(40402)**

SEIDENBERG, A.B., E.N. ORLAN, M.J. TRAVERS, AND E.L. SUTFIN. "Air Quality and Presence of Air Ventilation Systems Inside Waterpipe Cafés in North Carolina." *Tobacco Control*, 28: 356–358, 2019 **(60410)**

SEIDENBERG, A.B., O.A. AYO-YUSUF, AND V.W. REES. "Characteristics of 'American Snus' and Swedish Snus Products for Sale in Massachusetts, USA." *Nicotine and Tobacco Research*, 20: 262–266, 2018 **(60702)**

SEITZ, C.M., AND Z. KABIR. "Burn Injuries Caused by e-Cigarette Explosions: A Systematic Review of Published Cases." *Tobacco Prevention and Cessation*, 4: 32 (8 pp), 2018 **(80209)**

SELMAR, D., U.H. ENGELHARDT, S. HÄNSEL, C. THRÄNE, M. NOWAK, AND M KLEINWÄCHTER. "Nicotine Uptake by Peppermint Plants as a Possible Source of Nicotine in Plant-Derived Products." *Agronomy for Sustainable Development*, 35: 1185–1190, 2015 **(10504)**

SEMPLE, S., H. SWEETING, E. DEMOU, G. LOGAN, R. O'DONNELL, AND K. HUNT. "Characterising the Exposure of Prison Staff to Second-Hand Tobacco Smoke." *Annals of Work Exposures and Health*, 61: 809–821, 2017 **(40630)**

SEO, A.D., D.C. KIM, H.J. YU, AND M.J. KANG. "Accidental Ingestion of E-Cigarette Liquid Nicotine in a 15-Month-Old Child: An Infant Mortality Case of Nicotine Intoxication." *Korean Journal of Pediatrics*, 59: 490–493, 2016 **(10422)**

SEYS, L.J.M., W. WIDAGDO, F.M. VERHAMME, A. KLEINJAN, W. JANSSENS, G.F. JOOS, K.R. BRACKE, B.L. HAAGMANS, AND G.G. BRUSSELLE. "DPP4, the Middle East Respiratory Syndrome Coronavirus Receptor, is Upregulated in Lungs of Smokers and

Chronic Obstructive Pulmonary Disease Patients." *Clinical Infectious Diseases*, 66: 45–53, 2018 **(80310)**

SHAHAB, L., E. MORTIMER, L. BAULD, J.A. MCGOWAN, A. MCNEILL, AND R.F. TYNDALE. "Characterising the Nicotine Metabolite Ratio and Its Association with Treatment Choice: A Cross Sectional Analysis of Stop Smoking Services in England." *Scientific Reports*, 7: 17613 (9 pp), 2017 **(70214)**

SHAIK, F.B., G. NAGAJOTHI, K. SWARNALATHA, C.S. KUMAR, AND N. MADDU. "Quantification of Nicotine and Cotinine in Plasma, Saliva, and Urine by HPLC Method in Chewing Tobacco Users." *Asian Pacific Journal of Cancer Prevention*, 20: 3617–3623, 2019 **(30103)**

SHAKLEYA, D.M., AND M.A. HUESTIS. "Optimization and Validation of a Liquid Chromatography-Tandem Mass Spectrometry Method for the Simultaneous Quantification of Nicotine, Cotinine, Trans-3'-Hydroxycotinine and Norcotinine in Human Oral Fluid." *Analytical and Bioanalytical Chemistry*, 395: 2349–2357, 2009 **(30401)**

SHARMA, P., N. SANE, S.D. ANAND, P. MARIMUTTHU, AND V. BENEGAL. "Assessment of Cotinine in Urine and Saliva of Smokers, Passive Smokers, and Nonsmokers: Method Validation Using Liquid Chromatography and Mass Spectrometry." *Indian Journal of Psychiatry*, 61: 270–276, 2019 **(90104)**

SHARMA, S., S.K. MISHRA, AND N. MITTAL. "Influence of Tobacco Dependence on Caries Development in Young Adult Males: A Cross-Sectional Study." *Journal of Conservative Dentistry*, 21: 597–601, 2018 **(80504)**

SHEEN, S.J. "Detection of Nicotine in Foods and Plant Materials." *Journal of Food Science*, 53: 1572–1573, 1988 **(10502)**

SHERIF, N.A., S.M. KAMEL, O.S. AL-ASHKAR, O.A. SHARAKI, E.A. SEIF, AND E.A. HEGAZY. "Detection of Cotinine in Neonate Meconium as a Marker for Nicotine Exposure in utero." *East Mediterranean Health Journal*, 10: 96–105, 2004 **(80704)**

SHEU, R., C. STÖNNER, J.C. DITTO, T. KLÜPFEL, J. WILLIAMS, AND D.R. GENTNER. "Human Transport of Thirdhand Tobacco Smoke: A Prominent Source of Hazardous Air Pollutants into Indoor Nonsmoking Environments." *Science Advances*, 6: (14 pp), 2020 **(40704)**

SHI, Y., S.E. CUMMINS, AND S-H. ZHU. "Use of Electronic Cigarettes in Smoke-Free Environments." *Tobacco Control*, 26: e19–e22, 2017 **(50506)**

SIDDIQI, K., S. SHAH, S.M. ABBAS, A. VIDYASAGARAN, M. JAWAD, O. DOGAR, AND A. SHEIKH. "Global Burden of Disease Due to Smokeless Tobacco Consumption in Adults: Analysis of Data from 113 Countries." *BMC Medicine*, 13: 194 (22 pp), 2015 **(60602)**

SIEGEL, D.A., T.C. JATLAOUI, E.H. KOUMANS, E.A. KIERNAN, M. LAYER, J.E. CATES, A. KIMBALL, D.N. WEISSMAN, E.E. PETERSEN, S. REAGAN-STEINER, S. GODFRED-CATO,

D. MOULIA, E. MORITZ, J.D. LEHNERT, J. MITCHKO, J. LONDON, S.R. ZAKI, B.A. KING, C.M. JONES, A. PATEL, D.M. DELMAN, AND R. KOPPAKA. "Update: Interim Guidance for Health Care Providers Evaluating and Caring for Patients with Suspected E-cigarette, or Vaping, Product Use Associated Lung Injury: United States, October 2019." *Morbidity and Mortality Weekly Report*, 68: 919–927, 2019 **(50438)**

SIKORSKA-JAROSZYŃSKA, M.H.J., M. MIELNIK-BŁASZCZAK, D. KRAWCZYK, A. NASIŁOWSKA-BARUD, AND J. BLASZCZAK. "Passive Smoking as an Environmental Risk Factor." *Annals of Agricultural and Environmental Medicine*, 19: 547–550, 2012 **(10103)**

SILVA, A.P., J. SCHOLZ, T.O. ABE, G.G. PINHEIRO, P.V. GAYA, A.C. PERIERA, AND P.C.J.L. SANTOS. "Influence of Smoking Cessation Drugs on Blood Pressure and Heart Rate in Patients with Cardiovascular Disease or High Risk Score: Real Life Setting." *BMC Cardiovascular Disorders*, 16: (6 pp), 2016 **(70228)**

SINGH, I., A. SINGH, R. KOUR, A. SINGH, R. SINGH, AND A. BALI. "Is Sodium Carbonate in Snuff a Causative Factor for Oral Mucosal Lesions: A Cross-Sectional Analysis." *Journal of International Society of Preventive and Community Dentistry*, 8: 339–342, 2018 **(60605)**

SINGH, K.P., G. LAWYER, T. MUTHUMALAGE, K.P. MAREMANDA, N.A. KHAN, S.R. MCDONOUGH, D. YE, S. MCINTOSH, AND I. RAHMAN. "Systematic Biomarkers in Electronic Cigarette Users: Implications for Noninvasive Assessment of Vaping-Associated Pulmonary Injuries." *European Respiratory Journal Open Research*, 5: 00182-2019 (12 pp), 2019 **(50433)**

SINGH, S., M. SOUMYA, A. SAINI, V. MITTAL, U.V. SINGH, AND V. SINGH. "Breath Carbon Monoxide Levels in Different Forms of Smoking." *Indian Journal of Chest Diseases and Allied Sciences*, 53: 25–28, 2011 **(60308)**

SINHA, D.N., P.C. GUPTA, A. KUMAR, D. BHARTIYA, N. AGARWAL, S. SHARMA, H. SINGH, M. PARASCANDOLA, AND R. MEHROTRA. "The Poorest of the Poor Suffer the Greatest Burden from Smokeless Tobacco Use: A Study from 140 Countries." *Nicotine and Tobacco Research*, 1529–1532, 2018 **(60603)**

SITAS, F., M. URBAN, D. BRADSHAW, D. KIELOWSKI, S. BAH, AND R. PETO. "Tobacco Attributable Deaths in South Africa." *Tobacco Control*, 13: 396–399, 2004 **(40512)**

SLADE, J., G.N. CONNOLLY, AND D. LYMPERIS. "Eclipse: Does It Live Up to its Health Claims?" *Tobacco Control*, 11: ii64–ii70, 2002 **(50107)**

SLAUGHTER, E., R.M. GERSBERG, K. WATANABE, J. RUDOLPH, C. STRANSKY, AND T.E. NOVOTNY. "Toxicity of Cigarette Butts, and Their Chemical Components, to Marine and Freshwater Fish." *Tobacco Control*, 20: i25–i29, 2011 **(40710)**

SLOMP, F.M., T.S. BARA, G.L. PICHARSKI, AND M.L. CORDEIRO. "Association of Cigarette Smoking With Anxiety, Depression, And Suicidal Ideation Among

Brazilian Adolescents." *Neuropsychiatric Disease and Treatment*, 15: 2799–2808, 2019 **(80101)**

SLOWIK, N., S. MA, J. HE, Y.Y. LIN, O.P. SOLDIN, R.A. ROBBINS, AND G.M. TURINO. "The Effect of Secondhand Smoke Exposure on Markers of Elastin Degradation." *Chest*, 140: 946–953, 2011 **(40609)**

SMITH, E.A., AND R.E. MALONE. "An Argument for Phasing Out Sales of Cigarettes." *Tobacco Control*, 29: 703–708, 2019 **(70318)**

SMITH, E.A., AND R.E. MALONE. "'Everywhere the Soldier Will Be': Wartime Tobacco Promotion in the US Military." *American Journal of Public Health*, 99: 1595–1602, 2009 **(40120)**

SMITH, K.C., K. WELDING, C. KLEB, C. WASHINGTON, AND J. COHEN. "English on Cigarette Packs from Six Non-Anglophone Low- and Middle- Income Countries." *International Journal of Public Health*, 63: 1071–1079, 2018 **(70313)**

SOKOL, N.A., R.D. KENNEDY, AND G.N. CONNOLLY. "The Role of Cocoa as a Cigarette Additive: Opportunities for Product Regulation." *Nicotine and Tobacco Research*, 16: 984–991, 2014 **(40403)**

SOLARINO, B., F. ROSENBAUM, B. RIESSELMANN, C.T. BUSCHMANN, AND M. TSOKOS. "Death Due to Ingestion of Nicotine-Containing Solution: Case Report and Review of the Literature." *Forensic Science International*, 195: e19–e22, 2010 **(30603)**

SOLDZ, S., D.J. HUYSER, AND E. DORSEY. "Characteristics of Users of Cigars, Bidis and Kreteks and the Relationship to Cigarette Use." *Preventive Medicine*, 37: 250–258, 2003 **(60512)**

SOMMERFELD, C.G., D.J. WEINER, A. NOWALK, AND A. LARKIN. "Hypersensitivity Pneumonitis and Acute Respiratory Distress Syndrome from E-Cigarette Use." *Pediatrics*, 141: (4 pp), 2018 **(50418)**

SONEJI, S., J.L. BARRINGTON-TRIMIS, T.A. WILLS, A.M. LEVENTHAL, J.B. UNGER, L.A. GIBSON, J. YANG, B.A. PRIMACK, J.A. ANDREWS, R.A. MIECH, T.R. SPINDLE, D.M. DICK, T. EISSENBERG, R.C. HORNIK, R. DANG, AND J.D. SARGENT. "Association Between Initial Use of E-Cigarettes and Subsequent Cigarette Smoking Among Adolescents and Young Adults: A Systematic Review and Meta-analysis." *JAMA Pediatrics*, 171: 788–797, 2017 **(50623)**

SONG, M-A., N.L. BENOWITZ, M. BERMAN, T.M. BRASKY, K.M. CUMMINGS, D.K. HATSUKAMI, C. MARIAN, R. O'CONNOR, V.W. REES, C. WOROSZYLO, AND P.G. SHIELDS. "Cigarette Filter Ventilation and its Relationship to Increasing Rates of Lung Adenocarcinoma." *Journal of the National Cancer Institute*, 109: (18 pp), 2017 **(40202)**

SOOD, N., AND N. SOOD. "A Rare Case of Vanishing Lung Syndrome." *Case Reports in Pulmonology*, 2011: (2 pp), 2011 **(40522)**

SOULE, E.K., T. LIPATO, AND T. EISSENBERG. "Waterpipe Tobacco Smoking: A New Smoking Epidemic Among the Young?" *Current Pulmonary Reports*, 4: 163–172, 2015 **(60402)**

SRKALOVIĆ, G. "Lung Cancer: Preventable Disease." *Acta Medica Academica*, 47: 39–49, 2018 **(70301)**

ST. HELEN, G., D.A. DEMPSEY, C.M. HAVEL, P. JACOB III, AND N.L. BENOWITZ. "Impact of E-Liquid Flavors on Nicotine Intake and Pharmacology of E-Cigarettes." *Drug and Alcohol Dependence*, 178: 391–398, 2017 **(20109)**

ST. HELEN, G., J.T. BERNET, D.B. HALL, C.S. SOSNOFF, Y. XIA, J.R. BALMES, J.E. VENA, J-S. WANG, N.T. HOLLAND, AND L.P. NAEHER. "Exposure to Secondhand Smoke Outside of a Bar and a Restaurant and Tobacco Exposure Biomarkers in Nonsmokers." *Environmental Health Perspectives*, 7: 1010–1016, 2012 **(40635)**

ST. HELEN, G., K.C. ROSS, D.A. DEMPSEY, C.M. HAVEL, P. JACOB III, AND N.L. BENOWITZ. "Nicotine Delivery and Vaping Behavior During *Ad Libitum* E-Cigarette Access." *Tobacco Regulatory Science*, 2: 363–376, 2016 **(50805)**

ST. HELEN, G., N.L. BENOWITZ, K.M. DAINS, C. HAVEL, M. PENG, AND P. JACOB III. "Nicotine and Carcinogen Exposure after Water Pipe Smoking in Hookah Bars." *Cancer Epidemiology Biomarkers and Prevention*, 23: 1055–1066, 2014 **(60401)**

ST. HELEN, G., P. JACOB III, N. NARDONE, AND N.L. BENOWITZ. "IQOS: Examination of Philip Morris International's Claim of Reduced Exposure." *Tobacco Control*, 27: s30–s36, 2018 **(60113)**

STAPLETON, M., A. HOWARD-THOMPSON, C. GEORGE, R.M. HOOVER, AND T.H. SELF. "Smoking and Asthma." *Journal of the American Board of Family Medicine*, 24: 313–322, 2011 **(40517)**

STAUDT, M.R., J. SALIT, R.J. KANER, C. HOLLMANN, AND R.G. CRYSTAL. "Altered Lung Biology of Healthy Never Smokers Following Acute Inhalation of E-Cigarettes." *Respiratory Research*, 19: 78 (10 pp), 2018 **(50415)**

STEIN, J.H., M. BUSHARA, K. BUSHARA, P.E. MCBRIDE, D.E. JORENBY, AND M.C. FIORE. "Smoking Cessation, but Not Smoking Reduction, Reduces Plasma Homocysteine Levels." *Clinical Cardiology*, 25: 23–26, 2002 **(40545)**

STEPANOV, I., A. KNEZEVICH, L. ZHANG, C.H. WATSON, D.K. HATSUKAMI, AND S.S. HECHT. "Carcinogenic Tobacco-Specific N-Nitrosamines in US Cigarettes: Three Decades of Remarkable Neglect by the Tobacco Industry." *Tobacco Control*, 21: 44–48, 2012 **(90207)**

STEPANOV, I., J. JENSEN, L. BIENER, R.L. BLISS, S.S. HECHT, AND D.K. HATSUKAMI. "Increased Pouch Sizes and Resulting Changes in the Amounts of Nicotine and Tobacco-Specific N-Nitrosamines in Single Pouches of Camel Snus and Marlboro Snus." *Nicotine and Tobacco Research*, 14: 1241–1245, 2012 **(60710)**

STERLING, K.L., B.A. MAJEED, A. NYMAN, AND M. ERIKSEN. "Risk Perceptions of Little Cigar and Cigarillo Smoking Among Adult Current Cigarette Smokers." *Nicotine and Tobacco Research*, 19: 1351–1358, 2017 **(60211)**

STERLING, K.L., C.S. FRYER, I. PAGANO, AND P. FAGAN. "Little Cigars and Cigarillos Use Among Young Adult Cigarette Smokers in the United States: Understanding Risks of Concomitant Use Subtypes." *Nicotine and Tobacco Research*, 18: 2234–2242, 2016 **(60216)**

STEVENSON, T., AND R.N. PROCTOR. "The Secret and Soul of Marlboro. Philip Morris and the Origins, Spread, and Denial of Nicotine Freebasing." *American Journal of Public Health*, 98: 1184–1194, 2008 **(40121)**

STEWART, P.M., AND J.R. CATTERALL. "Chronic Nicotine Ingestion and Atrial Fibrillation." *British Heart Journal*, 54: 222–223, 1985 **(10306)**

STÖRK, T., H. EICHSTÄDT, M. MÖCKEL, R. BORTFELDT, R. MÜLLER, AND H. HOCHREIN. "Changes of Diastolic Function Induced by Cigarette Smoking: An Echocardiographic Study in Patients with Coronary Artery Disease." *Clinical Cardiology*, 15: 80–86, 1992 **(40534)**

SU, C-P., G. SYAMLAL, S. TAMERS, J. LI, AND S.E. LUCKHAUPT. "Workplace Secondhand Tobacco Smoke Exposure Among U.S. Nonsmoking Workers, 2015." *Morbidity and Mortality Weekly Report*, 68: 604–607, 2019 **(40624)**

SUBRAMONEY, S., AND P.C. GUPTA. "Anemia in Pregnant Women who Use Smokeless Tobacco." *Nicotine and Tobacco Research*, 10: 917–920, 2008 **(60627)**

SUFFREDINI, D.A., AND R.M. REED. "At the Twisted Heart of Nicotine Addiction." *BMJ Case Reports* (4 pp), 2012 **(40533)**

SUHLING, H., T. WELTE, AND T. FUEHNER. "Three Patients with Acute Pulmonary Damage Following Use of E-Cigarettes. A Case Series." *Deutsches Ärzteblatt International*, 117: 177–182, 2020 **(50430)**

SUMARTININGSIH, S., H-F. LIN, AND J-C. LIN. "Cigarette Smoking Blunts Exercise-Induced Heart Rate Response Among Young Adult Male Smokers." *International Journal of Environmental Research and Public Health*, 16: 1032 (9 pp), 2019 **(81007)**

SUNG, H-W., Y. WANG, T. YAO, J. LIGHTWOOD, AND W. MAX. "Polytobacco Use and Nicotine Dependence Symptoms Among US Adults, 2012–2014." *Nicotine and Tobacco Research*, 20: S88–S98, 2018 **(60809)**

SWAN, G.E., C.N. LESSOV-SCHLAGGAR. "The Effects of Tobacco Smoke and Nicotine on Cognition and the Brain." *Neuropsychology Review*, 17: 259–273, 2007 **(10320)**

SWAYAMPAKALA, K., J.F. THRASHER, D. HAMMOND, H-H. YONG, M. BANSAL-TRAVERS, D. KRUGMAN, A. BROWN, R. BORLAND, AND J. HARDIN. "Pictorial Health Warning Label Content and Smokers' Understanding of Smoking-Related

Risks—A Cross Country Comparison." *Health Education Research*, 30: 35–45, 2015 **(70307)**

T

TABUCHI, T., S. GALLUS, T. SHINOZAKI, T. NAKAYA, N. KUNUGITA, AND B. COLWELL. "Heat-not-Burn Tobacco Product Use in Japan: Its Prevalence, Predictors and Perceived Symptoms from Exposure to Secondhand Heat-not-Burn Tobacco Aerosol." *Tobacco Control*, 27: e25–e33, 2018 **(60106)**

TALHOUT, R., P.A. RICHTER, I. STEPANOV, C.V. WATSON, AND C.H. WATSON. "Cigarette Design Features: Effects on Emission Levels, User Perception, and Behavior." *Tobacco Regulatory Science*, 4: 592–604, 2018 **(40213)**

TANNER, J-A., M. NOVALEN, P. JATLOW, M.A. HUESTIS, S.E. MURPHY, J. KAPRIO, A. KANKAANPÄÄ, L. GALANTI, C. STEFAN, T.P. GEORGE, L. BENOWITZ, C. LERMAN, AND R.F. TYNDALE. "Nicotine Metabolite Ratio (3-hydroxycontinine/cotinine) in Plasma and Urine by Different Analytical Methods and Laboratories: Implications for Clinical Implementation." *Cancer Epidemiology, Biomarkers & Prevention*, 24: 1239–1246, 2015 **(30101)**

TAS, D., H. SEVKETBEYOGLU, A.F. AYDIN, K. CELIK, AND M.A. KARACA. "The Relationship Between Nicotine Dependence Level and Community-Acquired Pneumonia in Young Soldiers: A Case Control Study." *Internal Medicine* 47: 2117–2120, 2008 **(10313)**

TASSEW, Z., AND B.S. CHANDRAVANSHI. "Levels of Nicotine in Ethiopian Tobacco Leaves." *Springer Plus*, 4: 649 (6 pp), 2015 **(10105)**

TENEGGI, V., L. SQUASSANTE, L. IAVARONE, S. MILLERI, A. BYE, AND R. GOMENI. "Correlation and Predictive Performances of Saliva and Plasma Nicotine Concentration on Tobacco-Withdrawal Induced Craving." *British Journal of Clinical Pharmacology*, 54: 407–414, 2002 **(30404)**

THIRIÓN-ROMERO, I., R. PÉREZ-PADILLA, G. ZABERT, AND I. BARRIENTIOS-GUTIÉRREZ. "Respiratory Impact of Electronic Cigarettes and 'Low Risk' Tobacco." *Revista De Investigacion Clinica*, 71: 17–27, 2019 **(50404)**

THOMSEN, S.F., AND L.T. SØRENSEN. "Smoking and Skin Disease." *Skin Therapy Letter*, 15: 4–7, 2010 **(40531)**

THORNTON, S.L., L. OLLER, AND T. SAWYER. "Fatal Intravenous Injection of Electronic Nicotine Delivery System Refilling Solution." *Journal of Medical Toxicology*, 10: 202–204, 2014 **(10416)**

THURTLE, N., R. ABOUCHEDID, J.R.H. ARCHER, J. HO, T. YAMAMOTO, P.I. DARGAN, AND D.M. WOOD. "Prevalence of Use of Electronic Nicotine Delivery Systems (ENDS) to Vape Recreational Drugs by Club Patrons in South London." *Journal of Medical Toxicology*, 13: 61–65, 2017 **(60918)**

TIDEY, J.W., L.R. PACEK, J.S. KOOPMEINERS, R. VANDREY, N. NARDONE, D.J. DROBES, N.L. BENOWITZ, S.S. DERMODY, A. LEMIEUX, R.L. DENLINGER, R. CASSIDY, M. AL'ABSI, D.K. HATSUKAMI, AND E.C. DONNY. "Effects of 6-Week Use of Reduced-Nicotine Content Cigarettes in Smokers With and Without Elevated Depressive Symptoms." *Nicotine and Tobacco Research*, 19: 59–67, 2017 **(80107)**

TIERNEY, P.A., C.D. KARPINSKI, J.E. BROWN, W. LUO, AND J.F. PANKOW. "Flavour Chemicals in Electronic Fluids." *Tobacco Control*, 25: e10–e15, 2016 **(50303)**

TIMBERLAKE, D.S., J. WU, AND W.K. AL-DELAIMY. "Tribal Casinos in California: The Last Vestige of Indoor Smoking." *BMC Public Health*, 12: 144 (8 pp), 2012 **(40634)**

TINDLE, H.A., S. SHIFFMAN, A.M. HARTMAN, AND J.E. BOST. "Switching to 'Lighter' Cigarettes and Quitting Smoking." *Tobacco Control*, 18: 485–490, 2009 **(40214)**

TIWARI, R.K., V. SHARMA, R.K. PANDEY, AND S.S. SHUKLA. "Nicotine Addiction: Neurobiology and Mechanism." *Journal of Pharmacopuncture*, 23: 1–7, 2020 **(10214)**

TJONCKE, J-A., R. GONCALVES, N. CASTAING, M. MOLIMARD, F. TOVAGLIARO, AND K. TITIER. "Death Related to Nicotine Replacement Therapy: A Case Report." *Forensic Science International*, 309: (9 pp), 2020 **(30604)**

TO, N., D.J. GRACIE, AND A.C. FORD. "Systematic Review with Meta-Analysis: The Adverse Effects of Tobacco Smoking on the Natural History of Crohn's Disease." *Alimentary Pharmacology and Therapeutics*, 43: 549–561, 2016 **(40527)**

TOABORELLI, M., M. MONTELLA, M. LIBRA, R. TEDESCHI, A. CRISPO, M. GRIMALDI, L. DAL MASO, D. SERRAONO, AND J. POLESEL. "The Dose-Response Relationship Between Tobacco Smoking and the Risk of Lymphomas: A Case-Control Study." *BMC Cancer*, 17: 421 (9 pp), 2017 **(40542)**

TOMAR, S.L., AND J.E. HENNINGFIELD. "Review of the Evidence that pH is a Determinant of Nicotine Dosage from Oral Use of Smokeless Tobacco." *Tobacco Control*, 6: 219–225, 1997 **(20110)**

TØNNESEN, P. "Smoking Cessation and COPD." *European Respiratory Review*, 22: 37–43, 2013 **(70223)**

TØNNESEN, P. "Which Drug to Be Used in Smoking Cessation?" *Polskie Archiwum Medycyny Wewnętrznej*, 118: 373–375, 2008 **(70224)**

TURNER, J.A., M.W. MCNICOL, AND R.W. SILLETT. "Distribution of Carboxyhaemoglobin Concentrations in Smokers and Non-Smokers." *Thorax*, 41: 25–27, 1986 **(60207)**

TUSHINGHAM, S., C.M. SNYDER, K.J. BROWNSTEIN, W.J. DAMITO, AND D.R. GANG. "Biomolecular Archeology Reveals Ancient Origins of Indigenous Tobacco Smoking in North American Plateau." *Proceedings of the National Academy of Sciences of the United States of America*, 115: 11742–11747, 2018 **(10102)**

TUTKA, P., J. MOSIEWICZ, AND M. WIELOSZ. "Pharmacokinetics and Metabolism of Nicotine." *Pharmacological Reports*: 57: 143–153, 2005 **(20102)**

TWEED, J.O., S.H. HSIA, K. LUTFY, AND T.C. FRIEDMAN. "The Endocrine Effects of Nicotine and Cigarette Smoke." *Trends in Endocrinology and Metabolism*, 23: 334–342, 2012 **(10304)**

U

UEMATSU, T., A. MIZUNO, S. NAGASHIMA, A. OSHIMA, AND M. NAKAMURA. "The Axial Distribution of Nicotine Content Along Hair Shaft as an Indicator of Changes in Smoking Behaviour: Evaluation in a Smoking-Cessation Programme with or Without the Aid of Nicotine Chewing Gum." *British Journal of Clinical Pharmacology*, 39: 665–669, 1995 **(30505)**

UNGER, M., AND D.E. UNGER. "E-Cigarettes/Electronic Nicotine Delivery Systems: A Word of Caution on Health and New Product Development." *Journal of Thoracic Disease*, 10: S2588–S2592, 2018 **(50104)**

V

VANSICKEL, A.R., A. SHIHADEH, AND T. EISSENBERG. "Waterpipe Tobacco Products: Nicotine Labelling Versus Nicotine Delivery." *Tobacco Control*, 21: 377–379, 2012 **(60405)**

VARDAVAS, C., C. GIRVALAKI, A. VARDAVAS, S. PAPADAKIS, M. TZATZARAKIS, P. BEHRAKIS, AND A. TSATSAKIS. "Respiratory Irritants in E-Cigarette Refill Liquids Across Nine European Countries: A Threat to Respiratory Health? [Research Letter]." *European Respiratory Journal*, 50: (3 pp), 2017 **(50305)**

VASILJEVIC, M., A.S.J. WALLIS, S. CODLING, D-L. COUTURIER, S. SUTTON, AND T.M. MARTEAU. "E-Cigarette Adverts and Children's Perceptions of Tobacco Smoking Harms: An Experimental Study and Meta-Analysis." *BMJ Open*, 8: e020247 (8 pp), 2018 **(50628)**

VELICER, C., G. ST. HELEN, AND S.A. GLANTZ. "Tobacco Papers and Tobacco Industry Ties in Regulator Toxicology and Pharmacology." *Journal of Public Health Policy*, 39: 34–48, 2018 **(40109)**

VILLANTI, A.C., L.K. COLLINS, R.S. NIAURA, S.Y. GAGOSIAN, AND D.B. ABRAMS. "Menthol Cigarettes and the Public Health Standard: A Systematic Review." *BMC Public Health*, 17: 983 (13 pp), 2017 **(40305)**

VISSER, W.E., W.N. KLERX, H.W.J.M. CREMERS, R. RAMLAL, P.L. SCHWILLENS, AND R. TALHOUT. "The Health Risks of Electronic Cigarette Use to Bystanders." *International Journal of Environmental Research and Public Health*, 16: 1525 (14 pp), 2019 **(50502)**

VIVEROS, M-P., E.M. MARCO, AND S.E. FILE. "Nicotine and Cannabinoids: Parallels, Contrasts and Interactions." *Neuroscience and Biobehavioral Reviews*, 30: 1161–1181, 2006 **(60902)**

VLACHOPOULOS, C., N. IOAKEIMIDIS, M. ABDELRASOUL, D. TERENTES-PRINTZIOS, C. GEORGAKOPOULOS, P. PIETRI, C. STEFANADIS, AND D. TOUSOULIS. "Electronic Cigarette Smoking Increases Aortic Stiffness and Blood Pressure in Young Smokers [Letter]." *Journal of the American College of Cardiology*, 67: 2802–2803, 2016 **(50413)**

VLAHOY, D., S. GALEA, H. RESNICK, J. AHERN, J.A. BOSCARINO, M. BUCUVALAS, J. GOLD, AND D. KILPATRICK. "Increased Use of Cigarettes, Alcohol, and Marijuana Among Manhattan, New York, Residents After the September 11th Terrorist Attacks." *American Journal of Epidemiology*, 155: 988–996, 2002 **(80507)**

VŇUKOVÁ, M., R. PTÁČEK, J. RABOCH, AND G.B. STEFANO. "Decreased Central Nervous System Grey Matter Volume (GMV) in Smokers Affects Cognitive Abilities: A Systematic Review." *Medical Science Monitor*, 23: 1907–1915, 2017 **(40515)**

VOGEL, E.A., J. CHO, R.S. MCCONNELL, J.L. BARRINGTON-TRIMIS, AND A.M. LEVENTHAL. "Prevalence of Electronic Cigarette Dependence Among Youth and Its Association with Future Use." *JAMA Network Open*, 3: (13 pp), 2020 **(50607)**

VOGIATZIS, I., E. TSIKRIKA, V. SACHPEKIDIS, S. PITTAS, AND A. KOTSANI. "Factors Affecting Smoking Resumption After Acute Coronary Syndromes." *Hellenic Journal of Cardiology*, 51: 294–300, 2010 **(70216)**

VON RAPPARD, J., M. SCHÖNENBERGER, AND L. BÄRLOCHER. "Carbon Monoxide Poisoning Following Use of a Water Pipe/Hookah." *Deutsches Arzteblatt International*, 111, 674–679, 2014 **(60422)**

VOZORIS, N.T., AND M.B. STANBROOK. "Smoking Prevalence, Behaviours, and Cessation Among Individuals with COPD or Asthma." *Respiratory Medicine*, 105: 477–484, 2011 **(70217)**

VUOLO, M., B.C. KELLY, AND J. KADOWAKI. "Impact of Total Vending Machine Restrictions on US Young Adult Smoking." *Nicotine and Tobacco Research*, 18: 2092–2099, 2016 **(70314)**

W

WACKOWSKI, O.A., J.M. SONTAG, D. HAMMOND, R.J. O'CONNOR, P.A. OHMAN-STRICKLAND, A.A. STRASSER, A.C. VILLANTI, AND C.D. DELNEVO. "The Impact of E-Cigarette Warnings, Warning Themes and Inclusion of Relative Harm Statements on Young Adult's E-Cigarette Perceptions and Use Intentions." *International Journal of Environmental Research and Public Health*, 16: 184 (14 pp), 2019 **(50629)**

WADA, K., T. KAWACHI, A. HORI, N. TAKEYAMA, S. TANABASHI, S. MATSUSHITA, N. TOKIMIOTSU, AND C. NAGATA. "Husband's Smoking Status and Breast Cancer Risk in Japan: From the Takayama Study." *Cancer Science*, 106: 455–460, 2015 **(40616)**

WADE, W.S., AND K. WHITE. "The Package as a Weapon of Influence: Changes to Cigarette Packaging Design as a Function of Regulatory Changes in Canada." *Tobacco Prevention and Cessation*, 6: (4 pp), 2020 **(70311)**

WADGAVE, U., AND L. NAGESH. "Nicotine Replacement Therapy: An Overview." *International Journal of Health Sciences*, 10: 425–435, 2016 **(70219)**

WAIN, A.A., AND J. MARTIN. "Can Transdermal Nicotine Patch Cause Acute Intoxication in a Child? A Case Report and Review of Literature." *The Ulster Medical Journal*, 73: 65–66, 2004 **(10406)**

WALD, N.J., M. IDLE, J. BOREHAM, AND A. BAILEY. "Inhaling Habits Among Smokers of Different Types of Cigarette." *Thorax*, 35: 925–928, 1980 **(40208)**

WALLEY, S.C., K.M. WILSON, J.P. WINICKOFF, AND J. GRONER. "A Public Health Crisis: Electronic Cigarettes, Vape, and JUUL." *Pediatrics*, 143: (11 pp), 2019 **(50202)**

WANG, B., AND B. ROSTRON. "Tobacco-Related Poison Events Involving Young Children in the US, 2001–2016." *Tobacco Regulatory Science*, 3: 525–535, 2017 **(10407)**

WANG, J.B., J.E. OLGIN, G. NAH, E. VITTINGHOFF, J.K. CATALDO, M.J. PLETCHER, AND G.M. MARCUS. "Cigarette and E-Cigarette Dual Use and Risk of Cardiopulmonary Symptoms in the Health eHeart Study." *Public Library of Science One*, 13: (14 pp), 2018 **(50710)**

WANG, T.W., A. GENTZKE, S. SHARAPOVA, K.A. CULLEN, B.K. AMBROSE, AND A. JAMAL. "Tobacco Product Use Among Middle and High School Students—United States, 2011–2017." *Mortality and Morbidity Weekly Report*, 67: 629–633, 2018 **(60805)**

WARD, K.A., P.J. SPOKES, AND J.M. MCANULTY. "Case-Control Study of Risk Factors for Hospitalization Caused by Pandemic (H1N1) 2009." *Emerging Infectious Diseases*, 17: 1409–1416, 2011 **(40514)**

WARD, K.D., S. AHN, F. MZAYEK, R.A. ALI, S. RASTAM, T. ASFAR, F. FOUAD, AND W. MAZIAK. "The Relationship Between Waterpipe Smoking and Body Weight: Population-Based Findings from Syria." *Nicotine and Tobacco Research*, 17: 34–40, 2015 **(60417)**

WARDA, L., M. TENEBEIN, AND M.E. MOFFATT. "House Fire Injury Prevention Update: Part I. A Review of Risk Factors for Fatal and Non-Fatal House Fire Injury." *Injury Prevention*, 5: 145–150, 1999 **(80203)**

WARE, J.J., J-A. TANNER, A.E. TAYLOR, Z. BIN, P. HAYCOCK, J. BOWDEN, P.J. ROGERS, G.D. SMITH, AND R.F. TYNDALE. "Does Coffee Consumption Impact on Heaviness of Smoking?" *Addiction*, 112: 1842–1853, 2017 **(70110)**

WARNAKULASURIYA, S., AND K. STRAIF. "Carcinogenicity of Smokeless Tobacco: Evidence from Studies in Humans & Experimental Animals." *Indian Journal of Medical Research*, 148: 681–686, 2018 **(60620)**

WATKINS, R.E., AND A.J. PLANT. "Does Smoking Explain Sex Differences in the Global Tuberculosis Epidemic?" *Epidemiology and Infection*, 134: 333–339, 2006 **(80306)**

WAYNE, G.F., AND G.N. CONNOLLY. "How Cigarette Design Can Affect Youth Initiation into Smoking: Camel Cigarettes 1983–93." *Tobacco Control*, 11: i32–i39, 2002 **(40805)**

WAYNE, G.F., G.N. CONNOLLY, AND J.E. HENNINGFIELD. "Brand Differences of Free-Base Nicotine Delivery in Cigarette Smoke: The View of the Tobacco Industry Documents." *Tobacco Control*, 15: 189–198, 2006 **(40103)**

WEI, B., M.L. GONIEWICZ, R.J. O'CONNOR, M.J. TRAVERS, AND A.J. HYLAND. "Urinary Metabolite Levels of Flame Retardants in Electronic Cigarette Users: A Study Using the Data from NHANES 2013–2014." *International Journal of Environmental Research and Public Health*, 15: 201, 2018 **(50403)**

WEINBURGER, A.H., AND M. SOFUOGLU. "The Impact of Cigarette Smoking on Stimulant Addiction." *American Journal of Drug and Alcohol Abuse*, 35: 12–17, 2009 **(60916)**

WEINGARTEN, T.N., S.M. MOESCHLER, A.E. PTASZYNSKI, W.M. HOOTEN, T.J. BEEBE, AND D.O. WARNER. "An Assessment of the Association Between Smoking Status, Pain Intensity, and Functional Interference in Patients with Chronic Pain." *Pain Physician*, 11: 643–653, 2008 **(80602)**

WEN, C.P., S.P. TSAI, T.Y. CHENG, H.T. CHAN, W.S.I. CHUNG, AND C.J. CHEN. "Excess Injury Mortality Among Smokers: A Neglected Tobacco Hazard." *Tobacco Control*, 14, i28–i32, 2005 **(80413)**

WHINCUP, P.H., J.A. GILG, J.R. EMBERSON, M.J. JARVIS, C. FEYERABEND, A. BRYANT, M. WALKER, AND D.G. COOK. "Passive Smoking and Risk of Coronary Heart Disease and Stroke: Prospective Study with Cotinine Measurement." *British Medical Journal*, 329: 200–205, 2004 **(40619)**

WHITEHEAD, T., C. METAYER, M.H. WARD, M.G. NISHIOKA, R. GUNIER, J.S. COLT, P. REYNOLDS, S. SELVIN, P. BUFFLER., AND S.M. RAPPAPORT. "Is House-Dust Nicotine a Good Surrogate for Household Smoking?" *American Journal of Epidemiology*, 169: 1113–1123, 2009 **(40705)**

WICKHAM, R.J. "How Menthol Alters Tobacco-Smoking Behavior: A Biological Perspective." *Yale Journal of Biology and Medicine*, 88: 279–287, 2015 **(40302)**

WICKHAM, R.J. "The Biological Impact of Menthol on Tobacco Dependence." *Nicotine and Tobacco Research*, 20: 1–9, 2019 **(40301)**

WIENCEK, J.R., E.A. GEHRIE, A.M. KEISER, P.C. SZKLARSKI, K.L. JOHNSON-DAVIS, AND G.S. BOOTH. "Detection of Nicotine and Nicotine Metabolites in Units of Banked Blood." *American Journal of Clinical Pathology*, 151: 516–521, 2019 **(30202)**

WILENS, T.E., M. VITULANO, H. UPADHYAYA, J. ADAMSON, R. SAWTELLE, L. UTZINGER, AND J. BIEDERMAN. "Cigarette Smoking Associated with Attention-Deficit Hyperactivity Disorder." *Journal of Pediatrics*, 153: 414–419, 2008 **(80815)**

WILLS, T.A., I. PAGANO, R.J. WILLIAMS, AND E.K. TAM. "E-Cigarette Use and Respiratory Disorder in an Adult Sample." *Drug and Alcohol Dependence*, 194: 363–370, 2019 **(50423)**

WU, C-F., N-H. FENG, I-W. CHONG, K-Y. WU, C-H. LEE, J-J. HWANG, C-T. HUANG, C-Y LEE, S-T. CHOU, D.C. CHRISTIANI, AND M-T. WU. "Second-Hand Smoke and Chronic Bronchitis in Taiwanese Women: A Health-Care Based Study." *BMC Public Health*, 10: 44 (10 pp), 2010 **(40612)**

WU, D., AND D.F. O'SHEA. "Potential for Release of Pulmonary Toxic Ketene from Vaping Pyrolysis of Vitamin E Acetate." *Proceedings of the National Academy of Sciences of the United States of America*, 117: 6349–6355, 2020 **(50804)**

WU, J., AND D.D. SIN. "Improved Patient Outcome with Smoking Cessation: When is it Too Late?" *International Journal of Chronic Obstructive Pulmonary Disease*, 6: 259–267, 2011 **(70210)**

WU, S-Y., F. XING, S. SHARMA, K. WU, A. TYAGI, Y. LIU, D. ZHAO, P.R. DESHPANDE, Y. SHIOZAWA, T. AHMED, W. ZHANG, M. CHAN, J. RUIZ, T.W. LYCAN, A. DOTHARD, AND K. WATABE. "Nicotine Promotes Brain Metastasis by Polarizing Microglia and Suppressing Innate Immune Function." *Journal of Experimental Medicine*, 217: (17 pp), e20191131, 2020 **(10317)**

WU, S.Y-D., M.P. WANG, W.H. LI, A.C. KWONG, V.W. LAI, AND T.H. LAM. "Does Electronic Cigarette Use Predict Abstinence from Conventional Cigarettes Among Smokers in Hong Kong?" *International Journal of Environmental Research and Public Health*, 15: 400 (10 pp), 2018 **(70218)**

WU, W., S. SONG, D.L. ASHLEY, AND C.H. WATSON. "Assessment of Tobacco-Specific Nitrosamines in the Tobacco and Mainstream Smoke of Bidi Cigarettes." *Carcinogenesis*, 25: 283–287, 2004 **(60501)**

X

XAVIER, R.F., D. RAMOS, J.T. ITO, ET AL. "Effects of Cigarette Smoking Intensity on the Mucociliary Clearance of Active Smokers." *Respiration*, 86: 479–485, 2013 **(80301)**

XIA B., B.C. BLOUNT, AND L. WANG. "Sensitive Quantification of Nicotine in Bronchoalveolar Lavage Fluid by Acetone Precipitation Combined with Isotope-Dilution Chromatography-Tandem Mass Spectrometry." *ACS Omega*, 6: 13962–13969, 2021 **(30105)**

XIE, X., Q. LIU, J. WU, AND M. WAKUI. "Impact of Cigarette Smoking in Type 2 Diabetes Development." *Acta Pharmacologica Sinica*, 6: 784–787, 2009 **(10314)**

XU, X., M.M. IBA, AND C.P. WEISEL. "Simultaneous and Sensitive Measurement of Anabasine, Nicotine, and Nicotine Metabolites in Human Urine by Liquid Chromatography–Tandem Mass Spectrometry." *Clinical Chemistry*, 50: 2323–2330, 2004 **(90203)**

Y

YANG, Y.T., AND S. GLANTZ. "San Francisco Voters End the Sale of Flavored Tobacco Products Despite Strong Industry Opposition." *Annals of Internal Medicine*, 169: 708–709, 2018 **(40312)**

YANKELEVITZ, D.F., C.I. HENSCHKE, R. YIP, P. BOFFETTA, J. SHEMESH, M.D. CHAM, J. NARULA, AND H.S. HECHT. "Second-Hand Tobacco Smoke in Never Smokers Is a Significant Risk Factor for Coronary Artery Calcification." *JACC: Cardiovascular Imaging*, 6: 651–657, 2013 **(40611)**

YERAMANENI, S., K.N. DIETRICH, K. YOLTON, P.J. PARSONS, K.M. ALDOUS, AND E.N. HAYNES. "Secondhand Tobacco Smoke Exposure and Neuromotor Function in Rural Children." *Journal of Pediatrics*, 167: 253–259, 2015 **(40606)**

YERGER, V.B. "Menthol's Potential Effects on Nicotine Dependence: A Tobacco Industry Perspective." *Tobacco Control*, 20: ii29–ii36, 2011 **(40303)**

YIENGPRUGSAWAN, V., J. BERECKI-GISOLF, C. BAIN, R. MCCLURE, S-A. SEUBSMAN, AND A.C. SLEIGH. "Predictors of Injury Mortality: Findings from a Large National Cohort in Thailand." *BMJ Open*, 4: e004668, (8 pp), 2014 **(80405)**

YINGST, J.M., J. FOULDS, S. VELDHEER, S. HRABOVSKY, N. TRUSHIN, T.T. EISSENBERG, J. WILLIAMS, J.P. RICHIE, T.T. NICHOLS, S.J. WILSON, AND A.L. HOBKIRK. "Nicotine Absorption During Electronic Cigarette Use Among Regular Users." *Public Library of Science One*, 14: e0220300 (15 pp), 2019 **(20107)**

YOO, S-J., S-J. PARK, B-S. KIM, K. LEE, H-S. LIM, J-S. KIM, AND I-S. KIM. "Airborne Nicotine Concentrations in the Workplaces of Tobacco Farmers." *Journal of Preventive Medicine and Public Health*, 47: 144–149, 2014 **(10127)**

YOSHIOKA, D., H. ISHII, Y. HATANO, K. UMEKI, N. SAKAMOTO, R. SHIRAI, S. FUJIWARA, AND J. KADOTA. "A Case of Eosinophilic Pneumonia in a Tobacco Harvester." *Allergology International*, 60: 551–554, 2011 **(10124)**

Z

ZANDONAI, T., E. TAM, P. BRUSEGHINI, F. PIZZOLATO, L. FRANCESCHI, M. BARALDO, C. CAPELLI, P. CESARI, AND C. CHIAMULERA. "The Effects of Oral Smokeless Tobacco Administration on Endurance Performance." *Journal of Sport and Health Science*, 7: 465–472, 2018 **(81004)**

ZHANG, X., J. KAHENDE, A.Z. FAN, L. BARKER, T.J. THOMPSON, A.H. MOKDAD, Y. LI, AND J.B. SAADDINE. "Smoking and Visual Impairment Among Older Adults with Age-Related Eye Diseases." *Preventing Chronic Disease*, 8: (8 pp), 2011 **(40529)**

ZHAO, R., Y. WU, F. ZHAO, Y. LV, D. HUANG, J. WEI, C. RUAN, M. HUANG, J. DENG, D. HUANG, AND X. QIU. "The Risk of Missed Abortion Associated with the Levels of Tobacco, Heavy Metals and Phthalate in Hair of Pregnant Women." *Medicine*, 96: (5 pp), 2017 **(80702)**

ZHAO, S., F. CHEN, D. WANG, H. WANG, W. HAN, AND Y. ZHANG. "Effect of Preoperative Smoking Cessation on Postoperative Pain Outcomes in Elderly Patients with High Nicotine Dependence." *Medicine*, 98: (6 pp), 2019 **(80604)**

ZHOU, S., M. WEITZMAN, R. VILCASSIM, J. WILSON, N. LEGRAND, E. SAUNDERS, M. TRAVERS, L-C. CHEN, R. PELTIER, AND T. GORDON. "Air Quality in New York City Hookah Bars." *Tobacco Control*, 24: e193–e198, 2015 **(60411)**

ZHOU, Z., D. BOHAC, AND R.G. BOYLE. "Continuous Weeklong Measurements of Indoor Particle Levels in a Minnesota Tribal Casino Resort." *BMC Public Health*, 16: 870 (6 pp), 2016 **(40633)**

ZIEDONIS, D., S. DAS, AND C. LARKIN. "Tobacco Use Disorder and Treatment: New Challenges and Opportunities." *Dialogues in Clinical Neuroscience*, 19: 271–280, 2017 **(70104)**

ZIMMERMANN-STENZEL, M., J. MANNUSS, S. SCHNEIDER, AND M. SCHILTENWOLF. "Smoking and Chronic Back Pain. Analyses of the German Telephone Health Survey 2003." *Deutsches Arzteblatt International*, 105: 441–448, 2008 **(80601)**

ZVOLENSKY, M.J., K. MCMILLAN, A. GONZALEZ, AND G.J.G. ASMUNDSON. "Chronic Pain and Cigarette Smoking and Nicotine Dependence Among a Representative Sample of Adults." *Nicotine and Tobacco Research*, 11: 1407–1414, 2009 **(80603)**

Subject Index

About the Author

James Wigmore worked as a senior forensic toxicologist in one of the foremost forensic laboratories in North America, The Centre of Forensic Sciences, in Toronto, Ontario, for over twenty-nine years. He is a designated analyst under section 254(1) of the *Criminal Code of Canada* and was a member of a national advisory board, the Alcohol Test Committee (which advises the federal Department of Justice on breath and blood alcohol testing), from 1993 to 2005. Wigmore has testified in over 700 criminal trials at various levels of court across Canada as well as in civil (personal injury) and coroner's cases. He has published over sixty scientific articles on the forensic aspects of alcohol and cannabis, and his publications have been cited by the Supreme Court of Canada and the High Court of South Africa. He taught at the Center of Studies for Law in Action (the Borkenstein course) at Indiana University in Bloomington, Indiana from 2004 to 2012. He was the recipient of the most prestigious award of the Canadian Society of Forensic Sciences, the Prix Derome Award, which recognizes outstanding contributions to the field of forensic science. He was a member of the Health Canada Expert Panel on "Your Cannabis Questions Answered, Get the Honest Facts" in April 2018, and he has appeared on numerous TV and Radio shows. *Wigmore on Nicotine* is his fourth book.